THE PHYSICIST'S CONCEPTION OF NATURE

The Physicist's Conception of Nature

Edited by

Jagdish Mehra

D. Reidel Publishing Company

Dordrecht-Holland / Boston-U.S.A.

1973

Library of Congress Catalog Card Number 73–75765

ISBN 90 277 0345 0

Published by D. Reidel Publishing Company,
P.O. Box 17, Dordrecht, Holland

Sold and distributed in the U.S.A., Canada, and Mexico
by D. Reidel Publishing Company, Inc.
306 Dartmouth Street, Boston,
Mass. 02116, U.S.A.

Printed in The Netherlands by D. Reidel, Dordrecht

Dedicated to

PAUL ADRIEN MAURICE DIRAC

on the occasion of his seventieth birthday

SYMPOSIUM ON THE DEVELOPMENT OF
THE PHYSICIST'S CONCEPTION OF NATURE
IN THE TWENTIETH CENTURY

Held at the International Centre for Theoretical Physics,
Miramare, Trieste, Italy, 18–25 September 1972

Sponsoring Committee:

Co-Chairmen: H. B. G. CASIMIR (Philips Research Laboratories, Eindhoven),
EUGENE P. WIGNER (Princeton University)

EDOARDO AMALDI (University of Rome),
PIERRE AUGER (Université de Paris),
BRIAN FLOWERS (U.K. Science Research Council, London),
RUDOLF E. PEIERLS (University of Oxford),
FRANCIS PERRIN (Collège de France),
ISIDOR I. RABI (Columbia University),
GUNNAR RANDERS (NATO, Brussels),
LEON ROSENFELD (NORDITA, Copenhagen),
ABDUS SALAM (Imperial College, London, and International Centre
for Theoretical Physics, Trieste)

Symposium Director:

JAGDISH MEHRA (The University of Texas at Austin)

Financial Sponsorship:

The Science Committee, North Atlantic Treaty Organization (Brussels, Belgium);
Minna–James–Heineman Stiftung (Hanover, Germany);
Consortia of the City and the Region of Trieste (Trieste, Italy);
The Lorentz Foundation (Haarlem, The Netherlands)

Contents

Preface

The fundamental conceptions of twentieth-century physics have profoundly influenced almost every field of modern thought and activity. Quantum Theory, Relativity, and the modern ideas on the Structure of Matter have contributed to a deeper understanding of Nature, and they will probably rank in history among the greatest intellectual achievements of all time. The purpose of our symposium was to review, in historical perspective, the current horizons of the major conceptual structures of the physics of this century.

Professors Abdus Salam and Hendrik Casimir, in their remarks at the opening of the symposium, have referred to its origin and planning. Our original plan was to hold a two-week symposium on the different aspects of five principal themes: 1. Space, Time and Geometry (including the structure of the universe and the theory of gravitation), 2. Quantum Theory (including the development of quantum mechanics and quantum field theory), 3. Statistical Description of Nature (including the discussion of equilibrium and non-equilibrium phenomena, and the application of these ideas to the evolution of biological structure), 4. The Structure of Matter (including the discussion, in a unified perspective, of atoms, molecules, nuclei, elementary particles, and the physics of condensed matter), and finally, 5. Physical Description and Epistemology (including the distinction between classical and quantum descriptions, and the epistemological and philosophical problems raised by them).

These themes, taken together, might be regarded as constituting almost the whole of modern physics, and as a programme for one symposium it would have seemed to be too ambitious. It was, however, not all that insane. Our intention was not to cover the *details* of all the problems that come under these themes. Rather we had hoped to discuss the historical development of the principal conceptions, and the current horizons which have arisen from them, which would emphasize the structural and conceptual unity of the body of physical knowledge about Nature. These would be presented by some of their most authentic exponents.

The Sponsoring Committee came to the conclusion that the programme, even for a two-week symposium, would be too heavy. They thought that it would be advisable to hold two symposia to cover the intended programme, and that it could be done without damaging the unity of the general theme. They decided that we should deal with four themes in the first symposium, leaving the Structure of Matter to the second. Our financial sponsors also thought that this was a good idea, that based upon the success of the first symposium the second one could be held with confidence a couple of years later. We thus finally adopted the programme which is reproduced in Appendix 1.

With a programme so rich in scientific and historical fare, and with many stars among the lecturers, every occasion at the symposium represented a highlight. Yet, there were several exceptional occasions which many of us will specially cherish.

Dirac's evening lecture on the 'Development of the Physicist's Conception of Nature' was such an occasion. In pin-drop silence, all ears attuned to his words, Dirac developed his theme. He talked about classical mechanics and relativity, about quantum mechanics and the quantum theory of fields, and about how the development of physics could be pictured as a rather steady development with many small steps, superposed on which were a number of big jumps, the latter consisting usually of overcoming prejudices. Dirac wondered about the present state of physical theory and invoked the twin principles of 'beauty and logic' as the guidelines for the developing architecture of physics. He declared his belief that the physicist's conception of Nature has not stopped in growing: it is at an interim state at present and fundamental developments will occur in the future that will change it.

As he spoke, his voice feeble but firm, the sentences perfectly formed as only Dirac's sentences can be, his spirit grew and filled the hall. The audience shrank in size if not in numbers. As he finished his talk, one felt the presence of only this frail man who had the strength of character to match his wisdom, a man who, with a few kindred spirits, had created the language of modern theoretical physics, and who now enjoined his audience of distinguished physicists to cast away prejudice and seek a deeper understanding of Nature.

In his beautiful historical talk, dealing with a period of a few weeks in March and April 1926 when the passage took place from matrix mechanics and wave mechanics to a unified quantum mechanics, B. L. van der Waerden discussed at length a letter from Pauli to Jordan, written in April 1926, in which Pauli had referred to Lanczos' work on the integral equations. Van der Waerden pointed out in delectable detail how Pauli and Schrödinger had imperfectly understood the work of Lanczos, and how the latter's integral equation was equivalent to Schrödinger's. At the end of this talk, which had shown how deep and original had been Lanczos' understanding of the mathematical structure of quantum mechanics, it was pointed out that Professor Cornelius Lanczos was in the lecture hall and had listened to this forceful vindication of his early work against Pauli's criticism of it. As Lanczos got up to acknowledge the announcement of his presence Van der Waerden, who did not know him personally, was shocked with pleasant surprise, and the audience broke into sustained applause. It was a thrilling moment for all.

Another occasion with enormous impact came on Wednesday morning, 20 September 1972. This was just the fact that the speakers that morning, one after the other, were Paul Dirac, Werner Heisenberg, and Eugene Wigner. An occasion with such a succession, in one session of a conference, of speakers who had participated in the creation of quantum theory had probably not occurred since the first and the fifth Solvay conferences, and the audience was conscious of a sense of history. This was the veritable celebration of the seventieth birthday of physics in the twentieth century.

An evening we will all remember and cherish was the banquet of the symposium in honour of Dirac. The gracefulness of the occasion was highlighted by both the company present and the remarks that were made. Lord Snow's address on 'The Classical Mind' underscored the joy and pride of the occasion which we were allowed to share. The impromptu telling of the 'Dirac stories' by several people, and Dirac's own story about David Hilbert, delighted all who were present. These remarks have been gathered in a chapter in this volume, and hopefully many a reader will enjoy them.

Many of those who attended the symposium have written to me how 'successful, beautiful, historic, unique, etc.' this symposium was. It gives me great pleasure to thank those who contributed to the uniqueness, beauty, and success of this event.

First of all, the members of the Sponsoring Committee, all of whom helped at one time or another during the organization. Professors Edoardo Amaldi, Pierre Auger, Brian Flowers, Rudolf Peierls, Francis Perrin, I. I. Rabi and Léon Rosenfeld gave much wise counsel and help. Professor Gunnar Randers' encouragement, cooperation, and assistance were invaluable throughout. Professors Hendrik Casimir, Abdus Salam, and Eugene Wigner encouraged, helped, and sustained me at every step of the way in the effort to organize the symposium, and it was their continuous support that made this task both possible and a joy. The association of all these distinguished scientists with the symposium had assured its success, and every expression of gratitude will be inadequate to acknowledge the significance of their sponsorship.

The distinguished lecturers were most cooperative. The subject of the lecture was not of their choice at times, yet they accepted their assignments graciously. They made wonderful contributions to the programme and it is what they delivered that gave content and meaning to the symposium. Their papers, often with the discussions that followed the talks, are reproduced here. Also reproduced in this volume are the papers, prepared for other occasions, of those who were invited to give talks but preferred not to do so (Wentzel), and of those whom circumstances prevented from attending (Kapitza, Tomonaga). We can be truly grateful for the treat which the contributors to this volume have offered us.

The chairmen of sessions preserved a relaxed atmosphere throughout. The participants in the symposium were just marvellous. They braved a heavy programme, participating fully in all sessions and contributing to the discussions. The lecturers, chairmen, and participants stayed throughout the symposium, which was also a unique experience, deserving of the gratitude of the organizers.

The holding of the symposium was made possible by the financial sponsorship of NATO's Scientific Committee. The Minna–James–Heineman Foundation of Hanover, Germany, also contributed funds. At the request of Professor Casimir, the Lorentz Foundation made a timely contribution. Several friends of the Symposium Director, who wish to remain anonymous, rendered financial help, and the Consortia of the City and the Region of Trieste gave invaluable support for the social events, as did the Prince of Torre e Tasso. The publishers of this volume have supported the symposium enormously, and have cooperated in meeting our wishes regarding the publication. We are grateful for all this wonderful support.

The symposium was held at the beautiful premises of the International Centre for

Theoretical Physics in Miramare, by courtesy of the Director, Professor Abdus Salam. Professor Paolo Budini organized crucial assistance when the holding of the symposium was threatened with disruption. The Administrative Officer, Dr André Hamende, and the staff of the Centre, gave every help and assistance most readily and cheerfully. My gratitude to them is everlasting.

This volume is dedicated to Paul Dirac. I hope it will give him, and others who will read it, some pleasure.

JAGDISH MEHRA

Contributors

J. S. BELL, Theory Division, C.E.R.N., Geneva, Switzerland.

H. B. G. CASIMIR, Philips Research Laboratories, Eindhoven, The Netherlands.

S. CHANDRASEKHAR, Laboratory for Astrophysics and Space Research, University of Chicago, Chicago, Illinois, U.S.A.

E. G. D. COHEN, The Rockefeller University, New York, New York, U.S.A.

LEON N. COOPER, Physics Department, Brown University, Providence, Rhode Island, U.S.A.

P. A. M. DIRAC, Department of Physics, Florida State University, Tallahassee, Florida, U.S.A.

JÜRGEN EHLERS, Max-Planck-Institut für Physik und Astrophysik, Munich, Fed. Rep. Germany.

MANFRED EIGEN, Max-Planck-Institut für Biophysikalische Chemie, Göttingen-Nikolausberg, Fed. Rep. Germany.

CHARLES P. ENZ, Institut de Physique Théorique, Université de Genève, Geneva, Switzerland.

BERNARD D'ESPAGNAT, Laboratoire de Physique Théorique et Particules Elémentaires, Université Paris-Sud, Orsay, France.

DAVID FINKELSTEIN, Department of Physics, Belfer Graduate School of Science, Yeshiva University, New York, New York, U.S.A.

R. HAAG, II. Institut für Theoretische Physik, Universität Hamburg, Hamburg, Fed. Rep. Germany.

WERNER HEISENBERG, Max-Planck-Institut für Physik und Astrophysik, Munich, Fed. Rep. Germany.

JOSEF M. JAUCH, Institut de Physique Théorique, Université de Genève, Geneva, Switzerland.

PASCUAL JORDAN, Universität Hamburg, Hamburg, Germany.

MARK KAC, The Rockefeller University, New York, New York, U.S.A.

P. L. KAPITZA, U.S.S.R. Academy of Sciences, Moscow, U.S.S.R.

WILLIS E. LAMB JR., Physics Department, Yale University, New Haven, Connecticut, U.S.A.

G. LUDWIG, Institut für Theoretische Physik, Universität Marburg, Marburg, Fed. Rep. Germany.

JAGDISH MEHRA, The University of Texas at Austin, Austin, Texas, U.S.A.

RUDOLF E. PEIERLS, Department of Theoretical Physics, University of Oxford, Oxford, England.

ILYA PRIGOGINE, Faculté des sciences, Université Libre de Bruxelles, Brussels, Belgium, and Center for Statistical Mechanics, The University of Texas at Austin, Austin, Texas, U.S.A.

FRITZ ROHRLICH, Department of Physics, Syracuse University, Syracuse, New York, U.S.A.

LEON ROSENFELD, NORDITA, Copenhagen, Denmark.

ABDUS SALAM, Imperial College of Science and Technology, London, and International Centre for Theoretical Physics, Miramare, Trieste, Italy.

JULIAN SCHWINGER, Physics Department, University of California, Los Angeles, California, U.S.A.

DENNIS W. SCIAMA, Department of Astrophysics, University of Oxford, Oxford, England.

C. P. SNOW, 85 Eaton Terrace, London S.W.1, England.

V. L. TELEGDI, Enrico Fermi Institute, University of Chicago, Chicago, Illinois, U.S.A.

SIN-ITIRO TOMONAGA, Tokyo University of Education, Tokyo, Japan.

ANDRZEJ TRAUTMAN, Institute of Theoretical Physics, University of Warsaw, Warsaw, Poland.

GEORGE E. UHLENBECK, The Rockefeller University, New York, New York, U.S.A.

B. L. VAN DER WAERDEN, Universität Zürich, Zurich, Switzerland.

C. F. VON WEIZSÄCKER, Max-Planck-Institut zur Erforschung der Lebensbedingungen der wissenschaftlich-technischen Welt, Starnberg, Fed. Rep. Germany.

GREGOR WENTZEL, 77 Via Collina, Ascona, Switzerland.

JOHN ARCHIBALD WHEELER, Joseph Henry Laboratories, Princeton University, Princeton, New Jersey U.S.A.

EUGENE P. WIGNER, Joseph Henry Laboratories, Princeton University, Princeton, New Jersey, U.S.A.

CHEN NING YANG, Department of Physics, State University of New York, Stony Brook, Long Island, New York, U.S.A.

Symposium Lecturers, Chairmen, and Members of the Sponsoring Committee

Sitting: C. F. von Weizsäcker, F. Perrin, G. E. Uhlenbeck, R. Haag, E. P. Wigner, W. Heisenberg, P. A. M. Dirac, C. N. Yang, C. P. Snow, S. Chandrasekhar, H. B. G. Casimir, M. Eigen, D. W. Sciama, C. Møller, J. Schwinger.

Standing: J. M. Jauch, W. E. Lamb, Jr., F. Rohrlich, J. S. Bell, B. L. van der Waerden, R. E. Peierls, C. P. Enz, J. Mehra, [V. L. Telegdi obscured], A. Salam, J. A. Wheeler, M. Kac, E. G. D. Cohen, G. Randers, A. Trautman, J. Ehlers.

Opening Remarks at the Symposium

Abdus Salam: Mr Mayor, Mr Rector, Professor Casimir, ladies and gentlemen. It is a very great privilege for me to welcome you on behalf of the International Centre for Theoretical Physics. This gathering of physicists is the most distinguished we have had in the eight years of the Centre's history, and might well perhaps be the most distinguished ever assembled. We have been privileged to welcome some of you here before and we are looking forward to making more friendships among those who have come here for the first time.

I would like to introduce the Centre to those who have not been here before. The Centre was set up in 1964 by the International Atomic Energy Agency at Vienna, with the cooperation of the government of Italy, the University of Trieste and the City of Trieste. As you know, the International Atomic Energy Agency is a part of the United Nations family of agencies. A few years later, the International Atomic Energy Agency was joined by Unesco, and the Centre is presently a joint enterprise of these two United Nations agencies. Besides furthering international cooperation and the pursuit of theoretical physics, one of the many aims of the Centre, and this is what makes it a unique organization within the United Nations family, is to help senior physicists from developing countries to remain active in research while still teaching and working in their own home countries. We recognize that a good theoretical physicist, in order to remain at the peak of his powers, must come in contact with his fellows and peers for at least a fraction of the working year. The Centre aims to provide financial support and other facilities for this fraction of the physicist's working year in a stimulating environment of research, and annually some five hundred physicists from developing countries, out of a total of about eight hundred, make use of the facilities of the Centre.

In order to keep the Centre's research environment as invigorating and stimulating as possible it is imperative that the great physicists of the world remain associated with it and identify themselves with the Centre's aims and ideals. This is one of the reasons why it gives us the greatest happiness to welcome this very distinguished gathering here today. We believe that the Centre has the first faculty of a future United Nations University, that will serve as a prototype for further academic enterprises of a similar kind for furthering international co-operation and devoted to the task of building up science in the so-called 'third world'.

The symposium we are holding today had its origin in April 1970. There was a discussion between Jagdish Mehra and myself about the idea of holding a series of symposia on the physicist's concept of nature throughout centuries. The Centre, at that time, was planning for an occasion to celebrate the seventieth birthday of one of the greatest physicists of all time, P. A. M. Dirac, who is with us today, and to whom all of us wish many happy returns of his birthday. Jagdish Mehra thought it would be a wonderful idea to combine the two occasions, of holding the symposium envisaged by him and saluting Dirac at the same time. He took it from there, and we are very happy that his idea has become a reality. A most exciting programme has been drawn up by the most distinguished Sponsoring Committee for the week ahead.

As part of the United Nations Organization, the Centre cannot officially identify itself with any organizations devoted exclusively to the West or the East. This political circumstance, however, in no way diminishes the warmth of the welcome the Centre extends to the symposium by placing its facilities at your fullest disposal, although the University of Trieste is the official host of the symposium in Trieste.

It now gives me the greatest pleasure to request one of the two very distinguished co-chairmen of the Sponsoring Committee, a great physicist and a very old friend of the Centre, Professor Casimir, to take over from here.

Hendrik Casimir: Ladies and gentlemen. I have been looking forward to this symposium all along as you also have probably. When Jagdish Mehra first approached me in this connection in September 1970 and explained the idea, I at once felt that this symposium could become an extremely interesting and an extremely important occasion. I think that my enthusiasm was shared by practically everyone we approached about this project. Why is that? First of all, it is stereotyped to say that there has been an enormous growth of physics, but it just happens to be true that we are living in an age with an enormous expansion of empirical and theoretical knowledge; an age of enormous specialization in various branches of our subject. Therefore, a meeting where one tries not so much to go into the fine details of all the techniques of calculation and measurement, but where one can review the fundamental concepts and their background would appear to be extremely useful. Second, there may also be for some of us, who are now of an older generation, a certain nostalgic element in a gathering of this kind, because it makes us think back to a period when there were fewer physicists around and they knew each other, when it was still possible to read most of the really important papers on basic theoretical physics, and even to understand them, something which is becoming increasingly difficult today, in any case, for older people. I don't know how the youngsters manage. For these and other reasons, our Sponsoring Committee believes, and all of us expect, that a programme of this kind will be extremely interesting, instructive, and rewarding.

I believe I should here say a few words about certain criticisms that have been, and are being directed against a symposium of this kind. They are of different natures at various levels. On the one hand, there is a sort of philosophical level that believes that it is meaningless to discuss physics in the abstract, and not within the compass of some sort of philosophy of mankind, of society, of political structure. Such ideas evidently are not new. They have occurred even long ago in the disputes between the Church and Galileo. We may remember the textbook of physics called *Die Deutsche Physik*, German Physics, written during the Nazi period in Germany, which happened to be quite a good textbook, if you took only those chapters which were not particularly German. We may recall discussions on genetics in Soviet Russia. I am sure we still believe, as physicists, that we can discuss the basic issues and the abstract ideas of our subject without going into political questions. So I would like to discard that criticism. That does not mean that all of us are not in some way or the other, formed and shaped in our ideas and our beliefs and our ways of expressing ourselves by the background we come from. Of course we are. A historian might find it interesting later on, from the proceedings of a symposium like this one, to try to distinguish the various backgrounds of the participants, but it does not have to do with the subject matter which we are going to treat.

A second criticism is more down to earth. It points out that in our world physics has had an enormous impact on the whole society. Whereas, perhaps, older technologies were empirical and craftsmanlike, the new technologies have increasingly been based on the basic science that preceded them, often preceded them many years before. This development has been so enormous and has changed the pattern of our lives to a great extent. It holds such great promise, on the one hand, and such great dangers on the other, that the questions of this impact are more urgent and more important to some people than the abstract problems of the structure of physics. One can have sympathy with this point of view, but in my opinion it does not mean that any and every gathering of physicists should deal with that particular matter. If people find that it is more interesting to discuss these social implications – well, by all means, one can organize meetings and discussions on these questions. But I feel that it is a part of liberty and democracy that physicists who want to get together to discuss certain aspects of their own subject should be free to do so, and it should not be necessary at all always to include these other questions which, needless to say, are very important questions of our age.

I should like to say that, after discussions of our Sponsoring Committee, we arrived at the conclusion that although we don't feel that we should modify in any way the programme of the meeting as it stands, unless a majority of those present here would desire such a change of programme. I would like to stress the point that although we believe that we have every right to go on with the programme as it stands, it doesn't mean that many of us would not be willing to discuss other questions, if it is requested and if it can be done in an orderly manner.

I should like, at this moment, to ask Jagdish Mehra to step forward and to say a few words about the programme, and get our work started.

Jagdish Mehra: Mr Mayor, Mr Rector, Professor Casimir, Professor Salam, honoured guests, fellow participants, ladies and gentlemen:

'Nature is written in mathematical language.' This phrase first occurred in Galileo's *Saggiatore* in 1623 and expressed a singularly revolutionary idea. 'By a stroke of the pen, Galileo had abolished the *Natura* of the ancients with its substances, forms and qualities. Nature had become the sum total of quantitative phenomena and the very purpose of scientific research was henceforth completely changed.'*

* *History of Science*, Vol. II, Edited by R. Taton, Basic Books, New York, 1963.

   Galileo himself was not able to put his maxim entirely into practice, but from Galileo and Newton to James Clerk Maxwell, to Planck, Einstein, Rutherford, and Bohr, to Werner Heisenberg, Paul Dirac, Eugene Wigner and their successors, physical science has created a conception of Nature which would belong to the permanent intellectual heritage of man. And what a heritage it already is.

   This symposium has been organized to discuss the current horizons of physical theory in the context of its historical development in the twentieth century. It is a unique development. It has affected our intellectual premises and modes of thought. If the theory of relativity completed the logical development of classical physics, the quantum theory made a break with the past. Paul Dirac initiated the possibility of their unification. But the current frustration about the lack of theoretical structures that would unite them in lasting harmony might well bear out Wolfgang Pauli's prophecy that 'no man shall join what God hath put asunder.'

   The approach to truth is at best asymptotic. John Maynard Keynes, in an address delivered to the Royal Society Club in 1942 [and read again by Geoffrey Keynes at the Newton Tercentenary Celebrations in July 1946] said that 'Newton was not the first of the age of reason. He was the last of the magicians, the last of the Babylonians and Sumerians, the last great mind which looked out on the visible and intellectual world with the same eyes as those who began to build our intellectual inheritance rather less than 10000 years ago.' It is quite possible that a future age will look back on Einstein, Heisenberg, and Dirac as magicians. But in a certain sense we recognize the quality of magic in their work even today. For after all, it is in the apprehension of a sense of mystery in the search for the understanding of Nature, in faulting attempts at its mathematical and empirical description, that science grows and is never final. As the alchemist said in Anatole France's novel *La Rôtisserie de la Reine Pédoque*, 'If the repast I am about to offer you is not well prepared, it is not the fault of my cook, but of chemistry which is in its state of infancy.' The characteristic of science, if not infancy, is perpetual adolescence. The approach to truth is asymptotic.

   Still, I believe that we have cause for celebration. On the occasion of the seventieth birthdays of men like Dirac, Heisenberg and Wigner, we can look back at what has been served before us. They are among the principal authors of the conceptions to which our symposium is devoted.

   In seeking to understand the achievements of scientists of their calibre, one might well wonder about the play of chance or the choice of destiny that brought fulfilment to their undertakings. A historian does think about the conditions in which exceptional growth of science took place in the past, and the choices and encounters that led great minds to their pursuits and achievements.

   It is quite possible that Heisenberg, who was already enamoured of Plato while in school, could have pursued philosophy and the classics. He may have pursued pure mathematics, if the mathematician Lindemann at Munich, whom Heisenberg first visited for advice about his studies at the university, had been less austere and just a bit more friendly. But good fortune, his no less than of physics, led him to visit Sommerfeld. Sommerfeld's kindly guidance and careful advice – such as his remark that 'when kings go abuilding, wagoners have to do the work' – put a harness on Heisenberg and liberated his genius.

One finds it quite remarkable that Niels Bohr, who had been invited by Hilbert and the Wolfskehl committee to lecture on atomic theory in Göttingen in June 1921, could not do so until the following year – just when Heisenberg's preparation was ripe enough to encounter him. Sommerfeld may not have encouraged him to go along with him to Göttingen the previous year, and one can only wonder who else would have accompanied Bohr for those Socratic dialogues on the Hainberg and questioned him with such tenacity as did Heisenberg a year later. The pilgrimage which began in 1922 at the Hainberg in Göttingen, has continued for fifty years, and several generations have followed Heisenberg in it and shall continue to do so.

Dirac could very well have pursued a career in electrical engineering, a field which he pursued at Bristol University much to the disappointment of the mathematicians there. On graduation in 1921 he looked for a job in engineering. He couldn't find one because there was a depression on at that time. He went back to Bristol to study mathematics. Dirac and a Miss Dent were the only two students in the honours course in mathematics. Miss Dent was quite determined to pursue applied mathematics, and in order that the mathematical faculty should not have to give two sets of courses, Dirac also decided on applied mathematics. This was his way back to science, and he never looked for a job in engineering again. Thanks to Miss Dent and the depression, the choice had been made for him. And just eleven years later he succeeded Sir Joseph Larmor as Lucasian Professor of Mathematics in Cambridge – a chair whose first occupant was Isaac Barrow, Newton's teacher, and the second, Newton himself.

Wigner's boyhood ambition was to do pure physics, but it did not show promise as a career, because there were only two chairs of physics in all of Hungary in those days, and his father was not sure whether he should pin his future to occupying one of them. The family business, a prosperous one, was a tanning factory, and Wigner appropriately studied chemical engineering in Budapest in order to be useful to the enterprise. Fortunately, as a young man he was not an expert on employer–employee relations in the factory, and the loss to leather-tanning and management was surely minimal compared to the gain to science when he left Budapest and business for Berlin and physics.

Since 1925, when Wigner wrote his thesis on 'Bildung und Zerfall von Molekülen' under Polanyi in Berlin, he has known two ways of doing physics: either by pronouncing the first word on a subject or the last, and at times both.

We can only be happy that these great masters elected to do physics in their careers, a science which owes many of its riches to their conceptions. We are delighted that they are attending this symposium.

It has been my good fortune to have been associated with the organization of this symposium. I have enjoyed the utmost encouragement, confidence, and help from distinguished friends and supporters, especially Professors Casimir, Wigner, Salam, Randers, and the other eminent members of the Sponsoring Committee. We have a programme that might well test your stamina. Welcome to our symposium, and I wish you a pleasant and memorable stay at the symposium and in the beautiful city of Trieste.

P. A. M. Dirac

"The physicist has to replace prejudices by something more precise,
leading to some entirely new conception of nature."

1

Development of the Physicist's Conception of Nature

P. A. M. Dirac

When one looks back over the development of physics, one sees that it can be pictured as a rather steady development with many small steps and superposed on that a number of big jumps. Of course it is these big jumps which are the most interesting feature of this development. The background of steady development is largely logical, people are working out the ideas which follow from the previous set-up according to standard methods. But then, when we have a big jump, it means that something entirely new has to be introduced.

These big jumps usually consist in overcoming a prejudice. We have had a prejudice from time immemorial; something which we have accepted without question, as it seems so obvious. And then a physicist finds that he has to question it, he has to replace this prejudice by something more precise, and leading to some entirely new conception of nature.

One of the best examples of these jumps is provided by special relativity, which shows that we have to get rid of the conception that there is an absolute meaning to simultaneity. It seemed obvious to physicists previously that if two events are simultaneous it has a precise meaning, an absolute sense. But then when people came to do accurate experiments involving the propagation of light, taking into account the finite speed of propagation of light, they found that they had to give up that idea. It was Einstein who really grasped the need for getting rid of this absolute concept of simultaneity, and replacing it by a new picture of the world in which time appears as a fourth dimension and space and time have to be considered together and are subject to transformations in which the direction of the time axis is liable to change. This was a very big step forward and it led to the need for a reformulation of pretty well the whole of physics.

Previously, we had got used to the concept of vectors and tensors in three-dimensional space. A tensor is a definite physical concept. It can be described mathematically only by specifying its components with reference to some coordinate system. It has the property that its components transform linearly when we make a transformation of the coordinate system. But we can think of a tensor as existing quite independently of any system of coordinates. It is just something which is imbedded in space in some way, and the coordinates are needed for its mathematical description.

With the arrival of special relativity, all our vectors and tensors in three-dimensional space had to be changed to corresponding quantities in the four dimensions of space

J. Mehra (ed.), The Physicist's Conception of Nature, 1-14. All Rights Reserved
Copyright © 1973 by D. Reidel Publishing Company, Dordrecht-Holland

and time. That meant that they had to have more components. A vector, which physicists thought of as imbedded in three-dimensional space and requiring three components to specify it, became now something imbedded in four-dimensional space, requiring four components. And all the concepts of physics had to be changed in that way.

If you take the momentum concept in three-dimensional space, that had to be extended by a fourth component and the fourth component is just the energy. Previously we had the law of conservation of momentum and we had the law of conservation of energy as an independent law. Well, these laws get unified. We have a unified concept of momentum and energy and there is just one unified conservation law applying to the whole concept.

Then we may consider more complicated quantities. The stress tensor, for instance, which we have in three dimensions, has to be expanded by bringing in a number of further components when we go to special relativity, and these components are of the nature of rate of flow of momentum and energy.

A bit of a problem arises in connection with angular momentum. Angular momentum in ordinary three-dimensional space is a vector, but it should be treated as an axial vector, which is the vector product of two ordinary vectors. Such a vector, when we move it to four dimensions, forms a part of a tensor of the second rank, which is anti-symmetrical between its two suffixes. That requires the addition of three new components and the question arises 'What are these three extra components?' Well, angular momentum itself is quite an important concept, but the extra components which have to be brought in for the four-dimensional description turn out to be not important, because they are of the nature of a moment about some axis in space-time, which is of a transient character. We have the law of conservation of angular momentum playing an important role in non-relativistic theory. The extra three components are also conserved but because of their transient character, with nothing permanent about them, they are not important. The only important applications of angular momentum are in a non-relativistic sense, where we have some special time axis which is significant.

Well, that is a discussion of the changes which were introduced by special relativity, all of them consisting of overcoming the prejudice of absolute time. There is a further change when we pass over to general relativity, which is rather similar. Here the prejudice which we have to overcome is that Euclidean space applies to the physical world. The axioms of Euclid, which were formulated many centuries ago, of course are good axioms if you just assume them and proceed to derive their consequences. These consequences form Euclidean geometry. But it is a question whether these axioms really apply to distances as measured by the physicist. People have always assumed that they do because observations show that at any rate they apply with very great accuracy and people have been prejudiced in favour of perfect accuracy.

But it turns out that that is a prejudice which we have to give up. The distances measured by the physicist are not those which conform to Euclidean geometry. How do we know this? The differences are extremely small. Too small to show up with direct observation. Maybe at some future time when people can make observations much more accurately they will be able to show the differences. We get the need for modifying Euclidean geometry in an indirect way.

This departure from Euclidean geometry involves supposing that space should be pictured as a curved space in a space of a higher number of dimensions. Here again, it was Einstein who led the way, and he was influenced by the need to fit in gravitation with relativity. Newton supposed that we had the inverse square law of force governing all masses, but it was criticized on the grounds that it involved action at a distance. Philosophers said that a body cannot act where it is not. Therefore, there had to be something wrong with Newton's law. That is not really a valid criticism. One can bring in the concept of fields.

We have a field in physics when we have some physical quantity located at all points in space and varying usually continuously from one point to a neighbouring point. Now it was found that Newton's law of gravitation could be formulated in an alternative manner with the help of a field, involving the Newtonian potential, and with this field formulation, we do not need action at a distance. We could have a particle moving under the influence only of the field in its neighbourhood.

This field formulation satisfies philosophers, but to physicists, the field formulation and the action at a distance formulation should be considered as entirely equivalent, because it needs only a mathematical transformation to pass from one to the other, and the two formulations always give the same results when applied to examples. However, there are other respects in which the field formulation turns out to be superior.

Let us go over to consider electrodynamics. The original laws, such as Coulomb's law, involved action at a distance. They were not accurate and had to be reformulated. The improved laws were given by Maxwell and involved a field formulation, which led to the possibility of electromagnetic waves. Here you see you have a development possible with fields which you could not have had with the action at a distance formulation, and the field formulation proves to be superior.

Now the field formulation turns out to be necessary for describing gravitation in accordance with Einstein's general relativity. New equations were set up involving field quantities, with the one potential of the Newtonian theory replaced by ten potentials. Well, I don't mean to go into details of that.

We have thus had quite a big development in our ideas of space. First of all, the passage from three dimensions to four dimensions. Then the idea of introducing curvature into space. The resulting space is a Riemannian space. People have sometimes wondered whether this process of the physicist modifying his ideas of space should stop at that stage. There is the electromagnetic field, which is in a good many ways similar to the gravitational field. They both involve long-range forces, and in that way they are to be distinguished from the other fields of physics which come into atomic theory.

People have imagined that there should be some unification between them, and since Einstein has shown that gravitational fields can be explained in terms of geometry, people have suspected that the electromagnetic field should also be explicable in terms of geometry. Some more general geometry than the Riemannian geometry which forms the basis for Einstein's gravitational theory would be needed. A great deal of work has been done on those lines but the results are not satisfactory. I don't want to refer to unsuccessful developments of the concepts of physics, but keep to the successful ones. Under those conditions one should keep Einstein's space, namely the four-dimensional Riemannian space, as being retained up to the present time and still forming the basic space of physics, in which all physical processes have to be considered as imbedded.

That forms one development of the concepts of physics. The other main development which has taken place in recent times is the quantum theory, or the theory of atomic structure.

Now if we are discussing this question I think the title of my lecture is a little unfortunate. The title is 'The Development of the Physicist's Conception of Nature'. That might imply that all physicists have the same idea about how their concepts have developed. And that is not really true for the quantum theory. There the new concepts are not so easily explained as in the case of relativity. They are more recondite and the question of which are the important ones, which are the fundamental ones can very well vary from one physicist to another. I think a better title for this talk would be 'The Development of *a* Physicist's Conception of Nature'. I must give my own point of view, but I shall say at the outset that I don't claim that this is the only reasonable point of view to take. There are other points of view which are permissible and which could be defended. I prefer this one because I have found it to be the most successful one in my own case.

The quantum theory has developed through several big steps. The first one, of course, was Planck's introduction of finite quanta of energy in the description of the electromagnetic field; something which was very hard for physicists to accept and which had to be forced on them by experimental evidence. Then there was a whole mass of data provided by spectroscopy which was very puzzling to begin with. No order was made out of it until we had Ritz's combination rule of spectra, showing that the frequency of each spectral line, could be expressed as the difference of two terms. That was just a sort of a dodge which did not really provide a theory until Bohr set up his model of the atom. This model of Bohr's was perhaps the greatest single step of all in the development of atomic theory, because it showed that one could apply the laws of classical mechanics inside the atom, to the electrons moving in the atom, provided one put on certain extra conditions, and made certain approximations. The approximations were to neglect radiation damping, and the extra conditions one had to impose were the quantum conditions which fixed Bohr's stationary states. These were really very big developments in the physicist's conception of nature, and they were perhaps too drastic to be described merely as overcoming prejudices.

After the introduction of Bohr's stationary states, the further development of physics showed the need for having quantities connected with two states. Ritz's combination law, which resulted in Bohr's frequency condition, showed that the frequencies of the spectral lines are each connected with two states. Then Einstein introduced coefficients for emission and absorption, each connected with two states. So this idea of quantities connected with two states got developed. There was the important Kramers–Heisenberg dispersion formula, built up entirely in terms of quantities connected with two states.

That led Heisenberg to his really masterful step forward, resulting in the new quantum mechanics. His idea was to build up a theory entirely in terms of quantities referring to two states. Quantities referring to two states can be written out as a matrix array, so that one had to consider a whole matrix of numbers. Now the really important contribution of Heisenberg consisted not merely in introducing all these matrix elements, but in realizing that the whole complex of the matrix was an important physical concept which corresponded to a dynamical variable.

That led to the result that dynamical variables are subject to an algebra like matrix algebra in which one can carry out the processes of addition and multiplication but multiplication is in general non-commutative. $a \times b$ is usually not equal to $b \times a$. One then had to deal with the concept of dynamic variables obeying this kind of algebra.

Right from the beginning, when I first saw Heisenberg's original paper bringing out these ideas, it seemed to me that the most important idea there was the fact that we have to deal with dynamical variables subject to non-commutative algebra. I introduced a new name for such dynamical variables. I called them q numbers, while the ordinary numbers in mathematics, when one wanted to distinguish them, I called c numbers. The q numbers became a new concept which physicists had to get used to. They replaced the dynamical variables which they had been working with previously. They were subject to a different algebra. To begin with, the q numbers seemed to me to be something very mysterious. I made several assumptions about them, just for the purpose of being able to develop a theory and apply it to examples, and these assumptions were often wrong. But still the notion of the q number did get developed.

To begin with the mathematical nature of a q number was something completely unknown to me. But it was clear from the connection with Heisenberg's formalism that q numbers could sometimes be replaced by matrices. Well it turned out later that q numbers could always be replaced by matrices. They can be represented by matrices, as one says nowadays. But the q numbers can be represented by matrices in various different ways. The matrix should be looked upon as a set of coordinates for the q number in just the same way as the components of a tensor form the coordinates of the tensor. One can think of a tensor as existing independently of any system of coordinates and in the same way one can think of the q number as existing independent of any system of matrices.

When it became clear that the q numbers could always be represented by matrices, of course their mathematical nature was no longer mysterious. One could work out all that one wanted to about them, and one could correct my early mistakes. All that was left of the q number was a concept of something which has a physical meaning, independent of any matrix representation, and thus to be counted on the same footing as a tensor.

The early work with q numbers consisted just of algebraic deductions using the suitable algebra with non-commutative multiplication, and of course the interpretation of the results of these deductions was very obscure. People made a guess at the interpretation in simple examples. They found that a certain interpretation would give the correct answer and the interpretation was gradually generalized and built up in that way.

The general interpretation of quantum mechanics was very much helped by another development, which is due mainly to Schrödinger, working from the ideas of de Broglie. This involved bringing into physics a new concept, the concept of a state in atomic theory.

We already had the idea of states in classical mechanics. When we think about a given classical system, there will be various possible states of motion coming from the various solutions of the equations in motion. But the peculiar thing is that a quantum state does not just correspond to a classical state. It corresponds to a whole set of classical states, what one may call a family of classical states, which are related to one another in a special mathematical way, which was discovered by Hamilton a hundred years before. Hamilton discovered this special relationship between classical states just by considerations of mathematical beauty, trying to get a powerful formulation of the equations. And this work of Hamilton's turns out to be just what is needed as a preparation for our understanding of quantum states. Each quantum state corresponds to one of Hamilton's families of states.

Then the surprising thing turns up that the quantum states have superposition relations between them. That means they are to be pictured as a kind of vector. They are something which can be added to quantities of the same nature to produce sums of the same nature again. Strictly one should say that the quantum state corresponds not to one of these vectors but to the direction of a vector, but that is a rather unessential point. These vectors which correspond to the physical states in quantum theory are usually in a space of an infinite number of dimensions and when one brings in suitable conditions of convergence they become Hilbert vectors.

We have here some new vectors which become of importance in physics. When there is a new concept becoming important, I like to introduce a new name for it, like the name q numbers. I introduced the name of a ket vector. I won't go into the reason for that notation. We have then these ket vectors corresponding to physical states and there is a relationship between the q numbers and the ket vectors. Any q number can be multiplied into a ket vector to give another ket vector, or if you like you can say the q number is a sort of linear operator which can operate on the ket vectors.

The physicists had to get used to these new ideas of q numbers and ket vectors.

The q numbers can be represented by matrices. When we do that, we get a corresponding representation for the ket vector, that is, we get a set of coordinates for each ket vector. These coordinates are what are usually called a wave function. The reason for that is that in the early examples, as first formulated by Schrödinger, these coordinates of the ket vector did just form a wave function in three-dimensional space, when the theory is applied to a single particle and we use a suitable representation.

People now use the terminology 'wave function' quite generally for the coordinates of a ket vector in any system of coordinates, even when the result is not connected with waves at all.

That is the way in which waves come into quantum theory. The waves are a way of representing the physical states. Now some physicists are inclined to think that the waves are perhaps the most fundamental thing in atomic theory and that the particles are less important. I don't accept that view myself. One can go a long way just describing things in terms of their waves, but I don't think it is convenient to try to go the whole way like that. And I like to have this picture which I have been presenting to you, that we have particles whose variables consist of q numbers, and the particles can be in various states and that the states are described by wave functions.

With these concepts, people were able to get a powerful physical interpretation for quantum mechanics. They were able to get general rules for calculating probabilities of certain dynamical variables having certain values for a particular state.

Now the important thing I have stated there is that we calculate probabilities, we don't calculate that something will happen from given initial conditions. It means that the interpretation is a statistical interpretation. We do not have the determinism of classical mechanics. That is another of the concepts which physicists have had to get used to. They have had to get away from the prejudice in favour of determinism and it was very hard to get away from this prejudice. Some people are hoping to reintroduce determinism in some way, perhaps by means of hidden variables or something like that, but it just doesn't work according to the accepted ideas. I might add that personally I still have this prejudice against indeterminacy in basic physics. I have to accept it because we cannot do anything better at the present time. It may be that in some future development we shall be able to return to determinism, but only at the expense of giving up something else, some other prejudice which we hold to very strongly at the present time.

However there is not much point in speculating about what the future will hold. I just want to say that if you feel uncomfortable about having indeterminacy in the basic laws of physics, you are not alone in that feeling. Very many people do. I do. Schrödinger and Einstein have been very much opposed to it all along. But one has to accept that it is the best that one can do in our present state of knowledge.

With the introduction of these ket vectors or wave functions, one big development that has followed has been in connection with the application of the theory to a number of similar particles. The wave function will involve the variables of all these particles and it will be possible to consider that the wave function is symmetrical between the particles. If the world started off in a state which is symmetrical between these particles, it will always remain symmetrical, and we get a law of nature that only symmetrical wave functions occur in nature. That is a new kind of law, quite independent of anything that one could think of in classical theory, and not something which one can look upon as overcoming a prejudice.

An alternative possibility would be to have only anti-symmetrical wave functions occurring in nature.

It does seem that nature is really constructed on those lines. There are some particles for which only symmetrical wave functions occur, which are called bosons, and other particles for which only anti-symmetrical wave functions occur, which are called fermions.

I would like to return now to the concept of q numbers and to say a little about how this concept can be developed. The q numbers were introduced in the first place as playing a role analogous to the variables of classical dynamics. We need commutation relations between them. If we are told that ab is not equal to ba, we need some assumption which will tell us what the difference $ab - ba$ is. It wasn't hard to guess what this quantity is, for those dynamical systems that have a classical analogue, namely the Poisson bracket. With this connection the formalism of quantum mechanics becomes a generalization of the classical formalism. If one has a particular system in classical mechanics, involving particular particles interacting with particular laws of force, one can talk about the corresponding system in quantum mechanics.

But then it became possible to develop the idea of q numbers to apply to any linear operators which can operate on the ket vectors. In that way one could introduce q numbers which don't have any analogue in classical mechanics, and one added very much to the power of the formalism of quantum mechanics.

I might mention an example. If we take a wave function applying to several similar particles, it will involve the variables $q_1, q_2, ..., q_n$, for n particles, where the single variable q denotes collectively all the variables needed for the description of one particle. Then we may consider a permutation operator which makes a particular permutation of these variables. For example it might just be interchanging q_1 and q_2. Now each of these permutation operators should be considered as a q number. It is subject to equations of motion just like the equations of motion of other dynamical variables in Heisenberg's theory. But such a q number is very different from anything in classical theory. It has no classical analogue.

Now if we are going to apply this theory to wave functions which are symmetrical, then any permutation operator applied to such a wave function will be equal to 1. That means we have a q number always equal to 1, and that is not very interesting. Similarly, if we apply it to an antisymmetrical wave function, we get an operator which is equal to ± 1 according to whether the permutation is even or odd, again, not very interesting. But we can get interesting results by considering a more general kind of permutation operator. Suppose we have particles with spins. We might consider a permutation operator which involves only the positions of the particles and doesn't permute the spins. Now such a permutation operator will not be ± 1, it will be something more general. We can build up a theory of these permutation operators, considering them as dynamical variables, and get interesting results.

The q numbers can be developed in that way to bring in dynamical variables which have no classical analogue at all.

Now another development I would like to refer to is that one can consider operators which change the number of particles. Maybe increase the number of particles by 1 or reduce the number by 1. Such operators are emission or absorption operators.

They can be brought into the theory. They are just like the q numbers which we have had before. They are again subject to equations of motion. Bringing these into the theory allows one to set up a formalism in which the number of particles can change. In classical mechanics we do not have that possibility, but there is no difficulty at all about bringing it into quantum theory.

What I've been saying shows you the development of quantum mechanics and you see that it is really a very powerful theory which works very well. It just has one serious defect, namely, it is not relativistic. We have Heisenberg's original equations of motion running all the way through, equations involving d/dt, referring to one particular time variable, and this is against relativity. Now I have talked about two developments of physics, relativity and quantum mechanics. How are we to join them together? Physics must be unified. We must have a single theory conforming to both the principles of relativity and the principles of quantum theory. How can we get such a theory?

Let us first take a simple example, that of just a single particle. We can set up a wave function for it, a wave function involving just the variables x, y, z, describing the position of the particle, and if we consider it varying with the time we have t coming in also. We have then a wave function involving these four variables and that is something which we can handle relativistically, supposing these variables to describe a point in space-time. We can try to set up a relativistic wave equation for this wave function, conforming to the general principles of quantum mechanics. Well it turns out that it is possible to do that, and the surprising thing is (it was a great surprise to me when it first turned up!) that the simplest solution of this problem occurs not for a particle without spin, but for a particle with a spin of half a quantum. The spin of half a quantum seems to occupy a special role in nature in that it is possible to set up a relativistic quantum theory for such a particle agreeing with all the requirements of special relativity, as well as the laws of quantum mechanics.

We have to bring in a spin of half a quantum. Now that involves a new concept which physicists had to get used to, a concept of something which is called a spinor, which is a generalization of the concept of a tensor. A tensor you think of as something imbedded in space or in space-time, and we can imagine rotation operators applied to it. The spinor is similarly imbedded and we can apply rotation operators to it. But a spinor is such that if we turn it round once about an axis, we finish up with something which is minus what we started with, instead of being just what we started with, as in the case of a tensor. Mathematically it turns out that such quantities are quite possible, quantities imbedded in space which change sign whenever we apply to them one rotation about an axis. It is quantities of this kind which have to be introduced in order to describe the half quantum of spin.

Now the half quantum of spin which we get appearing naturally in the relativistic wave equation turns out to be very useful, because many of the elementary particles in nature do have just that spin; particularly the electron and the proton. So we have here a theory which is specially appropriate for the electron.

But there was a difficulty occurring right at the outset, namely, that when we apply the usual rules of interpretation for the wave function, we find that it allows states of negative energy as well as states of positive energy. That was quite a stumbling block to begin with, but it turned out that one could get over that difficulty in a very neat way at the expense of changing one's concept of the vacuum.

The physicist had always previously thought of the vacuum as a region where there is nothing at all, but that was a prejudice which we have to overcome. A better definition of a vacuum would be the state of lowest energy. Now if there are possibilities of electrons having negative energies, we should want to have as many of these electrons as possible in order to get the lowest energy. Electrons obey the Fermi statistic corresponding to anti-symmetrical wave functions. They satisfy Pauli's exclusion principle, which means that not more than one electron can be in any state. We thus get the state of the lowest energy for a region of space when all the negative energy electron states are occupied with one electron in each. That is the most that we can have of negative energy electrons.

That is a reasonable picture to set up for the vacuum when one gets over the prejudice that the vacuum contains nothing at all, and it did turn out to be satisfactory. It led to the possibility of our constructing states which depart from the vacuum in two ways, either by having electrons in positive energy states or by having holes among the negative energy states. And the holes among the negative energy states appeared as particles with a positive energy and charge, which were later interpreted as positrons.

With the new picture of the vacuum, we had the possibility of matter being created from radiant energy. If an electron jumps up from a negative energy to a positive energy state, then we have an electron, an ordinary electron, and a positron appearing, and the energy needed for this jump has been transformed into a material form.

There is also the possibility of polarization of the vacuum. We can disturb the vacuum distribution of negative-energy electrons by an electric field or a magnetic field, and we then have a sort of polarization of the vacuum occurring. We have all these developments occurring from the relativistic wave equation for the electron.

Now I said that just fitting in relativity with quantum theory led us to the spin of half a quantum. Now many particles are known with spins which are not half a quantum. Of special interest, of course, is the photon, with a spin one. Whatever are we to do with these other particles? There is a serious difficulty.

We can set up a quantum theory for them just by considering states referring to one particular time axis in space-time and considering how the states change when we change the direction of the time axis. But we have the difficulty that these changes are non-local. For an assembly of the particles we can set up field quantities which do change in a local way, but when we interpret them in terms of probabilities of particles, we get again something which is non-local.

Well, I feel that it is rather against the spirit of relativity to have quantities which transform in a non-local way. It means that we have a certain quantity in space-time referred to one direction of the time axis, and when we change the direction of the time axis the new quantity doesn't refer to the conditions in the neighbourhood of the point where the original quantity was, but refers to physical conditions some way away. That is rather like an action at a distance theory. It is against the spirit of relativity, but it is the best we can do at the present time.

This difficulty occurs also when we have several particles interacting with each other. The only theory which we can formulate at the present is a non-local one, and of course one is not satisfied with such a theory. I think one ought to say that the problem of reconciling quantum theory and relativity is not solved. The concepts which physicists are using at the present time are not adequate. They become very artificial when one just applies them in a formal way.

The difficulty becomes most apparent when one takes into account the interaction between, let us say, electrons and the electromagnetic field. If one supposes this interaction to arise from a point model of the electron, one gets infinities occurring in the equation. These infinities, of course, are not to be tolerated. One has to remove them is some way, and the natural way to remove them is to say that the electron is not a point charge, but the charge is distributed over a certain region.

Many physicists assume point charges and remove the difficulty of the infinities just by means of working rules. They say, let us depart from ordinary mathematics. Let us neglect infinities occurring in our equations when we don't want them. This formalism does sometimes lead to results is good agreement with observation, and many physicists are happy with this state of affairs. But I am most unhappy about it. I feel that we do not have definite physical concepts at all if we just apply working mathematical rules; that's not what the physicist should be satisfied with.

I might mention one idea which is very much talked about these days, the concept of renormalization. Already this concept occurred in classical theory with Lorentz' model of the electron. The electron, according to Lorentz, has a field around it, essentially the Coulomb field with some modification if the electron is moving. This field would have inertia, which would be added on to the mass of the electron, so that the mass of the electron should be considered as partly arising from the mass associated with the Coulomb field around it. It could be that the whole of the mass arises in this way.

Again we have to get away from the point model of the electron, because in that case the mass of the Coulomb field around it would be infinitely great.

We have there the original mass getting modified by the field which the particle produces. That effect persists in quantum mechanics, and results in the original mass of the electron with which we start in our equation becoming different from the observed mass. The observed mass is called the renormalized mass. The trouble there is that the renormalization effect is infinitely great, if we work with a point electron. There is similarly a charge renormalization coming from the fact that whenever we have a charge, it produces a polarization of the vacuum around it, which to some extent neutralizes the charge. Again, the effect is infinite for a point electron.

That shows the development of quantum theory as far as it has gone. It shows that our present position is far from satisfactory, because of the failure to fit quantum theory with relativity.

Many of the developments which have been occurring in physics in recent times have been concerned with the introduction of new particles. There again we see that the physicist had to overcome a prejudice. Up until about 1930 physicists thought there were only two fundamental particles, the electron and the proton. The reason for that was that there are two kinds of electricity, positive and negative, and a particle is needed for each of them, and that is all that one needs. There was very strong reluctance to postulate new particles up until about that time.

That led me astray when I first worked out the ideas about the holes in the vacuum distribution of electrons. I felt that the holes must represent protons because they certainly had a positive charge. I thought right at the beginning I would expect them to be symmetrical with the electrons and to have the same mass as the electrons. But it was rather inconceivable to me that there should be a new particle with a positive charge and the mass of the electron. I reasoned that if such particles did exist, the experimentalists would certainly have seen them.

Why did the experimentalists not see them? Because they were prejudiced against them. The experimentalists had been doing lots of experiments where particles were moving along curved tracks in a magnetic field. The curvature indicates which way the particle is moving along the track if one knows the sign of its charge. The experimentalists regularly saw electrons coming out from a source and having the appropriate curvature in the tracks. But they sometimes saw the opposite curvature, and interpreted the tracks as electrons which happened to be moving into the source, instead of the positively charged particles coming out. That was the general feeling. People were so prejudiced against new particles that they never examined the statistics of these particles entering the source to see that there were really too many of them.

Since about 1930, the climate of opinion about new particles has changed completely. Many new particles have been discovered and experimenters and theoretical people are both very willing to postulate new particles on the flimsiest evidence.

There was also a prejudice against fractions of the electronic charge occurring in nature. Right from the time of Millikan's very accurate observations of the charge of the electron, physicists assumed that all the charges in nature would be integral multiples of the electronic charge. Now in recent times theories have been set up which would involve some new particles called quarks, which have a charge of a fraction of the electronic charge. The usual kind of quark would contain a charge 2/3 of the electronic charge.

I remember meeting Ehrenhaft, who did a lot of work measuring charges on small particles, and he kept on claiming that he had discovered what he called the sub-electron, something with a charge less than the charge of the electron. Everybody thought that he was a crank, nobody took him seriously, and he was very much complaining about that. But it occurred to me recently that now that the climate of opinion has changed with regard to fractions of the electronic charge, it would be worthwhile looking at Ehrenhaft's results again. Ehrenhaft's early work was completely inaccurate, giving all sorts of values for the charge, but later on it did become rather more definite. But his work was always very inaccurate and careless compared to Millikan's work.

I looked up his last paper dealing with the electronic charge, published in 1941 in

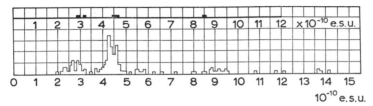

Fig. 1.

Philosophy of Science, Vol. 8, p. 403. There he has a diagram showing the results of a lot of experiments, as shown in Fig. 1. Horizontally, there is plotted the charge which he observed, and vertically the number of particles. He did experiments similar to Millikan's oil drop experiments, but using solid spheres instead of oil drops. The experiments which are referred to here were all done with small spheres of red selenium. You see that there are quite definitely two peaks on that curve. There is a main peak at somewhat less than the correct value, 4.7, and there is quite definitely a second peak which is roughly 2/3 of the main peak. I found that result rather interesting but could it mean that Ehrenhaft really had quarks? The disturbing feature is that he has so many of these sub-electrons. I find that very disconcerting, and also Ehrenhaft was not a good experimenter. You can see by the broadness of the peaks that his results were certainly not accurate.

Furthermore you will see some black squares on the top there. Those black squares represent results where he gives the size of the particles as well as the charge on them. For those results there is quite a definite correlation between the size and the charge. The smaller particles have the smaller charge. I thought that was very much against Ehrenhaft and that he must have had some systematic error which resulted in the smaller particles appearing to have smaller charges. But I am afraid that we shall never know the reason for those two peaks, because all the details of the experiments leading to them were lost when Hitler's army marched into Vienna.

Some six weeks ago, there was a summer school in Varenna dealing with the history of physics, and there was present Dr Holton, who talked about the old measurements determining the charge of the electron. He talked about Millikan's very precise work and also about the work of Ehrenhaft, very unprecise and very much discredited at the time. Dr Holton told me about a remark occurring in one of Millikan's papers which I find very interesting, and I would like to quote it. The historian of science does sometimes dig up something of special interest which is buried in the literature. This remark of Millikan's occurs in a paper which he published in 1910, describing experiments which led to the precise determination of the charge on the electron. I quote Millikan: 'I have discarded one uncertain and unduplicated observation apparently upon a singly charged drop which gave a value of the charge of the drop some 30% lower than the final value of *e*.'

From this remark one can infer two things. First, Millikan's great scientific honesty. Many experimentalists, when they get a result which is against what they are trying to establish and against the whole mass of their other observations would say simply 'My apparatus was out of order on that day. I don't know just what was wrong, but I cannot duplicate the result, so it is not of any interest to think of it further and it is not worth mentioning it when I come to publish my work.' Well, Millikan was not like that. He was scrupulously honest and had to refer to experiments which were discordant with what he was trying to prove. The second thing that one can infer was that Millikan had only one drop which gave a discordant result. It does set one wondering whether he didn't have a quark on that drop. I must leave that with you as an unanswered question.

Well, I would like finally to say a few words about the future. The development of the physicist's conception of nature is something which of course hasn't stopped now. So it is a mistake to attach too much importance to the present concepts. It is just an interim stage and we must expect future developments which will be of a fundamental character. I think the future developments will be just as fundamental as the passage from Bohr orbits to the Heisenberg quantum mechanics. I don't know how long we shall have to wait before these future developments occur. But they certainly must occur, and as I mentioned before they may cast new light on the question of determinacy versus indeterminacy.

People have often tried to figure out ways of getting these new concepts. Some people work on the idea of the axiomatic formulation of the present quantum mechanics. I don't think that will help at all. If you imagine people having worked on the axiomatic formulation of the Bohr orbit theory, they would never have been led to Heisenberg's quantum mechanics. They would never have thought of non-commutative multiplication as one of their axioms which could be challenged. In the same way, any future development must involve changing something which people have never challenged up to the present, and which will not be shown up by an axiomatic formulation.

People have often wondered whether one should change one's concept of space and time, perhaps introducing a discreteness into it. But so far there is no mathematical replacement for the standard space-time which is used by physics. I feel it is not much use speculating any further on these questions and I will conclude at this stage.

Part I

Space, Time, and Geometry

2

The Universe as a Whole

Dennis W. Sciama

INTRODUCTION

The foundations of our present understanding of the universe as a whole, both theoretical and observational, were laid down in the period 1917–1929. The last few years have seen developments of almost equal importance, again both theoretical and observational. In these circumstances I prefer not to emphasize the historical context, but to concentrate on the leading features of our present conception of the universe, with historical aspects entering only incidentally. I shall also avoid giving a detailed critical account of the relevant observational data since that would not be in the spirit of this conference. However it must be recognized that observational astronomy is far less reliable than experimental physics, and that, with one notable exception, cosmological data are amongst the least reliable in astronomy. Indeed at the present moment we are passing through a particularly controversial period, when even the standard interpretation of red shifts is being seriously challenged. To avoid prolixity I shall confine myself to the conventional attitude on these controversial questions, but I give a general warning that we lack any crucially convincing evidence. It seems likely that such evidence will be discovered when satellite astronomy gets into its stride.

It is convenient to classify the leading features of our present conception of the universe under the following heads:

1. Existence,
2. Evolution,
3. Symmetry,
4. Singularities.

1. EXISTENCE

The existence of the universe is clearly its most important characteristic, but I am referring here to the stronger idea that it is meaningful to talk of the universe as a whole as a single well-defined concept. This idea is one of the most important, perhaps the most important, scientific discovery of the twentieth century. Having made this grandiose claim, let me immediately introduce a number of reservations.

J. Mehra (ed.), The Physicist's Conception of Nature, 17-33. All Rights Reserved
Copyright © 1973 by D. Reidel Publishing Company, Dordrecht-Holland

(a)   *Other levels of structure may exist.*

To anticipate slightly, we conventionally represent the universe as a whole by one of the Robertson–Walker models of general relativity. However some cosmologists argue that this level of structure may not be the ultimate one. According to this view it is legitimate to regard our universe as a member of an ensemble of universes. Some of these other universes may have a different structure or even be governed by different physical laws than our own universe. The formation of galaxies, stars, planets and ultimately intelligent life may depend critically on the structure of the universe and on the physical laws governing it, so that it may be no accident that we observe the particular kind of universe that we do. This argument would provide an explanation of why the structure and the laws are as observed, but would regard the other universes as in some sense existing. This point of view seems very plausible to me, but it still has to be worked out in detail.[1] In the meantime it would perhaps be more conservative to take the opposite view and to suppose that:

(b)   *The universe is unique.*

This hypothesis deserves a heading of its own because it leads to a serious difficulty of principle which weakens our claim that it is meaningful to regard the whole universe as a well-defined concept. The difficulty arises when we attempt to apply the laws of physics to the whole universe. These laws were originally devised to deal with the situation where one has available many samples of a physical system. The properties which these samples have in common are regarded as 'essential', while their differences are 'accidental'. The common properties are then enshrined in laws of nature, usually in the form of differential equations, while the accidental properties are relegated to initial or boundary conditions that are more or less arbitrary. This procedure is obviously appropriate for the analysis of localized systems, such as projectiles at the surface of the earth. It is, however, inappropriate for a single unique physical system, where the distinction between the essential and the accidental breaks down. How then can we apply the laws of physics to the whole universe?

In some respects we may nevertheless regard the universe simply as a very large system subject to the laws of physics, although clearly we are thereby neglecting an essential aspect of the situation. To make partial progress therefore we assume that:

(c)   *The locally determined laws of physics apply to the universe as a whole.*

We are not yet at the end of our difficulties because we must now take account of the fact that the locally determined laws need to be known with great precision before they can be reliably applied to the whole universe. A change in the laws which may be small locally can sometimes be decisive cosmologically. Two well-known examples of this are the introduction of a cosmical term into Einstein's field equations (which Einstein later described as the greatest mistake of his life) and the hypothesis that some of the fundamental constants of nature change with time on a cosmological time-scale. This hypothesis will be discussed later on at this conference by Dirac and by Jordan, so I shall not consider it here, but confine myself to orthodox general relativity (with or without the cosmical term).

This is still a very significant step to take because general relativity provides us for the first time with both a physical and a mathematical apparatus which can be applied rigorously to a whole, even infinite, system. It is true that 'Newtonian' models of the universe have been constructed, but in these models inertial frames centred on different points are in relative acceleration. It is perhaps a semantic question rather than a scientific one, but it seems to me that one of the essential features of Newtonian theory is that different inertial frames should be in uniform motion relative to one another, so that a single inertial frame has world-wide significance.

In general relativity an infinite system can be both dynamically self-consistent and mathematically adequately described by Riemannian differential geometry. We owe this discovery to Einstein, de Sitter, Weyl and Friedmann, whose basic papers belong to the period 1917–1922. The modern form of this procedure was laid down in the nineteen-thirties by Milne, Robertson and Walker. One begins by assuming that the model universe is exactly symmetric both as regards isotropy and homogeneity. How good these assumptions are we shall consider later. One then determines the most general form of the metric compatible with these symmetry assumptions and so obtains the well-known Robertson–Walker metrics, which depend on one arbitrary function of time and one constant which can be chosen to be ± 1 or zero. One can now apply the field equations to find relations between this function and constant on the one hand and the physical energy momentum tensor on the other. One consequence of this procedure is the occurrence of physical singularities in the models. We shall return to this question later, but we may note here that these singularities limit the validity of our claim that general relativity provides us with a selfconsistent treatment of the whole universe. However, this difficulty is not a consequence of attempting to analyse a whole system, since similar singularities occur in collapsing stars. It seems that general relativity itself must break down in sufficiently strong gravitational fields.

So far our discussion has been entirely theoretical. We must, however, remember that the universe has also been discovered observationally in the twentieth century, and indeed in the same period as the development of relativistic cosmology. Thus it was in 1924 that Hubble showed conclusively that the Andromeda nebula lies outside the Milky Way, so putting to an end a vigorous controversy. In 1929 Hubble announced his linear law relating the red shift of a galaxy to its distance (a relation that had already been derived theoretically for a homogeneous universe 7 years earlier by Friedmann). Our observational understanding of the expanding universe dates from this time. To the observer a galaxy (or rather a cluster of galaxies) appears to be the basic building-brick of the universe. It is true that some astronomers believe in the reality of clusters of clusters of galaxies (superclusters), and others believe that there may be more matter between the galaxies than there is in the galaxies. But these are fine points. The essential thing is that with our telescopes we seem to have determined some of the main structural features of the universe, which repeat themselves fairly uniformly as far as the telescopes can penetrate.

We conclude that both theoretically and observationally the universe (probably) exists.

2. EVOLUTION

The most important single property of the universe, after its existence, is that it is a structure which is changing systematically with time. This is a very remarkable discovery since one might be tempted to think that, while evolution could proceed locally, the whole universe, the background against which local change takes place, would not itself be subject to evolution. This point of view (amongst others) seems to have motivated Bondi and Gold[2] and Hoyle[3] in their famous proposal of 1948 that the universe might be in a steady state, the continual creation of new matter compensating for the expansion. This steady state theory is very beautiful, but it is now in serious conflict with observation[4]. While the question is not settled as decisively as in laboratory physics and echoes of the controversy linger on, it seems very unlikely that the steady state theory can be saved.

Of course the most immediate evidence for the evolution of the universe (assuming that matter is conserved) is:

(a) *The red shifts of the galaxies.*
These red shifts are observed to be proportional to the distances of the galaxies (Hubble law), with a precision indicated in Fig. 1. In this figure the ordinate is the logarithm of the red shift $z(=\delta\lambda/\lambda_{rest})$ in units such that the maximum red shift represented is about 0.5. The abscissa is the measured optical brightness of the galaxies involved, which are selected in the hope that they have a standard intrinsic brightness so that their apparent brightness is a measure of their distance. According to Sandage[5], to whom Fig. 1* is due, the best selection procedure is to choose either the brightest galaxy in a cluster or strong radio galaxies. The small scatter around the straight line in the figure is a measure of the success of this procedure. The observed slope of this line then leads to the Hubble law

$$z \propto r,$$

and the intercept to the value of the Hubble constant τ $(z=r/\tau)$, which is about 10^{10} years.

Hubble originally formulated his law in 1929, and the observations he used fall into the region in Fig. 1 marked by the small black rectangle. Clearly his law extrapolates well to larger red shifts. However, even in uniform cosmological models (with one particular exception), we would expect the red shift-apparent brightness relation to become non-linear at large red shifts because of relativistic effects and of the fact that galaxies of large red shift are being observed as they were at much

---

* The six figures in this article are from the book *Modern Cosmology* by D. W. Sciama, by kind permission of Cambridge University Press.

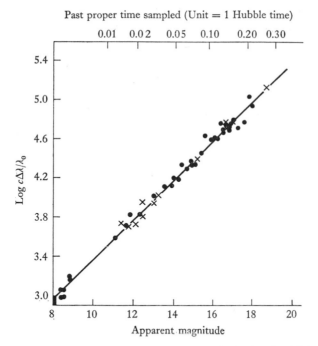

Fig. 1. The relation between red shift and apparent magnitude for the brightest galaxies in 42 clusters: ×, radio; ●, non-radio. Hubble's original (1929) linear relation is represented in this diagram (which is due to A. R. Sandage, *Observatory* **88**, 91 (1968)) by the black rectangle in the bottom left-hand corner.

earlier cosmic epochs, when the rate of expansion of the universe presumably differed appreciably from its present value. A measurement of the non-linearity, when combined with Einstein's field equations, would determine the type of universe we live in, and in particular whether the expansion will continue for ever or be converted by self-gravitation into a contraction. Unfortunately galaxies of large red shift are very faint so that reliable measurements cannot be made on them. We still do not know observationally even the sense of the non-linearity. It was at one time hoped that the quasars would solve this problem since their intrinsic optical brightness is much greater than that of even the brightest galaxies (if their red shift is a measure of their distance), and their observed red shifts range up to 3.40. Unfortunately quasars are far from being 'standard candles', and we know of no selection procedure to replace Sandage's criteria. The future of the universe thus remains undetermined.

Since the steady state theory would be compatible with the Hubble law (at the cost of abandoning or greatly modifying general relativity) we need more direct evidence that the universe is evolving. This is provided by:

(b) The radio source counts.
The first attempt to use these counts to draw cosmological conclusions was made in

1955 by Ryle and Scheuer[6], shortly after it was realized that some at least of the radio sources then catalogued lay outside our Galaxy. Both this and later attempts by Ryle and his colleagues at Cambridge have been subject to considerable criticism. In my opinion some of the criticism of the early attempts was justified, though I should declare my bias by admitting that I was one of those critics. On the other hand most of the criticisms of the later work are not so convincing, as has been clearly shown by Longair and Rees[7] and by Schmidt[8]. The most recent Cambridge results (due to Pooley and Ryle[9]) are shown in Fig. 2, in which the logarithm of the

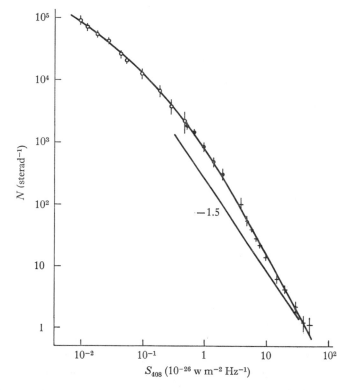

Fig. 2. Counts of radio sources derived by Pooley and Ryle. N is the number of sources per unit solid angle whose flux density at 408 MHz exceeds S_{408}. (From G. G. Pooley and M. Ryle, *Mon. Not. Roy. Astron. Soc.* **139**, 515 (1968).)

number N of sources per steradian brighter than the measured radio brightness S at 408 MHz is plotted against the logarithm of S. A uniform distribution of sources at rest would yield a straight line with slope -1.5, while the red shift corrections in both evolutionary and steady state theories would produce a flatter slope. The observed slope, however, is steeper.

Ryle's explanation of this discrepancy is that there are too many faint sources and that, since the faint sources are on the whole more distant and so are being observed as they were in the past, this means that *the sources must be evolving with cosmic epoch.* This conclusion, if valid, is very important, because it would represent the first occasion on which some average property of a population of objects in the universe has been found to change with cosmic epoch. In particular it would imply that the steady state theory is wrong. Ryle's arguments are summarized in a survey paper[10], but the papers mentioned earlier[7, 8] should be referred to for a rebuttal of the most recent criticisms.

Another important feature of Fig. 2, to which Pooley and Ryle draw attention, is that for the faintest sources the slope does flatten down below the static value of −1.5. They attribute this to the red shift corrections eventually dominating the evolutionary corrections, and point out that this behaviour is in any case required, since otherwise the *integrated* radiation from the radio sources would exceed the observed diffuse extragalactic background at the same frequency (408 MHz). In fact the sources shown in Fig. 2 account for about half of the extragalactic background, while a natural extrapolation of the curve would provide the other half.

One point that should be emphasized (and which is responsible for much of the criticism) is that most of the sources involved in the counts have not yet been optically identified. In particular we do not know (for most of the sources) which are radio galaxies and which are quasars. This ambiguity has led people to make a direct study of identified quasars, using their optical as well as their radio properties. This study has again yielded an excess of faint sources, as has been discussed most recently by Schmidt[8].

What one would like to do is to use these results to determine the correct model of the universe, and not simply to dispose of the steady state model. Unfortunately attempts to do this have not been successful. Too little is known about the expected physical evolution of the sources to separate out these physical effects from purely cosmological effects. Indeed any reasonable evolutionary model of the universe together with the source counts would lead to a required physical evolution which, in the present state of our knowledge, is also reasonable. There is not much prospect of solving this problem in the foreseeable future.

We now come to the most striking evidence in favour of an evolving universe. It is also the greatest discovery in cosmology since that of the expansion of the universe itself. This is:

(c) The microwave background.

The microwave background was discovered accidentally in 1965 by Penzias and Wilson[11] and immediately interpreted by Dicke and his colleagues[12] as black body radiation left over from the hot big bang origin of the universe. The present state of the observations is shown in Fig. 3. At wavelengths less than 100 cm, where radiation from the Galaxy is unimportant, the observed points fit quite well to a black body spectrum at a temperature of 2.7 °K. Unfortunately most of these points occur on the Rayleigh–Jeans part of the spectrum, and the data concerning the region at and beyond the peak ($\lambda \lesssim \sim 1$ mm), which have to be obtained from above the

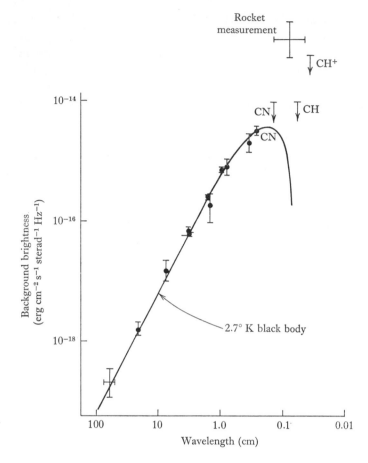

Fig. 3.   Measurements of the isotropic background at radio wavelengths, compared with a 2.7°K black body spectrum.

Earth's atmosphere, are confused and to some extent contradictory. The most recent data in this region tend to favour the black body hypothesis, but definitive results will probably only be obtained from satellite observations.

In view of this uncertainty, it is important to distinguish between deductions that can be made only if the black body hypothesis is basically correct, and those that require merely that the microwave background has an extragalactic origin. The high degree of isotropy of the background provides strong evidence that an extragalactic origin is indeed involved, and we discuss the implications of this in Section 3.

If we accept provisionally that the background does have a black body spectrum, then we can immediately deduce that the universe is probably evolving. The reason is that in its present state the universe is far too dilute to be able to thermalize radiation in the time available (10^{10} years). Since it is difficult to see how the radiation could have been produced already thermalized, we conclude that at sometime in the past the universe must have been sufficiently dense to thermalize radiation in the time-scale then prevailing. According to the standard cosmological models this would require a universal density of at least 10^{-14} gm cm^{-3} (that is about 10^{15} times larger than the present mean density). Gamow[13] had already suggested in 1948 that thermalization occurred close to the singularity at $t=0$, but an alternative possibility[14] is that dissipation processes occurred later to produce the necessary heat radiation. These processes must essentially have ceased by the time thermalization became impossible[15], the black body radiation existing at that time remaining black body radiation thereafter as in an ordinary adiabatic expansion. In either case we can conclude that the evolution of the universe goes back at least to the time when its density was about 10^{15} times greater than its present density.

(d) The cosmic abundance of helium.

A variety of observations and theoretical arguments indicate that the abundance of helium relative to hydrogen in the Galaxy is rather uniform[16], the number ratio being close to 0.1. This result creates an immediate problem, because it appears that the formation of helium out of hydrogen in stars (or massive objects) can account for

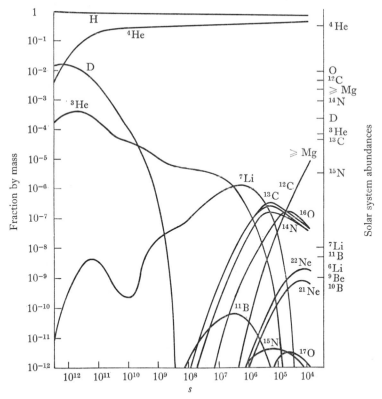

Fig. 4.   Element production in the hot big bang, according to Wagoner, Fowler and Hoyle. The fractional abundance by mass is shown as a function of the entropy per particle. (After *Astrophys. J.* **148**, 21 (1967).)

only about ten per cent of the observed helium (although similar processes for heaver elements do not face the same difficulty). Following the original proposals of Gamow[13], and of Alpher, Bethe and Gamow[17], we can resolve the problem by appealing to thermonuclear processes occurring about 100 seconds after the big bang, when the temperature was about $10^9 \,^\circ K$ and the material density about 10^{-3} gm cm^{-3}. The first realistic calculation of the resulting helium abundance was made by Hayashi[18], while more accurate results are due to Peebles[19] and to Wagoner, Fowler and Hoyle[20]. These latter authors also calculated the abundances of heavier elements, and their results are shown in Fig. 4. In this figure the ordinate is the logarithm of the abundance of helium by mass relative to hydrogen (the observed value by mass being 29 per cent). The abscissa is the logarithm of the present mean density of the universe. The abundances shown are calculated on the assumption that the present temperature of the black body radiation is $3 \,^\circ K$, and that this temperature is to be extrapolated backwards in time according to the usual adiabatic relations of relativistic cosmology.

Since the present mean density of the universe is believed to lie between about 10^{-31} gm cm^{-3} (the value based on galaxies alone) and 10^{-29} gm cm^{-3} (the maximum allowable intergalactic gas density), we see that the calculated value of the helium abundance lies close to the observed value. By contrast, with the possible exception of some of the lighter elements such as D, He^3 and Li^7, the remaining elements are produced in quantities much smaller than observed. These results fit very neatly with the stellar calculations, which broadly speaking can account for the formation of the heavier elements but not for He (the light elements Li^6, Be, B form a special case which may be understood in terms of cosmic ray spallation processes[21]).

The observations of the helium abundance in the Galaxy will undoubtedly improve over the years, but already we may claim rather impressive agreement between them and the hot big bang calculations. If we accept the obvious implications of this agreement it follows that we can extrapolate the evolution of the universe back to a time 100 seconds after the big bang. This is in fact the farthest back we can extrapolate on the basis of direct observations. However we shall see in Section 4 that according to general relativity we have to extrapolate the universe back to a singularity.

3. SYMMETRY

The second most important property of the universe is its large-scale symmetry. Of course on the scale of galaxies and clusters of galaxies (and perhaps superclusters) the universe is irregular, but on a larger scale the distribution of galaxies does seem to be fairly uniform. To be more precise, this distribution is roughly isotropic about us and also homogeneous. Strictly speaking the homogeneity should refer to one cosmic epoch (except in the steady state theory), but in practice the time interval involved in observing the galaxies is very much smaller than the age of the universe.

As we have already seen, this isotropy and homogeneity was assumed from the beginning in order to construct simple exact cosmological solutions of Einstein's field equations. Nevertheless it has recently been discovered that these symmetry assumptions are actually much more accurate than was envisaged 40 years ago. This discovery has come from detailed observations of the *isotropy* of the microwave background. To understand the significance of these observations it is helpful to regard this background as coming to us from a spherical surface situated one scattering mean free path away. The main contribution to this scattering comes from the Thomson interaction between the microwave radiation and intergalactic (or pre-galactic) free electrons. The red shift z_s of this last-scattering surface depends on the cosmological model and the density and ionization state of the intergalactic medium. Its possible values range from about 7, corresponding to a high-density model with an ionized intergalactic gas of density $\sim 10^{-29}$ gm cm^{-3}, to about 1000 for a low density model (at this red shift the radiation temperature $\sim 3000\,^\circ$K and the pre-galactic gas would be collisionally ionized). The temperature of the last scattering surface would be $2.7\,(1+z_s)\,^\circ$K, so that we are in effect surrounded by a glowing surface at this temperature.

We are interested in the precision to which the background is isotropic, that is, to which its temperature T_0 is isotropic (assuming for the moment that it has a black body spectrum in each direction). A measurement of this precision would also set limits to any possible variation of $T_s/(1+z_s)$ over the last-scattering surface. It would also set limits on possible large-scale inhomogeneities in the distribution of matter in the universe, since the gravitational effects of such an inhomogeneity would influence the temperature of radiation passing through it.

Fortunately the measurement of the background temperature as a function of direction can be carried out much more accurately than an absolute measurement in any one direction. Two sets of experimental results are shown in Figs. 5 and 6, which are due to Conklin and Bracewell[22] and to Partridge and Wilkinson[23] respectively. The fluctuations shown are not regarded as significant, and the net result of all these measurements is that on a variety of angular scales there is no observable anisotropy to a precision of about one part in a thousand.

This is by far the most accurate measurement in cosmology, and it has profound implications for our conception of nature. Unless there is a correlation between the values of T_s and z_s at different points on the last-scattering surface, it follows that

(*a*) T_s is constant over the surface to 1 in 10^3.

(*b*) z_s is constant over the surface to 1 in 10^3, that is, the expansion rate of the universe is isotropic to 1 in 10^3.

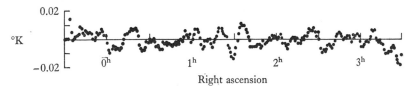

Fig. 5.   The distribution of background temperature along the celestial equator, according to Conklin and Bracewell, *Phys. Rev. Lett.* **18**, 614 (1967).

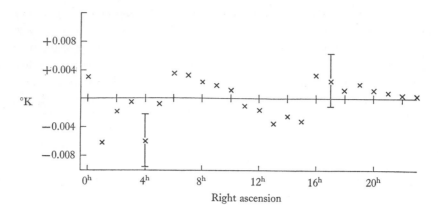

Fig. 6. The distribution of background temperature along the celestial equator, according to Partridge and Wilkinson. More than two years of data, obtained at Princeton and Yuma, have been averaged. The fluctuations about the mean are not regarded as significant. (After R. B. Partridge, *American Scientist* **57**, 37 (1969).)

(*c*) There can be no large-scale inhomogeneities with a density contrast of order unity and a length scale of order one-tenth of the radius of the universe[24]. This gives one a measure of the degree of homogeneity of the universe.

The rather strict isotropy referred to in (a) and (b) raises the important question, why is the universe so isotropic? This is one of the major unsolved problems of cosmology. The main difficulty of this problem arises from the existence of particle horizons[25] in these symmetric models, in other words, of galaxies which have not yet had time to communicate with us. Put more generally we can say that in these models there exist events which have never been under each other's physical influence. For example, consider two points on the last-scattering surface which are more than 30° apart. Then no physical communication between these points can have taken place. Yet their temperatures are the same to one part in a thousand. Why? Must we rely on the initial conditions in the universe being isotropic to this precision without further explanation being possible, or could causal influences be at work after all? There have been attempts to construct models in which the early universe was so anisotropic that particle horizons are eliminated[26]. Dissipation processes can then be envisaged which isotropize the universe. However these attempts have met with their own difficulties[27], and the problem is still quite open. An interesting suggestion[28] is that while the models which isotropize form a set of measure zero amongst all models, nevertheless the models in which galaxies can form may belong to the same set of measure zero. In that case we could understand why human beings would find their universe to be isotropic.

There is a further important consequence of the high observed isotropy of the microwave background. We have seen that we are at the centre of (our own) last-scattering surface. In addition, however, we must also be nearly at rest at this point. Otherwise the resulting Doppler effect would lead to an increase of temperature in the direction of our motion, a decrease of temperature behind us, and generally a dipole distribution in any plane. We therefore conclude that:

(d)   Our velocity relative to the last-scattering surface is less than 300 km sec^{-1}.

This conclusion has two important implications. The first is perhaps of more direct astronomical than conceptual interest. It concerns the compatibility of our motion through the microwave background with the hierarchy of localized motions to which the earth is subject,[30, 31] namely its motion round the sun, the motion of the sun round the Galaxy, of the Galaxy relative to the Local Group of galaxies, of the Local Group in the possible supercluster centred on the Virgo cluster of galaxies etc. Some of these individual motions are believed to amount to several hundred kilometres per second, so it is already of great interest that the resultant of all these motions is less than 300 km sec^{-1}. Moreover, experiments now under way are expected to reveal our motion through the microwave background if it exceeds 30 km sec^{-1} (in which case it should be possible to detect the motion of the Earth round the Sun).

However, from the conceptual point of view the most important consequence of (d) relates to what is historically the oldest problem in our subject, which goes back to Newton, and that is the problem of absolute versus relative rotation.

We can summarize this problem by the question: are inertial frames of reference determined by absolute space or by the matter in the universe? Even in general relativity it could be that the material universe is a small perturbation on absolute space. What Einstein called Mach's principle[32] would require there to be no absolute space, its role being taken over by the material universe. There would then be no such thing as absolute rotation, only rotation relative to the material universe.

If absolute space exists, we would expect that in practice the universe would have some rotation relative to it, although of course this could not be a rigorous requirement. On the other hand, if there is no such thing as absolute space then the universe must provide the standard of non-rotation, and could not itself rotate.

It is of interest to examine this question empirically, that is, to attempt to measure the rotation of the universe. If the period of this rotation turned out to be much greater than 10^{10} years, we might be tempted to argue that this was evidence in favour of effectively zero rotation. A full analysis of this problem would require us to study the detailed dependence of rotation on cosmic epoch, but this would take us beyond the broad survey we are attempting to give in this article.

Before the discovery of the microwave background the best upper limit on the rotation of the universe came from the observation that our Galaxy rotates *kinematically* relative to the universe outside it, and also shows the expected *dynamical* flattening. The rotation period T of the universe must thus exceed the rotation period of the Galaxy, or

$$T > 2.5 \times 10^8 \text{ years.}$$

From the point of view of ruling out the rotation of the universe, this limit is much too weak.

Now if the universe is rotating the Earth must be moving relative to the last-scattering surface of the microwave background. The observed upper limit of 300 km sec^{-1} for this latter motion can then be used to obtain a lower limit for the rotation period of the universe. This question has been studied by Hawking[33], who finds that the result depends somewhat on the cosmological model adopted. The broad result is the following inequality:

$$T > 2.5 \times 10^{14} \text{ years.}$$

The resulting limit is a million times more powerful than the previous one, and 2.5×10^4 longer than the usual time-scale associated with the universe. This is prima facie evidence that absolute space does not exist.

This tentative conclusion raises again in urgent form the old problem of giving rigorous content to the statement that absolute space does not exist. Recent work suggests that it may be possible to solve this problem within the context of orthodox general relativity. It has been found possible[34, 35, 36] to express the metric tensor at a point in space-time in terms of a Kirchhoff-type (hyper) volume integral over the energy-momentum tensor in a volume surrounding the point, together with a surface integral involving boundary data over the (hyper) surface. At first sight this would seem to be impossible since Einstein's equations are non-linear, and integration is a linear process. This difficulty can be avoided if one permits the kernel, or Green's function, of the integral representation to be a *functional* of the metric. Armed with this integral representation, one might be tempted to assert that Mach's principle corresponds to the requirement that the surface integral should vanish in the limit when the (hyper) volume includes the whole of space-time. This does in fact work for Robertson–Walker metrics[37], but in general it is too stringent a condition, because the boundary data are not freely disposable, but must satisfy elliptic constraint equations. Expressed more physically, the Green's function vanishes outside the light-cone of the field point, but we would in general expect galaxies outside the particle horizon of the point to make their 'Newtonian' contribution to the metric at the point. Only in isotropic universes would the total contribution from outside the particle horizon vanish (by symmetry).

A way around this difficulty has recently been found by Raine[38]. He generalizes the idea of an integral representation by expressing the metric tensor in terms of the Riemann curvature tensor, and the Riemann tensor in terms of the Ricci (or equivalently the energy-momentum) tensor. This enables Raine to formulate two conditions which must be satisfied in Machian universes, and he is able to show that a homogeneous Machian universe must be non-rotating. Thus a rigorous statement leading to the abolition of absolute space has at last been achieved.

4. SINGULARITIES

The occurrence of physical singularities in collapsing stars and expanding universes is the most serious problem faced by general relativity and cosmology today. The problem originally arose in the exactly isotropic Robertson–Walker models in which the density becomes infinite at a finite proper time in the past. This singularity used to be disregarded on the grounds that it would be eliminated by the introduction of reliastic irregularities, which would 'defocus' the cosmological model. We now know that in general this argument is not correct, owing essentially to the strength of self-gravitation in Einstein's theory. This result was proved for collapsing stars in 1965 in a remarkable paper by Penrose[39]. Hawking[40] then adapted and modified Penrose's (topological) methods and applied them to the universe as a whole. Further contributions to this problem were made by Ellis[41] and by Geroch[42], culminating in the paper of Hawking and Penrose[43] which contains the most powerful and comprehensive versions of the singularity theorems. This work has been fully described in an important book[44] by Hawking and Ellis, entitled *The Large Scale Structure of Space-Time*.

In this article we would like to give a brief account of one of these theorems. We have selected one which depends essentially on the observations of the microwave background. There are two reasons for this choice. The first is to show how the background can be used to draw still further conclusions beyond those of Section 3. The second is that some versions of the singularity theorems are based on assumptions about the universe which, though plausible, are difficult or impossible to verify in practice (for instance that the universe should admit a well-defined global Cauchy surface, that is, a space-like surface on which initial-value data can be prescribed). The most economical theorem in this respect uses the microwave background in the following way.

Theorem: Space-time is not singularity-free if the following conditions hold:

(*a*)   Einstein's field equations,

(*b*)   The energy condition ($T_{ab}W^a W^b \geqslant \frac{1}{2}T$ for all time-like unit vectors W^a),

(*c*)   Strong causality (every neighbourhood of a point contains a neighbourhood of that point which no non-spacelike curve intersects more than once),

(*d*)   There exists a point P such that all past-directed time-like geodesics through P start converging again within a compact region in the past of P.

Comment.   (*d*) is a precise statement of the idea that the gravitation due to the material in the universe is sufficiently attractive to produce a singularity.

This theorem has been proved by Hawking. The idea is now to use observations of the microwave background to show[45] that (*d*) is true in the actual universe, the point P corresponding to the Earth ((*b*) is a very weak and plausible condition on the energy-momentum tensor T_{ab}, while a violation of (*c*) would perhaps be worse than a singularity, being a global rather than a local breakdown of our ordinary physical concepts).

The procedure is to use the microwave observations to show that the actual universe is sufficiently like an exactly isotropic one, where the re-convergence certainly does occur. Now these observations imply that the actual universe is isotropic to one part in a thousand back to the last-scattering surface. There are then two possibilities:

(i)   The red shift z_s of this surface is small (~ 7). This would require a relatively large amount of intergalactic scattering material, and direct calculation then shows that the gravitational effect of this material would produce the required reconvergence before the red shift z_s is reached.

(ii) z_s is appreciably larger than 7. The influence of intergalactic matter may now itself be unable to produce re-convergence. On the other hand the isotropy of the universe would remain close to ideal out to z_s, that is, to a greater red shift than 7. In this case one can show that the energy density of the microwave background alone is enough to cause re-convergence. It follows that if general relativity is correct, and the other (weak and desirable) conditions of the theorem hold, then there is a physical singularity in our past.

We therefore face a crisis in theoretical physics. Either classical general relativity breaks down, or effectively negative energy densities can exist, or causality breaks down, or singularities exist in nature. One's first thought is that quantizing general relativity might resolve the crisis, but at the moment this remains only a hope. In any case we need to understand the behaviour of matter under conditions of high density and temperature when general relativity, quantum mechanics and elementary particle effects are simultaneously important. In this connection I am reminded somewhat wistfully of one of Professor Dirac's sayings: 'To make progress in physics you should separate the difficulties and solve them one at a time'. This seems to be a case where the solution of the problem may depend essentially on the interaction between the difficulties, and no doubt on the introduction of much new physics besides. Hard as this task may be, if we wish to develop further the physicist's conception of nature, I believe that we must tackle it.

REFERENCES

1.   M. J. Rees, *Comments Astrophys. Space Phys.* **4**, 179 (1972).
2.   H. Bondi and T. Gold, *Mon. Not. Roy. Astron. Soc.* **108**, 252 (1948).
3.   F. Hoyle, *Mon. Not. Roy. Astron. Soc.* **108**, 372 (1948).
4.   D. W. Sciama, *Modern Cosmology*, Cambridge University Press (1971).
5.   A. Sandage, *Astrophys. J.* **152**, 149 (1968); *Observatory* **88**, 91, (1968).
6.   M. Ryle and P. A. G. Scheuer, *Proc. Roy. Soc.* A. **230**, 448 (1955).
7.   M. S. Longair and M. J. Rees, *Comments Astrophys. Space Phys.* **4**, 79 (1972).
8.   M. Schmidt, *Astrophys. J.* **176**, 273, 289, 303 (1972).
9.   G. G. Pooley and M. Ryle, *Mon. Not. Roy. Astron. Soc.* **139**, 515 (1968).
10. M. Ryle, *Ann. Rev. Astron. Astrophys.* **6**, 249 (1968).
11. A. A. Penzias and R. W. Wilson, *Astrophys. J.* **142**, 419 (1965).
12. R. H. Dicke, P. J. Peebles, P. G. Roll, and D. T. Wilkinson, *Astrophys. J.* **142**, 414 (1965).
13. G. Gamow, *Rev. Mod. Phys.* **21**, 367 (1949).
14. C. W. Misner, *Nature* **214**, 40 (1967).
15. M. J. Rees, *Phys. Rev. Lett.* **28**, 1669 (1972).
16. L. Searle and W. L. W. Sargent, *Comments Astrophys. Space Phys.* **4**, 35 (1972).
17. R. A. Alpher, H. A. Bethe, and G. Gamow, *Phys. Rev.* **73**, 803 (1948).
18. C. Hayashi, *Prog. Theor. Phys.* **5**, 224 (1950).
19. P. J. E. Peebles, *Astrophys. J.* **146**, 542 (1966).
20. R. W. Wagoner, W. A. Fowler and F. Hoyle, *Astrophys. J.* **148**, 3 (1967).
21. M. Meneguzzi, J. Audouze and H. Reeves, *Astron. Astrophys.* **15**, 337 (1971).
22. E. K. Conklin and R. N. Bracewell, *Nature* **216**, 777 (1968).
23. R. B. Partridge and D. T. Wilkinson, *Phys. Rev. Lett.* **18**, 557 (1967).

24. M. J. Rees and D. W. Sciama, *Nature* **217**, 1511 (1968).
25. W. Rindler, *Mon. Not. Roy. Astron. Soc.* **116**, 663 (1956).
26. C. W. Misner, *Astrophys. J.* **151**, 431 (1968).
27. M. A. H. MacCallum, *Nature*, *Phys. Sci.* **230**, 112 (1971).
28. C. B. Collins and S. W. Hawking, *Astrophys. J.* **180**, 317 (1973).
29. D. W. Sciama, *Phys. Rev. Lett.* **18**, 1065 (1967).
30. J. M. Stewart and D. W. Sciama, *Nature* **216**, 748 (1967).
31. G. de Vaucouleurs and W. L. Peters, *Nature* **220**, 868 (1968).
32. D. W. Sciama, *The Physical Foundations of General Relativity,* Science Study Series No. 37 (1969).
33. S. W. Hawking, *Mon. Not. Roy. Astron. Soc.* **142**, 129 (1969).
34. B. L. Altshuler, *JETP,* **51**, 1143, (1966).
35. D. Lynden-Bell, *Mon. Not. Roy. Astron. Soc.* **135**, 413 (1967).
36. D. W. Sciama, P. C. Waylen and R. C. Gilman, *Phys. Rev.* **187**, 1762 (1969).
37. R. C. Gilman, *Phys. Rev.* D. **2**, 1400 (1970).
38. D. J. Raine, to be published.
39. R. Penrose, *Phys. Rev. Lett.* **14**, 57 (1965).
40. S. W. Hawking, *Proc. Roy. Soc.* A. **294**, 511 (1966); **295**, 490 (1966); **300**, 187 (1967).
41. S. W. Hawking and G. F. R. Ellis, *Phys. Letters* **17**, 246 (1965).
42. R. P. Geroch, *Phys. Rev. Lett.* **17**, 445 (1966).
43. S. W. Hawking and R. Penrose, *Proc. Roy. Soc.* A. **314**, 529 (1970).
44. S. W. Hawking and G. F. R. Ellis, *The Large Scale Structure of Space-Time,* Cambridge (1973).
45. S. W. Hawking and G. F. R. Ellis, *Astrophys. J.* **152**, 25 (1968).

3

A Chapter in the Astrophysicist's View of the Universe

S. Chandrasekhar

At the outset, I must state quite frankly that I have neither the knowledge nor the competence to undertake a coherent and a reasoned account of the developments that have led to our current ideas concerning the 'astrophysical universe'. But it is safe to assume that there are many facets to the astrophysical universe, and I shall therefore limit myself to only one facet, namely, the developments that have led to our present ideas concerning gravitational collapse and black holes. Even here I cannot be fair to all aspects though I can perhaps claim, by virtue at least of longevity, that I have been personally associated with related problems for a longer period than anyone else in this audience.[1]

In the first instance, one may be surprised that a problem of such current lively interest should have been a matter for serious discussion some fifty years ago; but the question concerning the last stages of stellar evolution occurs to one even if one has no precise knowledge of the physical processes that are responsible for the energy of the stars. Indeed, the question occurs to one almost inevitably: for no matter what the source of energy is, it must be exhausted sooner or later; and sooner or later the question must be confronted. The question was in fact formulated by Eddington in 1926 in one of his famous aphorisms: '*A star will need energy to cool.*' Let me rephrase the question in a less oracular fashion.

By 1925 it was known through the work of W. Adams on the binary system of Sirius that stellar masses with densities in the range of 10^5 to 10^6 grams cm^{-3} exist. And Eddington estimated that the central temperatures of such stars should be 10^7 to 10^8 degrees, on the assumption that ordinary perfect-gas laws hold and that matter is ionized to bare nuclei and electrons. On these same assumptions let E_V denote the negative electrostatic energy per gram. And let E_K denote the kinetic energy of the particles per gram. If such matter were released of the pressure to which it is subject, then it could resume the state of ordinary un-ionized matter only if $E_K > E_V$. Eddington's paradox[2] was this (though Eddington at a later time disclaimed to this particular formulation): if the quantities E_K and E_V are evaluated for the densities and temperatures expected in the interiors of the white dwarfs, then one finds that E_K(perfect gas) $< E_V$. In other words, on these premises, the matter will not be able to resume its ordinary state – an extraordinary situation if true. R. H. Fowler's resolution of this paradox in 1927 in terms of the then very new statistical mechanics of Fermi and Dirac is, in my opinion, one of the great landmarks in the development of our ideas on stellar structure and stellar evolution. May I spend a few minutes on it.

J. Mehra (ed.), The Physicist's Conception of Nature, 34-44. All Rights Reserved
Copyright © 1973 by D. Reidel Publishing Company, Dordrecht-Holland

Dirac's paper, which contains the derivation of what has since come to be called the 'Fermi–Dirac distribution', was communicated by Fowler to the Royal Society on 26 August 1926.[3] On 3 November Fowler communicated a paper of his own in which the application of the laws of the 'new quantum theory' to the statistical mechanics of assemblies consisting of similar particles is systematically developed and incorporated into the general scheme of the Darwin–Fowler method. And by 10 December (i.e., before a month had elapsed) his paper entitled 'Dense Matter' was read before the Royal Astronomical Society. In this paper Fowler drew attention to the fact that the electron gas in matter as dense as in the companion of Sirius, must be degenerate in the sense of the Fermi–Dirac statistics. Thus, to Fowler belongs the credit for first recognizing a field of application for the then 'very new' statistics of Fermi and Dirac, though among physicists this credit is generally given to Pauli for his explanation of the paramagnetic susceptibility of the alkali metals.

Fowler's resolution of Eddington's paradox is simply this: since at the temperature and densities in the white dwarfs the electron assembly will be highly degenerate, E_K should be evaluated by using the formulae appropriate in this limit. He showed that when E_K is so evaluated, it is indeed much greater than E_V. And Fowler concluded his paper[4] with the following statement.

The black-dwarf material is best likened to a single gigantic molecule in its lowest quantum state. On the Fermi–Dirac statistics, its high density can be achieved in one and only one way, in virtue of a correspondingly great energy content. But this energy can no more be expended in radiation than the energy of a normal atom or molecule. The only difference between black-dwarf matter and a normal molecule is that the molecule can exist in a free state while the black-dwarf matter can only so exist under very high external pressure.

The limiting form of the equation of state which Fowler used in his paper is the standard one which is familiar in solid-state physics, namely,

$$p = k_1 (n_e)^{5/3},$$

where k_1 is a constant and n_e is the concentration of the electron. The energy ($=3/2$ pV) associated with this pressure is zero-point energy; and the point of Fowler's paper is that this zero-point energy is so great that under normal circumstances a star could be expected to settle down into a state in which all of its internal energy is in this form. And it appeared for a time that all stars will have the necessary energy to cool. Fowler's arguments can be stated more precisely in the following manner.

On the basis of the equation of state, one can readily determine the structure which a configuration of an assigned mass M will assume when in equilibrium under its own gravity. One finds[5] that an *equilibrium state* is possible for any assigned mass: one finds in fact a mass-radius relation of the form

$$M = \text{constant } R^{-3}.$$

Accordingly, the larger the mass, the smaller is its radius. Also the mean densities of these configurations are found to be in the range of 10^5 to 10^6 grams cm^{-3} when the mass is of solar magnitude. These masses and densities are of the order one meets in the so-called white-dwarf stars. And it seemed for a time that the white-dwarf stage (or rather the 'black-dwarf' stage as Fowler described it) represented the last stage of stellar evolution for *all* stars. Since a finite state seemed possible for any assigned mass, one could rest with the comfortable assurance that all stars will have the 'necessary energy to cool'. But this assurance was soon broken when it was realized that the electrons in the centres of degenerate masses begin to have momenta comparable to *mc* where *m* is the mass of the electron. Accordingly, one must allow for the effects of special relativity. These effects can be readily allowed for and look harmless enough in the first instance: the correct equation of state, while it approximates to that by the equation given earlier for low enough electron concentrations, it tends to

$$p = k_2 (n_e)^{4/3}$$

as the electron concentration increases indefinitely. (In the foregoing equation, k_2 is another atomic constant.) This limiting form of the equation of state has a dramatic effect on the predicted mass-radius relation; instead of predicting a finite radius for all masses, the theory now predicts that the radius must tend to zero as a certain limiting mass is reached. The value of this limiting mass is

$$M_{\text{limit}} = 5.76\mu_e^{-2} \odot ,$$

where μ_e denotes the mean molecular weight per electron and $\odot$ stands for the solar mass. For the expected value $\mu_e = 2$,

$$M_{\text{limit}} = 1.44 \odot .$$

These results were obtained in 1930 and were published early in 1931.[6,7]

The existence of this limiting mass means that a white-dwarf state does not exist for stars that are more massive. In other words, 'the massive stars do not have sufficient energy to cool.'

But to return to 1931. Once the existence of the critical mass was established, the question that was puzzling was how one was to relate its existence to the evolution of stars which start as gaseous masses. If the stars had masses less than M_{limit} then the assumption that they would eventually settle down to a white-dwarf stage seemed natural enough. But what if their masses were greater?

By a simple comparison of the limiting forms of the equations of state, it can be readily shown that if at a given p, ϱ, and T, the fraction

$$1 - \beta = \frac{\frac{1}{3}aT^4 (= \text{radiation pressure})}{\frac{1}{3}aT^4 + \Re\varrho T(= \text{total pressure})} > 0.092,$$

then matter cannot become degenerate.

It is not difficult to convince oneself by fairly rigorous arguments that for perfect-gas stars $\langle 1 - \beta \rangle_{\text{average}}$ increases with mass. In fact, one can estimate that for $M > 6.6\mu_e^{-2} \odot$, matter can never become degenerate. And this gave rise to the question formulated already in 1932[8]:

Given an enclosure containing electrons and atomic nuclei (total charge zero) what happens if we go on compressing the material indefinitely?

But these questions so clearly raised in 1932 failed to attract any attention.

In 1934 the exact mass-radius relations for degenerate configurations were derived. [9] Fig. 1 exhibits the mass-radius relation that was deduced on the basis of the exact equation of state.

At this time the significance of existence of the limiting mass was so clear that to draw attention to the result as emphatically as possible, the conclusion was formulated in the following terms. [9a]

The life history of a star of small mass must be essentially different from the history life of a star of large mass. For a star of small mass the natural white-dwarf stage is an initial step towards complete extinction. A star of large mass cannot pass into the white-dwarf stage and one is left speculating on other possibilities.

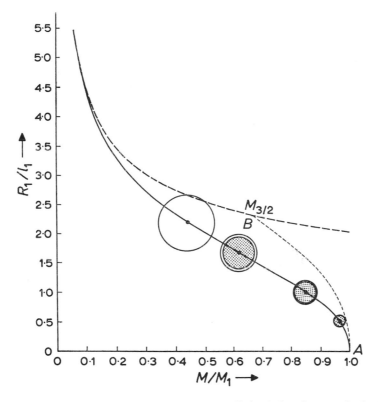

Fig. 1.   The full-line curve represents the exact (mass–radius)-relation for completely degenerate configurations. The mass, along the abscissa, is measured in units of the limiting mass (denoted by M_3) and the radius, along the ordinate, is measured in the unit $l_1 = 7.72\mu_e^{-1} \times 10^8$ cm. The dashed curve represents the relation (2) that follows from the equations of state (1); at the point B along this curve, the threshold momentum p_0 of the electrons at the centre of the configuration is exactly equal to mc. Along the exact curve, at the point where a full circle (with no shaded part) is drawn, p_0 (at the centre) is again equal to mc; the shaded parts of the other circles represent the regions in these configurations where the electrons may be considered to be relativistic ($p_0 \geqslant mc$). (This illustration is reproduced from S. Chandrasekhar, *Mon. Not. Roy. Astron. Soc.* **95**, 219 (1935).)

Statements very similar to the one I have just quoted from a paper written thirty-eight years ago frequently occur in current literature. But why, it may be asked, were these conclusions not accepted forty years ago? The answer is that they did not meet with the approval of the stalwarts of the day. Thus Eddington[10] commenting on the foregoing conclusion stated:

The star apparently has to go on radiating and radiating and contracting and contracting until, I suppose, it gets down to a few kilometres radius when gravity becomes strong enough to hold the radiation and the star can at last find peace.

If Eddington had stopped at that point, we should now be giving him credit for having been the first to predict the occurrence of black holes – a topic to which I shall return presently. But alas! he continued to say:

I felt driven to the conclusion that this was almost a *reductio ad absurdum* of the relativistic degeneracy formula. Various accidents may intervene to save the star, but I want more protection than that. I think that there should be a law of nature to prevent the star from behaving in this absurd way.

In spite of the then prevalent opposition, it seemed to me likely that a massive star, once it had exhausted its nuclear sources of energy, will contract and in the process eject a large fraction of its mass; and further that if by this process, it reduced its mass sufficiently, it could find a state in which to settle.

A theoretical advance in a different direction suggested another possibility. It is that as we approach the limiting mass along the white-dwarf sequence, we must reach a point where the protons at the centre of the configuration become unstable with respect to electron capture. The situation is this. Under normal conditions, the neutron is β-active and unstable while the proton is a stable nucleon. But if in the environment in which the neutron finds itself (as it will in the centre of degenerate configurations near the limiting mass), all the electron states with energies less than or equal to the maximum energy of the β-ray spectrum of the neutron are occupied, then Pauli's principle will prevent the decay of the neutron. Under these circumstances the proton will be unstable and the neutron will be stable. At these high densities, the equilibrium that will obtain will be one at which, consistent with charge neutrality, there will be just exactly the right number of electrons, protons, and neutrons with appropriate threshold energies that none of the existing protons or neutrons decay. At these densities the neutrons will begin to outnumber the protons and electrons by large factors. In any event it is clear that once neutrons begin to form, the configuration essentially collapses to such small dimensions that the mean density will approach that of nuclear matter and in the range 10^{13} to 10^{15} grams cm^{-3}. These are the neutron stars that were first studied in detail by Oppenheimer and Volkoff[11] in 1939 though their possible occurrence had been suggested by Zwicky some five years earlier.

Based on the work of Oppenheimer and Volkoff it appeared likely that a massive star during the course of its evolution could collapse to form a neutron star if during the process of contracting it had reduced its mass sufficiently. The process would clearly be cataclysmic, and it seemed likely that the result would be a supernova phenomenon. If the degenerate cores attain sufficiently high densities (as is possible for these stars) the protons and electrons will combine to form neutrons. This would cause a sudden diminution of pressure resulting in the collapse of the star onto a neutron core giving rise to an enormous liberation of gravitational energy. This may be the origin of the supernova phenomenon. But the formation of a neutron star, as the result of the collapse, will depend on whether a star, initially more massive than the limiting mass for the white-dwarf stars, ejects just the right amount of mass in order that what remains is in the permissible range of masses for stable neutron stars.

While the question of the ultimate fate of massive stars with all its implications was not faced until recently, the theory of the white-dwarf stars, based on the relativistic equation of state for degenerate matter, gained gradual acceptance during the forties and fifties. The principal astronomical reasons for this acceptance were twofold. First, the number of the known white-dwarf stars had, in the meantime, increased very substantially, largely through the efforts of Luyten; and the study of their spectra, particularly by Greenstein, confirmed the adequacy and in some cases even the necessity of the theoretically deduced mass-radius relation. And second, since a time scale of the order of ten million years, for the exhaustion of the nuclear sources of energy of the massive stars, requires the continual formation of these stars, one should be able to distinguish a population of young stars from a population of old stars. And spectroscopic studies provided evidence that the chemical composition of the young stars differs systematically from the chemical composition of old stars; and in fact the difference is in the sense that the young stars appear to have been formed from matter that has been cycled through nuclear reactions. And this last fact is consistent with the picture that during the course of the evolution of the massive stars a large fraction of their masses is returned to interstellar space. And it also seemed likely that this returning of processed matter to the interstellar space was via the supernova phenomenon.

While all these ideas became a part of common belief, it remained only as belief. Their full implications were not seriously explored before the discovery of the pulsars. The discovery, in particular, of a pulsar (with the shortest known period) at the centre of the Crab nebula added much credence to the views that I have described since the Crab nebula is itself the remnant of the supernova explosion. The discovery of the association of further pulsars (of longer periods) with what are believed to be the remnants of more ancient supernova explosions strengthens one's conviction. The story of the pulsars and their identification with neutron stars are matters of common knowledge, but I should like to refer to only one aspect that is related to the theory that I have been describing.

As we have seen, the construction of degenerate configurations on the Newtonian theory of gravitation leads to a limiting mass for these configurations; and at the limit the radius literally tends to zero unless, that is, no other factor intervenes. And what intervenes is general relativity. Mentioning general relativity I should like to briefly recall the following incident.

During my first year in Cambridge (1930–31), R. H. Fowler, who was my official supervisor, was absent in the United States. On this account, there were occasions when I consulted Dirac. My interests at that time were in the theory of white dwarfs.

While Dirac did evince encouraging interest in these matters, he nevertheless told me that if he should be interested in astrophysics he should rather be working on general relativity and cosmology. That was in 1930. It has taken me well over thirty years to take the first part of his advice!

A natural problem for someone with training in classical astrophysics is the radial pulsation of stars and the related question of their dynamical stability. In other words, the re-examination of the pulsation problem that was Eddington's first consideration in the modern theory of the internal constitution of the stars that he initiated.

The discussion of the radial pulsations of stars in the framework of general relativity happens to be a particularly simple problem. Indeed, Eddington could have solved it in 1918. The principal result of general relativity here is that stars that would be considered stable in the framework of the Newtonian theory can become unstable in general relativity. More precisely, this instability of relativistic origin would set in whenever the ratio of the specific heats is close to four-thirds.[12] This is the case for degenerate configurations near the limiting mass; and the application of the relativistic criterion shows that they become dynamically unstable before the limiting mass is reached. Precisely what happens is the following. On the Newtonian theory, it can be shown that the period of radial pulsation decreases monotonically to zero as we approach the limiting mass; but in the framework of general relativity, because of the instability it causes, the period attains a minimum and then tends to infinity prior to the limiting mass, where the sequence becomes unstable. In other words, while general relativity does not modify to any appreciable extent the structure derived from the Newtonian theory, it changes qualitatively the period mass relation; it exhibits a minimum period that was absent in the Newtonian theory. And this minimum period is about seven-tenths of a second. Since pulsars are known to have periods much shorter than this minimum value, the possibility of their being white-dwarf configurations was ruled out.

The principal conclusions that follow from these theoretical and observational studies can be summarized very simply.

Massive stars in the course of their evolution must collapse to dimensions of the order of ten to twenty kilometres once they have exhausted their nuclear source of energy. In this process of collapse, a substantial fraction of the mass will be returned (as processed matter) to the interstellar space. If the mass ejected is such that what remains is in the permissible range of masses for stable neutron stars, then a pulsar will be formed. The exact specification of the permissible range of masses for stable neutron stars is subject to uncertainties in the equation of state for neutron matter; but it is definite that the range *is* narrow: the current estimate is between 0.3 to 1.0 solar mass. While the formation of a stable neutron star could be expected in some cases, it is clear that their formation is subject to vicissitudes. It is not in fact an *a priori* likely event that a star initially having a mass of say ten solar masses, ejects during an explosion, subject to violent fluctuations, an amount of mass just sufficient to leave behind a residue in a specified narrow range of masses. It is more likely that the star ejects an amount of mass that is either too large or too little. In such cases the residue will not be able to settle into a finite state; and the process of collapse must continue indefinitely till the gravitational force becomes so strong that what Eddington concluded is a *reductio ad absurdum* must in fact happen: 'the gravity becomes strong enough to hold the radiation.' In other words, a black hole must form; and it is to these that I now turn.

Let me be more precise as to what one means by a black hole. One says that a black hole is formed when the gravitational forces on the surface become so strong that light cannot escape from it. That such a contingency can arise was surmised, already, by Laplace in 1798[13]. Laplace argued as follows. For a particle to escape from the surface of a spherical body of mass M and radius R, it must be projected with a velocity such that

$$\tfrac{1}{2}v^2 > GM/R;$$

and it cannot escape, if

$$v^2 < 2GM/R.$$

On the basis of this last inequality, Laplace concluded that if

$$R < 2GM/c^2 = R_S \quad (\text{say}),$$

where c denotes the velocity of light, then light will not be able to escape from such a body and we should not be able to see it.

By a curious coincidence, the limit R_S discovered by Laplace is *exactly* the same that general relativity gives for the occurrence of a *trapped surface* around a spherical mass. (A trapped surface is one from which light cannot escape to infinity.) While the formula for R looks the same, the radial coordinate r (in general relativity) is so defined that $4\pi r^2$ is the area of the 3-surface of constant r; it is not the proper radial distance from the centre.

That for a radial coordinate $r = R_S$, the character of space-time changes is manifest from the standard from the Schwarzschild metric that describes the geometry of space-time external to a spherical distribution of a total (inertial) mass M located at the centre. The metric is given by

$$ds^2 = -c^2\left(1 - \frac{2GM}{c^2 r}\right)dt^2 + \frac{dr^2}{1 - 2GM/c^2 r} + r^2\left(d\theta^2 + \sin^2\theta\, d\phi^2\right).$$

For a mass equal to the solar mass, the Schwarzschild radius R_S has the value

$$R_S = 2.5 \text{ km}.$$

At one time, the thought that a mass as large as that of the sun could be compressed to a radius as small as 2.5 km would have seemed absurd. One no longer thinks so: neutron stars have comparable masses and radii. Also one must remember that one need not associate high density with the occurrence of trapped surfaces: for $M = 4 \times 10^8\, \odot$, for example, $R_S = 10^{14}$ cm and the mean density of the enclosed matter is only $10^{-1/2}$ grams cm^{-3}.

From what I have said, collapse of the kind I have described must be of frequent occurrence in the galaxy; and black holes must be present in numbers comparable to, if not exceeding, the pulsars. While the black holes will not be visible to external observers, they can nevertheless interact with one another and with the outside world through their external fields. But one important generalization is necessary and essential.

It is known that most stars rotate. And during the collapse of such rotating stars, we may expect the angular momentum to be retained except for that part of it which may be radiated away in gravitational waves. The question now arises as to the end result of the collapse of such rotating stars.

One might have thought that the inclusion of angular momentum would make the problem excessively complicated. But if the current ideas are confirmed, the expected end result is not only simple in all essentials, it also provides n possibilities for the astrophysicist.

In 1963, Kerr[14] discovered the following solution of Einstein's equations for the vacuum which has two parameters M and a and which is also asymptotically flat:

$$ds^2 = -\frac{\Delta}{\varrho^2}(dt - a \sin^2 \theta \, d\phi)^2 + \frac{\sin^2 \theta}{\varrho^2}[(r^2 + a^2) \, d\phi - a \, dt]^2$$
$$+ \frac{\varrho^2}{\Delta} dr^2 + \varrho^2 \, d\theta^2,$$

where

$$\varrho^2 = r^2 + a^2 \cos^2 \theta \quad \text{and} \quad \Delta = r^2 - 2Mr + a^2.$$

(The solution is written in units in which $c=G=1$; and in a system of coordinates introduced by Boyer and Lundquist.)

Kerr's solution has *rotational symmetry* about the axis $\theta=0$: none of the metric coefficients depend on the cyclic coordinate ϕ. It is, moreover, *stationary*: none of the metric coefficients depend on the coordinate t which is time for an observer at infinity. And Kerr's solution reduces to Schwarzschild's solution when $a=0$.

A test particle describing a geodesic in Kerr's metric at a large distance from the centre will describe its motion as in the gravitational field of a body having a mass M and an angular momentum aM (as deduced from the Lens–Thirring effect).

It is now believed that the end result of the collapse of a massive rotating star is a black hole with an external metric that will eventually be Kerr's, all the asymmetrics having been radiated away. I shall not attempt to explain the reasons for this belief except to say that they derive, principally, from a theorem of Carter[15] (see also Chandrasekhar and Friedman[16]) which essentially states that *sequences of axisymmetric metrics, external to black holes, must be disjoint*, i.e. have no members in common.

The Kerr metric, like Schwarzschild's, has an event horizon: it occurs at

$$r = \frac{G}{c^2}[M + (M^2 - a^2)^{1/2}]. \tag{E}$$

In writing this formula, I have assumed that $a < M$; if this should not be the case, there will be no event horizon and we shall have a '*naked* singularity', i.e. a singularity that will be visible and communicable to the outside world. For the present, I shall restrict myself to the case $a < M$.

Trajectories, time-like or null, can cross the event horizon from the outside; but they cannot emerge from the inside. In this respect also the Kerr black hole is like the Schwarzschild black hole. But unlike the Schwarzschild metric, the Kerr metric defines another surface external to the event horizon whose equation is given by

$$r = \frac{G}{c^2} [M + (M^2 - a^2 \cos^2 \theta)^{1/2}]. \tag{S}$$

This surface touches the horizon at the poles; and it intersects the equator $(\theta = \pi/2)$ on a circle whose radius $(= 2GM/c^2)$ is larger than that of the horizon.

On the surface, an observer who considers himself as staying in the same place must travel with the local velocity of light: like Alice, he must run as fast as he can to stay exactly where he is! Light emitted by such an observer must accordingly appear as infinitely red-shifted to one stationed at infinity.

The occurrence of the two separate surfaces (E) and (S) in the Kerr geometry gives rise to unexpected possibilities. These possibilities derive from the fact that in the space between the two surfaces – termed the *ergosphere* by Wheeler and Ruffini – the coordinate r, which is time-like external to the surface (S), becomes space-like. Therefore, the component of the four-momentum in the t-direction, which is the conserved energy for an observer at infinity, becomes space-like in the ergosphere; it can accordingly assume here negative values. In view of these circumstances, we can contemplate a process in which an element of matter enters the ergosphere from infinity and splits here (in the ergosphere) into two parts in such a way that one part, as judged by an observer at infinity, has a negative energy. Conservation of energy requires that the other part acquire an energy that is in excess of that of the original element. If the part with the excess energy escapes along a geodesic to infinity while the other part crosses the event horizon and is swallowed up by the black hole, then we should have extracted some of the rotational energy of the black hole by reducing its angular momentum. The possibility that such processes can be realized was first pointed out by Penrose[17].

The fact that energy can be extracted from a Kerr black hole raises the question of its secular stability since it is known that rotating systems can become unstable in the presence of dissipative mechanisms. In the case of the Kerr geometry, dissipation by the emission of gravitational waves is clearly a possibility.

The only case of secular instability that is fully understood is that which occurs along the Maclaurin sequence of rotating homogeneous masses. Along this sequence secular instability by viscous dissipation occurs at the point where the Jacobian sequence bifurcates. In the absence of viscous dissipation, no instability occurs at this point; but if viscous dissipation is present, then instability occurs (with an e-folding time depending on the magnitude of the prevailing viscosity) by the neutral mode which transforms the Maclaurin spheroid into a Jacobian ellipsoid at the point of bifurcation[18]. When this same problem is considered by allowing for radiation-reaction terms that result from the emission of gravitational waves, one finds[19] that the Maclaurin spheroid does become secularly unstable for the same value of the eccentricity (for which viscosity causes instability in the Newtonian framework) but the mode by which the instability sets in is *not* the one which leads to the Jacobian sequence: instead, it is the one which leads to the Dedekind sequence[20]. (The Dedekind ellipsoid, in contrast to the Jacobi ellipsoid, is stationary in the inertial frame and derives its ellipsoidal shape from internal motions of uniform vorticity.) One may therefore ask whether along the Kerr sequence a Dedekind-like point of bifurcation occurs. A criterion for such occurrence has recently been established[21]. If by the application of this criterion it can be shown that there is a point along the Kerr sequence where secular instability does set in then, the astrophysical consequences will be immense in view of the existence of mechanisms (such as those of Penrose) by which energy of far larger amounts than by nuclear processes can be extracted from the rotational energy of the black hole.

REFERENCES

1. For accounts somewhat similar to the present one but with differing emphasis, see S. Chandrasekhar, *Amer. J. Phys.* **37**, 577 (1969) and *Observatory* **92**, 160 (1972).
2. A. S. Eddington, *Internal Constitution of the Stars*, Cambridge University Press (1926), p. 173.
3. R. H. Fowler, *Proc. Roy. Soc. A*, **113**, 432 (1926).
4. R. H. Fowler, *Mon. Not. Roy Astron. Soc.* **87**, 114 (1926).
5. S. Chandrasekhar, *Phil. Mag.* **11**, 592, 594 (1931).
6. S. Chandrasekhar, *Astrophys. J.* **74**, 81 (1931).
7. S. Chandrasekhar, *Mon. Not. Roy. Astron. Soc.* **91**, 456 (1931).
8. S. Chandrasekhar, *Z. Astrophys.* **5**, 321 (1932).
9. S. Chandrasekhar, *Mon. Not. Roy. Astron. Soc.* **95**, 207 (1935).
9a. S. Chandrasekhar, *Observatory* **57**, 373, 377 (1934).
10. A. S. Eddington, *Observatory* **58**, 38 (1935).
11. J. R. Oppenheimer and G. M. Volkoff, *Phys. Rev.* **55**, 374 (1939).
12. S. Chandrasekhar, *Astrophys. J.* **140**, 417 (1964).
13. P. S. Laplace, *Système de Monde*, Book 5, Chapter VI (as quoted by A. S. Eddington in *Internal Constitution of the Stars*, Cambridge University Press (1926); p. 6).
14. R. P. Kerr, *Phys. Rev. Letters*, **11**, 237 (1963).
15. B. Carter, *Phys. Rev. Letters*, **26**, 331 (1971).
16. S. Chandrasekhar and John L. Friedman, *Astrophys. J.* **177**, 745 (1972).
17. R. Penrose, *Nuovo Cimento*, Serie I, **1**, 252 (1969) and *Nature*, **236**, 377 (1972).
18. S. Chandrasekhar, *Ellipsoidal Figures of Equilibrium*, Yale University Press (1969), p. 98.
19. S. Chandrasekhar, *Astrophys. J.* **161**, 561 (1970).
20. S. Chandrasekhar, *Ellipsoidal Figures of Equilibrium*, Yale University Press (1969), p. 124.
21. S. Chandrasekhar and John L. Friedman, *Astrophys. J.* (1973) in press.

4

Fundamental Constants and Their Development in Time

P. A. M. Dirac

THE LARGE NUMBERS HYPOTHESIS

The information that experimentalists obtain provides us with a number of constants. These constants usually have dimensions, and then of course, they depend on what units one uses, whether centimetres or inches. Then they are not of theoretical interest. However, one can combine these constants and get from them some which are dimensionless, quite independent of the system of units one uses. It is only the dimensionless ones that are of interest in our theoretical discussion, and it is only these that I shall be dealing with today.

Some of the important ones are the reciprocal of fine structure constant, $\hbar c/e^2$ which is about 137, and the ratio of the mass of the proton to the mass of the electron, m_p/m_e, which is somewhere around 1800. Unless one goes into detail, these numbers can be counted as near to one. Now physicists believe that there is an explanation for them. At present, we do not know why they should have the values they have, but still one feels that there must be some explanation for them, and when our science is developed sufficiently, we shall be able to calculate them. We shall then have them turning up as certain combinations of 4π and other factors which occur in our equations.

Now there is another number which is also of fundamental importance, that is the ratio of the electric to the gravitational force between the electron and the proton. The electric force is e^2/r^2. The gravitational force is $\gamma m_p m_e/r^2$, where γ is constant. Their ratio is independent of r, namely $e^2/\gamma m_p m_e$. One can calculate the value of this dimensionless number and it turns out to be somewhere of the order of 10^{39}. It is an extremely large number, quite on a different footing from the others that we had. The question arises 'How can we explain it?' How can we possibly have such an enormous number occurring as a result of any theory? That provides a serious problem.

Now another enormous number is provided by cosmology. We have evidence that the universe is expanding, because the spiral nebulae are receding. At a certain time in the past, all the matter in the universe was collected very close together. This forms a sort of origin of time, and we can refer our present time to this origin. We get, in that way, an age for the universe. According to the latest estimates, it is about 2×10^{10} years. We can express this age, which is usually called the epoch, in terms of atomic units, instead of rather artificial things like years. Atomic units would be measured by some kind of atomic clock, it doesn't matter very much which particular atomic clock

J. Mehra (ed.), The Physicist's Conception of Nature, 45-54. *All Rights Reserved*
Copyright © 1973 by D. Reidel Publishing Company, Dordrecht-Holland

we use. When we write the epoch in these units, we again get a number somewhere near 10^{39}. It may differ by a few factors of ten, depending on which particular atomic clock we use, but still it is an enormous number, about the same as the one coming from the ratio of the electric to the gravitational fields.

Now, this connection, I believe, should be considered not as an accident. There must be some reason for it. We don't know what that reason is, but the more existence of the connection allows us to understand this number here, $\gamma m_p m_e / e^2$. It is just the epoch multiplied by a factor close to one.

If we accept this point of view, then we are led to a rather strange conclusion. The epoch is not constant. It gets bigger as the universe gets older. And if $\gamma m_p m_e / e^2$ is connected with it, it also must not be constant. It must get bigger as the epoch increases and be proportional to the epoch. We come to the conclusion that the gravitational constant γ, measured in atomic units, is not constant, but varies with the epoch t, in proportion to t^{-1}. Gravitational forces are getting weaker compared with electric forces.

That is the starting point of the theory which I am going to present to you. It involves the fundamental assumption that these enormous numbers are connected with each other. The assumption should be extended to assert that, whenever we have an enormous number turning up in nature, it should be connected to the epoch and should, therefore, vary as t varies. I will call this the Large Numbers Hypothesis.

It has important consequences for cosmology. Various theories of cosmology have been proposed. One may assume, as a reasonable basis, that Einstein's theory of relativity holds. If one then applies Einstein's equations to a universe which is assumed to be roughly homogeneous, one gets various cosmological models, which have been worked out by Friedmann. Some of these models involve the universe starting out very small, expanding up to a maximum size, then contracting again. For such a model to be acceptable we must suppose that we are at present in the expanding stage, and are observing the expansion through the recession of the galaxies. But we see that such a model contradicts our Large Numbers Hypothesis. It involves a maximum size of the universe. If we measure that maximum size in terms of atomic units, we get a very large number, and it will be a constant. Now the existence of a very large number which is a constant is just what our Large Numbers Hypothesis excludes.

THE LAW OF EXPANSION

Let us try and figure out the law of expansion of the universe.

Consider a neighbouring galaxy and express its distance from us in terms of atomic units. We take the velocity of light to be unity, so that atomic clocks, which provide a standard of time, will also provide a standard of distance. The distance of the neighbouring galaxy may then be written $kf(t)$, where k is a constant depending on which neighbouring galaxy we take and the function $f(t)$ expresses the law of expansion of the universe.

If we are not interested in the start of the universe, then we are concerned only with the asymptotic behaviour of $f(t)$ for large t. According to the Large Numbers Hypothesis, the asymptotic expression for $f(t)$ cannot contain any very large constants. It must therefore be of the form t^n er possibly log t. The velocity of recession of the neighbouring galaxy is then knt^{n-1} or k/t. To determine the law of expansion we must decide on the value of n or whether we have the log case. If n is equal to one, we have uniform expansion. If n is greater than one, we have an accelerating expansion. If n is less than one or if we have the log, there is a decelerating expansion.

Now there is a fairly simple argument for deciding between these possibilities. If we take a galaxy which is close to our own, its velocity of recession is somewhere of the order 10^{-3}, taking the velocity of light to be unity. Now if the present velocity of recession is 10^{-3} and we take a value of n greater than one, this velocity of recession will increase and will eventually become greater than one, and that's something which we cannot tolerate. Thus an accelerating expansion is not in agreement with our Large Numbers Hypothesis.

How about the other possibility, n less than one? It would mean that the velocity of recession was greater in the past than it is now, and if we go back in time a sufficiently long way, we shall come to an epoch when this velocity was greater than one. Now clearly the whole character of the universe must have been different at that time. How long ago was that time? Let us take an example, $n=3/4$. Then knt^{n-1}, if it is now 10^{-3}, was equal to 1 at the epoch $t=10^{27}$. At that epoch, even the closest neighbouring galaxy was receding with the velocity of light. There must have been some fundamental change then in our model of the universe. This particular epoch would be something which is very characteristic, and it involves a large number, not quite as big as the 10^{39}, but still too big to be allowed by our Large Numbers Hypothesis. So this case is again to be ruled out. Any value of n appreciably less than 1 and also the log case is thus to be ruled out. We are left only with the possibility $n=1$ or very close to 1. This is a preliminary theory, and we are not concerned with fine points, so we just take $n=1$. That means the expansion is uniform, and the velocity of recession is constant.

I first worked on this idea before the war, but I did not develop it correctly and there are mistakes in the early work. What I am giving now is different from my pre-war version, and I hope is an improvement on it.

CONTINUOUS CREATION OF MATTER

I want now to introduce another concept, the total amount of matter in the universe. We can express this in terms of the mass of the proton or neutron and get a dimensionless number. A number has to be dimensionless for this discussion. Now the value of this number is rather uncertain because we don't know how much invisible matter there is in the form of balck holes or interstellar gas, but one can make reasonable assumptions. If one assumes the amount of invisible matter is not so very much greater than the amount of visible matter, one gets for the total amount of matter in the universe a number somewhere around 10^{78} in proton masses, thus a number near the square of an epoch.

According to our Large Numbers Hypothesis, this number should be equal to the square of the epoch multiplied by some factor close to one, and therefore it must be increasing proportionally to the square of the epoch. That means that we have continuous creation of matter.

Now there has been a good deal of talk by cosmologists about continuous creation of matter, but that was in connection with another theory – steady state theory. What I am talking about is not the steady state theory; it is the evolutionary theory, where there is a definite start for the universe. This evolutionary theory, or the Big Bang Theory as it is often called, also requires continuous creation of matter when developed in accordance with the Large Numbers Hypothesis. I can see no escape from this requirement. It is just as forced upon us as the variation of the gravitational constant in the first place.

Now if we have continuous creation of matter, the problem immediately arises – Where is this matter created? There would seem to be three possible assumptions which one might make. *Assumption 1.* We may suppose that matter is created uniformly throughout space. *Assumption 2.* Matter is created where it already exists in proportion to the amount existing there. *Assumption 3.* Matter is created in special places such as the centres of galaxies, or something like that. I don't know of any decisive argument for deciding between those three possibilities, but one may have a preference on general considerations.

First of all, this assumption 3 I don't like at all. I don't see how one can proceed to set up any dynamical theory if one has matter sometimes being continuously created, and in other places not being continuously created. So I will rule out assumption 3.

We still have assumptions 1 and 2. What would they mean? Assumption 1 would mean that most of the matter that is created is created in intergalactic space, space which is very close to being a vacuum. It would seem that the vacuum has a sort of spontaneous radioactivity, being able to create matter, and this process occurs all the time and everywhere.

Assumption 2 could also be looked upon as a kind of radioactivity. Matter anywhere is multiplying itself. If we are not to get disagreement with a large body of established fact, we must assume with hypothesis 2 that the newly created matter consists of atoms of the same nature as those already existing there. Thus where there is hydrogen, there would be new hydrogen created; where there is iron, there would be new iron atoms created, and so on.

There is no really decisive argument for deciding between the assumptions, but I prefer assumption 2 because it can be fitted in with Einstein's theory of gravitation. Einstein's theory requires that matter should be conserved on the large scale. If matter is continually being created throughout space, that would provide a very direct contradiction with Einstein's theory. We would have to modify Einstein's theory in some way, and it is not at all clear how that modification should be made.

The assumption 2, we can get to agree with Einstein's theory by saying that Einstein's theory refers to a unit of mass which is varying with respect to the atomic units. According to the assumption 2, the number of protons or neutrons in a piece of matter should be increasing proportionally to t^2. Now if we assume that Einstein's theory refers to a unit of mass which, with respect to the atomic units, is varying according to t^{-2}, then for these Einstein units we shall have the mass staying constant. With assumption 2, we can thus get agreement with Einstein's theory by using a suitable unit of mass. We have here the possibility of developing our cosmology to agree both with the Large Numbers Hypothesis and with Einstein's theory.

For this reason I prefer assumption 2, but I see that there are difficulties in connection with it. It would require that matter everywhere shall be multiplying. The rate of multiplication is really quite slow. Using the value 2×10^{10} years for the epoch, it would mean that each atom has a chance of creating a new atom nearby of the order 10^{-10} per year. It is a very slow rate of increase, but still it would have important consequences, applied to the earth, when one considers the passage of geological ages. I think that Jordan will later on be talking about the effect of an assumption like this on the structure of the earth. I would just like to call attention to one particular difficulty. We have very old rocks which may show a crystalline structure. These rocks have been existing for hundreds of millions of years, and during that time there has been quite an appreciable creation of new matter. This creation of new matter must have occurred in some way which does not disrupt the crystal structure. Somehow the new atoms which the crystal forms have to be displaced to the surface of the crystal so as not to disrupt the whole crystal structure. I don't know just what physical process can give that result, but in any case, whichever hypothesis we are using for the creation of matter, we have to depart from standard physics.

THE TWO METRICS

Consider the development of our cosmology with assumption 2, keeping to both the Large Numbers Hypothesis and Einstein's theory of gravitation. An immediate consequence is that when we consider the element ds connecting two neighbouring points in space-time, there are two expressions for it which are of interest to the physicist. One of them is the ds occurring in the Einstein theory; let me call that one ds_E. The other is ds_A, measured by atomic clocks. These two ds's are not the same, because for one thing the unit of mass for one picture is different from the unit of mass for the other. We must try and get the connection between ds_E and ds_A. A simple argument will lead to the result.

Let us consider the motion of the earth in its orbit around the sun, neglecting the eccentricity. Let us just use Newton's approximation. We have then $\gamma M = v^2 r$, γ the gravitational constant multiplied by M the mass of the sun is equal to v^2, the velocity squared of the earth in its orbit, times r the radius of the orbit. Now this equation, one of the fundamental equations of mechanics, must hold for all systems of units, the Einstein units or the atomic units. The velocity v is the same with either system of units, because it is just a certain fraction of the velocity of light. Now for Einstein units, γ of course is a constant, in fact it is 1. M, V, γ, and r are all constant in Einstein units.

Now how about atomic units? The number of all nucleons in the sun must be multiplying according to the law t^2, which is the general law for the multiplication of nucleons everywhere. Thus M is proportional to t^2. Also γ is proportional to t^{-1}. This is what we had right at the beginning, and v is constant again, so r is proportional to t. Referred to atomic units, the earth's orbit has to be increasing proportionally to the epoch. But referred to Einstein units it is a constant. Now this example of the radius of the earth's orbit applies to a general ds. ds in Einstein units is thus connected with ds in atomic units by the equation: $ds_A = t\,ds_E$. Here we have the connection between the two metrics which are needed by physicists.

If we use Einstein units, we can measure the epoch in terms of those units and get a quantity, let us call it τ, for the epoch in Einstein units. If we connect it with t, which is the epoch in atomic units, we get $dt = t\,d\tau$, leading to $\tau = \log t$. It means that the Einstein units provide us with a different picture, in which the universe has existed throughout all time instead of having a definite start, $\tau = -\infty$ for $t = 0$.

I might mention as a point of history that the idea of two measures for the epoch connected in this way is not a new one. It was put forward by Milne* before the war, and he worked out a lot about it. But Milne's arguments were quite different from mine. He did not use the Large Numbers Hypothesis, and the details of his equations are different.

We have now two systems of units. We use the Einstein units for classical mechanics for working out mechanical problems. We use the atomic units for quantum theory. This means of course that there is some kind of departure from the standard correspondence principle connecting classical mechanics and quantum mechanics. But I don't think there is any objection to that, because the connection between the two units is one which varies only extremely slowly for laboratory purposes.

When we use Einstein units, all the atomic constants become variable. One can quite easily work out the way in which they vary. One finds, for instance, that the charge on the electron in Einstein units is proportional to $t^{-3/2}$ and h is proportional to t^{-3}; so e^2/h is constant. One might think that e^2/h could vary in a logarithmic way with the epoch. In fact Gamow has investigated that possibility, but there is observational evidence against it. So I am sticking to the view that e^2/h remains constant.

What is the picture of the universe that we have when we refer it to Einstein units? If we take the distance from our galaxy to a neighbouring galaxy and measure it in atomic units, it increases proportional to t. Thus ds_A is proportional to t. So ds_E is constant. Thus referred to Einstein units, the galaxies are not receding at all. We get therefore the Einstein picture of a closed universe with constant radius, the original Einstein model.

How is the recession of the nebulae to be understood with reference to this model? With respect to Einstein units, atomic clocks are continually speeding up, because with ds_A constant, ds_E is proportional to t^{-1}. Now if we think of light emitted from a distant nebula a long time ago, the wave length of that light was referred to atomic

* See Milne's books, *Relativity, Gravitation and World Structure* and *Kinematic Relativity*, Clarendon Press.

clocks at the time when it was emitted. The wave length of the light remains constant during its long journey to us with respect to the Einstein units. But when this light reaches us, it is to be referred to atomic clocks of the present time which are running faster. And thus the wave length of this light from a distant galaxy appears to be red-shifted. We have this picture in terms of the Einstein units which is, of course, mathematically equivalent to the previous picture and explains the red-shift equally well.

ASTROPHYSICAL CONSEQUENCES

Let us consider the question of how the temperature of the earth has varied during geological times. Teller worked out a theory for the variation of the luminosity of a star like the sun, according to which the luminosity varies in proportion to $\gamma^7 M^5$. Now if γ and M are both constant, as they are according to ordinary ideas, and the radius of the earth's orbit is also constant, we get the standard theory which leads to the earth's temperature being fairly constant. According to our present theory, γ is proportionate to t^{-1} and M to t^2, so the luminosity of the sun should be increasing according to the law t^3. The radius of the earth's orbit is also increasing proportional to t. So the intensity of heat which we receive from the sun is proportional to t. Thus in the geological past the earth was not receiving so much heat from the sun and would have been cooler than it is now. There may have been some compensation due to radioactivity in the earth. But in any case the temperature would not have varied rapidly, say according to the law $t^{1/4}$.

There is evidence that life has existed for something like 3×10^{10} years. During the passage of that period of time, the temperature of the earth would not have changed very much. The temperature would not have been unreasonable for life to have existed at that early age. There is no discrepancy there.

Gamow has worked on this question taking into account the variation of γ, but not the variation of M coming from continuous creation of matter. He obtained vastly different results, which were hard to reconcile with the existence of life such a long time ago, and he concluded that γ could not vary.

I would like to refer to some more drastic consequences of this theory. I talked about continuous creation of matter, but this requires also continuous creation of any form of mass in order to have consistency with Einstein's theory. If we just take a beam of light, the number of photons in it has to be increasing just like the number of nucleons in matter is increasing. That means that when we consider light coming to us from a distant galaxy, the number of photons has increased. The light is red-shifted and the energy gets less for that reason, but still the number of photons is increased and the apparent brightness is thus greater than it would otherwise be.

If we had some way of observing the absolute brightness of a distant galaxy, then we should find it very easy to check this result. But we don't have any way of determining the absolute brightness of a distant object. All we can see is its apparent brightness. If there is this factor of photon multiplication which increases the apparent brightness, that won't immediately give us any result in contradiction with observation.

If we had some other information about absolute brightness, then we could check it. We might assume that the absolute brightness of a distant galaxy is the same as that of a near one, on the average. The assumption is not really a reliable one, because we are observing the galaxies a long time ago, and things may have changed in the meantime. But if we do make that assumption, the present theory would lead to the conclusion that the number of galaxies with a certain apparent brightness is substantially greater for the smaller values of the apparent brightness. Sciama, in his talk, did refer to the observational evidence that there is a rather excessive number of distant galaxies with smaller values of apparent brightness, and this might be an explanation for it. But, as I mentioned, it is not altogether reliable because it does involve the assumption that absolute brightness doesn't change very much as the age of the universe changes.

There is another problem, in connection with the microwave radiation. The number of photons in the microwave radiation must also be increasing. Now if the microwave radiation is really a black body radiation, that would provide difficulty for this theory. Black body radiation is such that, expanding with the expanding universe according to the standard ideas with the number of photons unchanged, it remains black body radiation, and just the temperature falls. With the assumption of multiplication of photons, if the radiation is initially grey, it would become less grey and would approach black with the passage of time. If the natural radiation of the present time just happens to be black, that would be a difficulty for the theory, because it would mean a coincidence of a kind which one would not like. As Sciama mentioned, the observations at present are not at all conclusive with regard to whether the microwave radiation is black body radiation or not.

UNIFIED FIELD THEORY

I would like now to leave these astrophysical questions and talk a little about another subject. This is the question of unified field theory. There are two fields in physics for which the forces are long range – the gravitational field and the electromagnetic field. Einstein has explained the gravitational field in terms of geometry. As soon as Einstein did that, people began to wonder whether there should not be a similar explanation for the electromagnetic field. People set to work to try and unify the gravitational and the electromagnetic fields into a single geometric scheme, involving a more general geometry than the Riemannian geometry which Einstein needs.

Very soon, a remarkable way of unifying the electromagnetic and the gravitational fields was proposed by Weyl. I would like to mention the main idea of Weyl's theory. According to Einstein, there is a definite ds expressing the interval between neighbouring points, which has an absolute value. According to Weyl, this ds is not a well defined quantity. It has to be referred to some local standard. We may make a displacement of ds by a process like parallel displacement and then refer it to the local standard at the new point, and that would give us the variation of ds with parallel displacement. When we move around a closed loop, according to Weyl, ds does not end up with its original value, and that is the reason why we need the local standards of ds.

This was a more general geometry than Einstein's, and the new variables which appear in it can be connected immediately with the potentials needed to describe the electromagnetic field. A variation of the local standards gives a variation of the electromagnetic potentials leaving the field quantities unchanged. Weyl's theory thus provided a very beautiful synthesis of the electromagnetic field and the gravitational field. But it was not acceptable to physicists. And the reason for that is that Weyl's theory contradicted quantum phenomena.

The quantum phenomena give an absolute standard for measuring ds. We may just take an atomic clock, for instance, and measure all our ds with respect to it. There is then just no room for the uncertainty which Weyl requires. So physicists were reluctantly compelled to abandon Weyl's theory and to search for other ways of unifying the electromagnetic field with the gravitational field. That search has continued for decades without leading to any satisfactory solution. People come up with a solution from time to time, but it is always rather artificial and complicated and is not generally accepted. That remains the position to the present day.

Now let us look at the question again from the point of view that there are two metrics ds_E and ds_A. We see at once that the objections to Weyl's theory disappear. If Weyl's theory is to be applied at all, it must apply to ds_E. It will require ds_E to depend on our local standard of distance and to be subject to variation when we change the electromagnetic potentials. On the other hand ds_A is an absolute thing which doesn't depend on the choice of the electromagnetic potentials. All that we need to reconcile these two results is that the connecting factor between ds_E and ds_A shall not be exactly equal to t. It must be something which is approximately equal to t, and subject to variation when the electromagnetic potentials are changed. There is then no objection to Weyl's theory. We can accept it, and get in that way a very beautiful synthesis of the electromagnetic field and the gravitational field.

BROKEN SYMMETRIES

There is one consequence that we can deduce from that fairly easily. Let us consider a single particle in space-time. Let us take a field point close to its world line, and let us take an element ds at this field point. Now let us shift the field point into the future, in a direction roughly parallel to the direction of the world line, and let us ask how this ds varies when we shift it. We compare the shifted ds to the new local standard at the new point.

Now let us suppose that the particle is charged, so that there is the Coulomb force around it. That will require that there is an electrostatic potential at the field point, a Coulomb potential. Now this electrostatic potential, according to Weyl's theory, just gives the change in ds, referred to the local standard.

If we have a certain sign of the charge, we shall have, let us say, ds increasing when we go into the future. If we take the opposite sign of the charge, ds will be decreasing. Now, there is no symmetry between a thing increasing and decreasing. In the case of very small changes, there is an approximate symmetry, but accurately there is no symmetry between a thing getting bigger and a thing getting smaller. The result is that we shall have no symmetry between positive and negative electric charges.

Also, there is no symmetry between past and future. If ds is increasing when we go into the future, it is decreasing when we go into the past. Again, this is no symmetry. If we apply both operations, changing the sign of the charge and interchanging past and future, we get back to the original situation.

Now these results can be described in the terms of modern physics, such as Yang was using yesterday. In this way, C the operator which changes the sign of the charge, is not invariant; T, the operator which interchanges past and future, is not invariant; but the product CT is invariant. There is also a parity operation P, which is used in connection with these operations, and in the present theory the parity operation is invariant. We have a theory, therefore, with P invariant and C and T not invariant, and the product CPT is invariant.

According to experimental evidence, each one of these quantities P, C, or T is not invariant, although the product CTP is invariant. It appears, therefore, that the lack of invariance of P must be ascribed to some quantum phenomenon which cannot be explained classically. On the other hand, the lack of invariance of C and T can be explained classically. We can say that the lack of invariance of C and T arises from the interaction of the gravitational and the electromagnetic fields in accordance with Weyl's geometry. We have there a start on the problem of why there should be these violations in the symmetry laws.

I must say these ideas are very new and I haven't had time to work on them very much. The calculation ought to be made, of course, to find the order of magnitude of the violation of C and T from Weyl's theory and to compare it with the order of magnitude which is observed experimentally. That is a calculation which hasn't been done yet.

Well, that concludes the things I want to talk about, and I would like to say that I would welcome criticism of these ideas. They involve a big departure from accepted ideas, and there is liable to be strong opposition to them. I haven't talked about these ideas previously except at the meeting of the Pontifical Academy last April, and on that occasion there was no opportunity for discussion, so I would be very glad to have discussion on this occasion. Thank you.

Comment on Dirac's Paper

C. F. von Weizsäcker

Since Professor Dirac asked for response or criticism, I should like to make a remark about his talk. Personally I feel that his arguments are very convincing. It will certainly be necessary to test them empirically, and I can say nothing about that. I should like to add that the formula that the number of particles should be proportional to the square of the radius of the world, as measured in atomic units, seems to contain a certain degree of consistency which can only be brought out if you discuss the matter in quantum theory.*

A QUANTUM COSMOLOGY

According to Sciama's paper the universe is homogeneous and isotropic in space but evolving in time. The singularity in the beginning of time might be attenuated by a continuous creation of matter, as proposed by Dirac and Jordan. The comparison of this theory with recent experience might be worth an effort.

Let R be the radius of the universe as measured in elementary lengths. At present $R \approx 10^{40}$. The number n of baryons in the universe is at present $\approx 10^{80} = R^2$. One would assume this relation to hold at any time.

I should like to point out that this relation $n \approx R^2$ is the consequence of an attempt at a quantum cosmology which I have mentioned briefly in Section 5.3 of my paper on classical and quantum descriptions at this symposium.

An empirical question admitting of exactly k answers, or, to put it differently, a quantity admitting exactly k different possible values, is quantum-mechanically described by a k-dimensional Hilbert space. All truly empirical questions admit only a finite number of answers, but it will, in general, be possible to ask more questions in the future. A k-fold question can be subdivided into a number of twofold ('simple') alternatives or bits, each admitting precisely two answers. If we treat the alternatives as quantum objects, no loss of generality is implied by assuming Bose statistics for them. Any k-dimensional Hilbert space can be built up from $k-1$ two-dimensional Hilbert spaces as a fully symmetrical tensor product.

If the present universe is finite, we may try to describe it in a finite-dimensional Hilbert space. This implies a cut-off at small distances. We may assume the cut-off

---

* The following remarks on a 'quantum cosmology' had been prepared before hearing Dirac's talk on 'Fundamental Constants and Their Development in Time', and may be construed as indicating agreement with Dirac's ideas.

to be at 10^{-13} cm. This does not exclude that lengths smaller than 10^{-13} cm can be measured in particular experiments using high energies. But it means that the available energy would not permit all lengths to be determined simultaneously with a Δx smaller than 10^{-13} cm. The assumption can also be expressed by dividing the whole universe into $R^3 = 10^{120}$ elementary cells and by asking for every cell whether it is occupied by a baryon or not. The existence of other particles may add a factor of the order of 10, which we neglect here.

We can consider the simple alternatives as ultimate objects of the world. This is not an assumption but trivially possible as long as we leave their law of interaction indeterminate. It becomes a non-trivial assumption if we postulate that the law of interaction between the ultimate alternatives has the same symmetry as a single alternative. Let me, for brevity, call an ultimate alternative an 'ur' (from German 'uralternative'). The two possible answers are indicated by a two-valued variable $r = 1$ or 2. The Hilbert space of the ur consists of the complex vectors u_r. Its symmetry group will consist of $U(2)$ and complex conjugation. Complex conjugation will be described by introducing anti-urs. Systems of urs will best be described by (spinor) functions on this symmetry group. Its part $SU(2)$ is a three-dimensional real spherical space. In this description it takes over the role of cosmic space. This means: A simple alternative does not depend on any space that was given *a priori*, but a system of simple alternatives with an $SU(2)$-symmetry defines a space of the topological structure of the $SU(2)$ as its natural representation space. The hypothesis is that this *is* what we know as cosmic space.

For a world containing only a finite number of simple alternatives, continuous space is a space of parameters, but there is no position operator with continuous eigenvalues. In order to define what we mean by spatial measurements we must first describe how particles are built up from urs. In detail that will not be easier than elementary particle theory happens to be, but a few simple estimates can be made immediately. Let us speak of one type of elementary particle only, and let us call them baryons. They will be fermions. If the world consists of N urs we will have to find how many different baryon states can be built from them and filled by one baryon, respectively. Consider a particular baryon state consisting of m urs. m urs will admit of $(m+1) \approx m$ different states. Hence we can build m different baryons, each of which consists of m urs. Thus m^2 urs will be used up in filling all possible baryon states that consist of precisely m urs. If we fill all baryon states up to an upper limit of M urs, we will use up $M^3/3$ urs. If we use all N urs of the world for this purpose we will have $M^3 = 3N$. In this way our baryons would be a fully degenerate gas, and M would indicate the surface of the Fermi sea. Only baryons on or above the surface are observable. Hence M would be the minimum number of urs per observable baryon. Ergo the number n of observable baryons must be smaller than $N/M \approx N^{2/3}$. This consideration would be a step towards a derivation of the Dirac–Jordan law. The exponent 3 is of course mathematically connected with the three-dimensionality of the $SU(2)$. We can describe the same estimate by saying: A single ur cannot be localized in cosmic space (i.e. on the $SU(2)$). A particle localized with a $\Delta x \approx 10^{-13}$ cm must consist of 10^{40} urs. In a world consisting of 10^{120} urs (a number derived from the very assumption that particles can be so localized) there can be 10^{80} such particles.

Let N_+ be the number of urs, N_- the number of anti-urs. Our number N is the absolute number of ultimate alternatives, i.e. $N = N_1 + N_2$. But the interaction of urs will conserve $N_1 - N_2$. If N is small at some time, it will increase indefinitely. This is the expression of the fact that only finite questions can be answered at any time but that more questions can be asked in the future. Since $N \approx R^3$, an increase of N with time means an expanding universe. It means that an increasing number of urs will permit an increasingly fine subdivision of cosmic space. I take the creation of an ur to be essentially what Finkelstein calls a 'chronon' (a discrete time-event). Gravitation will be *defined* as the curvature of space. Its decrease with time is a consequence of the model. This theory fits into the general interpretation of time and irreversibility I give in my paper.

Just as the $SU(2)$ symmetry gives a reason why cosmic space is three-dimensional, the Lorentz symmetry of the empirical world must follow from the model. A fixed finite number of urs permits only non-unitary representations of the Lorentz group, but an indefinitely increasing number offers a basis for unitary representations. The general matrix of $U(2)$ can be written:

$$A = \begin{pmatrix} w + iz & x + iy \\ -x + iy & w - iz \end{pmatrix} (c + is) \tag{1}$$

with real parameters obeying the conditions

$$w^2 + x^2 + y^2 + z^2 = c^2 + s^2 = 1. \tag{2}$$

We call w, x, y, z cosmological space parameters, and c, s cosmological time parameters. In this description, space and time appear closed. Our actual measurement of space and time will be in atomic units. The relationship between atomic and cosmological measures may on average be given by a universal factor R. We define atomic co-ordinates by

$$\begin{aligned} x_0 &= Rw & x_4 &= Rs \\ x_1 &= Rx & x_5 &= Rc. \\ x_2 &= Ry \\ x_3 &= Rz \end{aligned} \tag{3}$$

The factor R must depend on time. But we will admit different definitions of atomic time, depending on the inertial motion of the observer. Different observers will consider different values of the co-ordinates as simultaneous. It is possible to define R as a function of atomic time such that there is a Lorentz transformation between different observers and that R depends on time only, for every observer. This is achieved by putting

$$R = c^{-1}. \tag{4}$$

From (2) we then get

$$x_0^2 + x_1^2 + x_2^2 + x_3^2 - x_4^2 = 1, \tag{5}$$

that is, a de Sitter universe. Yet values of the atomic co-ordinates, which are not very large compared with 1, are meaningless since there is no position operator. Thus we can use Equation (5) only on one side of the bottleneck, say for $x_4 \gg 1$. The universe is not empty, and one would have to devise equations of gravitation with a gravitational constant declining with increasing R that would admit this model as a solution. In such a theory the scalar R will be replaced by a tensor expressing local variations.

DISCUSSION

J. A. Wheeler: Dirac's starting point was the number 10^{39} for the ratio of the constants that measure the electric attraction between two particles, compared to the constants that measure the gravitational attraction between two particles. This number was compared with another number which had to do with the age of the universe, and everything flowed out of this – all sorts of things, the creation of particles, the change in the luminosity of the sun, and so on.

Could this be compared with the situation of the lady who had the bathtub in her house with the faucet on? The faucet was on, but the bathtub outlet was stopped up, and the bathtub started to run over. She got all the children to collect the water in all the teacups and in all the flower vases, and finally in all the suitcases and so on, to take care of this problem. Then somebody came along and said, well why not just turn off the water faucet?

My question, similarly, here has to do with the number 10^{39} which was identified with the *changing time* from the beginning of the universe to any particular epoch. Could it not equally well have been identified with the time from the start of the universe to the phase of maximum expansion, a number of the same order of magnitude, of 10^{39}, but a number which *does not change with time*? Then one would have eliminated all these issues that brought in such strange kind of physics.

P. A. M. Dirac: One would still have the difficulty of explaining the reason for this very big number. I just don't want to have any very big number which cannot be explained in some way or other. And I don't think you can explain it as being built up from 4 π's and so on.

J. A. Wheeler: How do you feel about the explanation of Brandon Carter that many cycles of the universe are possible and the constants in this particular cycle are such as will permit life?

P. A. M. Dirac: That doesn't get over the difficulty that you have to explain this very big number.

F. Perrin: I would like to ask Professor Dirac how he considers, besides the creation of matter, the possibility of creation of antimatter. Would he consider that in the hypothesis, which he prefers, matter would be created near in the neighbourhood of matter, and antimatter in the neighbourhood of antimatter, if such antimatter existed somewhere in the universe?

P. A. M. Dirac: I think that is very likely to be the case. I would agree with that.

J. Ehlers: I would like to ask Professor Dirac whether he could make a comment about how the Einstein type of metric quantities could be measured. You made the distinction between the atomic units and the Einstein units. Now I have an idea what it means to measure something in atomic units, I wonder whether you have an explicit prescription for what it would mean to measure the Einstein ds?

P. A. M. Dirac: I think it would be rather difficult to give an answer to that, because I am in favour of the Weyl geometry and the ds there is associated with some arbitrary standard of length. I would like to work a good deal more on the Weyl geometry before trying to answer that question. These ideas are all very new and I haven't had time to think about them very much.

I. Roxberg: May I address a question to Professor Dirac and possibly answer Professor Ehlers' question? There is, as Professor Dirac pointed out, quite a strong similarity between the theory he is putting forward here and the theory that was put forward by Milne some years ago. In Milne's theory one concluded that in the metric measured by atomic clocks and light signals, particles do not follow geodesics. The question to Professor Dirac is: do you not think that if you are thinking along the right lines it will be necessary to modify the field equations of general relativity in order to deduce something like particles moving along other worldlines and geodesics?

P. A. M. Dirac: My own theory is really very different from Milne's, and I don't accept Milne's arguments at all. For instance, the question of how *h* goes with the time is quite different in Milne's theory and in mine. All the details are different. I don't want to depart from Einstein's ideas about particles moving along geodesics except in so far as there is an electromagnetic field which will deflect them.

5

The Expanding Earth

Pascual Jordan

Since 1937 I have been deeply impressed and really fascinated by a certain idea of Dirac. He gave in his article of 1937 arguments in favour of the hypothesis that the so-called gravitational constant $\kappa = (8\pi/c^2)G$ might be in reality not a constant, but a function of the age of the universe. Thinking in the frame work of a Friedmann cosmological model we take the curvature of space and the average density of matter to be approximately constant in the whole space, but varying with time; and in the same manner the 'gravitational scalar' κ may be approximately spatially constant but decreasing in time, according to Dirac, in a first approximation inversely proportional to the age of the universe. Many discussions have been caused by this idea of Dirac, but the majority of participants in these discussions were inclined to deny its possibility. But I think that all objections made against this hypothesis are at least not better than the hypothesis itself.

I myself have always been inclined to think that this hypothesis may be probably correct. Today I believe that the logarithmic time derivative $-\dot{\kappa}/\kappa$ may have the order of 2×10^{-10} per year. Schapiro, from his highly admirable measurements of planetary radio echos, concluded that the absolute value $|\dot{\kappa}/\kappa|$ must be smaller than 4×10^{-10} per year.

If this idea of Dirac is correct, then we have to expect considerable consequences concerning our planet. For the earth is old enough to be affected considerably by such a variation of the gravitational scalar during the several billions of years of its history. Naturally a steady decrease of κ, if having taken place, must have caused a certain expansion of the earth. And it seemed to me highly interesting to see whether there are known empirical facts showing this consequence to be really false, so that one must conclude that also Dirac's hypothesis cannot be correct.

Naturally one cannot decide such a question by going to any geologist or geophysicist and asking him: 'Do you know something about an expansion of the earth during its history?' The answer would be: 'No, I never heard of such an idea.' And there exists a great diversity between the mentalities of physicists and of geologists. Physicists are eager to learn about new facts and new ideas caused by new facts. But many geologists are inclined to a more conservative style of thinking. This fact became visible in their reaction to the well known hypothesis of Alfred Wegener that Africa and South America might in the geological past have been situated together.

When Wegener put forward his hypothesis of continental drift, the reaction of geologists was not only mostly doubt about the possibility of such an idea; but these doubts, partially founded by impressive arguments, were connected with strong emotional disapproval and reproach. Today there is scarcely any serious author who would still doubt that Wegener correctly interpreted Africa and South America as two fragments of what formerly was one single great continent; but more than forty years elapsed before acknowledgement of Wegener's idea became general.

Considering this little fact of scientific history, and considering the psychological situation in those branches of science devoted to our planet and its development, I came to the decision not to ask other specialists about the compatibility of Dirac's hypothesis with empirical facts, but to try to learn myself what really are the proven facts, and to see whether they lead to real contradictions against Dirac's hypothesis. As a summary of what I intend to explain more carefully I can say that I did not find what might be called empirical evidence against this hypothesis.

The heading of this report about the geophysical consequences of Dirac's hypothesis coincides with the title of a book of mine, in its English translation: *The Expanding Earth* [Translated by A. Beer, Pergamon Press, Oxford (1971)].

Jeffreys after the publication of Wegener's idea made two severe objections:

(*a*) The similarity between the coastlines of Africa and South America is not sufficient to exclude a mere chance without deeper meaning.

(*b*) In order to move the floating continental blocks relative to the interior a huge

The Origin of the Oceans

Fig. 1. The joining of Africa and South America (after Carey).

resistance of friction must be overcome; therefore huge horizontal forces would be needed in order to cause such a movement.

Concerning point (*a*), Fig. 1 indicates the real situation as shown several years ago by Carey. The real boundaries of the continental blocks are not identical with the coast lines; one has to regard also the 'shelves', the areas of shallow sea, as parts of the continental blocks; these shelves are separated from the deep sea by the well defined 'continental slope' (compare Fig. 2). Correcting the boundaries of the

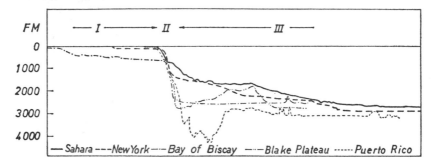

Fig. 2.   Profiles of the continental slope (after Heezen).

two continents accordingly Carey found a perfect fit between Africa and South America, leaving no doubt that mere chance is excluded convincingly.

But how about point (*b*)? We shall see it more clearly later on. But recent empirical facts show that in reality no movement of Africa and South America relative to a deeper layer has been performed. But the formation of the Atlantic ocean between them has been accomplished by a process of ocean spreading. This has been shown by recent discoveries of oceanography – I am inclined to think that the beauty of these geophysical discoveries is comparable to that of the quantum physical discoveries of our time.

It began with the detection, by Ewing, Heezen and Tharp, of a world-wide system of great oceanic rifts, continued on the continents in the form of great 'Grabenbrüche'. This is a word often used by German geologists; the English term 'rift valley' seems to correspond to it. Famous examples of rift valleys are found in Palestine, and in East Africa (Fig. 3) and along the upper Rhine. They all are parts or extensions of the mentioned system of oceanic rifts (Fig. 4).

Heezen explained the theory that these oceanic rifts are progressively broadening – their tow shores are retreating from each other; and magmatic masses from below are rising in the rifts. This theory found a wonderful confirmation by the detection of a palaeomagnetic strip pattern along the oceanic rifts. A picture of these stripes, according to Vine, is shown in Fig. 5. To understand its meaning, one must know that the two poles of the magnetic field of the Earth – in an erratic manner – change their positions. During the last four millions of years nine such field-reversals have taken place (Fig. 6). Elsasser, who successfully gave a theoretical explanation of

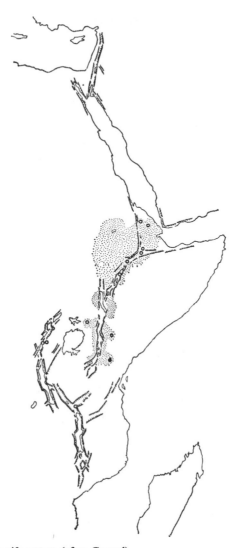

Fig. 3. The East African rift system (after Gognel).

the phenomenon of Earth magnetism, told me, that also these changes are theoretically well understandable.

The magmatic masses, rising from the Atlantic rift, are magnetized by the earth magnetic field before they are cooled. Therefore at each point near the Atlantic rift one can determine whether at the time of the formation of the sea bottom at this point the situation of the poles was as today or reversed. Making a palaeomagnetic chart of the ocean bottom along the rift the two possible situations of the poles show themselves in strips parallel and symmetric to the rift. (In the Red Sea, the rising magmatic masses heat large amounts of the deeper layers of the sea.)

The separation of Africa and South America is now to be understood as the result of progressive broadening of what originally was only a long rift valley through the old 'Gondwana' continent. The present line of the Atlantic rift still fits with the boundaries of the two now separated continents (Fig. 7).

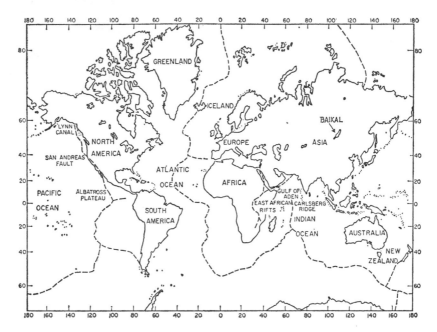

Fig. 4.   The system of rifts (after Ewing, Heezen and Tharp).

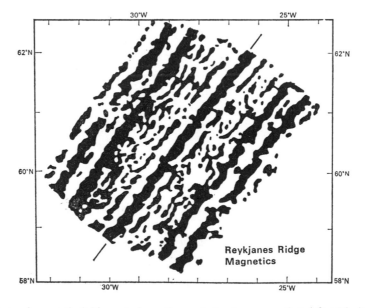

Fig. 5.   Effects of magnetic-field-reversals on the sea-bed palaeomagnetism (after Vine).

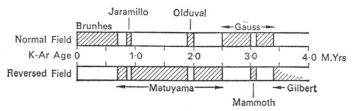

Fig. 6.   Magnetic field-reversals over the last 4 million years (after Cox, Doell and Dalrymple).

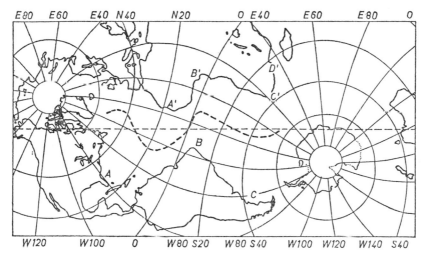

Fig. 7.   The Atlantic Rise as the former line of contact between Africa and South America (after Carey).

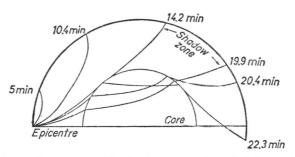

Fig. 8.   The Earth's mantle and core with seismic sound paths and their travel times.

From this we see that the increase of the distance between the two separating continents has not been caused by movements of the two blocks relative to their underground, but by a spreading process of the deep sea bottom. This 'spreading of oceans',
as it has been called, is a clear empirical fact; and the above-mentioned palaeomagnetic strips allow us also to determine the velocity of this spreading. This velocity of
the broadening of the Atlantic rift amounts to several cm per year. That is a rather
astonishing fact.

From seismological study the interior of the earth is physically well known to a considerable amount. The basic seismological measurements show the time elapsing between an earthquake and the beginning of its effects produced in any observational station having a certain distance from the centre of the seismic event. From these time differences one can deduce mathematically the path of the sound rays going through the earth; and from these rays one can calculate density and compressibility in the earth as functions of the radial coordinate (see Fig. 8 and Fig. 9).

Using this knowledge Dicke and his collaborators deduced the amount of earth's expansion that would arise if this expansion were a purely elastic response of the earth

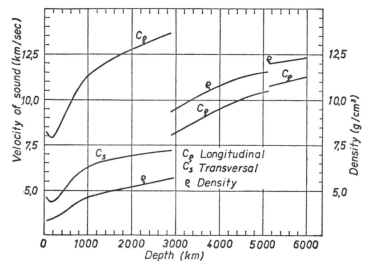

Fig. 9. The velocity of sound in the interior of the Earth (after Gutenberg; reproduced from MacDonald).

to the decrease of the gravitational scalar; concerning the radius R of the Earth in this case one would get approximately

$$\frac{\dot{R}}{R} = -0.1 \, \frac{\dot{\kappa}}{\kappa}.$$

The shadow-zone of seismic effects indicates qualitively that there is – below the 'mantle' of the earth, about 2900 km deep under the surface – a fluid core in which only longitudinal waves are propagated. There are strong arguments to believe that this core consists of iron, probably mixed with nickel.

Recent progress of evaluating the seismic data showed also that the mantle of the earth is chemically homogeneous only to the depth of 400 km. Below this so-called 'Byerly surface', 400 km deep, the chemical composition of the mantle is radially varying. This point seems to me to be highly important.

For from the mentioned result of Dicke concerning $\dot{R}/R$ in the case of an elastic response to the decrease of the gravitational scalar the following fact is to be learned. If we accept the interpretation of the factual spreading of oceans as a direct indication of a steady expansion of the earth – and this indeed seems to me to be the natural interpretation, which can be avoided only by the hypothetical and improbable invention of compensating effects giving a compensating decrease of the areas of deep sea oceans – then we cannot assume this expansion to be an elastic one. What we know about the velocity of broadening of the oceanic rifts means that the spreading of the oceans, if uncompensated, shows an increase of the earth radius of about one cm per year; and that cannot be only an *elastic* response of the earth's material to the decrease of κ.

Now there is the further fact of radial chemical inhomogeneity in the deep layers of the mantle of the earth; and this fact strongly suggests that a considerable part of the material of the core, originally held there in solution, must have crystallized out in a slow process and have formed what today are the deepest layers of the mantle. It seems that this process contributed essentially to the expansion of the earth.

From these considerations it becomes possible to give a very natural interpretation of a fundamental fact concerning the structure of the surface of the earth. We have at this surface two quite different hypsographic levels: the average level of the continental areas, and the level of the deep sea, about 5 km below zero. An older diagram of Wegener and Bucher (Fig. 10) shows this fact impressively, but probably with a certain degree of exaggeration; more modern diagrams show less overwhelmingly, but also clearly that this existence of two distinct hypsographic levels is an empirical fact. Several quite fantastic hypotheses have been forward to expain this fact – one of them assuming, that a great part of the so-called sialic layer (the material of the continental blocks) has been separated from the earth in the process of a separation of the moon from the earth. The other one of these two hypotheses tries to convice us that the material of the continents did not come from the earth itself, but from the

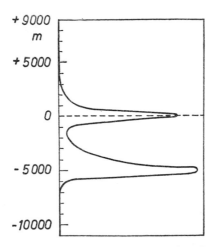

Fig. 10.   Frequency distribution of altitudes (after Wegener and Bucher).

moon, when it was 'captured' by the earth and broken down partially. The theory of earth's expansion helps us to discard these quite fantastic ideas, which certainly are in full contrast to physical possibilities and to the empirical facts as we know them.

My own concept is the following one: In a very early stage of earth's development the relatively light material of what today is the 'sial' layer, forming the continents, assembled itself as the outermost layer of the earth – as a layer covering the whole earth in a spatially constant thickness. After cooling and solidifying this outer skin of the earth was not able to maintain its connection. The expansion of the inner masses of the planet provoked a rupture of the outer skin, a tearing up into separated piece which are drawn to increasing distances from each other by the process of spreading of oceans between them.

Indeed many authors contributed to the result that the process of separation of Africa and South America has not been an exceptional one – there are other examples of similar situations – and a few authors discussed also the radical idea that the system of all continental blocks of the present earth would fit quite well into a reconstruction uniting them to a full covering of a former earth which had a surface which was not greater than the sum of all continental areas today.

The necessity, caused by the process of expansion, that the remaining great continental blocks had to perform a transformation allowing adaption to a smaller curvature of the surface of the earth, seems to give a sufficient theoretical basis for understanding the processes of mountain folding – thought in older times to be proof of a *contraction* of the earth.

My former younger co-worker Binge gave from the concept of the earth's expansion and Dirac's hypothesis an explanation for the phenomenon of volcanism. Probably many students of physics and especially geophysics are inclined to think that there is scarcely any difficult problem in the fact of volcanism: locally there may occur heating in the upper mantle of the earth, caused perhaps by radioactivity. Local melting may be the result of such heating, and the melted masses may find occasionally an outflow. But such an idea would be a total misunderstanding.

The basic fact, an explanation of which is needed, lies in that everywhere on the earth where there occurs in a certain area an unburdening of pressure, there comes – one can say, with certainty – volcanic activity into play. There is obviously all over the surface of the earth a tendency to explosions in layers under the surface. Two prominent specialists, Kuhn and Rittmann – feeling the enigmatical situation shown by this fact – came to the conviction, that a fundamental change in our picture of the interior of the earth must be introduced, in order to explain this decisive fact. They concluded that the interior of the earth – apart from a quite thin solid outer layer – might be in the state of gas – they called it 'solar matter'. Therefore the deeper layers below the surface contain, though being themselves solid, high percentages of gas, and the release of pressure allows this matter to undergo a process of separation of gas and solid matter. This process can result in the formation of fluid and heated substance, an increase of volume and explosion.

Surely this theory must be discarded because it is in total disagreement with what we have learned from seismology. But I think that Kuhn and Rittmann correctly felt that the fact of the universal tendency to explosions in the deeper layers below the surface of the earth necessitates an explanation which cannot be won without leaving the frame work of older concepts.

The second fundamental point seen by Binge very clearly is the fact, that the primary effect in volcanic activity is the occurrence of explosions – these cannot be explained as a consequence of mere heating and melting, but heating and melting must be understood as secondary effects of the explosions. Binge also saw that the concepts of Dirac's hypothesis and of earth's expansion provide us with both the possibility of explosions and of a permanent tendency to perform such explosions as soon as pressure is removed or decreases – for if the expansion theory is correct, then the deeper layers of the crust of the earth are exposed to a pressure which is decreasing continuously in the course of geologic history. Therefore in the mantle of the earth continuously high pressure phases of the mantle material must cease to be thermodynamically stable, and they must be transformed to phases of lower pressure. But this phase transition will not be going on without delay – in many cases the high pressure phases will at first continue to exist, but become unstable. Therefore occasion is given for numerous processes going on explosively.

This theoretical interpretation of terrestrial volcanism has consequences also concerning the moon. Since nearly a whole century specialists of the moon have been debating the question whether the so-called craters of the moon are generated by volcanic processes or by collisions of great meteoritic bodies with the moon. The fact that this question could remain a controversial one through many decades, shows already that it is really difficult to decide it. But from careful study of the literature I came to the conviction that the following answer to this controversial question is in accord with the empirical facts as they are seen by the best specialists of this field: all great craters on the moon are indeed caused by impacts; only some of the small craters are of volcanic nature, and these are results of mere gas eruptions.

This conclusion, drawn from the inspection of the empirical material and the numerous discussions about it, is in in best accord with what must be inferred from Binge's interpretation of the terrestrial volcanism. For the moon, though showing a certain amount of expansion by the existence of numerous analogues of rift valleys, certainly can be expanding only quite weakly in comparision with the earth. Therefore terrestrial volcanism, if really a consequence of earth's expansion, can have only a very weak analogue at the moon. Examples of rift valleys seem to exist also on Mars and seem to have been photographed by rockets.

Comment by P. G. Bergmann

Theories with a fifteenth field variable are a possible generalization of Kaluza's five-dimensional approach.[1] Einstein and I considered, what today is rightly known as the Jordan–Thiry theories, in the late thirties. We did not publish this attempt, as it did not achieve Einstein's objective, to yield a classical model of elementary particles. I presented a summary of our work much later.[2]

To analyze the physical significance of the new scalar, I studied the ponderomotive laws of their type of theory.[3] The results were: (a) There are conserved quantities playing the role of mass-linear momentum; (b) the particle will in general couple to the gradient of an 'incident' scalar field.

If we are to construct a theory with secularly changeable particle masses, we must somehow avoid the implications of Noether's theorem. I can see two possibilities: construct a theory without action principle; or, make the Lagrangian form dependent on the age of the universe, i.e. break the Poincaré–Einstein symmetry. The latter way can be defended on grounds of physical plausibility.

REFERENCES

1. T. Kaluza, *Preuss. Akad. Wiss. Ber.* **2**, 966 (1921).
2. P. Bergmann, *Ann. Math.* **49**, 255 (1948).
3. P. Bergmann, *Intl. J. Theor. Phys.* **1**, 25 (1968).

6

The Nature and Structure of Spacetime

Jürgen Ehlers

1. INTRODUCTION

Space and time and the even more basic notion of spacetime, and the structures assigned to them, belong to the most fundamental concepts of science. So far, every physical theory of some generality and scope, whether it is a classical or a quantum theory, a particle or a field theory, presupposes for the formulation or its laws and for its interpretation some spacetime geometry, and the choice of this geometry predetermines to some extent the laws which are supposed to govern the behaviour of matter, the laws of primary concern to physics. Thus Galileo's assertion[1] still applies: 'He who undertakes to deal with questions of natural sciences without the help of geometry is attempting the unfeasible.'

To substantiate the preceding remarks it suffices to recall a few characteristic examples. Newton's law of gravitation is based on (among other things) the assumption of an absolute simultaneity relation between events and a Euclidean distance function on the space sections $t =$ const. of spacetime. The classification of elementary particles according to their masses and spins rests, according to Wigner[2] (1939), on the structure of the Poincaré group, i.e., the group of isometries of the flat, pseudo-Riemannian spacetime of Einstein[3] (1905) and Minkowski[4] (1908). A satisfactory account of the extremely well established universal proportionality of inertial and gravitational mass[5] requires the assignment of a non-integrable, i.e., path-dependent linear connection (law of parallel displacement) with non-vanishing curvature to spacetime, as recognized by Einstein[6] between 1907 and 1915.

Spacetime geometry cannot adequately be considered in isolation from other parts of physics; its concepts and laws are inextricably interwoven with those of mechanics, electrodynamics etc. This insight has been emphasized particularly by Weyl[7], who said in this context that 'the individual laws of physics no more than those of geometry admit of an experiential check if each is considered by itself, but a constructive theory can only be put to the test as a whole'. This recognition implies that the 'truth' of geometry is to be judged on the basis of the consistency, simplicity and empirical adequacy of the system of physics of which it is a part, and not otherwise. This viewpoint is deeper and more fruitful than Poincaré's[8] conventionalistic attitude towards geometry, as has been amply demonstrated by Einstein.

The structure to be given to spacetime, and the reasons for this assignment, have been topics of major concern to physicists, mathematicians, philosophers and theolog-

J. Mehra (ed.), The Physicist's Conception of Nature, 71-91. All Rights Reserved

ians throughout the history of science[9]. Famous examples from the history of space-time research are the controversy between Leibniz and Clarke (and Newton) about absolute space, Mach's criticism of that same concept, Kant's claim concerning the aprioristic, transcendental character of Euclidian geometry[10], and Clifford's anticipation of Wheeler's geometrodynamics[11] based on Riemann's generalization of Gauss's intrinsic geometry of curved surfaces.

In this century the physicist's conception of spacetime underwent two major revisions.

The first was Einstein's relativization of time[3], the conceptually decisive step which transformed the various results and suggestions of Poincaré, Lorentz and others into a transparent, coherent theory, the special theory of relativity. Einstein's examination of the verifiable, concrete meaning of the notion of time and his elimination of apparent contradictions from the electrodynamics and optics of moving bodies which resulted almost without effort from the substitution of an operationally meaningful time concept for a dogmatically postulated absolute time provided the model for a critical, empirically oriented re-examination of physical concepts in general, and was thus of great methodical importance for the evolution of physics, in particular, quantum theory.

The second, even more surprising and profound idea was Einstein's gradual recognition[6], between 1907 and 1915, that gravity is not a force field existing besides the inertia-determining world-geometry, but should itself be considered as an aspect of the metrical and affine structure of spacetime, indicating in fact the curvature of the spacetime continuum and thereby furnishing the physical basis of that structure Riemann had speculated about at the end of his celebrated inaugural lecture[12] of 1854. With this second step, Einstein transformed the geometrical structure of spacetime from a rigidly given, never changing, absolute entity into a variable, dynamical field interacting with matter. He thereby removed a disparity between geometry and physics which had been criticized, in particular, by Mach. The not-yet-achieved incorporation of this fundamental idea of a dynamical spacetime geometry into a quantum theory of matter is one of the central open problems of physics whose solution may well require another radical change in the physicist's conception of nature. 'I do not believe that a real understanding of the nature of elementary particles can ever be achieved without a simultaneous deeper understanding of the nature of spacetime itself', says R. Penrose[13] in his profound lectures on the structure of space-time.

The purpose of this review is to describe the spacetime models which are currently important in physics, namely the special (flat) and general (curved) nonrelativistic spacetimes and the corresponding relativistic ones; to discuss reasons for their adoption or rejection for certain domains of experience; to sketch some recent work related to spacetime structure; and to hint at weaknesses and possible future developments. The emphasis is placed on the present understanding of spacetime structure; the historical development of ideas will be indicated only. Accordingly, even old ideas will be described in the language of present-day mathematics and physics.

2. NONRELATIVISTIC THEORIES OF SPACETIME

'The objects of our perception invariably include places and times in combination. Nobody has ever noticed a place except at a time, or a time except at a place' (Minkowski[4]). The primary extensive medium in which physical processes are imagined to occur (or, more abstractly, the common domain of definition of those fields which represent – at least at the classical, macroscopic level – the observable quantities) is therefore taken to be spacetime, a set M whose elements $p, q, \ldots$ are called events. Whether and how M can be decomposed into a 3-dimensional space and a 1-dimensional time is already a question about the structure of M whose answer is subject to empirical test.

According to Newton, there is an absolute time and an absolute space, recognizable by observations of mechanical phenomena such as the curvature of the surface of water in an 'absolutely rotating' pail. This means that for any two events p, q it is regarded as objectively decidable whether they are simultaneous and also whether they occur at the same place. Hence the set M of events is the cartesian product of the set T of all instants (of time) and the set S of all space points,

$$M = S \times T. \tag{2.1}$$

Moreover, S is assumed to be a Euclidean 3-space whose metric dl^2 is measurable by means of rulers, and T is taken to be a one-dimensional Euclidean space whose natural coordinate t (defined up to linear transformations $t \to at+b$) is given by standard clocks. This Newtonian spacetime is illustrated in Fig. 1.

The stratification of M given by the maximal subsets of simultaneous events can be interpreted as the causal structure of M, an idea apparently due to Leibniz[14]. The hyperplane $t=t(e)$ passing through an event e separates the causal future, or the domain of influence $(t>t(e))$ of e, from its causal past. The circumstance that future and past of e have a common boundary, the present, expresses the hypothesis implicit

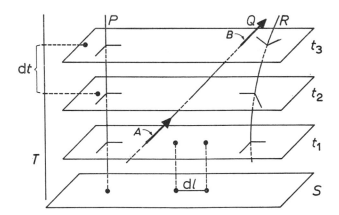

Fig. 1.   Newtonian Spacetime. Particle P is at absolute rest and carries non-rotating axes, Q moves uniformly, and R is accelerated and carries rotating axes. A and B are parallel.

in Newtonian Physics that there are arbitrarily fast signals, realizable by means of (strictly or arbitrarily nearly) rigid bodies or instantaneous action at a distance.

The stratification is also the formal counterpart of the ontological idea that the external world evolves in time: not only for any particular observer, but objectively the present state of the world is supposed to consist of the distribution of matter in the hyperplane 'now' existing, and the succession of the configurations of bodies in these hyperplanes represents the history of the material universe.

The group of transformations which preserve the structure of Newtonian spacetime – the product structure (2.1), and spatial and temporal congruence – is the direct product of the group of dilations, rotations and translations of S with the affine group of T; following Weyl[15] we call it the elementary group $\mathfrak{E}$.

Whereas the spatial and temporal metrics of M have a sound empirical foundation and the corresponding causal structure was an acceptable idealization as long as there was no clear evidence against instantaneous transmission of influences, the postulated absolute standards of rest and of no rotation have been rightly questioned by Berkeley, Huyghens and Leibniz[16] on the basis of the relativity of motion. If one drops these last two assumptions, one is led to (say) Leibniz's spacetime which is illustrated in Fig. 2.

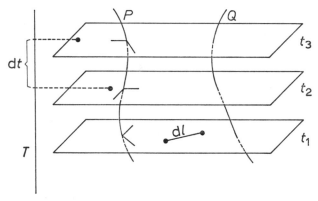

Fig. 2. Leibniz's Spacetime. There are no preferred motions (no 'straight' lines), and there is no parallelism except the one within a single stratum $t = $ const.

Leibniz's spacetime, the spacetime of nonrelativistic kinematics, has less structure than Newton's; in it there are neither preferred motions nor preferred 'nonrotating' spatial axes. Whereas Newtonian spacetime is an affine 4-space – since it is the product of Euclidean, hence affine spaces – Leibniz's spacetime has no affine structure; i.e. in the former one can speak of parallel 4-vectors, in the latter that is meaningless. Accordingly, the group of automorphisms of Leibniz's spacetime, called the kinematical group $\mathfrak{K}$, is much larger than the elementary group $\mathfrak{E}$. $\mathfrak{E}$ is a 9-dimensional Lie group, $\mathfrak{K}$ is not a Lie group since its elements require for their specification not only 3 parameters but also 6 arbitrary real functions of time (an angular velocity and a translation velocity).

Newtonian spacetime is obtained from Leibniz's spacetime by adding to the causal and metric structures of the latter the fibration which defines the state of absolute rest.

It is clear that a decision between these two spacetime models requires dynamical arguments. Newton's famous discussion of the spinning water-bucket (or Foucault's pendulum) can serve to justify the assumption that, dynamically, rotation has an absolute meaning, so that one aspect of the Newtonian spacetime structure, the parallel transport of spatial axes along timelike worldlines, is physically acceptable. In this respect dynamical facts decide in favour of Newton and against his relativistic opponents Huyghens and Leibniz. On the other hand, to identify in nature the state of absolute rest, Newton resorts to the statement 'that the centre of the system of the world is at rest', and this assertion has no observationally testable content.

Thus, both Newton and Leibniz are correct in their mutual criticisms, but the spacetime geometries corresponding to their positions are both dynamically inadequate. Leibniz's kinematical spacetime has not enough, Newton's dynamical spacetime has too much structure[29] (Equation (2.1)!). It is a remarkable historical fact that the resulting lack of clarity in the foundations of dynamics, although it was felt by many scientists, notably Euler, persisted until Lange[17] in 1885 recognized that what is needed besides the causal and metric structures is (in modern terms) the assumption that spacetime is an affine 4-space whose timelike straight lines (i.e. those not contained in a hyperplane $t = $ const.) represent free motions. This axiom is a precise formulation of the law of inertia (see Fig. 3), which emphasizes its intrinsic, coordinate-independent content.[18]

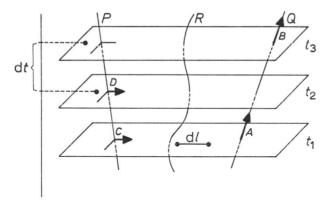

Fig. 3. The spacetime of classical dynamics. Uniform motion, exemplified by P and Q, is considered as absolutely preferred against accelerated motion, R; but no absolute rest (verticals) is defined. Parallelity of 4-vectors is meaningful; $A \parallel B, C \parallel D$.

It has often been argued that Newton's first law should not be postulated as a separate axiom since it is implied by the second law. That is of course true, but far more inportant than this trivial implication is that the law of inertia alone serves to define the affine structure of spacetime. The subsequent laws of dynamics presuppose that structure but do not restrict or enrich the spacetime geometry further.

Accepting the law of inertia one obtains the well-known spacetime of classical dynamics which we call 'special' nonrelativistic spacetime, for reasons to be explained later. Its group of automorphisms, the Galilei group $\mathfrak{G}$, is intermediate between $\mathfrak{E}$ and $\mathfrak{K}$,

$$\mathfrak{E} \subset \mathfrak{G} \subset \mathfrak{K}. \qquad (2.2)$$

The three groups $\mathfrak{E}$, $\mathfrak{G}$, $\mathfrak{K}$ characterize the geometrical or, perhaps better, chrono-geometrical structure of the corresponding spacetimes precisely in the sense of Klein's Erlanger programme (1872)[19].

$\mathfrak{G}$ is not only the symmetry group of the spacetime of classical nonrelativistic dynamics, but also the invariance group of the laws of mechanics which govern isolated systems of particles (or extended bodies). This statement is the essential content of the Galilean principle of relativity, which can be rephrased as the equivalence of all inertial reference frames for the description of dynamical phenomena. Within the spacetime so obtained also the nonrelativistic quantum theories of particles and fields can be formulated. The structure of $\mathfrak{G}$ plays, via its unitary ray representations, an essential role in constructing from first principles the quantum theory of free particles[20].

At this stage we depart from the (anyhow largely simplified) course of history. Whereas in fact the next two important steps in the evolution of spacetime concepts were taken by Einstein in 1905 and 1915, we should like to describe here first a natural extension of the special nonrelativistic spacetime due to Cartan[21] and Friedrichs[22] and elaborated further by Havas[23] and Trautman[24]. This theory was actually formulated only after and under the influence of Einstein's special and general theories, but from a systematic point of view it should be considered prior to these theories. Its merit is to show already at the nonrelativistic level that a satisfactory incorporation of gravity into the system of spacetime geometry and mechanics requires, because of the well established universal proportionality of inertial and (passive) gravitational mass, a change in the affine structure of spacetime. This step is independent of the relativization of time, since the phenomena in question do not necessarily involve large speeds or energies and thus are not relativistic.

Cartan's theory can be motivated by the following considerations.[25] The transition from kinematics to dynamics as sketched above consists in singling out a preferred class of motions the members of which define everywhere a standard of 'no acceleration'; the law which characterizes these motions is called the first law of dynamics. Once that has been done – and only then – forces are introduced via Newton's second law of dynamics; the acceleration of an arbitrary motion is to be judged relative to a preferred motion with the same instantaneous velocity passing (nearly) through the same event as the arbitrary motion. (This formulation presupposes a metric of spacetime, but avoids the use of dynamically preferred frames of reference and is purely local; it is meaningful at the level of the group $\mathfrak{K}$.) The traditional way of carrying out this general programme is to choose as the first law the law of inertia which leads to the spacetime geometry and dynamics belonging to the group $\mathfrak{G}$, as reviewed above. There is a grave objection to this procedure, however. Because of the experimentally extremely well established composition–independence of the ratio passive gravitational mass / inertial mass (see Dicke et al.[5] and Braginski et al.[5]) of macroscopic test bodies the actually available, unique candidate for the preferred class of motions is the class of free falls of (neutral, spherically symmetric, non-rotating) test bodies. These motions do not permit an observationally meaningful distinction between inertial forces and gravitational forces (and neither do any other known phenomena), and they do not satisfy the law of inertia since they exhibit relative accelerations. This insight suggests strongly to abandon the law of inertia and to reconstruct dynamics by taking as the first law the following characterization of free falls: With respect to suitable (so-called 'non-rotating') frames of reference, free falls obey the law

$$\ddot{\mathbf{x}} = - \nabla\phi, \tag{2.3}$$

in which $\phi(\mathbf{x}, t)$ is a frame-dependent real function called the gravitational potential relatively to that frame. This formulation is not completely satisfactory because it describes the structure provided by the class of free fall motions not directly, but with the help of preferred frames which are themselves defined by the free falls only. However, Equation (2.3) can be rewritten in the form

$$\frac{d^2x^a}{dt^2} + \Gamma^a_{bc} \frac{dx^b}{dt} \frac{dx^c}{dt} = 0,$$

$$\Gamma^\lambda_{44} = \phi_{,\lambda} (a, b, c = 1, ..., 4; \quad t = x^4; \quad \lambda = 1, 2, 3). \tag{2.4}$$

Hence, there exists on spacetime a unique symmetric, linear connection Γ the geodesics

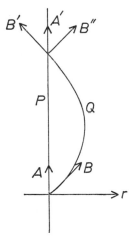

Fig. 4. A tunnel T, drilled along a diameter through a massive spherical body B_0, contains two freely falling particles P, Q. P rests at the centre of T and B_0, and Q oscillates in T. The spacetime diagram shows the world lines of P and Q. With respect to the gravitational-inertial connection Γ defined in the text, one has obviously: Parallel transport along P transfers A to A', B to B'', and parallel transport along Q transforms B into B'. Since $B' \neq B''$, the Γ-parallel transport is not integrable. P and Q form a geodesic 'diangle'.

of which represent the world lines of free fall. This statement is the core of Cartan's theory. The connection Γ together with the absolute time t and the spatial metric dl^2 can be characterized by some axioms[24] which ensure that in suitable coordinates Γ^a_{bc} can be expressed as in (2.4); Poisson's equation relating the gravitational field to its material sources turns out to be expressible as a relation between the contracted curvature tensor of Γ, the density of matter and the gradient of t; and the second law of dynamics can be formulated as in general relativity theory in terms of covariant derivatives with respect to Γ.

Whereas the parallel displacement of 4-vectors defined by Γ is path-independent if applied to spacelike vectors (which, by definition, are tangent to the space sections and are nothing but the familiar 3-vectors), it is in general path-dependent if applied to timelike vectors, as can be inferred from the curvature tensor associated with (2.4) or from dynamical thought-experiments like the one shown in Fig. 4 due to Heckmann and Schücking[26]. The non-integrability of the gravitational-inertial connection implies that geodesics are curved relatively to each other, or in physical terms that freely falling particles exhibit relative accelerations. The quantitative measure of this space-time curvature or gravitational tidal field is the curvature tensor associated with Γ.

The elementary mathematical fact that one can always introduce new coordinates $(\mathbf{x}, t)$ such that, at an event or even along a free-fall-worldline, the components of Γ vanish, corresponds to the physical fact, recognized by Einstein, that locally any gravitational field can be 'transformed away' (elevator experiment).

Whereas formally the local laws of Cartan's theory of spacetime, gravity and dynamics (as completed by the authors mentioned above), if expressed with respect to non-rotating coordinate-frames, do not differ from those of the standard Newtonian theory as given in the textbooks, conceptually it embodies an important advance by denying the separate existence of an integrable affine connection representing the inertial field and a vector field representing gravitation, and introducing instead of these two structures a single non-integrable connection representing both inertia and gravity. An empirically unjustifiable, fictitious distinction has thereby been removed, and the true nature of gravity as a connection has been recognized. The introduction of this concept (which is due to Levi-Civita (1917) and Schouten (1918)[27]) into physics, which resulted from the work of Einstein[6], Weyl[15], Cartan[21] and others is comparable to the introduction of vector fields for the description of electromagnetism by Maxwell. Connections, besides vector and spinor fields, form another type of mathematical entities suitable to represent physical objects, an insight exploited in the theory of gauge fields.[28]

A spacetime M with a nonrelativistic metric (t, dl^2) and a connection Γ according to (2.4) will be called a 'general-nonrelativistic' spacetime. In it, there are no exact, global inertial frames, but only local inertial frames, and these exhibit relative translational, though no rotational accelerations. The group $\mathfrak{N}$ relating non-rotating coordinate systems, consisting of those transformations which leave (t, dl^2) and the form of the law (2.3) unchanged, is larger than the Galilean group $\mathfrak{G}$, but smaller than the kinematical group $\mathfrak{K}$. $\mathfrak{N}$ contains arbitrarily time-dependent translations, but only time-independent rotations. We may extend (2.2) to

$$\mathfrak{E} \subset \mathfrak{G} \subset \mathfrak{N} \subset \mathfrak{K}, \tag{2.5}$$

a relation which indicates in a condensed form the evolution of the spacetime concepts at the nonrelativistic level. Whereas the transition from $\mathfrak{E}$ to $\mathfrak{G}$ represents the preliminary compromise between the absolutist Newton ($\mathfrak{E}$) and the relativist Leibniz ($\mathfrak{K}$), the step from $\mathfrak{G}$ to $\mathfrak{N}$ – or from a flat to a curved connection – is a (somewhat delayed) response to Mach's criticism[30] of the unfounded distinction between inertia and gravity.

All of nonrelativistic physics including quantum mechanics can be reformulated without difficulty within the framework of general-nonrelativistic spacetime[31]. (Thus, e.g., the change of the gravitational potential associated with a transformation of $\mathfrak{N}$ is accompanied by a phase change of a Schrödinger wave function to ensure form-invariance of the Schrödinger equation.) All the non-gravitational local laws have, in local inertial frames, the same form as in the gravity-free, special spacetime. Thus Einstein's strong principle of equivalence[32] is incorporated satisfactorily (as far as slow motion, low energy phenomena are concerned) into nonrelativistic physics. The 'special' theory based on $\mathfrak{G}$ now appears as a local approximation to the 'general' theory, valid as long as inhomogeneities of the gravitational field can be neglected.

One principal advantage of the generalized version of Newtonian mechanics is that it permits the treatment of unbounded and, in particular, spatially homogeneous selfgravitating systems as used in cosmology[33].

In the transition from the special to the general nonrelativistic spacetime the status of the connection has been changed from that of an absolute element[34], given once and for all, to a dynamical quantity depending on the physical state. The metric, however, is still treated as an absolute element. This is possible since a Galilean metric (t, dl^2) does not determine a unique connection (2.4), in contrast to the situation in relativity theory.

3. RELATIVISTIC THEORIES OF SPACETIME

Whereas nonrelativistic physics describes satisfactorily slow-motion phenomena at all scales including cosmology, its laws are wrong for fast motion, high energy processes. Especially it is incapable of accounting for the behaviour of massless fields such as electromagnetism.

After Rømer's discovery of the finiteness of the speed of light in 1676 and Bradley's discovery of aberration in 1728 it was natural to assume that light propagates in vacuo with the speed $c \approx 3 \times 10^{10}$ cm sec^{-1} along straight lines with respect to some non-rotating frame of reference which coincides, at least roughly, with the centre-of-mass frame of the solar system. This hypothesis singles out a preferred ether frame, since the only transformations of the kinematical group $\mathfrak{K}$ which preserve the assumed law of light propagation are those of the elementary group $\mathfrak{E}$. One is thus led back to the original Newtonian spacetime. It is well known that neither this theory of a rigid

Maxwell–Lorentz ether nor theories with deformable ethers nor emission theories of the Ritz type have been able to give a satisfactory account of the many phenomena of optics and electrodynamics of moving bodies. (For relevant experiments, see, e.g., refs. 35, 36.)

In his famous paper of 1905 on the electrodynamics of moving bodies[3] Einstein showed how the difficulties can be overcome by discarding the empirically unfounded assumption of an absolute time, and adopting instead a principle of relativity for mechanical and electromagnetic processes and by assuming the independence of the velocity of light on the velocity of the source (now experimentally established with an accuracy of 10^{-4}, see ref. 35). Three years later his former teacher Minkowski found a geometrical characterization of the new kinematics[4]. According to him, spacetime is a pseudo-Euclidean, four-dimensional space whose metric tensor η_{ab} has signature $(+ + + -)$. The null cones defined by η_{ab} describe light propagation (in vacuo), the timelike straight lines represent the world lines of free particles, and the arc length

$$\int \sqrt{-\eta_{ab}\,dx^a\,dx^b} = \int \sqrt{1 - v^2}\,dt \qquad (3.1)$$

$(x^4 = t,\, c = 1)$ of a timelike curve L gives the proper time measured by a standard clock carried by a particle with world line L. (See Fig. 5.)

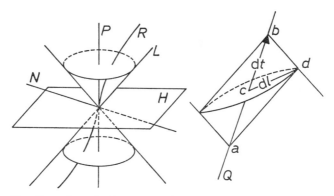

Fig. 5. Minkowskian Spacetime. P, Q represent free particles; R, an accelerated particle; L, a light ray; N a spacelike straight line. The events in H are simultaneous for P, but c and d are simultaneous for Q. The affine structure is as in Fig. 3, but the metric is different. $dl = c\,dt$, and b and d have 'distance' zero.

The spacetime geometry of the special theory of relativity can be derived from the assumption that there exist coordinates x^a with the following two properties:

(a) Two events x, y can be connected by a light signal in empty space if and only if

$$(\mathbf{x} - \mathbf{y})^2 = (x^4 - y^4)^2 . \qquad (3.2)$$

(b) Free particles obey the law of inertia

$$\frac{d^2 x^\lambda}{(dx^4)^2} = 0 . \qquad (3.3)$$

The transformations leaving these two laws invariant are precisely those which map the differential form $\eta_{ab}\, dx^a\, dx^b = dx^2 - (dx^4)^2$ into a constant multiple of itself; they form the Poincaré group $\mathfrak{P}$, augmented by dilations. Postulate (*a*) assigns a conformal structure (a field of null cones) to spacetime, and (*b*) gives it a projective structure (a family of 'straight' lines). These two primitive structures together define a Minkowskian geometry (Weyl[37] 1923, see also Fock[38]). This characterization of Minkowski space is a local one; (3.2) and (3.3) need to hold only in finite coordinate domains. If one assumes the coordinates x^a to range over the whole space $\mathbb{R}^4$ one can even dispense with (*b*), but such a global requirement seems physically unreasonable. Many other approaches are known, see ref. 36.

The metric of Minkowskian spacetime is simpler than that of the nonrelativistic spacetimes since it is specified by a single tensor η_{ab} rather than by two quantities t, dl^2.

The most important difference between the nonrelativistic spacetimes and that of special relativity lies in their causal structures. In Minkowski's spacetime the causal future (past) of an event e is bounded by the future (past) null cone, and thus there is a four-dimensional region whose events are causally disconnected with e, in contrast to the situation in nonrelativistic spacetimes. A bijection of Minkowski spacetime onto itself which preserves the causal order is the product of a dilation and an orthochronous Poincaré transformation (Zeeman, 1964[39]).

The (coordinate) topology of special relativistic spacetime can easily be obtained from its chronological order. Let b be called later than a, for a, $b \in M$, if b is contained in the interior of the future null cone of a, written $a < b$. Then the sets $\{x \,|\, a < x < b,\ a,\, b \in M\}$ generate the topology of M. This way of introducing the topology of M, due to Alexandrow[40], is physically very satisfactory since it says that an event x is close to y if there is a particle P through a and a 'short' time interval on P containing a within which P can 'communicate' with b. (See Fig. 6.)

The corresponding construction does not work in nonrelativistic spacetime, since it would lead to a non-Hausdorff topology. This illustrates the fundamental difference between the causal structures of relativistic and nonrelativistic spacetimes.

The absence of an observer-independent, transitive simultaneity relation between events in special (and general) relativity theory (spacelike separation is not transitive!) implies that the ontological conception of an external world evolving 'in time' has no

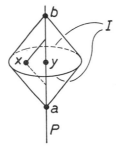

Fig. 6.   Causal topology. I is the causal interval between a and b. It is a neighbourhood of y containing events like x causally related to a particle like P.

formal counterpart in the laws of the theory. This recognition poses philosophical questions concerning the nature of time. (Weyl says[41]: The objective world simply *is*, it does not *happen*.)

Apart from the classical optical and electrodynamical effects which led to the special theory of relativity there are numerous other kinds of experimental results which confirm it. Examples are the time-dilation, now established by means of measurements of lifetimes of muons in storage rings with an accuracy of 2×10^{-2} (ref. 42), which supports directly the validity of Equation (3.1) and therefore the existence of a pseudo-Euclidean metric, and the multitude of data on high energy collisions which could hardly be ordered reasonably without use of the metric η_{ab} in 4-momentum space. Less direct, but at least equally convincing successes of the theory resulted from its combination with quantum theory. Examples are Dirac's electron-positron theory, the spin-statistics theorem, the CPT-theorem, the (already mentioned) classification of particles, and the Lamb shift.

There can be no doubt, therefore, that the Einstein–Minkowski spacetime theory is very nearly correct at the laboratory, atomic and nuclear scale. More precisely, one can say that the existence of a Minkowskian metric at each spacetime point is well supported empirically. On the other hand, the facts referred to do not demand the existence of global inertial frames[43]; they do not even permit one to decide whether a frame attached to the Earth's surface or one attached to a freely falling test particle is a better candidate for such a frame. Hence, the question arises how this uncertainty can be removed on observational grounds.

A related objection to special relativity theory is that its foundations involve in an essential way the law of inertia – its linear structure is based on it, and so the arguments concerned with the inseparability of inertia and gravity discussed in Section 2 all apply. These former considerations suggest that special relativity may be correct only approximately, as long as inhomogeneous gravitational fields are disregarded, and that the inclusion of gravity requires a modification of Minkowskian geometry similar to the one which led from special-nonrelativistic spacetime to general-nonrelativistic spacetime. Thus one is led to look for a theory which agrees locally (approximately) with special relativity, but has, instead of the integrable, affine connection of Minkowski spacetime, a non-integrable linear connection capable of representing, in the manner discussed above, the combined inertial-gravitational field.

In order to satisfy the first requirement, the metric should be related to the connection such that local inertial coordinate systems exist in which, at an event,

(a) the components of the metric tensor have their special relativistic standard values $\eta_{ab} = $ diag. $(1, 1, 1, -1)$,

(b) the first derivatives of the metric components vanish,

(c) the components of the connection vanish.

These requirements lead uniquely to the conclusion: Spacetime is a smooth 4-manifold with a pseudo-Riemannian metric g_{ab} of signature $(+ + + -)$. Its null geodesics represent light rays, the timelike geodesics of the Riemannian connection Γ^a_{bc} associated with g_{ab} represent worldlines of freely falling test particles, and the arc length along timelike lines measures the proper time shown by a standard clock. (See Fig. 7.) These assertions form the kinematical basis of Einstein's general theory of relativity[6]. The spacetime thus obtained incorporates the empirically supported structural elements of the 'Newtonian' gravity theory (non-integrable connection, curvature ≡ tidal field) and of special relativity (null cones, Minkowskian inner product of 4-vectors) and it does not contain the ill-founded, absolute, too special structures

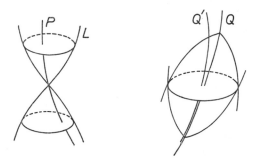

Fig. 7.   Curved pseudo-Riemannian spacetime. P, Q, Q' represent freely falling particles; L, a light ray. The relative acceleration between Q and Q' indicates curvature. The affine structure is qualitatively like that in Fig. 4, the metric (causal) structure is infinitesimally as in Fig. 5. Smooth deformation of Fig. 5 gives something like Fig. 6, though different details are shown.

(absolute time, integrable connection) of the earlier theories. The theory admits local inertial frames as defined above, and hence it is meaningful – as in the corresponding nonrelativistic case considered in Section 2 – to apply within its framework the following strong principle of equivalence[44]: For each class of physical phenomena except gravity a set of basic, local laws exists which, if expressed in terms of inertial coordinates at an event e, take on some standard form, independent of the spacetime location and the gravitational field gradient (the curvature). This principle is somewhat ambiguous like the correspondence principle between classical and quantum mechanics, since it does not tell which laws remain unchanged, but it is nevertheless useful for generalizing tentatively laws known in the absence of gravity to the case where gravity is present. (The ambiguity arises whenever second or higher derivatives occur, since the corresponding covariant derivatives do not commute – this again is formally similar to the factor ordering ambiguity in the quantum case.) It is the gravitational analog of the principle of minimal electromagnetic coupling.

In spite of the fundamental nature of the weak principle of equivalence it is desirable to give additional empirical reasons for abandoning flat spacetime in favour of a curved one. Two such reasons are:

(1) The terrestrial redshift measurements by Pound, Rebka and Snider (see ref. 35) are incompatible with the assumption that nuclear clocks show Minkowskian proper time, as shown convincingly by Schild[45]. Also, these experiments show that the frames of reference in which Maxwell's equations for the propagation of photons hold locally coincide, at least to within 1%, with frames falling freely towards the Earth[35]. These

frames, however, are relatively accelerated, and hence cannot be considered as strict inertial frames in the sense of special relativity. (In addition, these experiments support the strong principle of equivalence, since they show that the mechanically preferred frames are also electromagnetically preferred.)

(2) The deflection of light by the solar gravitational field, whose value is now established with about 10% accuracy[35], is incompatible with a Minkowskian light cone structure. It excludes conformally flat spacetime metrics (like that of Nordström's theory). The same conclusion can be drawn from the radar time delay measurements, the 'fourth test' of general relativity[35].

The preceding arguments demonstrate: If a pseudo-Riemannian spacetime metric g_{ab} is assumed to exist which is observable either by means of proper times as measured by atomic or nuclear clocks, or by means of its timelike geodesics as free fall orbits of test particles, or through its null geodesics as light rays, then the curvature associated with that metric does not vanish and provides an observer-independent measure of the gravitational tidal field. Moreover, a single such metric g_{ab} (together with its associated linear connection and curvature) accounts, in connection with Einstein's field equation, for the various observable phenomena; this is significant since, at least in principle, a metric is already determined by proper time measurements or by observations of geodesics separately. In view of these facts, the assertion that spacetime is 'really' curved, which was already fairly well established shortly after Einstein had proposed his theory, can now, in view of the recent experimental work[35] and further theoretical analyses, be considered as well established, and in the author's opinion the phenomenological foundation for the assignment of a curved, pseudo-Riemannian structure to spacetime is as firm as those of other fundamental theoretical conceptions of physics. (The precise form of the gravitational field equation is not as firmly established empirically. This question will not be discussed here, since it concerns not so much the structure of spacetime itself, but rather the detailed nature of the coupling of the gravitational field to its sources.)

In view of the successes of special relativity in elementary particle physics one may nevertheless ask whether it is possible to incorporate also gravity into it, at the unavoidable cost of sacrificing the observability of the flat metric. If one attempts to describe gravity as a Poincaré-invariant field which has to contain a massless spin two part as an essential ingredient because of its macroscopic, long range, attractive character and its universal coupling to all other fields, then according to Kraichnan 1955, Feynman 1956, Thirring 1959, Wyss 1965, Deser 1970 (for references see Deser[46]) the resulting theory can be reformulated in such a way that only a curved metric, and not the originally postulated flat metric, occurs in the laws of the theory, which in the pure spin two case turns out to be identical to Einstein's general theory, at least at the classical level. (A consistent quantum version is not known.) The originally postulated Poincaré invariance thus turns out to be physically meaningless in the theory finally obtained, just like the flat metric which is not only unobservable, but cannot even be uniquely computed from observable quantities and does not play any useful role in the theory. (The experimentally required approximate local validity of special relativity is guaranteed by the Riemannian nature of the observable metric, and has nothing to do with the initially postulated flat metric.) In the opinion of the author these remarkable results indicate strongly that there is no satisfactory flat space theory of gravity, and they strengthen the conclusion reached in the preceding paragraph. To interpret these results as showing that Einstein's theory may as well be considered as a somewhat peculiar Poincaré invariant theory with a complicated gauge group seems (to me) inappropriate and misleading. The usefulness of the formally Poincaré invariant description of Einstein's and similar theories of spacetime and gravity for making special relativistic techniques available to them, for comparing Einstein's with other theories, and for relating it to quantum field theories is an entirely different matter not to be confused with the issue with which we are concerned here. The physicist's conception of spacetime has been changed profoundly in the transition from special relativity to general relativity, and a return to the earlier, narrower scheme is as improbable as a return from quantum to classical mechanics.

4. REMARKS ABOUT RECENT WORK ON SPACETIME STRUCTURE; PROBLEMS

Since the curved, pseudo-Riemannian manifold of Einstein seems to be the most adequate and comprehensive model of spacetime presently available, one may wish to give a physically plausible axiomatics for it. The axiomatic construction should in particular clarify why a Riemannian rather than a different kind of geometry (e.g., a Finsler geometry or a Kähler manifold) is adopted, and it should enable one to understand why the same functions g_{ab} which describe the clock readings also determine the paths of test particles and light rays.

Such an approach has recently been elaborated by Pirani, Schild and the present author [47]. This work is closely related in spirit and partly inspired by papers or remarks due to Castagnino [48], Geroch, Hoffmann [51], Kronheimer [48], Kundt [51], Marzke [50], Penrose [48], Reichenbach [48], Synge [48], Trautman [48], Weyl [48], Wheeler [50], Woodhouse [48] and others. The main ideas of this approach, without technical details, will now be reviewed.

Neither rods nor clocks are used as primitive concepts. Instead, light rays and freely falling test particles are considered as the basic tools for setting up the spacetime geometry. Accordingly, the construction starts with a set M, the set of events, and two families $\mathfrak{L}$ and $\mathfrak{P}$ of subsets of M. $\mathfrak{L}$ represents the collection of all (possible) light rays, and $\mathfrak{P}$ that of all (possible) free fall world lines, briefly called particles hereafter.

The axioms about $(M, \mathfrak{L}, \mathfrak{P})$ express essentially the following. The events constituting a particle can be distinguished by means of a real parameter which can be thought of as a nonmetric, but smoothly varying time, determined only up to smooth and smoothly invertible transformations. Events which are (intuitively speaking) close to a particle are required to be connectible to the particle by precisely two light rays. Relatively to two particles (which may also be interpreted as observers carrying arbitrary clocks), events in their vicinity can be localized and coordinatized by means of

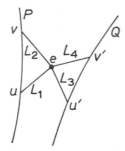

Fig. 8.   Radar coordinates based on two particles P, Q carrying clocks. L_1, L_2, L_3, L_4 are light rays.

the four 'times' u, v, u', v' at which the light rays connecting the particles with the event are emitted or received (see Fig. 8); and it is postulated that such 'radar coordinate systems' assign to M the structure of a differentiable manifold. Moreover, it is assumed that the set v_e of events contained in light rays through e consists, at least in an infinitesimal neighbourhood of e, of two (topologically) disconnected parts (which ultimately turn out to be the future and past light half cones); and that, again infinitesimally, v_e separates the vicinity of e into three parts containing, respectively, the events lying in the past, the present and the future of e.

A particle is assumed to be uniquely determined by an event and a direction (initial velocity) at that event, and the path-structure thus defined on M, which represents the combined inertial-gravitational field or, in Weyl's suggestive terminology, the guiding field, is quantitatively specified by the requirement that the law of inertia holds infinitesimally, which amounts to the assignment of a projective structure [49] to M. In this way the weak principle of equivalence is built into the theory.

In order to relate light propagation and free fall in accordance with experience, the set of all (possible) particles through an event e is assumed to cover, again locally, the interior of the timelike region bounded by v_e (see Fig. 9).

From these 'qualitative' assumptions about light propagation and free fall which form mathematical idealizations of well-established facts and which appear to be minimal requirements for the local validity of special relativity – any departure from these would seem to be a major change of the kinematical basis of physics – it follows that there exists a unique Lorentzian conformal structure (i.e., a field of null cones

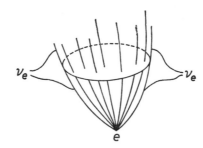

Fig. 9.   The world lines of freely falling particles through e fill (cover) the interior of e's lightcone, v_e.

derivable from a pseudo-Riemannian metric of the standard signature) on M whose null geodesics are identical with the light rays. Moreover, it follows that a unique symmetric linear connection Γ is determined by the following two requirements:

(*a*) the timelike geodesics of Γ coincide with the free fall world lines, and

(*b*) the parallel transport of vectors defined by Γ preserves the causal character of vectors, i.e. their being timelike, spacelike or null.

The affine parameters defined on particles by Γ provide preferred time coordinates; appropriate clocks can be constructed by means of particles and light rays using the method of Marzke[50] or Kundt and Hoffmann[51].

The geometry thus obtained – a conformal structure and a symmetric linear connection compatible with it (in the sense of property (*b*) above) – is nothing but a Weyl geometry, invented (from a different point of view) in connection with the first unified theory of gravity and electromagnetism[52]. In the approach outlined here, this geometry appears as the natural framework for the kinematics of light rays and free particles independently of and more basic than a metric. This geometry is, in general, not (pseudo) Riemannian; in it the transport of time intervals (or distances) will in general be path-dependent. Only if one adds the assumption that time-transport be path-independent – which one has to do if one identifies gravitational time (as given by Γ) with atomic or nuclear time, since the latter is integrably transported because of the indistinguishability of particles of a particular kind – then one obtains a metric and thus arrives at the standard spacetime structure of relativity theory.[53] By adding field equations relating the curvature of Γ to material sources, or by strengthening the infinitesimal law of inertia to the traditional, global law of inertia, one can finally specialize the theory to the general or the special theory of relativity, respectively. As discussed earlier, experience clearly decides in favour of the first possibility.

This approach shows how quantitative measures of time, angle and distance, and a procedure of parallel displacement (and hence covariant differentiation needed for formulating field equations) can be obtained constructively from 'geometry-free' assumptions about light-rays and freely falling particles; pseudo-Riemannian (or Weylian) geometry is recognized even more clearly than before as the appropriate language for a generalized kinematics which allows for the unavoidable and ever-present 'distortions' called gravitational fields.

A completely different, rather abstract approach to spacetime structure, the spinor approach of R. Penrose[54], will now be considered briefly. It is based on the observation that the covering group $SL_2(C)$ of the (homogeneous) Lorentz group $\mathfrak{L}$ is algebraically a much simpler object than $\mathfrak{L}$ itself, and that the simplest building blocks out of which the values of all tensor and spinor fields of standard field theory can be constructed are two-component spinors. Therefore, it is suggested to consider spacetime primarily as the carrier of such spinor fields, and to infer its structure from this its role.

Translated into mathematical language, this means that spacetime M is the base of a fibre bundle B whose typical fibre consists of a pair $(S, \bar{S})$ of complex, 2-dimensional vector spaces each equipped with a symplectic form (spin 'metric'), and related by an anti-isomorphism $S \to \bar{S}$. Such a pair of spin spaces determines, as is well known, a real, 4d-imensional vector space $\Re(S \otimes \bar{S})$ consisting of the hermitean spinors of the kind $t^{A\dot{A}}$ (in van der Waerden's notation), and this is the lowest-dimensional real vector space which can be built from $(S, \bar{S})$. Therefore, if and only if M is 4-dimensional, it is possible to tie the fibres of B to the base M in a simple manner, viz., by identifying smoothly the spaces $\Re(S \otimes \bar{S})$ with the tangent spaces. If this is done, M acquires a pseudo-Riemannian structure; thus one obtains not only Einstein's spacetime, but in addition a spinor structure – which is anyhow needed since there are Fermions in nature – and hence a time and space orientation[55], in accordance with the symmetry violations observed in weak interactions.[56]

The strength of this approach is that it is adapted specifically to dim $M = 4$ and signature $(g_{ab}) = (+ + + -)$ (or $(- - - +)$, which is equivalent), rather than to another type of semi-Riemannian manifold, and that it exploits the power and simplicity of spinor calculus for the analysis of spacetime. On the other hand, one would perhaps wish a more detailed physical motivation of the choice of the ingredients $SL_2(C)$; S, $\bar{S}$; and $\sigma: (S \otimes \bar{S}) \to T(M)$ on which this construction is based.

Whereas the preceding two approaches are conservative in that they analyze or reconstruct the pseudo-Riemannian spacetime of standard general relativity theory, there have also been many attempts to change that structure. One attractive possibility, proposed by Cartan[21] and elaborated by several authors, enriches the geometry by a torsion tensor coupled to the spin density of matter[57]. Other theories, based, e.g., on asymmetric connections or 'metrics', mostly created with the aim of obtaining unified field theories[58], will not be reviewed here. It appears that they have not led to significant physical insights, and they did not influence the main stream of physical thinking.

The great open question related to spacetime structure is its role in microphysics. In atomic, nuclear and particle physics the classical, nonrelativistic or relativistic, special spacetime structure has so far been used as if it were a universal classical external field; it even appears that the definitions of particles and fields presuppose a classical geometry with an isometry group such as $\mathfrak{G}$ or $\mathfrak{P}$. This does not fit with the claim of Einstein's general theory that the quantities defining the geometry, like g_{ab} or Γ^a_{bc}, are dynamical fields, an idea which in itself is overwhelmingly convincing. Why should 'geometry' stand apart from the rest of physics? 'The metric' is just a particular long-range field coupled in a universal way to all carriers of energy-momentum, and it so happens that this field governs at the laboratory scale, in conjunction with electrodynamic and quantum laws, those properties of solids which we use to describe in terms of distance, congruence etc., i.e. in terms of geometry; and at large scales this same field shows up in the phenomena called gravitation. This conception would seem to be much more satisfactory from the viewpoint of the unity of physics than one which treats 'geometry' differently from 'physical' structures – if only this view could be carried out in a consistent theory. So far, it has not; and it seems to the author that at present there is little hope that it will be, in the foreseeable future.

A crucial question related to this deep problem is: At which scale (if at all) does the concept of a smooth spacetime manifold cease to be adequate in the small? Does this happen at 10^{-13} cm, at 10^{-33} cm, or when? That it does happen at some scale is to be expected, in view of the (from a physicist's standpoint) highly artificial, non-constructive nature of the continuum of real numbers[59]; and because of a breakdown of the localizability of particles connected with creation processes[60]; or in view of violent quantum fluctuations of g_{ab} and Γ^a_{bc} at small scales[11]; it is also indicated by a number of singularity theorems established during the last several years.[61]

What kind of 'space' should take the place of the smooth manifold? A geometry of lumps without smallest elements (points), a statistical geometry as suggested by Menger[62], or a collection of classical spacetimes or three-spaces, weighted by probability amplitudes (see Wheeler[11]), or something else? 'The difficulty is that all our present theoretical work is based on a microscopic continuum and one is faced by the rather formidable problem of re-doing all physics in a continuum-free manner', says Finkelstein in this context.[63]. One generalization of ordinary spacetimes is provided by the causal spaces of Kronheimer and Penrose[48], another one is the 'quantum-computer-geometry' of Finkelstein[63], but so far little is known about their physical potentialities. Here we have arrived at the border of spacetime knowledge, and are left with questions only.

REFERENCES AND NOTES

1. G. Galilei, *Dialogo, Opere*, VII, p. 299, Edizione nazionale, Florence (1890–1909).
2. E. P. Wigner, *Annals Math.* **40**, 139 (1939).
3. A. Einstein, *Annalen der Physik* **17**, 891 (1905).
4. H. Minkowski, *Göttinger Nachr.* (1908), p. 53; see also his Address delivered at the 80th Assembly of German Natural Scientists and Physicians, at Cologne, September 1908, reprinted in the well-known Dover paperback *The Principle of Relativity*. The quotation on page 73 is taken from this address.
5. P. G. Roll, R. Krotkov, and R. H. Dicke, *Ann. Phys.* (*N.Y.*) **26**, 442 (1964); V. B. Braginsky and V. J. Panov, *Zh. Eksp. Teor. Fiz.* **61**, 875 (1971).
6. A. Einstein, *Jb. Radioakt. Electronik* **4**, 411 (1907); *Z. Math. Phys.* **62**, 225 (1913); *Sber. preuss. Akad. Wiss.* 844 (1915); *Annalen der Physik* **49**, 769 (1916).
   The statement in the text emphasizes what is now considered to be the essential content of Einstein's analysis. Historically, the role of the connection as an important structure independently of the metric was recognized only in 1917 and the following years. See ref. 27 and the remarks on page 78. For a detailed account of the history including more references see the paper by J. Mehra in this volume.
7. H. Weyl, *Philosophy of Mathematics and Natural Science*, Princeton (1949). The quotation is taken from section 18, part C.
8. See, e.g., H. Poincaré, *Revue de Métaphysique et de Morale*, vol. 3 (1895).
9. See, e.g., M. Jammer, *Concepts of Space*, Cambridge/U.S.A., (1954). This book contains many references to original sources as well as to historical accounts. Many interesting remarks about the development of spacetime concepts, from a modern point of view, are found in ref. 7.
10. For a new version of this attitude concerning the status of physical geometry see P. Lorenzen, *Philosophia naturalis* **6**, 415 (1961).
11. W. K. Clifford, in *The Common Sense of the Exact Sciences* (ed. J. R. Newman), New York (1946), p. 202; J. A. Wheeler, *Geometrodynamics*, New York (1962).
12. B. Riemann, *Über die Hypothesen, welche der Geometrie zugrunde liegen* (1854); *Gesammelte Mathematische Werke* (1876).
13. R. Penrose, *Structure of Spacetime*, in Battelle Rencontres (ed. C. M. De Witt and J. A. Wheeler), New York (1968), p. 121.

14. G. W. Leibniz, *Math. Schriften* **7**, p. 18 (ed. Gerhardt), Berlin (1849). The causal interpretation of time order and simultaneity is also used by Kant.
15. H. Weyl, *Raum, Zeit, Materie*, 5th ed., Berlin (1923). See in particular p. 142.
16. See, e.g., ref. 9, ch. IV.
17. L. Lange, 'Über die wissenschaftliche Fassung des Galileischen Beharrungsgesetzes', *Ber. kgl. Ges. Wiss., Math.-Phys. Kl.* (1885), p. 333.
18. See ref. 15, section 20.
19. F. Klein, *Vergleichende Betrachtungen über neuere geometrische Forschungen*, Erlangen (1972).
20. V. Bargmann, *Ann. Math.* **59**, 1 (1954). See also J. M. Jauch, *Foundations of Quantum Mechanics*, Reading/Mass. (1968), ch. 13.
21. E. Cartan, *Ann. Ec. Norm. Sup.* **40**, 325 (1923); **41**, 1 (1924).
22. K. Friedrichs, *Math. Ann.* **98**, 566 (1927).
23. P. Havas, *Rev. Mod. Phys.* **36**, 938 (1964).
24. A. Trautman, in *Perspectives in Geometry and Relativity* (ed. B. Hoffmann), Bloomington (1966), p. 413.
25. The following reasoning follows largely A. Trautman; see ref. 24 and his Brandeis lectures (1964). It is worth noticing that the section in which Schrödinger introduces gravitational fields – on pages 56–60 of his beautiful book *Space-Time Structure*, Cambridge (1950) – can be read verbatim as an introduction to Cartan's geometrical version of Newton's theory of gravity, although they are meant to apply to relativity (and do, of course).
26. O. H. L. Heckmann and E. Schücking, 'Newtonsche und Einsteinsche Kosmologie', in *Encycl. of Physics* (ed. S. Flügge), **53**, 489 (1959).
27. T. Levi-Civita, *Rend. Palermo* **42**, 173 (1917); J. A. Schouten, *Verh. Kon. Akad. Amsterdam* **12**, 95 (1918).
28. C. N. Yang and R. L. Mills, *Phys. Rev.* **96**, 191 (1954); R. Hermann, *Vector Bundles in Mathematical Physics*, vol. 1, New York (1970), ch. 3.
29. G. Holland, Ph.D-thesis, University of Hamburg (1965).
30. E. Mach, *Die Mechanik in ihrer Entwicklung*, 7th ed., Leipzig (1912).
31. See G. Rosen, *Am. Journ. Phys.* **40**, 683 (1972) and the references given there.
32. The distinction between a 'weak' and a 'strong' principle of equivalence is due to R. H. Dicke; see *The Theoretical Significance of Experimental Relativity*, New York (1964).
33. For a review and detailed references, see ref. 26.
34. J. L. Anderson, *Principles of Relativity Physics*, New York (1967), section 4.3.
35. K. S. Thorne, C. M. Will and W. T. Ni, in *Proceedings of the Conference on Experimental Tests of Gravitational Theories* (ed. R. W. Davies), NASA-JPL Technical Memorandum 33-499. See also Will's lectures presented at the 56th Fermi summer school, Varenna 1972.
36. E. Breitenberger, *Nuovo Cimento* ser. 11, **1B**, 1 (1971).
37. H. Weyl, *Mathematische Analyse des Raumproblems*, Berlin (1923), especially lecture 3.
38. V. A. Fock, *Theory of Space, Time and Gravitation*, Moscow 1955, Section 8 and Appendix A.
39. E. C. Zeeman, *J. Math. Phys.* **5**, 490 (1964).
40. A. D. Alexandrov, *Commun. of the State University of Leningrad*, No. **8**, 103 (1953).
41. Ref. 7, section 17.
42. F. J. M. Farley, J. Bailey, R. C. A. Brown, M. Giesch, H. Jöstlein, S. van der Meer, E. Picasso and M. Tannenbaum, *Nuovo Cimento* **45**, 281 (1966).
43. More precisely: elementary-particle experiments do not determine the variability of the metric field. This fact has been emphasized by Thorne and Will, see ref. 35.
44. Our formulation of the principle differs only slightly from that of Dicke given in ref. 32. The present formulation avoids inconsistencies which arise from the noncommutativity of covariant derivatives if one speaks of 'all' rather than of 'some' local laws.
45. A. Schild, ch. 1 in *Relativity Theory and Astrophysics* (ed. J. Ehlers); *Lectures in Applied Mathematics* **8**, Providence (1967).
46. S. Deser, *General Relativity and Gravitation* **1**, 9 (1970).
47. J. Ehlers, F. A. E. Pirani, A. Schild, ch. 4 in *General Relativity* (ed. L. O'Raifeartaigh), Oxford (1972), p. 63. See also the forthcoming lectures by J. Ehlers given at the Banff Summer School on Relativity and Gravitation, August 1972, to be published by the Canadian Physical Society.
48. M. Castagnino, *J. Math. Phys.* **12**, 2203 (1971); E. Kronheimer and R. Penrose, *Proc. Camb. Phil. Soc.* **63**, 481 (1967); H. Reichenbach, *Axiomatik der relativistischen Raum-Zeit-Lehre*,

Braunschweig (1924); J. L. Synge, *Relativity: the special theory*, Amsterdam (1956), *Relativity: the general theory*, Amsterdam (1964); A. Trautman, 'The general theory of relativity', *Soviet Physics Usp.* **9**, 319 (1966), see also ref. 24; H. Weyl, see refs. 15 and 37; N. M. J. Woodhouse, *J. Math. Phys.* **14**, 495 (1973).

49. A projective structure can be defined as an equivalence class of symmetric linear connections all having the same geodesics except for parametrization. For a direct definition (avoiding reference to connections) see ref. 47.

50. R. F. Marzke and J. A. Wheeler, in *Gravitation and Relativity* (ed. H. Y. Chiu and W. F. Hoffman), New York (1964).

51. B. Hoffmann and W. Kundt, in *Recent Developments in General Relativity*, p. 303, Warsaw (1962).

52. See, in particular, H. Weyl, *Physik. Z.* **22**, 473 (1921).

53. Instead of obtaining a metric by adding to a conformal structure a particular kind of projective structure, as indicated in the text and elaborated in ref. 47, one can also get a metric by supplementing the conformal (or causal) structure by a measure. Whereas this last procedure may lend itself more readily to generalizations at the microscopic level, as advocated by D. Finkelstein (ref. 62), there appears to be no phenomenological motivation for introducing a spacetime measure (4-volume) as a primitive concept.

54. See ref. 13, sections 4 and 5 for an excellent exposition.

55. These notions have been defined precisely by L. Markus, *Ann. Math.* **62**, 411 (1955).

56. The connection between spacetime orientations and elementary symmetry violations has been investigated by R. Geroch, Ph. D. Thesis, Princeton 1967; and by Ya. B. Zeldovich and I. D. Novikov, *Pisma V Red. Zh. E.T.F.* **6**, 772 (1967).

57. This theory is outlined in A. Trautman's contribution to this volume, where relevant references will be found.

58. See, e.g., Schrödinger's book mentioned in note 25, and M. A. Tonnelat, *Les théories unitaires l'électromagnétisme et de la gravitation*, Paris 1965.

59. R. Penrose, *An Analysis of the Structure of Spacetime*, Adams Prize essay, 1966, and ref. 13.

60. G. F. Chew, *Sci. Progr.* **51**, 529 (1963).

61. See S. W. Hawking and G. F. R. Ellis, *The Large Scale Structure of Space-Time*, Cambridge 1973, and the references therein.

62. See K. Menger's contribution to the well-known book *Albert Einstein*, (ed. A. Schilpp), Evanston, Ill., U.S.A. 1949; also *Topology without points*, Rice Institute Pamphlet, 1932.

63. D. Finkelstein, *Phys. Rev.* **184**, 1261 (1969); also *Coral Gables Conference on Fundamental Interactions at High Energy*, New York 1969, p. 338. The quotation is taken from the last reference.

7

Einstein, Hilbert, and the Theory of Gravitation*

Jagdish Mehra

Abstract. From 1907 to 1925, Albert Einstein dominated the development of the general relativity theory of gravitation. His own work on the theory of gravitation from 1912 to 1916 had the drama of high adventure. David Hilbert, fascinated by Einstein's work on relativity and Gustav Mie's work on electrodynamics, decided to construct a unified field theory of matter. In two communications to the Göttingen Academy (on 20 November 1915 and 23 December 1916), Hilbert developed his theory of the foundations of physics. In the first of these communications, he derived the field equations of gravitation and the conditions governing them. Hilbert's work, although inspired by Einstein, was independent of and simultaneous with Einstein's derivation of the field equations and, in certain essential respects, went beyond Einstein's. Einstein, at that time, was very critical of the efforts of Mie, Hilbert and Weyl to construct a unified field theory of gravitation and electromagnetism. After 1925, such a programme became his primary mission.

In this paper, we have given an account of the intellectual struggles of this fascinating period when the modern theory of gravitation was created.

CONTENTS

*This investigation developed from a question which Professor Eugene Wigner asked me in November 1971 about Hilbert's contribution to the equations of general relativity, and my reply to him. See Acknowledgements and Appendix.

1. INTRODUCTION

The modern theory of gravitation, more than any other theory in physics, is connected with the name of a single personality, that of Albert Einstein. Almost immediately after he had published his work on the theory of special relativity, Einstein restlessly moved on to incorporate the effects of gravitation in a consistent theoretical framework. Indeed this problem occupied him so much that he essentially abandoned quantum theory for many years, a field to which he had made singular contributions. A letter to Sommerfeld, dated 29 October 1912, gives evidence of the situation. Einstein wrote:

Dear Colleague: Your kind note only adds more to my predicament. But I assure you that with respect to the quantum [theory] I have nothing new to say that should arouse interest.... I am now exclusively occupied with the problem of gravitation and hope, with the help of a local mathematician friend [Marcel Grossmann], to overcome all the difficulties. One thing is certain, however, that never in life have I been quite so tormented. A great respect for mathematics has been instilled within me, the subtler aspects of which, in my stupidity, I regarded until now as pure luxury.* Against this prob-

---

* Writing to the physicist Paul Hertz at about this time, Einstein said that he had been 'a mathematical ignoramus until I reached this shore'. [I am grateful to Rudolf Hertz, the son of Paul Hertz, for sending me a photocopy of this letter.]

lem [of gravitation], the original problem of the theory if relativity is child's play. Abraham's new theory*, so far as I can see, is indeed logically correct, but otherwise a monstrous embarrassment. The hitherto existing relativity theory is certainly not as wrong as Abraham seems to think.[1]

Almost all the physicists of the first rank were absorbed at that time in an effort to understand the quantum problem which the work of Einstein, more than even Planck's, had brought to the forefront. Einstein had pushed the quantum problem into the background in his own work, having 'nothing new to say' about it for the time being, as he wrote to Sommerfeld. He was fully absorbed in 'generalizing' the relativity theory to include gravitation. Still, he did not develop his ideas on gravitation in isolation or in a vacuum of response from eminent physicists. Contrary to the lore which has become fashionable among the 'relativists' nowadays, Einstein had the full benefit of the cut and thrust of scientific debate with some first-rate minds in the development of his theory of gravitation. Indeed there were even those who made valuable, even crucial, contributions to this theory.

It is convenient, for the sake of discussion, to single out three distinct stages in the early development of general relativity, each of them brief in its duration. During the first period, from 1907 to 1914, Einstein attempted to formulate and sharpen the principle of equivalence, and engaged in numerous discussions, especially with Abraham and Nordström. The second period, from 1912 to 1914 which overlapped with the first, was one in which Einstein and his friend Grossmann incorporated into the new theory differential geometry and other great mathematical structures of the nineteenth century. During the third period, 1915–16, the field equations of gravitation were formulated by Einstein, were independently derived by Hilbert, and the mathematical structure of the equations was clarified. We shall discuss this lively development which culminated in the establishment of the general relativity theory of gravitation.

2. STEPS TOWARDS A NEW THEORY OF GRAVITATION

2.1 THE PRINCIPLES OF EQUIVALENCE AND RELATIVITY

As soon as the so-called theory of special relativity was formulated, Einstein sought for ways of 'generalizing' it. In his article on 'The Principle of Relativity and Its Consequences', which he wrote in 1907 for the *Jahrbuch der Radioaktivität und Elektronik*, he already suggested the equivalence between a gravitational field and an accelerated frame of reference.[2] Two points from this article are of great interest in the present context. First, the strong expression which Einstein gave to the assumption of proportionality between the inertial and gravitational masses of a body.

The proportionality between the inertial and gravitational masses holds for all bodies without exception, with the [experimental] accuracy achieved thus far, so that we may assume its general validity until proved otherwise.[3]

He then took a stand with respect to an extension of the special theory:

* We shall discuss Abraham's work later on in its proper context.

Is it conceivable, that the principle of relativity also holds for systems which accelerate relative to each other? This is, of course, not the place for a detailed discussion of this question. But since this question must occur to anyone who has followed the applications of the relativity principle made hitherto, I shall not fail to mention it here.[4]

Einstein considered two systems Σ_1 and Σ_2, one at rest in a field of constant gravity, and the other undergoing a uniform acceleration with respect to the first. The laws of physics then do not allow one to make any distinction between the two systems. He stated:

Hence, on the basis of our actual experience, we have no reason to suppose that the systems Σ_1 and Σ_2 can be distinguished from each other in any way. We shall therefore assume complete physical equivalence between the gravitational field and the corresponding acceleration of the reference system. This assumption extends the principle of relativity to the case of the uniformly accelerated translational motion of the reference system. The heuristic value of the assumption lies in the fact that it allows a homogeneous gravitational field to be replaced by a uniformly accelerated reference system, a fact which, to a certain extent, makes it amenable to theoretical treatment.[5]

Einstein then considered the properties of space and time in a uniformly accelerated frame of reference and found corrections only of the second order to the spatial shapes of bodies, but 'the process which occurs in watches – indeed any process, in general – takes place faster the larger is the gravitational potential present at the space point in question.'[6] From this he derived the result 'that the light coming from the surface of the sun has a wavelength which is larger by two parts in a million than the one produced by the same atoms on earth.'[7] Einstein further concluded that light beams, which do not travel along the axis of the acceleration, are bent by the gravitational field, leading to a decrease of the velocity of light which is proportional to the gravitational potential.[8]

Einstein's paper of 1907 may be regarded, certainly in the published literature, as the explicit starting point of the theory of general relativity, in which several of the assumptions and results to be justified later on were already present. With respect to the two central points of his theory – the equivalence of the inertial and gravitational masses and the principle of general relativity of uniformly accelerated frames of reference – Einstein would later on turn out to be rigid and uncompromising in all discussions with his competitors, and both of these ideas were present in his memoir of 1907.[9]

At this time Einstein's hypothesis of the light quantum was barely two years old, and he was deeply involved in the problems of radiation and the quantum theory. These problems would occupy him fully for another four years before he would return to work on general relativity in 1911.[10]

In a lecture to the Naturforschende Gesellschaft on 16 January 1911, shortly before he left Zurich to take the chair of physics at the German University of Prague, Einstein presented a synopsis of his ideas on special relativity.[11] The chair of physics in Prague, which had been occupied before him by Ernst Mach, proved to be very stimulating for the progress of general relativity. In June, 1911, Einstein submitted a paper to the *Annalen der Physik*, improving his earlier statements of 1907 on the action of gravitation on the propagation of light.[12]

What was new in this derivation in which he again denied the existence of an *absolute* acceleration? Einstein showed that, from the equivalence of an accelerated frame of reference with a corresponding rest frame including a static homogeneous field of gravitation, the identity of inertial and gravitational masses can be *derived*. Again the velocity of light turns out to be

$$c(\phi) = c\left(1 + \frac{\phi}{c^2}\right),\tag{1}$$

where ϕ is the (*negative*) gravitational potential. Hence 'the principle of the constancy of the speed of light, according to this theory, is not valid in the same manner as it serves to establish the ordinary [special] theory of relativity.'[13] With the help of Huygens' principle and his theory of gravitation Einstein derived the important consequence, that 'a light ray passing close to the sun would suffer a bending of 0.83 seconds of arc.'[14]

2.2 ABRAHAM'S OBJECTIONS

It was already at this early stage in the development of general relativity that Max Abraham entered into the discussion.[15] In a series of papers, Abraham formulated a theory of gravitation, starting from Einstein's consideration of a velocity of light dependent on the gravitational potential.[16] He assumed the existence of a potential ϕ [on the (four-dimensional) gradient of which the ten components of the (symmetric) gravitational tensor would depend].[17] In his calculation, the dependence of the velocity of light c on the gravitational potential turned out to be,

$$\frac{c^2}{2} - \frac{c_0^2}{2} = \phi - \phi_0,\tag{2}$$

which agreed with Einstein's expression (1) in the first order.

On considering the attraction of two masses, Abraham found that the gain of kinetic energy when the masses approached each other was only one-half of the loss in potential energy, leading to the conclusion that the gravitational field also contains energy.[18] He derived an expression for this field energy density,

$$\varepsilon = \frac{1}{8\pi\gamma}\left\{\left(\frac{\partial\phi}{\partial x}\right)^2 + \left(\frac{\partial\phi}{\partial y}\right)^2 + \left(\frac{\partial\phi}{\partial z}\right)^2 + \left(\frac{\partial\phi}{\partial ct}\right)^2\right\},\tag{3}$$

γ being the constant of gravitation. This expression is essentially positive definite, due to the fact that the plus sign appears in the last term within the curly brackets. The reason for this change of sign lay in a new *Ansatz* for the 'hydrostatic pressure' of the gravitational field.[19] The non-covariance under Lorentz transformations did not bother Abraham, since he found that invariance was missing even in infinitesimal regions.[20] One of the consequences of this theory was the dependence of the rest mass m_0 on the velocity of light

$$m_0(\phi) = \frac{M}{c(\phi)}\tag{4}$$

which agreed with Einstein's energy-mass formula in the first approximation.[21] Newton's law of gravitation did not, however, follow strictly from this theory, and had to be replaced by

$$K = \frac{A}{r^2} - \frac{B}{r^3},$$

where

$$\frac{B}{A} = \frac{\gamma m}{2c^2}$$

then, in general[22]

$$\frac{B}{Ar} \ll 1. \tag{5}$$

Abraham's theory had certain peculiarities which were quite unusual from the 'relativistic' point of view. First, the absolute value of the potential entered through Equation (2).[23] Second, his theory did not throw any light on the problem of matter, emphasizing rather the essential disparity between the dynamics of the electron and the theory of the gravitation.[24]

These features brought Abraham immediately into conflict with Einstein whose physical reasoning and mental attitude were against them. Abraham had produced a theory which, in Einstein's view [as he noted in his letter to Sommerfeld quoted earlier[1]], was 'logically correct, but otherwise a monstrous embarrassment'. In a paper on 'The Velocity of Light and the Statics of the Gravitational Field', Einstein criticized Abraham's theory and summarized his own point of view.[25] His main argument was that 'Abraham's system of equations cannot be brought into accord with the equivalence hypothesis, quite apart from the fact that his treatment of space and time could not be upheld from a purely formal mathematical point of view.'[26] Einstein again dealt with the equivalence between the static gravitational field and a uniformly accelerated frame of reference, suggesting the following relationship between the second derivatives of the velocity of light and the gravitational source term,

$$\Delta c = \gamma c \varrho, \tag{6}$$

where γ is the universal gravitational constant, $\gamma = 4\pi k/c^2$, and ϱ the mass density.[27] He improved upon this equation in a later work, replacing it by

$$\Delta c = \gamma \left\{ c\varrho + \frac{1}{2\gamma} \frac{\text{grad}^2 c}{c} \right\} \tag{6'}$$

where the second term expresses the energy density of the gravitational field.[28] In the case of small velocities, $v \ll c$, Δc may be replaced by $\Delta\phi$, ϕ being the gravitational potential. The usual gravitational constant γ is therefore no longer a universal constant.[29]

Having deduced the conservation of energy and momentum for a particle in the static gravitational field, Einstein commented on the space-time problem:

The space-time problem seems to me to be as follows. If one restricts oneself to a region of constant gravitational potential, the laws of nature would then assume an especially simple and invariant form, provided one refers them to that manifold of space-time systems which are related to each other by the Lorentz transformations of constant c. If one does not restrict oneself to regions of constant c, then the manifold of equivalent systems as well as the manifold of transformations which leave the laws of nature unaltered would become larger, and because of that the laws would become more complicated.[30]

In his next paper, Einstein discussed the influence of a static gravitational field on electromagnetic and thermal processes.[28] Similar to Abraham, he found [see Equation (4), for example] that some of the conventional expressions have to be multiplied by the varying light velocity c to form 'scalars'. For instance, the energy density of the electromagnetic field is given by

$$\tfrac{1}{2}(\mathscr{E}^2 + \mathscr{H}^2) \rightarrow \frac{c}{2}(\mathscr{E}^2 + \mathscr{H}^2), \tag{7}$$

or the genuine temperature T is given by

$$T = cT^*, \tag{8}$$

T^* being the usual temperature.

Einstein's clarification gave rise to a strong and lively discussion between him and Abraham.[31] In his note on 'Relativity and Gravitation, Reply to an Observation of Mr A. Einstein', Abraham insisted on the validity of the Lorentz invariance in infinitesimal regions, emphasizing the existence of an *absolute* system in which the gravitational field is quasi-static.[32] Einstein gave further clarification of the theory of relativity, since Abraham had concluded the 'breakdown' of that theory.[33] Abraham had also rejected any relativistic space-time concept, including the general space-time transformations.[34] These criticisms had to be answered, and Einstein did so very clearly by stating the two assumptions of special relativity, namely (i) the principle of relativity, claiming that no absolute frame of reference should exist, and (ii) the constancy of the velocity of light. The latter has to be given up if non-inertial frames are admitted, in agreement with the first principle.

Abraham was not very impressed by Einstein's answer, and he repeated his doubts in a second note.[35] He accepted the equivalence of inertial and gravitational masses, as established by the experiments of Eötvös[36], but did not wish to imprison this result into the equivalence hypothesis concerning the frames of reference. As for Einstein he proceeded to establish his claims within a few years.[37]

2.3 NORDSTRÖM'S THEORY

Abraham had certainly stimulated Einstein to strengthen his efforts. In the same year, 1912, a series of papers on the relativity principle and gravitation was published, which led to a more friendly and fruitful discussion with Einstein. The author of these papers was Gunnar Nordström of Helsingfors, who proposed in them the first consistent theory of gravitation.[38] Nordström started from the question whether one could construct a consistent theory of gravitation in which the principle of constancy of the velocity of light would be maintained, since the Abraham–Einstein discussions seemed to have shown the difficulties of the whole idea of relativity if one allowed for a variable velocity of light.[39] With Abraham, Nordström introduced a gravitational potential

$$\Box \phi = 4\pi k \varrho_0 \tag{9}$$

where k denotes the gravitational 'constant' and ϱ_0 the rest-density of matter. Then the constancy of the velocity of light c can be maintained provided one allows for a change of mass with the gravitational potential,

$$m = m_0 e^{g\phi/c^2}, \qquad g = 4\pi k. \tag{10}$$

In his second note, Nordström supplemented Equation (9) with equations for the gravitational force,

$$K_\mu = - g\varrho_0 \frac{\partial \phi}{\partial x_\mu}, \quad (\mu = 0, 1, 2, 3). \tag{11}$$

The system of Equations (9) and (11) would thus form a complete basis for the theory of gravitation. The gravitational factor g, however, does not turn out to be constant if the principle of equivalence of gravitational and inertial masses is to be preserved.[40] The rest density, according to Laue, is defined as

$$\varrho_0 = - \frac{1}{c^2} \operatorname{Tr} T_{\mu\nu} \tag{12}$$

where $T_{\mu\nu}$ = the matter tensor. The dependence of g on the gravitational potential is then given by

$$g(\phi) = \frac{g(\phi_0)}{1 + \dfrac{g(\phi_0)}{c^2}(\phi - \phi_0)}, \tag{13}$$

which guarantees the numerical equivalence between the inertial mass

$$m = m_0 \left[1 + \frac{g_0}{c^2}(\phi - \phi_0) \right] \tag{14}$$

and the gravitational mass

$$M_g = \int g(\phi) \varrho_0 \, d^3x, \tag{15}$$

to a certain extent, which we shall discuss later. The role of the gravitational potential ϕ_0 is to define the scale between gravitation and the other laws of nature.

Nordström's theory was the first attempt to provide a consistent approach to the problem of gravitation, a fact which was acknowledged by Einstein himself.[41] In a lecture on 'The Present Status of the Problem of Gravitation', delivered at the Eighty-fifth Conference of Natural Scientists in Vienna, in 1913, Einstein discussed the hypotheses and theories of gravitation. He listed the general postulates which may be valid in a theory of gravitation[42]: (i) energy-momentum conservation is satisfied; (ii) gravitational and inertial masses of closed systems are equal; (iii) special relativity is valid; (iv) the laws of nature do not depend on the absolute value of the gravitational potential. In discussing these, Einstein remarked that postulate (i) should be generally accepted, while postulate (iii) might not be valid except in regions of constant gravitational potential. The second postulate, he thought, ought to be included because of the experimental evidence, while postulate (iv) could only be justified by the simplicity which the laws of nature obtain once it is applied. Nordström's theory is derived from a scalar potential and satisfies these four postulates.[43] Among the theories of gravitation which should be regarded as the 'most natural', Einstein mentioned Nordström's theory and his own new theory worked out in collaboration with Grossmann. He derived Nordström's theory by assuming Hamilton's principle for a massive (point) particle and invariance under Lorentz transformations, and discussed the two consequences which follow from this theory: (a) light does not get deflected in a gravitational field; (b) the inertia of a particle is determined by the scalar potential ϕ, or the larger the masses which are accumulated in the vicinity of a particle the smaller becomes its inertia.

Nordström's theory had already been discussed in critical detail by Einstein and Fokker, especially its basis on the scalar potential.[44] It had been established that the line element ds depends on ϕ according to the relation

$$ds^2 = \phi^2 (dx^2 + dy^2 + dz^2 - c^2 dt^2), \tag{16}$$

the differential equation for the gravitational potential taking the form

$$\phi \,\square\, \phi = k \operatorname{Tr} T_{\mu\nu}, \tag{17}$$

where k is a constant and $T_{\mu\nu}$ is the general energy(-momentum) tensor. In his lecture at Vienna, Einstein summarized the situation as follows[45]:

We can sum up by saying that Nordström's scalar theory adheres to the postulate of the constancy of the velocity of light and satisfies all the requirements which can be placed upon a theory of gravitation, based upon the present experimental knowledge. The only unsatisfactory thing is the fact that according to this theory the inertia of a body seems to be *affected*, although not *produced*, by other bodies. Because, according to this theory, the inertia of a body increases the farther the other bodies are removed from it.[46, 47]

Einstein's main point was that he wished to stick to his four postulates, mentioned earlier, the more so because, with Grossmann, he was on the way to developing another consistent theory of gravitation, using the invariant theory of the absolute differential calculus. He had already stated in the work with Fokker:

We note that only the application of the invariant theory of the absolute differential calculus would provide us a clear insight into the formal content of Nordström's theory. Furthermore, this method makes it possible for us to determine the influence of the gravitational field on arbitrary physical phenomena, as is to be expected from Nordström's theory, without invoking new hypotheses. The relationship of the Nordström theory to that of Einstein-Grossmann thus becomes completely clear.

Finally, the role which the Riemann–Christoffel differential tensor plays in this [present] investigation suggests the idea, that it might also open the way for deriving the Einstein-Grossmann equations for gravitation independent of physical assumptions. The proof of the existence or non-existence of such a connection would signify an important theoretical advance.[48]

In other words, Einstein was now seeking to become a mathematician for the sake of the theory of gravitation.

2.4 EINSTEIN AND GROSSMANN: SEARCH FOR THE MATHEMATICAL TOOLS OF GENERAL RELATIVITY

Einstein returned from Prague to Zurich in the summer of 1912. He had by then already formulated the fundamental physical principles of the general relativity theory of gravitation, and was now searching for their mathematical structure. At the E.T.H. [the Swiss Federal Institute of Technology in Zurich], where he now returned as professor of theoretical physics, Einstein met again his old friend and former fellow student Marcel Grossmann, who was now a professor of mathematics and his colleague. With Grossmann, and under his guidance, Einstein studied the mathematical literature, especially the theory of invariants and the absolute differential calculus of Christoffel, Ricci, Levi-Civita and others. Einstein developed the mathematical structure of his theory jointly with Grossmann, and in his celebrated paper on the general theory of relativity in 1916 he acknowledged the help which his friend had given him. It was Grossmann's help which had, Einstein said, 'spared me not only the study of the relevant mathematical literature, but who [Grossmann] also assisted me in searching for the field equations of gravitation.'[49] This study of the mathematical literature and the search for the proper mathematical tools led to several joint papers with Grossmann during Einstein's all too brief stay in Zurich. These papers contained the first attempts toward a generalized theory of relativity, using the new mathematical tools, and gave full expression to Einstein's earlier physical insights. Although the actual outcome of these studies, which went into the final theory, seems to be small, their 'psychological value', to use Einstein's own phrase, was most important.[50]

Three papers resulted from the collaboration of Einstein and Grossmann, and Einstein reported on their results in his lecture at Vienna in the fall of 1913.[51] In a fourth paper[52], written by Einstein alone more than a year after he had moved from Zurich to his position at the Prussian Academy in Berlin, Einstein classified and summarized the achievements of the search up to that time.[53] The work of Einstein and Grossmann seems to have been a highly technical struggle with the formal possibilities offered by the absolute differential calculus. It did not have the simplicity and persuasive elegance of the later theory, mainly because of the fact that Einstein had to find the way of translating the physical postulates into mathematics. Another fact which made Einstein's search a sort of a comedy of errors was his great ability to give, for any mathematical result, a good looking physical 'reason', and he was very quick in providing such 'explanations'. This capacity of Einstein's to invent physical explanations, as we shall see later on, was the main reason why he missed the right solution during the period between 1912 and 1914, a fact which was lamented by him later on.[50] And the evidence suggests that it was not Einstein's collaborator, the mathematician Grossmann, who was responsible for the delay in the creation of a consistent theory of gravitation.

The first two joint papers of Einstein and Grossmann, and Einstein's lecture at Vienna, came at about the same time, and we shall consider them together.[54] After stating the physical postulates, which we have discussed above, Einstein went on to discuss the consequences which would follow from the general invariance of the line element ds, given by

$$ds^2 = g_{\mu\nu}\,dx^\mu\,dx^\nu, \tag{18}$$

[where summation is understood over the repeated indices (Einstein convention)] where the symmetric tensor $g_{\mu\nu}$ represents the metric dependent on space-time. The ten quantities $g_{\mu\nu}$ therefore characterize the gravitational field, replacing the scalar potential ϕ of the Newtonian theory and determining the behaviour of clocks and rods.[55] Now the line element, defined by Equation (18), also determines a Riemannian geometry, and the question arises about the relation between its invariants and co-variants, on the one hand, and the mathematical and the physical quantities, on the other. Grossmann, who searched through the mathematical literature in order to assist Einstein, also discussed the first of these questions.[56]

In dealing with tensor algebra and analysis, Grossmann introduced the notions of covariant and contravariant tensors of any rank, according to Ricci and Levi-Civita, and of divergence and gradients, using the Christoffel symbols,

$$\Gamma_{\mu,\,\nu\varrho} = \tfrac{1}{2}\left(\partial g_{\mu\nu}/\partial x^\varrho + \partial g_{\mu\nu}/\partial x^\nu - \partial g_{\nu\varrho}/\partial x^\mu\right) \quad \text{and} \quad \Gamma^\mu_{\nu\varrho} = g^{\mu\sigma}\Gamma_{\sigma,\,\nu\varrho}. \tag{19}$$

The so-called Riemann–Christoffel tensor of the fourth rank plays an important role in differential geometry, and is written out as

$$R^\mu_{\nu\varrho\sigma} = \partial\Gamma^\mu_{\nu\varrho}/\partial x^\sigma - \partial\Gamma^\mu_{\nu\sigma}/\partial x^\varrho + \Gamma^\mu_{\alpha\sigma}\Gamma^\alpha_{\nu\varrho} - \Gamma^\mu_{\alpha\varrho}\Gamma^\alpha_{\nu\sigma}. \tag{20}$$

Grossmann commented upon the importance of tensors as follows:

The overriding significance of these structures [tensors] for the differential geometry of a manifold, defined by its line element, makes it *a priori* probable, that these general differential tensors could also be important for the problem of the differential equations of a gravitational field. [The identical vanishing of the tensor $R^\mu_{\nu\varrho\sigma}$ provides the necessary and sufficient condition that the differential form can be transformed to the form $\Sigma_\mu \pm (dx^\mu)^2$, where the choice of signs depends on the signature of the manifold.] Indeed, it is possible to specify a covariant differential tensor $R_{\mu\nu}$, of second rank and second order, which might enter those equations, namely

$$R_{\mu\nu} = R^\varrho_{\mu\varrho\nu}. \tag{21}$$

This in itself shows that this tensor, in the special case of an infinitely weak gravitational field, would not reduce to the expression $\Delta\phi$. We must therefore leave open the question as to what extent the general theory of the differential tensors linked with the gravitational field is related to the problem of the equations of gravitation. Such a connection must exist, insofar as the equations of gravitation allow arbitrary substitutions, because already in this case it seems to be excluded to obtain differential equations of the second order. If, on the other hand, it could be established that the equations of gravitation only admit a certain group of transformations, then one could understand why the differential tensors yielded by the general theory would not be acceptable. As mentioned in the physical part [written by Einstein], as yet we are not in a position to discuss this question.[57]

Einstein continued in the addendum to their joint paper:

In writing up this paper, we have regarded it as a shortcoming that it has not been possible to write down equations for the gravitational field that are covariant generally, i.e. for arbitrary substitutions. Subsequently, I found, however, that equations which determine the $g_{\mu\nu}$ uniquely from $\theta_{\mu\nu}$ [the energy tensor], and which are uniquely covariant, cannot exist at all, as the following proof shows.[58]

These quotations from the 'workshop' of general relativity are important because they indicate that Grossmann had stressed the importance of the Riemann–Christoffel tensor to Einstein, but the latter failed, on account of his 'physical reasoning', to use it as the fundamental concept. From the procedure employed by Einstein, one can see how this mistake came about. He started from the generalized line element, Equation (18), and studied the motion of a material point in the field of gravitation according to the equation

$$\delta\left\{\int ds\right\} = 0, \tag{22}$$

from which the equation of motion follows

$$\frac{d^2 x^\mu}{ds^2} + \Gamma^\mu_{\varrho\sigma} \frac{dx^\varrho}{ds} \frac{dx^\sigma}{ds} = 0. \tag{23}$$

Einstein could now reformulate the laws of physics in the presence of a gravitational field, given by the metric $g_{\mu\nu}$. The energy tensor plays an important role in what he called the 'laws of material phenomena'.[59] Now, on generalization of the energy tensor (the special relativistic quantity) in the presence of a gravitation field in which the $g_{\mu\nu}$ are not constant, the newly formed quantity $T_{\mu\nu}$ is not conserved, and one finds

$$\frac{\partial (T^\nu_\sigma + t^\nu_\sigma)}{\partial x^\nu} = 0. \tag{24}$$

Here the new t^ν_σ represents the energy tensor of the gravitational field, and Einstein motivated its existence by saying: 'It thus represents the physical fact that in a gravitational field the impulse and energy of a material system vary with the time, in that the gravitational field transports energy and momentum to the material system.'[59] The fact that a gravitational energy-momentum tensor exists, has important consequences: 'Since we demand the validity of the conservation laws, we restrict the reference frame to the corresponding extent, and thereby renounce the setting up of the equations of gravitation in general covariant form.'[60] Thus it seemed that the whole programme of general relativity had broken down, as was correctly pointed out to Einstein after his Vienna lecture by the probings of Mie and Reissner.[61] What error had really crept into the framework of the theory?

In generalizing the Poisson equation (9), which can be written as a tensor equation by a generalization of Newton's theory, Einstein looked for an expression of the form

$$\gamma\theta_{\mu\nu} = \Gamma_{\mu\nu}, \quad (\text{where } \theta_{\mu\nu} = T_{\mu\nu}), \tag{25}$$

where $\Gamma_{\mu\nu}$ is a covariant tensor of the second rank, γ a constant and $\theta_{\mu\nu}$ or $T_{\mu\nu}$ the covariant energy tensor. In analogy with the Newton–Poisson law, one postulates with Einstein that Equations (25) are of the second order.[62, 63] The difficult question now arose about the tensor generalization of ϕ. Einstein appended the following remarks to the addendum in his joint paper with Grossmann.[64]

It must, however, be emphasized that it seems to be impossible, under this assumption [that the equations are of second order] to find a differential expression $\Gamma_{\mu\nu}$ which is a generalization of $\Delta\phi$ and turns out to be a *tensor* under *arbitrary* transformations. It cannot, however, be denied *a priori* that the final exact equations of gravitation can be of higher than the second order. Thus there always exists the possibility that perfectly exact differential equations of gravitation might be covariant with respect to *arbitrary* substitutions. The attempt at a discussion of such possibilities would, however, be premature in view of the present status of our knowledge about the physical properties of the gravitational field. We are thus restricted to [the equations of] the second order, and must therefore abstain from setting up equations of gravitation which would turn out to be covariant with respect to arbitrary transformations. Moreover, it must be emphasized that we have no clue regarding a general covariance of the equations of gravitation.

The idea was the following: If the metric $g_{\mu\nu}$ were determined entirely by the energy tensor where $\theta_{\mu\nu}$ or $T_{\mu\nu}$, and all general transformations were admitted, then a contradiction would arise. The minimum transformation group, under which the quantity $\Gamma_{\mu\nu}$ is permitted to be covariant, is a group of linear transformations. Einstein and Grossmann proposed that

$$\Gamma^{\mu\nu} = \frac{1}{\sqrt{-g}} \frac{\partial}{\partial x^\alpha} \left(g^{\alpha\beta} \sqrt{-g} \frac{\partial g^{\mu\nu}}{\partial x^\beta} \right) - g^{\alpha\beta} g_{\tau\varrho} \frac{\partial g^{\mu\tau}}{\partial x^\alpha} \frac{\partial g^{\nu\varrho}}{\partial x^\beta} +$$
$$+ \tfrac{1}{2} g^{\alpha\mu} g^{\beta\nu} \frac{\partial g_{\tau\varrho}}{\partial x^\alpha} \frac{\partial g^{\tau\varrho}}{\partial x^\beta} - \tfrac{1}{4} g^{\mu\nu} g^{\alpha\beta} \frac{\partial g_{\tau\varrho}}{\partial x^\alpha} \frac{\partial g^{\tau\varrho}}{\partial x^\beta}. \tag{26}$$

The equations which determine the gravitational field could then be simplified to

$$\Delta^{\mu\nu}(g) = \gamma \left(\theta^{\mu\nu} + \Lambda^{\mu\nu} \right), \tag{27}$$

with

$$\Delta^{\mu\nu}(g) = \frac{1}{\sqrt{-g}} \frac{\partial}{\partial x^\alpha} \left(g^{\alpha\beta} \sqrt{-g} \frac{\partial g^{\mu\nu}}{\partial x^\beta} \right) - g^{\alpha\beta} g_{\tau\varrho} \frac{\partial g^{\mu\tau}}{\partial x^\alpha} \frac{\partial g^{\nu\varrho}}{\partial x^\beta}$$

and

$$- 2\gamma \Lambda^{\mu\nu} = \left(g^{\alpha\mu} g^{\beta\nu} \frac{\partial g_{\tau\varrho}}{\partial x^\alpha} \frac{\partial g^{\tau\varrho}}{\partial x^\beta} - \tfrac{1}{2} g^{\mu\nu} g^{\alpha\beta} \frac{\partial g_{\tau\varrho}}{\partial x^\alpha} \frac{\partial g^{\tau\varrho}}{\partial x^\beta} \right).$$

The sum of the 'energy tensors' of matter and that of the gravitational field satisfy the conservation law, Equation (24).

Einstein and Grossmann, in their third paper of 1914, drew consequences from the existence of the energy tensor of the gravitational field, $t_{\mu\nu}$ $(t^{\mu\nu} \equiv \Lambda^{\mu\nu})$.[51] They 'showed' that one could not have generally covariant equations (25), which determine the characteristic metric $g_{\mu\nu}$, thereby improving their work of 1913. Taking the derivative of $\Delta_{\sigma\nu}$, and summing over ν, one obtains four conditions

$$B_\sigma \equiv \frac{\partial^2}{\partial x^\nu \partial x^\alpha} \left(\sqrt{-g}\, g^{\alpha\beta} g_{\sigma\mu} \frac{\partial g^{\mu\nu}}{\partial x^\beta} \right) = 0. \tag{28}$$

The equations of gravitation (27) are covariant with respect to all 'justified' transformations of the coordinate system, i.e. those which satisfy the conditions (28). In order to interpret these conditions, one considers a variational principle,

$$\delta J = \delta \int H \, d\tau = 0, \tag{29}$$

with

$$H = \sqrt{-g} \, g^{\alpha\beta} \frac{\partial g^{\tau\varrho}}{\partial x^\alpha} \frac{\partial g^{\tau\varrho}}{\partial x^\beta},$$

where H can be regarded as the Hamiltonian of the gravitational field. The allowed coordinate system is then 'chosen such that, for the fixed boundary values of the coordinates and their first derivatives (considered in an arbitrary coordinate system), the integral J approaches a maximum.'[65] One could, therefore, say that 'since the conditions $B_\sigma = 0$, with the help of which we have restricted the coordinate systems, are the immediate consequence of equations of gravitation, our considerations show that the covariance of the equations is as far-reaching as imaginable.'[66]

The fourth and final paper of the series in 1914, dealing with the results of the Einstein–Grossmann collaboration[52], which Einstein communicated (in his own name alone) to the Prussian Academy, did not add anything new to the theory except some mathematical clarifications.[67] Einstein, by that time, had mastered the mathematical tools himself, and he did not need his friend any more.[68] Yet, the theory of Einstein and Grossmann was not Einstein's final theory.

2.5 THE FOUNDATION OF GENERAL RELATIVITY

In the spring of 1914 Einstein moved from Zurich to Berlin to take up his appointment as a member of the Prussian Academy and director of the research institute of physics at the Kaiser-Wilhelm-Gesellschaft.[69] The new position provided him with full freedom for his research work, an activity for which Planck and his colleagues had estimated him so highly. Having succeeded sufficiently in his first attempt to generalize relativity, or at least in making a plan of its generalization, Einstein was not so sure if he could do much more at that time, and remarked to his friend and former fellow student Louis Kollros on the occasion of his farewell from Zurich: 'The Germans are betting on me as a prize hen; I am myself not sure whether I am going to lay another egg.'[70] Lay a few golden eggs he still would in Berlin in both the fields of quantum theory and general relativity. The latter was completed in Berlin soon after his arrival there, as far as the problem of gravitation was concerned.[71]

Before Einstein went to Berlin, what he and Grossmann had achieved was an *Entwurf* – a plan or outline of the shape which, they thought, the theory of general relativity would take. In this initial sketch, the mathematical ideas had played a smaller role than one would have expected. 'The general theory of invariants acted as an obstacle. The direct way turned out to be the only one practicable,' Einstein wrote to his friend Michele Besso in March 1914.[72]

During most of the year 1914 Einstein was almost content with the results of his

labours with Grossmann, achieved in 1913. Early in 1915, however, he was again at
work, starting by drawing the experimental consequences which the theory might
have. On 12 February 1915, he wrote to Michele Besso[73]:

I am able to inform you about two beautiful scientific matters: (1) Gravitation. Red-shift of spectral
lines – the spectroscopic double stars have the same average velocity at the radius of vision. The
masses of the stars are obtained from the periodic Doppler fluctuation of the lines. The component
due to the heavier star should have a mean red-shift of the spectral lines relative to the other. *This
has been verified.* Since it is possible to estimate the radii of the stars (from their spectral type), it
has even provided a satisfactory quantitative verification of the theory.' [The other scientific matter
which Einstein reported to Besso was the experimental verification of his Ampere-molecular current
hypothesis.]

At three consecutive sessions of the Prussian Academy – on 4, 11, and 18 November
1915 – Einstein presented his new theory. Opening his first contribution on the 'general theory of relativity', he remarked:

During the past [several] years, I have tried hard to construct a general relativity theory on the assumption of the relativity [also] of non-uniform motions. I actually though that I had discovered the only
law of gravitation corresponding analogously to the postulate of general relativity, and sought to
demonstrate the necessity of exactly this solution in a memoir published last year in these proceedings.
[See *Sitzungsberichte*, 1914, pp. 1066–1077, Ref. 59.]
    A fresh examination showed me that the necessity [of the solution] from the method proposed
there cannot be proved at all; that this seemed to be so was due to an error. The postulate of relativity,
to the extent that I have invoked it there, is always satisfied if one takes Hamilton's principle as the
basis; in fact, however, it does not provide any means of determining the Hamiltonian function H of
the gravitational field. It actually affects the equation restricting the choice of H [Equation (28)] in no
other way, except that H should be invariant with respect to linear transformations, a requirement
which is of no use in accomplishing the relativity of accelerated [motions]. Moreover, the choice [for
the Hamiltonian] indicated [by Equation (29)] is in no way determined [by Equation (28)].
    For these reasons I lost all faith in the field equations which I had set up, and began to look for a
path which restrict the possibilities in a natural manner. Thus I reverted to the requirement of the
general covariance of the field equations, from which I had departed with a heavy heart in the first
place three years ago when I worked together with my friend Grossmann. In fact, at that time we had
already come quite close to the solution of the problem given in the following.
    Just as the special relativity theory is founded upon the postulate that its equations must be co-
variant with respect to linear, orthogonal transformations, the theory developed here is based on the
postulate of the *covariance of all systems of equations with respect to the transformations of the substitution determinant 1.*
    The fascination of this theory would hardly leave anybody who has really grasped it. It represents
a real triumph of the method of the general differential calculus founded by Gauss, Riemann,
Christoffel, Ricci and Levi-Civita.[74]

The restriction on general transformations, with the substitution determinant 1,
simplifies the formal foundations of the theory of general relativity considerably. For
instance, it allows one to relate covariant and contravariant indices much more easily.
The most important change in attitude, however, was to express the problem of gravitation not by the artificial forms introduced earlier, but by the Riemann–Christoffel
tensor as suggested by Grossmann already in 1913.[75] Gravitation is connected with
a tensor of the second rank, and Einstein, therefore, now chose the reduced Riemann–
Christoffel tensor

$$R_{\mu\nu} = R^{\varrho}{}_{\mu\varrho\nu}.\tag{30}$$

He split this tensor into two terms,

$$R_{\mu\nu} = R'_{\mu\nu} + R''_{\mu\nu}$$

where

$$R_{\mu\nu} = -\frac{\partial \Gamma^{\varrho}_{\mu\nu}}{\partial x^{\varrho}} + \Gamma^{\varrho}_{\sigma\nu}\Gamma^{\sigma}_{\mu\varrho} \tag{31}$$

and $R''_{\mu\nu}$ vanishes wherever $\sqrt{-g} = 1$.

In considering the conservation law for the energy, Einstein had earlier assumed the components of the gravitational field to be described by

$$g^{\tau\mu}\frac{\partial g_{\mu\nu}}{\partial x^{\sigma}}, \tag{32}$$

but now he found the Christoffel symbol $\Gamma^{\mu}_{\nu\varrho}$ to be a more adequate expression for this purpose.

The most important question concerning the generalization of the Poisson equation had now been answered by the *Ansatz*[76]

$$R_{\mu\nu} = -\gamma T_{\mu\nu}. \tag{33}$$

These equations can be written in the Hamiltonian form with

$$\mathscr{L} = g^{\tau\sigma}\Gamma^{\alpha}_{\sigma\beta}\Gamma^{\beta}_{\tau\alpha}, \tag{34}$$

and the conservation of energy is satisfied with

$$\frac{\partial}{\partial x^{\lambda}}(T^{\lambda}_{\sigma} + t^{\lambda}_{\sigma}) = 0 \tag{35}$$

where

$$t^{\lambda}_{\sigma} = \frac{1}{2\gamma}\left(\mathscr{L}\delta^{\lambda}_{\sigma} - g^{\mu\nu}_{\sigma}\frac{\partial \mathscr{L}}{\partial g^{\mu\nu}_{\lambda}}\right).$$

Here again, the 'energy tensor' of the gravitational field (which, however, is a tensor only with respect to linear transformations) occurs.[77] The reason for its existence is the curvature of space, which does not allow for a conservation law of T^{λ}_{σ} alone.

Now the *Ansatz* (33) brings a problem along with it, since it demands that

$$T^{\sigma}_{\sigma} = 0, \tag{36}$$

for $\sqrt{-g} = 1$. Hence there seem to exist conditions which restrict the *allowed* coordinate frames.[78] Already in a previous note, in fact, Einstein proposed to take $T^{\sigma}_{\sigma} = 0$, since the electromagnetic energy-momentum tensor shows this property.[79] He stated explicitly:

We must remember, however, that according to our knowledge, *matter* is not to be conceived as (primitively) given and physically simple [entity]. And it is not just a few who hope that [the conception of] matter can be reduced to purely electromagnetic phenomena which, however, would follow a more complete theory than Maxwell's electrodynamics. Let us suppose for a moment that in such a complete electrodynamics the scalar of the energy tensor would also disappear! Would then the result just mentioned above prove that matter cannot be constructed with the help of this theory? I expect to be able to contradict this. Because it is indeed quite possible that gravitational fields constitute an essential part of *matter* to which pertains the expression given above. Hence T^σ_σ can be *apparently* positive for the entire structure (matter + gravitational field), while in reality only ($T^\sigma_\sigma +$ $+ t^\sigma_\sigma$) is positive and T^μ_μ vanishes everywhere. *In the following, we shall assume that the condition $T^\sigma_\sigma = 0$ is actually fulfilled in general.*[80]

One can then fix the coordinates by assuming that $\sqrt{-g} = 1$. In his contribution of 25 November 1915,[81] his fourth on the new approach during that November, Einstein went another step further, realizing that the assumption of the zero energy scalar is not necessary, if the equations of gravitation would be changed to

$$R_{\mu v} = - \gamma (T_{\mu v} - \tfrac{1}{2} g_{\mu v} T), \tag{37}$$

where $T = T^\sigma_\sigma$.

The reason for this new *Ansatz* is that now the tensors of energy, matter and that of the gravitational field occur always in the same way in all equations. Einstein, therefore, concluded[81]:

With that, the general relativity theory is finally completed as a logical structure. In its most general formulation, which turns the space-time coordinates into parameters devoid of physical significance, the principle of relativity leads, with conclusive necessity, to quite a definite theory of gravitation which can explain the perihelion motion of Mercury.[82] On the other hand, the postulate of general relativity reveals to us nothing more about the character of the other natural phenomena than what the theory of special relativity had already done. The opinion which I recently expressed in this regard [in these Proceedings] was wrong.[82] Every physical theory conforming to special relativity can, with the help of the absolute differential calculus, be presented in the framework of the theory of general relativity, without the latter offering a criterion for the validity of that theory.[83]

Einstein was filled with the joy and excitement of achievement, when he wrote to Besso again on 10 December 1915[73]:

The boldest dreams have been realized. *General* covariance. Perihelion motion of Mercury, wonderfully exact. The latter is perfectly determined from the astronomical point of view, because the determination of the masses of the inner planets by Newcomb was based upon *periodic* perturbations (not the secular ones). This time [the most natural mathematical assumptions] at hand turned out to be correct. Grossmann and I had thought that the conservation laws would not be satisfied, and Newton's law would not emerge in the first approximation....

Einstein sent his new papers to his friend Michele Besso, and wanted him to read them.[84] The winter of 1915 and the spring of 1916 was perhaps the period of greatest scientific contentment and happiness in Einstein's life.[85]

Two questions arise immediately with respect to Einstein's derivation of the equations of gravitation. First, how cogent and necessary is the derivation of Equation (37)? Second, how independent of gravitation are the other laws of nature? Hilbert took a stand with respect to both of these questions. First, he derived the equations of gravitation, Equation (37), from a set of axioms. Second, he showed that the same set of axioms also leads to conditions for the laws of nature, in particular electromagnetism, which cause matter to exist.

3. PHYSICS AND GEOMETRY

3.1 THE ROLE OF AXIOMATIZATION IN HILBERT'S WORK

To David Hilbert, one of the greatest mathematicians of the twentieth century, axiomatization signified a guiding principle of the structure of mathematics and the
mathematical laws of physics. That Hilbert's approach towards the foundations of
mathematics was too optimistic, and essentially unfeasible, was established by Gödel's
theorem more than a decade before Hilbert died. In his own work, Hilbert systematized and unified numerous branches of mathematics: theory of invariants (1885–
1893), theory of algebraic number fields (1893–1899), theory of analytic functions and
variational problems (1899–1909), and the theory of linear integral equations (1902–
1912). Hilbert's work on linear integral equations extended the pioneering work of
Fredholm, and led to the introduction of the celebrated 'Hilbert space', which
would provide the fundamental basis of the mathematical formulation of quantum
mechanics.[86]

The prelude which determined the second half of Hilbert's mathematical life, cre-
tainly from 1912 onwards, was stated in his famous book *Die Grundlagen der Geometrie*, which first appeared in 1899.[87] Hilbert worked on the axiomatic foundations
of geometry almost until 1909. Much later, after 1922, he turned to a systematic
axiomatization of mathematics, including mathematical logic. In between Hilbert
sought to apply his axiomatization methods to selected fields at the frontiers of physics,
having declared that 'physics is much too difficult for the physicists.'[87a]

The study of minimum principles on which physics could be based had occupied
Hilbert even before he formally undertook the axiomatization of physics in 1912,
as his earlier papers indicate. Hilbert had lectured on physics in Göttingen since
1902. After the death of his friend Hermann Minkowski in 1909, he lectured more
regularly on physics from 1910 to 1918. Minkowski had introduced him to several fields in physics.[88] Minkowski's death in January 1909 left him very lonely and
sad – and the axiomatization of physics seemed to be an active and loving tribute
to his memory. Hilbert's efforts picked up slowly, and beginning in 1912 they really
took on.

Hilbert had always been interested in the fundamental problems of physics, but now
he was really inspired. In 1913 he organized the 'Göttinger Gastwoche' which brought
together Planck, Debye, Nernst, Smoluchowski, Sommerfeld, Lorentz, and several
others to lecture on and discuss the problems of the kinetic and quantum theories.[89]
Another indication of his strong and continuing interest in physics was the fact that
he had already established, several years before the First World War, a special position
for an assistant who would work with him in physics, a position which he normally
offered to Sommerfeld's students.[90] Starting in 1912, numerous students wrote dissertations on various problems of theoretical physics under Hilbert.[91] The subjects which
Hilbert chose for his work in physics were characteristic of his mathematical temper-
ment. In his early lectures he dealt with the *foundations of mechanics*, using Gauss'
minimal principle as the fundamental axiom.[92] Other than mechanics, Hilbert found

it quite desirable to pursue the axiomatization of thermodynamics and kinetic theory, radiation theory, electricity and gravitation (including their unification), and later on quantum theory. Under Hilbert's inspiration at Göttingen, Carathéodory sought to axiomatize thermodynamics [93], and Paul Hertz discussed its 'mechanical' foundations. [94]

One of the important problems in the programme of axiomatization of physics was the examination of the consistency of the axioms and their independence, and Hilbert's fields of research were determined by it. Hilbert first turned to the mathematical foundations of kinetic gas theory. [95] He then dealt with the general formulation of Kirchhoff's laws of radiation without referring either to the classical wave theory or the quantum theory. [96] Inspired by the success of the special theory of relativity, and the progress of Einstein's work on general relativity and the theory of gravitation, Hilbert also entered upon research in this frontier of physics, making original and important contributions. [97]

It is well known that Hilbert vigorously stimulated research on the mathematical foundations of quantum mechanics, in particular through his collaborator Von Neumann. The new atomic mechanics had its first breakthrough (matrix mechanics) in Göttingen, and Heisenberg lectured on it in Hilbert's seminar already in the fall of 1925. [98] Hilbert not only guided Jordan in his work on the transformation theory [99], but entered the field himself, first lecturing on the foundations of quantum mechanics at Göttingen, and then publishing a paper jointly with Von Neumann and Nordheim. [100] This paper also initiated Von Neumann's work on the foundations of quantum mechanics, leading to his famous book on this subject. [101]

For the purpose of this study, we shall deal only with Hilbert's search for and foundation of a unified theory of natural phenomena, based upon a unification of electricity and gravitation in a single mathematical scheme.

Three principal ingredients determined Hilbert's research on this subject: (i) the mathematical inheritance of the nineteenth century, in particular the theories of space and the concept of invariance in geometry and dynamics; (ii) the ideas of Gustav Mie concerning the theory of matter; and finally, (iii) Einstein's formulation of the theory of gravitation. A review of the status of these 'ingredients' is of interest before discussing Hilbert's contributions to the 'foundations of physics'. [97] Einstein was strongly critical of Hilbert's work but others, such as Weyl, tried to follow it and improve upon it by bringing in other concepts of their own. Ironically, it was Einstein, who had pooh-poohed Hilbert's programme of a unified theory of matter, who would himself spend the last three decades of his life on this search along lines similar to those of Hilbert's – and that too during a period when discoveries of new particles and forces had rendered the entire approach doubtful.

3.2 CONCEPTS OF SPACE AND GEOMETRY

The concept of space has played an important role since the earliest times. The Greek philosophers and mathematicians, the medieval Christian thinkers and the men of science of the Renaissance had their notions of space. Newton finally decided upon 'absolute space' as the theatre of natural phenomena, and Kant made it into an *a priori* of experience.[102] The decisive progress in the ideas of space and geometry was, however, initiated by the great mathematicians of the early 19th century, especially Gauss, Lobatchevski and Bolyai, who first discovered the existence of non-Euclidean geometries.[103] In his survey, known as the 'Hannoversche Gradmessung' [the Hanover Survey] Gauss even attempted to determine as to which geometry of space would be valid in nature. Although his attempt was not successful, in that it did not prove (nor did it disprove) that the geometry of space was compatible with Euclidean geometry, he wrote to Olbers in 1817:

I became more and more convinced that the necessity of our geometry [i.e., Euclidean] cannot be demonstrated, at least neither by, nor for, the human intellect. In some future life, perhaps, we may have other ideas about the nature of space which, at present, are inaccessible to us. Geometry, therefore, has to be ranked until such time not with arithmetic, which is of purely aprioristic nature, but with mechanics.[104]

It was Riemann who developed Gauss' ideas. He introduced his conception in his famous 'trial' lecture [Probevorlesung] '*On the hypotheses which lie at the basis of geometry*'[105], which he delivered at Göttingen on 10 June 1854. He took note of the fact that he could not base the concepts of a new geometry on any available wealth of the existing mathematical literature, but could only refer to Gauss' 'second memoir on *Biquadratic Residues* in the *Göttingen Gelehrte Anzeiger*, and in his Jubilee-book, and some philosophical researches of Herbart'.[105a] Riemann introduced the general *line element* d*s*,

$$ds^2 = \sum_{\mu, \nu = 1}^{n} g_{\mu\nu} \, dx^\mu \, dx^\nu \tag{38}*$$

in the *n*-dimensional space, and the concept of the *geodesic* in curved space. His investigations led to the so-called Riemannian geometry which can be characterized by two axioms:[106]

Axiom I. The differential coefficient d*s*/d*t* at a given point on the curve depends only on the derivatives d*x*$^\mu$/d*t* at this point and not on higher derivatives or on the behaviour of the curve elsewhere.

Axiom II. d*s*/d*t* is the square root of a quadratic form of the d*x*$^\mu$/d*t*:

$$\frac{ds}{dt} = \sqrt{g_{\mu\nu} \frac{dx^\mu}{dt} \frac{dx^\nu}{dt}}. \tag{39}$$

Axiom II, in fact, implies Axiom I. The important parameters $g_{\mu\nu}$ are called the metric of the space, i.e. they determine whether it is Euclidean or not. Riemann concluded his lecture with the remark: 'The question of the validity of the hypotheses of geometry in the infinitely small is bound up with the question of the ground of the metric relations of space... This leads us into the domain of another science, of physics, into which the object of this work does not allow us to go today.'[107]

---

* In modern notation, Equation (38) is written as Equation (18).

Clifford not only translated Riemann's paper into English, but also tried to extend his ideas to a 'Space Theory of Matter'.[108] He proposed the thesis that the 'variation of curvature of space is what really happens in that phenomenon which we call the *motion of matter*, whether ponderable or ethereal'. Other mathematicians followed much less the physical speculations of Riemann, than seeking to work out the scheme of his geometry.

Another line of mathematical development by men like Beltrami, Christoffel, and Lipschitz, placed emphasis on certain aspects of differential geometry, making connection with invariance problems. Christoffel influenced Bianchi and Ricci in their development of the *absolute differential calculus* for the theory of invariants of quadratic differential forms. Levi-Civita, a disciple of Ricci's, carried on the programme further, and in 1917 introduced the remarkably useful concept of parallel displacement. The French school later on contributed to differential geometry through Élie Cartan. Rather than going into the details of these mathematical developments, let us mention an important extension of the geometry of space, namely, the introduction of the time as another coordinate of physical space by Hermann Minkowski.

On 21 December 1907 Minkowski submitted a lengthy memoir entitled 'Die Grundgleichungen für die elektromagnetischen Vorgänge in bewegten Körpern' to the Göttingen Academy.[110] He discussed a new formulation of the laws of physics, especially mechanics and electrodynamics, on the basis of the work of Lorentz, Poincaré and, more particularly, his former Zurich student Einstein who had 'most acutely' expressed the 'new concept of time.'[111] The important point of Minkowski's considerations was that he treated the time on an equal footing with the space coordinates and invented the notion of the *four-dimensional space-time continuum*.[112] Minkowski's new formal and intuitive space-time geometry opened the path for the later researches of Einstein, leading to the creation of general relativity on the basis of Riemannian geometry.

It is remarkable that at about the same time as Minkowski, Hargreaves quite independently suggested the use of space-time vectors.[113] His concept of using space-time as a unified geometrical concept was, however, only indirect in that by organizing space and time derivatives together the corresponding equations gained a much simpler form. In fact, both Hargreaves and Poincaré had used the equivalence of space and time coordinates in the mathematical forms which they had employed[113], without giving it the full physical and intuitive emphasis that characterized the approach of Minkowski. It may be of interest to note, however, that the first scientist, who suggested a connection between the space and time dimensions in modern science, seems to have been the physiologist Fechner in the 1840's.[114]

3.3 SPACE-TIME AND INVARIANTS

On 3 June 1868, Hermann von Helmholtz submitted to the Göttingen Academy of Sciences an extended version of his lecture 'On the facts which lie at the basis of geometry'.[115] He introduced his memoir with the remark:

My researches on spatial perception in the field of view have motivated me to undertake the investigation of the problem of the origin and nature of our perceptions of space in general. In this regard, the question which impressed upon me, and which obviously belongs in the domain of the exact sciences, was chiefly this: How much objectively valid meaning do the axioms of geometry have? How much, on the other hand, depends only on definitions, or deductions from definitions, or on the form of the representation? [116]

Helmholtz, who had worked, independently of Riemann on similar ideas, now extended the latter's conception.[117] He discussed in particular that hypothesis of Riemann, according to which the line element is equal to the square root of a homogeneous function of the second degree, as in Equation (39). This hypothesis, Helmholtz claimed, could be derived on the basis of more general assumptions. He started by considering the motions of a solid body in space. The 'translation invariance' yields $\frac{1}{2}n(n-1)$ equations in an n-dimensional space. In addition there might be the possibility of *congruence* under rotations, leading to further equations. All of these equations can be formulated as differential expressions if applied to infinitesimally small coordinate differences. The important point, which he could demonstrate, was that there exists a quantity of second degree in the differentials which is invariant under all the symmetry transformations. Together with the two other postulates, that space should have three dimensions and that it should extend to infinity, Helmholtz regarded the foundations of geometry as being fully laid.

The ideas of Helmholtz were both conservative and progressive at the same time. On the one hand, he limited himself to infinite spaces of three dimensions; on the other, he stressed the use of symmetry transformations and invariants of a geometry which became a turning point for the later development.

The concept of an *invariant* had been introduced by Sylvester in 1851.[118] The theory of invariants was systematically developed on the basis of Grassman's theory of 'Stufen', from which the tensor calculus also arose.[119] The next step was to determine the invariants of quadratic forms and their classifications, to which Jacobi, in particular, contributed. Felix Klein created a new approach to geometry when he announced his 'Erlanger Programm' in 1872.[120] In this programme he considered a geometry as being given by a 'manifold on which a transformation group is defined,' and he looked for the invariant theory with respect to the group'.[121] Beginning in 1872 Sophus Lie, who had influenced Klein, developed the theory of continuous groups. In his lectures in Göttingen on the development of mathematics in the nineteenth century, Klein declared that it was only a small step from his Erlangen Programme to the theory of general relativity.[122]

The theory of invariants of a symmetry group, which was so very much present already in Helmholtz' lecture, became familiar to the physicists and mathematicians through the work of Klein and Lie. Jacobi had in fact already found a connection between the Euclidean invariance of the mechanical Lagrangian and the conservation laws for linear and angular momenta.[123]

Finally Hamel, unaware of the work of his predecessors, proceeded systematically to exploit the connection between symmetry transformations and conserved quantities in mechanics.[124] Hamel extensively used Lie's work on continuous groups,[125] deriving the result that the coefficients in the Lagrange equations are constants if the related infinitesimal transformations generate a symmetry group, and if the coefficients are constants then the conservation laws are satisfied.[126]

A generalization of Hamel's work, without making any reference to it, however, was made by Herglotz in 1911.[127] He discussed the ten-parameter Poincaré group[128] of 'motions' in the four-dimensional space, and related its symmetries to ten general integrals, among which the most familiar are the energy, momentum and angular momentum. The other three conserved quantities correspond to a generalization of the center-of-mass integrals in classical mechanics.

As we have already mentioned, Hilbert himself knew much about invariants. Indeed, by solving the fundamental problems of the theory of invariants Hilbert had killed the subject.[129] In addition, he had with him a young expert from the mathematical centre of the theory of invariants by the name of Emmy Noether. Emmy, daughter of Max Noether, came from Erlangen where Klein had developed his famous 'programme', where Clebsch had worked and Gordan had been the 'king of invariants.' In Göttingen around 1915, where Hilbert, at this stage in his distinguished career, was buried in his studies on the foundations of physics and general relativity, where Klein whose great mathematical programme was being carried forward by the 'relativists' (and who, in his latter years, had again become interested in current research), Emmy Noether's work became very important.[129a]

Even before Klein knew about Noether's work, he had requested Engel, the old collaborator of Sophus Lie, to derive all the ten known integrals of the non-relativistic mechanical n-body system 'in the spirit of Lie'.[130] But the most general and systematic treatment was already available in Emmy Noether's work, which Klein communicated to the Göttingen Academy only in 1918, and which became celebrated later on.[131] The important generalization in her work lay in the fact that she studied both finite continuous groups, i.e. those that depend continuously on ϱ parameters ε, as well as infinite continuous groups depending on ϱ continuous functions. Noether also dealt with mixed groups, in the sense defined, and established two theorems in her fundamental paper:[132]

(1) If the variational integral is invariant under a finite continuous group G_ϱ, then ϱ linearly independent combinations of the Lagrangian expressions become divergences, and vice versa. This theorem is also valid in the limit of infinitely many parameters.

(2) If the variational integral is invariant under an infinite continuous group $G_{\infty\varrho}$, in which arbitrary functions and their derivatives up to the σ-th order appear, then ϱ identical relations are satisfied between the Lagrangian expressions and their derivatives up to the order σ. The inverse of this theorem is also true.

Emmy Noether established the most general equivalence between the invariance under symmetry transformations of the classical equations of motion of the action integral, from which they can be derived, and the conservation of certain quantities which generate the group in the sense of Lie. Her work was carried on, or explained better, in terms of physics by Bessel-Hagen.[133] Noether's theorem obtains a much simpler form in quantum theory due to the fact that the 'complicated' Poisson bracket of classical theory is replaced by the 'simple' commutator.[134]

3.4 MIE'S FIELD THEORY OF MATTER

Gustav Mie made the first attempt to construct a complete theory of matter in the twentieth century.[135] In particular, he wanted 'to explain the existence of an indivisible electron and to relate the phenomenon of gravitation to the existence of matter'.[136] Three assumptions formed the basis of his theory:

(i) Electric and magnetic fields exist *inside* the electrons as well.[137]

(ii) The principle of (special) relativity is valid throughout.

(iii) 'The hitherto known states of the ether, namely the electric field, the magnetic field, the electric charge, and the charge current are entirely sufficient to describe all phenomena in the material world.'[138]

Concerning these fundamental assumptions, Mie remarked that the second one is valid *beyond doubt* (!), and the third can only be justified by its success. As regards his concept of matter, Mie thought that the atoms consist of a number of electrons bound together by a dilute positive charge as in the Thomson model. Moreover, the atoms are surrounded by *atmospheres* which, since the whole atom is neutral, do not cause electric but gravitational fields. Mie thought that the enormous intensity of the [electromagnetic] field creates spatial concentrations of charge which we signify as electrons, to which the ordinary Maxwell equations do not apply. He further remarked:

The electron in my theory does not occupy a well-defined portion of space in ether. Rather, it consists of a 'nucleus' which goes over continuously into an 'atmosphere' of electric charge. The 'atmosphere' extends to infinity, becoming extremely thin in the neighbourhood of the 'nucleus', so that one cannot detect it experimentally in any way.[139]

From his three assumptions, Mie was led to a generalization of the equation for the ether. This extension of the Maxwell–Lorentz theory was determined on the basis of the validity of the principle of energy conservation and the existence of a *localizable energy*. Mie did not quite realize how much freedom was left to him, and how many more assumptions were built into his derivation than was necessary.

The Hamiltonian H played an essential role in Mie's theory, where H depends on what he called 'quantities of magnitude', $\mathscr{D}$ (electric displacement), $\mathscr{H}$ (magnetic field), ϱ (charge density) and j (electric current). From this function H, one obtains the 'quantities of intensity', E (the electric field), $\mathscr{B}$ (the magnetic induction), and the corresponding 'charge' e and 'current' i, by taking the derivative with respect to the correlated 'quantities of magnitude'. Or, inversely, one may start with a Hamiltonian Φ of the 'quantities of intensity' and derive the 'quantities of magnitude' by proper differentiation. Apart from this general structure, the Hamiltonian H should be invariant under Lorentz transformations, consistent with the assumption (ii). In addition, Mie simplified the form of the Hamiltonian by using the hypothesis:

$$H = H(\sigma, p),$$

with

$$\sigma = \sqrt{\varrho^2 - j^2}, \quad \text{and} \quad p = \mathscr{D}^2 - \mathscr{H}^2. \tag{40}$$

The detailed results which Mie obtained on the basis of his assumptions are both complicated and incomplete, and we shall only indicate certain features of his conclusions. He claimed, in particular, to have shown that it was possible to have stable electrons in the sense in which he had conceived them, with the electromagnetic and gravitational fields having the following properties:

(i) In the interior of an electron $\mathscr{E}$ and $\mathscr{D}$, as well as ϕ and ϱ, must have opposite signs.

(ii) The gravitational field must be represented by a four-vector.

(iii) The inertial and gravitational masses of a particle are identical only if there are no motions in the interior of the particle. The mass ratio and the gravitational constant thus turn out to be functions of the temperature, decreasing with increasing values of the latter.

Mie's theory obviously ran into serious difficulties. Some of these are connected with the fact that no one has succeeded in deriving solutions for static electrons in which the charge is 'quantized'. In fact, all the trial solutions are consistent with an arbitrary choice of the total charge. Even more serious is the fact that Mie's theory depends on an absolute potential. The equations of motion do *not* remain unchanged if one replaces the potential ϕ by $\phi + \text{const.}$, due to the fact that the skew symmetric tensor,

$$H^{ik} = \frac{\partial A^k}{\partial x_i} - \frac{\partial A^i}{\partial x_k}, \tag{41}$$

in the interior of the electron depends on the potential A^k explicitly. A material particle will therefore not be able to exist in a constant external potential field.[140] Mie's theory suffers from having been made much too early. Thus it could not properly recognize the existence of the energy quantum or forces other than the electromagnetic and gravitational ones.[141] Moreover, his approach to gravitation was very 'conventional' and would soon be overthrown by the new revolutionary ideas of Einstein and Hilbert.

3.5 Hilbert's Communication to the Göttingen Academy

What we have discussed in the previous sections represents, briefly, the state of affairs when Hilbert, in 1915, wrote his first note on the foundations of physics.

During the first World War Hilbert remained fully absorbed in physics. In the summer of 1914 Debye had joined the faculty of the University of Göttingen and organized, at Hilbert's request, a seminar on the structure of matter. Landé was Hilbert's assistant for physics at that time, and he kept Hilbert informed about the developments in quantum theory, but he left in December 1914 to join the Red Cross.[142] At this time, when only a few students attended the Hilbert–Debye seminars, Hilbert became fascinated with the new ideas of Gustav Mie and Einstein, and he sought to find a connection between them. Einstein visited Göttingen at Hilbert's invitation and presented his ideas there. In a letter to Arnold Sommerfeld, dated 15 July 1915, Einstein wrote: 'I had the great joy of seeing in Göttingen that everything [about the theory of relativity] is understood to the last detail. With Hilbert I am just enraptured. An important man!'[143] Hilbert was no less fascinated by his encounter with Einstein, and he devoted the following several months to the study of gravitation and electromagnetism in the search for a unified field theory of matter.

At about the same time when Einstein presented [on 4, 11, 18 and 25 November 1915] a series of communications on the general relativity theory of gravitation to the Prussian Academy, Hilbert presented [on 20 November 1915] a derivation of Einstein's equations of gravitation to the Royal Academy of Sciences in Göttingen. Hilbert's communication appeared in the third issue of the Proceedings of the Göttingen Academy for 1915, and in his published version he referred to all of Einstein's communications of November 1915 to the Prussian Academy. There can be no doubt that Hilbert's derivation was entirely independent. The dates of the papers involved, and remarks made in these papers by both Einstein and Hilbert, are quite clear. In the first three contributions, up to November 18, Einstein still used the gravitational equation (33) without the term involving the *trace*. Only in the last paper, read on November 25, Einstein gave the form of Equation (37) to the equations, saying that 'the field equations for vacuum, on which I have based the explanation of the perihelion motion of Mercury, remain unaffected by this modification.'[144] This would indicate that he had found the new improved equations between 18 November, when he gave the 'explanation' of the perihelion motion of Mercury, and 25 November, when he presented his new note on 'The Field Equations of Gravitation'. Hilbert reported his own derivation of the equations of gravitation on 20 November, a few days ahead of Einstein, certainly in the public domain. In his published paper Hilbert, realizing this situation, pointed out: 'It seems to me that the differential equations of gravitation so realized [by me] are in agreement with the beautiful theory of general relativity proposed by Einstein in his later [25 November 1915] memoir.'[145]

Hilbert had indicated quite correctly that he shared, at the very least, with Einstein, the derivation of the field equations of gravitation. Indeed, Hilbert's was more like a derivation than Einstein's, which was based more on the simplicity of the *Ansatz* than on a full grasp of the implications of the structure of the equations. Hilbert's achievement in deriving the field equations of gravitation is not generally known, even among 'general relativists', and seldom cited in the literature.[146] Of course, there never was any quarrel over priority between Hilbert and Einstein, who admired one another deeply.[147] Hilbert once remarked: 'Every boy in the streets of Göttingen understands more about four-dimensional geometry than Einstein. Yet, in spite of that, Einstein did the work and not the mathematicians.'[148] Or in a public lecture: 'Do you know why Einstein said the most original and profound things about space and time in our generation? Because he learned nothing at all about the philosophy and mathematics of time and space.'[149] It was also Hilbert who recommended the award of the third Bolyai Prize in 1915 to Einstein 'for the high mathematical spirit behind all his achievements'.[150]

3.6 AXIOMS OF HILBERT'S THEORY OF MATTER

Hilbert had studied the work of Abraham, Nordström, Mie and Einstein very thoroughly. He tried now, using his powerful mathematical tools including the variational calculus and the theory of invariants, to construct a consistent theory of matter on axiomatic grounds. In particular he was attracted by the 'giant propositions' of Einstein and the 'deep ideas and original concepts 'on which Mie had built his electrodynamics.[151] He thus proposed: 'Following the axiomatic method, in fact from two simple axioms, I would like to propose a new system of the basic equations of physics. They are of ideal beauty and, I believe, they contain the solution of the problems of Einstein and Mie at the same time.'[152] Starting from four 'world parameters' $x^\mu (\mu = 1, ..., 4)$, characterizing the world points in the space-time continuum, Hilbert assumed that the following quantities determine the processes at x^μ (*zeroth axiom*): (i) the ten gravitational potentials $g_{\mu\nu}$ of Einstein, exhibiting the form of a symmetric tensor under general transformations of x^μ; (ii) the four electromagnetic potentials A_μ, forming a vector with respect to general transformations.

Now the physical processes are determined by two axioms[153]:

Axiom I (Mie's axiom of the world-function): The laws of physics are determined by the world-function H, containing the arguments

$$g_{\mu\nu}, \; g_{\mu\nu,\varrho} = \frac{\partial g_{\mu\nu}}{\partial x^\varrho}, \qquad g_{\mu\nu,\varrho\sigma} = \frac{\partial^2 g_{\mu\nu}}{\partial x^\varrho \partial x^\sigma}$$

and

$$A_\mu, \; A_{\mu,\varrho} = \frac{\partial A_\mu}{\partial x^\varrho}, \quad (\varrho, \sigma = 1, ..., 4), \tag{42}$$

such that the variation of the integral

$$\int H \sqrt{g} \; \mathrm{d}^4 x, \quad (g = \det |g_{\mu\nu}|, \quad \mathrm{d}^4 x = \mathrm{d}x^0 \, \mathrm{d}x^1 \, \mathrm{d}x^2 \, \mathrm{d}x^3) \tag{43}$$

vanishes for any of the 14 potentials $g_{\mu\nu}$ and A_μ.

Axiom II (axiom of general invariance): The world-function H is an invariant with respect to an arbitrary transformation of the coordinates x^μ.

Concerning this axiom Hilbert remarked:

Mie had already stated the requirement of orthogonal invariance. In the above-mentioned Axiom II, Einstein's fundamental and basic idea of general invariance finds its simplest expression, although with Einstein the Hamilton principle plays only an auxiliary role. His H functions are in no sense general invariants, nor do they contain the electric potentials.[154]

Hilbert correctly noted the fact that Einstein's first publication of November 1915 still contained restrictions upon the allowed transformations. It seems that the necessity of the general invariance of *all* physical laws first became clearer to Hilbert than to Einstein. Of course, Hilbert had good reasons for that. Besides being a mathematician who was always interested in utmost generality and greatest possible elegance, he had an important argument for such an assumption; he had discovered a mathematical theorem which made the connection between the laws of gravitation and of electrodynamics possible.[155]

3.7 Hilbert's Fundamental Theorem, Energy Conservation, and the World Function

(i) Hilbert's Theorem

Hilbert formulated his fundamental theorem, on which he based his other derivations, in the following form:

Let J be an invariant under arbitrary transformations of the four world-parameter, comprising n quantities and their derivatives. From the variation

$$\delta \int J \sqrt{g} \, d^4x = 0, \tag{44}$$

with respect to the n quantities, n Lagrange (variational) equations can be formed, such that in this invariant system of n differential equations for the n quantities four of them are always a consequence of the remaining $n-4$, in the sense that between the n differential equations and their total derivatives four combinations, independent from each other, are always identically satisfied.[156]

This theorem of Hilbert's is a consequence of the second part of Noether's theorem. In fact, Emmy Noether had been in close contact with Hilbert in those days.[157] The consequences of the application of this theorem, together with the axioms I and II, are of great interest. From Hamilton's principle, and the form of H in accordance with axiom I, Hilbert derived 14 Lagrangian equations for the 14 potentials. Ten of these equations contain the variations with respect to the gravitational potentials $g_{\mu\nu}$, and are called the equations of gravitation; while four stem from the variation with respect to the electrodynamic potential A_μ, and give rise to the generalized Maxwell equations. However, Hilbert said:

Because of the theorem stated, the four equations [which describe the generalized Maxwell equations] can be regarded as a consequence of the equations [of gravitation]. On account of this mathematical theorem, we can therefore immediately make the assertion that, in the sense indicated, *the electrodynamic phenomena are the consequences of gravitation*. In the recognition of this fact, I discern the simple and very surprising solution of the problem of Riemann, who was the first to investigate theoretically the connection between gravitation and light.[158]

Hilbert's theorem thus led him to prove the connection between the laws of gravitation and the laws of electrodynamics. This result was not quite in agreement with the spirit of Einstein, who had explicitly stated that general relativity does not impose any restriction on the laws of physics, gravitation excluded. But even Einstein agreed in his paper of 1916 in the *Annalen der Physik* that the field equations of gravitation 'completely provide the equations of material processes if the latter can be characterized by four independent differential equations.'[159] More light was thrown on this problem by Hermann Weyl who stressed the concept of gauge groups, of which the general transformations were an example.

(ii) Energy Conservation

From axioms I and II Hilbert derived the energy-momentum vector in the form

$$e^\lambda = H p^\lambda - a^\lambda - b^\lambda - c^\lambda - d^\lambda, \tag{45}$$

with

$$a^\lambda = \frac{\partial H}{\partial g^{\mu\nu}_{\kappa\lambda}} A^{\mu\nu}_\kappa$$

$$A^{\mu\nu}_\kappa = p^{\mu\nu}_\kappa + (\Gamma^\mu_{\kappa\varrho} p^{\varrho\nu} + \Gamma^\nu_{\kappa\varrho} p^{\varrho\mu})$$

and

$$p^{\mu\nu}_\kappa = \frac{\partial p^{\mu\nu}}{\partial x^\kappa}, \qquad p^{\mu\nu} = (g^{\mu\nu}_\sigma p^\sigma - g^{\mu\sigma} p^\nu_\sigma - g^{\nu\sigma} p^\mu_\sigma),$$

and

$$b^\lambda = B^\lambda_{\mu\nu} p^{\mu\nu},$$

$$B^\lambda_{\mu\nu} = \frac{\partial H}{\partial g^{\mu\nu}_{,\lambda}} - \frac{\partial}{\partial x^\lambda} \frac{\partial H}{\partial g^{\mu\nu}_{,\lambda\kappa}} - \frac{\partial H}{\partial g^{\varrho\nu}_{,\lambda\kappa}} \Gamma^\varrho_{\kappa\mu} - \frac{\partial H}{\partial g^{\mu\varrho}_{,\lambda\kappa}} \Gamma^\varrho_{\kappa\nu},$$

and

$$c^\lambda = \frac{\partial H}{\partial A_{\kappa,\lambda}} p_\kappa,$$

and

$$d^\lambda = \frac{1}{2\sqrt{g}} \left(\frac{\partial}{\partial x^\mu} \frac{\partial \sqrt{g}\, H}{\partial A_{\mu,\nu}} - \frac{\partial \sqrt{g}\, H}{\partial A_{\mu,\nu}} p^\sigma A_\sigma \right).$$

This energy vector, although it still depends on an arbitrary vector p^σ, satisfies identically the *invariant energy conservation equation*

$$\frac{\partial \sqrt{g}\, e^\lambda}{\partial x^\lambda} = 0. \tag{46}$$

Hilbert's derivation clearly used the fact that his world function H was a general invariant. The question now was, how the energy theorem, Equation (46), had anything to do with the conservation of energy in Einstein's theory? Hilbert's approach was equivalent to Einstein's theory [as presented on 25 November 1915 (Ref. 83)], and Felix Klein would later on take up the crucial question of energy conservation in Einstein's theory.

(iii) *The Form of the World Function*

Hilbert took over Einstein's postulate that the equations of gravitation should depend only on the second derivatives of the potentials (in axiom I). He then concluded (*Axiom* III) that H must be of the form

$$H = R + L, \tag{47}$$

where

$$R = g^{\mu\nu} R_{\mu\nu}$$

is the invariant deriving from the Riemann tensor, the curvature, and where

$$L \equiv L \left(g^{\mu\nu}, \frac{\partial g^{\mu\nu}}{\partial x^\lambda}, A_\mu, \frac{\partial A_\mu}{\partial x^\kappa} \right). \tag{48}$$

$R_{\mu\nu}$ is the contracted Riemann–Christoffel tensor of Equation (21). For the invariant L Hilbert proved the fact, that

$$\frac{\partial L}{\partial A_{\mu,\,\nu}} + \frac{\partial L}{\partial A_{\nu,\,\mu}} = 0, \tag{49}$$

leading to the result that the derivatives of the electrodynamic vector A_μ occur only in the form

$$F_{\mu\nu} = A_{\mu,\,\nu} - A_{\nu,\,\mu} \tag{50}$$

which means that L depends on $g_{\mu\nu}$, A_μ and curl A_μ only.

Hilbert concluded:

This result, on which the character of Maxwell's equations depends, follows here essentially as a consequence of general invariance, i.e. from axiom II.[160]

He found that in the special case of a Minkowski metric,

the electromagnetic energy tensor of Mie is nothing but the generally invariant tensor obtained by differentiating the invariant L with respect to the gravitational potentials $g^{\mu\nu}$ and taking the limits $[g_{\mu\nu}=0$ for $\mu \neq \nu,\ g_{\mu\mu}=1] - a$ fact which first led me to the logically close connection between Einstein's theory of general relativity and Mie's electrodynamics, and gave me the conviction of the correctness of the theory developed here.[161]

By taking the variation of H and using the Ansatz of Equation (47), Hilbert derived two sets of equations. The first one read

$$\sqrt{g}\,(R_{\mu\nu} - \tfrac{1}{2}Rg_{\mu\nu}) = - \frac{\partial\sqrt{g}\,L}{\partial g^{\mu\nu}}. \tag{51}*$$

Under the condition that the energy momentum tensor of matter is completely given by the electrodynamic function L, that is

$$T_{\mu\nu} = \frac{\partial L}{\partial g^{\mu\nu}}, \tag{52}$$

Equations (51) turn out to be identical with Einstein's equations (37). In his published

---

* Hilbert wrote his equations (51) in the equivalent form

$$R_{\mu\nu} - \tfrac{1}{2}Rg_{\mu\nu} = - T_{\mu\nu}, \tag{1}$$

thus defining $T_{\mu\nu}$ by the equation

$$\frac{\partial\sqrt{g}L}{\partial g^{\mu\nu}} = \sqrt{g}\,T_{\mu\nu}.$$

Now multiplying (1) by $g^{\mu\nu}$ and summing over μ and ν, one obtains

$$R - 2R = -T \quad \text{or} \quad R = T. \tag{2}$$

Substituting this into (1), one obtains

$$R_{\mu\nu} = -T_{\mu\nu} + \tfrac{1}{2}Tg_{\mu\nu} \tag{3}$$

which is Einstein's equation (37) for $\gamma = 1$.

paper on 'The Foundations of Physics', Hilbert noted the agreement between his equations and those formulated by Einstein in his theory.[145]

The importance of Hilbert's result lies in the fact that the equations of electrodynamics immediately follow from the variational principle, by taking the variation with respect to the electromagnetic potentials, or symbolically,

$$[\sqrt{g}\, L]_h = 0,   \tag{53}$$

once the equations of gravitation (51) are given. This is the consequence of the Hilbert–Noether theorem. As Hilbert noted, '*This is the precise mathematical expression of the general assertion made above about the character [-istic feature] of electrodynamics as a consequence of gravitation.*'[162]

In fact, Hilbert regarded his derivation of Einstein's equations of gravitation, and generalization of the equations of electrodynamics of Mie, as a triumph of the axiomatic method. He remarked:

As one can see, the few simple assumptions stated in the axioms I and II are sufficient, by analogous interpretation, for the construction of the theory. With [the help of] these [axioms], not only are our conceptions of space, time and motion reconstructed in a fundamental way, in the sense put forward by Einstein, but I am also convinced that through the basic equations set forth here the hitherto concealed phenomena inside the atom would find the closest explanation. In particular the reduction, in general, of all physical constants to mathematical constants must be possible, because it is only then that the possibility would come closer that, in principle, from physics would arise a science of the type of geometry – surely [it would be] the greatest glory of the axiomatic method which, as we have seen here, is the most powerful tool that analysis, namely variational calculus and invariant theory, can employ.[163]

Hilbert was obviously fascinated by the possibility that a unified theory of gravitation and electromagnetism would explain the existence of elementary quanta [of matter and radiation], a possibility which Einstein would pursue the rest of his life. But in 1916, and during the following several years, Einstein did not think much of such a possibility. To Rudolf Humm, a student from Hilbert's circle in Göttingen, had gone to Berlin in the spring of 1917 to hear Einstein's lectures and to talk to him, Einstein confided some of his thoughts at that time. On a visit which he paid to Einstein in May 1917, Humm noted:

He [Einstein] is a bad calculator, he said; he rather works conceptually. He does not seem to believe to be right what we are doing in Göttingen. He himself has never thought so formalistically. His imagination is firmly tied to reality. He is very careful, and entirely a physicist. He does not rush immediately to generalize as we do in Göttingen. He explains this [attitude] by saying that he had to rid himself of his prejudices very deliberately. That's why he did not grasp straight away how a general invariance could exist. Rather, he had to come to this view step by step, which subsequently seemed to be very plausible indeed. But before that, he had real aversion to it because the quantities employed there – the curvature tensors – had seemed to him very unclear.

I [Humm] suggested to him that he had taken recourse to the quantum theory in order to modify the theory of gravitation, while Hilbert on the other hand wanted to derive the quantum theory from the theory of gravitation. At this he made a roguish grin. This would not do, he said, even though the theory of gravitation was more general. But [in his view] the ideas of relativity could not yield anything more than gravitation.[164]

Two questions are now important. First, the response of Einstein to Hilbert's first memoir on the foundations of physics. Einstein made only a few direct or indirect comments on this, referring to it in particular in his celebrated paper on the theory of general relativity in the *Annalen der Physik* in the early spring of 1916, which we shall discuss next. We shall then turn to Hilbert's second memoir on the foundations of physics, in which he went beyond Einstein in the physical interpretation of the new theory, before we deal with the second vital question: What had Hilbert's theory to do with Einstein's? Were they in any sense equivalent?

4. COMPLETION OF THE THEORY OF GRAVITATION

On 20 March 1916, Albert Einstein's paper on 'The foundations of the theory of general relativity' arrived at the *Annalen der Physik*.[165] A long paper of more than fifty printed pages, it established the theory of gravitation within the framework of *general relativity*, and gave a complete account of the fundamental principles and the most important consequences.

4.1 GENERAL COVARIANCE AND THE EQUIVALENCE PRINCIPLE

General relativity signified a generalization of Einstein's special theory of 1905, because it now postulated that 'the laws of physics must be so formulated that they remain valid with respect to frames of reference in arbitrary motion.'[166] [Einstein thought that this generalization would satisfy Mach's critique of mechanics.] An important consequence of this postulate, Einstein noted, would be that 'In the theory of ralativity, the magnitudes of space and time cannot be defined such that the differences of the space coordinates can be measured directly with a measuring rod, and that of time with a normal clock.'[167]

The introduction of a frame of reference only simplifies the description of coincidences, on which the laws of physics can be built. Of course, one has to say how the experiments which deal with events in space and time have to be described. This connection, in Einstein's theory, was made through the line element in which the metric $g_{\mu\nu}$ entered, and the important generalization became:

If the new space-time coordinates $x^1 \ldots x^4$ are introduced by means of an arbitrary substitution, then in the new system the $g_{\mu\nu}$ are no longer constants but functions of space-time. At the same time, the motion of a free mass point in the new coordinates becomes curvilinear, rather than uniform, whereby the law of motion becomes independent of the nature of the moving mass point. We shall interpret such motion as taking place under the influence of a gravitational field. We note that the appearance of a gravitational field is tied together with the variation of $g_{\mu\nu}$ with respect to space-time. Also in the general case, where we cannot ensure the validity of special relativity in a finite region by an appropriate choice of the coordinates, we will have to hold on to the view that the $g_{\sigma\tau}$ describe the gravitational field.

In the theory of general relativity, gravitation plays an exceptional role above all other forces, in particular the electromagnetic. This is so because the ten functions $g_{\sigma\tau}$ which represent the gravitational field also determine the metrical properties of the four-dimensional space.[168]

Although the last remark might seem to pinpoint the difference in the points of view of Einstein and Hilbert, it is quite consistent with Hilbert's results. In fact Hilbert had clearly stated that once the equations of gravitation are given, and thus also the metric, important consequences for the laws of electromagnetism would follow. That Einstein did not like this conclusion, he made amply clear later on.

It is perhaps worthwhile at this point to mention the discussions about the physical meaning of general relativity, the principle of equivalence in particular, which Einstein had with Friedrich Kottler and Erich Kretschmann, soon after the publication of his paper on general relativity in 1916. Kottler raised the question whether Einstein, in his final formulation of general relativity, had given up the real principle of equivalence.[169] Einstein insisted that this was not so and, in any case, it depended on what one understood by the principle of equivalence. Insofar as he was concerned, this principle was the very 'basis of my theory'. [170] The essential point about the equivalence principle, in Einstein's view, is its dynamical nature.

By no means can one assert that the gravitational field can be thought of as being purely kinematical. A 'kinematic, non-dynamical formulation of gravitation' is not possible. By merely transforming from a Galilean system to another through the acceleration transformations, we do not learn about *arbitrary* gravitational fields, rather only about those particular ones which must also satisfy the same law as all gravitational fields. This is again only another formulation of the equivalence principle (especially, in its application to gravitation).
A theory of gravitation violates the equivalence principle in a sense, as I understand it, only if the equations of gravitation are not satisfied in *any* reference frame which moves non-uniformly relative to a Galilean frame of reference. It is quite evident that this criticism cannot be leveled against my theory with the *general* covariant equations, for in this case the equations are valid with respect to every frame of reference. *The requirement of general covariance of the equations includes that of the equivalence principle as a special case.*[171]

Kottler's other questions referred to how fundamental were the various concepts that had been used in Einstein's theory, in particular the 'energy tensor of the gravitational field'. These, Einstein replied, were convenient names rather than deep concepts; the only real concept in general relativity, according to him, was general covariance.

Kretschmann's attempts were directed towards another question related to the equivalence principle.[172] He started from the observation that all physical laws could be brought into a general covariant form by mathematical re-arrangement. He thought:

It must be possible, however, to ascribe some significance to the postulates of realtivity other than the purely formal mathematical one. For only when such a significance is present could one explain, for instance, the obvious and generally recognized impossibility of adapting the notion of the rigid body (which, more than anything else, is so easy to characterize by means of purely topological features) to Einstein's original theory of relativity.[173]

Kretschmann sought to sharpen the physical and conceptual content of relativity by considering an example in four-dimensional geometry and attempting to bring it into agreement with a generally valid formulation of an arbitrary postulate of relativity.[173] He emphasized that:

A system of physical laws fulfills the relativity postulate of a group (G) of invariant transformations if, and only if, for every representation of arbitrary form of precisely these laws of the system, the physically valid (and thereby distinguishable, in principle, from all others by observation) inertial systems required for its representation form such a large set (under all the possible physical conditions permitted by the laws), that the set of transformations relating them comprises the group G in some fashion or is identical with it.[174]

According to this statement, the validity of the relativity postulate for a system of physical laws is independent of the form of its mathematical expression, and is determined entirely by its physical content. Einstein's relativity postulate was weaker than this, Kretschmann insisted, and only stated that all coordinate frames were possible. The theory of general relativity, he said, did not satisfy any sharply defined principle. 'In a physical sense, Einstein's theory does not satisfy any postulate of relativity at all; in its content it is a perfectly "absolutist" theory.'[175] In his reply, re-stating the principles of the general relativity theory, Einstein stressed the heuristic value of general covariance.[176] He maintained that the theory did in fact lead to a new equation for the gravitational field, and that was significant.

4.2 THEORY OF THE GRAVITATIONAL FIELD

After a lengthy section on a treatment of the mathematical tools in which, for instance, he introduced the sum convention [i.e., sum always over two equal indices in different positions[177]], Einstein presented the theory of the gravitational field in his famous paper of 1916. He had already, in the mathematical section, argued that g could not vanish:

Should $\sqrt{-g}$ vanish at a spot in the four-dimensional continuum, then it signifies that a finite coordinate volume there corresponds to an infinitesimal 'natural' volume. This, however, may not be so anywhere, and therefore the sign of g cannot change. Following special relativity, we sall assume that g always has a finite negative value. This represents a hypothesis about the physical nature of the continuum under consideration and at the same time a determination of the choice of coordinates. If, however, $\sqrt{-g}$ is always positive and finite, then it is obvious that *a posteriori* the choice of coordinates can be made such that this quantity is equal to 1.[178]

The equations then become much more simplified.

Einstein developed the theory in two steps. First, he set up the equations of gravitation without the presence of matter.

We distinguish in the following between the 'gravitational field' and matter in the sense that everything outside the gravitational field is regarded as 'matter' – i.c. not only 'matter' in the ordinary sense, but also [including] the electromagnetic field.[179]

In the case of special relativity, i.e. $g_{\mu\nu}$ being constants, the Riemann–Christoffel tensor given by Equation (20) is identically zero. The generalization of this situation clearly requires

$$R_{\mu\nu} = R^{\varrho}_{\mu\varrho\nu} = 0. \tag{54}$$

Einstein stated:

It must be pointed out that only a minimum arbitrariness is connected with these equations. For, other than $R_{\mu\nu}$, there is no tensor of the second rank connected with it which can be constructed from the $g_{\mu\nu}$ and their derivatives, which does not contain higher than second order derivatives, and is linear in them.[180]

And he added,

That these equations, which follow from the requirement of general relativity in a strictly mathematical way and, together with the equations of motion, yield Newton's law of attraction in the first approximation and the explanation of the excess (i.e. the excess remaining after the perturbation correction has been included) perihelion motion of Mercury (discovered by Leverrier) in the second approximation, should in my opinion be convincing regarding the correctness of the theory.[181]

Einstein then derived the field equations from a Hamiltonian principle, using the Hamilton function

$$H = g^{\mu\nu}\Gamma^{\alpha}_{\mu\beta}\Gamma^{\beta}_{\nu\alpha}. \tag{55}$$

The energy of the gravitational field is thus conserved, or

$$\frac{\partial t^{\alpha}_{\sigma}}{\partial x^{\alpha}} = 0 \tag{56}$$

with

$$\gamma t^{\alpha}_{\sigma} = \tfrac{1}{2}\delta^{\alpha}_{\sigma}g^{\mu\nu}\Gamma^{\alpha}_{\mu\beta}\Gamma^{\beta}_{\nu\alpha} - g^{\mu\nu}\Gamma^{\alpha}_{\mu\beta}\Gamma^{\beta}_{\nu\alpha}.$$

Although t^{α}_{σ} is not a tensor, the conservation equation is valid in all systems with $\sqrt{-g}=1$, and therefore the energy and momentum of the gravitational field are conserved.

The generalization of the equations of gravitation (54) starts with the *assumption* of a symmetric energy tensor $T_{\alpha\beta}$.[182] Since Equation (54) can be written as

$$\frac{\partial}{\partial x^{\alpha}}(g^{\sigma\beta}\Gamma^{\alpha}_{\mu\beta}) = -\gamma(t^{\sigma}_{\mu} - \tfrac{1}{2}\delta^{\sigma}_{\mu}t),$$

with

$$t = t^{\mu}_{\mu}, \tag{57}$$

one immediately obtains the equations of gravitation in the presence of matter by replacing it through

$$\frac{\partial}{\partial x^{\alpha}}(g^{\sigma\beta}\Gamma^{\alpha}_{\mu\beta}) = -\gamma[(t^{\sigma}_{\mu} + T^{\sigma}_{\mu}) - \tfrac{1}{2}\delta^{\sigma}_{\mu}(t + T)]. \tag{58}$$

Equation (58) is equivalent to the Hilbert–Einstein equation (37).

Concerning this 'derivation', Einstein remarked:

We must admit that the introduction of the energy tensor of matter by means of the postulate of relativity alone is not justified. For this reason, we have derived it above from the requirement that the energy of the gravitational field should act [i.e. gravitate] in the same way as every other form of energy. The strongest reason for the choice of the above equations [Equations (58)] is that they yield the conservation relations for the components of the total energy (energy and momentum) as a consequence.[183]

The energy momentum conservation can thus be obtained directly from the field equations (58). The occurrence of the gravitational energy term in the conservation equation is equivalent to

$$\frac{\partial T^{\alpha}_{\sigma}}{\partial x^{\alpha}} = -\Gamma^{\alpha}_{\sigma\beta}T^{\beta}_{\alpha}. \tag{59}$$

This condition expresses the influence of the energy of the gravitational field, and Einstein remarked, referring to Hilbert's four auxiliary conditions:

Thus the field equations of gravitation contain four simultaneous conditions, which material processes must satisfy. They yield the equations of a material process completely, provided the latter can be characterized by four independent differential equations.[184]

Immediately afterwards, however, Einstein stated that all laws of physics could be generalized into a covariant form without restricting the possibilities offered by special covariance.

This fact explains why it is not necessary to introduce special assumptions about the nature of matter (in a restricted sense). In particular, the question may remain open whether or not the theory of the electromagnetic and gravitational fields together provides a sufficient basis for the theory of matter. The postulate of general relativity, in principle, cannot teach us anything about it. Only the further development of the theory will show whether the principles of electromagnetism and gravitation together can accomplish what the former [electromagnetism] alone cannot.[185]

4.3 CONSEQUENCES AND APPLICATIONS OF THE THEORY OF GRAVITATION

Einstein generalized Euler's equations of hydrodynamics and Maxwell's equations of electrodynamics, without changing the structure of the latter. His treatment stands in sharp contrast to Hilbert's, for whom his own derivation of Mie's equations of electrodynamics from the equations of gravitation, in the limiting case of special relativity, meant a particular triumph. For this, Einstein criticized Hilbert rather sharply. In a postcard to Ehrenfest he wrote:

I do not like Hilbert's formulation. It [Hilbert's formulation] is needlessly specialized and, as far as matter is concerned, unnecessarily complicated. It is not honest (= Gaussian) in design, [and reflects] the pretension of a superman by a camouflage of techniques.'[187]

Einstein's triumph was not the field theory of matter but the correct theory of gravitation, which led to Newton's theory and the bending of light in the first approximation, and to the rotation of the perihelion of Mercury in the second. The approximation which reproduced Newton's law of gravitation was that the metric $g_{\mu\nu}$ becomes Minkowskian in the limit of spatial infinity, and that the term dx^4/ds is large compared to the other quantities $[dx^1/ds,$ etc.$]$. The equation of motion of a massive particle, in a static gravitational field, thus becomes

$$\frac{d^2x^i}{dt^2} = -\frac{1}{2}\frac{\partial g_{00}}{\partial x^i}, \qquad (i = 1, 2, 3), \qquad (60)$$

in agreement with Newton's result. Only the potential g_{00} plays a role in this approximation and the Poisson equation becomes

$$\Delta g_{00} = \gamma \varrho, \qquad (61)$$
$$(\gamma = 8\pi k/c^2 = 1.87 \times 10^{-27} \text{ c.g.s. units}).$$

For the bending of light by the sun, Einstein found 1.7 seconds of arc.[188]

The calculation for the rotation of the perihelion of Mercury was based upon the vacuum equations (54). Making an *Ansatz* of radial symmetry, one finds in the first order

$$g_{k\ell} = -\delta_{k\ell} - \alpha \frac{x_k x_\ell}{r^3}, \quad (k, \ell = 1, 2, 3),$$

$$g_{k0} = g_{0k} = 0, \quad (k = 1, 2, 3) \qquad (62)$$

$$g_{00} = 1 - \frac{\alpha}{r}.$$

However, in the second order, the motion of a planet of small mass deviates from the Kepler–Newton laws. The orbit obtains, in the direction of the motion, a small rotation of magnitude

$$\varepsilon = 24\pi^3 \frac{a^2}{T^2 c^2 (1 - e^2)},$$ (63)

where a is the semi-major axis, e the eccentricity of the orbit, T the period of the planet, and c the velocity of light.[188]

Soon after the publication of his memoir of 1916 on the foundations of the theory of general relativity, Einstein began to seek new conceptual developments of the theory. One of the first such developments was the introduction, in a note on the approximation integration method for the field equations of gravitation, of the concept of gravitational waves.[189] From the *Ansatz*

$$g_{\mu\nu} = -\eta_{\mu\nu} + h_{\mu\nu},$$ (64)

Einstein stated that

$$\eta_{11} = \eta_{22} = \eta_{33} = -1 = \eta_{44}; \quad \eta_{\mu\nu} = 0 \quad \text{for} \quad \mu \neq \nu$$

where, in first approximation, $h_{\mu\nu}$ is small compared to 1. According to a proposal of de Sitter, the first order calculation is simplified by not choosing coordinate systems with $\sqrt{-g} = 1$. Choosing new small parameters defined by,

$$\gamma_{\mu\nu} = h_{\mu\nu} - \tfrac{1}{2}\eta_{\mu\nu}\eta^{\varrho\sigma} h_{\varrho\sigma},$$ (65)

the equations of gravitation become

$$\frac{\partial^2 \gamma_{\mu\nu}}{\partial x^{\alpha^2}} - \frac{\partial^2 \gamma_{\mu\alpha}}{\partial x^\nu \, \partial x^\alpha} - \frac{\partial^2 \gamma_{\nu\alpha}}{\partial x^\mu \, \partial x^\alpha} + \delta_{\mu\nu} \frac{\partial^2 \gamma_{\alpha\beta}}{\partial x^\alpha \, \partial x^\beta} = 2\gamma T_{\mu\nu},$$ (66)

and reduce to

$$\eta^{\sigma\beta} \frac{\partial^2 \gamma_{\mu\nu}}{\partial x^\sigma \, \partial x^\beta} = 2\gamma T_{\mu\nu}$$ (67)

if

$$\frac{\partial \gamma_{\mu\alpha}}{\partial x^\sigma} = 0.$$ (68)

The latter restriction can be imposed because of general covariance.[190] This reasoning of Einstein is exactly the same as Hilbert's.

From Equation (67), one finds that gravitational fields propagate with the velocity of light, and the potentials $h_{\mu\nu}$ or $\gamma_{\mu\nu}$ can be calculated from the $T_{\mu\nu}$,

$$\gamma_{\mu\nu} = -\frac{\gamma}{2\pi} \int \mathrm{d}V_0 \frac{T_{\mu\nu}(x_0, y_0, z_0, t - r)}{r},$$ (69)

where $\mathrm{d}V_0 = \mathrm{d}x_0 \, \mathrm{d}y_0 \, \mathrm{d}z_0$. Since $T_{\mu\nu}$ is of the second order in $h_{\mu\nu}$, using a *wave Ansatz* and evaluating carefully, one finds

$$\gamma_{\mu\nu} = \alpha_{\mu\nu} f (x' \mp x^0),$$

where

$$\alpha_{\varrho_1} \mp i\alpha_{\varrho_0} = 0, \tag{70}$$

and one obtains a solution of the problem with

$$t_1 = \frac{1}{4\gamma} f^2 \left[\frac{(\alpha_{22} - \alpha_{33})^2}{2} + a_{23}^2 \right]. \tag{71}$$

Waves with $\alpha_{22} - \alpha_{23}$ and α_{23}^2 equal to zero are spurious.

Considering the total emission of *gravitational* radiation from a body, Einstein found that it is always finite, and in conclusion he remarked that, 'Already in the earlier treatment, we had emphasized that the final result (requiring an energy loss from bodies) of this consideration must arouse doubts about the general validity of the theory. It appears that an improved quantum theory would also bring about a modification of the theory of gravitation.'[191]

4.4 The Cosmological Term

Einstein's second extension of the theory of gravitation was his introduction of the so-called cosmological term or constant.[192] Starting from the consideration that Poisson's equation does not entirely replace Newton's theory of gravitation if one does not add the condition that the potential ϕ becomes finite in the limit of spatial infinity, Einstein looked for similar conditions in the theory of relativity. This condition, however, leads to difficulties even in Newton's theory, because the density of matter then has to be zero at infinity and stars could escape into the space beyond. A possibility of avoiding this conclusion would be to assume a very high value of the potential at infinite distances. 'This would be a feasible way, if the form of the gravitational potential were not determined by the celestial bodies themselves,' Einstein thought.[193] If equilibrium is assumed, with the presence of a finite potential difference between the centre and the infinity of space, the ratio of the densities must be finite; given a zero density at infinity, there must follow a zero density everywhere else. Einstein then noted:

These difficulties would hardly be overcome within the framework of Newtonian theory. One might ask whether they could be avoided by a modification of Newton's theory. Here we shall propose a way for which no claim is made to be taken seriously by itself, but it serves to bring the following into better focus. Replace Poisson's equation by,

$$\Delta\phi - \lambda\phi = 4\pi k\varrho, \tag{72}$$

where λ denotes a universal constant. Let ϱ_0 be the (uniform) mass density, then

$$\phi_0 = -\frac{4\pi k}{\lambda} \varrho_0 \tag{73}$$

is a solution [of Equation (72)].[194]

Einstein thus introduced a finite average density of matter. The real density distribution leads to accumulations of ϱ, and over and above the constant potential ϕ_0 a Newtonian potential due to $4\pi k\varrho$ is added. Space thus constructed could extend to infinity, but this would not be true in general relativity if the equations of gravitation were modified by the 'cosmological term'. Einstein's reasoning was that, 'In a logically consistent theory of relativity, there can be no inertia *relative to space*, but only a *mutual* inertia of masses. Hence if a mass is removed spatially far enough from all other masses of the universe, then its inertia must go to zero.'[195] An infinite space does not meet this Machian condition, Einstein concluded, after computations of models with Grommer, because one cannot find suitable boundary values at infinity. But

if the universe could be imagined as a *spatially closed but unbounded continuum*, boundary conditions of that kind would be totally unnecessary. I will show below that both the postulate of general relativity, and the fact that the velocities of fixed stars are insignificant, are consistent with the hypothesis of a spatially closed, unbounded universe. However, in order to carry through this fundamental idea, it is necessary to make a general modification of the field equation.[196]

Einstein considered a spherical space with constant curvature and radius a, leading to the metric

$$g_{k\ell} = -\delta_{k\ell} + \frac{x_k x_\ell}{a^2 - (x_1^2 + x_2^2 + x_3^2)}, \quad (k, \ell = 1, 2, 3) \tag{74}$$

$$g_{00} = 1 \quad \text{and} \quad g_{k0} = 0, \quad (k = 1, 2, 3).$$

This metric is compatible, not with the equations of gravitation (37), but with the generalized equations [i.e., those including the cosmological term],

$$R_{\mu\nu} - \lambda g_{\mu\nu} = -\gamma (T_{\mu\nu} - \tfrac{1}{2} g_{\mu\nu} T). \tag{75}$$

For sufficiently small λ, this field equation is in any case also consistent with the empirical facts pertaining to the solar system. It also satisfies the conservation of momentum and energy, because one obtains Equation [(75)] instead of Equation [(37)] if, in Hamilton's principle, in the place of the scalar of Riemann's tensor one uses this scalar together with an additional universal constant. In this way [i.e. using Hamilton's principle] the conservation laws are assured. That the field equations [75] are consistent with our notions concerning field and matter will be shown below....[197]

In Hilbert's formulation, the field equations (75) could be obtained very easily, just by replacing R by $R + \lambda$ in Equation (47).

For the cosmological term λ, Einstein obtained the expression

$$\lambda = \frac{1}{a^2}. \tag{76}$$

He was now ready to sum up these considerations:

The theoretical conception of the real world corresponding to our ideas would therefore be the following. The curvature of space varies with the temporal and spatial distribution of matter, but it can be roughly approximated by a sphere. In any case, this notion is logically consistent, and from the standpoint of general relativity the most obvious; whether it is also valid in the light of current astronomical knowledge will not be discussed here. In order to achieve this logically consistent formulation, we had to introduce a new generalization of the gravitational field equations, which however is still not justified by our actual knowledge of gravitation. It should be emphasized, however, that a positive curvature of space by the presence in it of matter also comes about if the additional term [universal constant] is not introduced. The latter is only necessary in order to allow for a quasi-static distribution of matter, which corresponds to the fact of the insignificant velocities of fixed stars.[198]

In any case, Einstein was very content with general relativity, and he felt really fulfilled as confirmations of his predictions were obtained, the most important perhaps being the measurements of the deflection of light from the sun which were carried out during the solar eclipse expedition of 1919 led by Eddington. When Eddington's cable with the results of the expedition arrived, Ilse Rosenthal-Schneider was with Einstein and he showed it to her. At her joy that the results agreed with his theory, 'He remarked totally without emotion, "I knew that the theory was correct." And when I told him what if there had been no verification of his prediction, he replied, "I would have felt sorry for the dear Lord! The theory is, of course, all right."'[199]

5. PHYSICAL INTERPRETATION OF GENERAL COVARIANCE

In his paper of 1916 on general relativity, Einstein sought to interpret the meaning of general covariance. An important postulate of all physical laws is their causal structure, and a causal 'law has significance for the empirical world only if observable facts alone appear as causes and effects'.[200] The postulate of general covariance takes away from space and time

the last vestige of physical objectivity.... All of our assertions concerning space-time always amount to the measurements of space-time coincidences. If, for instance, the physical process is only the motion of a material point, then in effect nothing more could be observed than the encounters of two or three such points. The results of our measurements are in fact nothing but the statement of such encounters of material points of our apparatus with other material points, or coincidences respectively between the pointers of clocks, dials, and other point-events under observation, occurring at the same place and time.[201]

Writing to Besso on 31 October 1916, after a visit to Holland, Einstein elaborated these points further:

The objective significance of space and time lies in the first instance in that the four-dimensional continuum is hyperbolic, in such a way that originating from every point there are spatial and temporal line elements, i.e. those for which $ds^2 > 0$ and $ds^2 < 0$ respectively. The coordinates x^μ per se do not have spatial or temporal character. In order to uphold usage, those systems are to be preferred for which

$$g_{00}(dx^0)^2 > 0, \qquad g_{11}(dx^1)^2 + 2g_{12}dx^1dx^2 + \ldots + g_{33}(dx^3)^2 < 0.$$

However, an objective justification of this kind is not involved in this choice. Hence the 'spatial' or 'temporal' character is real. But nature has not made one coordinate temporal and other spatial.[202]

It was exactly on this basis that Hilbert, in his axiomatic approach to the foundations of physics, developed his own physical interpretation of covariance.[203]

5.1 AXIOM OF PHYSICAL INTERPRETABILITY

In his second memoir on 'The foundations of physics', presented to the Göttingen Academy on 23 December 1916, Hilbert developed further the axiomatic foundations of general relativity by emphasizing 'some more general questions of a logical and physical character.'[204] He considered the Riemannian metric $g_{\mu\nu}$ in x–t space and formulated, what he later on called his Axiom IV, the space-time axiom. He asserted:

'The quadratic form

$$G\left(x^1, x^2, x^3, x^4\right) = \sum_{\mu, \nu} g_{\mu\nu}x^\mu x^\nu$$

is such that, in its representation as a sum of four linear quadric forms of x^μ, there always appear three square [terms] with the positive and one square [term] with the negative sign.'[205]

The quadratic form, given above, therefore gives the metric of a pseudo-geometry in the four-dimensional world of x^μ. Hilbert then defined the space- and time-lines, calling the segment of a curve $x^\mu = x^\mu(p)$, for which

$$G\left(\frac{dx}{dp}\right) > 0, \tag{77}$$

a 'space-line', with the length

$$\lambda = \int dp \sqrt{G\left(\frac{dx^\mu}{dp}\right)}, \tag{78}$$

and a segment of the curve, where

$$G\left(\frac{dx^\mu}{dp}\right) < 0 \tag{79}$$

a 'time-line' with the proper time [i.e. the time as measured by an observer moving on this world-line],

$$\tau = \int dp \sqrt{-G\left(\frac{dx^\mu}{dp}\right)}. \tag{80}$$

For a 'null line', therefore,

$$G\left(\frac{dx^\mu}{dp}\right) = 0. \tag{81}$$

He remarked:

In order to elucidate this notion of pseudo-geometry, let us imagine two ideal measuring devices: the *measuring string*, by means of which we can measure the length λ of any path; and, second, the *light-clock* by means of which we can determine the proper time of any time line. The *measuring string* reads null and the *light-clock* stands still along every null line; whereas the string cannot be used along the time line, the clock cannot be used along the space line.[206]

With either of these devices it should be possible to compute the values of $g_{\mu\nu}$ as a function of x^μ, if only one introduces a certain space-time frame of reference. It is, moreover, straightforward to axiomatize the pseudo-geometry suitably.

5.2 RESTRICTIONS DUE TO CAUSALITY

The above considerations have important consequences for the problem of causality, as Hilbert pointed out:

Until now we have considered all coordinate systems $x^{\mu'}$, which arise from any arbitrary transformation, as equivalent. This arbitrariness must be restricted as soon as we insist that two world points on the same time-line can be related to each other by cause and effect, and therefore it should be impossible to transform such world points into simultaneous world points. By designating x^0 as the *proper* time coordinate, we define: For a *proper* space-time coordinate system, not only $g \neq 0$ always, but the four inequalities

$$g_{11} > 0, \qquad \begin{vmatrix} g_{11} & g_{12} \\ g_{21} & g_{22} \end{vmatrix} > 0, \qquad \begin{vmatrix} g_{11} & g_{12} & g_{13} \\ g_{21} & g_{22} & g_{23} \\ g_{31} & g_{32} & g_{33} \end{vmatrix} > 0, \qquad g_{00} < 0 \tag{82}$$

are also always satisfied. A transformation which takes one such space-time coordinate system into another space-time coordinate system is called a *proper* space-time transformation.[207]

Hilbert's treatment was so straightforward and classical that Pauli took over an entire section of this paper in his review article on the theory of relativity.[208] Reichenbach did the same in his axiomatization of general relativity.[209] In his memoir, Hilbert went on to discuss the usual principle of causality, namely the postulate that from the knowledge of the physical quantities and their appropriate derivatives in the present, they can also be determined in the future. In general relativity this principle seemed to be violated, but it could be restored in the following way: 'From a knowledge of the 14 potentials $g_{\mu\nu}$, A_μ at a given moment, one can necessarily and without ambiguity infer all facts of interest about them at a later time, if only they have a physical meaning.'[210]

By introducing physically meaningful results Hilbert sharpened the formulation of the causality principle, thereby completing the axiomatic structure of the new physics:

As I have pointed out, physics is a four-dimensional pseudo-geometry, whose metric $g_{\mu\nu}$ is related to the electromagnetic quantities, that is matter, by the basic equations (4) and (5) of my first communication. With this realization, a long-standing geometrical problem can be readily resolved: namely, the problem whether, and in which sense, Euclidean geometry (which is only known from mathematics to be a logically consistent structure) has validity in reality.

The old physics, with its notion of absolute time, assumed the axioms of Euclidean geometry and made them the basis of every physical theory. Even Gauss hardly differed from this [approach]. He constructed hypothetically a non-Euclidean physics, retaining the concept of an absolute time, and waiving only the axiom of parallels of Euclidean geometry. The measurement of the [inner] angles of a triangle of large dimensions convinced him of the invalidity of this non-Euclidean physics.

The new physics of Einstein's general relativity principle takes an entirely different approach to geometry. It does not rely upon either the Euclidean or any other geometry in order to deduce the actual laws of physics. Instead, the new theory of physics at once yields, from one and the same Hamilton principle, the geometrical and physical laws [i.e. the fundamental equation (4) and (5) of Hilbert's first memoir]. These equations show how the metric $g_{\mu\nu}$, which also gives the mathematical expression for the physical phenomenon of gravitation, is connected with the values of the electrodynamic potentials A_μ.

The Euclidean geometry is *as foreign to modern physics as action-at-a-distance*. Insofar as the theory of relativity rejects it [Euclidean geometry] as the general basis of physics, it [the theory of relativity] teaches us that geometry and physics are equal in rank, comprising *one* science resting upon a common foundation.[211]

5.3 HILBERT'S CONCLUSION

Hilbert concluded that the only solution of the equations of matter (i.e. the equations of gravitation and electricity), which is consistent with Minkowskian geometry, is one in which all sources of matter and electricity are removed, because from the equations of gravitation in this metric it follows that

$$\frac{\partial \sqrt{g} L}{\partial g^{\mu\nu}} = 0. \tag{83}$$

This is both a necessary and a sufficient condition.

It is impossible to obtain a regular metric, which at the same time is not pseudo-Euclidean and yet corresponds to a universe without electricity, by the variation of the metric of pseudo-Euclidean geometry under the assumptions I and II. [The first assumption being that the variation from the pseudo-Euclidean geometry does not depend on the time variable, and the second that the variations should behave regularly at infinity.][212]

Central to Hilbert's conclusion is the assumption that the metric $g_{\mu\nu}$ remains regular. Hilbert emphasized his belief that 'it is only the regular solutions of the fundamental equations of physics that represent reality immediately.'[213] He also believed that the solutions which were not regular did provide an important mathematical tool in *approximating* characteristic regular solutions, citing the Schwarzschild solution of Einstein's equations, which are singular at $r = 0$, as an example.[214] With these considerations Hilbert concluded his treatment of the foundations of physics, a work whose significance in the development of the general relativity theory of gravitation has seldom been appreciated.

6. HAMILTON'S PRINCIPLE AND THE CONSERVATION LAWS

In his celebrated Encyclopaedia article on the theory of relativity, Pauli explicitly acknowledged Hilbert's formulation of the law of gravitation, Equation (37), and remarked:

At the same time as Einstein, and independently, Hilbert formulated the generally covariant field equations. His presentation, though, would not seem to be acceptable to physicists, for two reasons. First, the existence of a variational principle is introduced as an axiom. Secondly, of more importance, the field equations are not derived for an arbitrary system of matter, but specifically based on Mie's theory of matter.[215]

Pauli's comments about Hilbert's formulation, in which he had introduced the variational principle as an axiom, merely reflected the prevailing taste of the day. During the late nineteen-tens and early twenties [Pauli's article was published in 1921], several prominent theoretical physicists shared a belief that a Hamiltonian principle should not necessarily be used to describe the laws of physics. The main reason for this feeling was that there existed great difficulties at that time with respect to the description of the laws of microscopic physics. Although many of these difficulties were removed after the creation in 1925 of a consistent theory of atomic phenomena, there has again been a resurgence of doubt whether the laws of high energy physics can be formulated by using a variational principle. In the macroscopic domain, however, with which, according to expert scientific opinion, general relativity seems to deal principally, Hilbert's procedure of assuming a Hamiltonian principle as a fundamental axiom is now regarded as being well justified.[216]

The question persisted, of course, whether the two theories, Hilbert's and Einstein's, were identical. Lorentz and Einstein dealt with this question initially, and the proof was finally completed by Klein.

6.1 LORENTZ AND THE HAMILTON PRINCIPLE IN EINSTEIN'S THEORY

In a letter to Lorentz, dated 16 August 1914, Einstein discussed several details of the theory of general relativity which he had developed with Grossmann, in particular the meaning of energy conservation.[217] Already in 1915, before the new developments of the theory were published by Einstein and Hilbert, Lorentz studied the problem of the conservation of energy in a note on 'Hamilton's principle in Einstein's theory of relativity', which he submitted to the Royal Dutch Academy of Sciences.[217a] Lorentz thought that

the discussion of some aspects of Einstein's theory of gravitation would perhaps gain in simplicity and clarity, if it were based on a principle similar to that of Hamilton's, so much so indeed that Hamilton's name might properly be attached to it. Now that Einstein's theory is available, we can easily determine how this variational principle should be formulated for different kinds of systems and for the gravitational field itself.[218]

Einstein had already shown that one could derive the equation of motion of a material point in the gravitational field from the Lagrangian

$$L = - m \frac{ds}{dt} = - m \sqrt{g_{\mu\nu} v^{\mu} v^{\nu}} \tag{84}$$

where

$$v^{\mu} = dx^{\mu}/dt.$$

Equation (84) could be generalized to describe continuous matter if mv^{μ} were replaced by ϱv^{μ}, and L expresses a Lagrangian density. Lorentz then derived the equations of motion under a given force, and finally turned to the motion of a charged particle in electromagnetic and gravitational fields. From the theory of electrons two sets of equations arise. Lorentz considered one of these sets as defining the mathematical description of the system, and derived the second set from Hamilton's principle. By varying the electromagnetic potentials, he obtained Maxwell's equations and the laws of conservation of energy and momentum. The equations of gravitation could be derived from the Hamiltonian, including Einstein's function H defined by Equation (29), and varying it with respect to the gravitational potentials. Lorentz found full agreement with the results of Einstein, including the fact that the total energy and momentum are only conserved in systems of adapted coordinates, as defined by Equation (28).

Shortly after the appearance of Hilbert's first memoir on the foundations of physics in November 1915, and Einstein's papers of November 1915 and his memoir on general relativity early in 1916, Lorentz published a second note on 'Einstein's theory of gravitation'.[219] He wrote:

In the pursuit of his important researches on gravitation Einstein has recently attained the goal which he had constantly kept in sight: he has succeeded in establishing equations whose form is not changed by an arbitrarily chosen change of the coordinate system. Shortly afterwards [sic], working out an idea that had already been expressed in one of Einstein's papers [sic], Hilbert has demonstrated the use that may be made of a variation law, which may be regarded as Hamilton's principle in a suitably generalized form. With these results, the theory of general relativity may be said to have taken a definitive form, although much remains to be done still in its further development and application to special problems. It would also be desirable to present the fundamental ideas in as simple a form as possible.[220]

Lorentz took the general function H, the 'principal function', to consist of 'three parts, of which the first relates to the material (mass) points, the second to the electromagnetic field, and the third to the gravitational field itself.'[221] For the first part, he used the function L, Equation (84), giving the kinetic term, and it turned out to be invariant under general transformations. Passing on to the second part, Lorentz remarked that 'the mathematical expression for this part was communicated to me by Einstein in our correspondence. It is also to be found in Hilbert's paper, where it is remarked that the quantity in question may be regarded as the measure of the *curvature* of the four-dimensional generalization to which [the expression for the line element] relates. Here we shall talk only about the interpretation of this quantity.'[222] Again the term K, which Hilbert had used, could be interpreted with the help of Gauss, and it did not change with coordinate transformations. That part of the principal function which is due to the curvature is given by

$$H_3 = + \frac{1}{\gamma} \int R \, d\Omega, \tag{85}$$

where γ is the gravitational constant, and $d\Omega$ a four-dimensional volume element. The derivation of the electromagnetic part took rather long, but again the generalization of the usual expression was obtained, containing the sum of $\mathscr{E}^2$ and $\mathscr{H}^2$, i.e. the scalars of the electric and the magnetic fields in suitable form.[223]

In his third and fourth communications on this subject, Lorentz derived the Hilbert–Einstein field equations, in particular Equation (37), by a variation of the gravitational potential for the two cases, namely the $T_{\mu\nu}$ being due to the electromagnetic or the mechanical part respectively. Altogether Lorentz had produced a complete proof of the equivalence of Einstein's inductive and Hilbert's deductive methods, treating all the delicate points clearly and in detail.

6.2 EINSTEIN AND THE HAMILTON PRINCIPLE

Nevertheless, Einstein returned to a discussion of this problem at the session of the Prussian Academy on 26 October 1916.[224] He obviously felt that he should give the connection between Hilbert's formulation and his own, a still simpler form, and also take a stand against Hilbert's attempt to incorporate Mie. He did not waste any time in guarding his own stake in this problem.

Lorentz and Hilbert [*note the order of these names*] have recently succeeded in presenting the theory of general relativity in an especially clear form, by deriving its equations from a single variational principle. We shall also follow the same method in this work. In doing so, I shall attempt to present the fundamental connections [between Hamilton's principle and general relativity] as clearly and generally as possible, as far as the standpoint of general relativity permits. In particular, especially in contrast to Hilbert, as few restrictive assumptions as possible about the constitution of matter will be made. On the other hand, contrary to my own previous paper, the choice of the coordinate system shall remain perfectly free.[225]

Einstein reformulated the Hamiltonian function for the variation by letting it depend only on the first derivatives of $g_{\mu\nu}$ and A_μ. Again, the Hamiltonian function had to be split into a gravitational part and all the rest, by assumption, otherwise the equations of gravitation, etc., would not follow. He easily derived the conservation of energy and momentum, including both the matter tensor $T_{\mu\nu}$ and the field energy tensor $t_{\mu\nu}$. From the latter equations, and the field equations of gravitation, there would follow the four equations

$$\frac{\partial T_\sigma^\nu}{\partial x^\nu} - \frac{1}{2}\frac{\partial g^{\mu\nu}}{\partial x} T_{\mu\nu} = 0 \tag{86}$$

for the energy-momentum tensor of matter. These are essentially Hilbert's four conditions, which Einstein did not want to link with Mie's equations of electrodynamics. First, he did not like the latter, and second, because he was not convinced that electrodynamics alone could provide a complete theory of matter. Einstein changed this attitude in his later years, but at the time when Hilbert and Weyl professed interest in the electromagnetic theory of matter, he was strongly opposed to it.

Clearly, Einstein's conservation law for the energy tensor and Hilbert's four conditions for the energy tensor of matter are closely linked. We shall pursue this connection in some detail in the next section, thereby seeking to make a little more evident the meaning of energy conservation in general relativity. Let us mention here only that the so-called energy conservation law to which Einstein and Lorentz addressed themselves appears more like the definition of the energy tensor of gravitation $t_{\mu\nu}$ than a physical law. The fact, that it was possible to find a $t_{\mu\nu}$ which depends only on the gravitational fields and its derivatives, might be considered important. But that is not really surprising, since the total principal function H had been assumed to consist of two parts, one of which contained entirely the gravitational term R.

7. GENERAL COVARIANCE AND CONSERVATION LAWS

After strenuous efforts, having followed several dead-ends, Einstein had succeeded in establishing a consistent theory of general relativity. The mathematicians, whose illustrious ancestors had laid its mathematical foundation, happily took up the development and further interpretation of the new theory. Some disagreement arose on a basic question concerning its physical interpretation. Einstein, in this case, stuck to a more formal approach, putting a formal energy conservation law in the forefront. The reason is to be found in his earlier procedure, in which full covariance was rather doubtful [see Section II.4]. Hilbert, on the other hand, tried to exploit covariance physically. Finally it was Klein, the author of the Erlangen Programme and grand old master of geometry, who, in his last important work, completed the synthesis of the views of Einstein and Hilbert – the synthesis, in a sense, of the theory of relativity and the theory of invariants. In 1917, he published two memoirs dealing with 'energy-momentum conservation' in general relativity. Shortly before this, in an exchange of letters with Hilbert, Klein clarified the meaning of the four conditions which Hilbert had discovered in this context. What escaped Klein was the fact that there existed an even more elegant connection with the equations of general geometries, the Bianchi identities.

7.1 ENERGY CONSERVATION AND THE THEORY OF INVARIANTS

In an extensive memoir [226] on 'the differential laws for the conservation of momentum and energy in Einstein's theory of gravitation', which Klein communicated to the Göttingen Academy on 19 July 1918, he attempted to unify the different forms of the differential laws for the conservation of energy and momentum that could be found in the papers of Einstein, Lorentz and Hilbert. The important tool which he used in doing this was the theory of invariants.[227]

Under the assumptions of Einstein, Hilbert, and Lorentz, the variational integral could be split into two terms, of which only one contains the curvature scalar R. Klein proceeded in three steps: first he considered R as a function of $g^{\mu\nu}$ and its first and second derivatives; then he drew conclusions from the fact that R is an invariant under generally covariant transformations; finally he used the fact that R is an invariant of specific form and content. From this procedure, there followed a set of differential relations which could be interpreted in physical terms.

From the fact that R is a function of $g^{\mu\nu}$ and its first and second derivatives, Klein derived the equation

$$\mathscr{R}_{\mu\nu}g_\tau^{\mu\nu} = 2\frac{\partial\mathscr{U}_\tau^\sigma}{\partial x^\sigma}, \quad (\tau = 1, ..., 4) \tag{87}$$

where

$$\mathscr{R}_{\mu\nu} = \sqrt{g}\,R_{\mu\nu} \quad \text{and} \quad \mathscr{U}_\tau^\sigma = \sqrt{g}\,U_\tau^\sigma, \quad \text{and} \quad g_\tau^{\mu\nu} = \frac{\partial g^{\mu\nu}}{\partial x^\tau}.$$

Here $\mathscr{U}_\tau^\sigma$ is a mixed tensor under linear transformations. But due to the fact that R is generally invariant, one finds that the divergence of $\mathscr{R} + \mathscr{U}$ is generally covariant. Altogether Klein deduced 152 equations containing $\mathscr{R}$ and $\mathscr{U}$ and their derivatives, but not all of them were independent. One specific set reproduced three vector equations,

$$\sum\frac{\partial\mathscr{T}_\tau^\sigma}{\partial x^\sigma} + \tfrac{1}{2}\mathscr{T}_{\mu\nu}g_\tau^{\mu\nu} = 0, \tag{88a}$$

and

$$\frac{\partial\left(\mathscr{T}_\tau^\sigma + \dfrac{1}{\gamma}\mathscr{U}_\tau^\sigma\right)}{\partial x^\sigma} = 0, \tag{88b}$$

and

$$\frac{\partial\left(\mathscr{T}_\tau^\sigma + \ell_\tau^\sigma\right)}{\partial x^\sigma} = 0, \tag{88c}$$

which, in this sequence, are identical with Hilbert's four conditions, and with Lorentz' and Einstein's conservation laws. The second set essentially reproduced Hilbert's energy conservation law, Equation (46). But it is physically equivalent to Lorentz' law (88b), and merely means that an equation for the gravitational field is introduced. Nevertheless, Klein preferred Hilbert's complicated energy law, Equation (46); it is complicated because the vector e^λ has such a tricky structure.

However, it [Hilbert's energy conservation law] has the advantage that it not only makes a simple [invariant-theoretic] assertion, but also the quantity e^λ appearing in it can be readily characterized invariantly. *It [Hilbert's energy conservation law] contains the auxiliary vector p^τ, and besides it is independent of the $T_{\mu\nu}$, $R_{\mu\nu}$ and their differential coefficients.*[228]

Felix Klein's work is closely connected with the studies of Emmy Noether concerning the relations between symmetry properties and conservation laws.[229] Conservation laws can be stated in two forms, i.e. the differential and integral forms. Klein turned to the latter in his second note on 'the integral form of the conservation laws and the theory of the spatially closed universe'.[230] Concerning the physics underlying his considerations, Klein referred to the papers of Einstein[192] and de Sitter.[231] Expressing his motivation, Klein noted:

First of all, I wish to express my viewpoint about the integral form of the conservation laws, which Einstein had preferred to formulate in differential form. In this connection, I shall treat Einstein's theory of the spatially closed but unbounded universe and its modification by de Sitter. My goal is to state the *mathematical* connections as clearly as posssible, while touching upon the physical question only briefly. I feel a certain satisfaction in this, since [it will be seen that my] old ideas of 1871–72 attain decisive validity. [Here Klein was referring to his Erlangen Programme.] The reader may decide for himself the extent to which progress has been made.[232]

In some general remarks, Klein stated the physical concepts of vectors etc., which have meaning only if one refers to a certain group:

There are as many different kinds of theories of relativity as there are [mathematical] groups. This statement is perhaps in contradiction to the much publicized current discussion about Einstein's general theories, but it is not significantly different from Einstein's own detailed arguments, to which I attach great importance. Einstein's papers, upon which I shall comment here, indeed demonstrate that at times, although not systematically, he [Einstein] resorts to the same conceptual freedom which I have proposed in my Erlangen programme.[233]

Klein then proceeded to discuss the generalizations of Einstein's closed spherical universe, namely elliptical universes, and their related symmetry groups. These universes are based on the use of cosmological constants λ, which change the original Einsteinian energy tensor of the gravitation field to

$$\bar{t}_\tau^\sigma = t_\tau^\sigma + \frac{\lambda}{\gamma} \delta_\tau^\sigma,$$

or

$$\bar{U}_\tau^\sigma = U_\tau^\sigma + \lambda \delta_\tau^\sigma. \tag{89}$$

But otherwise the considerations leading to Equations (88) remain unchanged. It further turned out that the closed Einstein universes would lead to an integral vector whose first three (spatial) components are zero. This is easily understood since a static universe of fixed radius had been chosen. The fourth component was interpreted as the total energy of the universe. Now the amazing result of an explicit calculation was that the two differential conservation laws, Equations (88b) and (88c), do not yield the same total 'gravitational' energy for a finite universe.

Klein's results indicated that one had to be very careful in applying the law of energy conservation in general relativity. One might again stress the view which was expressed by Hilbert, that energy conservation is not necessarily a characteristic of general relativity. By energy conservation, Hilbert obviously meant well defined terms based upon physical considerations. This brings us to a discussion of the point whether one should prefer Hilbert's formulation based on constraints, or Einstein's concept of energy conservation, in general relativity.

7.2 IDENTITIES FROM GENERAL COVARIANCE

Felix Klein's considerations, which we have discussed in the foregoing, were rather general, resulting in the conclusion that both Hilbert's constraint conditions for the energy tensor of matter and Einstein's conservation of energy momentum followed from general covariance. An earlier note of Klein, which was both complete and general, already contained a specific discussion of the main points.[234] Discussing the variation of Hilbert's two Hamiltonian integrals

$$I_1 = \int R \, d\Omega, \quad \text{and} \quad I_2 = \int L \, d\Omega, \tag{90}$$

where

$$d\Omega = \sqrt{g} \, dx^1 \ldots dx^0, \quad (d\Omega = d^4x),$$

Klein derived fourteen field equations, namely the ten equations (37) for the gravitational field, and the four equations for the 'matter field'. The latter equations followed from the former in the case of general covariance.[235] Klein then discussed Hilbert's energy-conservation law, Equation (46), and concluded that it would follow from the fourteen field equations as a mathematical identity. Hence, it 'can obviously not be considered analogous to the law of conservation of energy, as it exists in ordinary mechanics. For, if in ordinary mechanics we write that

$$\frac{d(T + U)}{dt} = 0, \tag{91}$$

then [our understanding is that] this relation is not a mathematical identity, but rather a consequence of the differential equations of mechanics.'[236] Klein also admitted that Einstein's conservation law could hardly be considered as a law in the usual sense.[237]

Hilbert very much agreed with Klein's conclusions, and in a letter to Klein, which Klein submitted with his own note to the Göttingen Academy, Hilbert expressed this agreement.[238] He asserted, moreover, that 'in *general* relativity, i.e. in the case of the *general* invariance of Hamilton's function, equations for the energy (which correspond to the energy equations in orthogonally invariant theories) do not exist at all. In fact, I regard this situation as typical for general relativity.'[239] After discussing energy conservation in orthogonal invariant theories, Hilbert shared the view that in a generally invariant theory one could not derive the conservation law in the same fashion. 'In general relativity, instead of the missing energy equations there are just the four overdetermined Lagrange equations, as expressed by the four identities. Conversely, the energy equations of the orthogonally invariant theories appear as the residue of those four identities of the theory of gravitation.'[240]

In his reply, Klein went further in the discussion of the identities.[241] Starting from the point of view that the two terms, R and L, in the Hamiltonian function of Hilbert, should be separated, and discussing the general covariance of both terms, Klein found two sets of four constraints. In addition to these eight equations, he found the fourteen Lagrangian equations which exist for the fields, say the gravitational and the electric potentials. Klein showed that from the four constraints governing the L term there still follow the conservation laws, but they do not preserve their separate (physical) meaning any more, since they can be reduced to the four (Hilbert) conditions of constraint for matter.

7.3 Bianchi Identities and Hilbert's Constraint Conditions

On the basis of the foregoing discussion, it may be justified to describe Klein's studies of the conservation laws and invariants in general relativity as the completion of the latter theory. They also stressed the importance of the conditions which Hilbert had placed on the energy tensor of matter. Since these conditions are obtained only with the help of the field equations of gravitation (37) [i.e. when the latter are satisfied], we might as well consider them as conditions for the curvature tensor $R^\sigma_{\mu\nu\lambda}$ or, more precisely, for the contracted tensor $R_{\mu\lambda}$ which enters the equations of gravitation. Now Hilbert's statement means simply this: once general covariance is satisfied, there exist only four differential identities satisfied by $R_{\mu\lambda}$. What are these identities? It is remarkable that great mathematicians and geometers like Hilbert and Klein did not see the connection between the four identities of Hilbert and the well-known identities of Bianchi.[242] The latter read

$$R^\sigma_{\mu\nu\lambda;\tau} + R^\sigma_{\mu\lambda\tau;\nu} + R^\sigma_{\mu\tau\nu;\lambda} = 0, \tag{94}$$

where ';' denotes covariant differentiation with respect to the coordinates. The Bianchi identities are valid not only for Riemannian, but also for more general affine geometries. Contracting the indices $\sigma = \nu$, one obtains

$$R_{\mu\lambda;\tau} - R^\sigma_{\mu\lambda\tau;\sigma} - R_{\mu\tau;\lambda} = 0. \tag{95}$$

If we multiply Equations (95) with $g^{\mu\tau}$ and contract again, we arrive at

$$2\left(R^\mu_\lambda - \tfrac{1}{2} g^\mu_\lambda R\right)_{;\mu} = 0. \tag{96}$$

These four equations may be interpreted either as the four identities of Hilbert or Einstein's energy conservation laws, if one connects them with the equations of gravitation (37). Thus Hilbert had actually discovered, with his identities, a part of the Bianchi identities which can be derived from purely geometrical considerations. It should also be worthwhile to relate some of the remaining 148 identities, which Klein found in his work of 1918, to the other consequences of the Bianchi identities. The overlap certainly is very large if not complete.[243]

We might actually venture the guess that we owe Hilbert's memoirs on the foundations of physics to the fact that he did not recognize his four identities as a consequence of the Bianchi identities. If he had been aware of the connection, he might have quite possibly slipped over their content without trying to extract the essential physical meaning. But it was the physical consequences which led Hilbert to his valuable contributions to the theory of general relativity. In a certain sense, we could apply to Hilbert himself the words which he used about Einstein: that the reason why he made progress in the theory of relativity was that he did not know everything about the mathematics of space and time before he set out to make discoveries in it.

8. EXTENSION OF GENERAL RELATIVITY

Einstein's theory of general relativity incorporated gravitation entirely. Concerning matter itself, it was less complete, as Einstein's controversy with Hilbert showed. With respect to electromagnetism, in particular, Einstein had left everything as it was, but the unsatisfactory status of his assertions was felt more and more by other scientists. The first, who stepped forward, already in 1918, was Hermann Weyl. Weyl proposed nothing less than a new geometry and a unified theory of gravitation and electricity. Another attempt was started by Kaluza in 1921, and revived later on by Oskar Klein in 1926. Finally Einstein, soon after completing the first phase of his discussions with Bohr about the new quantum theory, moved into the direction of a unified field theory of matter, and spent the rest of his life trying to cope with it – entertaining various interesting, if not entirely successful, attempts. The underlying mathematics, which generalized differential geometry, was developed by Levi-Civita in 1917.

8.1 GENERALIZATION OF RIEMANNIAN GEOMETRY

Riemannian geometry rests essentially on the consequences derived from the line element ds^2, in which the metric tensor appears. From the metrical tensor is constructed the Riemann–Christoffel tensor, and from the latter the invariants which play such an important role in general relativity. But one could even go further in geometry, endowing the three-index Christoffel symbol Γ with full meaning. This was done by Levi-Civita, and extended by Weyl who, at the starting point of his study, stated that:

Inspired by the weighty inferences of Einstein's theory to examine the mathematical foundations anew, the present writer made the discovery that Riemann's geometry goes only half way towards attaining the ideal of a pure infinitesimal geometry. It still remains to eradicate the last element of geometry 'at-a-distance', a remnant of its Euclidean past. Riemann assumes that it is possible to compare the lengths of two line elements at *different* points of space, too; *it is not permissible to use comparisons at a distance in an 'infinitely near' geometry.* One principle alone is allowable; by this a division of length is transferable from one point to that infinitely adjacent to it.[244]

The initiative of developing the mathematical extensions was provided by Einstein's theory of gravitation itself. Levi-Civita responded first.[245] Then Hessenberg followed, who considered space as formed by a great number of small elements cemented together by parallel transports.[246]

The idea behind these attempts was to make the notion of covariant differentiation equivalent to a parallel transport of vectors in curved space. This can be done with the help of the geodesic components $\Gamma^{\mu}_{\varrho v}$ as, for instance, in the case of a vector a^{μ} or b_{μ}

$$a^{\mu}_{;v} = \frac{\partial a^{\mu}}{\partial x^{v}} + \Gamma^{\mu}_{\varrho v} a^{\varrho},$$

$$b_{\mu;v} = \frac{\partial b^{\mu}}{\partial x^{v}} - \Gamma^{\varrho}_{\mu v} b_{\varrho}.$$

(97)

The Christoffel symbol then expresses the infinitesimal transport of the line element in an 'affinely connected manifold'. This theory was completed in Weyl's dissertation on a 'pure infinitesimal geometry'.[247]

In a pure infinitesimal geometry of the n-dimensional continuum each set of coordinates $x^{1} \dots x^{n}$ refers to a different point. The connection between the different points is expressed by continuous functions possessing at least continuous first derivatives and a non-zero Jacobian. One can now define an infinitesimal displacement for a given *point* in space. Since the notion of vectors, tensors, etc., is *a priori* meaningful only at a point in a consistent differential geometry, one has to look for its meaning in the entire space. This can be done if one assumes that the space forms an 'affinely related' manifold, or the fact that for any point P there exists a neighbourhood such that every vector at P can be transformed by a parallel displacement into a vector at P' (where P' is in that neighbourhood). The coefficient in front of this displacement is the Christoffel symbol Γ. A 'metrical manifold' is finally obtained only if the line element at any point P can be compared with respect to the lengths.

8.2 Weyl's Unified Theory

What has been achieved by this generalization of geometry? Obviously, it is now possible only to compare lengths measured at one and the same point. Moreover, it is no longer the $g_{\mu v}$ that make sense now or can be determined by measurements, but rather their ratios or the relations between them. If we denote the length of a short rod by l, then at different points the measuring rod is changed infinitesimally as

$$\frac{dl}{dt} = - l \frac{d\phi}{dt},$$

(98)

where the parameter t appears in a curve connecting the infinitesimally close points P and P'. For the affinely related manifolds Weyl obtained,

$$d\phi = \phi_{\mu} dx^{\mu}.$$

(99)

Riemannian geometry is obtained if $\phi_{\mu} = 0$. The Christoffel symbols in the generalized theory read

$$\Gamma_{\mu, \varrho\sigma} = \Gamma^{\text{Riemann}}_{\mu, \varrho\sigma} + \tfrac{1}{2} (g_{\mu\varrho}\phi_{\sigma} + g_{\mu\sigma}\phi_{\varrho} - g_{\varrho\sigma}\phi_{\mu}).$$

(100)

Here the absolute values of the $g_{\mu\nu}$ seem to be fixed, which need not always be the case. The invariance of all geometrical relations must exist under the *gauge* transformation

$$g'_{\mu\nu} = \lambda g_{\mu\nu} \quad \text{and} \quad \phi'_\mu = \phi \ - \frac{1}{\lambda}\frac{\partial\lambda}{\partial x^\mu}. \tag{101}$$

Weyl could now proceed to interpret the metric vector ϕ_μ, which determines the lengths in accordance with Equation (99), as being identical with the electromagnetic four-potential, apart from a numerical factor.[248] In the presence of the electromagnetic fields $F_{\mu\nu}$, this geometry will therefore deviate from Riemann's.

This idea of Weyl's was very original and attractive, because it made electrodynamics as much a consequence of geometry as Einstein had made gravitation. But some difficulties arose immediately, and Einstein pointed out that in Weyl's theory spectral lines with definite frequencies could not exist.[249] To this, Weyl replied that since the $g_{\mu\nu}$ and ϕ_μ were not measurable, it made the connection between ideal mathematical quantities and the physical quantities rather loose but suggestive. One should, however, note that the existence of the quantum of action might very well be at the basis of this failure of Weyl's theory, just as its consequences for Einstein's theory of gravitation have not yet been ascertained. In any case, Einstein greatly admired Weyl for his mathematics, though not for his physics, as he repeatedly expressed in letters.[250] A little later on, Eddington considered even more general geometries than Weyl's. Even Einstein considered these geometries for a while, but left the field very soon.[251] After quantum mechanics was created, Weyl noted that his own theory had really no physical basis.

8.3 FIVE-DIMENSIONAL RELATIVITY

Whereas Weyl had introduced the variation of the measure ds^2 and linked it to the existence of electromagnetic potentials, Kaluza remained in the realm of metrical geometries and sought to include the electromagnetic field by extending the dimensions of the universe.[252] In the five-dimensional geometry the field theories of gravitation and electrodynamics could be unified in a very simple way. The line element could now be written as

$$d\sigma = \sqrt{\Gamma_{ik}\,dx^i\,dx^k}, \tag{102}$$

where $x^0, \ldots, x^4$ are the space coordinates, and Γ_{ik} are the 15 covariant components of a five-dimensional symmetric tensor related to the usual four-dimensional ones and the electromagnetic potentials, according to the fact that $x^1, \ldots, x^4$ are the conventional space coordinates and the *metric* should not depend on x^0. Hence we put

$$\Gamma_{00} = \alpha,$$

and

$$d\Lambda = dx^0 + \frac{\Gamma_{0i}}{\Gamma_{00}}\,dx^i; \qquad ds^2 = \left(\Gamma_{ik} - \frac{\Gamma_{0i}\Gamma_{0k}}{\Gamma_{00}}\right)dx^i\,dx^k \tag{104}$$

become the infinitesimal invariants, the sum of which gives the invariant $d\sigma^2$. One finds that the four Γ_{0i} transform like the covariant components of a four-vector, and the *Ansatz* that

$$\Gamma_{0i} = \alpha\beta\phi_i, \quad (i = 1, 2, 3, 4) \tag{105}$$

where ϕ_i is the electrodynamic potential, can therefore be used.

Einstein's equations of gravitation as well as the equations of electrodynamics (in covariant form) can be derived from the variational principle

$$\delta J = \delta \int dx^0 \, dx^1 \, dx^2 \, dx^3 \, dx^4 \, \sqrt{-\Gamma} \, P = 0 \tag{106}$$

where P is the curvature corresponding to the metric tensor Γ_{ik}, and

$$\tfrac{1}{2}\alpha\beta^2 = \gamma. \tag{107}$$

One might have sought to generalize the covariant or invariant expression for P in order to obtain more general equations, but Oscar Klein, who extended Kaluza's work, was not primarily interested in this problem; he rather hoped to incorporate quantum effects in the five-dimensional theory.[253] Since the motions in the fifth dimension are not perceptible in ordinary experiments, one could just as well average over the entire motion; the quantum of action could thus follow from a periodicity in the fifth dimension, and the conjugate momentum p could be fixed to have the two values $+e$ and $-e$. Then the *free* Hamiltonian of a charged particle, say an electron in an electromagnetic field, would read

$$H = \frac{1}{2m} \, \Gamma^{ik} p_i p_k, \tag{108}$$

and the usual Hamiltonian formalism would apply.

Klein attempted in various ways to relate the quantities of the five-dimensional theory to the wave function in quantum theory. One such attempt was to write the electrodynamical potential Φ of an electron as

$$\Phi = -x^0 + S(x^1, x^2, x^3, x^4), \tag{109}$$

and to obtain the periodicity of Φ from a periodicity in the fifth coordinate; this was possible because one could relate $-\Phi$ to the transformed coordinate x^0.[254] Another possibility which Klein proposed was to try 'to relate the fifteenth quantity Γ_{00} with the wave function U, which characterizes matter, in order to achieve a formal unity between matter and field.'[255]

Several years later, Veblen and Hoffmann would mathematically replace the Kaluza–Klein five-dimensional theory by a four-dimensional projective geometry.[256]

9. 'EIN BLEIBENDER KERN' – AN ENDURING CORE

The creation of the theory of relativity in the first two decades of this century drastically changed the inherited physical concepts of space and time. The vision of a unified theory of matter first appeared in the work of Mie, Hilbert, Weyl and Kaluza. But, at the same time, another revision of classical theory had begun which also affected the understanding of the notions of space and time; it led in the middle twenties to a consistent theory of the microscopic processes. Oskar Klein's early attempts to unite the two lines of development failed, and we are not much closer now to a unified theory of matter. In the late nineteen-twenties already, certainly since Dirac's theory of the electron, it became clear that electromagnetism was not the only force which had to be united with gravitation. In quick succession new forces and fields were discovered, and the existence of a massive particle with no charge at all destroyed the hopes of ever achieving an electromagnetic theory of matter.

It seems that Einstein's remarks concerning Hilbert's theory had been prophetic. He had said:

To me Hilbert's *Ansatz* about matter appears to be childish, just like an infant who is unaware of the pitfalls of the real world.... In any case, one cannot accept the mixture [i.e. Hilbert's] of well-founded considerations arising from the postulate of general relativity and unfounded, risky hypotheses about the structure of the electron [or matter in general]. I am the first to admit that the discovery of the proper hypothesis, or the Hamilton function, of the structure of the electron is one of the most important tasks of the current theory. The 'axiomatic method', however, can be of little use [in this].[257]

But strangely enough, Hilbert did not seem to be very impressed by Einstein's negative judgement. In 1924, he published an almost unaltered united version of his two memoirs on the foundations of physics, including the notes he had written to Felix Klein.[258] He again stressed the fact that Mie had first demonstrated the manner in which one could describe matter by a unified field theory. Now, in 1924, he wrote:

As I have pointed out in my first memoir, the tremendous problems and the structure of ideas of general relativity, developed by Einstein, find their simplest and most natural expression as well as, in a formal sense, a systematic generalization and completion, in the method proposed by Mie. Since the publication of my first memoir, significant publications have appeared on this subject. Let me mention only the brilliant and profound investigations of Weyl and the always new formulations and thoughtful contributions of Einstein. However, on the one hand, Weyl in the development of his ideas is finally led to the equations proposed by me; and on the other, even Einstein, in spite of his repeated efforts to start from various different premises, in his latest publications finally returns directly to the equations proposed in my theory.

I am quite sure that the theory as developed here by me contains an enduring core, and provides a framework within which there is sufficient margin for the future development of physics, pursuing the ideal of a unified field theory. In any case, it is also of epistemological interest to note in which way the few and simple assumptions as expressed in axioms I, II, III and IV, formulated by me, are sufficient for the construction of the complete theory.

Whether in fact the ideal of a pure unified field theory is a definitive one, or if certain extensions and modifications of it will be necessary, in order to obtain in particular the theoretical justification for the existence of the electron and the proton, and the consistent formulation of the laws valid in the interior of an atom – to answer this is the task of the future.[259]

Ironically, it was exactly this hope – the realization of the ideal of a unified field theory – the hopes of Mie, Hilbert, and Weyl, which he had brushed aside with a confidence bordering on contempt, that Einstein, in outspoken contrast to the developments in quantum theory, entertained later on. Such hope was already latent in a note on 'Do gravitational fields play an essential role in the structure of elementary particles?', which he communicated to the Prussian Academy on 10 April 1919.[260] He acknowledged that the theories of Mie, Hilbert and Weyl were deep and beautiful mathematically, even though they led to unsatisfactory physical results. He thought that he knew the first steps which might lead to a solution of the problem.[261] Starting from the observation that if gravitational fields were essential for the structure of matter, the field equations (37) could not be accepted in the limit of constant $g_{\mu\nu}$, he said that, 'Material particles could not exist, if $g_{\mu\nu}$ were constant. If, therefore, one considers the possibility that gravitation, among other forces, plays a part in the structure of corpuscles, then Equation [(37)] could not be regarded as valid.'[262]

Only if one would stick to Equation (37) would one have to use Mie's theory.[263] However, the inclusion of the cosmological constant λ would already change the equations of gravitation (37). As a new *Ansatz*, instead of these equations, he proposed the new equations

$$R_{\mu\nu} - \tfrac{1}{4}g_{\mu\nu}R = -\gamma T_{\mu\nu}; \; (T_{\mu\nu} = \tfrac{1}{4}g_{\mu\nu}F_{\alpha\beta}F^{\alpha\beta} - F_{\mu\alpha}F_{\nu\beta}g^{\alpha\beta}),\qquad(110)$$

with the electromagnetic energy tensor $T_{\mu\nu}$. Equation (37) multiplied by $g^{\mu\nu}$, would then hold, because $g^{\mu\nu}T_{\mu\nu}=0$. In addition, Einstein assumed the four equations

$$\frac{\partial \Phi_{\mu\nu}}{\partial x^{\varrho}} + \frac{\partial \Phi_{\nu\varrho}}{\partial x^{\mu}} + \frac{\partial \Phi_{\varrho\mu}}{\partial x^{\sigma}} = 0,\qquad(111)$$

getting 12 independent equations for 16 variables $g_{\mu\nu}$ and $\Phi_{\mu\nu}$. He then rewrote Equations (110) as

$$(R_{\mu\nu} - \tfrac{1}{4}g_{\mu\nu}R) = -\gamma\left[T_{\mu\nu} - \frac{1}{4\gamma}g_{\mu\nu}(R - R_0)\right].\qquad(112)$$

Since R is related to λ, the cosmological constant appeared only as a constant of integration in the new theory. He concluded by saying that:

The above-mentioned considerations indicate the possibility of a theoretical explanation on the basis of the gravitational and electromagnetic fields alone, without the introduction of additional hypothetical terms as in Mie's theory. This possibility appears to be particularly promising, since it makes it unnecessary to introduce the special constant λ, which would otherwise be needed to solve the cosmological problem. On the other hand, there remains a peculiar difficulty. Namely, if we consider Equation [(110)] in the spherically symmetric, static case, then we obtain one equation too few for the determination of $g_{\mu\nu}$ and $F_{\mu\nu}$, so that *every spherically symmetric distribution of electricity* can exist in equilibrium. Thus the problem of the structure of elementary quanta cannot be solved only on the basis of the equations given above.[264]

Again and again, Einstein returned to the question whether one could obtain the quantum of action from an *over determination* of the field variables.[265] This was also the motivation behind his later moves to abolish the quanta by means of the *unified field theory*, on which he had started to work seriously beginning in 1928, after his efforts to disprove the consistency of quantum mechanics had failed. On 5 January 1929, he wrote to his friend Michele Besso:

But the most wonderful thing is that, what I have brooded on and calculated all these days and nights, is now completed and in front of me, condensed to a mere seven pages under the title 'unified field theory'. This might appear old-fashioned, and my dear colleagues as well as you, my friend, will right away stick their tongues out as far as possible. Because, Planck's constant h does not appear in these equations. However, as soon as the limitations of the [current] statistical craze [i.e. quantum mechanics] are clearly seen, one will retreat repentantly to a space-time approach and then these equations [of mine] will become the starting point [of the new theory]. I have discovered a geometry which is characterized, not only by a Riemannian metric, but also by distant-parallelism. The latter was thus far generally felt to be a characteristic of Euclidean [geometry], and the simplest field equations of such a manifold lead to the well-known laws of electricity and gravitation. Even the equations $R_{ik} = 0$, in spite of their success, now belong to the junk pile. I shall not forget to send you the reprints. If [on reading] you don't stick out your tongue, then you must be a hypocrite, because I know you.[266]

Michele Besso did not, we can be sure, 'stick his tongue out' at what Einstein wrote to him. He was Einstein's old friend and admirer, and his sounding-board for scientific ideas ever since the early days of special relativity in Berne, Switzerland. Einstein continued to develop various aspects and formulations of his unified field theory to the end of his life, without ever really coming close to his great goal. In these efforts he was often alone, and the scientific problems on which he worked were no longer central to the interests of a more pragmatic age. He continued to publish his reflections in respected scientific journals, and at times they were reported in newspapers. No one ever stuck his tongue out at what Einstein had to say in scientific matters, although his hopes and attempts to create a unified theory of physics caused a few raised eyebrows and aroused some melancholy about the lost and better-to-be-forgotten visions of heroic times.

Dirac has called Einstein's theory of gravitation 'probably the greatest scientific discovery that was ever made'.[267] Discussing the difficulty of understanding the origin and meaning of this discovery, Chandrasekhar has remarked that, 'It appears that only by a mixture of physical reasonableness, mathematical simplicity, and aesthetic sensibility can one arrive at Einstein's field equations. The general theory of relativity is in fact an example of "the power of speculative thought".'[268] Hilbert had played a decisive role in determining the form of the equations of the general relativity theory of gravitation and the strict conditions to which they are subject. In the Einstein–Hilbert equations of 1915, there shall forever remain an enduring essence, 'a permanent core', which gave such satisfaction to Hilbert and brought fulfilment to Einstein.

The unification of gravitation and electricity into a single field theory of all physical phenomena, however, did not succeed. As Pauli said of these attempts, 'No man shall join what God hath put asunder.'[269]

REFERENCES AND NOTES

1. From *Einstein–Sommerfeld Briefwechsel*, edited by A. Hermann, Schwabe and Co., Basel/ Stuttgart 1968, p. 26. Writing to Hilbert on 1 November 1912 Sommerfeld said, 'My writing to Einstein was in vain.... Einstein is obviously so deeply immersed in [the theory of] gravitation, that he is deaf against everything else.' (p. 27).
2. A. Einstein, 'Über das Relativitätsprinzip und die aus demselben gezogenen Folgerungen', *Jahrbuch der Radioaktivität und Elektronik* **4**, 411–461 (1907); **5** (1908). Actually, Einstein never considered the principle of relativity as confined to inertial frames alone, certainly not

since he had derived the dependence of the inertial mass of a system on its motion or its energy in his paper 'Ist die Trägheit eines Körpers von seinem Energieinhalt abhängig?', in *Ann. Phys. (Leipzig)* **18**, 639–641 (1905). In May 1906, he submitted another paper on this subject, 'Das Prinzip von der Erhaltung der Schwerpunktsbewegung und die Trägheit der Energie', *Ann. Phys. (Leipzig)* **20**, 627–633 (1906), in which he proved the theorem that 'either one must give up [Newton's] first law of mechanics, according to which a body originally at rest, and not subject to external forces, cannot have a translational motion, or assume that the inertia of a body, according to the laws proposed [by us], depends on its energy content.' (*l.c.*, p. 633).

Einstein analyzed his equivalence postulate of energy and inertial mass in a paper entitled 'Über die von Relativitätsprinzip geforderte Trägheit der Energie', *Ann. Phys. (Leipzig)* **23**, 371–384 (1907). To whichever (simple) system he applied it, the results were consistent with the principles of mechanics such as energy conservation, etc.

This was the state of affairs when he wrote, at the request of J. Stark, the article in the *Jahrbuch*. In this article, Einstein dealt with the optical prerequisites, the principle of the constancy of the velocity of light and its kinematic consequences. In part II of the article, he dealt with the transformations in electrodynamics. The mechanics of the electron and the possible tests of the experimental consequences, in particular Kaufmann's results (*Ann. Phys.* **19** (1906)) were discussed in part III. In the following section (IV), devoted to the mechanics and thermodynamics of systems, he stated the important result, that 'the law of the conservation of mass for an individual system, according to our result, is valid only if its energy remains constant; it is then equivalent to the law of conservation of energy. However, the changes to which the mass of a physical system is subject, in ordinary physical phenomena, remains undetectably small. For instance, the decrease in the mass of a system, which loses 1000 calories (c.g.s.) of its energy, amounts to 4.6×10^{-4} gm. In the radioactive decay of a substance tremendous amounts of energy are set free; even so the change of mass involved in such a process is not large enough to be observed.' (*Jahrbuch der Radioaktivität und Elektronik* **4** (1907), p. 442). As Planck had noted, in the case of radium the loss in weight was beyond the reach of experimental accuracy which could be obtained in 1906. Finally, in section V, Einstein discussed the relation of the relativity principle to gravitation.

3. A. Einstein, *Jahrbuch der Radioaktivität und Elektronik* **4** (1907), pp. 443–444.
4. A. Einstein, *Jahrbuch* **4** (1907), Ref. (2), p. 454.
5. A. Einstein, *Jahrbuch* **4** (1907), Ref. (2), p. 454.
6. A. Einstein, *Jahrbuch* **4** (1907), Ref. (2), pp. 458–459.
7. A. Einstein, *Jahrbuch* **4** (1907), Ref. (2), p. 459.
8. A. Einstein, *Jahrbuch* **4** (1907), Ref. (2), p. 461.
9. Another development had started with the electromagnetic considerations of Henri Poincaré as early as 1905. In his remarkable paper 'Sur la dynamique de l'électron' which was received by the editor of *Rendiconti del Circolo Mathematico di Palermo* on 23 July 1905, and was published in Vol. **21**, pp. 129–176 (1906), Poincaré not only studied, almost at the same time as Einstein [whose article 'Zur Elektrodynamik bewegter Körper' arrived at the *Annalen der Physik* in Leipzig on 30 June 1905] the consequences of the Lorentz contraction, but he also referred to the problem of gravitation. Poincaré had first used the phrase 'principle of relativity' in his lecture on 'The present and future of mathematical physics' at the International Congress of Arts and Sciences in St Louis in 1904 (see H. Poincaré, *Oeuvres*, Vol. **9**, p. 248, Gauthier-Villars, Paris, 1954); and it occurred again on p. 129 of the communication under discussion. In the introduction to this paper, he wrote, 'If we accept the *postulate of relativity*, we shall find that among the laws of gravitation and the laws of electromagnetics there exists a common number. It is the velocity of light. We shall find that it occurs in all forces, of whatever origin, and it can only be explained in two ways: (1) Either there exists nothing in the universe that is not of electromagnetic origin; (2) or, this quantity, which is common to all physical phenomena, appears only because it relates to our methods of measurements.' (H. Poincaré, *Rendiconti*, 1906, pp. 131–132).

Poincaré considered Lorentz contraction, and tried to derive the Lorentz covariant action principles. Let us also mention here that it was Poincaré who first gave the name *Lorentz group*. In Section 9 of his paper (pp. 166–175), he finally dealt with the 'hypotheses about gravitation'. He stated: 'Lorentz was thus obliged to complete his hypothesis by supposing that *forces of all possible origins, in particular gravitation, are subject to a translation* (or, if one prefers, subject to the Lorentz transformation) *in the same way as electromagnetic forces.*' (*l.c.*, p. 166) Poincaré then studied the consequences of a Lorentz invariant theory of gravitation, finding a correction to Newton's law of gravitation, which is of the order of v^2/c^2.

The next paper in this connection, R. Hargreaves' 'Integral forms and their connection with physical equations' (*Trans. Camb. Phil. Soc.* **21**, 107–122 (1908)), did not refer to gravitation specifically, but to the 'four-dimensional formulation' of the laws of physics; 'We are concerned here with the variation of integral forms, or more specially with their invariance, under the action of an operator which is an extension of the hydrodynamical operator in Euler's equation. In the integral forms temporal terms are admitted, i.e. terms containing dt as well as the differentials of coordinates, and it appears that the forms have special properties when the temporal terms are associated with the non-temporal in a definite way. These associated terms are significant quantities which include vector and scalar products as particular cases.' [Ref. 113, *l.c.*, p. 107] The importance of this paper partly lay in the fact that Hargreaves' result was connected with the studies of H. Bateman. On 11 March 1909, Bateman read a paper on 'The Transformation of the Electrodynamical Equations' before the London Mathematical Society (*Proc. Lond. Math. Soc.* (2) **8**, 223–264 (1910)). He said: 'The object of the present paper is to find all the transformations for which the electrodynamical equation are invariant. In the case of the simpler equations of the theory of electrons, it is proved that the transformations belong to a certain group which is isomorphic with the group of conformal transformations of a space of four dimensions. It is assumed, however, that the transformation is such that the total charge on a system of particles is unaltered.' (*l.c.*, p. 224). What Bateman required was that the *equation*

$$\mathrm{d}s^2 = 0$$

should be invariant, rather than just the line element. From the study of the propagation of luminous disturbances, described by the above equations, he concluded that a more general *Ansatz* for ds^2 had to be chosen, than Minkowski's assumption that the coefficients $g_{\mu\nu}$ depend on the space-time components, since light does not travel with uniform velocity in media with varying dispersion. Although Bateman's motivation was entirely different from Einstein's, and moreover he dealt with the electromagnetic problem alone, he was led to a conclusion similar to Einstein's. Einstein obviously did not know about Bateman's work of 1912. (Also see Bateman's later paper 'The Electromagnetic Vectors', *Phys. Rev.* **12**, 459–481 (1918).)

10. There exist two papers dealing with the electrodynamics of moving bodies, which appeared almost simultaneously with Minkowski's work, Ref. (110), which Einstein wrote with his first student J. Laub: 'Über die elektromagnetischen Grundgleichungen für bewegte Körper', *Ann. Phys. (Leipzig)* **26**, 532–540 (1908), with errata in *Ann. Phys. (Leipzig)* **27**, 232 (1908) and **28**, 445–447 (1909); 'Über die im elektromagnetischen Felde auf ruhende Körper ausgeübten ponderomotischen Kräfte', *Ann. Phys. (Leipzig)* **26**, 541–550 (1908). Einstein wrote another note on this topic, 'Bemerkung zu der Arbeit von D. Mirimanoff "Über die Grundgleichungen…"', *Ann. Phys. (Leipzig)* **28**, 885–888 (1909).

11. A. Einstein, 'Die Relativitätstheorie', *Vierteljahresschr. Naturforsch. Ges. Zürich*, **56**, 1–14 (1911).

12. A. Einstein, 'Über den Einfluss der Schwerkraft auf die Ausbreitung des Lichtes', *Ann. Phys.* **35**, 898–908 (1911). He said, 'In a paper published already three years ago, I tried to answer the question whether the propagation of light is affected by gravity. [See Ref. 2] Here I shall return to this question again for two reasons: first, because I am no longer satisfied with my [earlier] treatment; second, and more important, because in retrospect I realize that one of the most significant consequences of this question can be tested experimentally. One finds that indeed, according to the theory which I shall present, light rays which pass close to the sun should be bent by its gravitational field giving rise to an apparent increase, of almost one second of arc, in the angular distance of a fixed star which appears to be in the vicinity of the sun.'

13. A. Einstein, Ref. (12), p. 906.

14. A. Einstein, Ref. (12), p. 908.

15. Max Abraham was a very gifted physicist, noted for his work in the field of electrodynamics, in particular electron theory. He had been a brilliant student of Max Planck, under whom he took his doctorate. In 1902 he published his first memoir on 'Dynamik des Elektrons', *Nachr. Ges. Wiss. Gött.* pp. 20–41 (1902), in which he developed the so-called 'rigid' electron model, by which he hoped to account for the stability of the electron. Abraham obtained a velocity dependence for the mass of the electron, different from the result of Lorentz and Einstein, but

consistent with the early experiments of Kaufmann. Another feature of Abraham's theory was that he stuck to an absolute coordinate system. (See e.g. *Phys. Z.* **5**, 576–579 (1904).) Only in the latter would the electromagnetic waves propagate with the velocity c in all directions.

16. M. Abraham, 'Zur Theorie der Gravitation', *Phys. Z.* **13**, 1–4, 176 (1912); 'Das Elementarge-setz der Gravitation', *Phys. Z.* **13**, 4–5, (1912); 'Der freie Fall', *Phys. Z.* **13**, 310–311 (1912); 'Die Erhaltung der Energie und der Materie im Schwerkraft Felde', *Phys. Z.* **13**, 311–314 (1912); 'Das Gravitationsfeld', *Phys. Z.* **13**, 793–797 (1912).

In the first paper of this series, he remarked, 'In a paper published recently, A. Einstein [Ref. 12] put forth the hypothesis that the velocity of light (c) depends on the gravitational potential (ϕ). In this note, I shall develop a theory of gravity, which satisfies the principle of relativity. Furthermore, I shall derive a relationship between c and ϕ, which in its first approximation, is equivalent to Einstein's. This [my] theory ascribes to the densities of energy and energy current values in the gravitational field that are different from those accepted at present.' (*Phys. Z.* **13**, 1.)

17. They actually do so on bilinear expressions in the gradients. (See M. Abraham, *Phys. Z.* **13**, 3.)

18. M. Abraham, Ref. (16), p. 3.

19. M. Abraham, Ref. (16), p. 4.

20. M. Abraham, *Phys. Z.* **13**, 793–794. 'I have just now resorted to the language of relativity. Still, we shall see that [the theory of relativity] is not in accord with the views concerning gravity as presented here, if only because the axiom of the constancy of light velocity is renounced here. In my previous papers on gravitation, I have attempted to preserve, at least in infinitesimal [domains], the invariance with respect to the Lorentz transformations. However, I am now convinced that my equations of motion for a material point can by no means be brought into accord with the principles of analytical mechanics. On the other hand, Einstein's equivalence hypothesis has been shown to be equally untenable. In order to develop the new theory of gravitation, I shall, therefore, prefer not to deal with the space-time problem here at all.'

21. M. Abraham, *Phys. Z.* **13**, p. 796.

22. M. Abraham, *Phys. Z.* **13**, p. 797.

23. Abraham derived from it an upper limit for the mass of a star which, however, turned out to be rather large ($m < 10^8 \, m_0$). (*Phys. Z.* **13**, 311.)

24. Abraham concluded his last paper (*Phys. Z.* **13**, 797 (1912)) with the remark: 'The theory of gravitation developed here is completely independent of the unresolved problems of the dynamics of the electron. Apart from the basic assumption that the gravitational field is completely determined by [specifying] the velocity of light [everywhere], our theory is based upon (i) the validity of the expressions for the gravitational tensor [(1a)–(1d) Ref. 16], (ii) the principles of mechanics, i.e. conservation of energy and momentum and the Lagrange equations, and finally, (iii) the hypothesis of the proportionality of gravity and energy.'

25. A. Einstein, *Ann. Phys. (Leipzig)* **38**, 355–369 (1912).

26. A. Einstein, Ref. (25), p. 355.

27. A. Einstein, Ref. (25), p. 360.

28. A. Einstein, 'Zur Theorie des statischen Gravitationfeldes', *Ann. Phys. (Leipzig)* **38**, 443–458 (1912), in particular p. 457.

29. A. Einstein, Ref. 25, p. 362.

30. A. Einstein, Ref. 25, pp. 368–369.

31. Abraham was very critical and argumentative, but he did strive for clarity. In their obituary of Max Abraham, Born and Von Laue wrote: 'Clarity was the essence of his nature, both in matters of mind and heart. Clarity was evident in all his writings and we believe that it meant more to him than the discovery of new phenomena. Whenever he found it wanting, he would always make a sharp, even exaggerated criticism.' [*Phys. Z.* **24**, 49–53 (1953).] This habit of sharp criticism got Abraham into many difficulties, and this former brilliant disciple of Max Planck did not gain a stable position for a long time. After eight years as Privatdozent in Göttingen (1900–1908), he went to the University of Illinois in 1908 as a professor of physics, and left after six months. In 1909 he accepted a professorship at the University of Milan, but during the war he was inducted into service and worked on telegraphy for the Telefunken Company. Finally, when he was appointed to the chair of theoretical mechanics at Aachen, Abraham became severely ill and, after several months of great suffering, died on 16 September 1922.

32. M. Abraham, *Ann. Phys. (Leipzig)* **38**, 1056–1058 (1912).

33. A. Einstein, *Ann. Phys. (Leipzig)* **38**, 1059–1064 (1912). Abraham wrote: 'Einstein's theory of relativity has exercised a fascinating influence, especially on the younger mathematical physicists, which threatens to retard a healthy development of theoretical physics. It was evident to the most sober critics that this theory could never lead to a complete worldview, if it cannot even integrate within its system the most important and omnipresent (natural) force of gravity. The failure of the efforts directed toward this has led to a crisis of relativity theory. Already a year ago, by assuming the dependence of the velocity of light on the gravitational potential, Einstein abandoned the essential postulate of his earlier theory, namely the constancy of the velocity of light. [See our Ref. 12]. Now, in a paper published recently [our Refs. 25, 28], he has also renounced the requirement of the invariance of the equations of motion under Lorentz transformations, thereby dealing the theory of relativity a *coup de grâce*. Anyone who, like this author, had warned against [listening to] the singing of the sirens of this theory, would greet with satisfaction [the fact] that the originator of this theory is himself convinced of its untenability.' [Ref. 32, p. 1056.]

34. Abraham continued: 'Indeed, any relativistic space-time formulation which is expressed through the relationships between the space-time parameters of [two coordinate systems] Σ and Σ' would be untenable. I shall certainly not consider such a space-time formulation. As I have pointed out elsewhere, it seems to me that an absolutist [rather than a relativistic] theory is needed. If one system is raised above all others, in which the gravitational field is static or quasi-static, then it is permissible to call a motion with respect to this system as being "absolute". In the old action-at-a-distance mechanics, it was possible to assume that the very remote masses of the fixed stars, or an imaginary [Neumann] "body a", would determine an inertial frame of reference. Such an idea is foreign to the new mechanics based on the concept of local interaction. However, it would be consistent with the new mechanics if the omnipresent field of gravity itself provides the absolute frame of reference for the motion of bodies. (If one so chooses, one may consider this as an argument for the "existence of the ether".)

'Einstein's starting point, in his recent investigation, is a heuristic hypothesis which he calls the equivalence principle, and which requires a space-time correspondence of two frames of reference in acceleration with respect to each other. But even in this new form, it does not seem to be possible to carry through a consistent relativistic space-time description. Thus even this latest theory of Einstein's rests on shaky foundations.' [Ref. 32, pp. 1057–1058.]

35. M. Abraham, 'Nochmals Relativität und Gravitation, Bemerkungen zu A. Einsteins Erwiderung', *Ann. Phys. (Leipzig)* **39**, 444–448 (1912). He repeated his conviction that the theory of relativity belonged to the past: 'The components of yesterday's [sic] relativity theory, which depend on the rigorous and general validity of its entire concept, are Einstein's kinematics and the space-time definition tied with it. It was just this relativistic concept of space and time which had endowed the relativity theory of the past [sic] with a certain philosophical lustre which is now fading. To be sure, in his reply [to all this], Mr. Einstein conjures up in the distance the *fata morgana* of a new all-embracing space-time definition of gravitation. He is already claiming credit [today] for a [non-existent] theory of relativity of the future, and is inviting the assistance of the experts for its discovery. Until this payment is received [i.e. this discovery is made], this claim [of Einstein's] will remain, in the account of relativity, on the debit side of the ledger.'

In a letter to L. Hopf, written in the autumn of 1913, Einstein remarked: 'It is proceeding splendidly with gravitation [his work with Grossman]. Unless I am entirely mistaken [and he was!], I have obtained the general equations. As you may have read recently in the *Physikalischen Zeitschrift*, Abraham has slaughtered me and relativity theory altogether in two fierce assaults, and has at the same time presented the only correct theory of gravitation (a 'nostrification' of my results) – a magnificent steed, which only lacks three legs. He also maintains that it was Robert Mayer who first recognized the mass of the energy.' (C. Seelig, *Albert Einstein*, Europa Verlag, Zürich (1954), p. 171.)

36. R. v. Eötvös, *Math. u. Natw. Ber. Ungarn*, **8**, 65 (1890); improved later on by R. Eötvös, D. Pekár, and E. Fekete, *Trans. XVI. Allgem. Konf. der int. Erdvermessung* (1909), and D. Pekár, *Naturwiss.* **7**, 327 (1919).

37. See his reply, in A. Einstein, *Ann. Phys. (Leipzig)* **39**, 704 (1912).

38. G. Nordström, 'Reiativitätsprinzip und Gravitation', *Phys. Z.* **13**, 1126–1129 (1912); 'Träge

und schwere Masse in der Relativitätsmechanik', *Ann. Phys. (Leipzig)* **40**, 857–878 (1912);
'Zur Theorie der Gravitation vom Standpunkt der Relativitätsprinzips', *Ann. Phys. (Leipzig)*
42, 533–554 (1913); 'Die Fallgesetze und Planetenbewegungen in der Relativitätstheorie',
Ann. Phys. (Leipzig) **43**, 1101–1110 (1914).

39. G. Nordström, *Phys. Z.* **13**, 1126. 'Einstein's hypothesis of the dependence of the velocity of
light *c* on the gravitational potential leads, as can be seen from the discussions between Einstein and Abraham, to considerable difficulties in connection with the principle of relativity.
Thus one wonders whether it might be possible to repalce Einstein's hypothesis by another,
which would leave *c* constant and yet modify the theory of gravitation in such a way that,
in accordance with the principle of relativity, gravitational and inertial masses would be
identical.'
In an addendum to the first paper, Nordström noted: 'From a letter from Professor Einstein
I have learned that he had already previously considered the possibility [proposed by me, as
quoted above] of treating gravitational phenomena in a simple way. However, he came to the
conclusion that the consequences of such a theory would not correspond to reality. With the
help of a simple example, he has shown that according to this theory a rotating system in a
gravitational field will be accelerated less than a non-rotating one. In fact, this conclusion is
really not questionable, because the difference [between the accelerations] is too insignificant
to contradict experience. Still, this conclusion shows that my theory is not consistent with
Einstein's equivalence hypothesis. According to the latter, an unaccelerated inertial system in
a homogeneous gravitational field is equivalent to an accelerated inertial system in gravity-free
space. However, in my view, this circumstance alone is not sufficient to abandon [my] theory.
Although Einstein's hypothesis is most ingenious, it does lead to great difficulties. It is therefore desirable to pursue other approaches as well to [the problem of] gravitation, and I would
like to contribute to these in my note.' (Ref. (39), p. 1129.)

40. In his third paper, Nordström placed certain conditions on the variability of the gravitational
factor *g*. He said, 'Every indefiniteness in the theory, noted above, can be removed by a very
plausible assumption, for which I am indebted to Laue and Einstein. Laue has shown that
Einstein's equivalence principle (although, not in its fully generality) can be maintained by
suitably defining the rest mass density of matter.... In addition, we shall see that Einstein's
equivalence principle requires a quite definite dependence of the gravitational factor *g* on the
gravitational potential *ϕ*....' [G. Nordström, *Ann. Phys.* **42**, 533–534.]

41. Nordström had gone to Zurich to meet Einstein (and Laue), and he submitted his third paper
(*Ann. Phys.* **42**, 533 (1913)) from there on 24 July 1913. At that time, therefore, he was in close
contact with Einstein.

42. A. Einstein, *Phys. Z.* **14**, 1249–1266 (1913), in particular p. 1250.

43. The postulate (ii), which states the equivalence of masses, is also satisfied if the gravitational
potential does not change significantly in the system considered.

44. A. Einstein and A. D. Fokker, 'Die Nordströmische Gravitationstheorie von Standpunkte des
absoluten Differentialkalkuls', *Ann. Phys. (Leipzig)* **44**, 321–328 (1914). In this paper, they
stated: 'The gravitational field is determined by the ten quantities $g_{\mu\nu}$. In the Einstein-Grossmann theory, ten formally similar equations are specified for these ten quantities. Nordström's
theory, on the other hand, is based on the assumption that it is possible to satisfy the principle
of the constancy of the velocity of light by suitably choosing the frame of reference. We shall
see [in this note] that this [Nordström's assumption] is equivalent to reducing the ten quantities
$g_{\mu\nu}$ to a single quantity ϕ^2 by an appropriate choice of the frame of reference.' (p. 324.)

45. A. Einstein, *Phys. Z.* **14**, 1254 (1913).

46. Nordström's theory has been revived during the past few years by the work of Brans and Dicke
in an attempt to give a certain scalar admixture to Einstein's tensor theory, which seems to be
consistent with recent experiments on the bending of light from the sun. C. A. Brans and
R. H. Dicke, *Phys. Rev.* **124**, 925 (1961).

47. More important, for the current situation, is the discussion following Einstein's lecture in
Vienna, in which Mie pointed out the connection between Abraham's theory and that of Nordström (*Phys. Z.* **14**, (1913) 1262). 'I would like to supplement Einstein's interesting comments
with some remarks on the historical development of the theory, since Einstein has referred to
them only in passing. Nordström's theory is tied together with the investigations of Abraham.
I consider it essential that it should be mentioned here that Abraham was the first who formu-

lated rather reasonable equations of gravitation. While previously one always tried (there are indeed many older theories of gravitation) to represent the gravitational field analogous to the electromagnetic field, Abraham discovered a new possibility. It is impossible to bring the older attempts into accord with the relativity principle; for if the principle of the equivalence of inertial and gravitational masses is sufficiently accurately satisfied, then the gravitational field cannot be represented by a six-vector. It is for this reason that Abraham first of all formulated a theory with a scalar gravitational potential. However, he made the mistake that he set ϱ [which is the density ϱ_0 in Equation (9)] identical with the density of inertial mass.... Naturally, the relativity principle cannot be satisfied in this way.

'Nordström then improved this theory by substituting for ϱ a quantity which is invariant with respect to the Lorentz transformation. At about the same time [as Nordström], I also formulated a theory of gravitation. However, my theory is embedded in an extensive work on the theory of matter, and that's why my investigations have escaped [the notice of] Mr Einstein.'

At the last remark, Einstein protested, 'No, no.' and replied, 'I have not spoken about Mr Mie's theory because the equivalence of the inertial and gravitational masses has not been worked through rigorously in it. It would, of course, have been illogical if, having proceeded on the basis of certain postulates, I had not held on to them. I agree that I have not studied Mie's theory as carefully as I should have, but it was the farthest from my thoughts to disparage it if I have not referred to it in this context.' (*Phys. Z.* **14**, 1263.)

Einstein then dealt with the connection between the ideas of Abraham and Nordström. 'As far as Nordström's theory is concerned, I cannot really say that it was Abraham who first proposed the way which has been suggested by Nordström. For Abraham's theory is based upon the fact that the velocity of light is variable and that, to a certain extent, it represents a measure of the gravitational potential. Nevertheless, he employs the form of the usual theory of relativity, which leads him to a contradictory, hybrid position. It is indeed so objectionable, that to me his theory seems to be quite untenable.'

Mie admitted these deficiencies [of Abraham's theory], but insisted that Nordström had started from Abraham's equations. Upon which Einstein remarked that, 'Yes, psychologically it is indeed so, but not logically, for Nordström's theory is fundamentally different from Abraham's.' (*l.c.*, p. 1263).

48.   A. Einstein, Ref. (44), p. 328.
49.   A. Einstein, *Ann. Phys. (Leipzig)* **49**, 769 (1916). Einstein's statements about the value of this collaboration differ at different places. Sommerfeld, for instance, had suggested to Einstein not to emphasize Grossmann's role too much, because 'he will claim half of this share [of credit] in the creation of general relativity'. To which Einstein replied, 'Grossmann will never raise a claim to be considered as the co-discover [of general relativity]. He only helped me in getting oriented with the mathematical literature, but he did not contribute [sic] anything material to the [ideas and] results.' (Letter from Einstein to Sommerfeld, dated 17 July 1915, quoted from *Einstein–Sommerfeld Briefwechsel*, (ed. by A. Hermann), Schwabe and Co., Basel and Stuttgart (1968), p. 30) [See our Ref. 1].
50.   In 1933, Einstein remarked about this collaboration with Grossmann: 'I worked on these problems from 1912 to 1914 with my friend Grossmann. We found that the mathematical methods for solving the first question [translation of the laws into general relativity] was already waiting for us in the absolute differential calculus of Ricci and Levi-Civita. As to the second problem [law of gravitation], we soon recognized that the methods for doing this had long ago been worked out by Riemann (curvature tensor). Already two years before the final publication of the general theory of relativity, we had considered the correct field equations of gravitation, but we failed to recognize that they were physically applicable.' (A. Einstein, 'The Origins of General Theory of Relativity', G. A. Gibson Foundation Lecture, University of Glasgow, 20 June 1933, Glasgow (1933), pp. 10–11.)
51.   A. Einstein and M. Grossmann, 'Entwurf einer verallgemeinerten Relativitätstheorie', *Z. Math. u. Phys.* **62**, 225–261 (1913), ('I. Physikalischer Tiel': A. Einstein; 'II. Mathematischer Tiel': M. Grossmann); A. Einstein, 'Physikalische Grundlagen einer Gravitationstheorie', M. Grossmann, 'Mathematische Begriffsbildungen zur Gravitationstheorie', *Vierteljahresschr. Naturforsch. Ges. Zürich* **58**, 284–290, 291–297 (1913); A. Einstein and M. Grossmann, 'Kovarianzeigenschaften der Feldgleichungen der auf die verallgemeinerte Relativitätstheorie gegründeten Gravitationstheorie', *Z. Math. u. Phys.* **63**, 215–225 (1914). See also our Ref. 47.

52. A. Einstein, 'Die formale Grundlage der allgemeinen Relativitätstheorie', *Sitz. Ber. Kgl. Preuss. Akad. Wiss., Math.-Phys. Kl.* (1914), 1030–1085, submitted on 29 October 1914, and read on 19 November 1914.

53. In the introduction Einstein wrote: 'During the past few years, I have perfected a generalization of the theory of relativity, in cooperation with my friend Grossmann in [certain] parts. A colourful mixture of physical and mathematical requirements has been employed as a heuristic device in these investigations. For this reason, it is not easy to have an overview of and to characterize this theory from a formal mathematical point of view. In this [new] work, I have attempted to fill this gap. In particular, we succeeded in obtaining the equations of the gravitational field in a strictly covariant-theoretic way. In order to assist the reader in obtaining a complete understanding of the theory without recourse to the study of other purely mathematical treatises, I have tried to give simple derivations of the basic principles of the absolute differential calculus – parts of which, I believe, are new. (Ref. (52), p. 1030.)

54. Einstein's lectures at the Naturforschende Gesellschaft, Zurich, were given on the occasion of the annual meeting of the Schweizerische naturforschende Gesellschaft, at Frauenfeld on 9 September 1913. The lecture at the Congress of Natural Scientists in Vienna was given in November 1913. The final version of the 'Entwurf' paper was written after his lectures at both of these conferences.

55. See the Frauenfeld lecture, 1913, the second paper in Ref. 51, p. 286.

56. In his lecture at Frauenfeld [the second paper in Ref. 51, p. 292]. Grossmann explained this point of departure. 'The fundamental mathematical concept of Einstein's theory of gravitation, which is to characterize a gravitational field by means of a quadratic differential form with variable coefficients, necessitates a generalization of the definitions and methods of vector analysis. Of fundamental significance for this purpose is Christoffel's famous treatise, *Über die Transformation der homogenen Differentialausdrücke zweiten Grades* [*J. fur Math.* 70 (1869)], as well as the Ricci and Levi-Civita's *Méthodes de Calcul différentiel absolu et leurs applications* [*Math. Ann.* 54 (1901)], which developed it further. In the latter treatise, the authors developed methods which allowed one to write the differential equations of mathematical physics in an invariant [i.e. independent of coordinate systems] form. The new developments in vector analysis shed even more light on the advantages of [employing] such a general invariant-theoretic treatment. Such a treatment at once provides a complete system of vector-analytical definitions which have been introduced [gradually] by Minkowski, Sommerfeld, Laue, and others, for the description of the four-dimensional world of the relativity theory.'

57. M. Grossmann, first paper in Ref. 51, Mathematischer Teil, p. 257.

58. A. Einstein, first paper in Ref. 51, p. 260.

59. A. Einstein, *S. B. Preuss. Akad. Wiss.* (1914), see pp. 1054–1057.

60. A. Einstein, *Phys. Z.* **14**, 1258 (1913).

61. *Phys. Z.* **14**, 1262–1266 (1913); see also *Phys. Z.* **15**, 108–110 (1914) and *Phys. Z.* **15**, 115, 169–176, 176–180 (1914). After Einstein's lecture, Mie remarked: 'It seems to be that in his work, Einstein has postulated the most intriguing principle of a general relativity. However, this principle has not been rigorously complied with in the theory which he has presented...' (*Phys. Z.* **14**, 1264). A little later, he went on to say that, 'The generalization of the relativity principle, in the form in which it has been expressed in Einstein's paper, applies only to linear transformations, and thus has absolutely nothing to do with accelerated motions.' (*Phys. Z.* **15**, 176 (1914).)

62. As Einstein remarked correctly, there is no justification for this assumption. (See e.g., *Z. Math. u. Phys.* **62**, 233 (1913).)

63. A. Einstein, the first paper in Ref. (51), *Z. Math. u. Phys.* **62** (1913), pp. 233–234.

64. First paper in Ref. (51).

65. A. Einstein and M. Grossmann, *Z. Math. u. Phys.* **63**, 221, (1914).

66. A. Einstein and M. Grossmann, *Z. Math. u. Phys.* **63** (1914), see p. 225.

67. We refer to various remarks in this paper concerning the meaning of the tensors appearing in the derivation and the formulation of the equations. [Ref. 52, 'Die formale Grundlage der allgemeinen Relativitätstheorie', *Sitz. Ber. Kgl. Preuss. Akad.*, Math. Phys. Kl. (1914), 1030–1085.]

  In an article entitled 'Die Neue Mechanik', M. Abraham discussed the Einstein–Grossmann theory. [*Rivista di Scienze* **15**, 8–27 (1914)]: 'With the help of the method of the "absolute differential calculus" developed by Ricci and Levi-Civita, the authors of the "Entwurf" have also succeeded in writing the electromagnetic and dynamical equations in a form which, at least in the infinitesimal case [i.e. differential form] satisfies the relativity requirement [of the general transformation]. However, it is the gravitational field, the integration of which into relativity theory seems to be the goal of the "Entwurf", does not yield to the proposed scheme. The differential equations of the gravitational field, proposed by the authors, are not invariant with respect to the general space-time transformations, i.e. a system of mutually gravitating masses in non-uniform or rotational motion is, in general, not equivalent to a system at rest.

'It is true that the dynamics becomes relativistic if one includes the forces of rotation along with gravitation. However, the gravitational field of a rotating system is then no longer identical with that of the non-rotating system. It is really of no avail to seek to develop a theory of gravitation which would satisfy the requirement of general relativity; even if such a mathematical theory exists, it would be physically meaningless. Every theory of relativity, be it the special theory of 1905 or the general theory of 1913, founders at the rock of gravity. The relativistic ideas are obviously not broad enough to provide a framework for a comprehensive worldview.' (M. Abraham, *l.c.*, p. 25.)

In writing to his friend Besso, Einstein showed some respect for Abraham's remarks. 'The community of physicists has shown little interest in [my] work on gravitation. Abraham probably understands it best. In the "Scienza", he has been screaming loudly against all relativity, but with a good grasp. I shall visit Lorentz in Spring, and expect to discuss all this with him. He takes great interest in these matters, as does Langevin. Laue does not quite seem to accept [my] principal ideas, nor does Planck, only perhaps Sommerfeld.' [A. Einstein to M. Besso, from Zurich, towards the end of 1913.]

Abraham wrote a sober review of the new gravitational theories in *Jahrbuch der Radioaktivität* **11**, 470 (1914), showing a deep understanding of the physical principles and mathematical structure.

[Since the completion of this paper in early September 1972, the correspondence between Einstein and Besso has been published: *Albert Einstein, Michele Besso, Correspondance, 1903–1955* (edited by Pierre Speziali), Hermann, Paris (1972). All the quotations from the letters of Einstein and Besso may be identified in this book from the dates which I have given in my references. The English translations of the quotations, used in this paper, are mine.]

68. With considerable confidence, Einstein wrote in this paper: [Ref. (52)]: 'The laws governing these differential forms have been obtained by Christoffel, Ricci and Levi-Civita. I would like to present here an especially simple derivation of these, which [also] appears to me to be new.

69. The invitation to the Prussian Academy was initiated by the leading Berlin physicists Max Planck, Walther Nernst, Heinrich Rubens, and Emil Warburg. In particular, they stressed his research on relativity in a petition to the Prussian Minister of Education: 'His name is primarily known through his famous treatise on "The Electrodynamics of Moving Bodies" (1905), on the Principle of Relativity, according to which the contradiction between the previously accepted Lorentz' theory of the light ether at rest and the subsequently (experimentally) established independence of the electrodynamic-optical phenomena in terrestrial bodies from the movement of the earth, is radically explained by the fact that an observer moving with the earth uses a time coordinate different from that of a stationary observer in a heliocentric system.

'The revolutionary consequences of this new interpretation of the concept of time have repercussions on the whole of physics, and above all on mechanics and epistemology. They [the consequences] later found, through the work of the mathematician Minkowski, a formulation which gives the whole system of physics a new unification inasmuch as the time dimension enters the stage on completely equal terms with the three space dimensions. Although this idea of Einstein's has proved itself so fundamental for the development of physical principles, the application of it still lies for the moment at the frontier of the measurable....' (See C. Seelig, *Albert Einstein* [Ref. 35], pp. 173–174.)

70. C. Seelig, Ref. (35), p. 178.

71. There are various accounts of the times which Einstein spent in Berlin. It is often asserted that Einstein was lonely in Berlin and did not receive the proper attention to which he was entitled. (See e.g., P. Frank, *Einstein: His Life and Times*, New York (1965), pp. 114–115) The creation of general relativity was surely something to which very few contributed, and in which the most important theoretician in Berlin, Max Planck, did not actively participate. Planck was much more concerned with the quantum problem, taking a point of view to which Einstein himself

would subscribe later on. Nevertheless, Einstein and his work received increasing attention in Berlin, and he was able to write to his old friend Michele Besso [after he had received and refused invitations to chairs of theoretical physics at Zürich University and E.T.H.]: 'I must say that it has been very hard for me to come to a decision about it. It goes without saying that the general conditions there [in Switzerland] have greater appeal for me. But if you could only see how beautiful are the relationships which have developed between my colleagues (particularly Planck) and myself, and how welcome all of them have made me feel, and continue to do so, and if you could just imagine how my work has progressed mainly on account of the understanding which I have found for it *here*, then you will realize that I could not really turn my back on this place.' [Letter, dated Berlin, 8 September, 1918. See *Correspondance*, Ref. 67.]

72. Letter to Besso, dated Zurich, March 1914. See *Correspondance*, Ref. (73).].
73. *Albert Einstein, Michele Besso, Correspondance, 1903–1955* (edited by Pierre Speziali), Hermann, Paris (1972).
74. A. Einstein, *Sitz. Ber. Preuss. Akad. Wiss.* pp. 778–786 (1915).
75. Einstein remarked explicitly: 'Mathematics teaches us that all of these covariants can be derived from the Riemann–Christoffel tensor of the fourth rank…. In the problem of gravitation, we are particularly interested in the tensors of the second rank, which can be constructed from these tensors of the fourth rank and the $g_{\mu\nu}$ by inner multiplication.' (Ref. (74), p. 781. Compare with our earlier quotations from Grossmann's paper.)
76. Einstein said that on the basis of his considerations 'the field equations of gravitation may be written in the form

$$R_{\mu\nu} = -\gamma T_{\mu\nu},$$

since we already know that these equations are covariant with respect to arbitrary transformations of the determinant 1.' (Ref. (74), p. 783.)
77. After his lecture at Vienna in 1913, Einstein had a discussion with Reissner, who asked for the meaning of this 'energy': 'Einstein has talked about the bending of electromagnetic radiation in a gravitational field. I wonder if Einstein would tell us something about the elementary question concerning the effect of gravity on its own mass, i.e. the mass of the energy of the gravitational field. How can one better understand, or show mathematically, that the static energy of a pure gravitational field, although it possesses inertia and weight, does not possess the other attributes of ponderable matter, i.e. its kinetic behaviour [such as free fall or collapse]?' (*Phys. Z.* **14**, 1265). It was a perceptive question, in response to which, Einstein made some comments, but Reissner (either then or afterwards) did not have the feeling that an answer had been given (*l.c.*, p. 1265). Actually Einstein had completely misunderstood Reissner's question. He returned to deal with it in considerable detail and quite satisfactorily in *Phys. Z.* **15**, 108–110 (1914).
78. A. Einstein, *Sitz. Ber. Preuss. Akad. Wiss.* p. 785 (1915).
79. A. Einstein, 'Zur allgemeinen Relativitätstheorie (Nachtrag)', *Sitz. Ber. Preuss. Akad. Wiss.* pp. 799–801 (1915).
80. A. Einstein, Ref. (79), pp. 799–800.
81. A. Einstein, 'Die Feldgleichungen der Gravitation', *Sitz. Ber. Preuss. Akad. Wiss.* pp. 844–847 (1915).
82. Einstein had explained this result, obtained empirically in his second note, submitted on 18 November 1915, entitled 'Erklärung der Perihelbewegung des Merkur aus der allgemeinen Relativitätstheorie', *Sitz. Ber. Preuss. Akad. Wiss.* pp. 831–839 (1915).
83. Ref. (81), p. 847.
84. He wrote to Besso again on 21 December 1915: 'Read the papers [i.e. Einstein's]! The final release from misery has been obtained. What pleases me most is the agreement with the perihelion motion of Mercury….Even Planck now begins to take the thing seriously, although he still resists it a bit. But he is a splendid human being.' [See *Correspondance*, Ref. 73.]
85. Writing again to Besso on 3 January 1916, he said: 'The great success with [the theory of] gravitation pleases me extraordinarily. I have seriously in mind to write a book in the near future on the special and general theory of relativity….' [See *Correspondance*, Ref. 73.]
86. See R. Courant and D. Hilbert, *Methods of Mathematical Physics*, Interscience, New York (1953).
87. D. Hilbert, *Die Grundlagen der Geometrie* (Leipzig, 1899; 9th edition, Stuttgart, 1962). 'This means that it is still being read, and obviously by more people than read Hilbert's other works. It has gradually been modernized, but few readers realize that foundations of geometry as a

field has developed more rapidly than *[Die] Grundlagen der Geometrie* as a sequence of re-editions, and that Hilbert's book is now a historical document rather than a basis of modern research or teaching.' – Hans Freudenthal in his biographical sketch of David Hilbert in the *Dictionary of Scientific Biography*, Charles Scribner's Sons, New York (1972).

87a. Hermann Weyl wrote: "Hilbert is the champion of axiomatics. The axiomatic attitude seemed to him one of universal significance, not only for mathematics but for all sciences. His investigations in the field of physics are conceived in the axiomatic spirit. In his lectures he liked to illustrate the method by examples taken from biology, economics, and so on. The modern epistemological interpretation of science has been profoundly influenced by him. Sometimes when he praised the axiomatic method he seemed to imply that it was destined to obliterate completely the constructive or genetic method." [See H. Weyl, *Gesammelte Abhandlungen*, Vol. 4, Springer Verlag, New York 1968, pp. 162–163.]

In his famous address, *Mathematische Probleme*, which Hilbert delivered before the international Congress of Mathematicians at Paris in 1900, he stressed the importance of great concrete fruitful *problems*. He posed and discussed twenty-three unsolved problems. His sixth problem dealt with the axiomatization of physics. [See D. Hilbert, Mathematische Probleme, *Gött. Nachr.* 1900; see also D. Hilbert, Axiomatisches Denken, *Math. Ann.* **36**, 405–415 (1918). For a discussion of Hilbert's twenty-three problems and their influence on the development of mathematics, see L. Bieberbach, *Die Naturwissenschaften* **18**, 1101–1111 (1930).

88. In his obituary of Minkowski, Hilbert wrote: 'During the period he was at Bonn [1892–1894], Minkowski would spend his vacations regularly in Königsberg, where his family lived, and where he met Hurwitz and me almost daily. Mostly we went for walks in the surroundings of Königsberg. Once, on Christmas 1890, Minkowski remained in Bonn. When I tried to persuade him to come to Königsberg, he described himself in a funny letter to me as one who was physically fully contaminated, and in need of a ten-day quarantine, before Hurwitz and I would allow him to go on our walks in the mathematically pure Königsberg.' [See D. Hilbert, *Gesammelte Abhandlungen*, Vol. 3, Berlin (1935), p. 355.] Hilbert continued, 'For more than six years [from fall 1902 to January 1909, when Minkowski died] we, his closest mathematical colleagues, took our "mathematical" walk every Thursday punctually at 3 o'clock on Hainberg, including the last Thursday before his death, when he especially spiritedly told us about the latest progress on his electrodynamic researches....' [D. Hilbert, *l.c.* p. 364.]

Hilbert's collaboration with Minkowski was noted by Blumenthal: 'It was with Minkowski that Hilbert first started intensive reading (something which he avoided) of classical theoretical physics.' (*Lebensgeschichte*, in D. Hilbert, *Gesammelte Abhandlungen*, Vol. 3, p. 417.)

89. See *Vorträge über die Kinetische Theorie der Materie und der Elektrizität*, B. G. Teubner, Berlin (1914); Hilbert wrote the foreword to these lectures.

90. O. Blumenthal, *Lebensgeschichte*, *l.c.*, p. 417 [see Ref. 88].

91. L. Föppl, 'Stabile Anordnungen von Elektronen im Atom' (1 March 1912), *J. reine angew. Math.* **141**, 251–302 (1912); H. Bolza, 'Anwendung der Theorie der Integralgleichungen auf die Elektronentheorie und die Theorie der verdünnten Gase' (2 July 1913); B. Baule, 'Theoretische Behandlung der Erscheinungen in verdünnten Gasen' (18 February 1914), *Ann. Phys. (Leipzig)* **44** (1914); K. Schellenberg, 'Anwendung der Integralgleichungen auf die Theorie der Elektrolyse' (24 June 1914), *Ann. Phys.* **47** (1915); H. Kneser, Untersuchungen zur Quantentheorie, (2 March 1921), *Math. Ann.* **84**, 277–302 (1921). (See D. Hilbert, *Gesammelte Abhandlungen*, Vol. 3, p. 433.)

92. G. Hamel, who had worked on his thesis, 'Über die Geometrien in welchen die Geraden die Kürzesten sind' (24 June 1901), under Hilbert, continued his attempts to establish, beginning in 1909, the axiomatic foundations of mechanics along the lines of Hilbert. See G. Hamel, 'Über die Grundlagen der Mechanik', *Math. Ann.* **66**, 350–397 (1909); 'Über Raum, Zeit und Kraft als apriorische Formen der Mechanik, *J. B. deutsch. Math. Ver.* **18**, 357–385 (1909); 'Über ein Prinzip der Befreiung bei Lagrange, *J. B. deutsch. Math. Ver.* **25**, 60–65 (1916); *Lehrbuch der elementaren Mechanik*, Leipzig (1912); 'Die Axiome der Mechanik', in H. Geiger and K. Scheel *Handbuch der Physik*, Vol. V, Berlin (1927). See also G. Hamel, 'Die Lagrange-Eulerschen Gleichungen der Mechanik', *Z. Math. u. Phys.* **50**, 1–57 (1904); and 'Über die virtuellen Verschiebungen in der Mechanik', *Math. Ann.* **59**, 416–434 (1905); and G. Herglotz, 'Über die Mechanik des deformierbaren Körpers von Standpunkte der Relativitätstheorie', *Ann. Phys. (Leipzig)* **36**, 493–533 (1911).

93. C. Carathéodory, 'Untersuchungen über die Grundlagen der Thermodynamik', *Math. Ann.* **67**, 355 (1909).

94. P. Hertz, 'Über die mechanischen Grundlagen der Thermodynamik', *Ann. Phys. (Leipzig)* **33**, 225–274, 537–552 (1910). See also Hertz' article on statistical mechanics in Weber and Gans *Repertorium der Physik*, Vol. I/2, Leipzig (1916). Hertz was close to Hilbert in 1912.

95. D. Hilbert, 'Begründung der Kinetischen Gastheorie', *Math. Ann.* **71**, 562 (1912).

96. D. Hilbert, 'Zur Begründung der elementaren Strahlungstheorie', I, II, III: *Nachr. Ges. Wiss. Gött.* (1912) p. 773 (*Phys. Z.* **13**, 1056 (1912); *Nachr. Ges. Wiss. Gött.* (1913), p. 409 (*Phys. Z.* **14**, 592 (1913)); *Nachr. Ges. Wiss. Gött.* (1914), p. 275 (*Phys. Z.* **15**, 878 (1914)).

97. D. Hilbert, 'Die Grundlagen der Physik I, II', *Nachr. Ges. Wiss. Gött.* (1915), p. 395; (1917), p. 201. He later on united these two communications into a single article with the same title, *Math. Ann.* **92**, 1 (1924).

98. W. Heisenberg, *Z. Phys.* **33**, 879–893 (1925); *Math. Ann.* **95**, 683–705 (1925/26).

99. P. Jordan, 'Über kanonische Transformationen in der Quantenmechanik', *Z. Phys.* **37**, 383–386 (1926); **38**, 513–517 (1926); 'Über eine neue Begründung der Quantenmechanik', *Z. Phys.* **40**, 809–838 (1927). Jordan proved that his formal transformation theory contains not only Schrödinger's wave mechanics, but also the Born–Wiener operator calculus and Dirac's *q*-number algorithm as special cases.

100. D. Hilbert, J. von Neumann, and L. Nordheim, *Math. Ann.* **98**, 1–30 (1928).

101. J. von Neumann, *Grundlagen der Quantenmechanik*, Springer, Berlin (1931) (English translation by R. T. Beyer, Princeton (1955)).

102. See, e.g., M. Jammer, *Concepts of Space*, Harvard University Press (1954); Harper Torchbooks, New York (1960).

103. See F. Klein's *Vorlesungen über die Entwicklung der Mathematik im 19. Jahrhundert*, Vol. I, Berlin (1926), reprinted by Chelsea Publ. Co., New York (1966), pp. 57–60. Also E. T. Bell, *The Development of Mathematics*, McGraw-Hill, New York (1945), 2nd edition, pp. 329–330.

    Gauss had thought about the possibility of non-Euclidean geometry since his student days, and he discussed the axiom of parallels with his friend Wolfgang Bolyai. In various letters he expressed the opinion that Euclidean geometry should not apply to nature. Two non-mathematicians, F. C. Schweikart and his nephew F. A. Taurinus, worked on this problem, and the former published a paper on 'astral geometry'. Gauss was aware of these attempts, and one knows that he had himself constructed a non-Euclidean geometry but not published it, because he feared, as he wrote to Bessel, 'the clamour of the Bœotians'. Then Wolfgang Bolyai's son Johann published his 'absolute geometry' in 1831, on which he had worked since 1820. Although Gauss acknowledges this 'genius of the first rank' privately, he neither discussed it publicly nor published anything about it, which made Johann Bolyai very bitter. At the same time, beginning in 1826, Nikolai Lobatchevski submitted his memoirs to the Kasan Academy. Beginning with his first paper on 'imaginary' or 'pan-geometry' in 1829, Lobatchevski continued publishing until 1855. In this case, Gauss praised the author openly; he had got to know the author and his work personally, and he had him elected to the Göttingen Academy in 1842.

104. Letter to H. W. M. Olbers, dated Göttingen, 28 April 1817, reprinted in *Werke*, Vol. 8, p. 177; translated in M. Jammer, Ref. 102, p. 145.

    Gauss had something to do with the method of triangulation ever since 1816, although the actual work was carried out in 1821–23 and 1828–1844. He reported on the results of the measurement of the record triangle with vertices on the mountain tops Hohenhagen, Brocken, and Inselberg, the distances being 69, 85, and 107 km respectively. [Gauss' 1827 paper on curved surfaces did not report on the results of the measurement of the triangle HBI. The results are to be found in his *Werke*, Vol. 4, p. 449. I am grateful to Prof. B. L. van der Waerden for this information.]

105. Printed in German, for the first time in *Abhand. Ges. d. Wiss. zu Göttingen* **13**, 1867, and then in *Riemanns Werke* (edited by R. Dedekind and H. Weber, 1st edition 1876, 2nd edition 1892): translated into English by W. K. Clifford, *Nature* **8**, 14–17, 36, 37, reprinted in *W. K. Clifford's Mathematical Papers*, London (1882), p. 54. Riemann had proposed, as he was required to do, three topics from which the faculty could choose any one and notify him about three days before the 'trial' lecture [*Probevorlesung*]. It was Gauss who, being especially interested in Riemann's ideas, selected the topic which Riemann had listed in the third and last place, almost as an afterthought. In fact, he was not really prepared to give this lecture, but this

disadvantage developed into a remarkably historic expansion of his ideas. For the mathematical and technical formulation, see his contribution for the Paris prize 1861, reprinted in *Riemann's Ges. Math. Werke*, Dover, New York, p. 391. Hermann Weyl published an edition of Riemann's *Habilitation* thesis with a detailed introduction, notes, and commentary. It is a most valuable guide to Riemann's geometrical thinking. See B. Riemann, *Über die Hypothesen, welche der Geometrie zu Grunde liegen*, neu herausgegeben und erläutert von H. Weyl (Julius Springer, Berlin, 1919)].

105a. See *Clifford's Mathematical Papers*, p. 55. [See Ref. 105] About Herbart's influence, Riemann was more explicit in remarks which were published later on in his *Werke*, pp. 507–508: 'My principal work consists in a new formulation of the known laws of nature, i.e. expressing them by means of other fundamental notions, whereby the use of experimental data concerning the interaction between heat, light, magnetism and electricity would make it possible to investigate their inter-connections. I was led to this by a study of the works of Newton and Euler, on the one hand, and Herbart, on the other. As for the latter, I could almost completely agree with the results of the earliest investigations of Herbart (which are presented in his doctoral and inaugural dissertations of 22 and 23 October 1802). I did, however, differ from the course of his later speculations concerning his natural philosophy and principles of psychology which refer to it [natural philosophy]. The author is a follower of Herbart in psychology and epistemology. However, he cannot subscribe to Herbart's natural philosophy and the metaphysical disciplines based upon it.'

We have checked the 'Habilitationsthesen' (reprinted in J. F. Herbart's *Sämtliche Werke*, Vol. I, Scientia Verlag, Aalen (1964), pp. 277–278) and quote thesis V (23 October 1802) (p. 278): 'Spatii et temporis cogitationem quod e mente nostra ejicere non possumus, hoc non probat, eas cogitationes natura nobis insitas esse. Qui in hac Kantianae rationis parte latet error, totum Aollit systema.' Another part of Herbart's writings which seems to bear on Riemann's interests, mentions the word 'quantum of the surface', which reappeared in Riemann's inaugural lecture, is contained in *Nachschrift zur zweyten Auflage, Pestalozzi's Idee eines ABC der Anschauung*, 1802 and 1804, reprinted in *Sämtliche Werke I* (see p. 254): 'Perhaps the content of a surface, considered in a pure sense, is a *totally* Platonic idea, which would destroy all forms including the perceptual. It is this notion which defines the *pure quantum* of the extent of the surface, quite apart from the fact whether this quantum may appear in a round or rectangular shape, whichever one pleases.' Concerning Gauss' paper, we refer to *New General Investigations of Curved Surfaces* (1825) and *General Investigations of Curved Surfaces* (1827); it is the latter memoir which Riemann had in mind. Both of these papers have been translated into English, and appeared as *General Investigations of Curved Surfaces*, New York (1965). [For Herbart's influence on Riemann's ideas, see Bertrand Russell, *An Essay on the Foundations of Geometry*, Dover edition, New York (1956), Chapter 1.]

106.   See W. Pauli, *Theory of Relativity*, Pergamon Press, Oxford (1958), p. 35.

107.   W. K. Clifford's translation, Ref. 105, p. 69.

108.   W. K. Clifford, *On the Space Theory of Matter*, read 21 February 1870; abstract printed in *Camb. Phil. Soc. Proc.* **2** (1876), reprinted in W. K. Clifford's *Mathematical Papers*, pp. 21–22.

109.   W. K. Clifford, *l.c.* p. 22. Other indications of Clifford's ideas may be found in his posthumously published book *The Common Sense of the Exact Sciences*, London (1885), in particular, p. 225: 'We may conceive our space to have everywhere a nearly uniform curvature, but that slight variations of the curvature may occur from point to point, and themselves vary with the time. These variations of the curvature with the time may produce effects which we not unnaturally attribute to physical causes independent of the geometry of our space. We may even go as far as to assign to this variation of the curvature "what really happens in that phenomenon which we term the motion of matter".'

110.   H. Minkowski, *Nachr. Ges. Wiss. Gött.*, pp. 53–112 (1908).

111.   H. Minkowski, Ref. (110), p. 55.

112.   H. Minkowski, Ref. 110, pp. 65–66. See also his lecture at the 80th Naturforscherversammlung (Congress of Natural Scientists) in Cologne on 'Raum und Zeit' printed in *Phys. Z.* **10**, 104–111 (1909), where he gave a most illuminating exposition of the concepts of space and time and introduced the notions of 'world vectors' and 'world postulate'.

113.   R. Hargreaves, 'Integral forms and their connection with physical equations', *Trans. Camb. Phil. Soc.* **21**, 107–122 (1908). In fact, H. Poincaré already used ict as the fourth coordinate x^4

in his memoir 'Sur la dynamique de l'électron', *Palermo Rend*, **21** (1906), which was submitted on 23 June 1905.

114. G. Fechner, 'Der Raum hat vier Dimensionen', in *Vier Paradora*, Leopold Voss, Leipzig (1846). Fechner wrote it under the pseudonym 'Dr Mises'.

Actually, D'Alembert suggested thinking of time as a fourth dimension in his article 'Dimension' in the *Encyclopédie*. Lagrange, in studying the reduction of quadratic forms to standard forms, casually introduced forms in *n* variables. He, too, used time as a fourth dimension in his *Mécanique analytique* (1788) and in his *Théorie des fonctions analytique* (1797). He says in the latter work, 'Thus we may regard mechanics as a geometry of four dimensions and analytical mechanics as an extension of analytical geometry.' Lagrange's work put the three spatial coordinates and the fourth one representing time on the same footing. Further, George Green in his paper of 1828 on potential theory did not hesitate to consider potential problems in *n* dimensions; he says of the theory, 'It is no longer confined, as it was, to the three dimensions of space.' Cauchy emphasized the concept of *n*-dimensional space [*Compt. Rend. Acad. Sci. Paris* **24**, 885–87 (1847)], and Grassmann considered the concept of *n*-dimensional geometry in a note published in 1845. [See M. Kline, *Mathematical Thought from Ancient to Modern Times*, Oxford University Press, New York (1972), pp. 1029–1030.]

115. H. von Helmholtz, 'Über die Tatsachen die der Geometrie zugrundeliegen', *Nachr. Ges. Wiss. Gött.*, pp. 193–221 (1863).

116. H. von Helmholtz, Ref. (115), p. 193.

117. H. von Helmholtz, Ref. (115), pp. 194–195. Riemann's 'trial' lecture [Probevorlesung] was published in the *Abhandl. kgl. Ges. Wiss. Gött.* **13** (1867), although it had appeared as a pamphlet in 1854.

118. J. J. Sylvester, *Collected Works*, Cambridge (1940), p. 198, originally printed as 'On the general theory of associated algebraical forms', *Cambridge and Dublin Math. J.* **4** (1851). In a further paper 'On the principles of the calculus of forms', *ibid.* **7** (1852), he introduced the expressions 'cogradient' and 'contragradient'.

Sylvester himself became interested in this field through Arthur Cayley, a London barrister, who had studied in Cambridge, and graduated with the highest distinction in 1841, and became professor in Cambridge only in 1863. Cayley, starting from the literature, in particular the contributions of Jacobi, 'created the algebraic geometry' (F. Klein, *Entw. d. Math.* I, 148, Ref. (103)). In 1858 Cayley introduced the concept of matrices which had been contained implicitly in the work of Grassmann. Cayley told Tait in 1894 as to what had led to matrices 'I certainly did not get the notion of a matrix in any way through the quaternions; it was either directly from that of a determinant, or as a convenient mode of expression of the equations

$$x' = ax + by$$
$$y' = cx + dy$$

(See E. T. Bell, *The Development of Mathematics*, McGraw-Hill, New York (1945), 2nd. ed., p. 205.) With G. Salmon and J. J. Sylvester, who went to Johns Hopkins University in 1876, and started the development of the science of pure mathematics in the United States, Cayley formed the very influential British school. He stimulated Hermite in France and cooperated with Clebsch and Gordan in Germany.

119. H. G. Grassmann, *Die lineare Ausdehnungslehre, ein neuer Zweig der Mathematik*, Berlin (1844); *Die Ausdehnungslehre, vollständig und in strenger Form bearbeitet* (1862). In 1872, R. F. A. Clebsch included the 'Stufen' of Grassmann in the theory of invariants in his paper 'Über eine Fundamentale Aufgabe der Invariantentheorie', *Abh. Gött. Ges. Wiss.* **17** (1872).

The word 'tensor' was first used by W. Voigt in 1898, in connection with the elasticity of crystals.

120. F. Klein, 'Vergleichende Betrachtungen über neuere geometrische Forschungen', Erlangen, appeared in December, 1872; reprinted in *Math. Ann.* **43** (1893) and in *Ges. Abh.*, Vol. I, p. 460; See F. Klein, *Vorlesungen*, Ref. 103, Vol. II, p. 28.

121. See p. 7 of Klein's Erlangen Programme, or Ref. 103, p. 28.

122. See Ref. 103, Vol. II, p. 28. In his Erlangen Programme, Klein had stated: 'We develop the theory of invariants with respect to groups.' Later on he remarked [Ref. 103, Vol. II, p. 28], 'Instead of this, one says: "the theory of the relationships which are invariant *relative to the*

group'; then from this to the phrase *theory of relativity* there is only one step, which the modern physicists, in their domain, set as a general goal.'

123. C. G. J. Jacobi, 'Vorlesungen über Dynamik', held in 1842–43, first edited in 1866 by Clebsch, and contained in the Supplement to 'Werke', Berlin (1884). Later J. R. Schütz, *Nachr. Ges. Wiss. Gött.* (1897), p. 110, derived the energy conservation from the symmetry principle. In fact, Jacobi was the first person who derived the 10 integrals of the mechanical equations of motion, (partly) by using the infinitesimal transformations contained in the Euclidean group. Jacobi worked in Königsberg from 1826 to 1843 as a great, powerful and influential teacher, contributing to nearly all fields of mathematics – elliptic functions, mechanics, differential equations and the theory of variations. His method was more inductive and impulsive than rigorous, as his famous statement indicates: 'Gentlemen: We have no time [or patience] for Gauss-like rigour.' [See F. Klein, *Entwicklung* I, Ref. 103, p. 114]. Jacobi's fundamental contribution to the theory of determinants, we have already mentioned above. His influence stretched far beyond Prussia and Germany. Hermite and Liouville acknowledged him as their teacher, and the same was true of Cayley in England.

124. G. Hamel, 'Die Lagrange–Eulerschen Gleichungen der Mechanik', *Z. Math. u. Phys.* **50**, 1–57 (1904) (inaugural lecture); See also 'Über die virtuellen Verschiebungen in der Mechanik', *Math. Ann.* **59**, 416–434 (1904).

125. S. Lie and F. Engel, *Theorie der Transformationsgruppen*, Teubner, Leipzig (1888–93); and S. Lie and G. Scheffers, *Kontinuierlichen Gruppen*, Teubner, Leipzig (1893). It should be mentioned that Sophus Lie, the friend and inspiring colleague of Felix Klein, did not possess the latter's systematic approach, and was only persuaded with effort, by Klein, with the help of Engel, to write the laborious masterpiece in three volumes. At a much later time, *after* Lie's work had been propagated by his book, Bell made the judgement: 'Engel's triumph in organizing Lie seems to have been a mistake. But it was not fatal. Three hundred leaden volumes could not have crushed and buried a talent like Lie's. His ideas had done their enduring work long before all the freshness and ferment were systematized out of them. As for Klein's part, in this comedy of well-intentioned errors, it is only fair to record that he was actuated by the least inexcusable of all excuses for meddling with another man's life: the sincere and unselfish desire to make a friend do something distasteful for his own supposed good.' (E. T. Bell, *The Development of Mathematics*, 2nd edition, New York (1945), p. 442.)

126. G. Hamel, Ref. 124, p. 12.

127. G. Herglotz, 'Über die Mechanik des deformierbaren Körpers von Standpunkte der Relativitätstheorie', *Ann. Phys. (Leipzig)* **36**, p. 1911.

128. The name 'Poincaré group' was first used by E. P. Wigner in: 'On Unitary Representations of the Inhomogeneous Lorentz Group', *Annals of Math.* **40**, 149 (1939).

129. "His papers on the theory of invariants had the unexpected effect of withering, as it were overnight, a discipline which so far had stood in full bloom." [H. Weyl, *Gesammelte Abhandlungen* Vol. 4, Springer-Verlag, New York 1968, p. 124.]

129a. The first mention of Emmy Noether in this context occurs in Felix Klein's 'Zu Hilberts erster Note uber den Grundlagen der Physik', *Nachr. Ges. Wiss. Gött., Math. Phys. kl.*, 469 (1917), especially on p. 476: 'When I spoke to Miss Noether recently about my result concerning your [Hibert's] energy vector, she informed me that she was able to derive the same result already a year ago on the basis of your note (i.e. not from the simplified calculation in me memoir Nr. 4). At that time she wrote a manuscript, which she showed me. She had brought it to the decisive stage as I recently showed before the Mathematical Society.' Klein's statements occur in a letter to Hilbert [which he submitted to the Göttingen Academy on January 25, 1918]. In his reply, part of which Klein also mentioned in his note, Hilbert said: 'Emmy Noether, whom I had invited already a year ago to assist me in the clarification of the analytical questions concerning my energy principle, found at that time that the energy components, which I had stated (as well as those of Einstein) formally by means of the Lagrangian differential equations [Equations (4) and (5) of my first communication], could be reformulated, without resorting to the Lagrangian equations, as expressions whose divergence vanishes identically.' (*l.c.*, p. 477.)

130. F. Engel, 'Über die zehn allgemeinen Integrale der klassischen Mechanik', *Nachr. Ges. Wiss. Gött.*, pp. 270–275 (1916), communicated by Klein on 11 November 1916. Engel stated in his introduction [*l.c.*, p. 270]: 'At our last meeting you brought to my attention a paper by Herg-

lotz [See Ref. 127]. He has treated a mechanical problem which is consistent with the ten-component Lorentz group, and has derived the ten known integrals of the problem from the infinitesimal transformations of this group (entirely in the sense of Lie). In this connection, you had remarked that from these integrals one can obtain the ten known integrals of the ordinary n-body problem, by taking the limit and letting the velocity of light go to infinity. It is now only left to show how one can obtain, in the sense of Lie, the latter integrals from the infinitesimal transformations of a ten-component group belonging to the n-body problem, which, unfortunately, has been called the Galilean group. It is because it deals only with the integral of the [kinetic energy] and its second moments, since the zero-th and first moments are quite obvious.'

131. E. Noether, 'Invariante Variationsprobleme', (dedicated to F. Klein on the fiftieth anniversary of his doctorate), *Nachr. Ges. Wiss. Gött., Math. Phys. Kl.* pp. 235–257 (1918), communicated on 26 July 1918. Emmy, daughter of the mathematician Max Noether from Erlangen, came to stay in Göttingen in 1916, although she had visited there earlier. The connections of Göttingen with Erlangen are obvious: Felix Klein had come to Göttingen from there and had kept up his contacts with Gordan and Max Noether (who became a professor of mathematics in Erlangen in 1875 and stayed there until his death in 1921). Max Noether 'played an important role in the development of the theory of algebraic functions as the chief representative of the algebraic geometrical school.' [H. Weyl, 'Emmy Noether', *Scripta Math.* **3**, 201–220 (1935), reprinted in *Gesammelte Abhandlungen* **3**, 425–444; see p. 425.]

When Gordan retired in 1910, Erhard Schmidt, one of Hilbert's students, brought his master's spirit to Erlangen, and so did Ernst Fischer who followed Schmidt there in 1912. At least since 1913, Weyl recalled, Emmy Noether had visited Göttingen occasionally, and both Hilbert and Klein welcomed her collaboration. In fact, Hilbert tried to get her through with her habilitation (i.e. installed as a Privatdozent in the University of Göttingen) already during World War I, but he did not succeed before 1919, because the philosophers and historians opposed her candidacy on the ground that she was a woman. The faculty members argued that since there existed no faculty rest rooms for ladies, this would cause difficulties for a woman lecturer. Upon which Hilbert remarked: 'I do not see that the sex of a candidate is an argument against her admission as Privatdozent. After all, we are a university and not a bathing establishment.' (H. Weyl, *l.c.*, p. 431). In 1922 she was given the courtesy title, without official salary, of an extraordinary (i.e. associate) professor. She stayed in Göttingen until 1933, where many worked with her. She had to leave Germany in 1933, and she became a professor of mathematics at Bryn Mawr College, also lecturing at times at the Institute for Advanced Study in Princeton. She died on 14 April 1935. See also B. L. van der Waerden, 'Nachruf auf Emmy Noether', *Math. Annalen*, pp. 469–476 (1935), and Auguste Dieck, 'Emmy Noether', Birkhäuser, Basel, 1970.

132. E. Noether, Ref. 131, pp. 238–239.
133. E. Bessel-Hagen, 'Über die Erhaltungssatze der Elektrodynamik', *Math. Ann.* **84**, 254 (1921).
134. The generalization of Noether's Theorem to quantum mechanics was made by E. P. Wigner in his paper 'Über die Erhaltungssätze in der Quantenmechanik', *Nachr. Ges. Wiss. Gött.*, p. 375 (1927).
135. Gustav Mie, 'Grundlagen einer Theorie der Materie (I)', *Ann. Phys. (Leipzig)* **37**, 511–534 (1912); (II), **39**, 1–40 (1912); (III), **40**, 1–66 (1913).
136. G. Mie, *l.c.*, (I), p. 511.
137. This is in contrast to the model, for instance, of Lorentz who thought of an electron, with its charge distributed on the surface of its shell. Poincaré had already pointed out that the Lorentz electron model suffered from many difficulties.
138. G. Mie, *l.c.*, (I), p. 513.
139. G. Mie, *l.c.*, (I), p. 512.
140. 'This, to us, seems to constitute a very weighty argument against Mie's theory', Pauli remarked in 1921. (W. Pauli, 'Relativitätstheorie' in *Encyklopädie der math. Wiss.* V/19, Leipzig (1921); *Theory of Relativity*, Pergamon Press, Oxford (1958), p. 192, see Ref. 106).

Although Mie did not succeed in discovering the appropriate 'world function' which could account for the existence, asymmetry, and stability of the proton and the electron, his investigations inspired the work of M. Born and L. Infeld on 'non-linear electrodynamics' which corresponded entirely with Mie's programme. [See M. Born's review article on the Born–Infeld non-linear electrodynamics in *Ann. Inst. H. Poincaré* **7**, 155 (1937), which gives an explicit discussion of the relation of this work to Mie.]

141. G. Mie, Section 35 of (III), in Ref. 135, entitled 'Das Plancksche Wirkungsquantum' does not represent a step in this direction.
142. Constance Reid, *Hilbert*, Springer, New York, (1970), pp. 140–141.
143. *Einstein-Sommerfeld Briefwechsel* (see Ref. 1), p. 30.
144. A. Einstein, *S. B. Preuss. Akad. Wiss.* (1915), p. 844.
145. D. Hilbert, 'Die Grundlagen der Physik', *Nachr. Kgl. Ges. Wiss. Gött.* pp. 395–407 (1915), especially p. 405.

In an article, 'Contribution to the History of Einstein's Geometry as a Branch of Physics', [Published in *Relativity*, Plenum Press (1970)], Eugene Guth wrote [p. 183]: 'Apparently, Weyl, in the first textbook of general relativity [H. Weyl, *Raum, Zeit, Materie*, Springer, Berlin (first edition 1918)], originated the story that the field equations of general relativity have been established also by Hilbert simultaneously and independently of Einstein. This remark was taken over also by Pauli in his well-known article on relativity [Ref. 140] and later on by several other textbook writers. *The remark of Weyl and Pauli does not correspond with the historical truth.* [My emphasis.] In the first place these two experts did not look up carefully the references in Hilbert's paper. Otherwise, they would have noticed that Hilbert quotes all of Einstein's communications of November 1915, in the Proceedings of the Berlin Academy. In particular, he quotes the paper on p. 844 which contains the final form of the GR [General Relativity] field equations.'

Guth goes on to say [pp. 183–184] that, 'It is interesting, however, how this "myth" has arisen. For the following remarks I am indebted to Professor P. P. Ewald who was, in 1915, Hilbert's "assistant for physics". His story is reinforced by the recently published exchange of letters between Einstein and Sommerfeld. [See our Ref. 1.] In a letter dated 15 July 1915, Einstein says about his visit in Göttingen that he had great joy there; everything he said was understood in detail. He was very enthusiastic about Hilbert. Clearly Einstein gave a talk in Göttingen. Hilbert, whose absentmindedness was legendary, started to think about the problem on the basis of what Einstein said, who at that time did not have yet the correct form of his field equations. I heard from Ewald that Hilbert, very likely in the fall of 1915, gave a talk in Göttingen presenting the correct equations without referring to Einstein. However, by that time, Einstein must have had also the correct equations, certainly before Hilbert did. Sommerfeld heard about Hilbert's talk in Göttingen and suggested that Hilbert write a letter apologizing to Einstein, which Hilbert, of course did. At any rate, he never claimed having been the independent discoverer of the field equations of G.R. [General Relativity].'

We have not been able to find any evidence that would substantiate these remarks.* There does not exist any trace of a letter of apology from Hilbert to Einstein. The reason for that is quite simple: Hilbert did not need to apologize at all. As for Weyl and Pauli, they had read the literature quite *carefully* and concluded that Hilbert had discovered the field equations of gravitation [Equations (37)] simultaneously with and *independently* of Einstein.

Hilbert was able to refer in his *published* paper in the *Göttinger Nachrichten* to all four of tingen Academy [his first communication on the foundations of physics] on 20 November 1915. [See the records of the meetings of the Königliche Gesellschaft der Wissenschaften, Göttingen.] Hilbert's derivation [of Equations (37)] was thus announced a few days before Einstein presented his formulation to the Prussian Academy in Berlin (on 25 November 1915). Hilbert was able to refer in his *published* paper in the *Göttinger Nachrichten* to all four of Einstein's communications of November 1915 because his paper, just as Einstein's communications, was published only several weeks later, and he could include these in the footnotes. [See footnotes on p. 395 and p. 405 of Ref. 145.]

Hilbert was clearly inspired by Einstein's work of 1913–14 on gravitation, and Mie's work of 1912–13 on electrodynamics, to develop his unified theory of electromagnetism, and acknowledged this fact clearly and unequivocally. But his derivation [of Equations (37)] was *completely independent* of Einstein's approach. He used an axiomatic method and employed the variational principle. In fact, Einstein did not like Hilbert's approach. [See our References 149, 187, 203, 242, and 257.]

---

* See Note 270 added in proof.

Instead of an apology for his first communication, Hilbert wrote a second memoir on the foundations of physics, and presented it to the Göttingen Academy on 23 December 1916. That Hilbert had been guided by the intuitive physical reasoning of Einstein, he repeated often enough. That he considered Einstein as the principal architect of general relativity, he confirmed by his often expressed admiration for Einstein and by his nomination of Einstein for the third Bolyai prize in 1915 'for the high mathematical spirit behind all his achievements'. [The first and second recipients of the Bolyai prize had been Poincaré and Hilbert respectively. Einstein did not receive this award.]

Hilbert was justly proud of his contribution to the foundations of physics and his derivation of the field equations of gravitation. By 1920, Einstein's dominance over the entire range of questions of general relativity was such that people were forgetting what Hilbert had contributed. That is why, in 1924, he again published a treatise which consolidated the principal contents of his communications of 1915 and 1916 with additional comments. He felt that his contributions contained 'an enduring core'. [See Refs. 147 and 258.]

146. Hilbert's contributions to the theory of gravitation are not mentioned in most textbooks or review articles on general relativity. There are, however, a few exceptions: e.g., H. Weyl, *Space, Time, Matter* (Dover, New York, 1952, based on the fourth German edition), H. Weyl, 'Zu David Hilberts siebzigstem Geburtstag, *Die Naturwissenschaften* **20**, 57–58 (1932), and H. Weyl, '50 Jahre Relativitätstheorie', *Die Naturwissenschaften* **38**, 73–83 (1950), W. Pauli, *Theory of Relativity*, p. 161, J. W. Anderson, *Principles of Relativity Physics*, Academic Press, New York (1967), p. 344. Only P. Jordan went so far as to call the equations of gravitation in the vacuum, the *Einstein–Hilbert equations*. (*Schwerkraft und Weltall*, Braunschweig (1952), p. 61.)

147. In an essay in honour of Hilbert's 60th birthday, Max Born wrote: 'This remarkable coincidence, however, never gave rise to any controversy about priority between the two men. Rather, their correspondence, which had originated in an exchange of scientific views, developed into more personal and friendly communications. Hilbert always remained aware of the fact that the great principal physical idea was Einstein's, and he expressed it in numerous lectures and memoirs....' (*Naturwiss.* **10**, 88–93 (1922), in particular p. 92, reprinted in M. Born, *Ausgewählte Abhandlungen*, Vol. II, pp. 584–598, see p. 595.)

148. C. Reid, *Hilbert*, Springer, New York (1970), p. 142.

149. Einstein once remarked, 'The people in Göttingen sometimes strike me, not as if they want to help one formulate something clearly, but as if they only want to show us physicists how much brighter they are than we.' (Reid, *l.c.*, p. 142.)

150. C. Reid, *Hilbert*, p. 142.

151. 'Einstein has posed immense problems, has brought forth profound thoughts and unique conceptions, and has invented ingenious methods for dealing with them. Mie was able to construct his electrodynamics [on the basis of these], and they have opened up new avenues for the investigation of the foundations of physics.' (D. Hilbert, see Ref. 97, 'Die Grundlagen der Physik', p. 395.)

152. D. Hilbert, Ref. (97), I, p. 395.

153. D. Hilbert, Ref. (97), I, p. 396.

154. D. Hilbert, Ref. (97), I, p. 396.

155. 'The leitmotiv for the construction of my theory has been the following mathematical theorem, the proof of which I shall give at another place.' (D. Hilbert, Ref. (97), I, p. 396.)

156. D. Hilbert, Ref. (97), I, p. 397.

157. See the remarks in Section III.3. Emmy Noether had given the proof in its most general form. Hilbert gave a complete proof for a special case (which he called Theorem 2) of the general theorem.

158. D. Hilbert, Ref. (97), I, pp. 397–398.

159. A. Einstein, *Ann. Phys. (Leipzig)* **49**, 769 (1916), p. 810.

160. D. Hilbert, Ref. (97), I, p. 403.

161. D. Hilbert, Ref. (97), I, p. 404.

162. D. Hilbert, Ref. (97), I, p. 406.

163. D. Hilbert, Ref. (97), I, p. 407.

164. C. Seelig, *Albert Einstein* [Ref. 35], p. 188.

165. A. Einstein, *Ann. Phys. (Leipzig)* **49**, 769–822 (1916).

166. A. Einstein, Ref. (165), p. 772.
167. A. Einstein, Ref. (165), p. 775.
168. A. Einstein, Ref. (165), p. 779.
169. F. Kottler, *Ann. Phys. (Leipzig)* **50**, 955 (1916). Kottler had already given, as early as 1912, the expression for the electromagnetic field equations in a generally covariant form (*S. B. Akad. Wiss. Wien* **121**, 1659 (1912)), and Einstein recognized his contribution. 'Among the papers which deal critically with the general theory of relativity, those of Kottler are particularly remarkable, because this expert has really penetrated into the spirit of the theory.' [A. Einstein, 'Über Friedrich Kottlers Abhandlung "Über Einsteins Äquivalenzhypothese und die Gravitation"', *Ann. Phys. (Leipzig)* **51**, 639–642 (1916).]
170. A. Einstein, *Ann. Phys. (Leipzig)* **51**, 639 (1916).
171. A. Einstein, *Ann. Phys. (Leipzig)* **51**, 639 (1916), esp. pp. 640–641.
172. E. Kretschmann 'Über den physikalischen Sinn der Relativitätspostulate, A. Einsteins neue und seine ursprüngliche Relativitätstheorie', *Ann. Phys. (Leipzig)* **53**, 575–614 (1917).
173. E. Kretschmann, Ref. (172), p. 576.
174. E. Kretschmann, Ref. (172), p. 584.
175. E. Kretschmann, Ref. (172), p. 610.
176. A. Einstein, *Ann. Phys. (Leipzig)* **55**, 241–244 (1918). He praised the author (Kretschmann) for his sharp wit, but did not find the sharper form of the equivalence principle either valuable or desirable.
177. A. Einstein, *Ann. Phys. (Leipzig)* **49**, 769 (1916), esp. p. 781.
178. A. Einstein, Ref. (177), p. 789.
179. A. Einstein, Ref. (177), pp. 802–803.
180. A. Einstein, Ref. (177), pp. 803–804. The formal proof of this statement was given by H. Vermeil, *Nachr. Ges. Wiss. Göttingen* (1917) 334; see also H. Vermeil, *Math. Ann.* **79** (1918) 289.
181. A. Einstein, Ref. (177), p. 804.
182. 'The special theory of relativity has led to the result that the inertial mass is nothing but the energy, which can be expressed mathematically by means of a symmetrical tensor of the second rank, the energy tensor. Hence, we shall have to introduce an energy tensor $T^\alpha{}_\sigma$ in general relativity also, having the same mixed character as the components $t^\alpha{}_\sigma$ of the gravitational field, but belonging to a mixed symmetrical covariant tensor.' (Ref. (177), p. 807.)
183. A. Einstein, Ref. (177), p. 808.
184. A. Einstein, Ref. (177), p. 810.
185. A. Einstein, Ref. (177), pp. 810–811.
186. Einstein had already done this in a paper entitled 'Eine neue formale Deutung der Maxwellschen Feldgleichungen der Elektrodynamik', *Sitz. Ber. Preuss. Akad. Wiss.* (1916), pp. 184–187. (Submitted and read on 3 February 1916.)
187. C. Seelig, Ref. (35), p. 199.
188. This value had already been calculated in his communication of 18 November 1915, dealing with the motion of the perihelion of Mercury.
189. A. Einstein, *Sitz. Ber. Preuss. Akad. Wiss.* (1916), pp. 688–696. See also the later paper, 'Über Gravitationswellen', *Sitz. Ber. Preuss. Akad. Wiss.* (1918), pp. 154–167, in which Einstein gave a more lucid presentation and corrected a previous error.
190. Einstein noted [similar to what Hilbert thought], 'At first sight, it seems strange that, in addition to the ten equations [Equations (66)] for the ten functions $\gamma_{\mu\nu}$, there should be four others which can be written arbitrarily along with them, without over-determination. The justification of this procedure is the following. The equations [of gravitation] are covariant with respect to arbitrary substitutions, i.e. they are satisfied for an arbitrary choice of the coordinate systems. If a new coordinate system is introduced, then the $g_{\mu\nu}$ of the new system would depend upon the four arbitrary functions which define the transformation of the coordinates. These four functions can be so chosen that the $g_{\mu\nu}$ of the new system would satisfy four arbitrarily prescribed relations. The latter are assumed to be chosen such that they transform into the equations [(68)] in the case of the approximation of interest to us. Thus the latter equations signify the condition chosen by us according to which the coordinate system has to be selected. By means of the equations [(68)], one can now obtain the simple equations [(67)] instead of [(66)].' (A. Einstein, Ref. (189), (1918), pp. 155–156.)
191. A. Einstein, Ref. (189), (1918), p. 164.

192. A. Einstein, 'Kosmologische Betrachtungen zur allgemeinen Relativitätstheorie', *Sitz. Ber. Preuss. Akad. Wiss.* (1917), pp. 142–152 (submitted and read on 8 February 1917).

In a letter dated 5 December 1916, Michele Besso thanked Einstein for the papers which he had sent him, and asked him rather vaguely about the role of gravitational energy. Einstein replied to him during the same month (December 1916), and said: 'Now concerning $\lambda = 1/R$. Quite apart from the fact whether it holds or not, the question here is not one of great scientific significance ... . If I choose the $g_{\mu\nu}$ to be Galilean at a certain location, and continue in a most appropriate manner, then [we ask ourselves] how do $g_{\mu\nu}$ depend on space and time at an extremely large distance? Is it possible to arrange things in such a way that the $g_{\mu\nu}$ shall be determined by matter alone, in accordance with the principle of relativity? ... . The permanence of the universe requires that motion (centrifugal forces) should oppose [gravitational] collapse. Such is the case with our solar system. But this is possible only if one lets the average density of matter suitably approach zero at infinity, or else infinite differences of potential would arise.

'Such a description would be unsatisfactory even in Newtonian theory. [The problem of the rarification of matter and the dissipation of energy at infinity.] It is even less satisfactory in relativity, because relativity does not satisfy inertia. The latter would be mainly determined at [spatial] infinity by the $g_{\mu\nu}$, and to a much lesser extent by the interaction with the other masses. I find such a description unbearable. The only solution which I can think of is the hypothesis of the *closedness of space*, the consistency of which I have already demonstrated.

'Of course, I do not really believe that the universe is in a statistical-mechanical equilibrium, even though I have [on occasion] treated it so. Because then, all the stars will coalesce together (if only a finite volume were at our disposal). However, a deeper examination shows that statistics can be used for the treatment of such questions which concern me. One can, however, also do without the statistical considerations. Certainly, the infinitely large differences of potential would give rise to very large stellar velocities, which should have been observable for a long time. Small potential difference, together with an infinite expansion of the universe, requires an empty space at infinity (constancy of the $g_{\mu\nu}$ at infinity with an appropriate choice of the coordinates), in contradiction to any sensible interpretation of relativity. Only the *closedness* of the universe can get rid of this dilemma, and in any case it is also suggested by the fact that *the curvature has the same sign everywhere because the energy density is nowhere negative, as we know.*

'The λ introduced here for the first time has nothing to do with the previous one. The fact that an additional term $+ \lambda g_{\mu\nu}$ on the left side of the field equation does not interfere with its tensor character, had escaped my notice. My new approach, however, seems to require a non-vanishing λ, which would give rise to a non-vanishing average density of matter ϱ_0. A count of fixed stars leads to an order of magnitude for ϱ_0 of 10^{-22} gm/cm^3, corresponding to a radius of the universe $R = 10^7$ light years, whereas the distance of the remotest fixed stars has been estimated 10^4 light years. You and Dällenbach should read [my] paper; you will have fun.'' [See *Correspondance*, Ref. (73.)]

193. A. Einstein, *S.B. Preuss. Akad. Wiss.* (1917), p. 143.
194. A. Einstein, Ref. (193), pp. 143–144.
195. A. Einstein, Ref. (193), p. 145.
196. A. Einstein, Ref. (193), p. 148.
197. A. Einstein, Ref. (193), p. 151.
198. A. Einstein, Ref. (193), p. 152. At first Einstein was very satisfied with his cosmological considerations. In a letter to Besso, dated 9 March 1917, he wrote: 'You will have received the ''cosmological considerations''. It provides at least a proof that my relativity [theory] can lead to a system without contradiction. Till now one had always to be afraid that the ''infinite'' would harbour inconsistencies which could not be resolved. Unfortunately there is little promise that these views can be checked in reality. If we take into consideration the astronomical results on the density of distribution of stars, we obtain an order of magnitude estimate of $R = 10^7$ light years, whereas visibility extends to $R = 10^4$ light years. Incidentally, the question arises whether we should not be able to see stars which lie close enough to our antipode. They should have a negative parallax. One must not forget, however, that the curvature of space is irregular, and light rays travel through an inhomogeneous medium.' [See Ref. 73.]

Writing to Besso again on 20 August 1918, he said: 'Either the universe has a centre, has a vanishing density everywhere, empty at infinity where all the thermal energy is gradually lost as radiation; or, all the points are equivalent on the average, and the mean density is everywhere the same. In either case, one needs a hypothetical constant λ, which specifies the particular mean density of matter consistent with equilibrium. One perceives at once that the second possibility is more satisfactory, especially since it implies a finite size for the universe. Since the universe is *unique*, there is no essential difference between considering λ as a constant which is peculiar to a law of nature or as a constant of integration.' [See Ref. (73).]

Later on, the cosmological constant had ups and downs, even in the esteem of Einstein. In the early nineteen-twenties he became interested in the mathematically more general affine geometry, close to the approaches of Weyl and Eddington, but he returned to his old equations in 1925. 'I had to reject my endeavour which followed Eddington's ideas. In any case, I am now convinced that unfortunately nothing is to be gained from the Weyl-Eddington complex of ideas. I regard the equation

$$R_{\mu\nu} - \tfrac{1}{2}g_{\mu\nu}R = -T_{\mu\nu}$$

as the best (electromagnetic) [equation] that we have today. There are nine equations for fourteen quantities $g_{\mu\nu}$ and $\gamma_{\mu\nu}$. New calculations seem to indicate that they yield the motion of the electrons. But it appears to be doubtful whether the quanta have a place in all this.' [Letter to Besso, dated 25 December 1925, *l.c.* in Ref. (73)]

199.   Ilse Rosenthal-Schneider, 'Erinnerung an Gespräche mit Einstein', manuscript dated 23 July 1957. [Einstein Archive, Princeton, N.J.]

Singularly lacking are any eye-witness accounts of Einstein's tremendous absorption in the problems of general relativity during 1915–16. A charming glimpse is contained, however, in Charles Chaplin's *My Autobiography* (The Bodley Head, London 1964).

Chaplin recalls a dinner at his home in California in 1926, at which Einstein, Mrs Einstein, and two other friends of Chaplin were present. At dinner Mrs Einstein 'told me the story of the morning he conceived the theory of relativity.' She related: 'The Doctor [i.e. Einstein] came down in his dressing-gown as usual for breakfast but he hardly touched a thing. I thought something was wrong, so I asked what was troubling him. "Darling," he said, "I have a wonderful idea." And after drinking his coffee, he went to the piano and started playing. Now and again he would stop, making a few notes then repeat: "I've got a wonderful idea, a marvellous idea!" I said: "Then for goodness' sake tell me what it is, don't keep me in suspense." He said: "It's difficult, I still have to work it out."'

Mrs Einstein told Mr Chaplin that Einstein continued playing the piano and making notes for about half an hour, then went upstairs to his study, telling her that he did not wish to be disturbed, and remained there for two weeks. 'Each day I sent him up his meals,' she said, 'and in the evening he would walk a little for exercise, then return to his work again.'

'Eventually,' Mrs Einstein said, 'he came down from his study looking very pale. "That's it," he told me, wearily putting two sheets of paper on the table. And that was his theory of relativity.'

200.   A. Einstein, *Ann. Phys. (Leipzig)* **49**, 771.
201.   A. Einstein, Ref. (200), p. 776.
202.   Letter from Einstein to M. Besso, dated Berlin, 31 October 1916. [See Ref. (73)]
203.   Einstein expressed his prejudice against Hilbert's axiomatic method. 'In this paper, it is not my goal to present the theory of general relativity as a system, logically as simple as possible, based on a minimum of axioms. Rather, my motivation is to develop this theory in such a way that the reader would sense the psychological simplicity of our approach, and the basic assumptions would be verified as much as possible by our empirical knowledge.' (A. Einstein, Ref. (200), p. 777.)
204.   D. Hilbert, *Nachr. Ges. Wiss. Gött., Math. Phys. Kl.* (1917), 53–76, in particular p. 53.
205.   D. Hilbert, 'Die Grundlagen der Physik', *Math. Ann.* **92**, 1–32 (1924). This is a slightly condensed version of Hilbert's two memoirs of 1915 and 1916, which were published in the Proceedings of the Göttingen Academy. The third axiom is the one which expresses H as a sum of a gravitational term K and an electrical term L. Axiom IV was stated in this paper (p. 11). [This paper was reprinted in Hilbert's *Gesammelte Abhandlungen*, Vol. 3, pp. 258–289.]
206.   D. Hilbert, *Nachr. Gött.* (1917), p. 54.
207.   D. Hilbert, *Nachr. Gött.* (1917), p. 57.

208. W. Pauli, *l.c.* in Ref. 106, Section 22, 'Reality relations', English edition, pp. 62–64.
209. H. Reichenbach, *Axiomatik der relativistischen Raum-Zeit-Lehre*, Vieweg, Braunschweig (1965); English translation by M. Reichenbach, University of California Press, Berkeley (1969), pp. 179–181. It seems to be quite remarkable that Reichenbach ignored Hilbert's system of axioms completely.
210. D. Hilbert, *Nachr. Gött.* (1917), p. 61.
211. D. Hilbert, *Nachr. Gött.* (1917), pp. 63–64.
212. D. Hilbert, *Nachr. Gött.* (1917), p. 66.
213. D. Hilbert, *Nachr. Gött.* (1917), p. 70.
214. Einstein oscillated between the opinions whether he should take the singular solution as the one corresponding to the real particle (electron, etc.) or not.
215. W. Pauli, *Theory of Relativity*, p. 145, footnote 277 (see Ref. 106).
216. Note, e.g., the statement in J. L. Anderson, *Principles of Relativity Physics:* 'Today most physicists would be not only willing to accept as axiomatic the existence of a variational principle but would also be loath to accept any dynamical equations that were not derivable from such a principle.' [J. L. Anderson, *l.c.*, p. 344, (see Ref. 146).]
217. Einstein-Lorentz correspondence, The Hague, The Netherlands.
217a. H. A. Lorentz, *Versl. K. Akad. Wetensch. Amsterdam* **23**, 1073 (1915); *Proc. K. Akad. Amsterdam* **19**, 751 (1915); reprinted in H. A. Lorentz, *Collected Papers*, Vol. 5, The Hague (1937), pp. 229–245.
218. H. A. Lorentz, Ref. (217a), p. 1073, or *Collected Papers*, Vol. 5, p. 229.
219. H. A. Lorentz, *Versl. K. Akad. Wetensch. Amsterdam* **24**, 1389, 1759; **25**, 468, 1380 (1916); *Proc. K. Akad. Amsterdam* **19**, 1341, 1354, **20**, 2, 20 (1916), reprinted in *Collected Papers*, Vol. 5, pp. 246–313.
220. H. A. Lorentz, *Collected Papers*, Vol. **5**, p. 246.
221. H. A. Lorentz, Ref. 220), p. 246.
222. H. A. Lorentz, Ref. (220), p. 251. In his remarks Lorentz seems to convey the impression that everything had been done by Einstein, which, as we have discussed, was not entirely so. The fact was that Einstein was in close contact with Leiden since the time his friend Paul Ehrenfest succeeded Lorentz there. Einstein always expressed the feeling that he had found the greatest understanding in Holland. For instance, in a letter to Besso, written either at the end of 1913 or early in 1914, Einstein remarked: 'I shall visit Lorentz in Spring in order to discuss all these matters with him [i.e. general relativity]. He has great interest in this, just as Langevin.' [See *Correspondance*, Ref. (73)].

Einstein also enjoyed continuous and warm personal contact with Lorentz during and after the first World War, at a time when relations between the scientists of Germany and those of other nations were rather limited. In the exchange of correspondence between the two men, several letters are concerned with the attitudes of Planck, Nernst, and others with whom Lorentz obviously did not have any exchange during that time. It was, therefore, quite natural, and further enhanced by the occasional visits of Einstein to Leiden, that Lorentz should regard almost everything in general relativity as coming from Einstein. One must also consider that Lorentz at that time, in 1915 and 1916, was beginning to depend much more on personal discussions for information about new developments rather than on his own study of the literature.
223. This part is contained in the second paper in H. A. Lorentz, *Versl. K. Akad. Wetensch. Amsterdam* **24**, 1759 (1916).
224. A. Einstein, 'Hamiltonsches Prinzip und allgemeine Relativitätstheorie', *Sitz. Ber. Preuss. Akad. Wiss.* (1916), pp. 1111–1116.
225. A. Einstein, Ref. (224), p. 1111.
226. F. Klein, *Nachr. Ges. Wiss. Gött., Math. Phys. Kl.* (1918), pp. 171–189.
227. He proudly noted: 'As one can see, there is nothing more to be really calculated in the integrals, except to make the obvious use of the elementary formulas of the classical variational calculation.' (F. Klein, Ref. (226), p. 172)
228. F. Klein, Ref. (226), p. 185.
229. In conclusion he noted: 'I must not forget to thank Miss Noether for her active assistance in my recent work. She has worked out in full generality the mathematical ideas which I have used in an attempt to relate the integral $I_1[= \int K \, dx_1 \dots dx_4]$ with the physical problems, and she will publish them soon in these Proceedings.' [F. Klein, Ref. (226), p. 189].

230. F. Klein, *Nachr. Ges. Wiss. Gött.* (1918), pp. 394–423 (submitted and read on 6 December 1918).

231. W. de Sitter, *Proc. Acad. Sci. Amsterdam* **19**, 1217; **20**, 229 (1917).

232. F. Klein, *Nacht. Gött.* (1918), Ref. (226), pp. 394–395.

233. F. Klein, Ref. (232), p. 399.

234. F. Klein, 'Zu Hilberts erster Note über die Grundlagen der Physik', *Nachr. Ges. Wiss. Gött* (1917), pp. 469–489 (read on 25 January 1917). Klein's aim was to simplify Hilbert's calculations of the variational problem and obtain 'a clearer insight into the significance of the conservation laws'. [F. Klein, *l.c.*, p. 469].

235. Klein also favoured Hilbert's generalization of Maxwell's equations. 'It is also clear now how the old theory of Maxwell's is related to the new theory as a limiting case. If Maxwell's theory is written in terms of the general curvilinear coordinates $x_1 \ldots x_4$, we always encounter ds^2, whose Riemannian curvature vanishes identically. On the other hand α is taken to be 0 [where α is proportional to the gravitational constant]. *With this, the ten equations [Equations (37)] are satisfied automatically; the $Q_{\mu\nu}$ of the electromagnetic field are no longer subject to any restrictions, and just Maxwell's equations remain.*' (Ref. (234), pp. 473–474).

236. F. Klein, Ref. (234), p. 475.

237. 'Because of all this I can hardly believe that it is appropriate to regard the arbitrarily chosen quantities $t^{\nu}{}_{\sigma}$ as the energy-components of the gravitational field.' (Ref. (234), p. 477.)

238. D. Hilbert, *Nachr. Gött.* (1917), p. 477–480.

239. D. Hilbert, Ref. (238), p. 477. Hilbert also mentioned that Emmy Noether, whose help he had requested in his study of the question of energy conservation over a year previously, had, in fact, studied the question thoroughly and arrived at Klein's conclusions. Klein had also seen Noether's proof after writing down his own derivation. (Ref. (234), p. 476.)

240. D. Hilbert, Ref. (238), p. 480.

241. F. Klein, Ref. (234), pp. 481–482.

242. L. Bianchi, *Mem. della Societa Italiana delle Scienze* (3a) **11**, (1897). Felix Klein quoted Bianchi in his *Development of Mathematics in the 19th Century*, Vol. II, on p. 149, [see Ref. 103] but he never referred to the Bianchi identities in the relevant papers.

The English translation of W. Pauli's Encyclopaedia article on relativity was published early in 1958. In the English edition, Pauli appended a large number of supplementary notes. I had opportunity of talking to Pauli in Spring 1958, among other things, about this book on relativity. I told him that the original article was already so perfect that it was surprising that he should have thought of adding supplementary notes. He replied with a wide grin, 'You know, it was not all that perfect. I had not even mentioned the Bianchi identities.' This he had done now in Note 7, and discussed the identities in detail pp. 212–214 [see Ref. 106].

It is remarkable that when Einstein fully saw the advantages of the derivation of the field equations from a variational principle, he no longer recognized that this was precisely why Hilbert had pursued this approach. Einstein wrote: 'The derivation of the field equations from a variational principle has the advantage that the compatibility of the resulting system of equations is assured and that the identities connected with the general covariance, the "Bianchi identities", as well as the conservation laws result in a systematic manner.' [A. Einstein, *The Meaning of Relativity*, Fifth Edition, Princeton (1955), Appendix II, p. 154.]

Although Einstein developed this remark to show how this could be done, he no longer remembered the fact that this is indeed what Hilbert had done in his first communication on the foundations of physics.

243. It might not be complete due to the fact that not all of Klein's 152 conditions or identities seem to be independent. Also, he derived them for a special matter system. On the other hand, the number of 256 Bianchi identities would surely be reduced in a Riemannian geometry. Finally one must ask whether the Bianchi identities express the fact of general covariance completely, or whether there are further consequences from it. Without going into these details, let us just note the fact that there is considerable overlap between the identities derived by Bianchi and those derived by Klein on invariant-theoretic grounds.

244. H. Weyl, *Raum-Zeit-Materie*, Leipzig (1918); the fourth edition (1921) was translated as *Space-Time-Matter*, Methuen, London (1922). Dover edition, New York (1952). The quotations are from the Dover edition, p. 102.

245. T. Levi-Civita, *Rend. del. Circ. Mat. di Palermo* **42**, 173 (1917).

246. G. Hessenberg, 'Vektorielle Begründung der Differentialgeometrie', *Math. Am.* **78**, 187 (1917).
247. H. Weyl, *Math. Z.* **2**, 384 (1918). See also Weyl, *Sitz. Ber. Preuss. Akad. Wiss.* (1918), p. 465.
248. This follows by integrating the Equation (101), and noting that *l* is a constant throughout the space if $\phi_\mu dx^\mu$ is a total differential, which means that the expressions

$$F_{\mu\nu} = \frac{\partial \phi_\nu}{\partial x^\mu} - \frac{\partial \phi_\mu}{\partial x^\nu}$$

vanish.
249. A. Einstein, *Sitz. Ber. Preuss. Akad. Wiss.* (1918), p. 478.
250. On receiving the proof sheets of Weyl's book (*Raum, Zeit, Materie*, Ref. (244)), Einstein wrote to him on 8 March 1918: 'It is like a classical symphony. Each word is related to the whole and the design of the work is grandiose. The magnificent use of the infinitesimal parallel displacement from vectors to the deduction of the Riemannian tensors! How naturally it all falls into place! And now you have given birth to something which I could never have expected: the construction of Maxwell's equations. I am naturally very delighted with you and with your handling of the subject... .' (C. Seelig, Ref. (35), p. 200.)

But already on 31 June 1918, Einstein was remonstrating with Weyl: 'Could one really accuse the Good Lord of being inconsequential if he rejected the opportunity discovered by you for harmonizing the physical world? I do not think so. If he had made the world to your specification, dear Weyl, I would have gone to Him and said reproachfully: "Dear Lord, if it did not lie within Thy power to give an objective meaning to the congruence of infinitely small rigid bodies, so that when they are removed from each other one cannot say if they are congruent or not, why hast Thou, O Inconceivable, not disdained to bequeath to the angle this property or that of similarity?"... . But since the Good Lord noticed long before the development of theoretical physics that he could not come to terms with the opinions of the world, He does as He likes.' (C. Seelig, Ref. (35), p. 201.)

Writing to Ehrenfest, he expressed himself again about Weyl's theory: 'A paper by Pauli on Weyl's theory already shows the consequence of the initial fallacy of this theory: in general, static solutions do not exist for non-vanishing potentials. It is quite incomprehensible to me that Weyl himself and all others cannot immediately sense that the theory is contrary to experience.' (Letter to Ehrenfest, dated 4 December 1919, reprinted in C. Seelig, Ref. (35), p. 202.)

And in a letter to Besso, dated 26 July 1920, he wrote: 'Of course, I have been totally convinced from the very beginning of the incorrectness of Weyl's theory. Most [experimental] facts are *against* it, and none in its favour.' [See *Correspondance*, Ref. (73)]
251. A. S. Eddington, *Proc. Roy. Soc. (London)* A**99**, 104 (1921); A. Einstein, *Sitz. Ber. Preuss. Akad. Wiss.* A**23**, 32, 76, 137.
252. T. Kaluza, *Sitz. Ber. Preuss. Akad. Wiss.* (1921), p. 27. This theory was extended by O. Klein in several papers: *Z. Phys.* **37**, 895 (1926); *Z. Phys.* **36**, 188 (1927), *J. Phys. Rad.* **8**, 242 (1927); *Ark. Mat. Ast. Fys.* **34**, No. 1 (1946).
253. O. Klein, *Z. Phys.* **37**, 903 (1926).
254. To conclude, he said: 'As in de Broglie's work, our approach is also based upon the motivation that the analogy between mechanics and optics (which is evident in the Hamiltonian method) leads to a deeper understanding of quantum phenomena. That this analogy has real physical significance is shown by the similarity of the conditions for the stationary states of atomic systems and the interference phenomena in optics. Concepts such as point charge and mass point now appear to be strange even in classical field physics. The hypothesis has often been used that material particles are special solutions of the field equations which govern the gravitational and electromagnetic fields. It is only a short step to relate the analogy mentioned above to this idea. For, according to this hypothesis, it does not seem at all strange that the motion of material particles shows similarities with the propagation of waves. The analogy in question is, however, incomplete so long as one considers wave propagation in four dimensions only. This becomes already evident in the varying velocities of material particles. If one imagines the observed motion as a projection on the space-time of a propagating wave (taking place in a five-dimensional space), then a complete analogy can be made. Mathematically, it means that the Hamilton–Jacobi equation cannot be understood as a set of characteristics of a four-dimensional wave equation, but rather of a five-dimensional one. We are led to Kaluza's theory in this way.

'Although the introduction of a fifth dimension into our physical picture might initially appear rather strange, such a radical modification of the geometrical basis of the field equations is suggested by quantum theory. For, as we know, it appears less and less likely, that quantum phenomena do not lend themselves to a unified space-time description. In contrast, the possibility of expressing these phenomena by a system of five-dimensional field equations should not be ruled out right away. (Footnote: Professor Bohr has made such remarks on numerous occasions, and they have had a decisive influence on our work.)' (Ref. (253), p. 906.)

525. O. Klein. *Z. Phys.* **46**, 207 (1927). Oscar Klein remarked immediately that his efforts were not successful. In a footnote in the same paper (pp. 190–191), he said: 'In the following I have given a much belated but complete exposition of my efforts on the five dimensional form of the theory of relativity which (since, in the meantime, various researchers have been occupied with this subject) hardly offers any new mathematical results. Still I hope that it would arouse some interest because it does differ from the other endeavours in this field, including the author's, both in form as well as the physical point of view. In particular, I consider it no longer possible, by introducing the fifth dimension, to be able to bring into harmony the deviations introduced by the quantum theory with the space-time description of the classical theory.'

256. O. Veblen and D. Hoffman, 'Projective Relativity', *Phys. Rev.* **36**, 810 (1930).

257. Letter to H. Weyl, on 23 November 1916 (quoted from C. Seelig, Ref. (35), p. 200).

258. D. Hilbert, 'Die Grundlagen der Physik', *Math. Ann.* **92**, 1–32 (1924), reprinted in D. Hilbert, *Gesammelte Abhandlungen*, Berlin (1935), pp. 258–289.

259. D. Hilbert, pp. 1–2 (*Ges. Abh.* 3, pp. 258–259).

260. A. Einstein, *Sitz. Ber. Preuss. Akad. Wiss.* (1919), pp. 348–356.

261. 'The theoreticians have tried very hard to devise a theory which would account for the electricity in equilibrium which constitutes an electron. In particular, G. Mie has investigated this question in great depth.... However beautifully wrought its formal construction by Mie, Hilbert, and Weyl may have been, this theory has not yielded satisfactory physical results. On the one hand, the abundance of possibilities is discouraging; on the other, the additional terms [in the theory] could not be put simply enough to lead to a satisfactory solution.' (Ref. (260), p. 349.)

262. A. Einstein, Ref. (260), p. 350.

263. A. Einstein, Ref. (260), p. 351.

264. A. Einstein, Ref. (260), pp. 355–356.

265. 'The idea which I am grappling with concerns the understanding of quantum phenomena and it is this: over-determination of the laws by having more differential equations than field variables. In this way the inarbitrariness of the initial conditions could be handled, without renouncing field theory. This approach could well be quite wrong, but it must be attempted and, anyhow, it is logically possible. The equation of motion of material points (electrons) would be given up completely, and the kinetic content of the latter would be determined along with the field laws. I shall send you the preliminary work on this problem as soon as it is printed. The mathematics [in it] is enormously difficult, and the connection with what can be experienced becomes ever more indirect. But it is still a logical *possibility*, to do justice to reality without *sacrificium intellectus*.' (Letter to M. Besso, dated 5 January 1924.) [See *Correspondance*, in Ref. (73).]

266. A. Einstein to M. Besso, dated Berlin, 5 January 1929 [see Ref. 73]. The paper in question was 'Einheitliche Feldtheorie', *Sitz. Ber. Preuss. Akad. Wiss.* (1929), pp. 2–7.

267. P. A. M. Dirac, Lecture on 'Methods in Theoretical Physics', in the series, *From A Life of Physics*, Trieste, Italy (June, 1968).

268. S. Chandrasekhar, 'On the Derivation of Einstein's Field Equations', *Am. J. Phys.* **40**, 224 (1972). Chandrasekhar has argued that no rigorous derivation of Einstein's equations exists. The basic assumption of Einstein's 1915 theory was that the space-time is a quasi-Riemannian manifold endowed with a symmetric metric. Assuming that the Riemann–Christoffel tensor vanishes identically in the absence of a gravitational field, the simplest generalization in the presence of matter are the Einstein–Hilbert equations (37). The constant γ is derived by taking the limit to Newton's theory, and the matter tensor is generalized from special relativity. This was Einstein's 'derivation'. The other derivation, followed more closely by Hilbert, started from a variational principle. The question is how to 'derive' the Lagrangian or the Hamiltonian. Assuming it to be essentially proportional to the scalar R, one finds the equations for the vacuum, which had been 'guessed' or 'assumed' by Einstein. If matter is present one has to add a

term γM, where M denotes the 'scalar' of matter, in order to obtain the Einstein–Hilbert equations by a variational procedure. It must still be argued whether one should replace R by $R + \lambda$ in the variation function, thus introducing the cosmological constant.

269. Professor Markus Fierz has pointed out to me that this remark originated with Hermann Weyl, who, more than Pauli, was used to quoting statements from the Bible or his own variations thereof. 'Pauli did enjoy repeating this remark without claiming authorship of it.' [Private communication]

270. *[Note added in proof. To be read in conjunction with Ref. (145).]* Ewald was Hilbert's assistant during 1912–13, not in 1915 as mentioned in the quotation in Ref. (145).

In a letter, dated 9 September 1973, Professor P. P. Ewald has informed me that, 'At the time the Hilbert and Einstein papers of 1915 appeared, I was with the German armies in Russia as a "Field-X-ray Mechanic" and I did not learn of the General Relativity developments before my return to civilian life in 1918. I then had so many problems before me in the rapidly developing field of X-ray crystallography that I took only a slight interest in the more general problems of physics. I cannot remember ever having had a conversation of the type ascribed to me [in Ref. (145)].'

I am very grateful to Professor Ewald for this clarification.

ACKNOWLEDGEMENTS

Not long ago I published a small piece* dealing with a charming little essay on 'the state of ether in magnetic fields', which the sixteen-year-old Einstein had written while he was awaiting admission to the E.T.H. in Zurich. This paper sought to trace the continuity between Einstein's early interest in electrodynamics and his later work on the special and general relativity theories. On reading this paper, Professor Eugene Wigner asked me whether David Hilbert had not independently discovered the field equations of gravitation.** His impression from his stay in Göttingen (where Wigner had been Hilbert's assistant for one year in the late nineteen-twenties) was that Hilbert had indeed done so, and he asked me if it was true. I replied to Professor Wigner about Hilbert's contribution to the theory of gravitation.† He kindly encouraged me to expand my account to deal with the intricate and exciting details of the early years in the formulation of the general relativity theory of gravitation. This is what I have sought to do in this study.

As work on this project developed, my friend Dr Helmut Rechenberg became interested in the arguments that were central to the investigation, and we discussed this subject together in great detail. He was collaborating with me in my work on the history of quantum theory, and we were both very interested in Hilbert's contributions to physics, especially his interest in atomic mechanics. Dr Rechenberg's assistance in an examination of primary sources, and in independently verifying my conclusions (such as Hilbert's rediscovery of the Bianchi identities), has been invaluable. I am grateful to him for his interest, support, and help in all aspects of the preliminary stages of this investigation.

Professor Lothar Frommhold went over the translations of numerous original German quotations with me, and I am indebted to him for this assistance.

* 'Albert Einstein's "First" Paper', *Science Today* (April 1971); 'Albert Einsteins erste wissenschaftliche Arbeit', *Physikalische Blätter* **9**, 385–391 (1971).
** See Appendix.
† See Appendix.

Several persons read a draft manuscript of this paper. I am grateful for the invaluable comments of Professors Markus Fierz, Peter Havas, Pascual Jordan and Andrzej Trautman. I am especially indebted to Professors Res Jost, B. L. van der Waerden, and Arthur Wightman whose painstaking reading and detailed comments have helped to improve this study at many places. Professors S. Chandrasekhar, Robert Dicke, P. A. M. Dirac, Ivar Waller, John Wheeler, and Eugene Wigner also read the manuscript and encouraged me throughout, and I am grateful for their support. For any errors of fact or interpretation that remain, I alone am responsible.

APPENDIX

I. LETTER FROM EUGENE WIGNER TO JAGDISH MEHRA:

8 Ober Road
Princeton, New Jersey   08540
November 29, 1971

Dear Jagdish:

Just a line to thank you for your article *Albert Einsteins erste wissenschaftliche Arbeit*. I enjoyed it very much indeed.

There is though one question in my mind which I would like to ask you. I was under the impression that, simultaneously with Einstein, Hilbert also found the now accepted equations of general relativity. Is this correct? If so, is there a reason no one seems to mention this now? I realize that the basic idea was due to Einstein but it is interesting that, even after the promulgation of the basic idea, it took a rather long time to find the correct equations incorporating that idea – even though both Einstein and Hilbert seem to have worked on it.

It was good to see you in Amherst and I wish to thank you again for your visit.

Sincerely,

Eugene (Wigner)

II. REPLY TO EUGENE WIGNER FROM JAGDISH MEHRA:

7606 Rustling Road
Austin, Texas   78712
12 December 1971

Dear Eugene:

Thank you very much for your kind note. I am happy to learn that you liked the article on Einstein's 'first' paper.

Einstein had embarked quite seriously, already in 1907, on the considerations that led to his general theory and the theory of gravitation, beginning with the equivalence of inertial and gravitational mass. After four years of serious meditation, he wrote (in 1911) on the influence of gravity on light. In 1912, he discussed the velocity of light in a gravitational field (*Ann. d. Phys.* **38**, 443 (1912)). With Marcel Grossmann (his fellow student from the early days at the E.T.H. in Zurich, who was now his colleague at the E.T.H. during the very brief period Einstein spent there after coming from Prague and before going to Berlin), Einstein summed up his ideas on the general theory of relativity and the theory of gravitation in 'Entwurf einer Verallgemeinerten Relativitäts-theorie und eine Theorie der Gravitation' (*Z. Math. u. Phys.* **62**, 225 (1913)). (By the way, it was Grossmann's father who had assisted Einstein in obtaining the job at the *Patentamt* in Bern. Later on, in discussing the origin of his 1913 paper with Marcel Grossmann, Einstein insisted that the entire inspiration and scheme of that paper were his own, and that Grossmann had merely helped him in carrying out the mathematical programme envisaged by him.)

In 1914, Einstein wrote on the formal foundations of general relativity theory (*Sitz. Ber. Preuss. Akad. Wiss.* 1030, (1914)), gave several lectures on the problem of gravitation and relativity, and published a paper with M. Grossmann on the general covariance properties of the field equations of the theory of gravitation (*Z. Math. u. Phys.* **63**, 215 (1914)). During 1915, he published new ideas on general relativity, especially the application to astronomy with the explanation of the perihelion of Mercury.

It was during 1915 that David Hilbert became actively interested in the general theory of relativity, aroused essentially by Einstein's statements of 1914 on its formal foundations. At the end of 1915, Hilbert presented to the *Göttinger Gesellschaft der Wissenschaften* his first *Mitteilung über die Grundlagen der Physik*, and followed it with his *Zweite Mitteilung* at the close of 1916. Both of these papers held great excitement for Felix Klein, who was led to a more compact formulation of Hilbert's calculations with the help of the old and trusted method of Lie's infinitesimal transformations. In 1924, Hilbert himself gave the final, simplified formulation of his two earlier Mitteilungen on the foundations of general relativity. (*Math. Ann.* **92**, 1–32 (1924).)

Einstein's theory of gravitation is founded upon the tensor equation

$$R_{\alpha\beta} - g_{\alpha\beta}R = 8\pi\,(G/c^4)\,T_{\alpha\beta}.$$

These ten equations have between them four identities, four equations that really put no condition at all upon the geometry because they are automatically fulfilled whether these (above) equations are satisfied:

$$(R^{\alpha\beta} - g^{\alpha\beta}R_{;\beta} = 8\,(G/c^4)\,T^{\alpha\beta}_{;\beta}.$$

This was the feature of the Einstein equations that was discussed by Hilbert (20 November 1915): Since the ten differential equations leave a freedom of four arbitrary functions in the solution, they cannot be entirely independent, but must have four inner relations. These relations are a consequence of the Bianchi identities for the full Riemannian tensor, and are generally(?) known in the form of the condition that the Ricci tensor be divergenceless.

Therefore, there are only six net equations for the ten quantities $g_{\alpha\beta}$. This circumstance at first drove Einstein away from the final form (25 November 1915) of the field equations (*Preuss. Akad. Wiss. Berlin*; *Sitzb.* 844 (1915)), so that his preliminary publication (*Preuss. Akad. Wiss. Berlin, Sitzb.* 778 (1915)) put forward equations of the form (11 November 1915),

$$R_{\mu\nu} = 4\pi \left(G/c^4 \right) T_{\mu\nu}.$$

Hilbert (*Gesell. Wiss. Göttingen, Nachrichten*, 395 (1915)) pointed out that the equations set up to determine the $g_{\alpha\beta}$ would be quite unacceptable if they were really expected to determine them. What is really relevant, and what should be determined is the geometry and curvature of space-time. What coordinates we use to describe the geometry should be immaterial. If the equations were completely to determine the ten $g_{\alpha\beta}$, they would tell us not only the geometry but also, unhappily, the coordinates in terms of which that geometry is expressed. However, we know that only the interval between one event and another (between one crossing of world lines and another, for example) has any real significance:

$$ds^2 = g_{\alpha\beta}\,dx^\alpha\,dx^\beta.$$

How one draws coordinate surfaces through space-time is a matter of paper work and bookkeeping, and has nothing to do with real physics.

In his celebrated paper on 'Die Grundlage der allgemeinen Relativitätstheorie' (*Ann. d. Phys.* **49**, 769 (1916)), Einstein mentioned (p. 810): 'Die Feldgleichungen der Gravitation enthalten also gleichzeitig vier Bedingungen, welchen der materielle Vorgang zu genügen hat. Sie liefern die Gleichungen des materiellen Vorganges vollständig, wenn letzterer durch vier voneinander unabhängige Differentialgleichungen charakterisierbar ist.' At the end of this sentence, without pursuing this point further in detail, Einstein gave reference to Hilbert's work in a footnote, referring especially (and only) to p. 3 of Hilbert's 1915 paper in the *Göttinger Nachrichten*.

Hilbert, in retrospect, could not have been satisfied by this weak reference to his work. In a sense, Einstein had 'appropriated' Hilbert's contribution to the gravitational field equations as a march of his own ideas – or so it would seem from the reading of his 1916 *Ann. d. Phys.* paper on the foundations of general relativity. In his article on 'Die Grundlagen der Physik' (*Math. Ann.* **92**, 1–32 (1924)), in which Hilbert brought together his two earlier communications on general relativity, he wrote: 'Die gewaltigen Problemstellungen und Gedankenbildungen der allgemeinen Relativitätstheorie von EINSTEIN finden nun, wie ich in meiner ersten Mitteilung ausgeführt habe, auf dem von MIE betretenen Wege ihren einfachsten und natürlichsten Ausdruck und zugleich in formaler Hinsicht eine systematische Ergänzung und Abrundung.

'Seit der Veröffentlichung meiner ersten Mitteilung sind bedeutsame Abhandlungen über diesen Gegenstand erschienen: ich erwähne nur die glänzenden und tiefsinnigen Untersuchungen von WEYL und die an immer neuen Ansätzen und Gedanken reichen Mitteilungen von EINSTEIN. Indes sowohl WEYL gibt späterhin seinem Entwicklungsgange eine solche Wendung, dass er auf die von mir aufgestellten Gleichungen ebenfalls gelangt, *und anderseits auch* EINSTEIN, obwohl wiederholt von abweichenden und unter sich verschiedenen Ansätzen ausgehend, *kehrt schliesslich in seinen letzten Publikationen geradenwegs zu den Gleichungen meiner Theorie zurück.*' [My emphasis.]

'Ich glaube sicher, dass die hier von mir entwickelte Theorie *einen bleibenden Kern* [my emphasis] enthält und einen Rahmen schafft, innerhalb dessen für den künftigen Aufbau der Physik im Sinne eines feldtheoretischen Einheitsideals genügender Spielraum da ist. Auch ist es auf jeden Fall von erkenntnis-theoretischem Interesse, zu sehen, wie die wenigen einfachen in den Axiomen (I–IV) von mir ausgesprochenen Annahmen der ganzen Theorie genügend sind.'*

Einstein did not entirely (or at all!) favour the developments of general relativity pursued by Weyl. By 1924 Hilbert had decided that, in spite of what Einstein and Weyl had wrought, his own contributions 'einen bleibenden Kern enthalten'. Hilbert's own goal, even in his earlier researches, had been to build a unified theory of electromagnetism and gravitation based upon the work of Mie. To this end, Hilbert made use of Hamilton's principle, applying it to an arbitrary function of the gravitational potential and an electromagnetic four-vector, invariant under an arbitrary transformation of the coordinates. On the basis of a remarkable property of the variational equations, four equations are obtained, connecting the electromagnetic and gravitational quantities.

The undeclared goal of Hilbert's two communications of 1915 and 1916 was to build a field theory of atomic nuclei and electrons, something in which he also did not succeed. What did still remain, of course, even as late as 1924 (or, for that matter, today) was 'ein bleibender Kern' – the fact that Hilbert had provided the form of the Einstein field equations for gravitation, as well as the inner conditions by which they were determined.

Your question is therefore quite to the point, and I hope I have provided the historical background. Hilbert had indeed provided the rigorous mathematical arguments for the form which the gravitational field equations were to obtain. Einstein absorbed these arguments in his great paper of 1916, just as he absorbed other things on several occasions. Hilbert had the vision of a unified field theory before Weyl and Einstein, and in this respect he was also the earlier failure. He had the dream of a universal law accounting both for the structure of the cosmos as a whole and of all atomic nuclei.

As for the form of the gravitational equations in which Einstein finally presented them in his 1916 *Ann. d. Phys.* paper on the foundations of general relativity, and which Hilbert had derived in his Göttingen Nachrichten Mitteilung, the dates for the priority are instructive. Einstein presented his communications of 1915 to the Berlin Academy on November 4, 11, 18, and 25 respectively, while Hilbert's first Mitteilung to the Göttingen Gesellschaft was on November 20, 1915. Hilbert's communication contained the result which Einstein used later.**

---

* Hilbert was very pleased with the geometrical abstraction and its beauty in Einstein's theory. He recommended Einstein in 1915 for the third Bolyai Prize 'for the high mathematical spirit behind all his achievements'. This prize had been awarded first to Poincaré and then to Hilbert.
** Einstein used Hilbert's result in his 1916 paper (Received 29 March 1916).

You ask: 'Is there a reason no one seems to mention this now?' – that is, Hilbert's work on the foundations of general relativity and the theory of gravitation. The only reason, probably, is that people do not read the literature. Another reason is the sociology of science, the question of the cat and the cream. Einstein was the big cat of relativity, and the whole saucer of its cream belonged to him by right and legend, or so most people assume! Still, I am very glad that you have raised the question.

By the way, the mathematical tradition of Göttingen from Gauss to Hilbert formed an important background of the work on quantum mechanics done there. It might perhaps be appropriate to discuss Hilbert's contributions to physics. I have got a copy of the notes of Hilbert's lectures on the mathematical foundations of quantum mechanics, and they make fascinating reading. I expect to use these to discuss Hilbert's interest in the structure of matter.

It might amuse you to know that the first problem of physics which Hilbert decided to work on seriously was the problem of the kinetic theory. He dealt with the problem of collisions in the Maxwell–Boltzmann distribution function, and this was the occasion to apply the theory of integral equations that he had developed to the problems of physics. A fact that is perhaps not known is that David Enskog attended Hilbert's lectures in Göttingen in 1912/13 and wrote his thesis in Uppsala in 1917. Chapman became familiar with Enskog's work only several years after he had done his own work independently. Enskog did not mention in his thesis that he had benefited from Hilbert's lectures, although the latter served as a point of departure for his work on the kinetic theory of non-uniform gases. Once it occurred to me, on the basis of the formal similarity between Hilbert's and Enskog's work, to explore whether Enskog had come in contact with Hilbert. I found Enskog's name, after some search, among those who had taken Hilbert's course on the kinetic theory.

With my best regards,

Sincerely yours,
Jagdish (Mehra)

Professor Eugene Wigner
8 Ober Road
Princeton, New Jersey   08540

8

Theory of Gravitation

Andrzej Trautman

1. INTRODUCTION

The work on, and the understanding of, gravitation greatly influenced not only the physicist's conception of nature but also the development of all exact sciences. Newton invented the method of fluxions, and thereby laid down the foundations of calculus, in connection with his research on the motion of bodies and on the law of universal attraction.[1] The calculus of variations, the theory of differential equations and the perturbation methods of solving them arose directly from the needs of mechanics and astronomy. Through the work of Poincaré[2], the consideration of global and stable properties of motions stimulated the birth of topology. The relativistic theory of gravitation of Einstein[3], and his search for a unified theory[4], enhanced the development of differential geometry. The notion of a superspace introduced recently by J. A. Wheeler[5] provides us with a concrete example of an infinite-dimensional manifold and leads to a number of difficult problems in the theory of Banach manifolds.

The theory of gravitation has had successes in all the fields where gravitational interactions are expected to play a dominant role. The laws of gravitation, very accurately checked within the solar system, seem to be applicable also on a much larger scale. It is amazing – and encouraging – that a simple theory of gravitation provides us with models of the entire Universe, some of which are at least in a qualitative agreement with the observations.

The achievements of the Newtonian theory of gravitation were later overshadowed by those of Maxwell's electromagnetic theory, by the discovery of the atomic nature of matter and by the development of quantum mechanics and relativity. The theory of general relativity although initially poor in experimentally verifiable predictions, greatly influenced our picture of the Universe and the understanding of space and time. It also gave rise to a hope – which is now believed to be false – of constructing a unified, geometric theory of electromagnetism and gravitation. In spite of its profound implications, for a long time Einstein's theory was being developed with little contact with the natural sciences. The situation has changed during the last years, thanks to the startling discoveries in astronomy, the progress in radio and radar measurements and the patient efforts to detect gravitational waves.[6] The theorists have followed suit and done relevant work on the process of collapse and formation of black holes, on new general relativistic effects, on the mechanisms of emission and absorption of

J. Mehra (ed.), The Physicist's Conception of Nature, 179–198. All Rights Reserved
Copyright © 1973 by D. Reidel Publishing Company, Dordrecht-Holland

gravitational radiation, and on the stability of relativistic, gravitating systems. The significance of the new discoveries and observations, as well as the role of general relativity in astrophysics, astronomy and cosmology, have been admirably presented in the lectures of S. Chandrasekhar and D. W. Sciama which are printed in the same volume.

Much excitement and justified interest surrounds the experiments performed with the purpose of measuring the flux of *gravitational waves* falling on the surface of the Earth. According to J. Weber, who initiated this field of research over a decade ago, there are sporadic pulses of radiation which seem to come from the centre of the Galaxy.[7] This result would constitute a beautiful confirmation of Einstein's predictions if it were not for the fact that its interpretation in terms of gravitational waves requires the existence of extremely powerful, hard-to-find sources of radiation. Although the issue is important and interesting, it is more relevant for astrophysics and cosmology than for the theory of gravitation as such. If Weber is right, then we are faced with the challenge to find the sources of the powerful radiation; if he is not, then this only confirms the earlier, conservative estimates of the amount of gravitational radiation in the Universe. In the latter case, more refined techniques than those available now will be needed to detect gravitational waves of cosmic origin. In either case, there does not seem to be any need for a change in the fundamental assumption of the general theory of relativity. Moreover, the recent accurate measurements of the time delay of radar signals passing near the surface of the Sun, and also those of the deflection of radio waves, seem to confirm the theory fairly well and favour the conventional theory rather than its modifications, such as those requiring an additional flat metric or a scalar field.[8]

An outstanding problem of theoretical physics is to build a *quantum theory of space, time, and gravitation*. For brevity, the problem is often formulated by stating that the gravitational field should be quantized. Such a description is not entirely adequate because it presupposes a quantum theory of gravity along the lines of quantum electrodynamics. Gravitation is so closely related to the structure of space-time that it is hard to conceive a profound modification of the description of the former without introducing drastic changes in the nature of the latter.

A pioneering work on the quantum theory of gravity was done in 1930 by L. Rosenfeld and a satisfactory Hamiltonian form of Einstein's equations was given by P. A. M. Dirac. Extensive research on various methods of quantizing general relativity and on possible quantum effects of gravitation has been carried out since 1950.[9, 10] Assuming that it is correct to describe gravitational interactions in terms of their quanta, the main quantitative result is that gravitons may produce observable effects only at extremely high energies, corresponding to the *Planck length*

$$(\hbar G/c^3)^{1/2} \approx 10^{-33} \text{ cm}. \tag{1}$$

For example, according to L. Parker[11], R. U. Sexl and H. K. Urbantke[12], Ya. B. Zeldovich and his co-workers[13], one can expect creation of pairs of particles by very strong gravitational fields, with curvatures of the order of 10^{33} cm^{-1}. It is presumed that such curvatures may occur during gravitational collapse, be it cosmological or local. However, as S. Hawking and R. Penrose[14] point out, and A. Salam predicts on the basis of his theory, extraordinary local effects should take place already at curvatures of the order of 10^{13} cm^{-1}. It is difficult to imagine how a particle such as an electron, whose radius is of the order of 10^{-13} cm, can survive in a space with a (local) radius of curvature 10^{20} times smaller. A different point of view, advanced by J. A. Wheeler, is to consider elementary particles as having a foam-like structure, the foam consisting of highly curved, quantum-fluctuating space-time with a characteristic length given by Equation (1).

The theoretical studies indicate the importance of the Planck length but so far there is no experimental evidence that this quantity is physically relevant in a similar sense as the fine structure constant, the classical radius of the electron, the Chandrasekhar mass or the gravitational radius of the Sun are known to be. In other words, can we be confident that nothing drastic happens when we consider the range of distances from 10^{-13} cm down to 10^{-33} cm (or rather the corresponding range of energies)? In this unexplored region there may occur completely new phenomena which will eventually mask over the quantum gravitational effects, as calculated from the present theory.

A short lecture on a broad subject cannot be comprehensive. It would not be appropriate to review here the fundamentals of the theory of gravitation. The article by J. Mehra, appearing in this volume, contains a lucid account of how the modern theory of gravitation was created, with an emphasis on the role played by Hilbert. In this lecture, I shall restrict myself to a few basic problems connected with the development of general relativity theory and to the Einstein–Cartan theory of gravitation.

2. THE PRINCIPLES

The principles which are associated with the theory of relativity and gravitation were the subject of many controversies and misunderstandings. One of the best known among them has been the discussion on the significance of the 'principle of general covariance', a polemic which started around 1917[15] and has been revived during the recent years by V. A. Fock.[16]

In part, the difficulties are due to a lack of clarity as to what is a principle in theoretical physics. We accept the following definition: a *principle* is a statement about physical theories, formulated on the basis of experiments or by extrapolation from known theories. If the principle is a true statement for any particular theory, then the theory is said to satisfy the principle. As a rule, a principle is meaningful (true or false) for a class of physical theories and not for all theories. In other words, a principle selects a set of theories, namely those for which it is true. These remarks may sound trivial but, if accepted, they show that such familiar arguments as 'it follows from the principle of equivalence alone that light propagating in a gravitational field changes its frequency' cannot stand good. It requires a definite physical theory to investigate the propagation of light. The principle of equivalence is satisfied by several theories, including the Newtonian theory of gravitation[17], which does not allow any reasonable description of electromagnetism.

To illustrate the general definition by a well-understood example, let us consider the *principle of (special) relativity*. It refers to the class of theories which assume an affine (flat) space as a model of space-time. Each theory is characterized by some additional non-dynamical structure (integrable linear connection, metric tensor, absolute time, ether, etc.). Moreover, free motions of point particles are described by a family of straight lines in the affine space, and this family is an open and non-void subset of the space of all straight lines. With every theory there is associated a group of automorphisms: it is the group of all these affine transformations which preserve the additional structure of space-time. The principle of relativity says that the group of automorphisms acts transitively in the family of free motions. Clearly, the principle of relativity, as defined here, is satisfied in Einstein's special theory and in Galilean physics, fails in pre-relativistic electrodynamics, and is meaningless in theories based on a curved space-time.

The *principle of equivalence* refers to classical theories of gravitation which assume an affinely connected space-time (i.e., a differentiable manifold with a linear connection). The principle says that, in the vacuum, the geometry of space-time defines locally only one linear connection. It implies that there is really no such thing as a gravitational force. Indeed, a force is described by a vector field which can be used to build a new linear connection from any given one, in contradiction with the principle of equivalence. Therefore, the equality of inertial and gravitational masses is a consequence of the principle.

I am tempted at this point to make the following remark which goes a little beyond the subject of my talk. As soon as one realizes that the gravitational force is not a correct concept, it becomes clear that any classical force, everything that can be legitimately put on the right-hand side of Newton's law of motion, is of electromagnetic origin. On the other hand, most of theoretical physics, except for general relativity but including quantum theory, is based on concepts such as the energy, a Hamiltonian or a Lagrangian, which all can be traced back to the notion of force. We are not so naive as to try to reduce all phenomena to electromagnetism, as were the nineteenth-century physicists with respect to 'mechanical forces' but we attempt to model all theories after electrodynamics, classical or quantum. It may be that this is one of the reasons of the slow progress in our understanding of the fundamental processes.

In order to formulate the *principle of general invariance* (or the principle of general relativity as it is sometimes called), it is desirable to distinguish between the dynamical and the absolute (non-dynamical) elements of a theory.[18, 19] The dynamical elements characterize the history of a physical system described by the theory and are subject to equations of motion. In any given theory, the absolute elements are the same for all histories. For example, the metric tensor is an absolute element in the theory of special relativity and acquires a dynamical character in Einstein's theory of gravitation. The automorphisms or symmetries of a theory are the transformations which preserve the absolute. According to the principle of general invariance, the automorphism group of a relativistic theory of gravitation consists of all diffeomorphisms of space-time. This is a highly non-trivial and strong statement; it has nothing to do with the possibility of going over to curvilinear coordinates.

The following is a list of the absolute elements in the known classical theories, supplemented by a conjecture about the future theory of space, time, and gravitation:

Theory of	Absolute elements		
	Time	Metric and flat linear connection	Topology and differential structure
Galilean mechanics	yes	yes	yes
special relativity	no	yes	yes
general relativity	no	no	yes
the future	no	no	no

The topological and differential structures of space-time do not seem to possess a well-defined operational meaning. Therefore, it is likely that they will have to be abandoned, or rather replaced by another structure which will be more closely related to, and influenced by, physical phenomena than the absolute, locally Euclidean manifold structure of space-time assumed in all current theories. In my opinion, a satisfactory quantum theory of space, time, and gravitation will have to do away with the notion of a differentiable manifold as a model of space-time.

The *principle of locality* in classical physics can also be precisely stated in the language of differential geometry. Roughly, it says that all fundamental laws of physics can be reduced to equations involving only local differential operators of finite order. The *principle of Mach* is a negation of the principle of locality. Of course, there is nothing unique or final about the formulation of any of the principles. The definitions given here should be considered as tentative examples of how the subject can be approached. In particular, Mach's principle ought to be sharpened to become significant.

3. CATEGORIES, FIBRE BUNDLES AND GAUGE INVARIANCE

The purpose of a physical theory is to construct mathematical models of nature, models that can be used to explain and predict physical phenomena and events. Any particular theory, perhaps with the exception of cosmology, provides us with many models, each of them adapted to a specific situation and giving a good description of events within a bounded region of space and time and with an accuracy characteristic of the theory. The details change from one model to another but all models of a theory have certain common features, determined by the basic assumptions of the theory. This remark leads at once to the idea that it should be possible to organize the mathematical models used in a physical theory into what is nowadays called a category.[20]

To establish the terminology and notation, let me recall that a *category* $\mathscr{A}$ consists of a class of *objects* $A, B, C, \ldots$, and a class of sets $\mathrm{Mor}(A, B)$, $\mathrm{Mor}(B, C), \ldots$ of elements called *morphisms* of $\mathscr{A}$. If $f \in \mathrm{Mor}(A, B)$ and $g \in \mathrm{Mor}(B, C)$, then there exists the composite morphism $g \circ f \in \mathrm{Mor}(A, C)$ and the composition of morphisms is associative. For any object B there is a morphism $1_B \in \mathrm{Mor}(B, B)$ such that

$$\text{if} \quad f \in \mathrm{Mor}(A, B) \quad \text{and} \quad g \in \mathrm{Mor}(B, C), \quad \text{then} \quad 1_B \circ f = f \quad \text{and} \quad g \circ 1_B = g.$$

In most cases, morphisms are certain mappings and one writes

$$f : A \to B \quad \text{instead of} \quad f \in \mathrm{Mor}(A, B).$$

A morphism $f : A \to B$ is called an *isomorphism* if there exists a morphism $f^{-1} : B \to A$, called its inverse, such that

$$f \circ f^{-1} = 1_B \quad \text{and} \quad f^{-1} \circ f = 1_A.$$

From any category $\mathscr{A}$ one can form the category $\mathscr{I}\mathscr{A}$ whose objects coincide with those of $\mathscr{A}$ and whose morphisms are isomorphisms of $\mathscr{A}$.

For example, the category of sets, $\mathscr{E}ns$, has sets as objects and mappings as morphisms; bijections are isomorphisms. The category of (real) vector spaces $\mathscr{V}ect$ has linear mappings as morphisms. In physics, of importance is the category $\mathscr{D}iff$ of finite-dimensional differential manifolds, with differentiable mappings as morphisms, and the category $\mathscr{A}ff$ of affine spaces, which may be considered as a subcategory of $\mathscr{D}iff$. Let $A_1 = (E_1, V_1, +)$ and $A_2 = (E_2, V_2, +)$ be two affine spaces, where E_1 and E_2 are the underlying sets, V_1 and V_2 are the associated vector spaces and $+$ in both cases denotes the transitive and free action of the additive groups V_1 and V_2 in E_1 and E_2, respectively. By definition, a morphism in $\mathscr{A}ff$ is a map $f : E_1 \to E_2$ such that there exists a linear map $\tau f : V_1 \to V_2$ and

$$f(v + p) = \tau f(v) + f(p)$$

for any $p \in E_1$ and $v \in V_1$. If $g : E_2 \to E_3$ is another affine morphism, then

$$\tau(g \circ f) = \tau g \circ \tau f.$$

We have here an example of a correspondence between categories, referred to as a *functor*.

More generally, if $\mathscr{A}$ and $\mathscr{B}$ are categories, a law τ associating to objects A, B and morphisms f, g of $\mathscr{A}$ certain objects and morphisms of $\mathscr{B}$, and such that

$$\text{if} \quad f : A \to B, \quad \text{then} \quad \tau(f) : \tau(A) \to \tau(B),$$
$$\tau(1_A) = 1_{\tau(A)}, \qquad \tau(g \circ f) = \tau(g) \circ \tau(f)$$

is called a covariant functor. (Here 1_A denotes the identity morphism of A.) A contravariant functor $\tau:\mathscr{A}\to\mathscr{B}$ is characterized by

$$\text{if } f:A\to B, \quad \text{then } \tau(f):\tau(B)\to\tau(A),$$
$$\tau(1_A)=1_{\tau(A)} \quad \text{and } \tau(g\circ f)=\tau(f)\circ\tau(g).$$

Clearly, $\tau:\mathscr{A}ff\to\mathscr{V}ect$ defined, in the notation of the previous paragraph, by

$$\tau(A)=E, \qquad \tau(f)=\tau f$$

is a covariant functor.

For any category $\mathscr{A}$ and any object C in $\mathscr{A}$ one defines the contravariant functor

$$\tau^C:\mathscr{A}\to\mathscr{E}ns$$

by

$$\tau^C(A)=\mathrm{Mor}\,(A,C)$$

and

$$(\tau^C f)\,(g)=g\circ f$$

for any $f:A\to B$ and $g:B\to C$. For example, if $\mathscr{A}$ is the category of vector spaces and $C=R$ then τ^C associates to any vector space V its dual V^*; this defines a contravariant functor $*:\mathscr{V}ect\to\mathscr{V}ect$.

One of the important applications of categories and functors is to define the concept of naturality. Given two categories, $\mathscr{A}$ and $\mathscr{B}$, and two functors of the same variance $\tau_1,\tau_2:\mathscr{A}\to\mathscr{B}$, a natural transformation N associates to each object A of $\mathscr{A}$ a morphism in $\mathscr{B}$, $N(A):\tau_1(A)\to\tau_2(A)$ such that the diagram

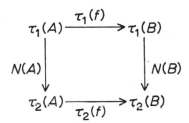

commutes for any $f:A\to B$. If $N(A)$ is an isomorphism for any A, one says that N establishes a *natural equivalence* of the functors τ_1 and τ_2. For example, let $\mathscr{A}=\mathscr{B}= =\mathscr{V}ect$ and $\tau_1=id$, $\tau_2=**$ (double dual). The mapping $N(V):V\to V^{**}$, defined for any vector space V by

$$\langle v^*, N(V)v\rangle = \langle v, v^*\rangle, \qquad v\in V, \quad v^*\in V^*,$$

is a natural transformation. It becomes a natural equivalence when restricted to the subcategory of finite-dimensional vector spaces.

Functors may be thought of as general constructions. The existence of a natural equivalence between a pair of functors means that these constructions lead to essentially the same result.

An important mathematical concept whose relevance for theoretical physics has been recently recognized is that of a *fibre bundle*.[21] Fibre bundles generalize the notion of a Cartesian product; locally, they can always be represented as Cartesian products. One may see the need for such a generalization by considering the development of our ideas of space and time.

According to the Ancient Greeks' picture of the world, space-time[†] E was a Cartesian product of time T and space S[22]: to any event one could associate an instant of time t and a location in space s; both time and space were absolute.

In Newtonian physics, space-time E may be represented as a product $T \times S$ in many ways; none of these representations are natural in a sense, which can be easily related to the notation of natural transformations. Space is relative because there is no absolute method of ascertaining whether or not two non-simultaneous events happen at the same place. In other words, there is no natural horizontal slicing of E; there is only a vertical fibring corresponding to the projection $\pi : E \to T$ which associates to any event $p \in E$ the corresponding instant of time $t = \pi(p)$; or, time is absolute.

The last example provides us with the essential set-theoretic ingredients of a *bundle* A: it consists of two sets, say M and E, called respectively the *bundle space* (or the total space) and the *base space*, and of a surjective map $\pi : M \to E$, called the *projection*; shortly $A = (M, E, \pi)$. The sets M and E usually have some additional structure, such as that of a topological space or a differential manifold, and π is then assumed to be compatible with these structures (i.e., to be a morphism in the corresponding category). In most cases, the spaces $\pi^{-1}(p)$, $p \in E$, are all alike, i.e., isomorphic (in that category) to a space F, called the *typical fibre*; the set $\pi^{-1}(p)$ is then called the *fibre* over p. For any $p \in E$, there is usually more than one isomorphism of $\pi^{-1}(p)$ onto F (otherwise M would admit a natural representation as $E \times F$); if f and f' are two such isomorphisms, then $f' \circ f^{-1}$ is an automorphism of F; the set of all automorphisms of F which are of this form constitutes a group, called the *structure group* G of the bundle.

In physics, we are most often interested in differential bundles: M, E and F are differential manifolds, π is differentiable and for any $p \in E$ there exists a neighbourhood U of p and a differential isomorphism (diffeomorphism) $h : \pi^{-1}(U) \to U \times F$ such that $pr_1 \circ h = \pi$.

The class of all differential bundles forms a category $\mathscr{B}un$; its morphisms are pairs (f, a) of mappings such that $\pi_2 \circ f = a \circ \pi_1$ with (M_1, E_1, π_1) and (M_2, E_2, π_2) denoting the bundles. A product bundle is $(E \times F, E, pr_1)$, where $pr_1(p, q) = p$. A bundle $\mathscr{B}un$-isomorphic to a product bundle is called *trival*.

Among differential bundles especially important are *vector bundles* and *principal bundles*. Roughly speaking, a vector bundle is a differential bundle with a vector space playing the role of the typical fibre. For example, the tangent bundle $T(E)$ and the

---

[†] Strictly speaking, one should distinguish between the physical space-time and models used to describe it. A sentence like 'according to Einstein, space-time is a Riemannian manifold' should really read 'in Einstein's theory a Riemannian manifold is used as a model of space-time'. We shall adhere, however, to the convenient abuses of language which prevail in the physical literature.

cotangent bundle $T^*(E)$ of a manifold E are vector bundles over E. There is a co-variant functor T from $\mathscr{Diff}$ into the category of vector bundles $\mathscr{VBun}$; $T(f)$ is the tangent mapping of the differentiable map f. A principal bundle has a Lie group as typical fibre and structure group at the same time; the group acts freely in the bundle and transitively on its fibres. For example, the bundle of frames $B(E)$ of a manifold E is a principal bundle with structure group $GL(n)$, where $n=\dim E$. There is a covariant functor B from $\mathscr{IDiff}$ into the category of principal bundles $\mathscr{PBun}$.

The functors $T, \tau : \mathscr{Aff} \to \mathscr{VBun}$, where T is the tangent functor and

$$\tau(A) = E \times V, \qquad A = (E, V, +)$$

are naturally equivalent to each other. The natural transformation

$$N(A): E \times V \to T(E) \tag{2}$$

associates to $(p, v) \in E \times V$ the vector $X \in T(E)$ tangent to the curve $t \mapsto tv + p$ at p.

We may now list a number of categories that are frequently used in physics. In most cases, it is possible to restrict the category by specifying the number of dimensions of its objects; this will not be done here because the dimensionality of space-time does not enter our elementary considerations. We shall only assume that all manifolds and vector spaces are finite dimensional.

I. The *Galilean category* $\mathscr{Gal}$ has Galilean spaces as objects. A Galilean space is an affine space $(E, V, +)$ endowed with a bilinear map

$$h: V^* \times V^* \to \mathbf{R}$$

which is (a) symmetric, (b) positive, and (c) of rank $n-1$, where $n = \dim V$. If $(E_1, V_1, +, h_1)$ and $(E_2, V_2, +, h_2)$ are two Galilean spaces, the affine morphism f is a Galilean morphism if $h_1 \circ (\tau f)^* = h_2$. A Galilean automorphism is called a Galilean transformation.

Let $\Sigma \subset V^*$ be the null space of h and $S \subset V$ the subspace of all vectors orthogonal to Σ. For any Galilean space $A = (E, V, +, h)$ the quotient space $T = E/S$ is called the absolute time of A. If

$$\pi: E \to T$$

is the canonical projection, then (E, T, π) is a fibre bundle with S as the typical fibre. The relation of absolute time to Galilean transformations is described by the following proposition: There is a covariant functor $\sigma: \mathscr{Gal} \to \mathscr{Bun}$ defined by

$$\sigma(A) = (E, T, \pi) \tag{3}$$

and

$$\sigma(f) = (f, a)$$

where $f: E_1 \to E_2$ is a Galilean morphism and $a: T_1 \to T_2$ is the unique map satisfying $\pi_2 \circ f = a \circ \pi_1$.

II. The *category of phase spaces*, $\mathscr{Phase}$, plays a role in classical mechanics. A phase space is a pair (M, β) consisting of an even-dimensional differential manifold M and a non-degenerate two-form field β on M. Morphisms of $\mathscr{P}hase$ are defined as differentiable maps carrying one two-form into another. Automorphisms of $\mathscr{P}hase$ are called canonical transformations. In classical mechanics, one makes frequent use of the existence of the following two functors:

$$\mathscr{I}\mathscr{D}iff \xrightarrow{T^*} \mathscr{P}hase \xrightarrow{\Pi} \mathscr{L}ie\,\mathscr{A}lg \tag{4}$$

T^* associates with a differential manifold E the phase space $(T^*(E), \beta_E)$, where $\beta_E = d\alpha_E$ and α_E is the canonical form on $T^*(E)$; Π is the Poisson functor mapping (M, β) into the Lie algebra of differentiable functions on M, with a bracket defined by β; the images of morphisms under T^* and Π are defined in an obvious way.

III.   The *classical theory of fields* has $\mathscr{V}\mathscr{B}un$ as the underlying category. Let σ denote a differentiable action of $GL(n)$ in R^m; $\sigma: GL(n) \rightarrow GL(m)$, $\sigma_{id} = id$ and $\sigma_a \circ \sigma_b = \sigma_{ab}$ for any $a, b \in GL(n)$. One defines the functor $\bar{\sigma}: \mathscr{I}\mathscr{D}iff_n \rightarrow \mathscr{V}\mathscr{B}un$ by introducing in the set

$$\bar{\sigma}(E) = (B(E) \times \mathbf{R}^m)/GL(n)$$

the structure of a vector bundle over the n-dimensional differential manifold E; the action of $GL(n)$ in $B(E) \times \mathbf{R}^m$ is defined by $(e, q) \mapsto (ea, \sigma_{a^{-1}}(q))$ for any $a \in GL(n)$, $e \in B(E)$ and $q \in \mathbf{R}^m$. A cross-section of $\bar{\sigma}(E)$ is called a field of quantities of type σ. For example, if σ is the obvious representation of $GL(n)$ in the space of k-forms, $\mathbf{R}^m = \Lambda^k \mathbf{R}^{n*}$, then $\bar{\sigma}(E) = \Lambda^{k*}(E)$ is the bundle of k-forms over E.

Let $\mathscr{V}\mathscr{B}un_n$ denote the category of vector bundles over n-dimensional differential manifolds. If (M_i, E_i, π_i), $i = 1, 2$, are two such bundles, then $f: M_1 \rightarrow M_2$ is a morphism of $\mathscr{V}\mathscr{B}un_n$ if it is differentiable, admits a differentiable isomorphism $a: E_1 \rightarrow E_2$ such that $\pi_2 \circ f = a \circ \pi_1$ and is linear on each fibre. For any integer k one defines the kth jet extension functor[23]

$$J^k: \mathscr{V}\mathscr{B}un_n \rightarrow \mathscr{V}\mathscr{B}un_n$$

which plays a basic role in the theory of partial differential equations and in particular for classical fields. A $\mathscr{V}\mathscr{B}un_E$-morphism

$$\lambda: J^k \circ \bar{\sigma}(E) \rightarrow \Lambda^{n*}(E)$$

is called a Lagrangian for the field of quantities of type σ. The fundamental relations between Lagrangians, Euler-Lagrange equations, invariant transformations and conservation laws may be given a natural and simple formulation in this framework.

IV.   The underlying category of *quantum physics* is that of complex Hilbert spaces, $\mathscr{H}ilb$. Its morphisms are unitary mappings. Let (E, μ) be a manifold E with a differentiable measure μ; one associates to it the Hilbert space $L^2(E, \mu)$ of square integrable complex functions, with a scalar product defined by

$$(f \mid g) = \int_E \bar{f}g\mu.$$

The class of all manifolds with differentiable measures constitutes a category $\mathscr{M}$; its morphisms are measure-preserving diffeomorphisms. There is a covariant functor $L^2 : \mathscr{M} \to \mathscr{H}i\ell\ell$ which assigns to an $\mathscr{M}$-morphism h the unitary transformation $L^2(h)$ defined by $L^2(h)(f) = f \circ h^{-1}$. Moreover, to a vector field X on E which preserves μ,

$$\mathscr{L}_X \mu = 0.$$

the functor L^2 associates the antihermitean operator $\mathscr{L}_X$ on $L^2(E, \mu)$, i.e., the Lie derivative with respect to X. In other words, L^2 gives rise to a functor $\mathscr{M} \to \mathscr{L}ie\,\mathscr{A}\ell g$. The study of its relation to the functor $\Pi \circ T^*$ defined by (4), is known as the problem of quantizing a mechanical system.

The principle of relativity as formulated in the preceding section, implies that there are no privileged inertial systems and this, in turn, may be interpreted to mean that Galilean space-time is not a Cartesian product of space and time (cf. Fig. 1). The principle of general invariance has a similar consequence. In the theory of special relativity, space-time is a flat (affine) space, i.e., a Riemannian space with an integrable linear connection. The existence of distant parallelism in this case implies that $B(E)$, the bundle of frames, is a Cartesian product. This is no longer so in general relativity where the result of transferring a vector from one point to another by parallel transport along a curve depends on that curve. One is tempted to say that the bundle of frames of space-time in general relativity is not trivial. However, according to the precise definition given above, $B(E)$ is trivial unless E has a non-Euclidean topology. Any global coordinate system on E induces an isomorphism of $B(E)$ onto $E \times GL(n)$. Nevertheless, the intuitive property of '$B(E)$ not being a product' may be given a precise formulation in terms of categories.

On the category $\mathscr{I}\mathscr{A}ff$, which is appropriate for both Galilean physics and special relativity, one can define two functors to $\mathscr{P}\mathscr{B}un$. One of them is the functor B associating to $E \in \mathscr{I}\mathscr{D}iff$, the bundle of frames $B(E)$, the second, C, is a functor of constructing the product bundle

$$C(A) = E \times B(V),$$

where $A = (E, V, +)$ and $B(V)$ denotes the set of all vector frames of V. These functors are naturally equivalent, as may be seen from the commutative diagram

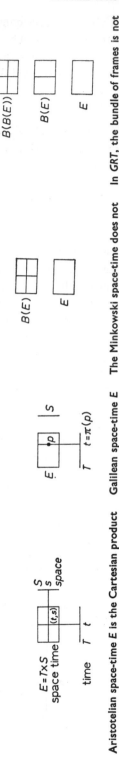

Aristotelian space-time E is the Cartesian product of time T and space S. Both space and time are absolute. This structure is assumed also in prerelativistic electrodynamics.

Galilean space-time E is a fibre bundle over the base T (time).

The Minkowski space-time does not even have a natural fibring. But the bundle of frames $B(E)$ is a product bundle.

In *GRT*, the bundle of frames is not a product but admits teleparallelism which turns $B(B(E))$ into a product.

Fig. 1.   Historical development of ideas on the structure of space and time.

where f is any affine isomorphism and $BN(A)$ is the $\mathscr{PB}un$-isomorphism induced from the $\mathscr{VB}un$-isomorphism $N(A)$ described in Equation (2). Nothing analogous exists for the larger category $\mathscr{IDiff}$; the bundles $B(E)$ and $E \times B(\mathbf{R}^n)$ may be isomorphic but there is no natural equivalence of the corresponding functors.

A similar analysis may be applied to show that, in the Galilean category, the functor σ occurring in Equation (3) is not naturally equivalent to the 'product functor' $\mathscr{G}al \to \mathscr{B}un$, associating with $A \in \mathscr{G}al$ the product bundle $(T \times S, T, pr_1)$.

Keeping in mind that all statements about spaces being or not being products should be understood as referring to the natural equivalence of appropriate functors, we may now *compare the meaning of the two principles of relativity*:

(a) the special principle implies that space-time E is not a product;

(b) the general principle implies that the bundle of frames $B(E)$ is not a product. These are analogous statements but they refer to different spaces and no wonder that this has led to numerous controversies in the past.

In the theory of general relativity, one can take as the underlying category that of differential manifolds with linear connections. A linear connection on E induces a privileged field of linear bases on $B(E)$ and thus turns $B(B(E))$ into a product.[25] One may speculate as to the existence of a theory of space-time in which $B(B(E))$ is not a product.† Even if it should turn out that this generalization is physically uninteresting, it is clear that fibre bundles provide us with a deep insight into the structure of space-time and the nature of its theories.

For a long time, it has been recognized that '*gauge-invariant theories*', such as electrodynamics, are conveniently described in terms of principal bundles with connections. If P is a principal bundle over the base E, with structure group G and a connection form ω, the group G may be interpreted as the group of gauge transformations of the first kind. For any cross-section $\phi: E \to P$, the form on E

$$A = \phi^* \omega$$

is the 'potential', whereas the two-form

$$F = \phi^* \Omega$$

$$\Omega = d\omega + \omega \wedge \omega$$

is the 'field' arising from gauge invariance. A change of the cross-section of P induces a change in A interpreted as a gauge transformation of the second kind. For $G = \mathbf{SO}(2)$ and $\mathbf{SO}(3)$ one obtains, in this way, the potentials and the fields of the Maxwell and Yang-Mills theories, respectively.[26]

An infinitesimal connection in P may be used to define covariant differentiation in vector bundles associated to P. In particular, if one constructs a complex vector

---

† D. D. Ivanenko suggested that the construction of such a theory should be referred to as the second relativization. It has been pointed out by F. A. E. Pirani that a theory of space-time based on conformal geometry is of this type.

bundle associated to the electromagnetic bundle by the homomorphism

$$\sigma(e^{i\alpha}) = e^{in\alpha}, \qquad \alpha \in \mathbf{R}, \quad n \in \mathbf{N},$$

he obtains for the covariant derivative, expressed in terms of local coordinates,

$$\nabla_k = \partial_k - inA_k, \quad (k = 1, 2, 3, 4),$$

and this may be used to justify the known form of the minimal electromagnetic coupling.

It is clear that the underlying category for the theory of fields connected with gauge invariance with respect to a Lie group G is that of principal bundles over space-time, with G as the structure group. A choice of a particular physical situation implies the choice of an object in the category, together with an infinitesimal connection; choosing a gauge is equivalent to picking up a cross-section of the bundle. All gauges are on the same footing (principle of relativity of gauges) because a bundle P over E is not naturally isomorphic to $E \times G$, in the sense explained above.

It is sometimes asserted that the general theory of relativity may also be obtained in this way, by taking G to be the Lorentz group or the Poincaré group.[27] This is not quite the case: the general-relativistic principal bundle has a structure richer than that of a bundle with the Lorentz group $\mathbf{O}(1, 3)$ as the typical fibre. In other words, the underlying category of general relativity is essentially narrower than that of principal bundles with $\mathbf{O}(1, 3)$ as the structure group. This is due to the following theorem: a principal fibre bundle P over an n-dimensional manifold E and with structure group $GL(n)$ is $\mathscr{PB}un$-isomorphic to $B(E)$ if and only if there exists an $\mathbf{R}^n$-valued one-form θ on P, such that

$$\theta(X) = 0 \Leftrightarrow T(\pi)X = 0,$$

$$\psi_a^*\theta = a^{-1}\theta,$$

where $\pi : P \to E$ is the projection and $\psi_a : P \to P$ denotes the action of $a \in GL(n)$ in P. The form θ is often called the 'soldering form' of P; the bundle $B(E)$ is 'soldered' to E rather than being loosely connected to E, as general principal bundles are. The covariant exterior differential of θ is the torsion form of the connection. Note also, that, for any manifold E, one can introduce the product bundle $E \times GL(n)$. In general, not only there is no natural isomorphism of $B(E)$ on $E \times GL(n)$ but no global isomorphism whatsoever (e.g., if E is a two-sphere).

A disadvantage of the gauge approach to electrodynamics is that it does not provide a natural method of deriving the other half of Maxwell equations (i.e., other than $dF = 0$). The full set of Maxwell equations is known to follow from a simple action principle in the Kaluza–Klein theory, or one of its modifications.[28] It is interesting to know that, in fact, there is a definite isomorphism between the theory based on an infinitesimal connection and the Kaluza–Klein five-dimensional theory. This isomorphism may be extended to a large class of theories with gauge invariant fields. In other words, for any such theory it is possible to construct a multidimensional,

Riemannian space which bears the same relation to that theory as the Kaluza–Klein space to electrodynamics.[29]

Let G be a Lie group, possessing an invariant metric h, i.e., a symmetric non-degenerate covariant tensor field of second order, defined on G and invariant with respect to both left and right translations. For example, if G is semi-simple, then one can define h by $h_e(A, B) = Tr(Ad_A \circ Ad_B)$ where $Ad_A(C) = [A, C]$, A, B, C belong to the Lie algebra of G and e is the unit of G. An abelian group, such as $SO(2)$, also has an invariant metric. Given a principal fibre bundle P with structure group G, over a base manifold E (space-time) with a Riemannian metric g, one can define a Riemannian metric γ on P as follows. Let X be a vector tangent to P at r and write $\gamma(X)$, $h(A)$, etc., instead of $\gamma(X, X)$, $h(A, A)$, etc. We put

$$\gamma_r(X) = g_{\pi(r)}(T(\pi)X) + h_e(\omega(X)).$$

It follows from the properties of h that γ is non-singular and invariant with respect to G,

$$\psi_a^* \gamma = \gamma \quad \text{for any} \quad a \in G.$$

In particular, if $G = SO(2)$, its Lie algebra can be identified with $\mathbf{R}$ and h may be taken to be the Euclidean metric on $\mathbf{R}$ (possibly, with a numerical coefficient), then γ on P is the Riemannian metric of the five-dimensional Kaluza–Klein theory. It is also clear how one can construct a principal fibre bundle from the Kaluza–Klein space.

This construction, when applied to the theory of a general field arising from gauge invariance, leads to the following possibility. One can consider the action integral $\int_{\pi^{-1}(\Omega)} K$ where K is the Ricci form, corresponding to the metric γ, and $\Omega \subset E$. By varying this action, with due care not to spoil the invariance of γ with respect to G, one obtains a set of field equations, analogous to the Einstein–Maxwell set that one gets in the Kaluza–Klein theory.

4. THE EINSTEIN–CARTAN THEORY

In 1922 Elie Cartan[30] proposed a slight modification of Einstein's theory of gravitation. According to Cartan, space-time corresponding to a distribution of matter with an intrinsic angular momentum should be represented by a curved manifold with *torsion*, the latter being *related to the density of spin*. This idea may be made plausible by the following considerations.

In the theory of special relativity, the group of inhomogeneous Lorentz transformations (the Poincaré group) plays a fundamental role in the description of elementary physical phenomena. In Cartesian coordinates (x^i), an infinitesimal Poincaré transformation is of the form

$$\delta x^i = \lambda^i{}_j x^j + \mu^i \quad (i, j = 1, 2, 3, 4) \tag{5a}$$

where

$$\lambda_{ij} + \lambda^i_j = 0. \tag{6a}$$

The Lie algebra of the Poincaré group has two basic invariants, interpreted physically as *mass* and *spin*. In Einstein's theory of general relativity, mass directly influences curvature but spin has no similar dynamical effect. On the other hand, curvature and torsion are related, respectively, to the groups of homogeneous transformations and of translations in the tangent spaces of a manifold endowed with a linear connection. Indeed, let (θ^i) be a field of co-frames, i.e. a set of four fields of one-forms which are linearly independent at each point of the manifold. If $(\omega^i{}_j)$ are the one-forms of the linear connection with respect to (θ^i), then the curvature and torsion two-forms are, respectively,

$$\Omega^i{}_j = d\omega^i{}_j + \omega^i{}_k \wedge \omega^k{}_j = \tfrac{1}{2}\theta^k \wedge R^i{}_{jk} = \tfrac{1}{2}R^i{}_{jkl}\theta^k \wedge \theta^l,$$

and

$$\Theta^i = d\theta^i + \omega^i{}_j \wedge \theta^j = \tfrac{1}{2}\theta^j \wedge Q^i{}_j = \tfrac{1}{2}Q^i{}_{jk}\theta^j \wedge \theta^k.$$

Denoting by D the exterior covariant derivative[25], one can define a radius-vector as a field (x^i) such that

$$Dx^i = \theta^i. \tag{7}$$

In a general curved space Equation (7) has no solutions but it can always be integrated along a curve. When this is done for a loop, one finds that the vector (x^i) does not return to its initial value. For an infinitesimal closed curve, the change in the radius-vector is

$$\delta x^i = (\Omega^i{}_j x^j - \Theta^i) \times \text{surface element}. \tag{5b}$$

If the linear connection is metric,

$$Dg_{ij} = 0, \tag{8}$$

then

$$\Omega_{ij} + \Omega_{ji} = 0. \tag{6b}$$

In other words, the curvature and torsion induce, respectively, a Lorentz transformation and a translation of the radius-vector field constructed along a closed curve in a space with a metric linear connection.

Cartan's basic idea has been developed by several authors.[31] The generalization due to Cartan constitutes a slight departure from Einsteins' theory: the field equations in empty space remain unchanged. In our opinion, the Einstein–Cartan theory is the simplest and the most natural modification of the original, Einstein's theory of gravitation. This modification deserves to be analyzed in detail, in precedence over the theories requiring an additional scalar field to describe gravitational phenomena.

The desirability of such an analysis may be related to recent discoveries in astronomy. It is conceivable that torsion may produce observable effects inside those objects which, as the neutron stars, have built-in strong magnetic fields, possibly accompanied by a substantial average value of the density of spin. One is tempted to speculate that the intrinsic angular momentum may influence – or even prevent – the occurrence of singularities in gravitational collapse and cosmology. A recent result of W. Kopczyński[32] supports this idea.

For a body with given values of spin and mass, the dimensionless numbers characterizing the order of magnitude of the effects of torsion and of curvature are, respectively,

$$\text{spin/(radius)}^2 \quad \text{and} \quad \text{mass/radius.}$$

(We use a system of units in which the gravitational constant and the velocity of light are equal to 1.) For an electron, the ratio of these two (very small) numbers is of the order of $1/\alpha \approx 137$; the influence of spin on geometry is larger than that of mass. This is no longer so for matter in bulk because mass is essentially additive whereas in most circumstances spins cancel out one another.

The *field equations* of the Einstein–Cartan theory can be derived from a variational principle,

$$\delta \int (K + L) = 0,$$

where L is the Lagrangian (four-form) of matter and K is the Ricci four-form,

$$16\pi K = \eta_k{}^l \wedge \Omega^k{}_l$$

and

$$\eta_{ij} = \tfrac{1}{2}\theta^k \wedge \eta_{ijk} = \tfrac{1}{2}\eta_{ijkl}\theta^k \wedge \theta^l,$$

$$\eta_{ijkl} = \eta_{[ijkl]}, \quad \eta_{1234} = |\det g_{mn}|^{1/2}.$$

If the sources of the gravitational field are described by a tensor or spinor field (φ_A), then, by varying with respect to θ^i, $\omega^k{}_l$ and φ_A, and assuming Equation (8), one arrives at the system of equations

$$\tfrac{1}{2}\eta_{ijk} \wedge \Omega^{jk} = -8\pi t_i, \tag{9}$$

$$\eta_{ijk} \wedge \Theta^k = 8\pi s_{ij}, \tag{10}$$

$$L_A = 0, \tag{11}$$

where t_i and s_{ij} are the densities (three-forms) of energy-momentum and of spin, respectively. In the absence of spin, the energy-momentum density is symmetric

$$\theta^k \wedge t^l = \theta^l \wedge t^k$$

and Equation (9) goes over into the Einstein equation. In the general case, there is a symmetric energy-momentum tensor

$$T^{ij} = \theta^j \wedge t^i - \tfrac{1}{2}Ds^{ij}.$$

In the approximation of special relativity there is a radius vector (x^i) subject to Equation (7) and the conservation law of energy-momentum, $Dt_i = 0$, together with the symmetry of T^{ij}, implies that the total angular momentum is conserved:

$$\text{if} \quad L_A = 0, \quad \text{then} \quad D(x^i t^j - x^j t^i + s^{ij}) = 0.$$

Similarly as in Einstein's theory, the *equations of motion* can be deduced, in simple cases, directly from the field Equations (9) and (10), without using Equation (11). The Bianchi identities for a curved space with torsion, applied to Equations (9) and

(10) yield the relations

$$Dt_j = Q^k_{\ j} \wedge t_k - \tfrac{1}{2} R^{kl}_{\ \ j} \wedge s_{kl}, \tag{12}$$

$$Ds_{kl} = \theta_l \wedge t_k - \theta_k \wedge t_l. \tag{13}$$

In the absence of spin, Equation (12) reduces to the usual covariant conservation law $Dt_j = 0$.

To derive the equations of motion of a spinning fluid or dust, it is convenient to introduce a 'particle derivative' defined as follows.[33] Let $u = v^i \eta_i$ be the three-form dual with respect to the velocity vector field (v^i), $\eta_i = \tfrac{1}{3} \theta^j \wedge \eta_{ij}$. The particle derivative of a tensor field (φ_A) with respect to (v^i) is given by the formula

$$\dot{\varphi}_A \eta = D(\varphi_A u),$$

where $\eta = \tfrac{1}{4} \theta^i \wedge \eta_i$ is the volume element in space-time. A *spinning dust* may be defined as a continuous medium characterized by its velocity (v^i), the density of energy and momentum (P_i), and the density of spin (S_{ij}). The three-forms of energy-momentum and of spin are

$$t_i = P_i u \quad \text{and} \quad s_{ij} = S_{ij} u,$$

respectively. From Equation (13) there follows the relation

$$P^i = \varrho v^i + v_k \dot{S}^{ki}, \tag{14}$$

where $\varrho = g_{ij} P^i v^j$ and $\dot{S}^{ij}$ is the particle derivative of S^{ij} with respect to (v^i). Equation (13) is equivalent to the system consisting of Equation (14) and the equation of motion of spin,

$$\dot{S}^{ij} = v^i v_k \dot{S}^{kj} - v^j v_k \dot{S}^{ki}.$$

The modified conservation law given by Equation (12) gives rise to the equation of *translatory motion*

$$\dot{P}_i = Q^k_{ij} v^j P_k + \tfrac{1}{2} R^{kl}_{\ \ ij} v^j S_{kl} \tag{15}$$

which is a generalization, to the Riemann–Cartan space of an equation derived by Mathisson[34] and Papapetrou[35] for point particles with an intrinsic angular momentum. If the dust has no spin, $S_{kl} = 0$, then there is no torsion, and the equations of motion are simply

$$\dot{\varrho} = 0 \quad \text{and} \quad \dot{P}_i = 0.$$

The Einstein–Cartan theory gives rise to a number of interesting possiblitities. According to F. Hehl[36] the new theory may contribute to an explanation of weak interactions. A more conservative attitude is to look for new macroscopic effects in regions with strong magnetic fields. M. A. Melvin[37] has suggested that torsion may play a significant role during the early stage of the development of the Universe.

REFERENCES

1. Sir Isaac Newton, *Mathematical Principles of Natural Philosophy* (translated into English by Andrew Motte), Cambridge University Press (1934).

2. H. Poincaré, *Les méthodes nouvelles de la mécanique céleste*, 3 vol., Gauthier-Villars, Paris (1892–99).
3. A. Einstein, 'Die Grundlage der allgemeinen Relativitätstheorie', *Ann. Physik* **49**, 769–822 (1916).
4. A. Einstein, *The Meaning of Relativity*, 5th ed., Princeton University Press (1955).
5. J. A. Wheeler, *Einstein's Vision*, Springer Verlag, Berlin (1968).
6. A short review of the recent theoretical and observational work on relativistic gravitation, together with references to literature, may be found in the author's *Summary of the 6th International Conference on GRG* (Copenhagen, July 1971), published in the *GRG Journal* **3**, 167 (1972).
7. J. Weber, 'Evidence for discovery of gravitational radiation', *Phys. Rev. Lett.* **22**, 1320 (1969) and 'Anisotropy and polarization in the gravitational radiation experiments', *Phys. Rev. Lett.* **25**, 180 (1970).
8. Various relativistic theories of gravitation have been recently reviewed by K. S. Thorne, C. M. Will, and their co-workers, cf. C. M. Will, *The Theoretical Tools of Experimental Gravitation*, Lectures presented at Course 56 of the International School of Physics 'Enrico Fermi', Varenna, July 1972.
9. A good review of the research on quantization of the gravitational field done prior to 1967, together with new results, and references to the literature of the subject, can be found in: B. S. DeWitt, 'Quantum theory of gravity', *Phys. Rev.* **160**, 1113 (1967); **162**, 1195 (1967); **162**, 1239 (1967).
10. A. Salam and J. Strathdee, 'Quantum gravity and infinities in quantum electrodynamics', *Lett. Nuovo Cimento* **4**, 101–108 (1970); C. I. Isham, A. Salam and J. Strathdee, 'Infinity suppression in gravity-modified quantum electrodynamics', *Phys. Rev.* **D3**, 1805 (1971) and Part II (ICTP preprint no. 14/71).
11. L. Parker, 'Quantized fields and particle creation in expanding universes', *Phys. Rev.* **183**, 1057 (1969).
12. R. U. Sexl and H. K. Urbantke, 'Production of particles by gravitational fields', *Phys. Rev.* **179**, 1247 (1969).
13. Ya. B. Zeldovich, 'Creation of particles in cosmology', *Pisma v Zh.E.T.F.* **12**, 443 (1970); Ya. B. Zeldovich and A. A. Starobinsky, 'Particle creation and vacuum polarization in the anisotropic gravitational field', a preprint of the Institute of Applied Mathematics, U.S.S.R. Academy of Sciences (in Russian); Ya. B. Zeldovich and I. D. Novikov, *Tieoria tiagotienia i evolutsyia zviezd*, Moscow (1971).
14. S. W. Hawking and R. Penrose, 'The singularities of gravitational collapse and cosmology', *Proc. Roy. Soc. Lond.* **A 314**, 529 (1970).
15. E. Kretschmann, 'Über den physikalischen Sinn der Relativitäts-postulate', *Ann. Physik* **53**, 575 (1917); A. Einstein, 'Prinzipielles zur allgemeinen Relativitätstheorie', *Ann. Physik* **55**, 241 (1918).
16. V. A. Fock, *Theory of Space, Time, and Gravitation*, Moscow (1955); and 'Principles of Galilean mechanics and Einstein's theory', *Uspekhi Fiz. Nauk* **83**, 577 (1964).
17. A. Trautman, 'Comparison of Newtonian and relativistic theories of space-time', in *Perspectives in Geometry and Relativity* (Essays in Honor of V. Hlavatý), Bloomington (1966).
18. A. Trautman, 'The General Theory of Relativity', *Uspekhi Fiz. Nauk* **89**, 3 (1966).
19. J. L. Anderson, *Principles of Relativity Physics*, Academic Press, New York (1967).
20. S. MacLane, 'Categorical algebra', *Bull. Amer. Math. Soc.* **71**, 40–106 (1965); I. Bucur and A. Deleanu, *Introduction to the Theory of Categories and Functors*, J. Wiley, New York (1969) A. Trautman, 'Relativity, categories, and gauge invariance', lecture given at the 5th International Conference on GRG, Tbilisi, 1968 (in print in the *Proceedings of the Conference*).
21. N. Steenrod, *The Topology of Fibre Bundles*, Princeton University Press (1951); A. Lichnerowicz, 'Espaces fibrés et espaces-temps', *GRG Journal* **1**, 235–245 (1971); R. Hermann, *Vector bundles in mathematical Physics*, vol. 1, Benjamin, New York (1970); A. Trautman, 'Fibre bundles associated with space-time', *Reports on Math. Phys.* (*Torun*) **1**, 29–62 (1970).
22. R. Penrose, 'Structure of Space-Time', Chapter VIII of the 1967 *Lectures in Mathematics and Physics* (Battelle Rencontres) (ed. by C. M. DeWitt and J. A. Wheeler), Benjamin, New York (1968).
23. R. Palais, *Foundations of Global Analysis*, Benjamin, New York (1968).
24. A. Trautman, 'Invariance of Lagrangian Systems', article in *General Relativity* (papers in honour of J. L. Synge), (ed. by L. O'Raifeartaigh), Clarendon Press, Oxford (1972).

25. S. Kobayashi and K. Nomizu, *Foundations of Differential Geometry*, vol. 1, Interscience, New York (1963).
26. C. N. Yang and R. L. Mills, 'Conservation of isotopic spin and isotopic gauge invariance', *Phys. Rev.* **96**, 191 (1954); T. D. Lee and C. N. Yang, 'Conservation of heavy particles and generalized gauge transformations', *Phys. Rev.* **98**, 1501 (1955).
27. R. Utiyama, 'Invariant theoretical interpretation of interaction', *Phys. Rev.* **101**, 1597 (1956); T. W. B. Kibble, 'Lorentz invariance and the gravitational field', *J. Math. Phys.* **2**, 212 (1961).
28. P. G. Bergmann, *Introduction to the Theory of Relativity*, Prentice-Hall, New York (1942).
29. A. Trautman, 'Riemannian Bundles', *Bull. Acad. Polon. Sci., sér. sci. math. et phys.* **18**, 667 (1970).
30. E. Cartan, 'Sur une généralisation de la notion de courbure de Riemann et les éspaces à torsion', *Compt. Rend. Acad. Sci. Paris* **174**, 593 (1922). E. Cartan, 'Sur les variétés à connection affine et la théorie de la relativité généralisée, I partie', *Ann. Ec. Norm.* **40**, 325 (1923) and **41**, 1 (1924).
31. Costa de Beauregard, 'Sur la dynamique des milieux doués d'une densité de moment cinétique propre', *Compt. Rend. Acad. Sci. Paris* **214**, 904 (1942); A. Papapetrou, 'Non-symmetric stress-energy-momentum tensor and spin-density', *Phil. Mag.* **40**, 937 (1949); D. W. Sciama, 'On a non-symmetric theory of the pure gravitational field', *Proc. Camb. Phil. Soc.* **54**, 72 (1958); V. I. Rodichev, 'Twisted space and non-linear field equations', *Zh. Eksper. Teor. Fiz.* **40**, 1469 (1961); F. Hehl and E. Kröner, 'Über den Spin in der allgemeinen Relativitätstheorie: Eine notwendige Erweiterung der Einsteinsches Feldgleichungen', *Z. Physik* **187**, 478 (1965) (this paper contains many references); K. Hayashi, 'Gauge theories of massive and massless tensor Fields', *Progr. Theor. Phys.* **29**,494–515 (1968); A. Trautman, 'On the Einstein–Cartan Equations I and II', *Bull. Acad. Polon. Sci., sér. sci. math. astr. et phys.* **20**, 185 and 503 (1972); K. Hayashi and A. Bregman, 'Poincaré gauge invariance and the dynamical role of spin in gravitational theory', *Ann. Phys. (N.Y.)*, (in print).
32. W. Kopczyński, 'A non-singular universe with torsion', *Phys. Lett.* **39A**, 219 (1972).
33. A. Trautman, 'On the Einstein–Cartan equations III', *Bull. Acad. Polon. Sci., sér. sci. math. astr. et Phys.* **20**, 895 (1972).
34. M. Mathisson, 'Neue Mechanik materieller Systeme', *Acta Phys. Polon.* **6**. 163 (1937).
35. A. Papapetrou, 'Spinning test-particles in general relativity', *Proc. Roy. Soc. (London)* **A209**, 249 (1951).
36. F. W. Hehl, 'Spin und Torsion in der allgemeinen Relativitäts-theorie oder die Riemann–Cartansche Geometrie der Welt', Habilitationsschrift (mimeographed), Tech. Univ. Clausthal (1970).
37. M. A. Melvin, private communication of 27 September 1972.

Comment on Trautman's paper

Remarks on Five-Dimensional Relativity

W. Thirring

This theory, as a possible generalization of relativity, had been initiated by Kaluza[1] and Klein some fifty years ago and has been further developed by many people. Several of them are in this distinguished audience, and I hope you will forgive me if not everybody is mentioned at the appropriate place.

The first observation was the miracle that the curvature R in a five-dimensional manifold with metric

$$
g = \quad
\begin{array}{c|c|c}
 & 1 & 4\ 5 \\
\hline
 & \bar{g} & fA \\
\hline
 & fA & -1
\end{array}
\tag{1}
$$

is simply*

$$
R = \bar{R} - \frac{f^2}{4} F_{ab} F^{ab}
\tag{2}
$$

if $\bar{g}$ and A do not depend on the fifth coordinate s. Here $\bar{R}$ is the four-dimensional curvature constructed with the $\bar{g}$, $F_{ab} = A_{a,\,b} - A_{b,\,a}$ and f, a constant to be adjusted later. Since $\sqrt{g} = \sqrt{-\bar{g}}$ one realizes that the Lagrangian $\sqrt{g}R$ unites gravitation and electricity. But why is everything independent of s?

Suppose that the manifold is periodic in s, so that we live on something like the surface of a cylinder with radius μ^{-1}.

In this case any field ϕ has an s-dependence $\sim \sum_n e^{is\mu n} c_n$, with integer n. Take for instance the Lagrangian of a scalar field, and substitute

$$
\phi(\mathbf{x}, t, s) = e^{is\mu n} \varphi(\mathbf{x}, t),
\tag{3}
$$

---

* Using the convention that the subscripts a, b go from 0 to 3 and u, v go from 0 to 4.

J. Mehra (ed.), The Physicist's Conception of Nature, 199–201. All Rights Reserved

$$g^{uv}\phi_{,u}\phi_{,v} = \bar{g}^{ab}(\varphi_{,a} + in\mu f A_a\varphi)(\varphi_{,b} + in\mu f A_b\varphi) - \varphi_{,s}^2. \tag{4}$$

Thus if μf is identified with the electric charge we get the Lagrangian for an n-fold charged particle. Therefore the theory
(1) unites gravitation with electricity,
(2) unites space-time with the interval space in field theory,
(3) tells us that charge is quantized.

Since f is the square root of the gravitational constant we calculate μ^{-1} to be $\sqrt{137} \times$ the Planck length $\sim 10^{-32}$ cm. This outrageous number comes from the fact that gravitation is so much weaker than electricity. It makes it clear why we don't see the fifth dimension. It just costs too much energy to produce s-excitations. Already the last term in (4) gives the ridiculous bare mass of 10^{18} gev ~ 1 microgram for singly charged particles. This difficulty has been partially overcome by Kerner[2] and others who considered manifolds of the type $G \times R^4$ where G is a compact Lie group.

The above theory corresponds to $G = U(1)$, and in the general case one gets, in addition to $\bar{R}$, the Yang–Mills Hamiltonian for the corresponding gauge fields. Thus for $G = SU(2)$ one obtains a charged particle without bare mass terms in the Lagrangian.

We are now asked to write down the Dirac equation in the five-dimensional manifold, and modern differential geometry tells us how to do that. If $\bar{g}$ is pseudo-Euclidean, one finds

$$m_0\Psi = \gamma^a(\psi_{,a} + fA_a\psi_{,s}) + \gamma^5\psi_{,s} + \gamma^5 f \frac{F_{ab}}{16}[\gamma^a, \gamma^b]\,\Psi. \tag{5}$$

Since there is an additional bare mass $\sim \gamma^5$ we put

$$\Psi(x, t, s) = e^{i(\varphi\gamma^5 + s\mu)}\psi(x, t) \tag{6}$$

and determine φ so as to get the usual free Dirac equation:

$$\left(\gamma^a(\partial_a + ieA_a) + M + \left(\gamma^5\frac{m_0}{M} - i\frac{\mu}{M}\right)f\frac{F_{ab}}{16}[\gamma^a, \gamma^b]\right)\psi = 0, \tag{7}$$

where

$$M = \sqrt{m_0^2 + \mu^2}\,.$$

Here we find the first genuine gravito-electric effect. Whereas in the scalar case, gravitation and electricity separated like oil and water, we now have our electric and magnetic moment proportional to f (not to e). The former may seem strange since it is known to violate P and T. In this theory there is, of course, a P-operation $\underline{P}$ which leaves the Lagrangian invariant[3]. But this does not commute with the rotation which leads from Ψ to ψ. Thus $\underline{P}$ is different from the usual parity. Charge conjugation C is just $s \to -s$ and agrees with the usual definition. Thus the usual CP-reflection is not conserved, and one is tempted to associate this gravito-electricity with the Fitch–

Cronin-effect. The size of the electric dipole moment depends on m_0 but it is never larger than 10^{-32}. Thus it is a few orders of magnitude too small to be the reason for the $K_L^0 \to 2\pi$ decay. However this may change if strong interactions are also included. In any case one still has the enormous bare mass M, and for a good theory the Lagrangian for the spinor field should also come out of R.

REFERENCES

1. Th. Kaluza, *Sitz. Preuss. Akad.*, p. 966 (1921).
2. R. Kerner, *Ann. Inst. H. Poincaré*, IX, 143 (1968).
3. W. Thirring, *Act. Phys. Austr. Suppl.* (1972).

9

From Relativity to Mutability

John Archibald Wheeler

'... the end of all our exploring
Will be to arrive where we started
And know the place for the first time.'
T. S. ELIOT[1]

1. INTRODUCTION

From Relativity to Gravitational Collapse; and from the Consequences of Collapse to the Principle that Nature Conserves Nothing

Relativity and the quantum principle constitute the two overarching concepts of 20th century physics. To review relativity here is to have an opportunity for something new. Casting an eye over what we have learned in this domain, can we discover out of it all some consideration that might guide us into tomorrow? The lesson need not be positive. It could be negative. In physics there are many negative principles. None is better known, and none has ever proved itself more powerful, than the original principle of relativity itself, saying that it is impossible to discover any difference in the laws of physics between two inertial reference frames that would distinguish the one frame from the other. If both special and general relativity were founded on negative principles, it is also true that some of the most remarkable consequences of relativity have a negative flavour, as for example these: (1) The universe cannot be static. (2) The volume of a closed model universe is not constant. (3) Total energy and total angular momentum cannot be defined for a closed universe. They are meaningless concepts. For such a system, no global law of conservation of energy and angular momentum has meaning or relevance. (4) Baryon number and lepton number are well defined quantities for a normal star; but when this star collapses to a black hole, the well established laws of conservation of particle number lose all applicability.

These and other interesting negative conclusions out of relativity have been recognized for some time. They are still startling enough to call for some review. However, after such a review, it is even more important for us to try to pull all these individual negatives together into a larger formulation of the way that nature acts. I have not been able to find any more reasonable way to state the situation than this: nature conserves nothing; there is no constant of physics that is not transcended; or, in one word, mutability is a law of nature.

J. Mehra (ed.), The Physicist's Conception of Nature, 202-247. All Rights Reserved

'... it would have been difficult to establish any laws of nature,' Wigner reminds us,[2] 'if these were not invariant with respect to displacements in space and time.' However, displacements in flat spacetime, or even in a curved spacetime that is asymptotically flat, make no sense in the closed universe of Einstein's general relativity. In that universe there is no global law of conservation of momentum and energy. More startling, such a universe undergoes gravitational collapse. In that collapse, classical space and time themselves come to an end. With their end, the framework falls down for everything that one has ever called a law of physics.

Nothing that relativity has ever predicted is more revolutionary than collapse, and nothing that collapse puts in question is more central than the very possibility of any enduring laws of physics.

The golden trail of science is surely not to end in nothingness. There may be no such thing as 'the glittering central mechanism of the universe' to be seen behind a glass wall at the end of the trail. Not machinery but magic may be the better description of the treasure that is waiting. Rather than Newtonian law it may resemble more the logic of relationships that Leibniz envisaged. But that is an issue for tomorrow. Today we look at the breakage that relativity has made among the laws of physics.

Not the slightest question is implied here about the everyday laws of physics under everyday conditions. No one will turn to relativity to learn something new about the physics or liquids or solids, unless he is concerned with a neutron star, or with conditions still more extreme.

Looking in on General Relativity through Six Windows

What relativity takes away, and what it gives, show nowhere better than on a tour about the structure that Einstein built, looking into it first through one window, then another. Framing each window is a different derivation of Einstein's law for the dynamic change of geometry with time. We scrutinize physics in turn through these six windows: (1) Einstein's original derivation, based on the principles of equivalence and correspondence; (2) Élie Cartan's derivation, resting on the fact that the boundary of a boundary is zero; (3) the most compact derivation one knows, based on the idea that density of mass-energy governs curvature; (4) the derivation of Hilbert and Palatini, founded upon the principle of least action; (5) the derivation of Hojman, Kuchař and Teitelboim, that introduces the group-theoretic concept of ' "group" of deformations of a spacelike hypersurface in spacetime'; and (6) the schematic derivation of Andrei Sakharov, founded upon the concept of 'the metric elasticity of space'. Another derivation starts from the theory of a spin-2 field in flat space, but it and other interesting derivations will not be touched upon here.

Several comments to be made here about these derivations come from the book of Charles W. Misner, Kip S. Thorne and myself, *Gravitation*,[3] now in the course of publication. Warm appreciation is expressed to these and other colleagues and not least to Paul Dirac himself for insights into the structure and consequences of Einstein's standard 1915 geometrodynamics.

'Tide-Producing Acceleration' or 'Riemann Curvature' as Local Measure of the Effect of Geometry on Motion

There is no derivation of the effect of a moving mass upon geometry that is not best prefaced by the effect of geometry upon the movement of a mass. That there appears at first to be no such effect comes as a shock to the beginning student of relativity. He expects to see the analogue of electromagnetism faithfully pursued, where the invariantly measured acceleration of a test charge is a direct measure of the electromagnetic field strength,

$$\frac{D^2 x^\mu}{D\tau^2} = \frac{e}{m} F^\mu{}_\nu \frac{dx^\nu}{d\tau}.$$ (1)

Instead, the neutral test particle of general relativity moves in a straight line with uniform velocity in the local Lorentz frame, a statement that expresses itself in an arbitrary curvilinear coordinate system in the form

$$\frac{D^2 x^\mu}{D\tau^2} = 0.$$ (2)

Gravitation seems to have disappeared. Has not Einstein gone too far, the beginning student may ask, in emphasizing that the only right description of a force is a local description? However, gravitation, at first apparently extinguished in this local description, springs into evidence again as a tide-producing force; that is, as a measure of the relative acceleration of two nearby test particles endowed with an initial separation η^α; thus,

$$\frac{D^2 \eta^\alpha}{D\tau^2} + R^\alpha{}_{\beta\gamma\delta} \frac{dx^\beta}{d\tau} \eta^\gamma \frac{dx^\delta}{d\tau} = 0.$$ (3)

The tide producing force or Riemann curvature $R^\alpha{}_{\beta\gamma\delta}$, as seen in its effect on a fleet of nearby test particles, is the central descriptor in Einstein's geometrical account of gravitation. At issue from this point onward is not the effect of curvature on mass but the back action of mass on curvature.

2. EINSTEIN'S ROUTE TO GENERAL RELATIVITY: ANTECEDENTS AND CONSEQUENCES

Riemann's 'Physical Geometry', Mach's Concept of Inertia, and Einstein's Equivalence Principle as Elements in Einstein's General Relativity

Einstein credits Riemann (Fig. 1) with one central idea of general relativity. Matter gets its moving order from geometry. In other words, geometry acts on matter. By the principle of action and reaction matter must therefore act on geometry. Thereupon geometry ceases to be a God-given Euclidean participant standing high above the battles of matter and energy. Geometry steps forward as a new participant in the world of physics.

A second idea that led him to relativity Einstein attributes to Mach. Acceleration can have no meaning unless there are objects with respect to which the acceleration takes place. Einstein could see consequences from this Mach principle. Thus inertia here must take its origin in mass-energy there. But gravitation here also arises from

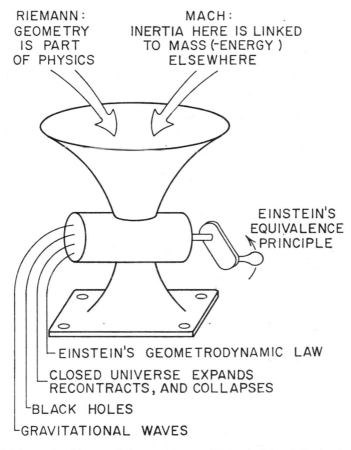

RIEMANN:
GEOMETRY
IS PART
OF PHYSICS

MACH:
INERTIA HERE IS LINKED
TO MASS (-ENERGY)
ELSEWHERE

EINSTEIN'S
EQUIVALENCE
PRINCIPLE

EINSTEIN'S GEOMETRODYNAMIC LAW
CLOSED UNIVERSE EXPANDS
RECONTRACTS, AND COLLAPSES
BLACK HOLES
GRAVITATIONAL WAVES

Fig. 1. Einstein's great achievement, to use his new 1907 principle of the local equivalence of 'gravitational' and 'propulsive' accelerations to bring together two currents of thought, going back to Riemann and Mach, and formulate (1907–1915) 'general relativity' or 'geometrodynamics', with all its consequences. In an unpublished essay of 1919[4] Einstein describes the equivalence principle (that came only two years after special relativity) as 'the happiest thought of my life': 'Thus, for an observer in free fall from the roof of a house there exists, during his fall, no gravitational field.' To Mach Einstein wrote enthusiastically from Zurich on 25 June 1913, more than two years before he had arrived at the final formulation of general relativity[5], 'If so [i.e., if the eclipse observations confirm the new theory], then your helpful investigations on the foundations of mechanics – Planck's unjustified criticisms notwithstanding – will receive a brilliant confirmation. For it necessarily turns out that *inertia* has its origin in a kind of interaction, entirely in accord with your considerations on the Newton pail experiment.' Einstein also gives warm testimony to the contribution of Riemann[6], '... space was still, for them [physicists], a rigid, homogeneous something, susceptible of no change or conditions. Only the genius of Riemann, solitary and uncomprehended, had already won its way by the middle of last century to a new conception of space, in which space was deprived of its rigidity and in which its power to take part in physical events was recognized as possible.'

mass-energy there. Therefore tentatively conclude that gravitation and inertia are transmitted by the same machinery.

That the machinery required to carry both gravitation and inertia is geometry was Einstein's great synthesis of the two currents of thought going back to Mach and Riemann. No consideration impelled him more directly to this synthesis than his 1909 principle of the local equivalence of gravitational and propulsive accelerations. Gravitation stops producing a curvilinear track in a flat spacetime. Motion becomes straight in every local Lorentz frame. Gravitation becomes the curvature encountered in passing from one local Lorentz tangent space to the next.

The principle of correspondence with the Newtonian theory of gravitation requires Einstein's conserved tensorial measure of curvature, $G_{\mu\nu}$, to agree (Einstein's papers[7] before the Berlin Academy, on 4, 18 and 25 November 1915) with 8π times the conserved measure of the density of mass-energy*, that is, the standard tensor of stress and density of momentum and energy, T; thus,

$$G_{\mu\nu} = 8\pi T_{\mu\nu}. \tag{4}$$

Schwarzschild Geometry as Source of Four Predictions

Fig. 2 illustrates the geometry calculated by Schwarzschild from Einstein's general relativity for the region within and around a centre of attraction such as the Sun. This Schwarzschild geometry leads directly to four well known predictions:

(1) the bending of light by the Sun.

$$\theta = 4M_\odot/R_\odot = 4 \times 1.47 \text{ km}/6.96 \times 10^5 \text{ km} = 1.75''; \tag{5}$$

(2) the redshift of light from the Sun,

$$z = \Delta\lambda/\lambda = M_\odot/R_\odot = 2.12 \times 10^{-6}, \tag{6}$$

and on earth,

$$z = g_{conv}h/c^2 = gh = h/0.92 \times 10^{18} \text{ cm}; \tag{7}$$

(3) the relativistic precession of the perihelion of Mercury about the Sun,

$$6\pi M_\odot/a\,(1 - e^2) \text{ radians per revolution or } 43''15 \text{ per century}; \tag{8}$$

(4) the $\sim 200\ \mu$sec delay experienced by a radar pulse on its trip from Earth to Venus and back, when it passes close to the Sun. But by far greater than any of these consequences of general relativity is the revolutionary prediction that the universe itself is dynamic.

---

* The units here are geometrical. The factor of conversion from the conventional unit of time to the geometrical unit of time is $c = 3.00 \times 10^{10}$ cm/sec; from the conventional unit of mass to the geometrical unit of mass is $G/c^2 = 0.742 \times 10^{-28}$ cm/g (Earth mass, 0.44 cm; Sun mass, 1.47×10^5 cm).

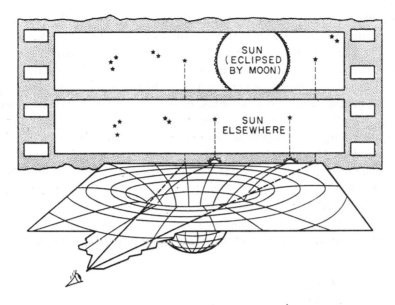

STARLIGHT BENT BY SUN'S
GRAVITATION (=GEOMETRY)

Fig. 2. Geometry within and around the Sun. Both inside and outside, the geometry departs from flatness (non-zero components of the 'tide producing acceleration' or Riemannian curvature $R^{\alpha}{}_{\beta\gamma\delta}$); but outside, 'Einstein's conserved tensorial measure of curvature', $G_{\mu\nu} = R^{\alpha}{}_{\mu\alpha\nu} - \frac{1}{2}g_{\mu\nu}g^{\beta\delta}R^{\alpha}{}_{\beta\alpha\delta}$, is zero. The analogy is close with electrostatics, where (1) the individual second derivatives $\partial^2\phi/\partial x^2$, $\partial^2\phi/\partial y^2$, $\partial^2\phi/\partial z^2$ of the electric potential have non-zero values both outside and inside a spherically symmetric cloud of electric charge, but (2) the combination of second derivatives, $\partial^2\phi/\partial x^2 + \partial^2\phi/\partial y^2 + \partial^2\phi/\partial z^2 = 4\pi\varrho_\varepsilon$, vanishes wherever there is no electric charge ($\varrho_\varepsilon = 0$).

Geometry as Dynamic New Participant in Physics

In the first days of general relativity geometry had been only the slave of matter. Matter here curved space here. Curvature here meant curvature there. Curvature there meant gravitation there. Thus Einstein's Riemannian geometry seemed to do nothing more than carry Newtonian pull from one mass to another. Then, before Einstein's eyes, geometry cast off its chains. It stepped onto the stage of physics as a participant in its own right. It asserted dynamic degrees of freedom of its own. Earlier, under Maxwell, the electromagnetic field had also won liberation and a position as an independent dynamic entity. However, geometry became all this and more; not only new dynamic entity, but also background and home for all other fields.

Cosmology and the Closure of the Universe

Nowhere did the new dynamics of geometry display itself more dramatically or more simply than in the predicted expansion and recontraction of a closed model universe, filled to effectively uniform density with a 'dust' of stars. The uniformity and the 'dust' (i.e., negligible pressure) were conveniences in the analysis; but the closure was to Einstein a matter of principle. This closure, moreover, owing to the advance of astrophysics, looks like someday being a testable prediction. For example, the apparent angular diameter of objects of standard size, that goes down forever with increase in distance in Euclidean geometry, is predicted in the Friedmann universe to go up again with distance at sufficiently great distances, owing to the lens-like action of the great curve of space itself.[8] Test or no test, I would be omitting an important point if I did not suggest that *Einstein's general relativity means today not only the set of differential equations that bear his name, but also the boundary condition of closure that marks solutions of these equations as interesting.* Closure was demanded in Einstein's eyes by Mach's principle[9]: '… this idea of Mach's corresponds only to a finite universe, bounded in space, and not to a quasi-Euclidean, infinite universe. From the standpoint of epistemology it is more satisfying to have the mechanical properties of space completely determined by matter, and this is the case only in a space-bounded universe.'

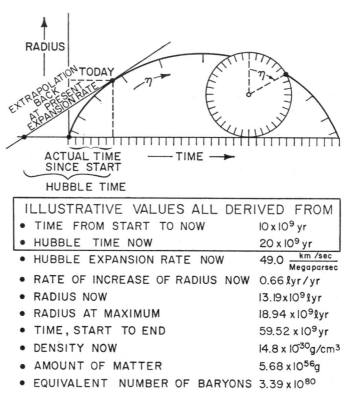

Fig. 3.   The dynamics of the Friedmann matter-dominated universe is spelled out by tying a paint brush to the rim of a wagon wheel and rolling the wheel along beside the side of the barn. Vertical coordinate gives radius of curvature of the 3-sphere (cm); horizontal coordinate gives time from the start of the expansion (in cm of light travel time). Illustrative numbers are adapted from ref. 3, Box 27.4. At very early times and very late time radiation dominates over matter in any model universe at all compatible with what one knows of the actual universe; but the resulting corrections to the cycloid curve and the listed numbers are small and, for simplicity, are not shown.

Elsewhere[10] he remarked, 'In my opinion the general theory of relativity can only solve this problem [of the origin of inertia] satisfactorily if it regards the world as spatially self enclosed.' Some able physicists disagree; but this is *Einstein*'s relativity.

No one can forget that it was the Russian meteorologist and physicist A. Friedmann who first worked out the dynamics of Einstein's simple closed model universe (Fig. 3). All follows from the decisive '00' or 'tt' component of the standard geometrodynamic law,

$$G_{00} = 8\pi\, T_{00}. \tag{9}$$

Specialize to a 3-sphere universe, of time dependent radius $a(t)$, filled with 'dust' of density ϱ given by some constant divided by the volume of the 3-sphere. Then this equation reads

$$\frac{3}{a^2}\left(\frac{da}{dt}\right)^2 + \frac{3}{a^2} = 8\pi\varrho = \frac{\text{constant}}{a^3}. \tag{10}$$

Multiply by $a^2/3$, rearrange and give the so-far-unspecified constant of proportionality (a measure of the total of the masses of the individual 'dust' grains or stars) the name a_0, to stand for the radius of the universe at the phase of maximum expansion; thus,

$$\left(\frac{da}{dt}\right)^2 - \frac{a_0}{a} = -1. \tag{11}$$

In what way does this Friedmann result differ from what one would expect for a compact cluster of rocks sitting out in space, suddenly driven apart by a blast of dynamite at the centre of the cluster? There the corresponding formula, with a now identified as the radius of the cluster, reads

$$\left(\frac{da}{dt}\right)^2 - \frac{a_0}{a} = \text{constant}. \tag{12}$$

The term $-a_0/a$ is a measure of the gravitational potential energy of binding of the cluster, and is always negative. The constant on the right, on the other hand, measures the excess of the energy of the dynamite over the original binding of the cluster. If the explosion is strong enough, the 'constant of energy' is positive, and the cluster flies apart for ever. If the explosion is weaker than a certain critical amount, the 'constant of energy' is negative, and eventually pulls the Newtonian cluster back together again.

Einstein's closed universe has no such option. There is no adjustable 'constant of energy' on the right hand side of the equation. The system is gravitation-dominated at all times. The radius rises to the maximum amount $a = a_0$ (proportional to the amount of matter present) and then recontracts and collapses to zero. That the Einstein geometrodynamics of a closed universe always ends in collapse has been proved in recent times without any appeal to spherical symmetry and under remarkably general conditions.[11, 12] Einstein's condition of closure is essential to this reasoning.

Old textbooks deal with closed model universes that expand, pause or nearly pause at a certain radius, and then start again to expand, at first slowly, then more and more rapidly. A special case of these now disfavoured models is a universe that stands forever in unstable equilibrium at a certain radius, like a pencil balanced for all time on its tip. Such models lie outside Einstein's 1915 (and today standard) general relativity. The radius as a function of time does not fulfil the normal geometrodynamic law. It satisfies another law, obtained by adding to Einstein's equation a so-called cosmological term. Without that ill-starred term relativity would have shouted out the greatest of predictions, the prediction that the universe itself is dynamic. Today, letting that term fade into the oblivion of the past, one can say that the 'would-have-been prediction' *is* the greatest prediction of all. It is a prediction almost too fantastic to be believed, and a prediction that is nevertheless dramatically confirmed by observation.

In 1915 one thought of the universe as enduring from everlasting to everlasting. Einstein could not believe the prediction that the universe is dynamic. He tried to escape it. But the considerations that lead to general relativity are compelling. He could find no natural way out. Therefore he took the least unnatural way out that he could find. He introduced the 'cosmological term'. Its whole purpose was to make possible a static universe. Then came 1927 and Hubble and the discovery that the universe is dynamic. Thereafter Einstein spoke of the cosmological term as[13] 'the biggest blunder of my life'.

Two Other Cycles of Doubt and Test

That was the first of the three great cycles of doubt and test of general relativity. The second came when the Hubble time, the 'extrapolated time' for galaxies to arrive at their present distances expanding at their present recession velocities, turned out to be shorter (of the order of 2 to 3×10^9 years) then the best estimates one could make of the actual age of the universe (of the order of 10×10^9 years). This meant that the expansion had been speeding up. In contrast, the Einstein–Friedmann predictions say it must slow down ('pull of gravitation'). So 'Give up general relativity,' more than one group said. Thus came the era of theories outside the framework of relativity and at variance with principle that the laws of physics are local in character: theories of the 'steady state expansion of the universe' and theories of 'the continuous creation of matter'. Then, thanks not least to Walter Baade[14], a revolution took place in one's understanding of the scale of astrophysical distances. Previously accepted distances to the galaxies in the Hubble catalogue had to be revised upward by a factor between 6 and 10. The linearly extrapolated time, H^{-1}, back to the start of the expansion rose by the same factor to a value now estimated to be not far from 20×10^9 years[15]. But galaxies actually got where they are in about half this time, according to more than one way of evaluating the time back to the beginning of all astrophysical processes ($\sim 10 \times 10^9$ years ago.) Therefore one believes today that one has clear evidence for the predicted slowing down of the expansion.

The third cycle of doubt and test began in 1958, when Jan Oort[16] gave 0.31×10^{-30} g/cm^3 as the best available figure for the averaged out density of matter present in the form of galaxies. In contrast, an amount of mass-energy of the order of 15×10^{-30}

g/cm^3 is required to curve space up into closure (Fig. 3). Einstein's geometrodynamics thus predicts that there is of the order of 10 to 100 times as much mass-energy in space as one sees in the form of galaxies. But where? And in what form? This 'mystery of the missing mass' is the central point of much present-day work.[17] Hydrogen gas unassembled into galaxies[18] will some day be detected by its ultraviolet absorption[19] or by its X-ray emission[20] if it is present in the required amount.

If cosmology once seemed a subject fit only for dreamers, today it is the heartland of observational astrophysics. For example, nothing did more to destroy the concept of a 'steady state expansion' of the universe than the observation of the 3 °K primordial cosmic fireball microwave radiation.[21, 22]

The Friedmann–Einstein prediction that the universe itself is dynamic, in the beginning too incredible for even Einstein himself to believe, has now become a central fact of modern physics. With the universe proved dynamic, one is the readier to accept three other ideas from Einstein's general relativity: (1) that other incredible prediction, that collapse is inevitable; and two prior ideas, (2) that the universe is closed, and (3) that geometry is a new dynamic participant on the stage of physics.

'Total Energy' and 'Total Momentum' as Concepts with No Meaning for a Closed Universe

The closure of Einstein's universe has a special consequence for energy. The law of conservation of energy connects the amount of mass-energy inside a closed surface with the value of a certain integral extended over that surface. Deform this surface of integration bit by bit in imagination at any one time so as to engulf more and more volume. At first the surface swells. Then it reaches a maximum extent. When it includes the entire volume, it has collapsed to nothingness. Thus the law of conservation of energy, applied to the complete closed system, degenerates to the trivial identity, 'zero equals zero'. The concept of 'total mass-energy' makes no sense for a closed universe. How could it? (1) There is no natural Lorentz frame in which to do the pointing and measuring off of a 4-vector of energy and momentum even if one had such a 4-vector. (2) There is no platform' on which to stand to measure the gravitational attraction of the closed system. (3) There is no place outside the system to put a planet into Keplerian orbit around it. It is satisfying that the mathematics kills at the start a concept that is bad physics. There is no such thing as the energy (or the angular momentum) of a closed universe.[23] *The dynamics of a closed geometry transcends the laws of conservation of angular momentum and energy.*

Gravitational Radiation from Gravitational Collapse

The dynamics of geometry, so central to these cosmological considerations, must also reveal itself in a testable way at a smaller scale in gravitational radiation, according to Einstein's standard general relativity. Moreover, the basic factor in the formula for gravitational radiation[24] is the very large number

$$P_{0,\,\text{grav}} = c^5/G = 3.6 \times 10^{59} \text{ erg/sec}. \qquad (13)$$

In other words, any system, big or small, that is highly asymmetric, and that changes its configuration in a time comparable to the time required for light to cross it, will give off gravitational radiation at a rate of the order of magnitude of $P_{0,\text{grav}}$.

Few events are more spectacular than a supernova, nor more relevant as a source of gravitational radiation. A normal star with slowly rotating white dwarf core develops gravitational instability in the course of its standard astrophysical evolution. The core collapses to a rapidly rotating neutron star. As the core implodes, it generates a powerful shock, in consequence of which the envelope explodes. This now accepted picture goes back for its beginnings to 1934 and Baade and Zwicky.[25] It received dramatic support in 1968 when Hewish and his collaborators[26] discovered the first few pulsars, among them one pulsing 30 time a second. It lies at that point in the Crab Nebula where Baade and Zwicky, 34 years before, had said the neutron star (from the July 1054 supernova) should be.

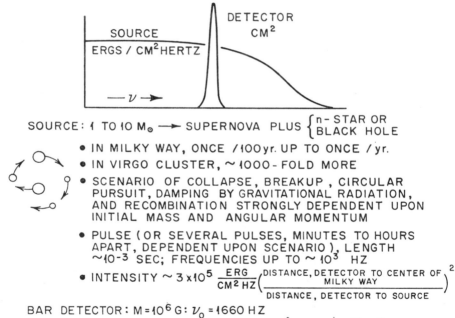

Fig. 4. Gravitational radiation, from source to detector.

To follow the internal dynamics of future supernovae, an optical telescope does not suffice, nor does X-ray or infrared astronomy, remarkable though the advances are today in all three methods of observation. Oceans of star stuff block all view of the Niagara Falls that pours its tumult inward at the centre. One signal nevertheless makes its way out, a pulse of gravitational radiation, with characteristic shape, yet to be calculated, dependent on the mass and angular momentum of the collapsing core. What has been calculated is the order of magnitude of the pulse. Press and Thorne conclude[27] that a Weber-bar detector,[28] built of a 100 kg monocrystal of quartz, cooled to $T = 3 \times 10^{-3}$ K, if it has a $(1/T)$-proportional damping time of $\sim 10^6$ sec (technology of late 1970's or early 1980's) should suffice to detect gravitational waves from a supernova in the Virgo cluster of galaxies (distance 3×10^7 light-years), if one can construct a sensor to measure changes in vibration amplitude $\lesssim 10^{-19}$ cm on a time scale of $\lesssim 0.1$ sec (Fig. 4). Supernovae flash out within that distance once a month or more.

Thanks to the initiative of Joseph Weber and the subsequent work of many other able investigators, at least twenty detectors of gravitational radiation have been constructed and exploited to give upper limits to the flux of energy streaming past the earth at selected frequencies. The time nears for a decisive test of one of the greatest of Einstein's predictions, energy-bearing waves in the geometry of spacetime itself.

Collapse of a too Massive Neutron Star to a Black Hole

Complete gravitational collapse of an overcritical mass, $M > M_{crit}$, is another great prediction. The precise value of the critical mass of a neutron star is uncertain but is believed to lie in the range $0.5 M_\odot < M_{crit} < 3 M_\odot$ (unless, as would seem possible at most for the first few days of its life, it is endowed with large amounts of differential rotation, in which case James Wilson gives a figure about 50% larger). When a neutron star of greater mass is formed by the gravitational collapse of the core of a star with white dwarf core, the collapse may slow down temporarily as neutron-star densities are reached (10^{14} to 10^{15} g/cm^3); but the collapse is then predicted to continue and to speed up, with the matter becoming more and more compact, until a horizon forms and a black hole comes into being. The proper circumference of the horizon divided by 2π, otherwise known as the Schwarzschild radius of the black hole, is $2M$ (cm) $= = 2(G/c^2)M_{conv}$ (g); that is, 3 km (roughly a tenth the size of a neutron star) for an object of solar mass, and 10 to 10^4 light seconds for a black hole of 10^6 to $10^9 M_\odot$, such as one may expect to find in a compact and highly evolved galactic nucleus.

In contrast to the 'dead' or Schwarzschild black hole of the traditional text, the object formed in the collapse of matter with any net spin angular momentum at all, $S \neq 0$, is a 'live' black hole, as first emphasized by Bardeen[29]; and it can give up energy to an external particle of field, as first pointed out by Penrose.[30] Hawking[31] showed that neither in the Penrose process, nor in any other process, can be surface area of the horizon of a black hole ever increase. Independently Christodoulou[32] showed that a black hole is characterized by an 'irreducible mass', M_{ir} (later shown to be connected with the area of the horizon by the formula $A = 16\pi M_{ir}^2$). The 'irreducible mass' is constant in any process that reversibly exchanges energy with a black hole, but it always rises in any irreversible process. Christodoulou, and Christodoulou and Ruffini[33], derived the wonderfully simple formula

$$M^2 = \left(M_{ir} + \frac{Q^2}{4M_{ir}} \right)^2 + \frac{S^2}{4M_{ir}^2} \tag{14}$$

for the mass-energy of a 'live' black hole in terms of its charge and spin.

Three processes offer themselves for the detection of a black hole: (1) the pulse of gravitational radiation given out at the time of formation; (2) the X-rays given out in the traffic-jam of matter accreting onto a black hole after formation, as analyzed by Zel'dovich and Novikov[34]; and (3) 'activity': activity arising from energy imparted to outside matter, or fields, or both, out of the stockpile of energy in a live black hole (see for example ref. 35 and 36). All three processes are being actively investigated, and have many interesting astrophysical consequences, most of which are reviewed in some detail in the 1972 Les Houches lecture series[37] and in ref. 3.

Roughly 50% of all stars are 'married'; and of such double star systems, roughly 40% are near-binaries, with periods of the order of a few days. When one component of a revolving double star system is a neutron star or black hole, it has a good chance to feed on the envelope of its companion, and in consequence become a powerful source of X-rays. When the compact component is a neutron star, its rotating off-axis magnetic field produces the normal pulsar phenomenon, but in a denser than normal plasma. Whether the compact component is a neutron star or a black hole, the inpouring gas, adiabatically compressed to 10^{10}–10^{11} K, emits far more in the X-ray region than in the visible. Only in this way has one been able to understand some of the spectacular eclipsing X-ray sources observed in recent months by Giacconi and his collaborators. Leach and Ruffini emphasize[38] the sharp division of these double-star X-ray sources into two classes. In one class the X-ray source flashes regularly like an optical pulsar. In this case, it is generally agreed, the compact (and optically invisible) component is to be identified with a neutron star. In the other class (two cases so far, Cygnus X$-$1, and the X-ray sources 2U1700–37) the X-ray intensity fluctuates, with the fluctuations amounting to as much as a factor of a hundred in a time as short as 50 msec. Ruffini reasons that this effect indicates (1) small size and (2) hydrodynamic instability of the flow of plasma into the black hole. In conformity with this reasoning, the mass of the compact component (as deduced from the period and range of Doppler velocities of the visible component) appears in the one case to be more than $8M_\odot$, and in the other case more than $4M_\odot$. If this object were a normal star, it would be far too bright (luminosity $\sim 8^3 L_\odot$ and $\sim 4^3 L_\odot$, respectively) to escape observation. It cannot be a white dwarf, because for these objects the critical mass limit is $M_{crit} \simeq 1.2 M_\odot$; and likewise it cannot be a neutron star if for such objects the critical mass is indeed $M_{crit} < 3.2 M_\odot$. Few see any alternative for these two X-ray sources except to conclude that the compact object is a black hole. Moreover, it is difficult to imagine how a neutron star continuously fed from a sufficiently massive companion can ever end up as anything except a black hole. Therefore it seems reasonable to conclude that science has now been launched, quietly but momentously, into the age of black hole astrophysics.

Black Hole as 'Experimental Model' for the Collapse of the Universe Itself

With black holes one has come full circle around the application of Einstein's geometrodynamics, past the traditional tests of general relativity, through the world of gravitational radiation, and into the world of gravitational collapse. The black hole of today is more than a black hole. It is symbol, 'experimental model', and provider of lessons for the collapse Einstein predicted in far later days for the universe itself.

If collapse is the most startling prediction that physics has ever made, it is also true that general relativity (except for the quantum principle) is the strangest edifice that physics has ever reared. Therefore it is appropriate to look into this structure from windows other than Einstein's original point of entry, aiming especially in the later derivations to enlarge one's view of what collapse is and what is means.

3. CARTAN'S DERIVATION: CONSERVATION OF THE SOURCE COMES ABOUT VIA THE PRINCIPLE THAT 'THE BOUNDARY OF A BOUNDARY IS ZERO'

Riemann Rotation or 'Tide-Producing Effect' Associated with Each Face of a Cube

The central point of electrodynamics is conservation of charge. The central point of geometrodynamics is conservation of mass-energy. Take the relevant field – the electromagnetic field in the one case, the Riemannian curvature or 'tide-producing acceleration' in the other – and *wire the field up' to the source in such a way that* this *conservation comes about automatically, through the principle that 'the boundary of a boundary is zero'*. These ideas go back for their origin to Cartan[39] (see note 39 and ref. 3 for a more complete exposition) and are illustrated in Fig. 5. Rather than look at all of spacetime, direct attention to an arbitrary 'simultaneity' or spacelike hypersurface Σ slicing through spacetime. Rather than examine all of Σ, focus (see enlarged view through magnifying lens) on a small cubical 3-dimensional element of volume located anywhere on Σ, and narrow attention to the 'front' face of this cube. Place a vector at the upper left hand corner (ULHC) of this face. Transport the vector parallel to itself around the periphery of this route, in the sense indicated by the arrow, ending up back at the original starting point. The vector undergoes a rotation. This rotation is proportional to (1) the size of the face and (2) the relevant component of the Riemann curvature of the 4-dimensional geometry. Repeat, taking the same vector on a tour from the same starting point and ending up at the same end point but this time around the top face of the cube. Repeat for all six faces of the elementary cube. Then the combined effect of all six rotation totals to zero. The cancellation of rotations occurs because each edge of the cube has been traversed as often in one direction as in the opposite direction. In other

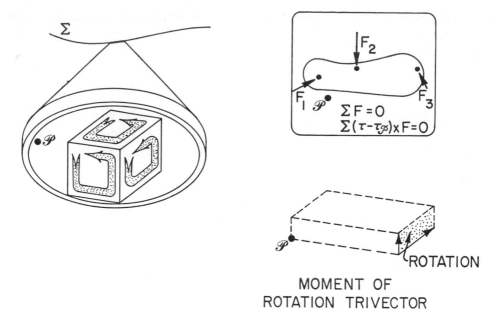

MOMENT OF
ROTATION TRIVECTOR

Fig. 5.   The analogy with mechanics as background for Cartan's 'moment of rotation'. The value of the moment of rotation, totalled over all six faces of the elementary cube, is independent of the location of the point $\mathscr{P}$. Likewise in mechanics the location of the point $\mathscr{P}$ makes no difference in the statement of the conditions for mechanical equilibrium.

words, the 3-cube has a boundary that is made of six 2-dimensional surfaces; and each surface has a boundary that is made of four 1-dimensional edges. However, each edge occurs twice. Thus, when due account is taken of sign, the contributions of all edges cancel. In brief, the 1-dimensional boundary of the 2-dimensional boundary of an elementary 3-dimensional volume, V, is automatically zero; or, in the symbolism of algebraic geometry,

$$\partial\partial V \equiv 0, \tag{15}$$

where ∂ stands for 'boundary of'. The resulting statement about the Riemannian curvature of spacetime, the so-called Bianchi identity, takes the form

$$\sum_{\substack{\text{all six} \\ \text{faces}}} \left(\begin{array}{l} \text{rotation associated} \\ \text{with each face} \end{array} \right) \equiv 0. \tag{16}$$

Moments in Mechanics and in Geometrodynamics

Compare geometry to mechanics. The body in the inset in Fig. 5 cannot be in equilibrium unless the forces all add to zero:

$$\sum_{\text{all forces}} \mathbf{F} = 0. \tag{17}$$

However, for equilibrium, another requirement must also be satisfied. The moments must add to zero:

$$\sum (\mathbf{r} - \mathbf{r}_{\mathscr{P}}) \times \mathbf{F} = 0 \tag{18}$$

About what point the moments are taken does not matter, by reason of the requirement $\sum \mathbf{F} = 0$ (cancellation of the multiplier or $r_{\mathscr{P}}$).

Turn from the idea of 'the moment of a force' in mechanics to the idea of 'the moment

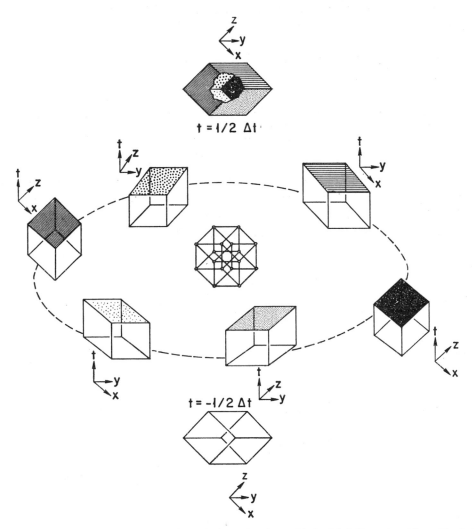

Fig. 6.   The principle that 'the boundary of a boundary is zero' in its 4-3-2 dimensional form. Exploded off the 4-cube at the centre of the figure are its eight 3-dimensional faces, every one a cube. Each of these cubes has six 2-dimensional faces. However, these 2-dimensional faces counterbalance each other in pairs; or, otherwise stated, and with due account of sign, the $8 \times 6 = 48$ faces 'add up to zero'. As example, the black face of the top cube nests against the black face of the right hand cube. Thus the 4-dimensional cube exposes no 2-dimensional face to the outside world; it is 'faceless'. The boundary of the boundary of the 4-dimensional cube is zero.

of a rotation' in geometrodynamics. It will not matter about what point one evaluates these moments. Therefore select the arbitrary point $\mathscr{P}$ shown in Fig. 5, both in 'the view through the lens', and repeated, for better seeing, at the lower right. Also shown at the lower right, depicted as a bivector, is the rotation (measure of Riemann curvature) associated with one of the faces of the cube. This bivector, together with the vector from $\mathscr{P}$ to the centre of the relevant face of the cube, defines a trivector. The value of this trivector depends upon the location of the point $\mathscr{P}$. However, the location of $\mathscr{P}$ drops out from, and has no influence on the value of, the sum of these trivectors, taken over all six faces of the cube:

$$\sum_{\substack{\text{all six} \\ \text{faces}}} (\mathbf{r} - \mathbf{r}_{\mathscr{P}}) \wedge \left(\begin{array}{c} \text{rotation associated} \\ \text{with each face} \end{array} \right) =$$

$$= \left(\begin{array}{c} \text{moment of rotation} \\ \text{trivector associated} \\ \text{with elementary cube} \end{array} \right) \rightarrow \left\{ \begin{array}{l} \text{identified by general relativity with } 8\pi \text{ times} \\ \text{the trivector representation (dual to an ordi-} \\ \text{nary vector) of the amount of energy and} \\ \text{momentum contained in this cube (= 'content} \\ \text{of source' in the cube)} \end{array} \right. \quad (19)$$

Identify this sum with 8π times the amount of energy-momentum contained in this elementary volume. Repeat this statement for all spacelike slices through the given region of spacetime, and for all regions of spacetime. Then one has stated the entire content of Einstein's 10-component field equation. This is relativity in brief!

ELECTRODYNAMICS

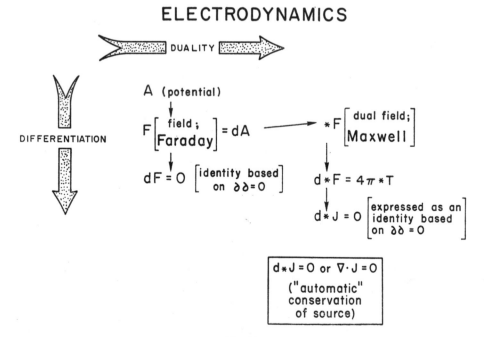

Fig. 7a.

Conservation of Energy-Momentum via the '2-Facelessness' of the 4-Cube

How does this 'wiring up' of the 'field' (geometry) to the 'source' (energy-momentum) guarantee the desired conservation of the source? How does it guarantee that, as time goes on, say from $t = -\frac{1}{2}\Delta t$ to $t = +\frac{1}{2}\Delta t$, no source is created in the element of 4-volume, $\Omega = V \Delta t$ (Fig. 6)? To have conservation means that the amount of source in the top cube (V at $t = \frac{1}{2}\Delta t$) must turn out to be equal to the amount of source in the bottom cube (V at $t = -\frac{1}{2}\Delta t$) plus the inflow of source during the time Δt (as described by the 'inflow' or 'content of source' in the six remaining cubes of Fig. 6); or means that the 'content of source' in all cubes together, with due account of sign, must add up to zero. But Equation (19) wires up the source to the field in such a way that the content of source in any one cube is given by the sum of (moments of rotation) associated with

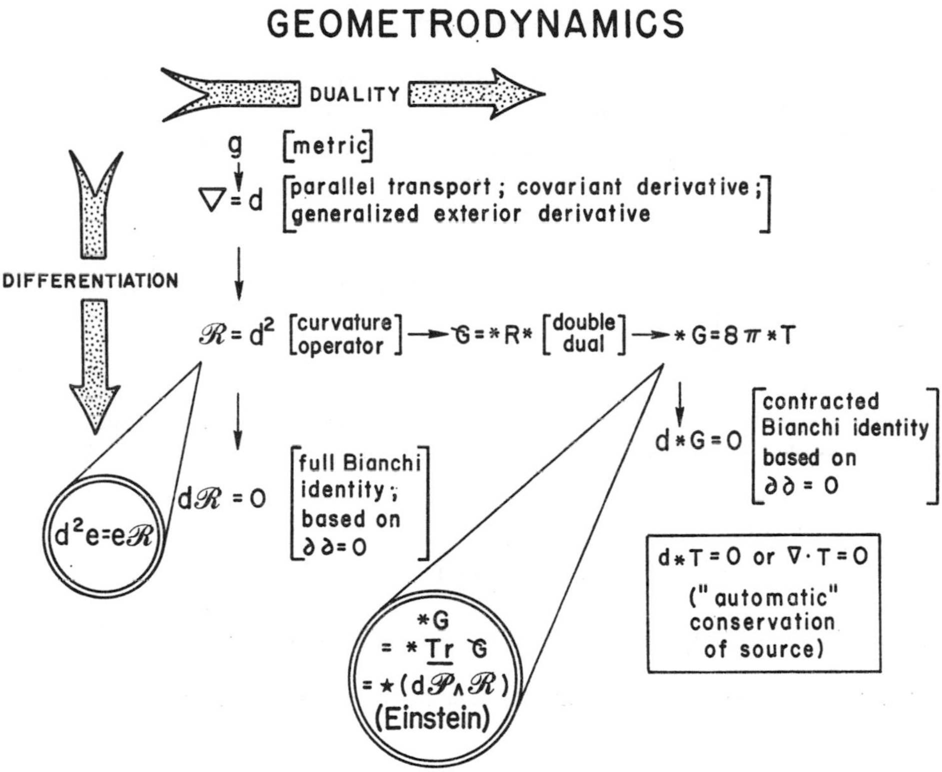

GEOMETRODYNAMICS

Fig. 7b.

Fig. 7. (*a*) The structure of electrodynamics compared to (*b*) the structure of geometrodynamics. In both diagrams the principle that 'the boundary of a boundary is zero' appears twice, once in the left hand column in its 3-2-1 dimensional form, and again in the right hand column, in its 4-3-2-dimensional form (diagrams adapted from Misner, Thorne and Wheeler[3]).

the faces of that cube; and the contributions of all 8×6 faces together cancel out identically; thus

$$\begin{pmatrix} \text{creation} \quad \text{of} \\ \text{source in } \Omega \end{pmatrix} = \sum_{\substack{\text{all eight} \\ \text{3-cubes} \\ \text{bounding } \Omega}} \begin{pmatrix} \text{content of source in} \\ \text{3-cube} \end{pmatrix} =$$

$$= \sum_{\substack{\text{all eight} \\ \text{3-cubes of given 3-cube}}} \sum_{\text{all six faces}} \frac{1}{8\pi} \begin{pmatrix} P\text{-dependent moment of} \\ \text{rotation associated with} \\ \text{that face} \end{pmatrix} \equiv 0 \tag{20}$$

$$\begin{pmatrix} \text{because the 48 faces cancel} \\ \text{out identically, in pairs} \end{pmatrix}$$

The same 'conservation via the principle $\partial\partial \equiv 0$' applies in electromagnetism, as one sees by comparing Figs. 7a and 7b.

Algebraic Geometry Rises above Dimensionality

One used to believe, and often still finds it useful to postulate, that the source comes first in the scheme of things, and the field second. However, one sees that today the possibility is open to think of the field as coming first. On this view the conservation of the source, and therefore in some sense even the existence of the source, is a consequence from and mere aspect of the existence of the field. Moreover, the principle of algebraic geometry ($\partial\partial \equiv 0$) that legislates and enforces 'conservation of the source' is a principle that rises above any particular dimensionality in its most general mathematical version. But the concepts of 'manifold' and 'dimensionality' are presupposed in the laws of physics as they look today. Can one look beyond and above existing statements of physics to a formulation that does not presuppose dimensionality? If so, the principle '$\partial\partial \equiv 0$' would seem an essential part of such a formulation. No principle reaches closer to the heart of general relativity.

4. THE MOST COMPACT FORMULATION OF GENERAL RELATIVITY

Intrinsic Curvature Plus Extrinsic Curvature Equals Energy Density

One knows no more compact statement of general relativity than this:

$$(\text{curvature}) = 8\pi \, (\text{density of mass-energy}). \tag{21}$$

More specifically, take any event P in spacetime, and any spacelike hypersurface Σ through P, and that local Lorentz frame at P in which Σ is a 'simultaneity'. Take the density ϱ (in cm^{-2}; cm of mass-energy per cm^3 of volume) in this frame, multiply it by 8π, and equate the product to the linear scalar measure of the 4-dimensional curvature projected on Σ; thus (after doubling)

$$\underbrace{^{(3)}R}_{\substack{\text{intrinsic}\\\text{curvature}}} + \underbrace{(\text{Tr }\mathbf{K})^2 \qquad\qquad - \qquad\qquad \text{Tr }\mathbf{K}^2}_{\substack{\text{'second invariant' of the extrinsic curvature; or, more}\\\text{briefly, 'extrinsic curvature'}}} = 16\pi\varrho \qquad (22)$$

$$\underbrace{}_{\substack{\text{twice the linear scalar measure of the}\\\text{4-dimensional curvature projected on }\Sigma}}$$

Make this demand for every inclination of the hypersurface through P, and for every choice of P, and have in this one demand the whole content of all ten components of Einstein's field equation.

In electrodynamics one similarly requires

$$\text{div }\mathbf{E} = 4\pi\varrho_e \qquad (23)$$

and imposes (covariance plus) this demand for every inclination of the hypersurface Σ through P and in this way recovers the other three Maxwell equations,

$$\text{curl }\mathbf{B} = \dot{\mathbf{E}} + 4\pi\mathbf{j}_e. \qquad (24)$$

In geometrodynamics, additional to the inclination of Σ, the curvature of Σ seems to matter, as evidenced not least in the appearance of the 3-dimensional scalar curvature invariant, $^{(3)}R$, in (22). However, the remaining two terms in (22) not only compensate for this curvature, but even follow uniquely[3] from the requirement that they should compensate for this curvature of Σ. Thus the left hand side of (22) is a measure of the 4-dimensional curvature. In this equation $\mathbf{K}$ (units cm^{-1}) is the so-called tensor of extrinsic curvature of the hypersurface Σ. It measures the fractional contraction of any local geometric object in Σ when all points of this object are projected forward a unit distance in time (cm) normal to Σ.

For another window into the content of general relativity we now turn from geometry to dynamics as the guiding idea.

5. FROM HILBERT'S DERIVATION TO SUPERSPACE

Hilbert's Principle of Least Action

In no branch of dynamics does a variational principle give a more comprehensive grip on the whole subject than in general relativity. David Hilbert recognized this point and presented the new variational principle to the Göttingen Academy[40] on 20 November, 1915. His step forward derived its guidance and inspiration from Einstein's earlier work. However, it based itself upon a principle of least action from the start. The resulting geometrodynamic law, independent of Einstein in its derivation, was nevertheless identical in form with what Einstein was to lay before the Berlin Academy only five days later.

The idea is simple. Give one spacelike hypersurface σ_0 and a second spacelike hypersurface σ and fill in between them a 4-geometry, $^{(4)}\mathscr{G}$. Try different 4-geometries. For each calculate the action integral*,

$$I = (1/16\pi) \int_{\sigma_0}^{\sigma} {}^{(4)}R \, d \, (4\text{-volume}).$$ (25)

That 4-geometry is allowed by classical physics that maximizes or minimizes or, more generally, extremizes this integral.

What is a 4-geometry? An automobile fender is a 2-geometry. Stretch a ruled transparent rubber sheet over the fender. In this way assign x and y coordinates to every tiny bump and pit in the metal surface. Now pull the rubber harder here and there and thus change the coordinates everywhere. Yet the fender continues to keep its 2-geometry. The difference between a Ford and a Fiat fender is invariant with respect to all changes in coordinatization. Hilbert understood well that the 4-geometry resulting from his variational principle is also invariant with respect to all changes in coordinates.

What is Fixed at the Boundaries Defines 'the Initial Value Problem'

To understand in addition and in coordinate-free geometrical terms what it is that one fixes on the two hypersurface σ_0 and σ is an achievement of recent times. It is also an important achievement. It permits one to state (1) what are appropriate initial value data for the classical dynamics and (2) on what the state function or probability amplitude function depends in quantum dynamics (as illustrated for the physics of a single particle in Fig. 8).

Arnowitt, Deser and Misner [41] turned away from any direct attempt to discover what was fixed at the boundaries, σ and σ_0, in Hilbert's action principle. They added a complete divergence to the Hilbert integrand. Such an addition affects in no way the resulting Einstein field equation, but does alter the quantities fixed at limits. The new quantities, expressed in coordinate-free geometrical form, turned out to be the 3-geometries, ${}^{(3)}\mathcal{G}_0$ and ${}^{(3)}\mathcal{G}$, of the bounding hypersurface, σ_0 and σ. Among other consequences of this result it follows [42] that there is a representation of quantum geometrodynamics in which the state function depends upon and is fixed by the 3-geometry:

$$\psi = \psi \, ({}^{(3)}\mathcal{G}).$$ (26)

The totality of all closed 3-geometries with positive definite signature is called superspace, and what has just been discussed is often known as the superspace representation of general relativity.

---

* Here the element of 4-volume, generalizing an expression like $r^2 \sin\theta \, dr \, d\theta \, d\phi$, is

$$d(4\text{-volume}) = (-g)^{1/2} \, d^4x.$$

The integrand is the 4-dimensional scalar curvature invariant, ${}^{(4)}R$, in a problem of pure geometrodynamics; or this supplemented by the Lagrangian of the other fields when other fields are present.

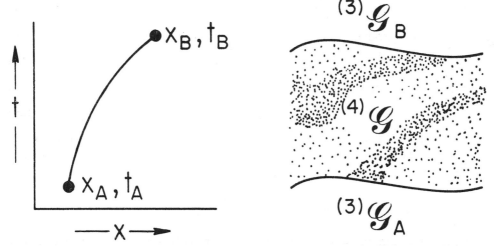

Fig. 8.   The classical history of particle *in* spacetime (world line; left) compared and contrasted with the classical history *of* space (4-geometry; right). In both cases the classical history is selected out from the other conceivable histories that connect initial configuration, A, and final configuration, B, by the circumstance that it extremizes the 'action integral' or 'dynamical path length', $I = I(A, B)$, from A to B. In quantum physics the 'wave function' or 'state function' or 'probability amplitude' depends upon the same variables that define the final state configuration, B; thus, $\psi = \psi(x, t)$ in particle dynamics. In geometrodynamics, one has $\psi = \psi(^{(3)}\mathscr{G})$ in the superspace representation; or, in the York representation (conformal part of the 3-geometry and local Hubble contraction rate specified), $\psi = \psi(^{(3)} <, \diagdown\!\!\!\diagup)$

York's Formulation of the Initial Value Data

In recent months James W. York, Jr., returning to the Hilbert action principle in its original form, has discovered[43] that it demands that one should specify at each point on σ_0 and on σ (1) the conformal part, $^{(3)}<$, of the 3-geometry and (2) the local extrinsic or Hubble time, τ, a concept first introduced by Karel Kuchař. To give the conformal part of a 3-geometry is to give for each point, not the absolute distance, but the relative distance, to every nearby point. In other words, angles are fixed, but not distances. Missing from the information that would be contained in a full 3-geometry at each space point is a scale factor; but in its place one has to specify at each space point something like the dynamical conjugate of this scale factor; namely, the rate at which this scale is decreasing with time, the local Hubble time τ, symbolized by the angular spread between two timelike vectors that stand perpendicular to the given spacelike hypersurface; thus, symbolically,

$$\diagdown\!\!\!\diagup \text{ represents } \tau. \tag{27}$$

In this mathematical representation, York, following earlier work of André Lichnerowicz[44] and Yvonne Choquet-Bruhat[45], has been able to show that one can determine the future from the given information by simple and elegant methods. The solution of an elliptic differential equation yields the unknown scale factor. Moreover, the solution always exists and is unique.

Wave Function, Wave Equation, and Hamilton–Jacobi Equation for Phase of the Wave

Quantum geometrodynamics in the York representation leads to a state function

$$\psi = \psi\left(^{(3)}<, \diagdown\!\diagup\right); \tag{28}$$

in the superspace representation, a state function

$$\psi = \psi\left(^{(3)}\mathscr{G}\right). \tag{29}$$

In neither case is the proper order of factors in the relevant wave equation quite free of all ambiguity, despite a most valuable analysis of this problem by Bryce DeWitt.[46] However, in the semiclassical approximation, one writes

$$\psi \simeq \left(\begin{array}{c}\text{slowly varying}\\ \text{amplitude factor}\end{array}\right) e^{iS/\hbar} \tag{30}$$

with the important physics showing up in the rapidly varying phase factor, $S/\hbar$. There is no ambiguity in the order of factors in the equation satisfied by the Hamilton–Jacobi function S. This definiteness follows not least because a value for S is directly given by the extremal value I of the action integral:

$$S(\sigma) = I_{\text{extremal}}(\sigma, \sigma_0). \tag{31}$$

Moreover, in the superspace representation, the equation for the dynamical evolution of this Hamilton–Jacobi function is a local equation. This equation was first written down by Peres.[47] It reads

$$(16\pi)^2 \left(\tfrac{1}{2}g^{-1/2}\right)\left(g_{ik}g_{jl} + g_{il}g_{jk} - g_{ij}g_{kl}\right)\left(\delta S/\delta g_{ij}\right)\left(\delta S/\delta g_{kl}\right) + g^{1/2}{}^{(3)}R = 0. \tag{32}$$

Here $S = S\left(^{(3)}\mathscr{G}\right)$ is, up to a factor, the phase of the wave function in superspace. Wave crests in superspace are described by surfaces of constant S. Three features of the geometry $^{(3)}\mathscr{G}$ put in an appearance in the Peres or 'Einstein–Hamilton–Jacobi' equation: its metric, $g_{ij}(x, y, z)$; the square root of the determinant of the metric tensor, $g^{1/2}(x, y, z)$; and the local value of the 3-dimensional scalar curvature invariant of the 3-geometry, $^{(3)}R(x, y, z)$.

Out of the law of propagation of wave crests in superspace one can deduce the law of propagation of a wave packet. In other words, one can discover how a 3-geometry evolves with time in the semi-classical approximation. In this way Ulrich Gerlach[48] has succeeded in deriving from the one Einstein–Hamilton–Jacobi equation all ten components of Einstein's standard geometrodynamic law.

Superspace as Arena for the Dynamics of Geometry

In no formulation of dynamics is the leap from the classical to the quantum outlook shorter than in Hamilton–Jacobi theory. Sharply intersecting wave crests reproduce the determinism of classical dynamics; waves of finite wavelength reproduce the finite wave packets and indeterminism of quantum dynamics. All this is familiar. What is new is superspace. It imposes itself on our attention exactly because we insist on analyzing the dynamics of geometry from the wave point of view. Demand Einstein geometrodynamics, demand the quantum principle, and end up with superspace.

What kind of an arena for dynamics is superspace? And what lessons does it teach? Fig. 9 illustrates at the left a smooth closed 2-geometry. One can approximate this

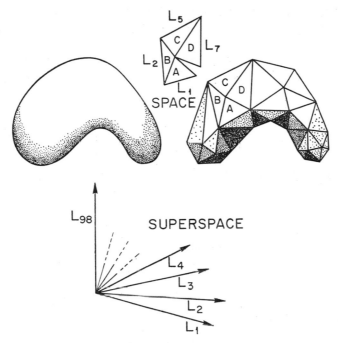

Fig. 9. A 2-geometry (upper left) is approximated by a skeleton 2-geometry (upper right). All the details of the shape of this skeleton 2-geometry are completely specified by giving (in this example) all 98 edge lengths, $L_1, L_2, \ldots, L_{98}$. This information is represented by a single point (lower diagram) in a 98-dimensional 'truncated superspace'.

2-geometry arbitrarily closely by a polyhedron or 'skeleton 2-geometry' (illustration at right) built of a sufficiently great number of faces. Euclidean geometry rules in each face. In this illustration the 98 edge lengths determine all the details of the shape of the polyhedron. Represent this information by a single point in a space of 98 dimensions. The projections of this point onto 98 coordinate axes give back all the original information about the 98 edge lengths. Move this 'representative point' slightly in the 98-dimensional space. Then all 98 coordinates of this point – and therefore all 98 edge lengths of the triangles in the polyhedron – also change slightly. The skeleton 2-geometry bends, twists, swells and otherwise changes in shape in obedience to the motion of the representative point. Take the analysis given here for skeleton 2-geometries built out of triangles and redo it[49] for skeleton 3-geometries built out of tetrahedrons. Also go from finite-dimensional or 'truncated' superspace to the limit where (1) the skeletonization is infinitely finegrained, (2) the edge lengths are infinitely numerous, and (3) superspace rises in dimensionality from the purely illustrative number of 98 to the actual number of infinity.

Dynamics of the Universe as a Leaf of History in Superspace

A *leaf of history* cuts through superspace. It describes the deterministic dynamic development of the geometry of space with time. Fig. 10 illustrates how. At the right is

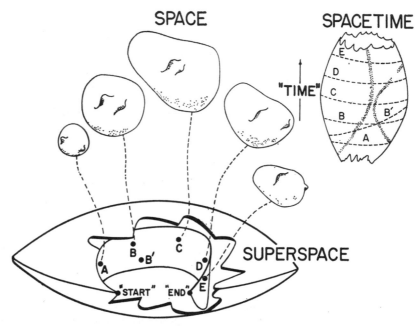

Fig. 10. Space (upper left), spacetime (upper right) and superspace (below). The 'leaf of history' that curves through superspace includes all the configuration (*A*, *B*, *B'*, ...) achieved by space in its classical dynamical evolution in time; that is, all spacelike slices through the given spacetime. A different spacetime (not shown); that is, a classical history of space when the dynamics of space is started off with different initial conditions, corresponds to a different leaf of history (also not shown) cutting through superspace.

spacetime, the usual deterministic classical picture of space evolving with time. Any spacelike slice through this spacetime, such as *A*, is a 3-geometry, a momentary configuration of space. It is represented in superspace by a single point, also denoted by *A*. Another slice *B* through the same spacetime provides another 3-geometry, and thus another point *B* in superspace. A one-parameter family of spacelike slices through spacetime thus 'generates' a one parameter family of points running through superspace: a line or curve. However, time in general relativity has a many fingered character. It bursts the bounds of anything so narrow as a one-parameter family of spacelike slices. The explorers of spacetime have full liberty to push ahead their exploration faster in one place than another. It is a perfectly legitimate action for them to measure up the 3-geometry of the spacelike slice B'. This 3-geometry is a new point in superspace. No line in superspace can accommodate all the points, all the 3-geometries, that one gets by making spacelike slices in all conceivable ways through a given spacetime. The region of superspace occupied by all these points is not a line; it is a leaf.

Given the spacetime, we have seen how we construct the leaf of history in superspace. Conversely, given the leaf of history in superspace, we obtain all the 3-geometries we need to reconstruct the spacetime. The procedure required, and used by Gerlach, but not spelled out here, reminds us in some ways of how we interlock together the disassembled wooden pieces of a Chinese-puzzle elephant to reconstitute the elephant.

New Features of Quantum Geometrodynamics

Quantum geometrodynamics differs drastically in principle from classical geometrodynamics. No longer is there the sharp yes–no difference between 3-geometries. The classical analysis clearly marked off the YES 3-geometries, that lie on a given leaf of history, from the NO 3-geometries, that do not. In the quantum analysis there is instead a probability amplitude $\psi(^{(3)}\mathscr{G})$ for this, that and the other 3-geometry. The 3-geometries with appreciable probability amplitude are far more numerous than can be accommodated in any one spacetime. There are too many wooden pieces to be fitted into one elephant. The concept of a deterministic classical spacetime has to be abandoned.

The idea has been discussed for many years that quantum effects smear out the local light cone.[50] A much more drastic conclusion emerges out of quantum geometrodynamics and displays itself before our eyes in the machinery of superspace: *there is no such thing as spacetime in the real world of quantum physics.* Spacetime is a classical concept. It is incompatible with the quantum principle. It has to be discarded in any deep-going analysis of the foundations of physics. It is an approximation idea, an extremely good approximation under most circumstances, but always only an approximation.

If we had a deterministic spacetime, we could take spacelike slices through it at two immediately succeeding instants, and thus find both a 3-geometry and a time rate of change of this 3-geometry. But complementarity forbids. It does not forbid our determining the 3-geometry alone an on initial spacelike hypersurface within arbitrarily narrow limits. However, the reciprocal uncertainty in the time rate of change of this 3-geometry is then arbitrarily great. This uncertainty deprives us of any possibility whatsoever to give any sharply defined meaning either to 'spacetime' or to 'the dynamical history of space'.

In summary, superspace leaves us space but not spacetime and therefore not time. With time gone the very ideas of 'before' and 'after' also lose their meaning.

Quantum Fluctuations in the Geometry of Space

These quantum effects show up in significant measure only at small distances. There is a convenient name for them – 'quantum fluctuations in the geometry'. They have nothing directly to do with particle physics. They are a property of all space.[51]

Analogous quantum fluctuations in the electromagnetic field are also a property of all space. To analyze these fluctuations, to calculate their effect upon the motion of the electron in the hydrogen atom, and to observe the resulting shifts in the spectral lines of hydrogen, together constitute one of the greatest triumphs of physics since World War II.[52] Thus today it is fully confirmed that the quantum fluctuations of the electric field in a region of extension L are of the order of magnitude

$$\Delta \mathscr{E} \sim (\hbar c)^{1/2}/L^2 \,. \tag{33}$$

Apply the same kind of analysis to the gravitational field and equally directly conclude[53,54] that the inescapable fluctuations in the metric $(-1, 1, 1, 1)$ are of the

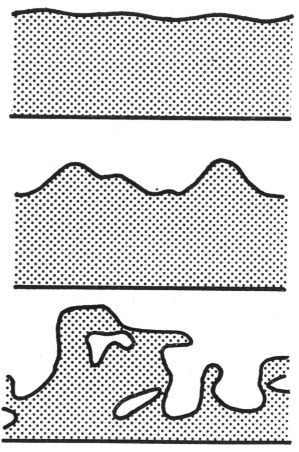

Fig. 11.   Symbolic representation of the quantum fluctuations that take place everywhere and all the time in the geometry of space: above, as evidenced at a scale of observation, L, far larger than the Planck length, L^*; middle, L only a little larger than L^*; below, L comparable to L^*. Compare with the view of a stormy ocean as seen by an aviator flying miles above it, flying a hundred metres above it, and tossing in a lifeboat on the surface.

order

$$\Delta g \sim L^*/L. \tag{34}$$

Here

$$L^* = (\hbar G/c^3)^{1/2} = 1.6 \times 10^{-33} \text{ cm} \tag{35}$$

is the Planck length.

These quantum fluctuations in the geometry of space are completely negligible at the scale of atoms and nuclei and elementary particles (L from 10^{-8} cm to 10^{-15} cm; Δg from 10^{-25} to 10^{-18}). In the domain of everyday physics space can be considered to be flat. Therefore, it is not surprising that no immediately measurable effect of the fluctuations, like the Lamb–Rutherford shift in hydrogen, has yet come to light. However, at smaller and smaller distances of observation the predicted fluctuations in the geometry become larger and larger, until at dimensions of the order of the Planck length one is open to believe that fluctuations take place even in the topology or connectivity itself (Fig. 11).

Quantum Fluctuations and Multiple Connectivity

Without any thought of quantum fluctuations, William Clifford[55], a century ago, had considered local changes in the connectivity of space as connected with the physics of particles. Again half a century ago, Hermann Weyl[56] pointed out that space, here and there, may be multiply connected in the small and, consequently, 'The argument that the charge of an electron must be spread over a finite region, because otherwise it would possess infinite inertial mass, has thus lost its force. One cannot all at say, here is charge, but only, this closed surface encloses charge.' The writer gave reasons[51] for the first time in 1957 out of fluctuation theory to consider 'wormholes' a property, not of particles, but of all space, and all electric charge as 'lines of electric force trapped in the topology of space'. In the same year Charles Misner[57] showed the beautiful ties that connect Maxwell's theory in a multiply connected space with the mathematics of differential forms and homology groups.

Today, reconsidering electric charge, we can turn around the order of history in our imagination. Deny the existence in nature of any such thing as a mystic magic electric jelly. Rule out also any point singularity in any solution of Maxwell's equations. Agree with Einstein that once one admits the possiblity of a singularity here, he has to admit it there, and therefore everywhere, and then he has destroyed the force of his field equation. Insist then that Maxwall's source free field equations hold everywhere without exception. Then electric charge becomes possible only if space is multiply connected. Therefore search nature for any evidence of electric charge. Find it – and conclude that space must, indeed, be multiply connected in the small. From this point of view, the existence of electric charge is the most compelling evidence we have today for Planck-scale fluctuations taking place in geometry and connectivity throughout all space. These are the fluctuations that say 'No!' to spacetime and to time at small distances.

Gravitational Collapse Reexamined within the Framework of Superspace

If Hilbert's variational principle leads to superspace, and superspace leads to fluctuations and two decisive negatives, may not superspace also lead to an important positive? It furnishes an arena in which to take a fresh look at gravitational collapse, the greatest crisis in the theoretical physics of our times (Fig. 12).

COLLAPSE	MATTER (1911)	SPACE (1970's)
DYNAMIC SYSTEM	e and + CHARGE	GEOMETRY (HUBBLE)
CLASSICAL PREDICTION	∞ KINETIC ENERGY IN A FINITE TIME	∞ COMPACTION OF MATTER AND GEOMETRY IN A FINITE TIME
ONE REJECTED SOLUTION	GIVE UP COULOMB LAW (10^{-8}cm, 5×10^{-13}cm)	GIVE UP EINSTEIN'S EQUATION
ANOTHER ATTEMPT AT A "CHEAP WAY OUT"	ABANDON IDEA THAT ACCELERATED CHARGE RADIATES	ABANDON IDEA THAT MATTER CAN BE COMPACTED INDEFINITELY
PRINCIPLE OF CAUSALITY RULES THIS OUT	J.J.THOMPSON E. PURCELL IN BERKELEY PHYS.	$v^2_{SOUND} = \dfrac{dp}{d\rho} > c^2$
IMPLICATION OF PLANCKS QUANTUM PRINCIPLE	QUANTUM SPREAD IN SPACE $\Delta p \sim \hbar / \Delta x$	QUANTUM SPREAD IN SUPERSPACE

Fig. 12.   Parallels between past and present crisis.

The electric collapse of matter, the great problem of the early 1910's, found its solution in the quantum principle. According to classical theory, the electron headed for the point centre of attraction arrived in a finite time at a condition of infinite kinetic energy. One had only to translate the classical Hamilton–Jacobi equation of motion of this particle to the Schrödinger wave equation to see deterministic collapse turned into probabilistic scattering (Fig. 13).

A classical leaf of history shows the universe expanding, reaching a maximum volume, recontracting, and finally collapsing in a finite proper time to a state of infinite compaction. Turn from classical determinism to a probability wave propagating in superspace. Can this wave not also undergo scattering at the point in superspace where otherwise collapse would have been expected? And if the electron scattered by the nucleus goes off on a quite new worldline, cannot the wave scattered in superspace go off on a quite new leaf of history?

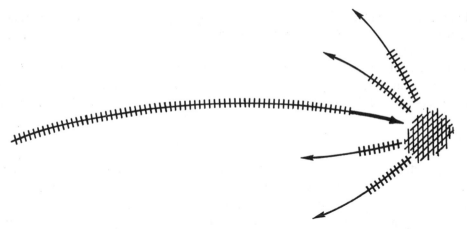

Fig. 13. Not deterministic collapse (in the cross-hatched 'zone of collapse') but probabilistic scattering, is the outcome of the encounter (motion in 3-space) of the negatively charged electron with the positively charged centre of attraction; and is also on outcome natural to consider seriously for the gravitational collapse of the universe itself (motion of representative point in superspace described, not by a deterministic leaf of history, but by a probability wave; and this wave undergoing scattering in superspace; not a deterministic new cycle of the universe, but a 'probability distribution' of new cycles of the universe).

In What Sense Do Other Leaves of History 'Coexist' with Our Own?

We have only to ask questions such as these to find ourselves facing a still deeper question. With two or more quite different leaves of history located in one and the same superspace, what strange kind of 'coexistence' of two universes are we confronting? Is it not absolute nonsense to speak of another universe coexisting with our own, no matter in how attenuated and ethereal a way we use the word 'coexist'? Almost a century ago Auguste Comte[58] also decried as absolute nonsense the idea of attributing a chemical composition to a distant star. It may have a sense to speak of the chemical composition of the Sun, he was willing to admit; but certainly not the composition of a star to which there is not the slightest possibility of anyone ever travelling. Of course, in the meantime, half a dozen ways have been found to get at the composition of a star, and many a satisfactory check has been made of one method against another. No one would think of dispensing with this concept.

There is also not the slightest possibility to travel to another leaf of history. Gravitational collapse places an impenetrable barrier between one leaf and another. Life cannot get through. Even such ideas as 'before' and 'after' lose their relevance in the final stage of collapse, thus altogether forbidding any direct comparison of time between one leaf of history and another.

Consider more closely this question of 'coexistence' of alternative histories of the universe. Quantum spread moves the representative point that describes the universe little way off one classical leaf in history in superspace. A larger movement takes it to another classical leaf of history. There is no difference of principle between the two.

There is only a difference of degree. No one can deny the 'coexistence' of alternative histories of the universe who accepts the existence of quantum fluctuations in the geometry of space.

One has only to recall the famous double slit electron interference experiment to see the same principle in a simpler context. The 'coexistence of two histories' of the electron is the very heart of the observed interference. No one has ever successfully contested it.

'Scattering in Superspace' as the Final Phase of Collapse

Between 'fluctuations' and 'scattering' there is a difference only in degree, not in kind. Do then the predicted final stages of collapse of the universe lead, not to the deterministic catastrophe of classical theory, but to a probabilistic scattering in superspace? If the electron, moving faster and faster towards the disaster, experiences scattering, not catastrophe, does the universe do the same? The arena for the dynamics of the electron is Minkowski spacetime; the arena for the dynamics of geometry is superspace; but is there otherwise any reason why scattering into a new history is not as truly the outcome in the one case as in the other?

Why collapse may not be final, why it may be followed by a new history or, rather, by a probability distribution of new histories, when the dimensions of the universe get down to a value governed by the Planck length, may be put in still other words. Already here and now, according to quantum geometrodynamics, violent fluctuations are going on in geometry as viewed at the Planck scale of distances. On such a worm's eye view a fluctuation is hardly distinguishable from the collapse of the universe itself. In effect, gravitational collapse of the 'local universe' is *already* over and over taking place and being undone. Moreover, this doing and undoing of collapse is going on everywhere in space and all the time without catastrophe. So why anticipate catastrophe from the collapse of the universe itself?

If one can foresee along these lines the answer to the paradox of collapse, why not work it out and demonstrate it by calculation? In the problem of the electron one goes easily from the classical Hamilton–Jacobi equation to the Schrödinger wave equation and from that to the Rutherford law for the probability distribution of scattering angles. Why not proceed similarly here, where one already has the Hamilton–Jacobi equation?

First, there are unsolved problems of factor ordering in translating the H–J equation into a Schrödinger equation. Second, both equations presume classical differential geometry. Classical differential geometry leaves no room for changes in topology. Yet it is an inescapable characteristic of quantum field theory that, in the phrase of John Klauder,[59] unruly configurations predominate. From unruly configurations of 3-geometry like those symbolized in Fig. 11 it is a small step in the imagination to go to a doubly connected 3-geometry, as would also seem to be required by the existence of electric charges. But classical differential geometry says 'No' to this step. If that mathematics applied to nuclear matter, it would also say 'No' to nuclear fission, with *its* change in connectivity. But the nucleus elongates and divides, all prohibitions of differential geometry notwithstanding. For the description of this change in connectivity today's nuclear physics has the right mathematical machinery. Today's geometrodynamics does not.

Lastly, no one can believe any purportedly quantitative treatment of the final stages of collapse as 'scattering in superspace' that assigns no role to Fermion fields and particles, has no explanation [60] for their spin, and pays no heed to their fate.

Scattering and Superspace as Waystations on the Road to Deeper Views

'Scattering in superspace' contains two concepts. One is 'scattering' as the terminal phase of collapse. The other is 'superspace' as the arena for that collapse. Both concepts, it is possible to believe, are way stations, useful way stations, but nevertheless only way stations, on the road to still deeper penetration. Therefore take a second look: at 'scattering', later; at superspace, now.

Superspace is a point of farthest advance in the understanding of relativity. In no arena does the dynamics of geometry express itself more compactly. From no vantage point do collapse and quantum fluctuations in metric appear more clearly as two aspects of the same geometrodynamics. Superspace is here to stay.

The mathematics of superspace nevertheless seems at first sight in two ways too frozen to expose to view any still deeper level of physics. (1) The dynamic law, the Einstein–Hamilton–Jacobi law (32) for the propagation of wavecrests in superspace, looks as if handed down from on high and beyond further explanation. Riemann [61] fought to make geometry a part of physics. How could he have counted it a victory to see the God-given geometry of Euclid replaced by the God-given geometry of superspace? [62] To have superspace instead of space is no advance towards the explanation of space. (2) The representative point in superspace is a 3-geometry. Three-geometry appears as the one and only dynamic entity. What about the rest of physics? Has one locked himself, unawares, into the view that particles and fields are all derivative, somehow constructed from geometry as from a 'magic building material'? Has one adopted the Clifford–Einstein 'space theory of matter'? Not at all.

Only to minimize detail has one limited attention to pure geometry: to gravitational waves, geons made out of gravitational waves, and black holes made by collapse of such geons, all in a universe curved up into closure by its content of black holes, geons and gravitational waves. How then does one give an account of electromagnetic fields and effects? One augments the variables that appear in the state functional, ψ. from $^{(3)}\mathcal{G}$ to $^{(3)}\mathcal{G}$ plus $\mathbf{B}$, where $\mathbf{B}$ is a divergence-free magnetic field defined everywhere throughout the manifold $^{(3)}\mathcal{G}$. Similarly for other fields: the field coordinate, or the field momentum, but not both, grace to complementarity, also appears in ψ: or, in the semiclassical approximation, appears in the Hamilton–Jacobi functional S. Accordingly 'augmented superspace', the configuration space of the dynamics, contains additional and non-geometrical coordinates.

Deeper questions do not arise. Are electromagnetism and particle fields a manifestation of pure geometry? Or is geometry a mere bookkeeping for relations between particles? Or are particles and geometry both primordial? Or have they both derived from something more primordial than either, call it pregeometry or call it what one will?

No immediate help does one get from the previous four derivations of relativity (Einstein, Cartan, compact, Hilbert) in penetrating deeper into such questions, either to understand why superspace has the special Hamilton–Jacobi structure (32), or to suggest what particles have to do with geometry. Guidance into these issues comes first from the final two derivations of relativity. Number five has for key idea that 'dynamically changing space must be imbeddable in spacetime'. Number six, epitomized, says 'space acquires its resistance to curvature from the curvature-dependence of the zero-point energy of particles and fields'. In penetrating to a stratum of ideas deeper than those encountered in previous derivations, these approches begin to recognize that geometry may be a derivative rather than a primordial concept.

Rejection of the View that Space is the Primordial Dynamic Entity

The directly contrary vision, going back to Clifford and Einstein, that geometry is the primordial entity, and everything else is derived or constructed from geometry, deserves its assessment before one turns to these final two derivations of general relativity.

As early as February 21, 1870, in a paper before the Cambridge Philosophical Society *On the Space Theory of Matter* [63] W. K. Clifford (1845–1879; Clifford algebras), inspired by the 1854 lecture of B. Riemann (1826–1866), had proposed that a particle is a 'hill' built out of the geometry of space rather than a foreign and physical object immersed in the geometry of space. Einstein himself was animated by the vision of a purely geometrical account of physics. Many a worker since who has occupied himself at all with general relativity has found himself little by little caught up in the same Clifford–Einstein vision. In such cases it is not rare to arrive at a little new understanding of Einstein's general relativity, a great appreciation of the crisis of gravitational collapse, and also, in the end, the conviction that the quantum principle is even more fundamental than geometrodynamics to the make-up of physics and the elucidation of collapse. A sample case history, for one of the many workers in the field, will illustrate this course of evolution of ideas:

1953: Accept gravitational collapse as a central issue. Simplify equation of state of collapsing object by taking radiation alone as the source of its mass-energy.

1954: Insist this radiation shall travel perpendicular, or nearly perpendicular to *r*. Arrive at a 'geon'. It holds itself together by its own gravitational attraction for a time long in comparison to periods of individual quanta. Attracts as a mass, moves as a mass, but nowhere contains any 'real' mass. Model for 'mass without mass'. A classical object. No direct relation whatsoever to a particle.

1955: 'Charge without charge': electricity as lines of force trapped in a multiply connected space. Existence of charge in nature taken as evidence that space in the small is multiply connected. 'Electromagnetism without electromagnetism' 2nd order Maxwell equations and 2nd order Einstein equations put together in 4th order Rainich equations dealing with geometry and nothing but geometry.

1956: A particle – that looks impressive – is as unimportant relative to the quantum fluctuation physics of the vacuum as a cloud – that looks impressive – is unimportant to the physics of the sky. Particle physics is not the right starting point for dealing with particle physics. Vacuum physics is. Space, owing to quantum fluctuations in geometry and connectivity at small distances, of necessity has a 'foamlike structure',

1964: Superspace: allows one to see the structure of geometrodynamics at a glance, and see collapse and quantum fluctuations of geometry within the same dynamic framework.

1968: When an orientable 3-geometry is multiply connected, (n handles or 'worm-holes') superspace has 2^n sheets. Each sheet corresponds to a topologically distinct continuous field of triads that can be laid down on the 3-geometry. There are 2^n distinct probability amplitudes associated with the same 3-geometry; or, per wormhole, one 'non-classical two-valuedness' or spinor degree of freedom ('spin without spin'). Question raised, can a particle be regarded as a 'geometrodynamical exciton'? And can neutrino fields, pion fields, hyperon fields and other fields likewise be interpreted in terms of 'modes of excitation' of multiply connected geometry?

1969: Continuing analysis of black hole physics.

1970: Outcome of gravitational collapse of the universe itself discussed in terms of 'scattering in superspace'.

1971: No dynamics of topological spin and no quantum fluctuations in topology – and therefore, one can believe, no proper treatment of collapse as scattering in superspace – without change in connectivity; but no continuous change in connectivity is allowed by differential geometry. Differential geometry presupposes a concept of 'point neighborhood' that cannot be a correct description of the physics at small distances, The thinning and breaking of a handle makes points that were near suddenly become far. Conversely, far away points have a potentiality for becoming immediately adjacent that is incompatible with the ideas of differential geometry. Even the concept of dimensionality cannot be applicable at small distances. With the failure of differential geometry, general relativity also fails: it cannot provide anything more than a crude approximation to what goes on at the smallest distances. Geometry 'is not crazy enough' to describe all of physics. But particle physics also does not provide any 'magic building material'. No account of particles that deals only with particles will ever explain particles. There must exist an entity ('pregeometry') more primordial than either geometry or particles on the foundation of which both are built. The nature of pregeometry will first become clear when one sees the quantum principle in all completeness, not as something strange and foreign imposed on the world, but as the central principle without which the world could not even come into being.

Out of a case history such as this, and many another, each with its pluses and minuses, what is the conclusion?

To those who have laboured in the garden of geometrodynamics, or watched its development, it has been a reward to see the blossoming of neutron-star astrophysics and the budding of black-hole astrophysics. It has been a satisfaction to observe that new dimension come to life that Einstein's theory gives to all of physics – *geometry*, from tidal acceleration as Riemannian curvature to superspace as the arena for geometrodynamics. It has been tantalizing that electricity lets itself be interpreted as lines of force trapped in a multi-wormhole geometry, with one spin $\frac{1}{2}$ tied to each wormhole. It has been both a disappointment and an inspiration to sense at last that one must look beyond geometry for the understanding of geometry – and of collapse.

The view that 'everything is geometry' has shown itself in the end a view 'too finalistic to be final'. The very surprise of the predictions of general relativity (expansion of the universe predicted, and predicted correctly, and predicted against all expectation; gravitational collapse; black hole) and the scope of its explanations (gravitation as a manifestation of geometry; conservation of mass-energy guaranteed by the principle that the boundary of a boundary is zero) created a new standard for the surprise of a prediction and for the scope of an explanation. The standard has meantime risen, not least because of the beautiful regularities uncovered in particle physics. General relativity has not kept up with the rise.

The 'Surface Geology' and 'Underground Geology' of the Vacuum

The student who first takes up geology finds no feature of the landscape more interesting than its topography. Later he sees cores drilled out from widely separated locations with identical strata. He comes to think of the stratum as the primary concept. Finally he begins to appreciate that underground strata and surface topography are manifestations of one and the same dynamic geology.

To the student who first learns about relativity no feature of the vaccum attests more clearly its power to take part in physics than its curvature. Then he sees energy slammed into the vacuum here, and discovers particles spray out with a characteristic spectrum of masses. He observes energy poured in at a remote point and finds that the same spectrum emerges from the vacuum. Particles come to look like the central feature of vacuum physics. But further study makes him believe that both the geometry, 'the surface geology of space', and the particles, 'the underground stata', are manifestations of a something more primordial than either. This is the point of view we adopt in looking into the structure of relativity through the last two windows.

6. THE STRUCTURE OF SPACETIME DERIVED FROM THE 'GROUP' OF DEFORMATIONS

General Relativity as Representation of the 'Group' of Deformations

The many-fingered time of Einstein's general relativity is a concept so simple that its sophistication does not immediately surface: its central presupposition that *space is imbedded in spacetime*. Let a band of observers explore the dynamics of geometry and other fields. Like a line of soldiers, they can advance faster on one front, slower on another, and later push ahead more rapidly in the second region, slower in the first, until they come to the preassigned 'river line', or spacelike hypersurface. What they find there must be the same whether the moving hypersurface surged ahead first on the left or on the right. The change in the physics from the initial simultaneity to the final simultaneity must be independent of the choice of simultaneities in between. In mathematical terms, the dynamics must provide a representation of the 'group' of deformations of a spacelike hypersurface. This requirement, Hojman, Kuchař and Teitelboim show[64], fixes the Hamiltonian of general relativity as of the form (32), up to an arbitrary canonical transformation, and up to the permitted addition of the cosmological term that Einstein first introduced and later rejected.

If the structure of Euclidean geometry ever seemed arbitrary, its general-relativity substitute, the law (32) of propagation of wave fronts in superspace, must have appeared as still more arbitrary; but it is not, one now sees. Superspace turns out to follow the only law that one can easily imagine, a law so simple in its principle that anything simpler could hardly be a law.

Were the Hamiltonian different, one would still have the geometrodynamical field coordinates, $g_{ij} = g_{ij}(x, y, z)$, and the geometrodynamical field momenta,

$$\pi^{ij} = \frac{\delta S}{\delta g_{ij}} ; \tag{36}$$

and an acceptable 'initial value set' of these $6 + 6 = 12$ functions of position would still determine an entire leaf of history in superspace. However, the 3-geometries making up this leaf of history would no longer fit into any one spacetime.

The band of observers would still have the freedom to push forward 'many-fingered time' with all the individual free choice that that term implies. However, these time increments would no longer let themselves be described as increments of a time coordinate in any manifold that in any way whatsoever constituted a spacetime.

Demand Hamiltonian theory in superspace and demand that that Hamiltonian theory shall yield spacetime, and automatically end up with the Einstein–Hamilton–Jacobi equation – that is the beautiful route to general relativity opened up by Hojman, Kuchař and Teitelboim. When a vector field is added, electromagnetism also emerges. When other fields are included, their dynamics similarly comes out of the condition of imbeddability.

Relativity Compared to Elasticity

The very austerity of 'relativity out of imbeddability' shows how little of a fundamental nature goes into the derivation of Einstein's law of gravity, and how little of the inner working of physics one really can read out of relativity. One is led to compare relativity with elasticity. The elastic energy-per-unit-volume of the small deformation,

$x^i \rightarrow x^i + \xi^i$, of a homogeneous isotropic solid, expressed in terms of the strain tensor $\mathbf{e}$ with components

$$e_{ij} = \tfrac{1}{2}\left(\frac{\partial \xi^i}{\partial x^j} + \frac{\partial \xi^j}{\partial x^i}\right) \tag{37}$$

is

$$c_1 (Tr\,\mathbf{e})^2 + c_2 Tr\,\mathbf{e}^2 \tag{38}$$

according to reasoning based upon considerations of symmetry and group theory alone. The binding of a mixed solid arises from bonds between a multitude of different atoms. Each bond has its own potential energy curve and resistance to bending. However, only the sums of the second derivatives of these many potential appear in the elastic constants c_1 and c_2. Not one hint do these two totals give about the size of the individual atoms, the composition of the solid, or the origin of a potential energy curve.

In general relativity there appears only the one constant, the Newtonian constant of gravity G. The existence of such a constant again follows, as shown by Hojman, Kuchař and Teitelboim, from group theory alone (the 'group' of deformation of a spacelike hypersurface). Nevertheless the origin and nature of any individual contributions to G are again totally concealed from view.

7. SAKHAROV'S DERIVATION: GRAVITATION AS THE 'METRIC ELASTICITY OF SPACE'

Nothing forces the student of elasticity to rely on measurement alone for values of the elastic constants of the solid. He can evaluate them from spectroscopic or calculated or estimated values of the stiffness parameters of the individual bonds. Sakharov[65] (see also Zel'dovich and Novikov[66]) similarly proposes to view the gravitation constant (1) as measuring the 'metric elasticity of space', and (2) as given by the sum of individual contributions, each of which in principle can be estimated. On this view space is like an empty sausage skin, which is 'floppy' and deprived of all resistance to bending until it has been filled with sausage meat. The 'sausage meat' is the zero-point energy of particles and fields.

In undeformed space the electromagnetic field, as an example, has per unit of volume a zero-point energy that is obtained by integrating the product of the following factors:

Number of independent modes in interval of circular
    wave numbers from k to $k + \mathrm{d}k$, $4\pi k^2 \mathrm{d}k/(2\pi)^3$
Number of states of polarization per mode,   2
Zero point energy per mode, $\hbar ck/2$

$$\text{Product,} \quad \hbar ck^3 \mathrm{d}k/2\pi^2 \tag{39}$$

The result diverges. It has to be renormalized to zero to be compatible with experience. The result is similar for other fields. Moreover, the result is qualitatively the same whether one deals with energy (one component of a 4-vector) or the Lagrangian (invariant density). However, when space is curved, correction terms arise in the renormalized invariant Lagrangian density for each field proportional to the 4-dimensional Riemann scalar curvature invariant:

$$\delta L_{\text{one field}} \sim \hbar c^{(4)} R \int k \, dk \qquad (40)$$

(see Berger, Gauduchon and Mazet[67] for more on the effect of curvature on the spectrum of standing waves). This integral is still divergent. Sakharov reasons that there is a highest circular wave number, $k = k_{\text{crit}}$, for which the calculation makes sense. Here k_{crit} for all fields alike, he proposes, is of the order of the reciprocal of the Planck length,

$$k_{\text{crit}} \sim (\hbar G / c^3)^{-\frac{1}{2}} \sim 10^{33} \text{ cm}^{-1}. \qquad (41)$$

It follows that the contribution to the Lagrangian of the vacuum from the curvature-dependent zero-point energies of all fields together has the same form and order of magnitude as the Lagrangian of Einstein's theory of gravity,

$$L_{\text{grav}} = (c^4 / 16\pi G)^{(4)} R \sim \hbar c \, k_{\text{crit}}^{2 \, (4)} R \sim \sum_{\text{fields}} \delta L_{\text{one field}}. \qquad (42)$$

This is the sense in which Sakharov considers gravitation to be the metric elasticity of the vacuum.

The constant of gravitation estimated in this way can be given almost any value one chooses by appropriate choice of the cutoff wave number k_{crit}. Sakharov tailors k_{crit} to give the known value of G. No one sees how to get k_{crit} from first principles. Nevertheless, Sakharov reasons, the proper order of ideas is not, first graviation and then fields and particles, but first fields and particles and then gravitation, as a derivative effect.

From Sakharov's 'particle first' point of view, gravitation is as much derivative from particle physics as elasticity is derivative from molecular physics. If one accepts his point of view one does wrong to try to build particles out of geometry. One does wrong whether one speaks of 1870 Clifford 'hills' in space or 1970 'geometrodynamic excitons'. One might as well try to build atoms out of elasticity! Atoms come first, and only then elasticity; particles first, and only then geometrodynamics.

8. MUTABILITY AND BEYOND

Pregeometry as More Primordial than Either Particles or Geometry

The last two derivations of relativity, different though they are, suggest that gravitation is as far removed as elasticity from being primordial. But does that mean that particles are primordial? Hardly. The derivative character of elasticity by no means implies that atoms are the primitive entities. On the contrary, it was the first and smallest advance in the study of solids to understand the two elastic constants in terms of scores of molecular potential energy curves, many of them not known in any detail.

Only when those scores of interactions found explanation in terms of a system of electrons and positively charged nuclei and Schrödinger's equations and nothing more did the decisive advance in understanding come. Likewise it may be only the first step forward to interpret the one 'constant of gravitation as the sum of the coefficients of curvature dependency of the vacuum energies of all the fields and particles of physics'. Yet to come would seem a second and far larger step: to see both geometry and all these fields and particles as manifestations of something more basic ('pre-geometry') than any of them.

Constants and Dynamic Laws not as Immutable but as 'Frozen in' in the First Stage of the Big Bang

What difference does it make if geometry and fields and particles are built up from something more primordial? Does not one then have to ask, when were conditions ever intense enough to form, and when were these conditions ever released fast enough to freeze, this structure of geometry and fields and particles into its present set of laws? When else than in the 'big bang'?

This piece of wood, this solid, is a 'fossil' from a photochemical reaction in a tree twenty years ago at a few hundred degrees Kelvin. One has only to subject it to higher temperatures to alter drastically its molecular constitution and switch it over to a new 'fossil'.

The iron nuclei in this steel pen nib are 'fossils' from a thermonuclear reaction in a star some billions of years ago at a temperature of some tens of millions degrees Kelvin. One has only to put these nuclei back into a star where conditions are sufficiently intense to transmute them into still heavier nuclei which, upon removal, rate as new 'fossils'.

Can particles themselves (and fields and geometry) be anything but 'fossils' from the most violent conditions of all, those encountered in the very earliest phase of the 'big bang', that mirror of gravitational collapse?

That there was a big bang (see for example the review of Peebles[68]) is evidenced not least by the recession of the galaxies, the proportion between primordial helium and hydrogen, and the primordial cosmic fireball radiation. The inevitability of gravitational collapse of every closed model universe, no matter how irregular, is by now a well established prediction of standard relativity.[69,70,71] Both at big bang and at collapse, calculated temperatures and pressures rise without limit. Between these times of conditions unprecedented in their extremity, physics is fossilized. No change with time has ever been found in the fine structure constant (see the impressive evidence adduced by Dyson[72]), in the mass of any particle, or in any other constant of physics.

One used to think of someday finding a 'theory' of the fine structure constant, of the basic constants of particle physics and of the 'big number scale',

$$\begin{pmatrix} \text{number of photons per} \\ \text{baryon in the universe} \end{pmatrix} \sim 10^{10}$$

$$\left(\begin{array}{l}\text{particle dimensions, } 10^{-13} \text{ cm,} \\ \text{relative to Planck length}\end{array}\right) \quad \sim 10^{20}$$

$$\left(\begin{array}{l}\text{estimated radius of universe} \\ \text{at full tide relative to} \\ \text{nuclear dimensions}\end{array}\right) \quad \sim 10^{40}$$

$$\left(\begin{array}{l}\text{electric force between} \\ \text{two particles relative to} \\ \text{gravitational force}\end{array}\right) \quad \sim 10^{40}$$

$$\left(\begin{array}{l}\text{estimated number of} \\ \text{baryons in universe}\end{array}\right) \quad \sim 10^{80}.$$

$$(42)$$

Today, forty years later, such a dream is as far from realization as ever. One is open to believe that one has been looking for the right answer to the wrong question. A century and a half ago Laplace dramatized the difference between initial conditions and dynamic law. The intervening decades have seen new laws uncovered, but not a single discovery about what fixes the initial conditions. The time has come to ask if the constants and the scale of the big numbers belong in the realm of law at all. Are they not more reasonably to be understood as initial conditions?

Mutability as Central Feature of Physics

'Constants' and laws alike 'frozen in' at the very earliest stage of the big bang, and rubbed out in the very last stage of gravitational collapse: that is the picture that one is led to examine seriously. On this picture physics is a staircase. Each tread registers a law (e.g., law of chemical valence). Each riser marks the transcendence of that law (e.g., temperatures and pressures so high that valence loses its significance). The staircase climbs from step to step: density, and density found alterable; valence law, and valence law melted away: conservation of net baryon and net lepton number, and these conservation laws transcended; conservation of energy and angular momentum, and these laws likewise overstepped; and then the top tread displaying all the key constants and basic dynamic laws – but above a final riser leading upward into nothingness. It bears a message: With the collapse of the universe, the framework falls down for every law of physics. There is no dynamic principle that does not require space and time for its formulation; but space and time collapse; and with their collapse every known dynamic principle collapses.

If the laws of conservation of particle number are transcended in black hole physics; if all dynamic laws are transcended in the collapse of the universe; if laws and constants of physics are first imprinted as initial conditions in the earliest phase of the big bang and erased in the final stage of gravitational collapse, then dimensionality itself can hardly be exempt from the universal mutability.

The review one by one of fixed points of physics has left not a single one unquestioned, neither 'constant' nor principle. It is difficult to find any other way to summarize

the situation as it now appears than this: 'There is no law except the law that
there is no law;' or more briefly, 'Ultimate MUTABILITY is the central feature of
physics'.

Beyond Mutability

One is led to think of a universe more ephemeral than would be admitted by any
'bootstrap particle model', or any model based upon a 'fundamental field', or any
model that considers geometry to be the 'magic building material' of existence.
Only by giving up almost everything, it would seem, can one be truly responsive to the
imperatives of collapse.

In all the marvellous history of physics nothing stands out more impressively than
the step-by-step transcendence of categories. 'Green' was adequate as description of
the colour of a mineral, but 'green' disappeared when one came to the motion of the
electron around the nucleus. The planetary circle of Copernicus faded from view
before the differential equation of Newton and Euler. Gravitation disappeared and
geometry took its place. The classical orbit made its exit when the wave of de Broglie
and Schrödinger made its entrance. The fantastic wealth of chemical fact and force
boiled down to electrons and nuclei and Schrödinger's equation. Each complication
of the evidence was not matched by a corresponding complication of principle.
The more one gave up the more one gained; and the more one gained the more one
gave up.

Dynamic Laws Transcended

If mutability demands the giving up of almost everything, what goes, what comes, and
what stays?

Superspace is the quintessence of relativity; and in the context of this arena one has
been led to think of the outcome of gravitational collapse as 'probabilistic scattering
in superspace'. On this view collapse is followed, not by a unique new cycle of big
bang, expansion, recontraction and collapse, but by a probability distribution of such
histories, each (because of transcendence of conservation laws) with its own new number of particles and own new time from big bang to collapse. This picture now
appears inadequate because it presumes, not too much to change, but too little.
When one began to consider particle number and particle masses and the dimensionless
constants of physics as altering from one cycle of the universe to the next, one also
started to view the dynamic laws themselves as like the laws of valence, wiped out by
conditions sufficiently extreme, and therefore extinguished in collapse. One had
already found it impossible to calculate his way through the quantum mechanics of
collapse within the context of superspace. Also one had already realized that the
superspace of general relativity is an incomplete arena. But to count relativity as
wiped out in collapse is to destroy superspace, and therefore take away the foundations
for any picture of collapse as 'scattering in superspace'. That 'scattering' is way

station to a larger picture, in which all the constants and dynamic laws get established only in the first stage of the big bang itself. Thus superspace goes and law goes. What comes?

Chaos Accepted, and Law Built on Chaos

If law goes, what can replace it but chaos? Chaos is not new for physics to encounter. Physics has mastered chaos before and translated it into the order of law. One can solve the two-body problem easily and the three- and four-body problem with greater and greater difficulty; but the N-body problem, with $N \geqslant 5$, is intractable. Nevertheless, when N grows and grows, the curtain rises to reveal temperature and entropy, new concepts unimagined and unimaginable at an earlier phase of physics. Moreover, the molecular chaos underneath in no way deprives the resulting laws of thermo-dynamics of the most impressive precision.

A 'pregeometry' that is primordial chaos, and law built upon this chaos: that is the vision of physics that we are led to examine.

How is one to find the key element of this underlying chaos or 'pregeometry'?

Nothing did one learn from a hundred years of elasticity about chemical forces; and a hundred years of chemistry unfolding all its wonderful regularities, provided not one clue to Schrödinger's equation. The order of understanding ran, not down, but up. One had to have quantum mechanics to understand chemical forces; and one had to know chemical forces to understand elasticity. Likewise a half century of gravita-tion – as – geometry has revealed nothing of the constitution of particles; and a half century of particle physics, laying open so many beautiful symmetries, has given not one hint of what lies beneath. Not down, but up; not down from particle physics or geometrodynamics, but up from the quantum principle would appear the right route to the primordial element, the 'pregeometry' that we visualize as chaos.

The Quantum Principle as the Only Principle

With law going and chaos arriving, one principle remains, the quantum principle. With all other laws of physics rated as mutable, it is the only principle. If no one ignorant of evolution has the first idea about the origin of life, it is also true that no one who is unacquainted with the quantum principle has the first idea how nature works. Physics without the quantum is medieval physics.

The quantum principle might almost be called the Merlin principle. Merlin the magician, on being pursued, changed first to a fox, then a rabbit, then a bird fluttering on one's shoulder. The quantum concept underwent still more spectacular changes in outward appearance: Mendeléev's[73] 'individuality amid uniformity'; Planck's law for the energy of an oscillator; the law of Rutherford and Einstein for radioactive decay and atomic transitions; Bohr's quantization of angular momentum; the non-com-muting observables of Heisenberg and Dirac; Heisenberg's uncertainty principle; Bohr's principle of complementarity; Feynman's principle of the democracy of all histories;

Everett's 'many-universes' formulation[74,75]; and the lattice of propositions of von Neumann and Birkhoff.[76]

Nothing is more surprising about quantum mechanics than this, that it still comes to us as a surprise. We have not yet discovered the most central consideration of all, the consideration that would tell us that the universe could not even have come into being had there been no quantum principle. We have no answer to the great question of Leibniz, 'Why is there something rather than nothing?'

Nothing is more important about the quantum principle than this, that it destroys the concept of the world as 'sitting out there', with the observer safely separated from it by a 20 centimeter slab of plate glass. Even to observe so miniscule an object as an electron, he must shatter the glass. He must reach in. He must install his chosen measuring equipment. It is up to him to decide whether he shall measure position or momentum. To install the equipment to measure the one prevents and excludes his installing the equipment to measure the other. Moreover, the measurement changes the state of the electron. The universe will never afterwards be the same. To describe what has happened, one has to cross out that old word 'observer' and put in its place the new word 'participator'. In some strange sense the universe is a participatory universe.

Is this instance of participation the tiny tip of a giant iceberg? Molecular chaos leads to concepts like temperature and entropy only when limitations are imposed, such as fixity of volume and total energy. Otherwise chaos is chaos. Does the chaos, the 'pregeometry', that we think of as underlying the universe, also fail to yield any law until it is analogously limited? Do we ourselves supply this limitation, we who have been forcibly elevated from observers to participators? Are we, in the words of Thomas Mann[77] 'actually bringing about what seems to be happening'? Are we destined to return to the deep conception of Parmenides[78], precursor of Socrates and Plato, that, 'what is, ..., is identical with the thought that recognizes it'?

Leibniz[79] reassures us that, 'Although the whole of this life were said to be nothing but a dream and the physical world nothing but a phantasm, I should call this dream or phantasm real enough if, using reason well, we were never deceived by it'. Never was the call to 'use reason well' more timely than today. Collapse and mutability make unprecedented demands on imagination and judgment. Now more than ever one is certain that no approach to physics that deals only with physics will ever explain physics.

No proud tower of human thought can remain unshaken by the greatest crisis one can name in the history of science: neither mathematics nor logic, neither philosophy nor physics. The budget officer may be able to parcel out money neatly to those areas of thought; but 'the good Lord' did not appreciate these fine distinctions and mixed them all up in the founding of the world. It will take the power of all of thought together if we are ever to understand why we have 'something rather than nothing'. We can believe that we will first understand how simple the universe is when we recognize how strange it is.

NOTES AND REFERENCES

1. T. S. Eliot, *Little Gidding*, Faber and Faber, London (1942).
2. E. P. Wigner, July 14, 1963 lecture, reprinted in *Symmetries and Reflections: Scientific Essays of Eugene P. Wigner*, Indiana University Press, Blommington, Indiana (1967) p. 32.
3. C. W. Misner, K. S. Thorne, and J. A. Wheeler, *Gravitation*, Freeman, San Francisco (1973).
4. A. Einstein as quoted in W. Sullivan, 'The Einstein papers', *New York Times* (March 28, 1972) p. 1
5. A. Einstein letter to E. Mach of June 25, 1913, Einstein archives, Institute for Advanced Study, Princeton, New Jersey, courtesy of the Einstein estate.
6. A. Einstein, *Essays in Science*, Philosophical Library, New York (1934) p. 68.
7. A. Einstein, *Preuss. Akad, der Wissenschaften*, Berlin, *Sitzungsberichte*, pp. 778–786 (session of Nov. 4, 1915, published Nov. 11); pp. 799–801 (session of Nov. 11, published Nov. 18); and (the field equation in its final form) pp. 844–847 (session of Nov. 25, published Dec. 2).
8. See for example ref. 3, Fig. 29.2.
9. A. Einstein, *The Meaning of Relativity*, Princeton University Press, Princeton, N.J., 3rd ed. (1950), p. 107.
10. A. Einstein, *Essays in Science*, Philosophical Library, New York (1934), p. 52. Einstein acknowledged a debt of parentage for his theory to Mach's principle. It is therefore only justice that Mach's principle should in turn owe its elucidation to Einstein's theory, in the sense that (a) the 'inertial properties of matter' are today understood in terms of, and as equivalent to, the straight lines of the local tangent space (local Lorentz geometry) of the curved spacetime manifold; (b) the 4-geometry of this manifold (and therefore these inertial properties of matter everywhere and at all times) is uniquely determined by appropriate arbitrarily chosen initial value data on an intial spacelike hypersurface of this manifold (for greatest convenience taken to be a hypersurface on which the locally defined local expansion rate or 'local Hubble time' has everywhere the same value); (c) specify this initial hypersurface arbitrarily but to this extent completely, that one gives everywhere the conformal part of its 3-geometry (otherwise one would not be able to specify 'where' one is putting the mass-energy that is to determine inertia); (d) give everywhere on this 3-geometry the appropriate measures of density and flow of mass-energy plus (e) the appropriate 'wave coordinates' and 'wave momenta' of the gravitational field (because the gravitational waves and fields are effectively a source of mass-energy, too); (f) solve the appropriate initial value equation of geometrodynamics, a non-linear second-order partial differential equation, for the conformal scale factor of the 3-geometry (the existence and uniqueness of this solution being assured by the work of N. O'Murchadha and J. W. York, Jr.); (g) thus have in hand complete initial value data that satisfy the initial value equations of general relativity; (h) solve the six dynamic equations of Einstein's theory for the entire 4-geometry; and in the end (i) have in this 4-geometry the full description of the inertial properties of every test particle, everywhere and at all times, as determined by the initial distribution of mass, as called for by Mach's principle. For details, see for example pp. 543–549 of ref. 3.
11. R. P. Geroch, *Singularities in the Spacetime of General Relativity: Their Definition, Existence, and Local Characterization*, doctoral dissertation, Princeton University, Princeton, N.J. (1967).
12. S. W. Hawking and R. Penrose, *Proc. Roy. Soc. London* A314, 529–548 (1969).
13. A. Einstein as quoted by G. Gamow, *My World Line*, Viking Press, New York (1970).
14. W. Baade, *Publ. Astron. Soc. Pacific* 28, 5–16 (1956).
15. A. Sandage in *Proceedings of the Symposium on the Galaxy and the Distance Scale*, Essex, England, in press, 1972.
16. J. H. Oort in *Onzième Conseil de Physique Solvay: La structure et l'évolution de l'univers*, Editions Stoops, Brussels (1958).
17. J. H. Oort, *Science*, 170, 1363–1370 (1970).
18. D. W. Sciama, *Modern Cosmology*, Cambridge University Press, Cambridge, England (1971), and earlier work cited there.
19. See for example ref. 18.
20. See for example ref. 18.
21. A. A. Penzias and R. W. Wilson, *Astrophys. J.* 142, 419–421 (1965).
22. R. H. Dicke, P. J. E. Peebles, P. G. Roll, and D. T. Wilkinson, *Astrophys. J.* 142, 414–419 (1965).
23. For a summary and literature, see for example ref. 3, p. 1214; pp. 457–459 and L. Landau and

E. Lifshitz, *The Classical Theory of Fields*, transl. by M. Hamermesh, Addison-Wesley, then Cambridge, now Reading, Mass. (1951), p. 320.

24. Theory reviewed in ref. 3, pp. 978–979.
25. W. Baade and F. Zwicky, *Proc. Nat. Acad. Sci. U.S.* **20**, 254–259 and 259–263 (1934) and *Phys. Rev.* **45**, 138 (1934) (abstract).
26. A. Hewish, S. J. Bell, J. D. H. Pilkington, P. F. Scott, and R. A. Collins, *Nature* **217**, 709–713 (1968).
27. W. H. Press and K. S. Thorne, 'Gravitational-Wave Astronomy', *Ann. Rev. Astron. Astrophys.* **10**, 335–374 (1972).
28. J. Weber, *Phys. Rev.* **117**, 307–313 (1960); also *Phys. Rev. Lett.* **24**, 276–279 (1970) and papers therein cited.
29. J. M. Bardeen, *Nature* **226**, 64–65 (1970).
30. R. Penrose, *Nuovo Cimento* **1**, special number, 252–276 (1969).
31. S. W. Hawking, *Phys. Rev. Lett.* **26**, 1344–1346 (1971) and *Commun. Math. Phys.* **25**, 152–166 (1972).
32. D. Christodoulou, *Phys. Rev. Lett.* **25**, 1596–1597 (1970).
33. D. Christodoulou and R. Ruffini, *Phys. Rev.* **D4**, 3552–3555 (1971).
34. Ya. B. Zel'dovich and I. D. Novikov, *Relativistic Astrophysics*, Vol. I: *Stars and Relativity*, University of Chicago Press, Chicago, Ill. (1971).
35. J. A. Wheeler, 'Mechanisms for Jets' in *Study Week on Nuclei of Galaxies*, *Pontificiae Acadameiae Scientarum Scripta Varia* no. 35, 539–567 (1971).
36. S. W. Hawking, *Commun. Math. Phys.* **25**, 152–166 (1972) and W. H. Press and S. Teukolsky, *Nature* **238**, 211–212 (1972).
37. C. DeWitt and B. S. DeWitt (eds.), *Black Holes*, Proceedings of 1972 session of École d'été de physique théorique, Gordon and Breach, New York (1973).
38. R. W. Leach and R. Ruffini, *Astrophys. J.* **180**, L15–L18 (1973).
39. É. Cartan, *Leçons sur la géometrie des espaces de Riemann*, Gauthier-Villars, Paris (1928 and 1946); for more details, see J. A. Wheeler, 'Gravitation as Geometry, II' in H.-Y. Chiu and W. F. Hoffman (eds.), *Gravitation and Relativity*, W. A. Benjamin, New York (1964) and C. W. Misner and J. A. Wheeler, 'Conservation Laws and the Boundary of a Boundary' in V. P. Shelest (ed.), *Gravitatsiya: problemi i perspectivi: pamyati Alakseya Zinovievicha Petrova posvashaetsya*, Naukova Dumka, Kiev (1972), and ref. 3, chapter 15.
40. D. Hilbert, *Konigl. Gesell. d. Wiss. Göttingen*, *Nachr.*, *Math.-Phys. Kl.*, 395–407 (1915; presented Nov. 20, 1915). [See J. Mehra's essay on Einstein, Hilbert and the Theory of Gravitation in this volume.]
41. R. Arnowitt, S. Deser and C. W. Misner, 'The Dynamics of General Relativity' in L. Witten (ed.), *Gravitation: An Introduction to Current Research*, Wiley, New York (1962), pp. 227–265; see also ref. 3, chapter 21.
42. J. A. Wheeler, 'Geometrodynamics and the Issue of the Final State' in C. DeWitt and B. S. DeWitt (eds.), *Relativity, Groups, and Topology*, Gordon and Breach, New York (1964); see also ref. 3, chapter 21.
43. J. W. York, Jr., *Phys. Rev. Lett.* **28**, 1082–1085 (1972) and *J. Math. Phys.*, **14**, 456–464 (1973).
44. A. Lichnerowicz, *J. Math. Pures et Appl.* **23**, 37–63 (1944).
45. Y. Choquet-Bruhat, *Acad. Sci. Paris*, *Comptes Rend.* **A274**, 682–684 (1972) and *Gen. Rel. and Grav.* (1973), in press.
46. B. S. DeWitt, *Phys. Rev.* **160**, 1113–1148: **162**, 1195–1239; and **162**, 1239–1256 (1967).
47. A. Peres, *Nuovo Cimento* **26**, 53–62 (1962).
48. U. Gerlach, *Phys. Rev.* **177**, 1929–1941 (1969).
49. T. Regge, *Nuovo Cimento*, **19**, 558–571 (1961).
50. W. Pauli, *Helv. Phys. Acta Suppl.* **4**, 69 (1956).
51. J. A. Wheeler, *Geometrodynamics*, Academic Press, New York (1962), pp. 75–83.
52. Summarized for example in F. J. Dyson, *Advanced Quantum Mechanics*, Cornell University, Ithaca, New York (1954) (mimeographed), p. 54.
53. Ref. 51, pp. 71–75.
54. K. Kuchař, *J. Math. Phys.* **11**, 3322–3334 (1970).
55. W. K. Clifford, *Lectures and Essays*, L. Stephen and F. Pollock (eds.), Macmillan, London, vol. 1,

p. 244 and p. 322, 1879; see also *Mathematical Papers*, R. Tucker (ed.), Macmillan, London (1882), p. 21.

56. H. Weyl, *Was ist Materie*, Springer, Berlin (1924), pp. 57–58.

57. C. W. Misner and J. A. Wheeler, *Ann. Phys.* **2**, 525–603 (1957).

58. A. Comte, *Cours de philosophie positive*, Paris (1835).

59. J. R. Klauder, *Ann. Phys.* **11**, 123 (1960).

60. Regarding the connection between topology, superspace, and spin, see for example J. A. Wheeler, 'Superspace and the nature of quantum geometrodynamics' in C. M. DeWitt and J. A. Wheeler (eds.), *Battelle rencontres: 1967 Lectures in Mathematics and Physics*, W. A. Benjamin, New York (1968), 242–307, esp. pp. 284–289.

61. B. Riemann, 1854, *Habilitationsvorlesung* in H. Weber (ed.), *B. Riemann: Gesammelte mathematische Werke*, 2nd ed., paperback reprint, Dover, New York (1953); transl. into English by W. K. Clifford, *Nature* **8**, 14 (1873).

62. Riemann himself was the first to introduce superspace. In his case it was a superspace built of conformal 2-geometries,[2]<. This superspace has since been the object of research of more than a hundred papers. Today it is known as Teichmüller space.

63. See ref. 55.

64. S. A. Hojman, K. Kuchař, and C. Teitelboim, 'New Approach to General Relativity', April 1973 preprint.

65. A. D. Sakharov, *Doklady Akad. Nauk S.S.R.* **177**, 70–71 (1967); English transl. in *Soviet Phys. Doklady* **12**, 1040–1041 (1968).

66. See ref. 34.

67. M. Berger, P. Gauduchon, and E. Mazet, *Le spectre d'une variété Riemannienne*, Springer, Berlin and New York (1971).

68. P. J. E. Peebles, *Physical Cosmology*, Princeton University Press, Princeton, N.J (1971).

69. A. Avez, *Acad. Sci., Paris, Comptes Rend.* **250**, 3583–3587 (1960).

70. See ref. 11.

71. See ref. 12.

72. F. J. Dyson, 'The Fundamental Constants and Their Time Variation' in A. Salam and E. P. Wigner (eds.), *Aspects of Quantum Theory*, Cambridge University Press, Cambridge, England (1972).

73. D. I. Mendeléev, selection of quotations from, in J. A. Wheeler, 'From Mendeléev's Atom to the Collapsing Star', 189–233 in *Atti del Convegno Mendeléeviano*, Accademia delle Scienze, Torino; reprinted in *Trans. New York Acad. Sci.* **33**, 745–779 (both in 1971).

74. H. Everett III, *Rev. Mod. Phys.* **29**, 454–462(1957).

75. B. S. DeWitt, 'The Many-Universes Interpretation of Quantum Mechanics', 211–262 in B. d'Espagnat (ed.), *Foundations of Quantum Mechanics*, Academic Press, New York (1971).

76. J. M. Jauch, *Foundations of Quantum Mechanics*, Addison-Wesley, Reading, Mass. (1968), especially chapters 5 and 8.

77. T. Mann, *Freud, Goethe, Wagner*, Knopf, New York (1937). Essay on Freud published originally as *Freud und die Zukunft (Vortrag gehalten in Wien am 8 Mai 1936 zur Feier von Sigmund Freuds 80 Geburtstag)*, Borman-Fischer, Wien (1936).

78. Parmenides of Elea, section on *Truth* in his poem, as paraphrased in *Encyclopedia Brittanica* article on Parmenides, p. 327, column 2, lines 9 and 16–17.

79. G. W. von Leibniz (1646–1716) as quoted by J. R. Newman, *The World of Mathematics*, Simon and Schuster, New York (1956), p. 1832.

Part II

Quantum Theory

10

The Wave-Particle Dilemma

Léon Rosenfeld

The dilemma between wave and particle aspects of the various constituents of matter and radiation arose after Planck's discovery of the quantum of action, and Einstein's quite independent proposal of a sort of revival of the corpuscular theory of light, again including the concept of the quantum of action in the form of energy or light quanta.

A famous confrontation of the two points of view took place in Salzburg in 1909 between their respective protagonists. The attitude of the two great masters, Planck and Einstein, is very characteristic of two tendencies that were then competing, and equally uncomfortably, in facing the quite unexpected dilemma that the theory of radiation presented. Planck took the conservative side. He took it openly by saying that, 'in introducing novelties in physics one must proceed in as conservative a manner as possible.' That was his outspoken principle: one must be very reluctant to take any new step which is not logically connected with the preceding one. In other words, Planck was willing and daring enough to overcome a prejudice, to use Dirac's pointed way of expressing the progress of physics, but he was uncomfortable about it.

Planck felt uncomfortable about abandoning the classical theory of radiation. And the discomfort, or hesitation, that he felt was an inhibiting feature in the progress of his own thinking. However, he had good reasons to feel uncomfortable and he expressed those reasons at Salzburg in very strong terms. He pointed out that the classical theory of radiation based on the field conception of Maxwell, or rather of Hertz to be more accurate, was supported by a great mass of experimental evidence about interference and diffraction phenomena, phenomena which were based on the superposition principle and which, therefore, could not, except by very forced assumptions, be incorporated in any corpuscular description of the propagation of light.

Einstein, on the contrary, represented the radical attitude. In fact, when he introduced the concept of the light quantum, he probably was not quite aware of the connection between his conception and Planck's introduction of the fundamental constant that bears his name. He thought that certain phenomena, like the photoelectric effect and the photochemical reactions, indicated that in the interaction between matter and radiation, the exchange of energy (and also, he added later, momentum) took place according to the laws of collisions of particles, so that radiation in such interactions acted as if it were composed of photons. However, when you treat those photons as independent particles you do not get Planck's radiation law; you get

J. Mehra (ed.), The Physicist's Conception of Nature, 251-263. All Rights Reserved
Copyright © 1973 by D. Reidel Publishing Company, Dordrecht-Holland

Wien's law, which is only a limiting case valid for high frequencies or for high values of the ratio $h\nu/kT$.

Very soon, however, Einstein realized that by abandoning the idea that photons composing a flow of radiation were independent, he could then recover Planck's law. His argument was essentially based on the consideration of fluctuations in the radiation field. He found the surprising result that the fluctuation in the radiation field, which is governed by Planck's law, was the sum of the fluctuations that would arise from a classical radiation field and those that would arise from an assembly of independent photons. As the outcome of this work, one had the very uncomfortable situation that the two aspects with which one had tried to describe radiation phenomena were now concurring in producing the result that was known to be the right one from Planck's analysis.

In Salzburg Einstein tried to defend nevertheless the radical assumption of the corpuscular structure of radiation by suggesting, although it could not be more than a qualitative suggestion, that the photons, or the light quanta as they were called then, were some kind of singularity, of concentration of energy and momentum inside a radiation field. The radiation field would so to speak guide the photons in such a way as to produce also the interference and diffraction phenomena that Planck wanted, with good reasons, to see reproduced by the radiation theory. Of course, none of these two opposing attitudes was satisfactory. Planck was perhaps too conservative and Einstein too radical.

Now one may ask at this stage what Bohr thought of the whole matter. Bohr was then still a student, but he was an uncommonly thoughtful student, and he was very eagerly following all those discussions. In this particular confrontation he took sides with Planck, but not entirely. He did not reject Einstein's conception of the light quantum, because he saw that this conception had its justification simply from the fact that it was a useful one, that it could form the basis of a method of analysis of the interaction between radiation and matter, which led to consistent results in a very simple way that the classical theory of radiation, of course, could not achieve. That was especially evident in the photoelectric effect. But why did he then not treat the two aspects of light on the same footing? Why did he give a sort of preference to Planck's view of the electromagnetic field as being in some sense more fundamental than the photon concept?

This is a point of view that Bohr never abandoned. The reason reveals the first germ in his mind of an attitude which was in the sequel extremely fruitful, and which he then called the correspondence argument. To put it in somewhat anachronistic terms, as we see it after the event, he attached reality, i.e. reality as it was defined by him, to those aspects that could be directly observed in certain limiting circumstances, by direct macroscopic observation. And, of course, in the case of radiation it is clear that direct observation in the limiting case of small values of $h\nu/kT$ gives the usual classical wave description of Hertz and Maxwell. As to the photon or the light quantum concept, introduced by Einstein, Bohr regarded it as a useful but an *auxiliary* concept, one which he later called symbolical, meaning thereby that it was not an aspect of the radiation phenomena which could be directly observed as such. It could only be observed through the bias of an application and verification of the conservation laws of energy and of momentum. Bohr always made this distinction between the two aspects of radiation.

By mentioning the word 'correspondence' in this context, at this early stage, I am of course being a bit unfaithful to Bohr's later idea of correspondence which was much deeper and also much bolder. In fact, the point that for low values of hv/kT one has to expect the classical theory of radiation to be valid had already been made by Lord Rayleigh in 1900, when he pointed out that for low frequencies the application of the equipartition theorem gave the law that bears his name. Rayleigh also suggested that it could not be valid generally as it led to a divergence, but should be valid at least for the modes of low frequency.

When I mentioned to Bohr that this was perhaps the first germ of the correspondence argument, he protested. He said emphatically that 'it is not the correspondence argument. The requirement that the quantum theory should go over to the classical description for low modes of frequency, is not at all a principle. It is an obvious requirement for the theory.' Nevertheless that was the first beginning of it.

Bohr also realized at that time the importance of the quantum of action as a stabilizing element for the Rutherford atom. As soon as he became acquainted with the Rutherford model of the atom, he realized that this model had a number of consequences of the greatest importance. There was no doubt that this model had a serious grain of truth in it, since after all it was introduced by Rutherford to explain results of the experiments on the scattering of alpha particles, which would not be understandable in any other way. He pointed out already at that early stage that this model would, for the first time, allow a clear distinction between an atom and a molecule, a molecule being characterized by the fact that it has a structure with more than one nucleus and the appropriate number of electrons. He also made at that time a large amount of calculations on the stability of certain ring structures of electrons which would bind the nuclei of a diatomic or polyatomic molecule together.

Bohr also realized that radioactive phenomena, on the other hand, would be nuclear phenomena. It was for the first time that such a clear distinction was made, and this led him to formulate what were later known as the displacement laws of radioactivity, which on the view of the Rutherford model, were rather visible consequences. When he presented those considerations to Rutherford, the latter gave him a severe lesson by condemning the impatience of youth in publishing too quickly speculations which could only be justified by some more experimental material than was then available. Later on, when this material accumulated within the next few months, the displacement laws were recognized by Fajans and Soddy. It is rather curious to note that Fajans, when he published those displacement laws, asserted that they were an argument *against* Rutherford's model. Well, this is an aside.

The main thing was that Bohr was convinced right from the start that the stability of Rutherford's model was, as he expressed himself, of 'non-mechanical' origin. Rutherford's atom was essentially unstable, and it could only be stabilized by the introduction of an element which was not part of classical physics, and it was obvious that it should be the quantum of action. It was known already, not only by Bohr but also by other people, that the orders of magnitude were just about right. Nicholson, especially, had made some very fancy speculations about the radiation of atoms being governed by Planck's quantum hypothesis.

I shall pass over rather rapidly on Bohr's formulation of the quantum postulates only to point out that the formulation of those postulates represented an extremely daring step and a typical example of overcoming prejudice. And the prejudice that was overcome in those postulates was the idea that the frequency of radiation emitted by a vibrating system of charges should be the same as the frequency of vibration of the charges, which is of course not the case according to the idea of transition between stationary states.

Einstein's reaction was very positive to the news that experiments had confirmed Bohr's ascription of the Pickering and Fowler series to helium. Hevesy wrote in a letter to Bohr that when he (Hevesy) told Einstein that it was established with certainty that the Pickering-Fowler spectrum belongs to helium, Einstein remarked: 'Then the frequency of the light does not depend at all on the frequency of the electron. And this is an *enormous achievement*. The theory of Bohr must then be right.' In a letter to Rutherford, Hevesy wrote: 'When I told him of the Fowler Spectrum the big eyes of Einstein looked still bigger and he told me "Then it is one of the greatest discoveries."'

In the derivation of Balmer's formula from the postulates which was Bohr's triumph at that time, he again made use of this argument which was only one part, and not the main part, of the correspondence argument. He argued, namely, that if you go to the orbits containing many quanta, the very excited orbits, then the difference of energy between neighbouring orbits becomes smaller and smaller, and the frequency of the transition between neighbouring orbits in that region approaches the mechanical frequency, so that you will recover the classical picture in the limit of high quantum numbers. That is, the phenomena involving many quanta are therefore not distinguishable from the classical description.

But again, this was not yet the correspondence argument. This was only a part of it which Bohr had described at that time. He had not yet used the word correspondence. He used the word analogy. He said that it was a beautiful analogy with classical electrodynamics. He was extremely concerned about deciding by very careful argumentation as to which parts of the classical theory he could retain and which parts he had to abandon. Of course, he had to abandon the classical mechanism of emission and absorption of radiation. These were described as transition processes, which were quite outside the scope of classical description. But the stationary states themselves, i.e. the calculation or the estimation of the energy for stationary states, he argued, were still amenable to classical theory, involving only the Coulomb field, the statical part of the electromagnetic field.

Bohr tried very hard to find ways of determining the energies of stationary states for given systems. His idea, which by the way was also developed independently at the same time by Ehrenfest under the name of adiabatic invariant theory, was simply to start from the only mechanical system for which the quantization was well-known and verified, i.e. the harmonic oscillator. Then he argued that if you transform the system, if you change the forces acting in the system in a very slow way, then no quantum transition was possible, and the system would gradually evolve into some other system. Whatever invariants belonged to the first system would keep the same value for the ultimate system. He called this the principle of mechanical transformability, by which he then tried to determine the stationary states of given systems of electrons and nuclei.

The success, and even the possibility, of formulating the postulates and of defining the idea of stationary state as he did, was only due to the weakness of the radiation coupling. If it had not been for that weakness there would not have been sharply defined, or reasonably sharply defined, energies of the excited states. They would have widths and that would have disturbed the picture. Bohr was, however, very conscious about it and he used repeatedly the weakness of the fine structure constant in order to make a sharp separation, in an ideal way of course, between those parts of the theory that were governed by the Coulomb field, by the static force only, and which he thought at that time could still be managed by classical physics, while the other parts which represented the pure radiation field were governed by the postulates and by transition processes which escape any possibility of classical analysis.

Even at that early stage the fundamental ideas of quantum mechanics were already contained in this picture, though they were not clearly expressed in the first publications; especially the fact that the radiation field is entirely outside any classical picture and must be treated by postulates, by the concept of transition processes, which are not analyzable in classical terms. Any attempt to analyze the mechanism of emission or absorption of radiation would involve a continuous succession of processes, each of which would involve an exchange, or modification of action less than the quantum, and therefore were quite excluded by this model. However, Bohr thought at that time that processes that depended only on static forces, were still amenable to classical analysis.

The next stage in the development of the theory was first of all accelerated in a very essential way by Sommerfeld's intervention, Sommerfeld's extension of Bohr's quantum conditions, derived from the principle of mechanical transformability or the use of adiabatic invariants, to more than one degree of freedom. Bohr had used until 1916, until Sommerfeld's work, the idea of electrons distributed in rings. When I asked him why he did that, why he used a rather improbable picture like that, he said, 'Well, that was what was used then, that was what we were used to.' It had come, of course, from Thomson's attempts to describe an atomic model with stable rings of electrons inside a distribution of positive charge.

Sommerfeld's paper, in which he introduced the quantization in three dimensions, made a deep impression on Bohr. Fortunately he had then the help of Kramers to refresh his mathematics and mechanics. Both of them undertook a deeper analysis of atomic structures based on the Hamiltonian scheme of equations and on the theory of multiply periodic systems. It is very important to remember that the whole framework of quantum mechanics has its origin in the consideration of multiply periodic systems, and this was quite essential.

The problem was then to analyze the spectroscopic data that had accumulated in the mean time. The spectroscopists were stimulated to obtain a deeper understanding of the regularities of the spectra that they had been so accurately measuring. But they did not quite know what to do with it. In the exciting environment of the German physics of that time, this was done in a very systematic way at different places, especially in Tübingen. The problem arose of accounting, not only for the series of stationary states called the spectroscopic terms that could be disentangled from the measurements of the frequencies of the spectra, but also for the intensities of the spectral lines, which seemed to be quite capriciously varying from one transition to the other.

It was here that Bohr developed the correspondence argument in an essential way. He said that the correspondence with classical physics, with the classical radiation theory, is valid not only in the classical limit, in the limit of many quanta, but is also approximately valid in a semi-qualitative way even in the domain, far away from the classical limit, in which the radiation processes are governed by the transition postulate.

In the classical theory of multiply periodic systems, the mechanical frequencies of the electrons are described by definite modes of frequencies and their harmonics, and then rules were established by trial and error, by just fumbling between those classical harmonics and the transitions between pairs of stationary states. This was not a simple correspondence, since it was a problem of associating a single number in the classical description, the number of the harmonic, with two numbers, or two sets of quantum numbers corresponding to the two states between which the transition was proceeding. A number of people concentrated their efforts on that problem, and it is amazing to think of the tremendous amount of work that was put into that in a few years' time. Reasonably clear understanding of optical spectra was obtained on the basis of the correspondence argument. I say, the '*correspondence argument*.' That is the phrase that Bohr liked to use rather than the 'correspondence principle', because he was a bit afraid of giving the name of principle to a rule the foundation of which was so shaky.

In the course of this work Bohr and Kramers were led more and more to the conviction that the classical picture, the classical model of atoms as systems of electrons revolving in orbits around the nucleus, orbits which were not simple Keplerian ellipses then, but which were turning around the nucleus in a complicated way, was breaking down in a very serious way. As we know now, after the event, the missing element was the spin, of course, that showed up by the necessity of introducing half integral quantum numbers, whereas there was no justification for such a half integral quantization. The feeling arose that those models were after all very unreliable, and that the only element on which finally the analysis of spectra was based was the system of empirical rules that were developed by various people for calculating the intensities of the various transitions in terms of the quantum numbers involved.

There was a curious divergence at that stage between Sommerfeld and Bohr, Sommerfeld keeping his faith in the literal validity of the models at a time when Bohr had already abandoned this idea. Bohr was always amused to recall that when he had sent reprints of his first papers of 1913 to Sommerfeld, he received in reply from Sommerfeld a very flattering comment on the content of the papers with the remark that 'I doubt very much in the reality of models.'

Now, at that time, it was in 1924, an incident occurred which induced Bohr and Kramers to crystallize their opinion. It was the arrival in Copenhagen of a young American, Slater, who had ideas of his own about the possible reconciliation of the two aspects of the radiation field. He had the idea that the classical electromagnetic wave would be a sort of guiding wave for the photons, and he realized that the connection between the emitting or absorbing atoms and the classical radiation field could not be upheld in the classical way. He had therefore conceived the idea that one had to replace the actual atoms by a system of oscillators, which could play an equivalent role in the mechanism of emission and absorption to the real atoms. Bohr and Kramers took up that idea, but only partially. It was quite in keeping with what I said before about Bohr's attitude towards the photon concept, that he rejected from Slater's proposal the photon altogether. He kept only the idea of the virtual oscillators creating a virtual radiation field, containing potentially all the possible frequencies that could be exchanged between the oscillators representing the atoms. Slater was very dissatisfied with that turn of events. He felt that what he considered as his main idea had been just left aside, and that of his own conception there remained only the grin of the Cheshire cat. And yet, he did nothing else than to let Kramers write up the paper and put his name together with those of Bohr and Kramers underneath. This paper of Bohr, Kramers and Slater represents the precursor stage to what emerged finally as the conceptual framework of quantum mechanics.

In the first place, it was a completely statistical description. The function of the radiation field was now reduced to transmitting interaction by the mechanism of absorption and emission in this particular model. But it was, as the name virtual field indicates, not a real field in the sense of being directly observed. It was a virtual field, a field containing in itself all the elements, all the frequencies, that could be absorbed or emitted by the various atoms. And the atoms themselves were also reduced to virtual systems of oscillators. Thus it was a rather abstract theory, eliminating all that remained of the classical description, even eliminating, and that was the culminating point, the validity of the conservation laws of energy and momentum. Since there was no element that could decide, after an atom, or a virtual oscillator, had emitted its corresponding radiation, when this radiation could be absorbed again, the conservation of energy and momentum could only be upheld in a statistical sense on the average.

Now this was a challenge in the first place to the experimentalists. Bothe and Geiger, within a very short time, and Compton and Simon in another way, tested this very revolutionary consequence of the statement of the famous paper of Bohr, Kramers and Slater, and found that it was not true in the case of the Compton effect. In the Compton effect they could show that there was a coincidence between the absorption and the emission of radiation and the recoil of the electron.

One would normally expect that Bohr and Kramers, and Slater also, would have abandoned those ideas in view of these experimental results, but Bohr's attitude was quite different. It did not influence him in the least. Not that he despised experimental results, far from it. But he said that those results only showed that we have not yet got to the depth, to the crux, of the matter, and that the difficulty, the dilemma of the two aspects of light, lies still deeper than we thought; and there was no cheap way of avoiding the dilemma, not even by renouncing a deterministic relation between the acts of absorption and emission. That did not mean that the fundamental conception of this representation of the phenomena by means of virtual oscillators and virtual radiation between them should not be a way, although very imperfect, and it ought to be pursued further. You may call it the resilience of Bohr, which was a very typical characteristic of his optimism in all matters.

At that time, even before the arrival of Slater in Copenhagen, Kramers had already started work on the theory of dispersion. Bohr's postulates coped with emission and absorption, and there remained the phenomena of dispersion, which were much harder nuts to crack from the point of view of the correspondence argument. Kramers made the first step in that direction, and then he and Bohr had the good fortune of Heisenberg's coming to their rescue. Heisenberg not only completed Kramers' work on the dispersion theory, but he was led from that to the crucial discovery of the actual mathematical formalism that was the correct expression for all those concepts that were only vaguely expressed in the paper of Bohr, Kramers and Slater. Heisenberg has himself testified that this paper made a great impression on him, and was very helpful in leading him in the right direction.

It is a bit embarrassing for me to talk about Heisenberg in his presence at this symposium, but I do that on good authority I think, i.e. on Heisenberg's own testimony in his writings. Heisenberg developed a matrix theory, operating only with observable quantities, the latter being the transition probabilities and the energies of the stationary states, and he firmly maintained that his theory was complete. It was true at least for multiply periodic systems in the framework of quantum mechanics. And then, another great pioneer who unfortunately is no longer among us, Pauli, a close friend of Heisenberg, was a bit more critical. He said, 'Well, after all the moon also revolves along a Keplerian orbit, and we have no difficulty in localizing the moon and defining its trajectory. So there is an element in the description which is still missing after all, in spite of the formal completion of the matrix scheme.' Who knows, perhaps Pauli and Heisenberg by their own exertions would have completed the scheme, but they were helped anyway, unwittingly, by Schrödinger's publication of wave mechanics.

Schrödinger's wave mechanics was a direct onslaught on quantum mechanics, on quantum theory altogether. He wanted to get rid of Bohr's postulates. He wanted to reinstall the classical description of radiation. He showed how, by means of the wave functions that he had determined as eigenfunctions of the eigenvalues of a mechanical system, he could construct charge and current densities, which were almost periodic functions with just the periodicities that would produce in a classical way the correct frequencies observed in the spectra. But Heisenberg immediately pointed out that that was an illusion, because any classical mechanism of emission and absorption of radiation of whatever frequencies would never lead to Planck's distribution law, but to the classical distribution.

As a result Heisenberg, at that time, was very sharply sensitized against Schrödinger's theory. Pauli, on the other hand, recognized that although Schrödinger's proposal was untenable, nevertheless Schrödinger's wave conception and the conception of a wave packet could just help to provide the missing element of a description, comprising all the elements of observation – not only those which had been contemplated by Heisenberg originally, but all the other elements of a spatial localization of charge and matter as well as a treatment of aperiodic phenomena, such as scattering processes. That programme was very quickly developed by Born, and the crowning piece to the edifice was provided, at about the same time, by Jordan and Dirac. That was the transformation theory, showing that you can pass by a canonical transformation from any mode of description, based on the choice of variables, to any other.

The very rapid sketch which I have given of the development shows that the impression that the pioneers of quantum mechanics, especially Heisenberg, found a formalism and then wondered about how to interpret it, is completely wrong. They were guided all the time by physical ideas, and at every step they were very clear about the physical implications of the symbols, operators, etc., that they were introducing. That was not the problem. The problem that arose very rapidly, and it was recognized by Heisenberg, was that this comprehensive description provided by the more flexible formalism arising from the fusion of the matrix idea, of the operator idea of Dirac, with the wave function, wave vector, state vector, or ket vector as Dirac called it, that this whole formalism implied limitations in the application of the classical concepts.

There was never any doubt in the minds of the people working on the elaboration of the theory that the symbols that they used had to be associated with classical concepts. That was quite clear at that time; although it does not seem to be so clear nowadays, I do not know why. But at that time nobody had any qualms about that. However, what appeared immediately, or very soon, was that this application of classical concepts was subject to limitations. And the real problem that Heisenberg had to face was to account for these limitations, to understand them. At that stage Bohr intervened again actively. Bohr was always, of course, in the background. Although he did not take part in the elaboration of the new theory directly, at this critical stage concerning the understanding of the limitations in the application of the classical concepts, Bohr intervened again very forcefully in collaboration with Heisenberg, and this intervention presented a very interesting and instructive feature.

Heisenberg argued that since his scheme was complete, it had to contain also the answer to the question on the origin of the limitations. Heisenberg tells us in his wonderful book *Der Teil und das Ganze* [Physics and Beyond], that when he was pondering about that problem he was reminded of a remark that Einstein had made to him when he presented to Einstein his ideas about using only observable quantities. Einstein said: 'What quantities are observable should not be our choice, but should be given, should be indicated to us by the theory.' That was a very wise remark which Einstein ought perhaps to have remembered later on. Be that as it may, it was extremely fruitful in this particular case and it led Heisenberg to the discovery of the uncertainty relations.

The question then was: What to do with those uncertainty relations? What was their significance? That was a matter for discussion, and Bohr fumbled very much for the right way of formulating the meaning of the uncertainty relations. And then, Bohr finally saw in these relations, in this peculiar situation, the solution of the dilemma that had hovered over the whole theory all the time. The dilemma of light, with which I began, had been doubled by the similar dilemma of matter, arising from Schrödinger's introduction of matter waves. Here again we have the two opposing aspects of particles and waves, applicable to matter as well as to radiation. Bohr reflected that the uncertainty relations were a consequence of the commutation rules, the only element in the theory which was not classical, in which Planck's constant was introduced with the coefficient i, the imaginary unit. Bohr was always fond of pointing out this feature which excluded any immediate classical interpretation of these relations. This situation was regarded by Bohr as the feature of the theory that opened up new prospects and eliminated all logical difficulties.

He did that by introducing the notion of complementarity as a logical relation of which we had so far no example in physics. This was a way of accommodating concepts, or phenomena as he called them later, which were mutually exclusive, but nevertheless were part of a comprehensive description of the phenomena. It is curious to point out in this connection that the situation between the complementary aspects of light, the wave aspect which is in direct correspondence with the classical description, and the concept of photon which is symbolical, which is only an expression for the exchange of energy and momentum between matter and radiation, is not paralleled, as de Broglie thought, by the two wave and particle aspects of matter, but it is just the other way around. For matter, the aspect which is in correspondence with classical observation is the particle aspect, of course; whereas the wave aspect is a symbolical one, and is the basis for the statistical description of the processes that can occur, and which Bohr called *individual* processes, meaning thereby that they are not divisible, that they are not analyzable in the classical way. Complementarity necessarily involves statistics. This is obvious, because if you want to accommodate two concepts which are mutually exclusive, if you want to use them together, you can only do that by renouncing the precise determination to a certain extent, so that you have to introduce a statistical element.

This was not a surprise for any of the Copenhagen group, because statistical causality was already contained in the paper by Bohr, Kramers and Slater, and was accepted as a natural form of causal relations in atomic physics, that was quite well adapted to the conditions of observation. There was no mystery about it. Here I may respectfully disagree with the statement made yesterday by Dirac, and I do not think that there is any reason to be uncomfortable about the use of statistical causality in physics. As a matter of fact, I do not believe that causality poses any problem at all in physics at any stage, for the simple reason that causality is not a matter of our choice. Causality is inherent in the form of the description that we develop, and we develop this manner of description in reference to the conditions of observation. If the conditions of observation are such that they allow for several possibilities, well, statistics is the indicated method. It is a method created for just such purposes. In fact, the first to introduce statistics in physics – I disregard of course Maxwell and statistical mechanics because that is another thing – but in fundamental atomic processes, were Rutherford and Soddy when they formulated the law of radioactive transformations. Now anyone who has seen Rutherford remembers a force of nature, and not a very deep, subtle or intricate thinker. As a matter of fact, he expressed his own character so gently in a famous lecture in which he said, 'I like simple pictures because I am a simple person myself.' So, if Rutherford introduced a statistical law for radioactive transformations, he did not have any deep epistemological qualms about it. He simply used statistics in the same way as an insurance agent uses statistics to calculate premiums. And that is the right way to do it; statistics is just made for that kind of problem. Moreover, it is not strictly true to say that quantum mechanics has eliminated determinism, because after all the laws of conservation of energy and momentum are just as deterministic as ever, and they have universal application.

Well, I must leave it at that. I hope to have described to you a very exciting period, one of the most exciting periods in the history of physics, which exhibits a remarkable continuity in the logical development of a trend of ideas. There has been very little of external influences disturbing the course of events. The only external influence was the advent of Schrödinger's wave mechanics, and that was a very helpful one, although Schrödinger himself was rather sorry for the use, rather the misuse, as he thought, that was made of his wave ideas.

DISCUSSION

F. Bopp: I have a question concerning the action integral. It seems to have been a curious development. The action integral appears in Bohr's famous theory of 1913. I came across with this integral in Sommerfeld's famous lecture in 1911 at Karlsruhe. Do you know the point to which the ideas were directed in the action integral up to 1910? There is the wave quantization of energy, and suddenly the action integral comes up. There is the possibility that perhaps Sommerfeld invented it and Bohr knew of this development. At the Solvay conference in 1911, when Lorentz put his question concerning adiabatic invariance, Einstein answered it. Was this the origin of this transformation from quantum energies to the connection of the action integral with Planck's constant?

L. Rosenfeld: I think that these two concepts, the action integral and the adiabatic invariance, were a bit earlier. The action integral was introduced for the first time by Planck in an effort to generalize the quantum condition to many degrees of freedom. He then concentrated on the statistical aspect of the quantum of action and tried to establish a connection with the cells in phase space. That was, I think, about 1910.

H. Rechenberg: That was already in 1906.

L. Rosenfeld: Already in 1906? Did he really do that in 1906 for n degrees of freedom? I don't think so. For one, yes, but then he pursued this line and it was by 1910 that he tried to introduce it in a general way.* Both Sommerfeld and Bohr knew Planck's work and were interested in it. The adiabatic invariants were introduced by Ehrenfest in connection with the statistical theory of radiation. It was a very brilliant piece of work, though it is not such a well-known paper unfortunately. Ehrenfest tried to see where the discontinuity introduced by Planck's constant occurred, and how necessary it was.

* See *Editor's Note* at the end of Discussion.

There were anticipations of other things in that paper, in particular the notion of adiabatic invariance. When Ehrenfest learned about Bohr's papers he was well prepared to think of this idea as a natural one. He knew that the phase integral in particular was an adiabatic invariant for periodic systems and so it was natural for him to try to use it to determine the stationary states of other systems.

F. Bopp: The most quoted paper of Ehrenfest was later than this.

L. Rosenfeld: Yes. Later on, Burgers, a pupil of Ehrenfest's had a thesis devoted to the theory of adiabatic invariance.

F. Hund: I believe that the introduction of the phase integral came about more gradually. It is difficult to say who at first had noticed its consequences. Planck, in his lectures of the winter term in 1905, which were essentially the same as his book of 1906 [*Theory of Heat Radiation*], had a footnote – he didn't emphasize it – that in a harmonic oscillator, the stationary state or the energy used, is expressed by the integral; but *only* for the harmonic oscillator. At the Solvay conference in 1911 it was somewhat more complicated. I think it was Nernst who asked the question as to how to quantize the rotator. Lorentz gave an answer, which was not quite correct, and Hasenöhrl then came up and pointed out the difference between harmonic oscillator and rotator. In a harmonic oscillator, equal steps of energy are also equal steps of the phase integral, but in the rotator this is not the case. Then he said that one has the choice of either taking equal steps of the energy or taking equal steps of the phase integral, and he [Hasenöhrl] thought that 'the second choice would be in the sense of Planck.' I think this was the first clear conception of a generalization, but some pages later he confused it totally for reasons that are not so important. Three weeks later, at the *Naturforscherversammlung* at Karlsruhe, Hasenöhrl used that integral as the essence of a quantum theory, but he gave an unhappy example afterwards. Again, at the Solvay conference in 1911, and in 1912, Planck took the phase integral as the central consideration for the harmonic oscillator, but only for the harmonic oscillator. In April 1913, at the so-called *Kinetischen Gas Kongress* at Göttingen [organized by David Hilbert], Debye gave the action integral as the real generalization of Planck's ideas without citing Hasenöhrl. He didn't know of him, I think. And, in the same year [1913], I think in May, Ehrenfest gave the right quantization of the rotator, but he touched upon the adiabatic principle in a few words only. So it was in 1913 that Debye and Ehrenfest have first understood that this was the sound foundation. It was, as I said, a somewhat gradual and complicated development.* [*Spontaneous and sustained applause from the audience.*]

L. Rosenfeld: Thank you very much. May I just say one word more about Ehrenfest. Bohr criticized Ehrenfest's application of the adiabatic principle in the case of the rotator, because if you try to go from an oscillation to a rotation, then you get a singularity when you pass to the rotation.

F. Hund: Ehrenfest tried another approach which was connected with that discontinuity, but it was afterwards.

B. L. van der Waerden: Rosenfeld quoted what Einstein said to Heisenberg about what is observable, but he quoted only the second half of what Einstein said. But in his book [*Physics and Beyond*], Heisenberg also told the beginning of the story, approximately as follows: Einstein expressed his doubts about his opinion that a theory should contain only statements about facts that are observable in principle. Heisenberg was astonished and said: 'But you yourself propagated this principle.' To which Einstein replied, 'Yes, maybe I have made use of this kind of philosophy, but still it is nonsense (Quatsch). *The theory alone decides what is observable and what not*'.

Editor's Note

At the first Solvay conference in 1911, Poincaré raised the question in a discussion as to how Planck's treatment of the harmonic oscillator and its quantum condition, according to which the equation $\iint dp\, dq = h$ determined the elementary region of a priori probability in the pq plane, should be extended to systems with more than one degree of freedom. In his reply, Planck expressed the belief that the

* See *Editor's Note* at the end of Discussion.

formulation of a quantum theory for systems with more than one degree of freedom would soon be possible. ['Une hypothèse des quanta pour plusieurs degrés de liberté n'a pas encore été formulée, mais je ne crois nullement impossible d'y parvenir.' *La Théorie du Rayonnement et les Quanta*, 1912, p. 120.]

Planck and Sommerfeld formulated the generalization to many degrees of freedom only *four years later*. [M. Planck, 'Die Quantenhypothese für Molekeln mit mehreren Freiheitsgraden', *Verh. Deut. Phys. Ges.* **17**, 407–418, 438–451 (1915); 'Die Physikalische Struktur des Phasenraumes', *Ann. Physik* **50**, 385– 418(1916); A. Sommerfeld, 'Zur Theorie der Balmerschen Serie', *Münchener Berichte*, 1915, 425–458, 'Die Feinstruktur der Wasserstoff- und wasserstoffsähnlichen Linien', *Münchener Berichte*, 1915, 1–94, 125–167 (1916).]

Planck considered dynamical systems of f degrees of freedom, the equations of motion of which admit f regular intervals. He generalized his treatment of the harmonic oscillator by dividing the phase space by means of surfaces $F(p_k, q_k) = $ constant, defined by the integrals, into regions of volume h^f. Planck postulated that the stationary states correspond to the f-dimensional intersections of these surfaces.

Paul Epstein showed later on [P. S. Epstein, 'Über die Struktur des Phasenraumes bedingt periodischer Systeme', *Berliner Berichte*, 1918, 435–446] that Planck's conditions for characterizing the stationary states were equivalent to the quantum conditions of Sommerfeld. Kneser, who wrote a thesis on quantum theory under Hilbert in 1921 [H. Kneser, 'Untersuchungen zur Quantentheorie', *Math. Ann.* **84**, 277–302 (1921)], also proved rigorously the equivalence of the quantum conditions of Planck and Sommerfeld for the characterization of stationary states.

Author's Note added in proof

In my reply to Bopp's question I gave (from memory) the correct date 1910 (anterior, of course, to the dates of publication) for Planck's non-trivial conception of the statistical meaning of the quantum of action related to the phase integral, and for Ehrenfest's first application of the adiabatic invariance of the phase integral for periodic systems (his paper was published in *Ann. Physik* **36**, 91 (1911)). For more details, see L. Rosenfeld, *Max-Planck Festschrift* (Berlin 1959), p. 203 and *Osiris* **2**, 149 (1936).

For the subject of the lecture, see also L. Rosenfeld, *Archive for Hist. of Exact Sc.* **7**, 69 (1971).

11

Development of Concepts in the History of Quantum Theory

Werner Heisenberg

In one of his lectures on the development of physics Max Planck said: 'In the history of science a new concept never springs up in its complete and final form as in the ancient Greek myth, Pallas Athene sprang up from the head of Zeus.' The history of physics is not only a sequence of experimental discoveries and observations, followed by their mathematical description; it is also a history of concepts. For an understanding of the phenomena the first condition is the introduction of adequate concepts. Only with the help of correct concepts can we really know what has been observed. When we enter a new field, very often new concepts are needed. As a rule, new concepts come up in a rather unclear and undeveloped form. Later they are modified, sometimes they are almost completely abandoned and are replaced by some better concepts which then, finally, are clear and well-defined. I would like to describe this development in three cases which have been important for my own work. First, the concept of the discrete stationary state, which obviously is a fundamental concept in quantum theory. Then, the concept of state, not necessarily stationary or discrete, which only could be understood after quantum mechanics and wave mechanics had been developed. And finally, closely connected with the first two, the concept of the elementary particle which is under discussion until now.

The concept of the discete stationary state was introduced by Niels Bohr in 1913. It was the central concept in his theory of the atom, the intention of which was described by Bohr in the sentence: 'It should be made clear that this theory is not intended to explain phenomena in the sense in which the word *explanation* has been used in earlier physics. It is intended to combine various phenomena, which seem not to be connected, and to show that they are connected.' Bohr stated that only after this connection had been established could one hope to give an explanation in the sense as explanations were meant in earlier physics. There were mainly three phenomena, which had to be connected. The first was the strange fact of the stability of the atom. An atom can be perturbed by chemical processes, by collisions, by radiation or anything else and still it always returns to its original state – its normal state. This was one fact which could not be explained satisfactorily in earlier physics. Then there were the spectral laws, especially the famous law of Ritz that the frequency of the lines in a spectrum could be written as a difference between terms and these terms had to be considered as characteristic properties of the atoms. And finally there were the experiments of Rutherford which had led him to his model of the atom.

So these three groups of facts had to be combined, and as you know the idea of the discrete stationary state was the starting point for their combination. First of all, one had to believe that the behaviour of the atom in the discrete stationary state could be explained by mechanics. This was necessary, because else there would be no connection with Rutherford's model, since Rutherford's experiments were based on classical mechanics. Then also one had to combine the discrete stationary states with the frequencies of the spectrum. There one had to apply the law found by Ritz which now was written in the form that h times the frequency of the line was equal to the difference between the energies of the initial and the final state. This law, however, could best be explained by an assumption which Bohr did not accept, namely by Einstein's idea of the light quantum. Bohr was for a long time not inclined to believe in light quanta, and he considered therefore his stationary states as stations during the motion of the electron which in its orbit around the nucleus loses energy by radiation. The assumption was that during this process of radiation the electron stops radiating at some stations called discrete stationary states. For some unknown reason it does not radiate in these stations, and the final station is the normal state of the atom. When the radiation takes place, the electron goes from one of thes tationary states to the next.

According to this picture the time in the stationary state was much longer than the time required for going from one state to another. But of course this ratio of time was never well defined.

What could be said about radiation itself? One could use the general ideas of Maxwell's theory. From this point of view the interaction between atom and radiation seemed to be the source of all trouble. In the stationary state there was no such interaction and therefore one could, so it seemed, apply classical mechanics. But could one use Maxwell's theory for radiation? I might mention that it was perhaps not necessary to take this point of view. One could have taken the light quanta more seriously. One could have said that the interference patterns which we see in light come about by some extra conditions on the motions of the light quanta. I remember vaguely a discussion with Wentzel in the old times where he explained to me the possibility that the motion of light quanta could be quantized and thereby possibly the interference patterns could be explained. But anyway this was not the point of view which Bohr took. Wherever one started one came into a lot of difficulties and I would like to describe these problems in some detail.

First of all, there were strong arguments in favour of the mechanical model of the stationary state. I have mentioned the experiments of Rutherford. Then the periodic orbits of the electrons in the atom could easily be connected with quantum conditions. So the idea of the stationary state could be combined with the idea of a specified elliptic orbit of the electron. In his earlier lectures, Bohr very frequently showed pictures of electrons moving in their orbits around the nucleus.

This model worked perfectly well in a number of interesting cases. First of all in the hydrogen spectrum. Then in the theory of the relativistic fine structure of the hydrogen lines by Sommerfeld, and in the so-called Stark effect, the splitting of lines in an electric field. So there was an enormous material which seemed to show that this connection of quantized electronic orbits with the discrete stationary states was correct.

On the other hand, there were other reasons for arguing that such a picture cannot be correct. I remember a conversation with Stern who told me that in 1913 when Bohr's first paper had appeared, he had said to a friend, 'If that nonsense is correct which Bohr has just published, then I will give up to be a physicist.'

I shall now point out the difficulties and errors of this model. The worst difficulty was perhaps the following. The electron made a periodic motion in the model defined by quantum conditions and therefore it moved around the nucleus with a certain frequency. However this frequency never turned up in the observations. You could never see it. What you saw were different frequencies which were determined by the energy differences in the transitions from one stationary state to another. Then there was a difficulty with the degeneracy. Sommerfeld had introduced the magnetic quantum number. When we have a magnetic field in some direction, then the angular momentum of the atom around this field should be one or zero or minus one according to this quantum condition. But then if you take a different field with a different direction, quantization has to be carried out with respect to this different direction. But one may have an extremely weak field first in one direction and after a short time in another direction. The field is too weak to turn the atom around. Hence the contradiction with the quantum conditions seems unavoidable.

My first discussion with Niels Bohr – which was just fifty years ago – circled around one of these difficult points. Bohr had given a lecture in Göttingen and had stated that in a constant electric field one can calculate the energy of the stationary states according to the quantum conditions, and that a recent calculation of Kramers on the quadratic Stark effect should probably give correct results because in other cases the method had worked so well. On the other hand, there is very little difference between a constant electric field and a slowly varying electric field. When we have an electric field varying not very slowly but say with a frequency which comes near to the orbital frequency, then we know that of course resonance takes place not when the frequency of the outer electric field coincides with the frequency of the orbit, but when it coincides with the frequency given by the transitions and observed in the spectrum.

When we discussed this problem at length Bohr tried to say that as soon as the electric field varies with time then forces of radiation come in and therefore it may not be possible to calculate the result in a classical way. But of course at the same time he saw that it is rather artificial to invoke the forces of the radiation at this point. Therefore we were soon inclined to say that there must be something wrong with the mechanical model of the discrete stationary state. There was one very decisive paper which has not been mentioned yet. It was a paper of Pauli on the H_2^+ ion. Pauli thought that we can possibly apply the Bohr–Sommerfeld quantization rules when we have a well defined model with periodic orbits, like in hydrogen; but perhaps not in a model which is so complicated like say the helium atom, where two electrons move around the nucleus, because there we would get into all the terrible mathematical difficulties and complications of the three-body problem. On the other hand, when we have two fixed centres, two hydrogen nuclei and one electron, then the motion of the electron is still a nice periodic motion and can be calculated. For the rest the model is already rather complicated; hence it can be used as a check as to whether the old rules really apply in such an intermediate case. Pauli did work out this model and found that he actually did not get the correct energy of the H_2^+ by his calculations. So the doubts against the use of classical mechanics for the calculation of discrete stationary states increased, and attention was moved over more and more to the transitions between the stationary states. One understood that in order to get the whole explanation of the phenomena it was not sufficient to calculate the energy. One also had to calculate transition probabilities. We knew from Einstein's paper of 1918 that the transition probabilities are defined as quantities referring to two states, initial and final state. Bohr had pointed out in his correspondence principle that these transition probabilities could be estimated by connecting them with the intensities of higher harmonics in the Fourier expansion of the electronic orbit. The idea was that every line corresponds to one Fourier component in the expansion of the electronic motion; from the square of this amplitude one can calculate the intensity. This intensity of course is not immediately connected with Einstein's transition probability, but it is related to it, so it allows some estimate of Einstein's quantities. In this way the attention gradually moved over from the energy of the stationary state to the transition probability between stationary states, and it was Kramers who started seriously to study the dispersion of an atom and to relate the behaviour of Bohr's model under radiation with the Einstein coefficients.

Kramers was guided by the idea of virtual harmonic oscillators in the atom corresponding to the harmonics, in writing down a dispersion formula. Then Kramers and I also discussed scattering phenomena where the frequency of the scattered light is different from the frequency of the incident light. Here the scattered light-quantum is different from the incoming quantum because during the scattering the atom makes a transition from one state to another. Such phenomena had just been discovered by Raman in the band spectra. When one tried to write down formulas for the dispersion in these cases one was forced not only to speak about the transition probabilities of Einstein but also to speak about transition amplitudes; one had to give phases to these amplitudes and one had to multiply two amplitudes – say the amplitude going from state m to state n with the amplitude going from n to the state k or so, and then to sum over the intermediate states n. Only when we did that did we get reasonable formulas for the dispersion.

So you see that by fixing the attention not to the energy of the stationary state but to the transition probabilities and to dispersion one eventually came into a new way of looking at things. Actually, as I just said, these sums of products which Kramers and I have written into our paper of the dispersion, were almost already products of matrices. So it was only a very small step from there to say: Well, let us abandon this whole idea of the electronic orbit and let us simply replace the Fourier components of the electronic orbit by the corresponding matrix elements. At that time I must confess I did not know what a matrix was and did not know the rules of matrix multiplication. But one could learn these operations from physics, and later it turned out that it was matrix multiplication, well known to the mathematicians.

By this time you see that the idea of an electronic orbit connected with the discrete stationary state had been practically abandoned. The concept of the discrete stationary state however had survived. This concept was necessary and had its basis in the observations. But the electronic orbit could not be connected with observations and therefore it had been abandoned and what had remained were these matrices for the coordinates.

I should perhaps mention that already before this happened in 1925, Born in his Göttingen seminary in 1924 had emphasized that it was wrong to put the blame for the difficulties of quantum theory only on the interaction between radiation and the mechanical system. He propagated the idea that mechanics had to be revised and to be replaced by some kind of quantum mechanics in order to supply the basis for an understanding of atomic phenomena. And then matrix multiplication was defined. Born and Jordan, and independently Dirac, discovered that those extra conditions which had been added to matrix multiplication in my first paper can actually be written in the elegant form $pq - qp = h/2\pi i$. Thereby they were able to establish a simple mathematical scheme for quantum mechanics.

But even then one could not say what this discrete stationary state really was, and therefore now I am coming to the second part of my talk – the concept of a 'state'. In 1925 one did have a method for calculating the discrete energy values of the atom. One also had at least in principle a method for calculating the transition probabilities. But what was this state of the atom? How could it be described? It could not be described by referring to an electronic orbit. So far it could be described only by stating an energy and transition probabilities; but there was no picture of the atom. Furthermore, it was clear that sometimes there are non-stationary states. The simplest example of a non-stationary state was an electron moving through a cloud chamber. So the question really was how to handle such a state which can occur in nature. Can such a phenomenon as the path of the electron through a cloud chamber be described in the abstract language of matrix mechanics?

Fortunately at that time wave mechanics had been developed by Schrödinger. And in wave mechanics things looked very different. There one could define a wave function for the discrete stationary state. For some time Schrödinger thought that the following picture of a discrete stationary state could be developed. One had a three-dimensional standing wave which can be written as the product of a function in space and a periodical $e^{i\omega t}$ of time, and the absolute square of this wave function meant the electric density. The frequency of this standing wave was to be identified with the term in the spectral law. This was the decisive new point in Schrödinger's idea. These terms did not necessarily mean energies; they just meant frequencies. And so Schrödinger arrived at a new 'classical' picture of the discrete stationary state which at first he believed could actually be applied in atomic theory. But then it soon turned out that even that was not possible. There were very heated discussions in Copenhagen in the summer of 1926. Schrödinger thought that the wave picture of the atom, with continuous matter spread out around the nucleus according to its wave function, could replace the older models of quantum theory. But the discussions with Bohr led to the conclusion that this picture could not even explain Planck's law. It was extremely important for the interpretation to say that the eigenvalues of the Schrödinger equation are not only frequencies – they are actually energies.

In this way of course one came back to the idea of quantum jumps from one stationary state to the other, and Schrödinger was very dissatisfied with this result of our discussions. But even when we knew this and accepted the quantum jumps, we did not know what the word 'state' could mean. One could of course try – and that was tried very soon – to see whether one can describe the path of the electron through a cloud chamber by means of Schrödinger's wave mechanics. It turned out that this was not possible. In its initial position the electron could be represented by a wave packet. This wave packet would move along and thereby one got something like the path of the electron through the cloud chamber. But the difficulty was that this wave packet would become bigger and bigger so that, if the electron just ran long enough, it might have a diameter of one centimetre or more. This is certainly not what we see in the experiments and so this picture again had to be abandoned. In this situation of course we had many discussions, difficult discussions, because we all felt that the mathematical scheme of quantum or wave mechanics was already final. It could not be changed and we would have to do all our calculations from this scheme. On the other hand, nobody knew how to represent in this scheme such a simple case as the path of an electron through a cloud chamber. Born had made a first step by calculating from Schrödinger's theory the probability for collision processes; he had introduced the notion that the square of the wave function was not a charge density as Schrödinger had believed, that it meant the probability to find the electron at the given place.

Then there came the transformation theory by Dirac and Jordan. In this scheme one could transform from $\psi(q)$ to for instance $\psi(p)$, and it was natural to assume that the square $|\psi(p)|^2$ would be the probability to find the electron with momentum p. So gradually one acquired the notion that the square of the wave function, which by the way was not the wave function in three-dimensional space but in configuration space, meant the probability for something. With this knowledge we returned to the electron in the cloud chamber. Could it be that we had asked the wrong question? I remembered Einstein telling me, 'It is always the theory which decides what can be observed.' And that meant, if it was taken seriously, that we should not ask: 'How can we represent the path of the electron in the cloud chamber?' We should ask instead: 'Is it not perhaps true that in nature only such situations occur which can be represented in quantum mechanics or wave mechanics?'

Turning around the question, one saw at once that this path of an electron in a cloud chamber was not an infinitely thin line with well-defined positions and velocities; actually the path in the cloud chamber was a sequence of points which were not too well-defined by the water droplets, and the velocities were not too well-defined either. So I simply asked the question: 'Well, if we want to know of a wave packet both its velocity and its position what is the best accuracy we can obtain, starting from the principle that only such situations are found in nature which can be represented in the mathematical scheme of quantum mechanics?' That was a simple mathematical task and the result was the principle of uncertainty which seemed to be compatible with the experimental situation. So finally one knew how to represent such a phenomenon as the path of the electron, but again at a very high price. Namely, this interpretation meant that the wave packet representing the electron is changed at every point of observation, that is at every water droplet in the cloud chamber. At every point we get new information about the state of the electron; therefore we have to replace the original wave packet by a new one, representing this new information.

The state of the electron thus represented does not allow us to ascribe to the electron in its orbit definite properties like coordinates, momentum and so on. What we can do is only to speak about the probability to find, under suitable experimental conditions, the electron at a certain point or to find a certain value for its velocity. So finally we have come to a definition of state which is much more abstract than the original electronic orbit. Mathematically we describe it by a vector in Hilbert space, and this vector determines probabilities for the results of any kind of experiments which can be carried out on this state. The state may change by every new information.

This definition of state made a very big change or, as Dirac has said, a big jump, in the description of natural phenomena, and I doubt whether the unwillingness of Einstein, Planck, von Laue and Schrödinger to accept it should be reduced simply to prejudices. The word 'prejudice' is too negative in this context and does not cover the situation. It is of course true that Einstein for instance thought it must necessarily be possible to give a kind of objective description of the state of affairs, the state of an atom, in the same sense as that had been possible in older physics. But it was indeed extremely difficult to give up this notion because all our language is bound up with this concept of objectivity. So all the words which we use in physics in describing experiments, like the words measurement or position or energy or temperature and so on, are based on classical physics and its idea of objectivity. The statement that such an objective description is not possible in the world of the atoms, that we can only define a state by a direction in Hilbert space – such a statement was indeed very revolutionary; and I think it is really not so strange that many physicists of that time simply were not willing to accept it.

I had a discussion with Einstein about this problem in 1954, a few months before his death. It was a very nice afternoon that I spent with Einstein but still when it came to the interpretation of quantum mechanics I could not convince him and he could not convince me. He always said 'Well, I agree that any experiment the results of which can be calculated by means of quantum mechanics will come out as you say, but still such a scheme cannot be a final description of Nature.'

Now we come to the third concept I wanted to discuss, the concept of the elementary particle. Before the year of 1928, every physicist knew what one meant by an elementary particle. The electron and the proton were the obvious examples, and at that time we would have liked simply to take them as point charges, infinitely small, just defined by their charge and their mass. We had to agree reluctantly that they must have a radius, since their electromagnetic energy should be finite. We did not like the idea that such objects should have properties like a radius but still we were happy that at least they seemed to be completely symmetrical like a sphere. But then the discovery of the electronic spin changed this picture considerably. The electron was not symmetrical. It had an axis and this result emphasized that perhaps such particles have more than one property and that they are not simple, not so elementary as we had thought before. The situation was again changed completely in 1928, when Dirac developed the relativistic theory of the electron and discovered the positron. A new idea cannot be quite clear from the beginning. Dirac thought first that the negative energy holes of his theory could be identified with the protons; but later it became clear that they should have the same mass as the electron and finally they were discovered in the experiments and were called positrons. I think that this discovery of antimatter was perhaps the biggest jump of all the big jumps in physics of our century. It was a discovery of utmost importance because it changed our whole picture of matter. I would like to explain this in more detail in the last part of my talk.

First Dirac suggested that such particles can be created by the process of pair production. A light quantum can lift a virtual electron from one of these negative energy states in the vacuum to a higher positive energy and that means that the light quantum has created a pair of electron and positron. But this meant at once that the number of particles was not a good quantum number anymore; there was no conservation law for the number of particles. For instance, according to Dirac's new idea one could say that hydrogen atom does not necessarily consist of proton and electron. It may also temporarily consist of one proton, two electrons, and one positron. And actually when one takes the finer details of quantum electrodynamics into account these possibilities do play some role.

In the case of interaction between radiation and electron such phenomena as pair production can happen. But then it was natural to assume that similar processes may occur in a much wider range of physics. We knew since 1932 that there are no electrons in the nucleus, that the nucleus consists of protons and neutrons. But then Pauli suggested that beta decay could be described by saying that an electron and a neutrino are being created in beta decay. This possibility was formulated by Fermi in his theory of beta decay. So you see that already at that time the law of conservation of particle number was completely abandoned. One understood that there are processes in which particles are created out of energy. The possibility for such processes was of course given already by the theory of special relativity, energy being transmuted into matter. But its reality occurred for the first time in connection with Dirac's discovery of antimatter and of pair-creation.

The theory of beta decay was published by Fermi in 1934. A few years later, in connection with cosmic radiation, we asked the question 'What happens if two elementary particles collide with very high energy?' The natural answer was that there was no good reason why one should not have many particles created in such an act. So actually the hypothesis of multiple production of particles in high energy collisions was a very natural assumption after Dirac's discovery. It was checked experimentally only fifteen years later when one studied very high energy phenomena and could observe such processes in the big machines. But when one knew that at very high energy collisions any number of particles can be treated under the only condition that the initial symmetry should be identical with the final symmetry, then one had also to assume that any particle was really a complicated compound system, because with some degree of truth one can say that any particle consists virtually of any number of other particles. Of course we would still agree that it may be a reasonable approximation to consider a pion composed only of nucleon and antinucleon and we should not consider higher compositions. But that is only an approximation, and if we have to speak rigorously then we should say that in any one pion we have a number of configurations of several particles up to an arbitrarily high number of particles if only the total symmetry is the same as the symmetry of the pion. So it was one of the most spectacular consequences of Dirac's discovery that the old concept of the elementary particle collapsed completely. The elementary particle was not elementary any more. It is actually a compound system, rather a complicated many-body system, and it has all the complications which a molecule or any other such object really has.

There was another consequence of Dirac's theory which is important. In the old theory, let us say in non-relativistic quantum theory, the ground state was an extremely simple state. It was just the vacuum, the empty world, nothing else, and had therefore the highest possible symmetry. In Dirac's theory the ground state was different. It was an object which was filled with particles of negative energy that could not be seen. Besides that, if the process of pair-production is introduced one should expect that the ground state must contain probably an infinite number of virtual pairs of positrons and electrons or of particles and antiparticles; so you see at once that the ground state is a complicated dynamical system. It is one of the eigensolutions defined by the underlying natural law. If the ground state is to be interpreted in this way, one can further see that it need not be symmetrical under the groups of the underlying natural law. In fact the most natural explanation of electrodynamics seems to be that the underlying natural law is completely invariant under the isospin group while the ground state is not. The assumption that accordingly the ground state is degenerate under rotations in isospace enforces the existence of long range forces or of particles with rest mass zero following a theorem of Goldstone. Coulomb interaction and photons should probably be interpreted in this way.

Finally Dirac – in consequence of his theory of holes – in his Bakerian lecture in 1941 propagated the idea that, in a relativistic field theory with interaction, use should be made of a Hilbert space with indefinite metric. It is still a controversial question, whether this extension of conventional quantum theory is really necessary. But after many discussions during the last decades, one cannot doubt that theories with indefinite metric can consistently be constructed and can lead to a reasonable physical interpretation.

So the final result at this point seems to be that Dirac's theory of the electron has changed the whole picture of atomic physics. After abandoning the old concept of the elementary particles, those objects which had been called 'elementary particles' have now to be considered as complicated compound systems and should be calculated some day from the underlying natural law, in the same way as the stationary states of complicated molecules should be calculated from quantum or wave mechanics. We have learned that energy becomes matter when it takes the form of elementary particles. The states called elementary particles are just as complicated as the states of atoms and molecules. Or to formulate it paradoxically: every particle consists of all other particles. Therefore we cannot hope that elementary particle physics can ever be simpler than quantum chemistry. This is an important point because even now many physicists hope that some day we might discover a very simple way to describe elementary particle physics like the hydrogen spectrum in the old times. This I think is not possible.

In conclusion I would like again to say a few words about what had been called the 'prejudices'. You may say that our belief in elementary particles was a prejudice. But again I think that would be a too negative statement, because all the language which we have used in atomic physics in the last 200 years is based directly or indirectly on the concept of the elementary particle. We have always asked the question: 'Of what does this object consist and what is the geometrical or dynamical configuration of the smaller particles in the bigger object?' Actually we have always gone back to this philosophy of Democritus; but I think we have now learned from Dirac that this was the wrong question. Still it is very difficult to avoid questions which are already part of our language. Therefore it is natural that even nowadays many experimental physicists – even some theoreticians – still look for *really* elementary particles. They hope for instance that the quarks, if they existed, could play this role.

I think that this is an error. It is an error because even if the quarks would exist we could not say that the proton consists of three quarks. We would have to say that it may temporarily consist of three quarks, it may also temporarily consist of four quarks and one antiquark, or five quarks and two antiquarks and so on. And all these configurations would be contained in the proton; and again one quark would be composed of two quarks and one antiquark and so on. So we cannot avoid this fundamental situation; but since we still have the questions from the old concepts, it is extremely difficult to stay away from them. Very many physicists have looked for the quarks and will probably do so in the future. There has been a very strong prejudice in favour of quarks in the last ten years, and I think they would have been found if they existed. But that is a matter to be decided by the experimental physicist.

There remains the question: What then has to replace the concept of a fundamental particle? I think we have to replace this concept by the concept of a fundamental symmetry. The fundamental symmetries define the underlying law which determines the spectrum of elementary particles. Now I will not go into a detailed discussion of these symmetries. From a careful analysis of the observations I would conclude, that besides the Lorentz group also SU_2, the scaling law, and the discrete transformations P, C, T are genuine symmetries. But I would not include SU_3 or higher symmetries of this type among the fundamental symmetries; they may be produced by the dynamics of the system as approximate symmetries.

But this is again a matter which should be decided by the experiments. I only wanted to say that what we have to look for are not fundamental particles, but fundamental symmetries. And when we have actually made this decisive change in concepts which came about by Dirac's discovery of antimatter, then I do not think that we need any further breakthrough to understand the elementary – or rather non-elementary – particles. We must only learn to work with this new and unfortunately rather abstract concept of the fundamental symmetries; but this may be bad enough.

DISCUSSION

P. A. M. Dirac: I would like to ask Heisenberg a question. I am in general agreement with this point of view about elementary particles that a concept really doesn't exist, but there is one reservation. I wonder whether the electron should not be considered as an elementary particle. It may be that I am prejudiced because I have had success with the electron and no success with other particles. I would like to hear Heisenberg's view on that.

W. Heisenberg: I cannot see that one could consider the electron as an elementary particle in the old sense, because an electron can produce light quanta. Light quanta can produce baryons. So actually the electron is connected with this world of baryons and hadrons and so on. So I don't see that you can separate it out. As soon as an electron has these interactions, then, of course, it is also surrounded by a cloud consisting of all these other things. Would you not agree?

E. P. Wigner: May I make a comment on that? What Heisenberg said is valid, and it is true for an electron, for a particle which is represented in Fock space, in the first Fock space. But if one looks at it from the point of view of the invariance, then the electron is the particle which is surrounded by those pairs and light quanta and what not, but it *is* a particle, and it transforms according to the irreducible representation of the Poincaré group and it obeys Dirac's equation. This is a matter of definition.

W. Heisenberg: Well, I can't see the problem. You are certainly right that an electron is a particle which has all these properties with respect to the Poincaré group, which Dirac has formulated, but I would say that every particle which exists, whether a compound system or not, will have symmetry properties which can be well-defined, symmetry with respect to the Lorentz group, $SU(2)$, and whatever else. So, I cannot see that this makes a fundamental difference.

E. P. Wigner: No, this is not the only criterion, but these are matters of definition, to some extent. This is not the only criterion for a particle to be elementary, that it transforms by an irreducible representation of the Poincaré group and so on. Another criterion is that it is orthogonal to the state of composite part of particles which are composed of others. Now, a deuteron transforms by an irreducible representation of the Poincaré group, but its state vector is not orthogonal to the joint state vector of proton plus neutron. In other words, I use two criteria for 'elementary'.

W. Heisenberg: You certainly would not claim that it is orthogonal to *any other* compound system which has the same symmetry as the electron.

E. P. Wigner: The electron state, I believe, is surely orthogonal to the only other state with spin one-half, namely the proton, and it is surely also orthogonal to the neutron. The neutron is surely not, with a certain probability, just a single electron.

W. Heisenberg: Is the electron orthogonal to helium ion which consists of a helium nucleus plus one electron? The helium ion then transforms like the electron, and I think these two things are not orthogonal to each other.

E. P. Wigner: Excuse me, I hate to argue with you, but the helium atom has four protons in it and therefore it has a baryon number four.

W. Heisenberg: Oh yes, my example was wrong. Now let me see.

E. P. Wigner: Yes, we should try a better example.

W. Heisenberg: I would suggest that you consider an object consisting of a muon, an anti-muon, and an electron. This is of course not a stable object. It will decay into an electron plus photon, but finally such an object might have exactly the same symmetry as an electron.

E. P. Wigner: I am sorry, but they don't belong to the same representation, because the mass is very different.

12

From Matrix Mechanics and Wave Mechanics
to Unified Quantum Mechanics

B. L. van der Waerden

The story I want to tell you begins in March 1926 and ends in April 1926. Early in March two separate theories existed: matrix mechanics and wave mechanics. At the end of April these two had merged into one theory, more powerful than the two parents taken separately.

Wave mechanics was based upon three fundamental hypotheses:

A. Stationary states are determined by complex-values wave functions $\psi(q)$, which remain finite everywhere in q-space.

B. The functions ψ satisfy a differential equation

$$H\psi = E\psi$$

in which the operator H is obtained from the classical Hamiltonian $H(p, q)$ by replacing every momentum p by

$$\frac{K}{i}\frac{\partial}{\partial q}, \qquad K = \frac{h}{2\pi}.$$

C. The eigenvalues E are the energy values.

To these three hypotheses, Schrödinger added Bohr's postulate:

D. $E_m - E_n = h\nu_{mn}$.

This theory was presented in Schrödinger's first and second communications on 'Quantisierung als Eigenwertproblem' in *Annalen der Physik* **79**. The first communication was received on 27 January, and the second on 23 February 1926.

On the other hand, *matrix mechanics* was invented by Heisenberg in June 1925, and presented in a fully developed form in Dirac's first paper on quantum mechanics (received 7 November 1925) and also in the famous 'three-men's paper' of Born, Heisenberg and Jordan (received 16 November 1925). This theory was based upon four mechanical hypotheses and two radiation hypotheses. The mechanical hypotheses are:

1. The behaviour of a mechanical system is determined by the matrices $\mathbf{p}$ and $\mathbf{q}$ (one matrix $\mathbf{q}$ for every coordinate q, and one $\mathbf{p}$ for every momentum p).

2. $\mathbf{pq} - \mathbf{qp} = (K/i)\mathbf{1}$, if p belongs to the same coordinate q, otherwise equal to 0.

3. $H(\mathbf{p}, \mathbf{q}) = W = $ diagonal matrix, having diagonal elements E_n, the energy values.

J. Mehra (ed.), The Physicist's Conception of Nature, 276-293. All Rights Reserved
Copyright © 1973 by D. Reidel Publishing Company, Dordrecht-Holland

4. Equations of motion:

$$\dot{\mathbf{p}} = -\frac{\partial H}{\partial \mathbf{q}}, \qquad \dot{\mathbf{q}} = \frac{\partial H}{\partial \mathbf{p}}.$$

These hypotheses imply:

$$p_{mn} = a_{mn}e^{2\pi i(v_m - v_n)t}$$
$$E_n = hv_n.$$

The radiation hypotheses determine the frequency and intensity of the radiation emitted or absorbed:

5. $E_m - E_n = hv_{mn}$.

6. The transition probabilities are proportional to the $|a_{mn}|^2$.

In his second communication, Schrödinger confesses that he did not succeed in finding a link between his own approach and Heisenberg's. This was written in February 1926, but in March he found the link. In his paper, 'Über das Verhältnis der Heisenberg–Born–Jordanschen Quantenmechanik zu der meinen', received 18 March 1926, Schrödinger writes: 'In what follows ... the *inner connection* between Heisenberg's Quantum Mechanics and my own will be made clear. From the formal mathematical standpoint one may even say that the two theories are *identical.*'

Now is this true? Are the two theories really equivalent in the formal mathematical sense?

Equivalence (or identity, as Schrödinger says) would mean

$$A, B, C, D \gtreqless 1, 2, 3, 4, 5, 6.$$

Now what Schrödinger actually proves, is

$$A, B, C \Rightarrow 1, 2, 3,$$

and of course

$$D \Rightarrow 5.$$

Moreover, if time-dependent functions ψ are allowed, satisfying Schrödinger's time-dependent differential equation, one can prove 4. However, hypothesis 6 can in no way be derived from Schrödinger's set of hypotheses.

The converse $\Leftarrow$, Schrödinger does not even attempt to prove. Yet he refers to his proof as '*Äquivalenz-Beweis*', and he asserts confidently:

'Die Aquivalenz besteht *wirklich*, sie besteht *auch in umgekehrter Richtung.*'

From the formal logical point of view, one may even say that it is impossible to derive A, B, C from 1, 2, 3, 4, 5, 6, because in hypothesis A the notion '*stationary state*' occurs, which does not occur in 1, 2, 3, 4, 5, 6.

After the publication of this paper, everybody accepted Schrödinger's conclusion that the two theories are 'equivalent'.

Everybody – except Pauli. He knew better.

On April 12, just after the publication of Schrödinger's first communication, but before his 'equivalence' paper came out, Pauli wrote a very remarkable letter to Jordan, in which he established the connection between wave and matrix mechanics, in a logically irreproachable way, independent of Schrödinger. He never published the contents of this letter, but he signed the carbon copy (which is quite unusual) and he kept the letter in a plastic cover until his death. I am indebted to his widow, Franca Pauli, for giving me her consent to publish this letter.

PAULI'S LETTER

[This letter was probably written and typed at Copenhagen]

12th April 1926

Dear Jordan,

Many thanks for your last letter and for your looking through the proof sheets. Today I want to write neither about my Handbuch-Article nor about multiple quanta; I will rather tell you the results of some considerations of mine connected with Schrödinger's paper 'Quantisierung als Eigenwertproblem' which just appeared in the *Annalen der Physik*. I feel that this paper is to be counted among the most important recent publications. Please read it carefully and with devotion.

Of course I have at once asked myself how his results are connected with those of the Göttingen Mechanics. I think I have now completely clarified this connection. I have found that the energy values resulting from Schrödinger's approach are always the same as those of the Göttingen Mechanics, and that from Schrödinger's functions ψ, which describe the eigenvibrations, one can in a quite simple and general way construct matrices satisfying the equations of the Göttingen Mechanics. Thus at the same time a rather deep connection between the Göttingen Mechanics and the Einstein–de Broglie Radiation Field is established.

To make this connection as clear as possible, I shall first expose Schrödinger's approach, styled a little differently. According to Einstein and de Broglie one can assign to any moving particle with energy E and momentum G [taking care of the relativity terms,

$$G = \frac{m_0 v}{\sqrt{1 - v^2/c^2}}, \qquad E = \frac{m_0 c^2}{\sqrt{1 - v^2/c^2}}$$

normed in such a way that the energy at rest is $= m_0 c^2$, hence $E^2 - c^2 G^2 = m_0^2 c^4$] an oscillation with frequency $v = E/h$ and wave length $\lambda = h/|G|$. (This assignment is invariant with respect to Lorentz transformations.) The phase velocity V is

$$V = \lambda v = \frac{E}{|G|},$$

hence the wave equation of de Broglie's radiation field

$$\Delta \psi - \frac{1}{V^2} \frac{\partial^2 \psi}{\partial t^2} = 0$$

assumes the form

$$\Delta \psi - \frac{G^2}{E^2} \frac{\partial^2 \psi}{\partial t^2} = 0. \tag{1}$$

Taking care of the relation

$$E^2 - c^2 G^2 = m_0^2 c^2$$

between energy and momentum, one obtains

$$\Delta \psi - \frac{E^2 - m_0^2 c^4}{c^2 E^2} \frac{\partial^2 \psi}{\partial t^2} = 0. \tag{2}$$

Now if we have a mass point moving in a field of force and if E_{pot} is its potential energy, the relation between energy and momentum becomes (taking care of the variability of the mass)

$$(E - E_{pot})^2 - c^2 G^2 = m_0^2 v^2$$

provided E is again so normed that for the mass point at rest $E - E_{pot} = m_0 c^2$. (For the hydrogen atom with relativity correction one obviously has to put $E_{pot} = -Ze^2/r$). Substituting this into (1) one obtains instead of (2)

$$\Delta \psi - \frac{[E - E_{pot}(x, y, z)]^2 - m_0^2 c^2}{c^2 E^2} \frac{\partial^2 \psi}{\partial t^2} = 0. \tag{3}$$

The phase velocity now depends on the place.

Schrödinger's approach is now as follows: *A quantum state of the system with energy E is only possible if a standing de Broglie-Wave without spatial singularities, depending on t like a sine function with frequency $v = E/h$, can exist in accordance with (3)*.

So one has to replace Ψ in (3) by a product of a new function $\bar{\psi}(x, yz)$ depending only on the location with the factor

$$e^{2\pi i v t} = e^{2\pi i (E/h) t}$$

thus obtaining

$$\psi = \bar{\psi} e^{2\pi i (E/h) t}$$

then

$$\frac{\partial^2 \psi}{\partial t^2} = -\frac{4\pi^2}{h^2} E^2 \psi$$

and one obtains

$$\Delta \bar{\psi} + \frac{[E - E_{pot}(x, y, z)]^2 - m_0^2 c^4}{c^2 K^2} \bar{\psi} = 0, \tag{4}$$

putting, as Schrödinger does, $K = h/2\pi$.

This is an eigenvalue problem for the possible values of $E = h v$. These v are enormously large, because in E the energy of the electron at rest is included. The Frequency Condition now says that the light waves can formally be considered as difference-oscillations of the de Broglie-radiation. Planck's constant enters the theory only at that point where one passes from the energy of the states to the frequency of the radiation of de Broglie.

Neglecting relativity corrections one obtains from (4) by putting $E = m_0 c^2 + \bar{E}$ and expanding according to powers of $1/c^2$:

$$\Delta \bar{\psi} + \frac{2m_0}{K^2} (\bar{E} - E_{\text{pot}}) \bar{\psi} = 0. \tag{5}$$

This equation is given in Schrödinger's paper, and he also shows how it can be derived from a Variation Principle.

   Here another remark for which I am indebted to Mr Klein. The difference between the general Quantum Theory of periodical systems and Schrödinger's Quantum Mechanics based upon Equation (5) is, from the point of view of the de Broglie-Radiation, the same as the difference between Geometrical Optics and Wave Optics. Namely if the wave length of the de Broglie-Radiation is small, one can put in (5), as is well-known

$$\bar{\psi} = e^{i(1/K)S}.$$

If S/K is large, one now obtains from (5), according to Debye, the Hamilton–Jacobi differential equation for S. In this case $\bar{\psi}$ becomes a univalued point function only if the moduli of periodicity of S/K are integer multiples of 2π. This leads to the usual condition $\int p\,\mathrm{d}q = nh$, which has been interpreted already by de Broglie from the point of view of the geometrical optics of his Radiation Field.

   In reality, however, S/K is not large generally, so one has to stick to (5) and to use the mathematics of Wave Theory to integrate this equation.

   Next comes my own contribution, namely the connection with the Göttingen Mechanics. For the sake of simplicity I shall consider a one-dimensional problem and use Cartesian coordinates (in the three-dimensional case and with arbitrary coordinates everything goes just so, also if gyroscopic terms are added). So let the wave-equation be given as

$$\frac{\mathrm{d}^2 \psi}{\mathrm{d}x^2} + \frac{2m}{K^2} [E - E_{\text{pot}}(x)]\,\psi = 0$$

(compare (5), the bars are omitted).

   Now let $E_1, E_2, ..., E_n, ...$ be the eigenvalues, $\psi_1, \psi_2, ..., \psi_n, ...$ a complete set of eigenfunctions. For these we have

$$\int_{-\infty}^{\infty} \psi_n \psi_m \,\mathrm{d}x = \begin{cases} 0 & \text{for} \quad n \neq m \\ 1 & \text{for} \quad n = m. \end{cases}$$

The first equation (orthogonality) follows from Green's formula, the second means a normation of the multiplicative constants in the ψ_n. Any arbitrary function of x can be expanded in a series with respect to the ψ_n. Now one considers in particular the expansion of $x\psi_n$:

$$x\psi_n(x) = \sum_m x_{nm}\psi_m(x);$$

$$x_{nm} = \int_{-\infty}^{\infty} x\psi_n\psi_m \,\mathrm{d}x. \tag{I}$$

One also puts

$$(p_x)_{nm} = iK \int_{-\infty}^{\infty} \frac{\partial \psi_n}{\partial x} \psi_m \, dx \; ; \qquad iK \frac{\partial \psi_n}{x} = \sum_m (p_x)_{nm} \psi_m(x) \qquad (II)$$

(i=imaginary unit, $K=h/2\pi$). Now $x_{nm}=x_{mn}$ is real, $(p_x)_{nm}=-(p_x)_{mn}$ purely imaginary. It can be shown without difficulty, that the matrices for x and p_x thus defined satisfy the equations of the Göttingen Mechanics. [Namely

$$\mathbf{p}_x \mathbf{x} - \mathbf{x} \mathbf{p}_x = -iK, \qquad \frac{1}{2m} \mathbf{p}_x^2 + \mathbf{E}_{pot}(x) = \mathbf{E} \quad \text{(Diagonal matrix)]}.$$

From the rule of multiplication it follows that the matrix belonging to any function $F(x)$ of x is just given by

$$F_{nm} = \int_{-\infty}^{\infty} F(x) \, \psi_n \psi_m \, dx.$$

I shall not write out the calculations in detail; you will be able to verify the assertion easily.

I have calculated the oscillator and rotator according to Schrödinger. Further the Hönl–Kronig-formulae for the intensity of the Zeeman components are easy consequences of the properties of the spherical harmonics. Perturbation theory can be carried over completely into the new theory, and the same thing holds for the transformation to principal axes, which in general is necessary if degenerations (multiple eigenvalues) are cancelled by external fields of force. At the moment I am occupying myself with the calculation of transition probabilities in hydrogen from the eigenfunctions calculated by Schrödinger. For the Balmer lines finite rational expressions seem to come out. For the continuous spectrum the situation is more complicated: the exact mathematical formulation is not yet quite clear to me.

As regards Lanczos, my considerations have only very few points of contact with his ideas. He considers a problem for which the eigenvalues are the reciprocal energy values, whereas here the eigenvalues are just the energy values. In his exposition certain functions depending, like Green's function, on two points, play an essential role; such functions are not used here. On the whole I feel that Lanczos' approach has not much value.

About the physical signification of the expressions (I) and (II) I do not know much. In any case they seem to be connected with the idea that the ordinary light waves are difference oscillations (beats) of de Broglie's radiation. The fact that in (I) and (II) no indefinite phases occur is due to the trivial reason that in passing from (3) to (4) the periodical factor depending on time has been suppressed. If this factor is taken into account, one obtains in x_{nm} and $(p_x)_{nm}$ besides $\exp[(2\pi i/h)(E_n-E_m)t]$ a phase factor $\exp[2\pi i(\delta_m-\delta_n)]$ in which δ_m and δ_n are the phases of eigenvibrations belonging to E_m and E_n. In principle, in the Göttingen theory as well as in de Broglie's statement of the quantum problem, no description of the motion of the electron in the atom in space and time is given. In the latter theory this is clear from the fact that outside the domain of validity of Geometrical Optics it is impossible to construct 'rays' in de Broglie's Wave System that can be considered as orbits of particles. The problem of the asymptotic linkage with the usual pictures in space and time for the limiting case of large quantum numbers remains unsolved. Yet it is a definite progress to be able to see the problems from two different sides. It seems one also sees now, how from the point of view of Quantum Mechanics the contradistinction between 'point' and 'set of waves' fades away in favour of something more general.

Cordial greetings for you and the other people at Göttingen (especially Born, in case he is back from America; please show him this letter).

Yours

(carbon copy signed:)   W. Pauli

COMMENTS ON PAULI'S LETTER

Pauli's wave equation (1) is called in the letter 'the wave equation of deBroglie's radiation field'. It is not given in de Broglie's thesis, but it is very easy to derive it from the given expressions for v and λ. It is valid for plane waves, i.e. for a free electron.

Equation (3) is essentially the Klein–Gordon equation. It is true that the magnetic terms are missing, but Pauli expressly says in the course of his letter that 'everything goes just so if gyroscopic terms are added', which shows that Pauli. who was at that time thinking very hard about the anomalous Zeeman effect, knew perfectly well how to handle magnetic fields. He omitted the magnetic terms only 'for the sake of simplicity'.

We know from Schrödinger's letters that he also tried the Klein–Gordon equation, but he gave it up because it did not yield the right fine-structure of the hydrogen atom.

The Klein–Gordon equation was discovered independently by Schrödinger, by Pauli, by Klein and Gordon and by at least two other people.*

To Pauli's orthogonality relations

$$\int_{-\infty}^{\infty} \psi_n \psi_m \, \mathrm{d}x = \begin{cases} 0 & n \neq m \\ 1 & n = m, \end{cases}$$

we may remark that in the one-dimensional case the eigenfunctions are single and real, so that complex conjugate factors ψ_n^* are not needed.

The paper of Lanczos, to which Pauli refers at the end of his letter, was published in *Zeitschrift für Physik* **35** (received 22 December 1925). I feel it has more value than the contemporaries suspected. Let us use our hindsight and start with Schrödinger's

---

* Jagdish Mehra has informed me that the other people were V. Fock (*Z. Phys.* **38**, 242 (1926); **39**, 226 (1926)), H. van Dungen and Th. de Donder (*Compt. Rend. Acad. Sci. Paris*, July 1926), and J. Kudar (*Ann. Physik* **81**, 632 (1926)).

equation, which I shall write as

$$- \Delta\psi + V\psi = E\psi, \tag{6}$$

leaving out all numerical factors. Lanczos considers a finite domain in q-space, so let us enclose our atom in a large sphere of radius R. As a boundary condition we may assume, $\psi = 0$ on the boundary. Since the zero-point on the energy scale is quite arbitrary, we may suppose that it lies below the lowest energy value. It follows that zero is not an eigenvalue.

Under these assumptions, the boundary value problem

$$\left. \begin{array}{l} - \Delta\psi + V\psi = u \\ \quad\psi = 0 \quad \text{on the boundary} \end{array} \right\}$$

can be solved by means of Green's function $K(P, Q)$ as follows:

$$\psi(P) = \int K(P, Q)\, u(Q)\, dQ. \tag{7}$$

Replacing $u(Q)$ by $E\psi(Q)$, and dividing by E, one obtains

$$\int K(P, Q)\, \psi(Q)\, dQ = \frac{1}{E}\psi(P). \tag{8}$$

This integral equation is equivalent to Schrödinger's equation (6). Its eigenvalues are just $1/E$, the reciprocal energy values.

Now this is just the kind of integral equations Lanczos considers. He does not specify what kind of function $K(P, Q)$ is, but he does say that the eigenvalues of his integral equation are the reciprocal energy values.

Now let us hear what Schrödinger says about the paper of Lanczos. In a footnote on p. 754 of his 'equivalence' paper he writes:

'Similar ideas are exposed in an interesting paper of Lanczos, which already contains the valuable insight that Heisenberg's atomic dynamics is capable of a continuous interpretation. For the rest, the paper of Lanczos has less points of contact with mine than one might think at first sight.

The determination of the system of formulae, which Lanczos leaves quite undetermined, *cannot* be found in the direction of identifying the symmetrical kernel $K(s, \sigma)$ with Green's function of our wave equation... *For this function of Green, if it exist, has as its eigenvalues the quantum levels themselves.*'

This is an error of Schrödinger, for which I have no explanation. We have just seen that the eigenvalues of Green's kernel $K(P, Q)$ are $1/E$ and not E.

Schrödinger continues: 'On the contrary, the kernel of Lanczos is required to have as its eigenvalues the *reciprocal* quantum levels.'

Schrödinger just missed the point.

If Lanczos' kernel $K(P, Q)$ is identified with the Green's function of Schrödinger's differential equation, its eigenfunctions φ_1, φ_2,..., are Schrödinger's eigenfunctions.

Besides the integral operator K defined by the kernel $K(s, \sigma)$:

$$K\psi(s) = \int K(s, \sigma)\,\psi(\sigma)\,d\sigma,$$

Lanczos introduces two more integral operators $\mathbf{p}$ and $\mathbf{q}$:

$$\mathbf{p}\psi(s) = \int p(s, \sigma)\,\psi(\sigma)\,d\sigma$$

$$\mathbf{q}\psi(s) = \int q(s, \sigma)\,\psi(\sigma)\,d\sigma$$

in such a way that

$$\mathbf{pq} - \mathbf{qp} = \frac{1}{2\pi i}\,\mathbf{1}. \tag{9}$$

Since p and q are supposed to be integral operators, the unit operator $\mathbf{1}$ must also be an integral operator

$$\mathbf{1}\psi(s) = \int E(s, \sigma)\,\psi(\sigma)\,d\sigma.$$

This implies, as Lanczos says, that $E(s, \sigma)$ is zero for $\sigma = s$, and that

$$\int\limits_{-\infty}^{\infty} E(s, \sigma)\,d\sigma$$

is equal to 1. Hence, Lanczos' function $E(s, \sigma)$ is just Dirac's function $\delta(s-\sigma)$.

Lanczos concludes that the functions $p(s, \sigma)$ and $q(s, \sigma)$ cannot be everywhere finite. In fact, if one wants to reach complete agreement between Lanczos, Schrödinger, and Pauli, one has to assume

$$q(s, \sigma) = s\cdot\delta(s - \sigma)$$

$$p(s, \sigma) = -\frac{h}{2\pi i}\,\delta'(s - \sigma).$$

Next, Lanczos defines the matrices corresponding to the operators $\mathbf{p}$ and $\mathbf{q}$:

$$p_{ik} = \int p(s, \sigma)\,\varphi_i(s)\,\varphi_k(\sigma)\,ds\,d\sigma$$

$$q_{ik} = \int q(s, \sigma)\,\varphi_i(s)\,\varphi_k(\sigma)\,ds\,d\sigma$$

and proves that the matrices $\mathbf{p}$ and $\mathbf{q}$ satisfy the Born–Jordan condition

$$\mathbf{pq} - \mathbf{qp} = \frac{h}{2\pi i}\,\mathbf{1}.$$

From this analysis we see that Lanczos' approach had more points of contact with the ideas of Schrödinger and Pauli than these two suspected. His weakness was that he was not able to specify his functions $K(s, \sigma)$, $p(s, \sigma)$ and $q(s, \sigma)$.

Let us now return to Pauli's letter. Pauli says at the end: 'The problem of the asymptotic linkage with the usual pictures in space and time for the limiting case of large quantum numbers remains unsolved.'

More light on this problem was shed by the study of the behaviour of wave packets. A most interesting contribution was a little-known paper of Ehrenfest, in which he proved that the centre of gravity of a wave packet moves according to the classical law: force = acceleration × mass, provided the force exerted upon the electron by the electromagnetic field is calculated by integrating the Lorentz force over the charge density $-e\psi^*\psi$. Another important contribution was, of course, Heisenberg's Uncertainty Principle, which was also derived from the study of the behaviour of wave packets in q-space and p-space.

A quite new point of view was Born's interpretation of $\psi^*\psi$ as a probability density, proposed in connection with his study of collisions. Dirac extended Born's probability interpretation to much more general measurements. However, in this lecture I wanted to restrict myself to what happened in March and April of 1926, so I shall stop here.

DISCUSSION

L. Rosenfeld: Thank you very much, Professor Van der Waerden. We are very happy that Professor Cornelius Lanczos is present here and has heard this very well deserved vindication of his contribution at that time. I remember having read your paper, Professor Lanczos, but I cannot claim that I understood it so well as Professor Van der Waerden has explained it.

B. L. van der Waerden: Oh, you are Lanczos?

C. Lanczos: Yes.

B. L. van der Waerden: Oh, that is marvellous. I didn't know that you were here at this symposium or that you would come to this lecture. [*Spontaneous and sustained applause from the audience at this dramatic moment when Professor Lanczos, unknown to Professor Van der Waerden and to most of the audience, just happened to be present. It was an historic encounter.*]

D. S. Saxon: My question is addressed to Van der Waerden. I think in Pauli's remarkable letter [to Jordan], which is remarkable not only for what it says but for how self-contained it is, there is one thing written down without any explanation which I find very interesting: namely, the differential characterization of the momentum operator. That is simply stated without any comment, more or less as a definition. Pauli doesn't say exactly where it came from. Do you have any comment on that?

B. L. van der Waerden: You have the text of the letter, and I don't know anything more about what was in Pauli's mind than is written in the letter.

D. S. Saxon: The wording is interesting because he says, 'one also puts', as if it were a matter of definition, but there is no comment about that very interesting identification.

L. Rosenfeld: I don't think there was anything about that in Schrödinger's paper either, in the paper in which he gave the equivalence. Schrödinger also gave the rule, the correspondence between momentum operator and the derivation, without any comment. I suppose it was a matter of trial and error, and it worked.

C. Lanczos, B. L. van der Waerden and L. Rosenfeld
"Oh, you are Lanczos?"

B. L. van der Waerden and C. Lanczos
"Pauli was a vicious man, as everybody knows. He could allow himself such viciousness because he
was such a great man."

D. S. Saxon: I was wondering about just that. Since the Heisenberg rules were known and the question of the matrix for the position operator was not a matter of definition, I wonder whether it might have been a matter of trial and error.

C. Lanczos: Let me make a remark which has nothing to do, by the way, with my paper, but it is something which is not well-known. Sommerfeld obtained the famous quantum conditions for separable systems, which are only possible if the Hamilton–Jacobi equation is separable, so that you can reduce the problem from *n* dimensions to a single dimension, but Einstein had an exceedingly ingenious idea which is not very well-known because I think it was only published once in the *Verhandlungen der Deutschen Physikalischen Gesellschaft.* Instead of these quantum conditions, $\int p \, dq = nh$, for each one of the separable pairs of variables, he took a sum, and he used only one single equation. Now the sum has an invariance significance, which has nothing to do with separating in coordinates, because the sum of the $\int p \, dq$ is a *S* function; so it is the change of the *S* function for a complete revolution. And he used the principle, not an equation, but a principle, and he didn't used the actual path but *any* path, and he said that along *any closed path* this integral should be an integer times *h*. This, in principle, would be applicable also for non-separable systems, although it has never been tried. But it is, in principle. The important thing is that Einstein gets away from the special coordinates in which the Hamilton–Jacobi equation can be separated, because the *S* function now has an invariance significance. It is true that the space to which he applied it is not an ordinarily connected space, but a so-called Riemannian space; not a Riemannian curved space, but a Riemannian surface. It was an exceedingly ingenious idea which has somehow not gone into the history books of physics, and that is the reason I though of mentioning this.

B. L. van der Waerden: Now that you have this instrument [the microphone] in your hand, may I ask you a question? Did you know all this to which I have referred in my paper? Were you aware of these connections?

C. Lanczos: You are absolutely right. You rehabilitated my work. Pauli was a vicious man, as everybody knows. [Laughter] Anything which didn't agree with his ideas was wrong, and anything was right only if he made it, if he discovered it, which is all right for such a great man. He could allow himself such viciousness, but I am very grateful to you for pointing out what you have.

I certainly was aware of these connections, but as you pointed out, the weakness was that I didn't specify the kernel function and the special functions. At that time, you know, after the matrix mechanics it looked as if you couldn't do anything with the continuum, and one would have to operate with the discontinuum. Everything is discontinuous in the matrices. Now I was very much interested at that time in integral equations, and actually the integral equation which I used was not the Schrödinger equation, but the inverse equation, because only a differential equation can be changed to an integral equation, as you pointed out. And that is the reason why the energies actually come in with a reciprocal, so that the Green's function which I had is basically the Schrödinger equation with the source of a delta function. If you have a source, and that source happens to be a delta function, then this function does have a physical significance, whereas it looks after Pauli's criticism that it has no physical significance. But it does have a physical significance.

B. L. van der Waerden: Yes, but after you had read Schrödinger's paper, did you realize that this was the case?

C. Lanczos: Afterwards, it was too trivial. Yes, afterwards it was trivial. I mean, it was no longer of interest, because Schrödinger came along and he did it. As it often happens, it is the second man who hits the nail on the head and not the first one. [Applause.]

R. Peierls: I would like to suggest an answer to Saxon's question, because I believe that in writing the differential operator for *p* it was obvious to Pauli from the de Broglie relation which he quoted, the relation between momentum and wavelength. This was probably the connection, because it is not a very long step, but that is only my guess.

While I am on my feet, may I make a small correction to a remark by Professor Van der Waerden. It is a very small point, but as Professor Van der Waerden is so careful with details, I think one should get this detail right. Ehrenfest *did not* prove that the centre of mass of a wave packet moves according to the classical trajectory, because that statement is not right. He proved merely that the acceleration of the centre of mass equals the average force, and that amounts to the other statement in the limit of a very small wave packet.

B. L. van der Waerden: Yes, that is what I meant to say, but it seems I did not say it clearly enough.

H. B. G. Casimir: I have two little remarks and one question. I remember Pauli referring once to this situation, not to this letter but to the situation, but only very casually. That was in 1932. And he said, 'Well, of course, when Schrödinger's first paper appeared, it was quite *obvious* that this was exactly the same as the matrix calculus.' So, he [Pauli] never published it, but he said to me, 'It was *at once clear* to me that this explained matric calculus when you thought of the system of orthogonal functions and so on.' That's the only time he referred to it in conversations that I had with him. But, as I said, this was in a casual way that he didn't consider it all that extraordinary sort of thing.

A second thing I just would mention for historical interest. Of course, you get the action function and in a certain limit you find back the solution *S* of the Hamilton–Jacobi equation, and we also know that these played a role in the equivalence of ordinary mechanics and geometrical optics. It is very curious that in experimental practice these things later on worked in the opposite direction. I remember being present at a meeting of the German Physical Society in 1932 in Berlin, which, in a way, gave birth to electron microscopy, where a number of physicists, partly from industry and partly from the Technische Hochschule, reported on the first work on electron microscopy. Several of them went again the other way round. They said, 'Well, we know about wave mechanics and the Schrödinger equation, and after all now the electron is a wave, and light is a wave, and therefore it tough to be possible to carry out geometrical optics with electrons.' This they proceeded to do. It seems to me to be a curious twist of history that just about six years after wave mechanics, they went in the opposite direction to make electron microscopes.

Then I have one question concerning adiabatic invariance. Is there in all the textbooks on theoretical mechanics any systematic study of, let's say, penduli where you slowly change the length of the pendulum and what remains constant in those cases? The only ancient application of the principle that I know of is the Hawaiian guitar, which you can only play because it is not the energy which remains constant when you slide up your finger over a vibrating string, but essentially the energy divided by frequency. Therefore when you twist the string and slide up your finger, the energy goes up, and that makes this higher pitch (to which you slide up on the guitar) still audible, otherwise it wouldn't be. If you play it the other way round you hear nothing at all. That is the only old practical application I know of adiabatic invariance, but I wonder whether in textbooks of mechanics there are others.

L. Rosenfeld: The notion of adiabatic invariance goes back to Boltzmann, who tried other applications, not so practical ones; for instance, to a mechanical model of Maxwell's theory. As a systematic textbook on adiabatic invariance, there is one by Levi-Civita in Italian, a very good one.

C. N. Yang: Since there are so many originators of the quantum conditions in this room, perhaps it is not inappropriate for me to add a footnote. We are familiar with the quantum condition $pq - qp = -i\hbar$. But, if you apply it to curvilinear coordinates, or to a system in which there is general constraint, this equation, as an operative equation, is not quite meaningful. I must confess that I got a shock when a graduate student at Stonybrook last year told me something that I totally did not know. Many people in the audience might be in the same boat as myself, so I thought that a footnote may not be out of place.

If you deal with a general dynamical system with curvilinear coordinates, whose Lagrangian is of the form $g_{ij}\dot{x}_i\dot{x}_j$, plus an appropriate potential energy term, where g_{ij} is a general function of x, what is the Schrödinger equation for a system like this? After the student convinced me that his result was correct, I tried to understand what would be the canonical quantization of this. My conclusion is that there is no general canonical quantization for a general system, if the g_{ij} represents a Riemannian geometry whose curvature is not Euclidean. However, what he discovered was that, in fact, if you use Feynman's path integral method for the formulation of quantum mechanics, this gives a unique Schrödinger equation. Up to this point, it is obvious that that is the generalization of the Laplacian, but what is strange is that there is an additional term with h^2 and R, where R is the total curvature of the system computed in Riemannian geometry out of the g_{ij}, treating the g_{ij} as the metric. Now, the product of this term cannot be argued out on grounds of the correspondence principle.

You may argue, as I tried to argue, that a confinement is necessary. If the space of g_{ij} has a curvature, there is a well-known embedding theorem which says that in higher dimensions, it can be embedded in Euclidean space. You might at first try to say, as we did, that in fact you should quantize it in the higher-dimensional Euclidean space so that you don't have this problem, and then ask whether the confinement to the subspace in that higher-dimensional Euclidean space would give you something like this. That is a very tantalizing idea, but after many attempts we finally concluded that that argument is not fruitful. In fact, we concluded that the canonical quantization rule does not give a unique equation and Feynman's path method does. This fact was apparently discussed already 20 years ago in an article in the *Review of Modern Physics*, but we found that this discussion was not fully satisfactory. The work to whaich I have referred will be published soon.

L. Rosenfeld: This reminds me of a similar phenomenon which has been pointed out recently by Jensen and collaborators, but in a quite different context, but which is perhaps connected with this. They asked how you can generalize D'Alembert's principle, the idea of constraints of classical mechanics, to the quantum case. They found that in the quantum case, without relativity, there is an extra term which also vanishes with h and which depends on the curvature. They treated the case of a point constrained to a given trajectory, and there arises a term depending on the curvature of this trajectory. Now if you think in terms of the wave packet, if the curvature is too strong the wave packet will oscillate within this curved part of the constrained trajectory and not propagate further.

APPENDIX

ORIGINAL GERMAN TEXT OF PAULI'S LETTER

(Nachlass Prof. W. Pauli)

12. April 1926

Lieber Jordan,

Vielen Dank für Ihren letzten Brief und die Durchsicht der Korrekturen. Heute will ich weder von meinem Handbuch-Artikel, noch von Mehrfachquanten schreiben, ich will Ihnen vielmehr die Resultate von einigen Ueberlegungen mitteilen, die ich in Zusammenhang mit der in den *Annalen der Physik* erschienenen Arbeit über 'Quantisierung als Eigenwertproblem' von Schrödinger angestellt habe. Ich glaube, dass diese Arbeit mit zu dem Bedeutendsten zählt, was in letzter Zeit geschreiben wurde. Lesen Sie sie sorfgältig und mit Andacht. Natürlich habe ich mich gleich gefragt, wie seine Resultate mit denen der Göttinger Mechanik zusammenhängen. Diesen Zusammenhang glaube ich jetzt vollständig klargestellt zu haben: Es hat sich ergeben, dass die aus dem Schrödingerschen Ansatz folgenden Energiewerte stets dieselben sind wie die der Göttinger Mechanik und dass aus den Schrödingerschen Funktionen ψ, welche die Eigenschwingungen beschreiben, in einfacher und allgemeiner Weise Matrizen hergestellt werden können, die den Gleichungen der Göttinger Mechanik genügen. Damit ist zugleich ein wohl ziemlich tiefliegender Zusammenhang zwischen der Göttingen Mecharik und dem Einstein–deBroglieschen Strahlungsfeld hergestellt.

Um diesen möglichst deutlich hervortreten zu lassen, will ich zunächts den Schrödingerschen Ansatz, ein wenig anders frisiert, darstellen. Nach Einstein und de Broglie kann einem bewegten Massenpunkt mit der Energie E und dem Impuls G [Berücksichtigung der Relativitätsglieder,

$$G = \frac{m_0 v}{\sqrt{1 - v^2/c^2}}, \qquad E = \frac{m_0 c^2}{\sqrt{1 - v^2/c^2}}$$

so normiert, dass Ruhenergie$=m_0c^2$ wird; $E^2-c^2G^2=m_0^2c^4$] ein Schwingungsvorgang mit der Schwingungszahl $\nu=E/h$ und der Wellenlänge $\lambda=h/|G|$ zugeordnet werden. (Diese Zuordnung ist invariant gegenüber Lorentz Transformationen). Die Phasengeschwindigkeit V ist gleich

$$V = \lambda\nu = \frac{E}{|G|}$$

daher nimmt die Wellengleichung des de Broglieschen Strahlungsfeldes

$$\Delta\psi - \frac{1}{V^2}\frac{\partial^2\psi}{\partial t^2} = 0$$

die Form an

$$\Delta\psi - \frac{G^2}{E^2}\frac{\partial^2\psi}{\partial t^2} = 0. \tag{1}$$

Mit Berücksichtigung der Beziehung

$$E^2 - c^2G^2 = m_0^2c^4$$

zwischen Energie und Impuls wird dies

$$\Delta\psi - \frac{E^2 - m_0^2c^4}{c^2E^2}\frac{\partial^2\psi}{\partial t^2} = 0. \tag{2}$$

Haben wir nun einen nicht kräftefrei bewegten Massenpunkt und ist E_{pot} seine potentielle Energie, so wird die Beziehung zwischen Energie und Impuls (bei Berücksichtigung der Massenveränderlichkeit)

$$(E - E_{\text{pot}})^2 - c^2G^2 = m_0^2c^4$$

wenn E wieder so normiert ist, dass für den ruhenden Massenpunkt $E-E_{\text{pot}}=m_0c^2$ (Für das Wasserstoffatom mit Relativitätskorrektion ist offenbar $E_{\text{pot}}=-Zc^2/r$ zu setzen). Dies in (1) eingesetzt gibt an Stelle von (2):

$$\Delta\Psi - \frac{[E - E_{\text{pot}}(x, y, z)]^2 - m_0^2c^4}{c^2E^2}\frac{\partial^2\psi}{\partial t^2} = 0. \tag{3}$$

Die Phasengeschwindigkeit hängt nun von Ort ab.

Der Schrödingersche Ansatz ist nun folgender: *Ein Quantenzustand des Systems mit der Energie E soll dann und nur dann möglich sein, wenn eine räumlich singularitätsfreie, zeitlich sinusförmige, stehende de Broglie-Welle mit der Frequenz $\nu=E/h$ gemäss (3) existenzfähig ist.*

Man ersetze also in (3) Ψ durch das Produkt einer neuen nur vom Ort abhängigen Funktion $\bar\psi(x, y, z)$ und dem Faktor

$$e^{2\pi i\nu t} = e^{2\pi i(E/h)t}$$

$$\psi = \bar\psi e^{2\pi i(E/h)t}$$

dann wird

$$\frac{\partial^2 \psi}{\partial t^2} = -\frac{4\pi^2}{h^2} E^2 \psi$$

und man erhält

$$\Delta \bar{\psi} + \frac{[E - E_{\text{pot}}(x, y, z)]^2 - m_0^2 c^4}{c^2 K^2} \bar{\psi} = 0, \tag{4}$$

wenn mit Schrödinger $K = h/2\pi$ gesetzt ist.

Dies ist ein Eigenwertproblem für die möglichen Werte von $E = h\nu$. Diese ν sind enorm gross, weill in E die Ruhenergie des Elektrons mit einbezogen ist. Die Frequenzbedingung sagt dann aus, dass die Lichtwellen formal als Differenztöne der de-Broglie-Strahlung aufzufassen sind. Das Wirkungsquantum h wurde nur an der Stelle in die Theorie hereingebracht, wo die Zuordnung von Energie der Zustände zu Frequenz der de-Broglie-Strahlung vorgenommen wird.

Bei Vernachlässigung der Relativitätskorrektionen erhält man aus (4) indem man $E = m_0 c^2 + \bar{E}$ setzt und nach $1/c^2$ entwickelt:

$$\Delta \bar{\psi} + \frac{2m_0}{K^2} (\bar{E} - E_{\text{pot}}) \bar{\psi} = 0. \tag{5}$$

Diese Gleichung steht in Schrödingers Arbeit und es ist dort auch gezeigt, wie sie aus einem Variationsprinzip gewonnen werden kann. Hier noch eine Anmerkung, die ich Herrn Klein verdanke. Der Unterschied zwischen der alten Quantentheorie der Periodizitätssysteme und der auf dem Ansatz (5) basierenden Schrödingerschen Quantenmechanik ist, vom Standpunkt der de-Broglie-Strahlung aus, derselbe wie der zwischen geometrischer Optik und Wellenoptik. Bei kleiner Wellenlänge der de-Broglie-Strahlung kann man nämlich in (5) in bekannter Weise den Ansatz machen

$$\bar{\psi} = e^{i(1/K) S}.$$

Ist S/K gross, so erhält man dann aus (5) nach Debye die Hamilton–Jacobische Differentialgleichung für S. Ueberdies wird in diesem Fall $\bar{\psi}$ nur dann eine eindeutige Ortsfunktion, wenn die Periodizitätsmoduln von S ganze Vielfache von 2π sind. Dies führt auf die bisher übliche $\int p \, dq = nh$ Bedingung, die ja schon von de Broglie selbst vom Standpunkte der geometrischen Optik seines Strahlungsfeldes aus interpretiert wurde.

In Wirklichkeit aber ist S/K im allgemeinen nicht gross, man muss bei (5) bleiben und die Mathematik der Wellenlehre zur Integration dieser Gleichung anwenden.

Nun kommt das, was ich zu Schrödinger hinzugefügt habe, nämlich die Verbindung mit der Göttinger Mechanik. Der Einfachheit halber möge ein eindimensionales Problem ins Auge gefasst und cartesische Koordinaten verwendet werden (Im dreidimensionalen und bei beliebigen Koordinaten geht es ganz analog, ebenso wenn noch gyroskopische Terme hinzutreten). Die Wellengleichung sei also gegeben durch

$$\frac{d^2\psi}{dx^2} + \frac{2m}{K^2} [E - E_{\text{pot}}(x)] \psi = 0$$

(vgl. (5), die Querstriche sind weggelassen).

Nun seien E_1, E_2,..., E_n,... die Eigenwerte, ψ_1, ψ_2, ..., ψ_n,... ein zugehöriges vollständiges System von Eigenfunktionen. Für diese gilt

$$\int_{-\infty}^{\infty} \psi_n \psi_m dx = \begin{cases} 0 & \text{für } n \neq m \\ 1 & \text{für } n = m. \end{cases}$$

Die erste Gleichung (Orthogonalität) folgt aus der Greenschen Formel, die zweite bedeutet eine Normierung der multiplikativen Konstanten in den ψ_n. Jede willkürliche Funktion von x ist nach den ψ_n entwickelbar. Nun betrachte man speziell die Entwickelung von $x\psi_n$ nach Eigenfunktionen:

$$x\psi_n(x) = \sum_m x_{nm}\psi_m(x); \quad x_{nm} = \int_{-\infty}^{\infty} x\psi_n\psi_m\, dx. \tag{I}$$

Ferner setze man

$$(p_x)_{nm} = iK \int_{-\infty}^{\infty} \frac{\partial\psi_n}{\partial x} \psi_m\, dx; \quad iK \frac{\partial\psi_n}{\partial x} = \sum_m (p_x)_{nm} \psi_m(x) \tag{II}$$

(i = imaginäre Einheit, $K = h/2\pi$). Es ist $x_{nm} = x_{mn}$ reell, $(p_x)_{nm} = -(p_x)_{mn}$ rein imaginär. Mann kann nun unschwer zeigen, dass die so definierten Matrizen für x und p_x den Gleichungen der Göttinger Mechanik genügen. Nämlich

$$\mathbf{p}_x\mathbf{x} - \mathbf{x}\mathbf{p}_x = -iK, \quad \frac{1}{2m} \mathbf{p}_x^2 + \mathbf{E}_{\text{pot}}(x) = \mathbf{E} \quad \text{(Diagonalmatrix)}.$$

Aus der Multiplikationsregel folgt übrigens, dass die zu irgend einer Funktion $F(x)$ von x allein gehörige Matrix $\mathbf{F}$ gerade durch

$$F_{nm} = \int_{-\infty}^{\infty} F(x) \psi_n\psi_m\, dx$$

gegeben ist.

Ich will die Rechnungen hier nicht im Einzelnen anführen, Sie werden die Behauptung unschwer verifizieren können. Den Oszillator und Rotator habe ich nach Schrödinger durchgerechnet. Ferner folgen die Hönl–Kronigschen Formeln für die Intensität der Zeemankomponenten in einfacher Weise aus den Eigenschaften der Kugelfunktionen. Die Störungstheorie lässt sich vollständig übertragen, ebenso die bei Aufhebung von Entartung (mehrfache Eigenwerte) durch äussere Kraftfelder im allgemeinen notwendige Hauptachsentransformation. Momentan bin ich damit beschäftigt, aus den von Schrödinger berechneten Eigenfunktionen beim Wasserstoff die Uebergangswahrscheinlichkeiten auszurechnen. Für die Balmerlinien scheinen sich endliche rationale Ausdrücke zu ergeben. Beim kontinuierlichen Spektrum ist die Sache aber komplizierter; die exakte mathematische Formulierung ist mir da noch nicht ganz klar. Was übrigens Lanczos betrifft, so haben meine Ueberlegungen nur ganz wenige Berührungspunkte mit den seinen. Er betrachtet ein Problem, dessen Eigenwerte die reziproken Energiewerte sind, während hier die Energiewerte selbst die Eigenwerte sind. Ferner spielen bei ihm Funktionen, die wie die Greensche Funktion von zwei Stellen abhängen, eine wesentliche Rolle; derartige Funktionen werden hier nicht herangezogen. Ich glaube überhaupt, dass der Ansatz von Lanczos nicht viel taugt.

Ueber den physikalischen Sinn der Ausdrücke (I) und (II) weiss ich nur wenig. Jedenfalls scheint er mit der Auffassung der gewöhnlichen Lichtwellen als Differenzschwingungen (Schwebungen) der de Broglieschen Strahlung zusammenzuhängen. Dass in (I) und (II) zunächst keine unbestimmten Phasen auftreten, hat nur den äusserlichen Grund, dass beim Üebergang von (3) zu (4) der in der Zeit periodische Faktor fortgelassen wurde. Fügt man ihn hinzu, so erhält man in x_{nm} und $(p_x)_{nm}$ neben $\exp\left[(2\pi i/h)(E_n - E_m)t\right]$ einen Phasenfaktor $\exp\left[2\pi i(\delta_m - \delta_n)\right]$, worin δ_m und δ_n die Phasen der zu E_m und E_n gehörigen de Broglieschen Eigenschwingungen bedeuten. Sowohl in der Göttinger, als auch in der Schrödingerschen Formulierung des Quantenproblems wird grundsätzlich keine raumzeitliche Beschreibung der Bewegung des Elektrons im Atom gegeben. In der letzteren äussert sich das darin, dass es ausserhalb des Gültigkeitsbereiches der geometrischen Optik nicht mehr möglich ist, im de-Broglieschen Wellensystem 'Strahlen' zu konstruieren, die als Bahnen vom Teilchen gedeutet werden können. Das Problem des asymptotischen Ausschlusses an die gewöhnlichen raum-zeitlichen Bilder im Grenzfalle grosser Quantenzahlen sowie der Bestimmung der Zeitpunkte der Uebergangsprozesse bleibt nach wie vor ungelöst. Aber es ist doch ein Fortschritt, die Probleme jetzt von zwei verschiedenen Seiten aus zu sehen. Man sieht wohl auch, wie von Standpunkt der Quantenmechanik aus der Gegensatz zwischen bewegtem Punkt und Wellensystem zu Gunsten von etwas Allgemeinerem verblasst.

Nun seien Sie und die anderen Göttinger (besonders Born; falls er schon aus Amerika zurück ist, zeigen Sie ihm bitte, diesen Brief) herzlichst gegrüsst,

von

Ihrem

gez. W. Pauli

13

Early Years of Quantum Mechanics: Some Reminiscences*

Pascual Jordan

In 1925 Heisenberg published his famous article 'Über quantentheoretische Umdeutung kinematischer und mechanischer Beziehungen', giving the fundamentals of what later has been called quantum mechanics. For the participants of the Symposium on the Physicist's Conception of Nature, no emphasizing of the singular character of that article is necessary; it may be more interesting to say that the ideas of this overwhelmingly fruitful publication were in close accord with the endeavours of many physicists who in those years tried to develop from Bohr's correspondence principle consequences giving little bits of quantitative knowledge about special cases of exact laws of quantum physics. Van Vleck gave a derivation of Einstein's laws of the relation between the probabilities of spontaneous emission and positive and negative absorption. This result of Einstein's had been looked upon for a long time in a sceptical manner by Niels Bohr; now it was highly interesting to see, just how from Bohr's preferred way of thinking, a derivation of Einstein's law could be given. Born and I performed a simplified mathematical derivation of the results of Van Vleck. Our article on this topic did not contain anything new apart from our simpler form of calculation, but by studying this topic we both came to a more intimate understanding of Bohr's leading idea.

Dutch physicists succeeded in finding empirically simple mathematical rules concerning the spectral intensities of multiplet and Zeeman components. Several authors – let me mention only Hönl and Kronig – found that these rules too could be derived from classical relations concerning the Fourier components of the simplest classical model for an atom, showing the coupling of rotations and Larmor precession. The phrase '*verschärfte Anwendung des Korrespondenzprinzips*' thereby came into fashion. Bohr himself had originally made only extremely simple applications of his idea of correspondence; for instance he had shown how the selection rules for the quantum numbers m, l, j could be understood as confirmations of the correspondence principle. But now one learned how to derive in this manner also the intensity relations for such cases, where several finite intensities, unequal to zero, could be compared. In the famous example of the harmonic oscillator one could see that the transition $n \rightarrow n-1$ must have a probability exactly proportional to n as a consequence of Kuhn's 'sum rule' for intensities, derived from the dispersion formula of Kramers, which was introduced by Heisenberg as a meaningful part of his new theory.

---

* At the request of Jagdish Mehra, I take great pleasure in writing some of my reminiscences of the early years in the development of quantum mechanics, as they are still vivid in my memory.

J. Mehra (ed.), The Physicist's Conception of Nature, 294-299. All Rights Reserved
Copyright © 1973 by D. Reidel Publishing Company, Dordrecht-Holland

Naturally this dispersion formula, and its extension to the Raman effect, worked out by Heisenberg and Kramers, were impressive further examples of the beginning of the detection of exact quantum physical laws under the guidance of the correspondence principle. Heisenberg wrote his above-mentioned decisive article – so far as I remember – partly at Göttingen and partly at Copenhagen.[1] He informed Born about his work. At first, he intended to treat the intensity relations in the Balmer spectrum, in a manner similar to what Kronig and other authors had done so successfully concerning the intensity rules in multiplets. But this problem was too difficult for a direct attack; it found its solution a few years later, after Schrödinger's wave mechanical treatment of the hydrogen atom. Instead of solving this very special and very complicated problem Heisenberg came to formulating the general principles of quantum mechanics, a name which had been invented by Born, and applied by him in the heading of another article already before Heisenberg's article of 1925.[2]

It was Born too, who saw that what Heisenberg defined as the product of two quantum mechanical observables was the mathematically well-known product of two matrices. Born himself had remarked already a little earlier that such a 'multiplication' could be performed – he told me about this remark of his, but at that time I was unable to understand the fruitfulness of this idea.

But Heisenberg's article of 1925 was a real eye-opener. Born and I began to build up, from the base given by Heisenberg, a systematic quantum mechanics for system of only one degree of freedom. The possibility of performing the multiplication and addition of matrices, and to differentiate them as functions of time, allowed a quantum formulation of canonical equations of motion. It had to be proved that as a mathematical consequence of these and of the commutation relation $pq - qp = h/2\pi i$, conservation of energy resulted, and the equality of the frequencies, multiplied by h, gave the pair-differences of the energy eigen values.

A somewhat hectic effort was needed in order to attain these results also for a quite general case. I remember that Born and I conducted a lively correspondence also during some holiday weeks and Born's stay in a sanatorium. However, the physician prevented Born from the continuation of this task which he thought to be dangerous for his health. Born had found that the above mentioned sum-rule of Kuhn means that in the difference $pq - qp$, the diagonal matrix elements equal $h/2\pi i$, and we found further that all the other elements of the matrix $pq - qp$ vanish.

Our correspondence with Heisenberg, who still remained at Copenhagen, led to the writing of a 'Dreimänner Arbeit', Born–Heisenberg–Jordan, containing the quantum mechanics of systems of any number of degrees of freedom together with a preliminary study of the angular momentum, from which we got our first results about the representations of the three-dimensional group of rotations. A decisive contribution, made by Heisenberg, was the formulation of the general canonical commutation relations:

$$[p_R, p_e] = [q_R, q_e] = 0 ;$$

$$[p_R, q_e] = \delta_{Re} \frac{h}{2\pi i}.$$

The most interesting special application of the quantum mechanics of systems of several degrees of freedom was made by Pauli who succeeded in deriving the Balmer formula from the new theory – still before Schrödinger, who by means of his wave mechanics found a new, second way to the exact laws of quantum theory.

Another piece in the 'Dreimänner Arbeit' gave a result, which I myself have been quite proud of: It was possible to show that the laws of fluctuations in a field of waves, from which Einstein derived the justification of the concept of corpuscular light quanta, can be understood also as consequences of an application of quantum mechanics to the wave field.

Soon after publication of the 'Dreimänner Arbeit', I saw a new article in the *Proceedings of the Royal Society*, containing a great part of what had been put forward in the article of Born and Jordan, and in the 'Dreimänner Arbeit'. The author of this new publication had been known to me only from one former article, treating in a wonderful manner the statistics of collision processes. His name was Dirac. Dirac, in other publications, made further astonishing contributions, especially a wonderful work deriving fully the formulas of Kronig concerning the intensity relations in multiplets.

The next point which I would like to mention is the publication of Schrödinger which gave the whole matter a new turn, showing the meaningful relation of de Broglie's idea to the problems of quantized atoms. Schrödinger, as one knows, in the first of a series of famous publications showed how the hydrogen atom was to be treated according to his new wave mechanical method. He himself, and at the same time Pauli, could show that the two methods of treatment for a quantum mechanical system are equivalent in their results, and how the Heisenberg matrices of any system could be calculated by a simple integration if one already knows the Schrödinger eigenfunctions of the same system. Pauli used this result in order to calculate from Schrödinger's eigenfunctions of the hydrogen atom the intensities in the Balmer spectrum, which Heisenberg had tried to determine directly from correspondence considerations, still before the systematic formulation of quantum mechanics or wave mechanics.

After Schrödinger's beautiful papers, I formulated what I liked to call the statistical transformation theory of quantum mechanical systems, answering generally the question concerning the probability of finding by measurement of the observable b the eigenvalue b', if a former measurement of another observable a had given the eigenvalue a'. The same answer in the same generality was developed, in a wonderful manner, by *Dirac*.

I have been extremely thankful to Dirac in another connection. My idea that the solution of the vexing problem of Einstein's light quanta might be given by applying quantum mechanics to the Maxwell field itself, aroused the doubt, scepticism, and criticism of several good friends. But one day when I visited Born, he was reading a new publication of Dirac, and he said: 'Look here, what Mr Dirac does now. He assumes the eigenfunctions of a particle to be non-commutative observables again.' I said: 'Naturally.' And Born said: 'How can you say "naturally"?' I said: 'Yes, that is, as I have asserted repeatedly, the method which leads from the one-particle problem to the many-body problem in the case of Bose statistics.'

Later on we enjoyed at Göttingen an extended visit by Dirac, during a time in which Oppenheimer was also there with us, and it was a highly interesting time. My friend Oppenheimer, by a short remark of his, also encouraged my conviction that it must be possible somehow to formulate the many-particle problem for the case of fermions instead of bosons by taking the wave function of the one-particle problem as a non-commutative observable. I tried to find this solution when I was at Copenhagen, but Dirac saw that my solution contained an error. I went back to Göttingen and talked with Eugene Wigner, and in a very short time we found the full solution of this problem.

It is a very pleasant memory of mine that Dirac had also the great amiability to write to me a small but meaningful letter in his nice handwriting when he had found the relativistic wave equation of the electron.

Just after Schrödinger's great discoveries on the topic of wave mechanics further rapid progress became possible, especially because application of wave mechanics or quantum mechanics to new examples of mechanical systems were now easy to perform. Born, who, several years before quantum mechanics had been busy in winning from the methods of astronomical perturbation theory new help for the mathematical attack against the problems of atomic models, now used similar methods in the questions of the wave mechanical approach to collision processes. In this connection, he developed his statistical interpretation of the Schrödinger field, in opposition to Schrödinger himself who was inclined to hope that wave mechanics might give a possibility of reducing all quantum effects to a purely classical function of waves. This statistical interpretation of the Schrödinger field given by Born was a great step in the direction taken later on by Dirac and by myself in what I have called above the statistical transformation theory of quantum mechanics.

Already in the 'Dreimänner Arbeit', Born–Heisenberg–Jordan, we saw that the eigenvalues of a component of the rotational impulse might be whole numbers or possibly half numbers $\pm 1/2$, $\pm 3/2$, ... It was from a study of Lucy Mensing that we learned, that from the model of the electron as a mass point only whole numbers 0, ± 1, ± 2 and so on could arise. Indeed Pauli came to this result also when he studied the hydrogen atom with the method of quantum mechanics. Therefore a new idea was needed in order to describe the concept of the spin of the electron – as concluded at that time from spectroscopic evidence by Uhlenbeck and Goudsmit. The English physicist Darwin found a possibility of formulating this concept by a modification of the Schrödinger equation. He showed that the introduction of a wave amplitude with two components was the necessary step. Pauli performed a translation of this theory of Darwin into the method of pure quantum mechanics, without explicitly using any wave function. And Heisenberg showed that the last, still existing, enigmas of the Zeeman effect could now be removed. Later on, Dirac made his new sensational progress by taking a wave function with even four components. New difficulties seemed to arise, but as we all know Dirac found a solution leading to the theoretical discovery of the positron, the first discovered example of antimatter.

Heisenberg in the meantime gave, by making an explicit use of the Schrödinger eigenfunctions, an explanation of the phenomenon of para-helium and ortho-helium. Dirac found the possibility of expressing Pauli's famous exclusion principle as the principle of antisymmetry of eigenfunctions in a system containing several electrons. But further difficult mathematical problems remained to be solved before a full understanding of the spectroscopic consequences of the Pauli exclusion principle became possible. This was the problem through which Eugene Wigner entered quantum mechanics, making use of I. Schur's theory of group representations.[3]

J. von Neumann, the famous mathematician, was also interested in quantum theory, and from him we learned to see in the mathematical conception of the Hilbert space the natural gackground of quantum mathematics. It may be mentioned that already shortly after the publication of the 'Dreimänner Arbeit', Born for one (or for half a) year went to the U.S.A. and together with the mathematician N. Wiener he wrote an article introducing the concept of operators into quantum mechanics, emphasizing in this manner the usefulness of a branch of mathematics at that time still scarcely known to the majority of physicists.

The famous mathematician Hilbert at Göttingen, who in former years had made fundamental contributions to Einstein's theory of general relativity, was also highly interested in quantum theory. Together with M. Born and the famous experimental physicist J. Franck, one of the liveliest and most competent admirers of Niels Bohr's ideas, he organized the seminar on the structure of matter. Apart from Schrödinger, nearly every essential contributor to quantum theory during those years delivered some lectures in that Seminar. Hilbert especially admired the fact that Schrödinger, with his wave equation of the hydrogen atom, had found a very simple example of an eigenvalue problem showing on the one hand discrete eigenvalues, and on the other hand a continuous interval of eigenvalues. Hilbert himself had known from his fundamental researches on integral equations and the quadratic forms of infinitely many variables that such occurrences must exist, but no mathematician had been able to find a simple example of this kind. Only Nature itself, as studied by the quantum physicists, held in store simple examples of this kind, at that time not yet detected mathematical fantasy. But it turned out that many systems studied already mathematically allow application to the most simple and the most fundamental problems of wave mechanics and quantum mechanics. Richard Courant's famous book about the mathematics of eigenvalue problems became one of the most valuable instruments for the work of quantum theoretical physicists, for it informed thoroughly about all those polynomes of Hermite and of Laguerre, and so on which now showed their applicability to physics. Oppenheimer showed his admirable ability of using difficult mathematics which became physical actuality.

In those eventful years it became possible also to understand, as consequences of quantum mechanics or wave mechanics, the basic chemical forces, formulated in an empirical manner by the concept of valence. Heitler and London were the leading figures in this endeavour. Only a few problems of atoms and molecules, such as superconductivity, remained unsolved still for later years.

In afterthoughts one could see that the inner logic of quantum laws is so stringent and so strikingly impressive that one should have been able to deduce essential parts of them as soon as one had found only a few of them. For instance, Duane had put forward an explanation of the interference properties of corpuscular light quanta by the remark, that the translation of an infinite spectroscopic lattice orthogonal to its lines has the character of a periodic motion, and therefore suggests application of the quantum rule, leading to the consequence that the translational impulse of the lattice can change only by quantum jumps. Taking this consequence seriously Duane would have been able to deduce already that, for instance, an *electron* colliding with the lattice must also show interference properties, wave properties, exactly the properties of its de Broglie wave.

Though the idea of application of quantum mechanics to wave fields met for several years with earnest doubts, as mentioned above, two further steps could be made during the few years after the 'Dreimänner Arbeit'. O. Klein and I found that Dirac's method of describing a bosonic many-particle system by taking the Schrödinger field of the single particle to be a field of quantized waves can be extended also to the case of interactions between the particles (for instance Coulomb interaction). W. Pauli and I showed how quantization of the Maxwell field (at least in the vacuum case) can be made to be Lorentz invariant. Together with the possibility already mentioned (Wigner–Jordan) of treating in a similar manner also the case of fermions instead of bosons, these steps allowed Heisenberg and Pauli to try to develop, with highly interesting new ideas, a complete quantum electrodynamics. As is well known, new difficulties arose. Further admirable investigations of different authors on the topic of 'renormalization' allowed one to remove, at least partially, these difficulties; but fundamental new ideas concerning the reduction of the manifold of all existent elementary particles to a unified basic law seem perhaps to favour the hope that all the difficulties met with in this wide field of theoretical investigation and speculation may lead to a still deeper understanding.

EDITOR'S NOTES

1. Heisenberg wrote his famous paper on the 'quantum-theoretical reinterpretation of kinematic and mechanical relations', to which Jordan has referred, entirely in Göttingen on his return from a vacation on the rocky island of Helgoland. This paper was published as Z. Phys. **33**, 879–893 (1925), and was signed from 'Göttingen, Institut für Theoretische Physik'.
2. Max Born's paper on 'quantum mechanics' was received on 13 June 1924, and published as Z. Phys. **26**, 379–395 (1924).
3. Wigner was the first physicist to introduce group theory into quantum mechanics. He saw the group-theoretical implications of the new quantum theory from its very beginnings. In his profound papers he contributed to the deeper understanding of Schrödinger's discovery by adding to it the group-theoretical viewpoint: see E. Wigner, Z. Phys. **40**, 492 (1926); **40**, 883 (1927); **43**, 624 (1927). Weyl delivered lectures in Zurich on related subjects during the academic year 1927–28; these were later expanded into his book.
    [See also H. Weyl, 'Quantenmechanik und Gruppentheorie', Z. Phys. **46**, 1–46 (1927).]

14

The Mathematical Structure
of Elementary Quantum Mechanics

Josef M. Jauch

INTRODUCTION

The discovery of quantum mechanics is one of the most exciting incidents in the intellectual history of man. It marks the synthesis of contributions from many brilliant men who assembled different facets into a unified and coherent theory. The result is more than the sum total of their individual contributions. It is also a new paradigm, a programme which has been the starting point of innumerable extensions into all areas of physics.

It was Dirac who more than anyone else perceived, as if in a vision, the mathematical structure of the new theory. It shall be my endeavour in this lecture to reveal some of the developments and flowering of the ideas which Dirac originated during the heroic times of quantum theory. I shall deal with elementary non-relativistic quantum mechanics only.

1. THE CANONICAL COMMUTATION RULES

One of the most beautiful results that emerged after the birth of quantum mechanics was the close formal analogy of the quantum dynamics with its classical counterpart. This analogy furnished the efficient and concise expression of Bohr's *correspondence principle*.

None other was more successful in revealing this hidden correspondence than Dirac who, by a stroke of genius, revealed the formal analogy between commutators of quantal observables and Poisson brackets for their classical analogues. Dirac expressed this analogy in the following manner:

If $F(q, p)$ and $G(q, p)$ are two classical observables considered as functions of the canonical variables q and p, then their Poisson bracket

$$\{F, G\} = \frac{\partial F}{\partial q}\frac{\partial G}{\partial p} - \frac{\partial F}{\partial p}\frac{\partial G}{\partial q} \tag{1}$$

corresponds to the commutator of their quantum analogues

$$[F, G] = FG - GF \tag{2}$$

J. Mehra (ed.), The Physicist's Conception of Nature, 300–319. All Rights Reserved
Copyright © 1973 by D. Reidel Publishing Company, Dordrecht-Holland

as indicated by the formula[1]

$$i\{F, G\} \leftrightarrow [F, G]. \tag{3}$$

In particular for the canonically conjugate variables P and Q one obtains in quantum mechanics the canonical commutation rules[2]

$$[Q, P] = i \cdot I. \tag{4}$$

This formal analogy suggests further that one may develop a transformation theory whereby the unitary transformations in quantum mechanics become the analogue of the contact transformations in classical mechanics[3].

Thus, if Q^*, P^* is another pair of canonical operators, functions of Q, P satisfying also

$$[Q^*, P^*] = i \cdot I,$$

then the hope was that every such transformation could be generated by a unitary transformation S $(S^* = S^{-1})$ such that

$$Q^* = SQS^{-1}, \qquad P^* = SPS^{-1}. \tag{5}$$

For this to be true one should be able to establish a uniqueness theorem which says essentially that every irreducible representation of the relation (4) is unitarily equivalent to every other one.

That such a theorem might be true was reinforced when Stone and von Neumann[4] proved such a theorem for a closely related but not identical problem:

Let $U_\alpha = e^{i\alpha P}$, then by formal manipulation of power series one proves easily

$$U_\alpha V_\beta = e^{i\alpha\beta} V_\beta U_\alpha. \tag{6}$$

These are the Weyl[5] form of the canonical commutations. Although seemingly less symmetrical than (4) they have the great advantage of introducing only unitary and hence bounded operators.

The first solutions of the canonical commutation rules (4) were given in matrix form[6], but it became clear soon after Schrödinger[7] had succeeded in establishing the equivalence of his wave-mechanics with the matrix-mechanics, that the problem should be considered as one concerning the theory of operators in a Hilbert space rather than matrices.[8]

It is then readily seen that there is a seemingly technical but nevertheless important difference between the two commutation rules (4) and (6). The latter involves only unitary, hence bounded operators, while it can be proved that at least one of the two operators P and Q must be unbounded[9].

Now an unbounded symmetrical operator can never be defined on the entire Hilbert space $\mathcal{H}$. It is at most defined on a dense linear subset D of $\mathcal{H}$. This fact makes an enormous difference in the representation problem so that for instance there exist actually infinitely many inequivalent representations of the commutation rules (4)[10].

The mathematical problem of the commutation rules splits therefore into two different ones each of which must be investigated with very different methods.

The commutation rules in the form (4) were studied from the mathematical point of view by several mathematicians. The problem reduces itself essentially to the question to find an additional property, such that when this property is satisfied the representation is unique up to unitary equivalence. Such additional properties were given by Rellich[11], Dixmier[12], Foiaş, Gehér, Sz.Nagy[13], Kilpi[14] and Tillmann[15]. Unfortunately most of these conditions fail to have any simple and obvious physical interpretation.

An important concept which replaces the notion of the bounded and hence continuous operators is that of *closedness*.

An operator A with dense domain $D \subset \mathcal{H}$ is called *closed* if the convergence of the two sequences $f_n \in D$ and $g_n = Af_n$ to the limits f and g implies $f \in D$ and $Af = g$.

An operator A with dense domain D is *symmetric* if for any pair of vectors $f, g \in D$ $(f, Ag) = (Af, g)$.

A symmetric operator can always be extended to a closed symmetric operator. But such an operator is not necessarily self-adjoint. This is only the case if the domain D_A of the operator A and the domain D_{A^*} of the operator A^* are the same. In that case we can write $A = A^*$, while in general for symmetric operators one has only $A \subset A^*$, that is A^* is an *extension* of A.

Von Neumann discovered that there exist closed symmetric operators which have no self-adjoint extension[16]. He also showed that the symmetric operators can be characterized by two positive integers (n_1, n_2) called the *defect indices*. An operator can be extended to self-adjoint ones if and only if $n_1 = n_2$.

Furthermore for a symmetric operator with defect indices $n_1 = n_2 = 0$ the extension is unique. In that case the operator is called *essentially self-adjoint*.

Symmetric operators with defect indices $\neq 0$ are by no means particularly freakish mathematical objects. One encounters them quite readily in actual physical problems. Thus, for instance in a spherically symmetrical problem the radial momentum operator, that is the operator canonically conjugate to the multiplication operator on the half line, has defect indices $(1, 0)$. Hence there exists no self-adjoint extension of it in the Hilbert space of physical state vectors. Similarly the symmetric operator which is canonically conjugate to the angle variable around a fixed axis in space has defect indices $n_1 = n_2 = 1$. Hence it admits self-adjoint extensions but they are not unique. One might have hoped that the canonical commutation rules are unique (that is up to unitary equivalence) if they are satisfied on a domain D on which they are essentially self-adjoint. However, it has not been possible to prove such a theorem. The closest to a physically interpretable theorem is Rellich and Dimier's which says that if (4) is true on a dense domain D and $P^2 + Q^2$ is essentially self-adjoint on this domain then the irreducible representation of P and Q is essentially unique.

These results can all be extended in a trivial manner to finite sets of canonical operators. In spite of this, the representation theory of the unbounded operator is even today still incomplete.

In contrast to this, the theory of the Weyl commutation rules can be considered as complete with the simplest possible imaginable result: There exists essentially only one irreducible representation of the commutation rules (6). These rules then define two self-adjoint operators P and Q defined as the generators of the associated one-parameter groups. Some corollaries of this result are the following:

(1) The spectra of both P and Q are the same and equal to $\mathbf{R}$, the set of all real numbers.

(2) P and Q are unitarily equivalent. In the Schrödinger representation the unitary transformation which establishes the equivalence is the Fourier transformation.

(3) The spectra of both operators are absolutely continuous.

(4) Unitary transformations leave these properties unchanged.

There is only one flaw in these results, one which would be noticed more readily by a physicist than by a mathematician. There is mathematically a complete symmetry between the two operators P and Q. Physically, on the other hand, the position operator Q plays an entirely different role from the momentum operator P. They are measured with different instruments and there is no obvious correlation between them.

It is therefore of interest to note that a less symmetrical formulation of the commutation rules offers a better physical interpretation and leads to a mathematical theorem of far reaching consequences.

To this end we consider the spectral resolution of the self-adjoint operator Q. It consists of a family of projection operators E_A called the *spectral measure* associated with the Borel sets $\Delta \in \mathfrak{B}(\mathbf{R})$ over the real numbers. The particular set $E_\lambda = E_{(-\infty, \lambda]}$ defines the spectral distribution function. It permits the definition of the functional calculus

$$u(Q) = \int_{-\infty}^{+\infty} u(\lambda)\, dE_\lambda$$

provided $u(\lambda)$ is measurable with respect to the spectral measure E_A. In particular

$$Q = \int_{-\infty}^{+\infty} \lambda\, dE_\lambda$$

and

$$V_\beta = e^{i\beta Q} = \int_{-\infty}^{+\infty} e^{i\lambda\beta}\, dE_\lambda.$$

From (6) one obtains by an elementary calculation

$$U_\alpha^{-1} E_\lambda U_\alpha = E_{\lambda+\alpha}$$

or more generally

$$U_\alpha^{-1} E_\Delta U_\alpha = E_{[\Delta]\alpha}$$

where $[\Delta]\alpha = \{q \mid q - \alpha \in \Delta\}$ is the set translated by the amount Δ to the right (for $\alpha > 0$).

We have therefore the following situation: There is a space $M = \mathbf{R}$ and a group G of operations α on M, which in this case are the translations by the amount α. We have furthermore a spectral measure E_Δ based on the Borel sets of M and a representation of the group G by unitary operators U_α such that

$$U_\alpha^{-1} E_\Delta U_\alpha = E_{[\Delta]\alpha}.$$

Although this formulation seems quite different at first, it is completely equivalent to (6).

The theorem of Stone and von Neumann says then that such a system of operators E_Δ, U_α is unique if it is irreducible.

There are two major advantages of formulating the canonical commutation rules in this manner. In the first place, they have a physical interpretation which illuminates a new physical aspect, that of homogeneity of space. In this interpretation, the projection operators E_Δ are the observables which locate the particle in the set Δ. The homogeneity of physical space is then expressed by the fact that displacements of these observables by a fixed amount α generate symmetry transformations. Secondly‘ this form is mathematically a special one of a profound mathematical theorem of great generality which has become the main tool for the representation theory of locally compact groups.

This theorem enables us in fact to construct all the Euclidean invariant localizable systems (particles) and all the irreducible representations of the Galilei and the Poincaré group (dynamics of one-particle systems). It also makes it possible to extend quantum mechanics to non-Euclidean spaces, something that is needed as soon as one wants to treat problems which involve constraints such as the rigid rotator, for example.

This generalization of the Stone–von Neumann theorem is the fundamental *imprimitivity theorem* of G. Mackey [17], which in a sense is the most important theorem for the elucidation of the mathematical structure of elementary quantum mechanics.

The theorem is rather long to express in the most general form, however its understanding is greatly facilitated if it is seen as a generalization of the Stone–von Neumann theorem quoted above.

It involves the following concepts: a homogeneous space M, a locally compact group G, a representation space $\mathscr{H}$, a unitary representation U_x of G, the little group G_0 and its representation L and finally the induced representation U_x^L of G.

For the canonical commutation rules M is $\mathbf{R}$, G the additive group of reals and $\mathscr{H}$ the Hilbert space, U_x is the operator U_α, the little group G_0 is the identity and U_x^L is $U_x = U_\alpha$.

Let us proceed to the definition of these concepts. The homogeneous space M is a locally compact topological space with a locally compact transformation group G which acts transitively and continuously on M. This means there exists a function from $M \times G$ to M denoted by $[q]x$ ($q \in M$, $x \in G$) with the following properties:

(*a*) For each fixed $x \in G$ the function $q \mapsto [q]x$ is a bijective continuous map of M into itself.

(b) $[[q]x_1]x_2 = [q]x_1x_2$ for all $x_1, x_2 \in G$ and all $q \in M$.

(c) $[q]e = q$.

(d) If $q_1, q_2 \in M$, then there exists an $x \in G$ such that $[q_1]x = q_2$ (Transitivity).

Next we assume that M is a measure space so that there exists a field $\mathscr{S}$ of measurable subsets of M. Furthermore, we assume that there exists a projection valued measure E_A on $\mathscr{S}$ with the properties

(a) $E_{\mathscr{S}} = 0$, $E_M = I$.

(b) for any disjoint sequence A_i ($A_i \perp A_k$ for $i \neq k$) $E_{\cup A_i} = \sum E_{A_i}$.

(c) $E_{A'} = I - E_A \equiv E_A'$

The projections E_A are linear operators in a Hilbert space $\mathscr{H}$. That same space is supposed to be the carrier of a unitary representation U_x of the group G, so that for all $x_1, x_2 \in G$, $U_{x_1} U_{x_2} = U_{x_1 x_2}$.

Finally the two structures E_A and G_x are linked together by the fundamental relation

$$U_x^{-1} E_A U_x = E_{[A]x}$$

where

$$[A]x = \{q| [q]x^{-1} \in A\}.$$

The pair $\{U, E\}$ consisting of the representation U_x and the projection valued measure E_A is called a *transitive system of imprimitivities* based on the space M.

There is an obvious way of constructing such systems. We begin with the simplest case. Suppose that there is defined a positive measure μ on the measurable sets $\mathscr{S}$ with the property that it is *quasi-invariant*. This means, for any $x \in G$ the measure μ_x defined by

$$\mu_x(A) = \mu([A]x)$$

has the same sets of measure zero as the measure μ. There exists then an essentially positive function, the Radon–Nikodym derivative written $d\mu_x/d\mu$ such that

$$\mu_x(A) = \int_A \frac{d\mu_x}{d\mu} \, d\mu.$$

We define the Hilbert space $\mathscr{H}$ by setting $\mathscr{H} = L^2(M, \mu)$ consisting of all complex valued square integrable functions $f(q)$ on M. The transformation U_x, defined by

$$(U_x f)(q) = \sqrt{\frac{d\mu_x}{d\mu}} \, f([q]x)$$

furnishes then a unitary representation of the group G. If we define the projection valued measure E_A by

$$(E_A f)(q) = 1_A(q) f(q)$$

where

$$1_A(q) = \begin{cases} 1 & \text{for } q \in A \\ 0 & \text{for } q \notin A \end{cases}$$

then the couple $\{U, E\}$ is the *canonical system* of *imprimitivities*. One easily verifies in fact that

$$(U_x^{-1} E_\Delta U_x f)(q) = (E_{[\Delta]x} f)(q).$$

This procedure can be generalized as follows:

Consider an arbitrary but fixed point $q_0 \in M$. The set of all $\xi \in G$ such that $[q_0]\xi = q_0$ is then a subgroup $G_0 \subset G$. It is called the *little group*. If q is an arbitrary point in M then it follows from $[q_0]x_1 = [q_0]x_2 = q$ that the points $q \in M$ are in one-to-one correspondence with the right cosets of the little group G_0. We may therefore identify the space M with the right cosets G/G_0 of G.

Suppose we are given an irreducible representation $\xi \to L_\xi$ of G_0 by unitary operators L_ξ in a Hilbert space $\mathscr{H}_0$ with scalar product $(\,,\,)_0$. We can then construct a new Hilbert space $\mathscr{H}^L$ in the following way. The elements of $\mathscr{H}^L$ are functions $f(x)$ on the group G with values in $\mathscr{H}_0$ satisfying the following conditions:

(a) For all $x \in G$ and $\xi \in G_0$ one has $f(\xi x) = L_\xi f(x)$,

(b) The function $\|f(x)\|_0^2$ is a measurable function on the space $M = G/G_0$,

$$\int_M \|f(x)\|_0^2 \, d\mu < \infty.$$

In condition (b) we have used the fact that if x_1 and x_2 are in the same right cosets then $x_1 = \xi x_2$ for some $\xi \in G_0$ and hence

$$\|f(x_1)\|_0^2 = \|f(\xi x_2)\|_0^2 = \|(L_\xi f)(x_2)\|_0^2 = \|f(x_2)\|_0^2.$$

The set of functions with these properties define the Hilbert space $\mathscr{H}^L$. In this space the unitary representation U_x^L of G is defined by

$$(U_x^L f)(y) = f(yx) \sqrt{\frac{d\mu_x}{d\mu}}.$$

It is called the representation *induced* by L.

Just as we did before, we can define a canonical system of imprimitivities by setting

$$(E_\Delta^L f)(x) = 1_\Delta(q) f(x)$$

and verify that this is a projection valued measure which satisfies

$$(U_x^L)^{-1} E_\Delta^L U_x^L = E_{[\Delta]x}^L.$$

We are now ready for the formulation of the imprimitivity theorem:

If $\Delta \to E_\Delta$, $x \to U_x$ is a transitive system of imprimitivities in a Hilbert space $\mathscr{H}$, then there exists an irreducible unitary representation L of the little group G_0 and a unitary map Ω which maps $\mathscr{H}$ onto $\mathscr{H}^L$ such that the image system of imprimitivities under this mapping is the canonical one.

The applications of this theorem in quantum mechanics are numerous. One may say that it is the theorem which replaces the role of canonical transformation theory of classical kinematics.

The theorem was applied by Wigner [18] in 1939, before it was proved, for the construction of all the unitary representations of the Poincaré group. Another application of the same theorem by Wightman [19] elucidated completely the notion of localizability for relativistic particles. This problem was actually raised and solved before by Newton and Wigner [20] with more elementary but less rigorous methods. The conclusion of this analysis is that all particles with restmass $m \neq 0$ are localizable. But systems which belong to representations with $m = 0$ and helicity $s \neq 0$ are not localizable.

This corresponds exactly to the known result that in the field theory for such such systems it is impossible to define a positive definite probability density [21].

It was later shown by Jauch, Piron [22], and Amrein [23] that the systems with $m = 0$ and $s > 0$ (this includes the Weyl neutrino and the photon, for example) are still localizable in a weaker sense. This sense can be made precise by changing the additivity of the spectral measure into subadditivity, so that for any set Δ one has in general only

$$E_\Delta + E_{\Delta'} < I.$$

The representation theory of such systems can again be solved completely by a combination of the imprimitivity theorem with the theorem by Neumark and Sz. Nagy [24] on positive operator-valued measures.

Another application of the same theorem enabled G. Emch [25] to construct all the representations of the Poincaré group in a quaternionic Hilbert space. The conclusion of this work was that from a quaternionic quantum mechanics one could not expect any new physical properties for relativistic localizable systems, than these obtained from the complex Hilbert space.

The validity of this method depends on the local compactness of M. As long as the system is finite-dimensional, M is locally compact and there is no problem. In fact, the formalism and the result can be taken over by a suitable and obvious generalization of the notation.

Thus, for instance, the Weyl commutation rules for an n-dimensional system could be written in the form

$$U_\alpha V_\beta = e^{i\alpha \cdot \beta} V_\beta U_\alpha \tag{7}$$

provided

$$\alpha = \{\alpha_1\, \alpha_2, ..., \alpha_n\}, \qquad \beta = \{\beta_1, \beta_2, ..., \beta_n\}$$

and

$$\alpha \cdot \beta = \alpha_1 \beta_1 + \alpha_2 \beta_2 + \cdots + \alpha_n \beta_n.$$

In field theory and in quantal systems in the thermodynamic limit one has to drop the condition $n < \infty$. The space M is then no longer locally compact and we are led to consider commutation rules of the kind

$$U(f) V(g) = e^{i(f, g)} V(g) U(f), \tag{8}$$

where

$$U(f) = e^{iP(f)}, \qquad V(g) = e^{iQ(g)}.$$

P, Q are now operator valued distribution on some real function space and (f, g) is the scalar product.

Unfortunately the imprimitivity theorem is no longer valid in this case and this is one of the main causes of the mathematical difficulties in the quantum theory of fields.

2. CONTINUOUS SPECTRA AND SCATTERING THEORY

Although quantum mechanics was first conceived as a matrix mechanics with a view to describe the stationary states of atomic systems and the radiative transitions between them, the need for a treatment of aperiodic, hence non-stationary, states was recognized from the very beginning. Some indication how such a theory would look like was already given at the end of the famous paper by Heisenberg, Born and Jordan[26].

The typical and most important case of aperiodic systems would be the scattering systems where incoming and outgoing particles evolve with a simple motion. Only during a short scattering time does their motion deviate from the simple one in a significant manner.

The first treatments of quantum scattering theory were based on Schrödinger's wave mechanics and they were strongly influenced by the problems of classical wave-scattering which existed in the well developed form of Rayleigh's theory of sound. Dirac was the first who showed that scattering theory could be given a more abstract form by treating the problem in momentum space[27]. However, this was not yet a general scattering theory and the problem resisted a rigorous and general treatment for a considerable time. The success of certain approximations, such as the Born approximation and the WKB-method, was very encouraging evidence that it should be possible to develop a complete and general scattering theory.

An important step in this direction was made by Heisenberg[28] in 1942 when he developed the theory of the S-matrix, thereby generalizing a concept which was introduced before by Wheeler[29] in connection with the theory of nuclear reactions. The motivation in Heisenberg's theory was similar to the one which led him to the conception of matrix mechanics. By eliminating from the theory concepts which had no relation to empirical facts he had hoped to develop a theory of the asymptotic properties of interacting elementary particles without relying on a detailed description of the evolution of the state at all times. This detailed description should be replaced by general physically motivated requirements on the S-matrix such as symmetries, unitarity, conservation laws and analyticity. The ultimate objective of such a theory is limited to the correlation of observable facts by S-matrix postulates.

While this programme, also called pure S-matrix theory, has not yet led to many tangible results, the definition of and the emphasis by Heisenberg on the S-matrix as the central object of any quantal scattering theory has greatly stimulated the subsequent research and led to an important mathematical theory.

In the idealized and most elementary form, the essential points of this scattering theory derive from a precise definition of the asymptotic behaviour of a scattering system. This behaviour is given by two unitary groups $U_t = e^{-i\mathring{H}t}$ and $V_t = e^{-iHt}$ describing respectively the free and the true evolution of the system. The first one is always known and has in fact a very simple form. The second is unknown but its properties are only needed in the asymptotic region $t \to \pm\infty$ in order to develop the S-matrix theory.

The asymptotic condition expresses the fact that in the remote past and the distant future these two evolutions are physically indistinguishable. The simplest and most natural way to express this in a mathematically rigorous fashion was to assume that $V_t^* U_t$ approaches a limit as $t \to \pm\infty$:

$$s\text{-}\lim_{t \to \pm\infty} V_t^* U_t = \Omega_\mp . \tag{9}$$

This led immediately to a precise definition of the so-called wave operators previously introduced by C. Møller[30]. The scattering operator S is then defined by

$$S = \Omega_-^* \Omega_+ . \tag{10}$$

The operators $\Omega_\pm$ are isometries. Each satisfies the relation $\Omega^*\Omega = I$. From this follows that $\Omega_+\Omega_+^* = F_+$ and $\Omega_-\Omega_-^* = F_-$ are projection operators. Furthermore the Møller operators satisfy the *intertwining relation*

$$\mathring{H}\Omega = \Omega H$$

or

$$V_t\Omega = \Omega U_t . \tag{11}$$

From these relations one obtains easily the conservation of energy in scattering processes

$$[\mathring{H}, S] = 0 .$$

In this abstract form of scattering theory the unitarity of the S operator cannot be established from (9) alone. An attempt of reconstructing an explicit form of such a proof on the basis of Dirac's and Heisenberg's explicit theory failed because, upon closer examination, it was found that the classical proofs of the unitarity of S[31] were based on a hidden assumption, which, when translated into the abstract version takes on the form

$$F_+ = F_- . \tag{12}$$

For a long time, this rather innocent looking hypothesis (which has a physical interpretation) was thought to be a consequence of the asymptotic condition, but a proof was not possible and was left as an unsolved problem until Kato[32] showed by a counterexample that condition (12) was independent of (9). Once (12) is assumed the proof of the unitarity of S is trivial (we write $F = F_+ = F_-$):

$$S^*S = \Omega_+^*\Omega_-\Omega_-^*\Omega_+ = \Omega_+^* F\Omega_+ = \Omega_+^*\Omega_+ = I$$
$$SS^* = \Omega_-^*\Omega_+\Omega_+^*\Omega_- = \Omega_-^* F\Omega_- = \Omega_-^*\Omega_- = I .$$

The superiority of the abstract scattering theory was clearly evident by the extreme simplicity of its basic hypotheses and the explicitness of the logical structure built on them. Its extension to the multichannel theory was accomplished in subsequent publications [33].

There were two limitations to this form of the scattering theory:

(1) It is a time-dependent theory. While this form is closer to the actual physical processes taking place in nature when a scattering experiment is performed, it does not conform to the practice of physicists to carry out their calculations in a stationary formalism. In fact, calculations are rather difficult in the time dependent form of the theory.

(2) The theory is not applicable to systems with long range or persistent interactions. For instance the Coulomb potential does not satisfy the asymptotic condition (9) [34].

Both of these limitations have been largely overcome. The relationship between the time-dependent and the stationary scattering theory was elucidated in a series of fundamental papers by Kato and Kuroda [35]. Another method for establishing the relation between the two forms of the theory was developed independently by three different groups [36, 37, 38]. It is based on an idea of Galindo [39] and uses a new mathematical tool of great power and simplicity, the Spectral Integral.

This is a generalization of the concept of the classical integral by extending it to operator valued functions with operator valued measures. With the help of this integral, it has been possible to establish a kind of dictionary between the formal and mathematically heuristic Lippman–Schwinger theory [40] and the correct stationary state theory, thereby justifying many of the results of this formalism and eliminating the mathematical inconsistencies which it contains. In particular one of the main results of this theory, contained in the heuristic formula

$$(S - I)_\lambda = 2\pi i \, (V - V R_{\lambda + i\varepsilon} V)_\lambda, \quad R_z = (H - z)^{-1}$$

(The index λ means the operator on the energy shell λ) could be reformulated and rigorously established as a consequence of the asymptotic condition [36].

Another method of stationary scattering theory, very popular with the physicists, is the eigenfunction expansion, particularly convenient for establishing the analytical properties of the S-matrix. Here, too, considerable progress was obtained especially through the work of Ikebe, Regge, Reijto, Newton and others [41]. In particular it was possible to establish conditions under which this expansion is valid.

The limitation of the short range interaction was overcome in an important 'break through the Coulomb barrier' by J. Dollard in a thesis at Princeton [34]. Dollard's method consisted in modifying the free evolution operator by a distorted evolution $\tilde{U}_t$ which is no longer a group but includes the long-range effects. One can then still show a modified asymptotic condition in the form

$$\Omega_\pm = \text{S-lim}_{t \to \mp \infty} V_t^* \tilde{U}_t \tag{$\tilde9$}$$

which replaces the old condition (9).

This technique opened the way to a much more general scattering theory which was found to be most conveniently formulated in algebraic form developed almost simultaneously by several workers[42] (Combes, Lavine, Martin, Amrein, Misra).

The transference of this method to the stationary theory is being carried through now and it gives hope that the imaginative but non-rigorous theory of van Hove for persistent interactions can be vindicated to a large extent.

One recent application of this theory is the establishment of upper bounds for the total cross-section at very high energy.

Another development of scattering theory followed more closely the outline sketched by Heisenberg in his original paper. The idea is to determine as much as possible of the S-matrix structure by imposing on it conditions which follow from very general principles. Among these principles are first of all those which follow from the relativistic invariance and the unitarity. They are supplemented by conservation laws, crossing symmetry and analyticity.

The ultimate objective of this approach is not a complete theory of the evolution of physical particle systems, but rather a correlation of empirical facts by S-matrix postulates based on physically plausible postulates. This approach has yielded some useful results in this direction, but it is doubtful whether such a meagre input basis is sufficient for obtaining the rich variety of empirical facts on elementary particles, as some of its adherents seem to believe[44].

3. WHY HILBERT SPACE?

An entirely different line of research concerning the mathematical structure of elementary quantum mechanics started with the question: 'Why Hilbert space?' In particular, why complex and infinite-dimensional Hilbert space?[45] More precisely, the question could be decomposed into three related ones:

(1) Why should the quantum mechanical formalism need a Hilbert space for the representation of the observables in the theory? How is it possible to construct the Hilbert space on the basis of these observable data by a chain of logical deductions?

(2) Why should this space be of infinite dimensions, since physical data are only finite in number?

(3) What is the empirical foundation for the choice of complex numbers for the coefficients of this space?

These questions can now largely be answered but the answers are much more involved than one might have expected and they give a deep insight into the mathematical structure of quantum mechanics[46].

The most successful approach to find a convincing answer was through an axiomatic reconstruction of the theory starting only from the most elementary but most characteristic experiences concerning quantal systems.

The road to this approach was laid out by B. Birkhoff and J. von Neumann in a paper entitled: 'Quantum Logic', a paper which marks the beginning of an important development in the axiomatic reconstruction of quantum theory[47].

The starting point of this development is the observation that the phase of the state-vector is not an observable quantity. Hence a theory, which uses the phase explicitly, uses a redundant element which conceivably could be eliminated from the theory. Mathematically this means that the theory does not involve vectors in Hilbert space and their linear combinations but rather the *rays* containing such vectors, or more generally the *sub-spaces* of the Hilbert space.

This situation recalls immediately the analogous situation in projective geometry. The geometrical elements at infinity can be treated on the same footing as the finite ones if one uses homogeneous coordinates defined only up to an arbitrary common numerical factor, that is sub-spaces of some vector space. It is well known that the structure of projective geometries can be completely described by the axioms referring to the intersections, unions and complements (duals) of such sub-spaces.

Thus quantum mechanics could equally well be formulated by the use of such operations alone and thereby establish incidentally that its mathematical structure is on the abstract level nothing else than a kind of projective geometry. This mathematical structure is called a *lattice* and its physical interpretation a *proposition system*.

The elements of this lattice are yes–no experiments, or rather equivalence classes of such experiments, and they are partially ordered. This order relation is denoted by $\subset$.

There are accordingly three lattice operations, the intersection (denoted by $\cap$) the union (denoted by $\cup$) and the orthocomplement (denoted by ').

These basic lattice theoretic operations are similar to those for the atomic Boolean lattices of subsets, except for the distributive law

$$a \cap (b \cup c) = (a \cap b) \cup (a \cap c)$$

which is false for the quantal lattices. It is replaced by a weaker one, which can be expressed in many different equivalent forms. One of the simplest way to do it is to require that if $a \subset b$ then the sublattice generated by (a, b, a', b') is Boolean. A lattice which satisfies this law is called a *weakly modular lattice* [48].

This more general property corresponds to the fact that in ordinary quantum mechanics there exist non-commuting observables. When observables commute they can be measured simultaneously and we may then say that they are *compatible*.

We define compatible propositions a and b in a lattice theory by the property that a and b generate a Boolean sublattice.

One denotes this relation by $a \leftrightarrow b$. The weakly modular lattices are then characterized by the property

$$a \subset b \Rightarrow a \leftrightarrow b.$$

In words: If a implies b then a and b are compatible. This formulation recalls the physical interpretation and makes this property quite plausible.

Once the structure of the lattice of propositions is established for quantal systems one may attack the problem: How does one pass from the lattice of propositions to Hilbert space? It is clear that the subspaces of the Hilbert space are a lattice which satisfies all these properties, but it is not clear why there should not exist other vector-spaces which also satisfy them, or why there should be any vector-space at all which furnishes a representation of a given lattice?

This question has been cleared up during the last ten years in a series of steps which I shall indicate:

(1) First it is necessary to define the notion of the *centre* ϕ of a lattice $\mathscr{L}$. It is the set of all propositions which are compatible with any other proposition. In formula

$$\phi = \{c \mid c \in \mathscr{L}, c \leftrightarrow x \; \forall x \in \mathscr{L}\}.$$

A lattice is called *irreducible* if the center ϕ is trivial. A lattice which is not irreducible can be decomposed into the union of two lattices $(\mathscr{L}_1, \mathscr{L}_2)$ consisting of pairs of elements, each in one of the two sublattices. The lattice operation of the union of two sublattices is obtained from the component wise operations of the component lattices. Thus if $x = (x_1, x_2)$, $y = (y_1, y_2)$, then

$$x \cap y = (x_1 \cap y_1, \; x_2 \cap y_2)$$
$$x \cup y = (x_1 \cup y_1, \; x_2 \cup y_2)$$
$$x' = (x'_1, x'_2).$$

If $\mathscr{L}_1$ and $\mathscr{L}_2$ are not irreducible then the decomposition can be continued until one arrives at a complete decomposition of a reducible lattice into irreducible ones. The decomposition is unique up to permutations.

(2) The next step is the proof that every irreducible lattice can be mapped one-to-one unto the closed linear subspaces of a Hilbert space over a field[49].

(3) It is possible to show that the field must be infinite[50].

(4) With a mild additional assumption one can then show that the field must contain the reals as a subfield[51].

(5) The reals and the quaternions can be excluded by separate considerations[52].

This leaves the complex numbers as the only possibility for the field and one has reconstructed the general mathematical frame for conventional quantum mechanics.

There remains one question to be settled: Why are observables represented by self-adjoint operators? The answer is intimately connected with the spectral theorem.

Every self-adjoint operator determines uniquely a spectral measure, that is a map E_A from the Borel sets $\mathscr{B}(\mathbf{R})$ on the real line to the projection operators.

In the lattice theory such a map becomes a σ-homomorphism of the Borel sets into the lattice. The physical interpretation is obvious: The lattice element a_A which corresponds to the set $A \in \mathscr{B}(\mathbf{R})$ represents the proposition: 'The value of the measured quantity lies in the set A'.

Because of the spectral theorem every such σ-homomorphism in Hilbert space is represented by exactly one self-adjoint operator. Thus the representation of observables by self-adjoint operators has ceased to be a seemingly arbitrary and often ill-motivated hypothesis of the theory, but it is simply a corollary of the spectral theorem for self-adjoint operators in Hilbert space.

With this result the reconstruction of conventional quantum mechanics from first principles based on an inductive phenomenological axiomatics is complete.

4. THE FORMALISM

As Dirac has said in his book, one must distinguish the abstract theory from its representation by numbers for the purpose of carrying out actual calculations. He compares this step with the choice of coordinate axes in geometry. The representations are not unique and they may be chosen in accord with considerations of simplicity and convenience.

One of the most important representations is the spectral representation with respect to a complete set of commuting observables (c.s.c.o.).

Such a set $\mathfrak{S}$ consists of a finite or countably infiite number of pairwise commuting self-adjoint operators which are *complete* in the following precise sense:

Let $\mathfrak{S}' = \mathfrak{A}'$ denote the set of all bounded operators which commute with every operator in $\mathfrak{S}$. It is a weakly closed *-algebra, also called a von Neumann algebra. The algebra $\mathfrak{A} = \mathfrak{A}''$ is the algebra generated by $\mathfrak{S}$. For any commuting set $\mathfrak{S}$ one has $\mathfrak{A} \subseteq \mathfrak{A}'$. The set $\mathfrak{S}$ is *complete* if $\mathfrak{A} = \mathfrak{A}'$. In that case the algebra $\mathfrak{A}$ is said to be *maximal abelian*.

Let $A_r \in \mathfrak{S}$ ($r = 1, 2, ...$) be the operators contained in $\mathfrak{S}$ (which we assume finite or at most countably infinite) and let $\Lambda_r = \mathrm{sp}\, A_r$ denote the spectrum of A_r. We define by $\Lambda = \Lambda_1 \times \Lambda_2 \times \cdots$ the Cartesian product of Λ_r and by $\lambda \in \Lambda$ an element in Λ.

There exists then a measure μ on the product space and a Hilbert space $L^2(\Lambda, \mu)$

$$(A_r f)(\lambda) = \lambda_r f(\lambda)$$

such that for $f(\lambda) \in L^2(\Lambda, \mu)$.

In this *spectral representation* the operators A_r appear as multiplication operators. They can be considered diagonalized [53].

The variety of different c.s.c.o. are reflected in the varieties of measures μ which can occur. This variety turns out to be much richer than physicists usually assume. Let us illustrate this by the case of one observable which is complete. In that case its spectrum is called simple.

Every self-adjoint operator A of this kind, bounded or not, decomposes the Hilbert space $\mathscr{H}$ into three orthogonal subspaces $\mathscr{H} = \mathscr{H}_p \oplus \mathscr{H}_{ac} \oplus \mathscr{H}_{sc}$. Let g be a cyclic vector so that $\{\overline{\mathfrak{A}g}\} = \mathscr{H}$. Then the spectral family E_λ defines a measure $\mu(\lambda) = (g, E_\lambda g)$ which contains a point spectrum, an absolutely continuous and singularly continuous part. There is no physical example known in quantum mechanics where the singularly continous part is different from zero.

It is convenient to introduce a notation which retains the advantages of Dirac's bra-ket notation but at the same time avoiding its pitfalls: If $f \in \mathscr{H}$ is an abstract vector in the Hilbert space and $\lambda \in \Lambda$ a point in the spectrum, we denote by $\langle \lambda | f \rangle$ a function which represents f with respect to some c.s.c.o. The following formulas are then obvious and they are basic for the calculus

$$\langle \lambda \,|\, f) + \langle \lambda \,|\, g) = \langle \lambda \,|\, f + g)$$
$$\langle \lambda \,|\, \alpha f) = \alpha \langle \lambda, f)$$
$$(f \,|\, \lambda\rangle = \langle \lambda \,|\, f)^{*}$$
$$(f, g) = \int (f \,|\, \lambda\rangle \, \mathrm{d}\mu(\lambda) \, \langle \lambda \,|\, g).$$

These formulas are generally valid, no matter what the nature of the measure μ may be. However, the linear operators may be represented as integral operators in the form

$$\langle \lambda \,|\, Af) = \int \langle \lambda| \, A \,|\lambda'\rangle \, \mathrm{d}\mu(\lambda') \, \langle \lambda' \,|\, f)$$

only in special cases. Hence neither $\langle \lambda |A| \lambda'\rangle$ nor $\langle \lambda \,|\, \lambda'\rangle$ can in general be given a meaning in this formalism.

With the theorem on the spectral representation a large part, in fact the essential part of Dirac's bra-ket formalism, can be mathematically based on a rigorous foundation without any special effort, thereby not only justifying a large part of Dirac's ingenious symbolic calculus, but also pointing out its deeper mathematical significance.

In contrast to Dirac's formalism the bras $\langle \lambda |$ and the kets $|\lambda\rangle$ never occur as independent quantities, only the spectral representations of the form $\langle \lambda \,|\, f)$ and $(f \,|\, \lambda\rangle$ have a precise mathematical meaning.

5. CONCLUSION

The mathematical structure of elementary quantum mechanics is now reasonably well understood although many technical problems are still awaiting solutions. Once again, as so often in the history of mathematical physics, this structure reveals the almost unreasonable effectiveness of the imaginative mathematical creations for the descriptions of the relations occurring in the real world of phenomena.

Dirac has brought to light many of these relations in a breathtaking succession of unparalleled successes. We, who have inherited the wealth of ideas from Dirac and the other founders of quantum mechanics, humbly acknowledge our debt for the inexhaustible treasure of scientific ideas which they have opened up for us.

NOTES

1. P. A. M. Dirac, *Principles of Quantum Mechanics*, 4th ed. (referred to as P. in the following), Oxford (1958). The idea of the correspondence between Poisson bracket and commutator appears already in Dirac's first publication on the new mechanics (*Proc. Roy. Soc.* A **109**, 642 (1925)).
2. We write for simplicity the equations only for systems with one degree of freedom. There is no difficulty in generalizing them for a finite number of degrees of freedom by distinguishing sets of variables with an index r $(r = 1, ..., n)$.
3. P. A. M. Dirac, *loc. cit.*, ref. 1, p. 106 ff.
4. J. von Neumann, *Math. Ann.* **104**, 570 (1931).

5. H. Weyl, *Gruppentheorie und Quantenmechanik*, S. Hirzel, Leipzig (1928). See also *Z. Phys.* **46**, 1 (1928).

6. M. Born and P. Jordan, *Z. Phys.* **34**, 858 (1925). They are

$$Q = \frac{1}{\sqrt{2}} \begin{pmatrix} 1 & 1 & 0 & 0 & . \\ 1 & 0 & \sqrt{2} & 0 & . \\ 0 & \sqrt{2} & 0 & \sqrt{3} & . \\ 0 & 0 & \sqrt{3} & 0 & . \\ . & . & . & . & \end{pmatrix} \qquad P = \frac{i}{\sqrt{2}} \begin{pmatrix} 0 & -1 & 0 & 0 & . \\ 1 & 0 & -\sqrt{2} & 0 & . \\ 0 & \sqrt{2} & 0 & -\sqrt{3} & . \\ 0 & 0 & \sqrt{3} & 0 & . \\ . & . & . & . & \end{pmatrix}.$$

7. E. Schrödinger, *Ann. Physik* **79**, 734 (1926).

8. The matrix-theoretic formulation of quantum mechanics motivated the work of A. Wintner, *Spektraltheorie der unendlichen Matrizen*, Hirzel, Leipzig (1929), which clearly showed that this problem was wrought with difficulties and obviously not yet ripe for a solution.

9. This important result was first proved by A. Wintner, *Math. Zeits.* **30**, 228 (1929). See also C. R. Putnam, *Commutation Properties of Hilbert Space Operators and Related Topics*, Springer (1967).

10. As an example we give two inequivalent matrix representations of P and Q, both of which are inequivalent with the standard one (ref. 6). They are defined on the domain $D \subset l^2$ (space of square summable complex sequences), consisting of all sequences x_n with a finite number of components subject to the condition $\Sigma_n x_n = 0$. We define

$$P = \begin{pmatrix} 0 & 0 & 0 & 0 & . \\ 0 & 1 & 0 & 0 & . \\ 0 & 0 & 2 & 0 & . \\ 0 & 0 & 0 & 3 & . \\ . & . & . & . & \end{pmatrix} \qquad Q = i \begin{pmatrix} 0 & -1 & -\frac{1}{2} & -\frac{1}{3} & . \\ 1 & 0 & -1 & -\frac{1}{2} & . \\ \frac{1}{2} & 1 & 0 & -1 & . \\ \frac{1}{3} & \frac{1}{2} & 1 & 0 & . \\ . & . & . & . & \end{pmatrix}.$$

A second representation on the same domain D is given by

$$P = \frac{i}{2} \begin{pmatrix} 0 & -1 & -1 & -1 & . \\ 1 & 0 & -1 & -1 & . \\ 1 & 1 & 0 & -1 & . \\ 1 & 1 & 1 & 0 & . \\ . & . & . & . & \end{pmatrix} \qquad Q = \begin{pmatrix} -1 & 1 & 0 & 0 & . \\ 1 & -3 & 2 & 0 & . \\ 0 & 2 & -5 & 3 & . \\ 0 & 0 & 3 & -7 & . \\ . & . & . & . & \end{pmatrix}.$$

11. F. Rellich, *Nachr. Akad. Wiss. Gött. Math. Phys. Klasse* 107 (1945).

12. J. Dixmier, *Comp. Math.* **13**, 263 (1958). Rellich and Dixmier proved the following theorem:
    Suppose P and Q are symmetric irreducible and closed on their respective domains D_P and D_Q and suppose their exists a dense domain $D \subset D_P \cap D_Q$ which is invariant under P and Q and on which the operator $P^2 + Q^2$ is essentially self-adjoint then P and Q are self-adjoint and they are unitarily equivalent to the Schrödinger operators.

13. C. Foiaş, L. Gehér and B. Sz.-Nagy, *Acta Sci. Math. (Szeged)* **21**, 78 (1960). The relevant portion of their theorem is the following:
    Suppose P and Q are both self-adjoint and satisfy the commutation rules (4) on a set $D \subset D_{[Q, P]}$ and that either $(P + iI)(Q + iI) D$ or $(Q + iI)(P + iI) D$ is dense in $\mathcal{H}$, then P and Q, if they are irreducible, are unitarily equivalent to the Schrödinger operators.

14. Y. Kilpi, *Ann. Acad. Sci. Fennicae* A **315**, 3 (1962).

15. H. G. Tillmann, *Acta Sci. Math. (Szeged)* **24**, 258 (1963); *ibid.*, **15**, 332 (1964.)

16. J. von Neumann, *Mathematische Grundlagen der Quantenmechanik*, Springer (1932), esp. Section II, 9, p. 75 ff.

17. G. Mackey, *Ann. of Math.* **55**, 101 (1952); **58**, 193 (1953); *Ann. J. Math.* **73**, 576 (1951). A convenient summary of this work is contained in G. Mackey, *Included Representation of Groups and Quantum Mechanics*, Benjamin (1968).

18. E. P. Wigner, *Ann. of Math.* **40**, 149 (1939).

19. A. S. Wightman, *Rev. Mod. Phys.* **34**, 845 (1962).

20. T. D. Newton and E. P. Wigner, *Rev. Mod. Phys.* **21**, 400 (1949).

21. This result was known a long time ago to Pauli. See also M. Fierz, *Helv. Phys. Acta* **12**, 3 (1939), esp. p. 4, and J. M. Jauch, Diploma Thesis, E.T.H. Zürich, 1939.

22. J. M. Jauch and C. Piron, *Helv. Phys. Acta* **42**, 559 (1967).

23. W. Amrein, *Helv. Phys. Acta* **42**, 149 (1969).
24. Neumark, *Doklady Acad. Sci. URSS* **41**, 359 (1961). B. Sz.-Nagy, *Functional Analysis*, Appendix, Ungar Publ. Co. New York (1960).
25. G. Emch, *Helv. Phys. Acta* **36**, 739; 770 (1963). See also J. M. Jauch, 'Projective Representations of the Poincaré Group in a Quaternionic Hilbert Space', in *Group Theory and its Applications*, Academic Press, New York (1968).
26. M. Born, W. Heisenberg and P. Jordan, 'Zur Quantenmechnik II', *Z. Phys.* **35**, 557 (1926).
27. P. A. M. Dirac, *Z. Phys.* **44**, 585 (1927).
28. W. Heisenberg, *Z. Phys.* **120**, 513; 673.
29. J. A. Wheeler, *Phys. Rev.* **52**, 1107 (1937).
30. C. Møller, *Del. Kgl. Danske Vid. Selskab* **23**, No. 1 (1945); **22**, No. 19 (1946).
31. The foundations of this theory were laid by J. M. Jauch, *Helv. Phys. Acta* **31**, 136 (1958). The first proof of the actual convergence of the asymptotic conditions was given by J. M. Cook, *J. Math. Phys.* **36**, 82 (1957). Later generalizations of these kinds of theorems were given by M. N. Hack, *Nuovo Cimento* **9**, (1958), and J. M. Jauch and I. I. Zinnes, *Nuovo Cimento* **11**, 553 (1959). The strongest theorems in this direction are due to T. Kato, *Perturbation Theory for Linear Operators*, Springer (1966), esp. Chapter X.
32. T. Kato and S. T. Kuroda, *Nuovo Cimento* **14**, 1102 (1959).
33. A selection of useful references in multichannel scattering theory are the following:
    W. Brenig and R. Haag, 'Allgemeine Quantentheorie der Streuprozesse', *Fortschr. Physik* **7**, 183 (1959). A general review giving a survey of the state of the art before the development of the abstract theory.
    H. Ekstein, *Phys. Rev.* **101**, 880 (1956). The first attempt of a general multichannel theory.
    J. M. Jauch, *Helv. Phys. Acta* **31**, 661 (1958).
    L. D. Fadeev, *Trudy Mat. Inst. Steklov.* **69**, (1963). A stationary state theory for the three-body problem using the techniques of Fredholm integral equations.
    K. Hepp, *Helv. Phys. Acta* **42**, 425 (1969). Generalization of Fadeev's theory to the *n*-body problem.
    J. M. Jauch and J.-P. Marchand, *Helv. Phys. Acta* **39**, 325 (1966). Proof of unitarity of *S*-matrix with overlapping channels.
    C. van Winter, 'Theory of Finite Systems of Particles' I and II, *Kgl. Danske Vid. Selsk. Mat.-Fys. Srifter* **2**, No. 8 and 10.
34. J. D. Dollard, *J. Math. Phys.* **5**, 729 (1964). A general account including many particle scattering with long range potentials is found in J. D. Dollard, 'Quantum Mechanical Scattering Theory for Short-Range and Coulomb Interactions' *Rocky Mountain J. of Math.* **1**, 4 (1971).
35. T. Kato and S. R. Kuroda, *Proceedings of Stone Conference*, University of Chicago (1969). S. T. Kuroda, *Proceedings of the Symposium on Advanced Topics in the Mathematical Foundations of Scattering Theory*, Institute for Theoretical Physics, Univ. of Geneva (1968), Ed. by W. Amrein.
36. W. O. Amrein, V. Georgescu and J. M. Jauch, *Helv. Phys. Acta* **44**, 407 (1971).
37. E. Prugovečki, *Nuovo Cimento* **63 B**, 569 (1969); *ibid.*, **4 B**, 124 (1971).
38. D. B. Pearson, *Nuovo Cimento* **2 A** 853 (1971).
39. A. Galindo, *Helv. Phys. Acta* **32**, 412 (1959).
40. c.f. R. G. Newton, *Scattering Theory of Waves and Particles*, McGraw Hill (1966), esp. Part III, ch. 6. and 7.
41. P. A. Reijto, *Comm. Pure Appl. Math.* **16**, 279 (1963), *ibid.*, **17**, 257 (1964).
42. W. O. Amrein, P. Martin and B. Misra, *Helv. Phys. Acta* **43**, 313 (1970); R. B. Lavine, *J. Funct. Anal.* **5**, 368 (1970); J. M. Combes, C.N.R.S. Marseilles, Preprint 1969.
43. J. M. Jauch and K. Sinha, Preprint Geneva, to be published in *Helv. Phys. Acta*. P. Martin and B. Misra have generalized the result of the preceding authors to the relativistic case in a paper to be published in *Helv. Phys. Acta* and they were able to show that all total cross-sections must be bounded in the limit of high energy λ by ln λ.
44. Some of the most important work on this programme is by
    G. F. Chew, *The Analylic S-Matrix,* Benjamin (1966). P. Landshoff, D. Olive, J. Polkinghome, *The Analytic S-Matrix*, Cambridge University Press (1966).
    J. M. Charap and S. Fubini, *Nuovo Cimento* **14**, 540 (1959); **15**, 73 (1959).
    S. Mandelstam, *Phys. Rev.* **112**, 1344 (1958).
    For a recent report on the status of the *S*-matrix philosophy see G. F. Ghew, *Phys. Rev.* **D 4**, 2330 (1972).

45. As far as I could make out, the question concerning the complex numbers was posed (probably for the first time in print) by P. Ehrenfest and an answer to Ehrenfest's question was given by W. Pauli, *Z. Physik* **80**, 573 (1933). Pauli's answer did not completely satisfy everybody and the question has come up again and again in numerous discussions on the foundations of quantum mechanics. In a series of papers, Stueckelberg and several collaborators established that one can in fact develop a quantum mechanics in a real Hilbert space. [E. C. G. Stueckelberg *et al.*, *Helv. Phys. Acta* **33**, 727 (1960); **34**, 621, 675 (1961); **35**, 673 (1962)]. However, it was found that in order to obtain an uncertainty relation between canonically conjugate variables such as position and momentum the observables must be subject to a super-selection rule of the following kind: All the observables commute with an antisymmetrical operator J, satisfying $J^{\sim} = -J, J^2 = -I$ ($\sim$ means the transposed operator). It is then easy to show by a suitable choice of coordinates that such an operator can always be represented by a matrix operator of the form

$$J = \begin{pmatrix} 0 & I \\ -I & 0 \end{pmatrix}$$

and that a Hilbert space with such a supplementary structure is mathematically completely equivalent with a complex space.

46. It is perhaps useful to recall at this point a now almost forgotten historical episode between Born and Wiener during the early days of quantum mechanics, an account of which can be found in M. Jammer, *The Conceptual Development of Quantum Mechanics*, McGraw Hill (1966), esp. p. 221–2. In retrospect it can now be seen that Wiener and Born came very close to establishing wave mechanics. Their work was published in *Z. Physik* **36**, 174 (1926). However, their function space is more general than Hilbert-space and their operators can only be expressed as ergodic limits of certain integral operators. This was a natural concept in the framework of Wiener's theory on generalized Fourier-transforms, but it was rather cumbersome for the establishment of a general transformation theory.

   Von Neumann showed later (cf. ref. 16) that separable Hilbert space furnishes a more natural mathematical frame for the establishment of the identity of matrix and wave mechanics because of the Riesz–Fischer theorem. This theorem establishes an isomorphism between the sequence space l^2 and the function spaces $L^2(\Lambda, \mu)$ and it is this isomorphism which permits the identification of the unique abstract mathematical object, the Hilbert space, which is behind all these different representations.

47. G. Birkhoff and J. von Neumann, *Ann. Math.* **37**, 823 (1936). This paper made at first little impression on theoretical physicists who at that time were busy with applying quantum mechanics to the ever-increasing domain of applications in atomic, molecular, nuclear, solid state and particle physics. After the war the paper was virtually unknown to theoretical physicists. It was D. Finkelstein who in a series of preprints, written at CERN in 1959 with Jauch and Speiser, insisted on the fundamental importance of this work for the deeper understanding of the mathematical structure of quantum mechanics.

48. Birkhoff and von Neumann favoured the *modular law* as the basic axiom which replaces the distributive law. In any lattice the following is true:

   If $x \subset z$ then $x \cup (y \cap z) \subseteq (x \cup y) \cap z$.

For Boolean lattices the two expressions on the right are identical (that is instead of $\subseteq$ one can write $=$). But there exists an important class of lattices which are not Boolean for which the identity still holds. These are the modular lattices and they are in a sense the weakest natural generalization of the Boolean lattices. In addition they have very nice mathematical properties. These observations have induced Birkhoff and von Neumann to favour modular lattices as the basic structures for the quantal proposition systems.

   They also pointed out that the closed linear subspaces of Hilbert space are *not* modular lattices which led to the conjecture that more general lattices, such as the projection lattices in a von Neumann algebra of type II (continuous geometries) might be a more suitable mathematical structure for the description of quantal systems.

   The conjecture was left in suspense until C. Piron [C. Piron, Ph.D. thesis, published in *Helv. Phys. Acta* **37**, 439 (1964)] proved that localizable systems cannot be described by a modular lattice. This means physically that a quantum theory based on a modular lattice could admit no particles. This seems to rule out modularity.

49. C. Piron, *Helv. Phys. Acta* **37**, 439 (1964), Theorem 22. Actually the proof of this theorem is incomplete. It was later completed by I. Amemiya and H. Araki, *Publ. Research Inst. Sc. Kyoto Univ.* A **2**, 423 (1967).

50. J.-P. Eckmann and Ph. Ch. Zabey, *Helv. Phys. Acta* **42**, 420 (1969).

51. C. Piron and S. Gudder, *J. Math. Phys.* **12**, 1583 (1971).

52. E. C. G. Stueckelberg, *loc. cit.*, ref. 25.

53. The proof of the existence of this spectral representation was given by J. M. Jauch and B. Misra, *Helv. Phys. Acta* **38**, 30 (1965). The proof includes the uniqueness of the equivalence class of the measure μ in case the number of the c.s.c.o. is finite. If the spectra are absolutely continuous this measure is equivalent to Lebesgue measure.

DISCUSSION

W. Thirring: It was mentioned in Professor Jauch's talk that in addition to the usual continuous spectrum and the point spcetrum, there is the so-called singular continuous spectrum, which is a continuous spectrum concentrated on a set of Lebesgue measure zero. It has been shown that if you have only Coulomb interactions, the singular continuous part of the spectrum of the Hamiltonian is absent, and this may be the reason why it has not been seen by physicists.

R. Peierls: If I understand the analysis that Professor Jauch has presented, it is to make respectable and consistent what otherwise would be rather like the formalism he referred to or just a recipe. The question I would like to ask him is: Does this analysis show whether nonrelativistic quantum mechanics, learned from Dirac and others, is in fact respectable? In order to understand a statement it is often illuminating to ask what the opposite would mean. Supposing the theory was unsuccessful and failed, or showed that the recipes that we are working with are not consistent and do not have a satisfactory mathematical background. What would then be the answer? Would it mean that we would have to modify or throw away the recipes, or would it mean that one has simply to look at more mathematics? If one looks more deeply probably something will be found that will justify the recipe. What is the answer?

J. M. Jauch: There are two aspects of this question. One of them is positive, and the other rather negative. The negative part is the one which Professor Peierls has asked, as to what would be the answer if it were not consistent. This is an interesting question, but it doesn't come into play because it *is* consistent. There is another aspect to it which is just as interesting. To what extent do we have more liberties than we thought we had? In other words, the phenomena, which are given to us from direct inspection, may permit more generalized theories which we did not expect, if we did not have a chance to examine the mathematical structure that is implied by phenomena. I am personally much more interested in this positive aspect of the investigation, and the answer to this question is not yet completely known.

A. Trautman: Under the heading, 'Why Hilbert Space?', Professor Jauch raised the question why it is infinite dimensional. Isn't it enough to argue that since all the unitary representations of the Poincaré group are infinite dimensional, one will need an infinite dimensional space in which to represent them, or do you have some other reason?

J. M. Jauch: This is a good argument, but it would no longer be applicable when one does not insist on unitary representations.

15

Relativistic Equations in Quantum Mechanics

Eugene P. Wigner

1. RECALLING THE DISCOVERY OF DIRAC'S ELECTRON EQUATION

Let me begin by recalling the way I learned about Dirac's relativistic equation of the electron. It was from a letter to Max Born who asked Dirac to review an article by a third person. The letter started with this review, pointing out that two of the chapters could be united and that this would render the article more concise, that is shorter, and also easier to read. It had a few more remarks on the article in question. Then, in a new paragraph of about ten lines he said that he had been working on the relativistic theory of the electron. He gave his four-component equation, including the interaction with an electromagnetic field.

Both Jordan, who also read it, and I were quite flabbergasted when reading it. We had tried to develop the relativistic equation for a spin 1/2 particle also, but all our attempts were directed toward a two-component equation – after all, the spin 1/2 particle has only two states. We were not satisfied with any of the equations we had found but were not yet ready to give up. The letter to Born changed all that. As Jordan put it 'Well, of course, it would have been better had we found the equation but the derivation is so beautiful, and the equation so concise, that we must be happy to have it.' I still have the letter in question, it is one of my most precious possessions.

2. INTRODUCTORY REMARKS

Actually, Dirac's electron equation was not the first relativistic wave equation: the so-called Klein–Gordon equation, for spin-less particles, was established before. However, the existence of two relativistic wave equations stimulated me to try to find all such equations – at least all which apply to single particles. When Jagdish Mehra suggested that I speak on this subject, I was afraid, first, that all I would be able to say about it would appear as 'old hat'. Worse than that, I was afraid that the subject is closed to such an extent that even Feynman's motto would not apply, the motto according to which the solution of every problem of physics brings, in its wake, a host of new and unsolved problems. However, a closer study of the problem made me realize how much I could learn from the existing literature and how many questions relating to it have remained unanswered. Hence, I am now glad that this is my subject. My discussion will be divided into two parts. The first part will deal with rela-

J. Mehra (ed.), The Physicist's Conception of Nature, 320-330. All Rights Reserved
Copyright © 1973 by D. Reidel Publishing Company, Dordrecht-Holland

tivistic equations and transformations in the absence of external fields. The effect of fields will be taken up in the second part.

3. EQUATIONS AND RELATIVISTIC TRANSFORMATIONS

As an introduction to this subject, I fear, it will be necessary to repeat very briefly the contents of a very old paper of mine.

The postulate that an equation be relativistically invariant implies that an arbitrary state, represented by a solution of the equations, be transformable to any other coordinate system obtainable by a Poincaré transformation from the coordinate system in which the original specification of the state was given. Since the Poincaré transformations contain the operation of time displacement, if one knows the behaviour of all states under all Poincaré transformations, one already has the relativistic equation of motion – this gives only the change of the state under time displacement. Conversely, however, it is also true that in order to prove that an equation is relativistically invariant, one must exhibit the behaviour of its solutions under all relativistic transformations. Actually, the situation is even worse: the wave equation, i.e. the time displacement operator, does not determine all the other operators of the Poincaré group uniquely. As a rule, though, only one set of such operators will appear natural – few of us would be inclined to interpret the Dirac equation as representing two scalar particles. However, mathematically, this is possible and the totality of the operators of the Poincaré group form a more complete description of the properties of the particle than the time displacement operator, i.e. the wave equation, alone. This is the reason for discussing in this section, at least superficially, the totality of the Poincaré transformations. On the other hand, when we try to describe the behaviour of the particles to the vacuum states of which these transformations apply, when we try to obtain their behaviour in external fields, the equations of motion will prove more informative. It seems that both have their importance, though in different contexts, but the former one will be discussed first.

It is easy to see that the transformations of the state vectors which correspond to the various transformations of the Poincaré group form a unitary representation of that group, but only up to a factor. If we denote the transformations of the Poincaré group by $P_1, P_2, \ldots$, the corresponding operations in Hilbert space by $D(P_1), D(P_2), \ldots$, must satisfy the equations

$$D(P_1)D(P_2) = \omega(P_1, P_2)D(P_1, P_2) \tag{1}$$

the $D(P)$ being unitary operators, the $\omega(P_1, P_2)$ numbers of modulus 1. The only real difficulty in solving these equations was to show that the $D(P)$ can be replaced by $D'(P) = \omega(P)D(P)$, the $\omega(P)$ being again numbers of modules 1, so that

$$D'(P_1)D'(P_2) = \pm D'(P_1 P_2) \tag{1a}$$

holds. Once this is done, the possible sets of operators are easily found – so easily that the argument leading to them can be very quickly reproduced at least in the most important case of a finite mass.

Since the displacements all commute, one can choose basic vectors $|p, \xi\rangle$ in such a way that the effect of a displacement T_a by the four-vector a is given by (we assume (la) but omit the primes on the D)

$$D(T_a)|p, \xi\rangle = e^{ip \cdot a}|p, \xi\rangle \tag{2}$$

p and a are four-vectors in the underlying Minkowski space, $p \cdot a$ is their Minkowski scalar product. The variable ξ can be discrete, it is introduced because there may be several states which transform according to (2) with the same p.

If we apply a Lorentz transformation L to the state $|p, \xi\rangle$, we obtain a state which still transforms by (2) but its momentum p is replaced by Lp.

This follows simply from the formula

$$D(L^{-1})D(T_a)D(L) = D(T_{L^{-1}a}) \tag{3}$$

which is the equation of the Poincaré group connecting Lorentz transformations with displacements. Indeed, it follows from (3) that

$$D(T_a)[D(L)|p, \xi\rangle] = D(T_a)D(L)|p, \xi\rangle = D(L)D(T_{L^{-1}a})|p, \xi\rangle =$$
$$= D(L)e^{ip \cdot L^{-1}a}|p, \xi\rangle = e^{iLp \cdot a}[D(L)|p, \xi\rangle]. \tag{3a}$$

The third member follows from (3), the fourth from (2) as applied to the displacement $L^{-1}a$, and the last one from the linear nature of $D(L)$ and the invariance of the Minkowski scalar product under the Lorentz transformation L. Equation (3a) shows that, indeed, $D(L)|p, \xi\rangle$ is a state with momentum Lp.

Equation (3a) is of basic significance inasmuch as it shows that the magnitude of the momentum is the same for all states which can be obtained from a $|p, \xi\rangle$ state by any Poincaré transformation. Indeed, such a transformation can be written as the product of a Lorentz transformation and a displacement. The latter, as evident from (2), does not change p at all – it only multiplies $|p, \xi\rangle$ by a number. As (3a) further shows, the Lorentz transformation L replaces p by Lp which has the same length as p because of $p \cdot p = Lp \cdot Lp$. One concludes, hence, that if the set of all states is irreducible, the momenta of all will have the same length because states with momenta of different lengths do not mix as a result of the application of Poincaré transformations.

If there are states $|p_0, \xi\rangle$ with a p_0 which is parallel to the time axis – this turns out to be the condition for the finiteness and real nature of the mass – and if R is a rotation, $D(R)$ applied to the states $|p_0, \xi\rangle$ will still give states with momentum $Rp_0 = p_0$. This means that the states $|p_0, \xi\rangle$ transform among themselves under rotations – they must transform by a representation of the group of rotations of space. If we assume that the whole set of operators $D(P)$ form an irreducible manifold, i.e. do not leave any subspace of the Hilbert space invariant, it becomes reasonable to choose the $|p_0, \xi\rangle$ in such a way that they transform by an irreducible representation R^j, with matrix-elements $R^j_{\xi', \xi}$, of the rotation group

$$D(R)\,|p_0, \xi\rangle = \sum_{\xi'} R^j_{\xi'\xi}|p_0, \xi'\rangle. \tag{4}$$

One recognizes this j as the spin of the particle, and the length of p_0 will be interpreted, of course, as its mass.

Equation (4) gives only the effect of rotations on $|p_0, \xi\rangle$. In order to obtain the effect of the general Lorentz transformations thereon, one decomposes this into a 'boost' B, that is a symmetric Lorentz transformation, and a rotation, i.e. one sets $L=BR$. The boost changes the momentum p_0 into Bp_0 and one can define $|Bp_0, \xi\rangle$ as

$$|Bp_0, \xi\rangle = D(B)\,|p_0, \xi\rangle. \tag{5}$$

This gives for $L=BR$

$$D(L)\,|p_0, \xi\rangle = D(B)D(R)\,|p_0, \xi\rangle = D(B)\sum_{\xi'} R^j_{\xi'\xi}|p_0, \xi'\rangle =$$
$$= \sum_{\xi'} R^j_{\xi'\xi}|Bp_0, \xi'\rangle. \tag{5a}$$

Finally, we have to apply the Lorentz transformation L not only to the states $|p_0, \xi\rangle$ the momentum vectors of which are parallel to the time axis, but to all states $|p, \xi\rangle$ though we can assume p to have the same Minkowski length as p_0. In order to do this, we note, first, that the boost Bp is uniquely defined by p and p_0 as a result of the equation $p=B_p p_0$. For the calculation of $D(L)|p, \xi\rangle$ we decompose, therefore, the Lorentz transformation LB_p into a boost and a rotation

$$LB_p = BR. \tag{6}$$

We can write, therefore,

$$D(L)\,|p, \xi\rangle = D(B)D(R)D(B_p^{-1})\,|p, \xi\rangle$$
$$= D(B)D(R)\,|p_0, \xi\rangle = \sum_{\xi'} R^j_{\xi'\xi}|Bp_0, \xi'\rangle. \tag{6a}$$

The second member is a consequence of (6), the third of (5), applied for B_p instead of B and multiplied by $D(B_p)^{-1}=D(B_p^{-1})$. The R and B in (6a) are given by (6).

This completes the determination of the transformations of all the state vectors under all Poincaré transformations. Every such transformation can be written as the product of a Lorentz transformation and a displacement $P=LT_a$ and the effect of $T(a)$, that is the operator $D(T_a)$, was given already by (2).

This completes the determination of the Poincaré transformations in the case that there is at least one time-like momentum vector, i.e. for finite real mass of the particle. Once (1a) is established – this was used when obtaining (4) – the calculation is indeed easy. The calculation is not significantly more difficult if the momenta are null-vectors, i.e. for light-like momenta. The same applies in the case of states – probably not realized in nature – in which the momenta are space-like ('tachyons'). The case that all four components of the momentum vanish is somewhat more difficult to deal with, and this has first been done by the Russian mathematicians, Gelfand, Neumark, and their collaborators. However, surely, there is no real particle all states of which are invariant under all displacements. Such a particle would be present always and everywhere with the same probability.

4. WHAT HAVE WE ACCOMPLISHED?

What we have accomplished by the determination of the Poincaré invariant manifolds in Hilbert space is the character of possible elementary particles. The emphasis is on possible, not actual. Just as the laws of physics do not tell us what phenomena actually take place but only what succession of events is possible, the invariance considerations do not tell us what particles actually exist but only give some of the properties of the particles which may exist. In particular, the invariance considerations do not tell us why the ratio of the masses of the two stable finite-mass particles is 1836.1, why their electric charges are so precisely oppositely equal, and so on. Before expressing reservations even with regard to what we seem to have accomplished, it may be worthwhile to give the reasons for doubting the existence of particles with imaginary mass, and of zero-mass particles with infinite spin. The reasons are semi-experimental.

If particles with imaginary mass, that is space-like momenta, existed, we could take any two of them, give them momenta with oppositely equal spatial components and extract any amount of energy from them. Their energy would become negative but the energy spectra of such particles would extend from $-\infty$ to ∞. If such particles existed, no thermal equilibrium would be possible. Their existence, or the possibility of their being produced, would give rise to phenomena which are fantastic and which would have been observed. This does not mean that equations with imaginary mass should not be used in calculations – Sudarshan, for instance, used them to advantage and so did Hadjioannou even more – but it does mean that one has to consider the existence of particles with space-like momenta most unlikely.

The same objection does not apply to particles with zero mass, that is of light-like momenta, and infinite spin. Their energy spectrum can be very naturally so chosen that it does not contain negative values. In fact, G. Mack here in Trieste, and Melvin at Drexel, have given serious consideration to the existence of such particles. It is true, nevertheless, that their virtual existence would give an infinite heat-capacity to a finite volume of space. This is not impossible – the doubling of the heat capacity due to the existence of neutrinos was hardly noted. Nevertheless, it is good to realize that the existence, or the possibility of production, of such particles would also render any true thermal equilibrium impossible. This is not in conflict with any experimental fact and, if we think of the origin of the world and the fact that it is still far from equilibrium, may even have attractive features. It is contrary to accepted views.

Let us now return to our principal question: are the results which we obtained concerning the limitations on the possible types of particles unquestionable. There are two questions in this connection the elucidation of which would be of interest. The first concerns the validity of the Poincaré group as the basic invariance group. It would seem that even if the space were finite, and the world had perhaps de Sitter character, the Poincaré group would be an adequate approximation, at least in microscopic physics. However, this is not a priori clear because the invariance group of de Sitter space, for instance, is very different from that of Minkowski space. In particular, the representations of the de Sitter group do not permit the definition of positive definite operators which could play the role of energy. The problems which an energy spectrum containing negative values would raise were mentioned before, at the discussion of the existence of tachyons. Nevertheless, Thirring, who has investigated this question most thoroughly, came to the conclusion that the conclusions derived from Poincaré invariance are valid to a very high approximation even in de Sitter space. It may be good to return to this question later, it seems to me that there are problems here which are not definitely and finally settled.

The second assumption which entered the considerations which were sketched in the preceding section is that the states are described by vectors, or rays, in a complex Hilbert space. What I have in mind in this regard when voicing reservations is not the replacement of the complex Hilbert space by a real one – such a space would be even more restrictive – but something like a quaternionic Hilbert space, or, more generally, a Hilbert space based on some other algebra but that of complex numbers. Such Hilbert spaces may permit other types of invariant manifolds, the existence of other types of particles. Jauch and his collaborators have given a good deal of thought to such possibilities and so did Gürsey but again, it seems to me, that the possibilities are not fully exhausted and that there remain problems of unquestionable mathematical interest and possibly of physical significance.

5. SOME FURTHER PROBLEMS

The manifolds of states, their various types of momenta which the transformation properties of their state vectors provide, are still a very meagre information. They determine the probabilities of the outcomes of very few measurements and we know that, at least in principle, the statements of quantum mechanics are formulated in terms of outcomes of measurements. One can say, of course, that the measurements should be described as operators in the Hilbert spaces which we found, but this requirement, though the equivalent of much repeated statements on the foundations of quantum mechanics, means very little in terms of the actual design of the measuring apparata.

We have learned, of course, from dispersion theorists, that the measurement of precisely defined operators is unnecessary, and one could content oneself with rather inaccurate position measurements. Goldrich and I have discussed this point most recently. It seems worthwhile, nevertheless, to discuss the possibility of accurate position measurements – the position is surely the most simple classical variable. In addition, a short discussion of a new 'observable' will be given even though its role is by no means fully clear.

There are two types of position operators which have been discussed. From an invariant-theoretic point of view, it is most natural to consider an operator, or a quartet of operators, which refer to a position in space-time. The meaning of a position in space-time was not always made clear by the proponents of such operators. Clearly, the position in question is not the position of a particle. If it were, and if one had observed the particle at position x, y, z, t, the probability would be zero of observing it at any point x', y', z', t' different from x, y, z, t. This is surely not so: even if we found the particle somewhere at time t, it will be somewhere also at time $t' \neq t$. In quantum mechanics, however, the states which are characteristic vectors of the quartet of position operators with characteristic values x, y, z, t are orthogonal to the characteristic vectors of any other set of characteristic values x', y', z', t' and this is true, for instance, as soon as $t' \neq t$.

The physical meaning of the space-time location was most clearly explained by A. Broyles. The localization refers not to a particle but to an event. The search for them constitutes a departure from the now accepted concepts of quantum mechanics! This does not contain observables referring to events. In fact, the greatest conceptual difficulty in the reconciliation of the quantum mechanics with the general theory of relativity is the basic difference in the observables they consider. This is the coincidence of two particles in general relativity, that is an event, whereas the quantum mechanical scattering and reaction theories do not attribute a space-time point to such an event. Hence, the interpretation of the space-time operators is still foreign to the present structure of quantum mechanics and it is, in fact, never specified what kind of events these space-time operators localize. A conceptual problem remains to be solved here – apparently, the mathematical problem is easier in this case than the interpretation of the formulae obtained. It may be added that, in addition to Broyles, several other colleagues have greatly contributed to this question.

The other concept of localization conforms with standard, old-fashioned ideas. It refers to the position of a particle at a definite time – definite time implying also a coordinate system in a definite state of motion. Newton and I determined these operators first but our ideas were made more precise by Wightman and given more elegant relativistic formulation by Fleming. He characterized these operators as functions of a vector in space-time, the vector being perpendicular to the space-like plane in which the particle's position is to be determined and ending at the position of the particle in that plane.

One may well have reservations concerning the basic nature of these position operators. First, the measurement on a space-like plane implies that one can receive the measurement signal instantaneously, in other words that the signal that the particle is at a given point in space reaches the observer with over-light velocity. This is the first reason for trying to replace these position measurements by others. The second reason is that though our postulates for the position operators could be satisfied for particles with finite mass, it is not possible to satisfy them for particles with zero mass and spin larger than $1/2$. In particular, they cannot be satisfied for light quanta. The most decisive efforts in this direction were made by Bertrand, and by Suttorp and de Groot, but their conclusions were very much the same as ours. In particular, it proved to be necessary for the definition of the position operators to extend the original

Hilbert space, referring only to positive energy states, to include also negative energy states. Even this constitutes significant progress – the usual position operators, that is multiplication with x, y, and z, also mixes positive and negative energy states. However, this is just what Newton and I wanted to avoid.

Conceptually, it would appear most natural to define the position of a particle on a light cone rather than on a space-like plane. It is possible to receive signals at the tip of the cone from any point thereof. It is perhaps not obvious, but it is true nevertheless, that the state vectors which correspond to states traversing a light cone at different points are orthogonal to each other. This means that the operators of which they are eigenfunctions, that is the operators of the position on the light cone, are self-adjoint. However, they do not seem to be very simple in terms of the operators commonly used. Nevertheless, it seems to me that it would be worthwhile to determine the characteristics of these operators more closely – as I consider altogether the light cone physics very promising.

The operators in the Hilbert space of single particles – not only the positions operators – are the first subject about which it would be good to know more. The second subject concerns the behaviour of such particles in external fields. This will naturally lead to the more customary forms of the relativistic wave equations, to their forms not as representations of the Poincaré group but as old-fashioned equations. There are even more unsolved problems in this connection than in the areas we have touched upon so far. However, before taking up this subject, it may be useful to take up the question under what conditions an outside effect can be described as an 'external field'.

6. WHEN CAN AN INTERACTION BE DESCRIBED BY AN EXTERNAL FIELD?

The question under what conditions the effect of a system on a particle can be described as that of an external field is a very general one but will be discussed only briefly. The condition includes, evidently, that the interaction leave the particle in a pure state, i.e. that no correlation become established between the state of the particle and the state of the system to the influence of which the particle is subjected. The field interaction is, therefore, the opposite extreme of the measurement interaction since the purpose and characteristic of the latter is just the establishment of a correlation between the states of the object, in our case the particle and the measuring device, i.e. the system with which it is to interact. Further, the interaction which is to be described by a field must leave the scalar product of any two possible initial states of the particle unchanged. Since these states are rather arbitrary, this means that the field-producing system must be left, in spite of its interaction with the particle, in the same state, no matter what the initial state of the particle was. This can be accomplished in a simple manner only if it remains in its initial state in spite of its having influenced the state of the particle by the interaction. This will be true for all states of the particle if it is true for a complete set of states thereof.

A complete set of states of our particle can be characterized by the momentum and helicity of the states. These will be changed by the field and this implies a change in the momentum and angular momentum of the field-producing system. This can leave the state of this system unchanged only if its initial state's momentum spectrum extended over a range considerably larger than the largest momentum change of the particle that can be expected and if the same is true of its angular momentum spectrum. This also must give significant probabilities to a wider range of angular momenta than the maximum angular momentum transfer can be expected to be, and the probabilities of the various angular momenta within that range must not show appreciable variations. These are only necessary conditions but their specification may suffice in the present context. They also suggest that strong interactions cannot be described by fields under any conditions and, indeed, the fields which appear in the Klein–Gordon and Dirac equations are only the electromagnetic fields.

7. EXTENSION OF THE HILBERT SPACE OF A FREE PARTICLE TO DESCRIBE THE EFFECT OF THE FIELD

It is important to remember that the general state vector of the free particle

$$\sum_{\xi} \int \frac{d^3 p}{p_0} \phi(p, \xi) |p, \xi\rangle \tag{7}$$

is a state vector in Heisenberg's sense, that is, in Wightman's words, sub specie aeternitatis. When we derive the various operators, such as the position operator, these will depend on the time of the measurement of the quantity in question. If these operators are the same as for the free particle, all the properties obtained will be the same as those of the free particle. In particular, if p is considered to be the momentum operator – this is independent of time – the momentum distribution will remain the same throughout time. Hence, if one wants to introduce the action of external fields on the particle, one has to make one of two possible modifications. One either has to change the operators which correspond to the various physical quantities, such as position, velocity, etc., the change to depend on the external field, or else one has to extend the Hilbert space and introduce state vectors different from (7). In the literature only the second alternative is considered and, also almost invariably, one works with the Fourier transform of (7), the wave function in Minkowski space.

The question naturally arises concerning the extent to which the Hilbert space is to be extended. In the case of the Dirac and Klein–Gordon equations, the restriction on the length of the momentum is abolished – this can range over all possible values. The restriction on the spin is, on the other hand, maintained. Furthermore, if the action of the field is of finite duration, the Fourier transform of the wave function will continue to contain a part which contains a Dirac delta function of the square of the momentum minus the mass square $\delta(p_t^2 - p_x^2 - p_y^2 - p_z^2 - m^2)$ so that both at $t = -\infty$ and at $t = \infty$ only the coefficient of this delta function will be relevant. But whereas one assumes that, before the interaction, the coefficient of the delta function at negative p_t was zero, this is no longer true after the interaction. The interpretation of this, in terms of the creation of antiparticles, was given of course by Dirac and Oppenheimer, and this interpretation is generally accepted now.

The point to which I wanted to call attention is different, though. It concerns particles with spins larger than 1/2. For these, no equations including the effect of a field have been proposed which would leave the spin unchanged. The extension of the Hilbert space to account for the effect of the field includes not only the whole four-dimensional momentum space but also all spin values, in integer steps, up to, and naturally including, the initial spin value. Furthermore, even if the interaction is of only finite duration, the lower spins do not disappear but remain just as the negative p_t part, caused by the electromagnetic interaction, remains.

It is not clear what this means. No two particles with identical masses but different spins are known. Does this mean that the interaction of particles with a field cannot be described for particles with spin 1 or more? Is this perhaps the seed of an explanation that no elementary particles with such spin exist? Or is it simply that we have not yet discovered all methods to describe the effects of fields? This is another puzzle worth adding to those mentioned before.

DISCUSSION

C. Fronsdal: Concerning Professor Wigner's remark about the instability of the spin for the higher wave equation, it is my understanding that the Fierz–Pauli theory was developed exactly to avoid this problem.

E. P. Wigner: The Fierz–Pauli theory, to my knowledge, has never been generalized to higher spins. Perhaps you are right that somebody will come around and find a very beautiful equation for particles with higher spin, and with interaction with the electromagnetic field, and perhaps even with other fields, but it looks very difficult.

C. Fronsdal: It has been done. Difficulties of a deeper nature have been pointed out, but the difficulty that you mentioned, I think, is not present.

E. P. Wigner: I do not know of equations which do what you indicate for particles with higher spin, but I hope I will learn from you about it.

A. Martin: I have a question about stable versus unstable particles. Professor Wigner mentioned that unstable particles are not orthogonal to the Hilbert space built out of stable particles and, of course, from a mathematical point of view it's certainly nice to build the Hilbert space exclusively with stable particles. However, I think, that from a physical point of view, we are really not in a satisfactory situation, because it appears that stability is a kind of dynamical accident, of which the most impressive example is the prediction of the omega-minus particle by Gell-Mann. The omega-minus particle is stable with respect to strong interactions, but it was predicted from the existence of resonances. So, somehow, the omega-minus particle would be more honourable than these resonances which were used to predict it, in spite of the fact that it is perfectly true to say that you can build a Hilbert space out of stable particles. It appears to me to be a very unsatisfactory situation.

E. P. Wigner: Dr. Martin is referring to an impromptu remark I made about unstable particles. The description of unstable particles as resonances is, from the point of view of present basic theory, the logical description. Surely, the neutron's state vector in Hilbert space is not orthogonal to all states of the trio of proton, electron, and antineutrino. It is, in fact, a linear combination of the state vectors of the triple. Nevertheless, since the neutron's properties are so very different from the ordinary states of the proton-electron-antineutrino triple, one is not used to describing the neutron as a resonant state of the triple. The situation is very different for the usual nuclear resonances the properties of which are expected to follow from the properties and interactions of the constituents. For unstable 'particles', such as the neutron or the omega-minus, this is not the case and it is natural, therefore, to disregard the basic resonance character of these and to treat them as independent particles. We have no theory which gives their properties in terms of those of the constituents. For this reason, I will at least temporarily and perhaps permanently, throw up my hands.

16

The Electron: Development of the First Elementary Particle Theory

Fritz Rohrlich

Section Headings: 1. Introductory Remarks; 2. Turmoil and Beginnings; 3. Thomson's Electron and the Lorentz Force; 4. The Models by Abraham and Lorentz; 5. The Saga of 4/3 and Poincaré Stresses; 6. Dirac's Classical Point Electron; 7. Recent Theory of the Point Electron; 8. Spin and Magnetic Moment; 9. The Classical Electron as a Limit of its Quantum Mechanical Description; 10. The Classical Electron Today.

1. INTRODUCTORY REMARKS

It is fair to say that the history of the electron lies in the very centre of the development of modern physics. It is therefore most appropriate to celebrate the 75th anniversary of the discovery of the electron (by Thomson) by reviewing its history and reporting on the present state of the classical theory.

But already before the experimental identification of the electron there existed a theory of electrons initiated by Lorentz (1892) which became the atomistic theory of all electrodynamic phenomena for a whole generation of physicists. The electron played the key role in identifying the photon as a particle, first in the photoelectric effect (1905) and then in the Compton effect (1922). The first particle of matter to show de Broglie wave behaviour was the electron (1927); the first particle of antimatter produced artificially was the positive electron (1932); and the first observable radiative correction was computed for the anomalous magnetic moment of the electron. Every physicist can easily add to this list.

The electron was the first elementary particle about which a detailed mathematical theory was constructed with the intention to describe its structure and its interaction. This classical theory of the electron eventually became a relativistic theory and enjoyed continued development and improvement long after the quantum mechanical theory had been started. Today, the classical theory of point charges has important applications such as particle accelerators and is still a stepping stone toward quantum electrodynamics in many investigations. And it will form a testing ground when a completely understood quantum electrodynamics will be found. For, the classical theory must emerge from it in a suitable limiting process.

We shall trace the development of this first elementary particle theory from its early beginnings to its present state. We shall then review to what extent this classical

J. Mehra (ed.), The Physicist's Conception of Nature, 331–369. All Rights Reserved
Copyright © 1973 by D. Reidel Publishing Company, Dordrecht-Holland

theory can be obtained as a limit of relativistic quantum mechanics which is of course still incomplete.

In this brief survey only the main points can be made and many important contributions to the subject will have to be omitted. Therefore, I cannot claim completeness in any sense of the word.

The literature is almost devoid of critical studies of the historical development of the theory of the electron. The most notable exceptions are the monumental work by Whittaker[1] and, (for the non-relativistic quantum mechanical aspect only) the excellent study by Jammer[2]. I am very indebted to these treatises, especially to the former, for many pointers and innumerable references to the original literature.

I shall not use the notation of the various authors quoted. This can only cause confusion, and it makes a comparison difficult since notations vary greatly and have changed over the years. I shall use Heaviside–Lorentz units; in the relativistic part I shall use a metric tensor for Minkowski space with trace $+2$. We shall also use units such that the velocity of light $c = 1$ whenever the dependence on c is not of interest.

2. TURMOIL AND BEGINNINGS

The natural philosophers of the 19th century were faced with descriptions of physical phenomena which were based on extreme opposites. There was the notion of a fundamental graininess of all things, of corpuscles, of atoms and of molecules; but there was also the notion of the continuum, the wave, the field as the basic entity. These are incompatibly different concepts, so that serious dilemmas arose when the same phenomenon could be explained on either basis. It is against this background of conceptual turmoil that the development of the classical theory of the electron must be viewed.

It started with optics. The corpuscular theory of Newton was the generally accepted explanation of optical phenomena. There was no greater scientist before him and his tremendous success in so many different fields from gravitation theory to differential calculus endowed his *Opticks* (1704) with tremendous prestige. Huygens, Newton's contemporary and adversary had successfully explained the discovery by Boyle and Hooke of 'Newton's Rings', but since those rings had also been explained by Newton's corpuscular theory, few scientists were willing to take seriously a wave theory of light. After all, light propagates in thin, straight lines through relatively small openings, very different from the propagation of waves, and very much like the notion of free particles. But the nineteenth century started with the first serious blow at the corpuscular theory when, in 1801, Thomas Young explained the colours of thin films by interference of waves. Thus was revived the old wave theory. Not long thereafter, the wave theory was led to its definitive triumph by Fresnel's explanation of diffraction, which was shown to dominate when the opening is made small enough, and of polarization (1816–18). The corpuscular theory of light, which was a corner stone in the attempt to reduce all of physics to mechanical terms, from then on died a quick death. History had moved from corpuscles to waves.

Then came electricity and magnetism. The electric attraction and repulsion of static charges used to be explained first by a two-fluid theory (Du Fay, 1730) and then by a one-fluid theory (Franklin, 1747). But then came the great discoveries by Ampère and by Faraday. Magnetic and electric phenomena became strangely linked to one another and those strange 'tubes of force' began to appear in the scientific literature by which Faraday seemed to be able to explain all the startling new phenomena he found. But the crowning achievement came when in 1864 Maxwell's 'On a Dynamical Theory of the Electromagnetic Field'[3] combined the extensive experimental knowledge of his day on this subject into a beautiful mathematical theory. The electromagnetic field thus won a permanent victory over the mechanical electric fluid of earlier days. History had moved from fluids to fields.

Lastly, came heat. The old phlogiston theory which thought heat to be a material fluid had been killed by Lavoisier in 1770. And there was no clear conception of the nature of heat for about one hundred years (except for a noteworthy but completely ignored paper by J. J. Waterston[4] in 1845). Luckily, the development in chemistry had put the molecular and atomic constitution of matter on a more and more secure foundation; it culminated in Mendeleyev's periodic table (1869). Thus, the time was ripe for a molecular theory of heat. Great contributions, especially by Maxwell, Boltzmann, and Gibbs established heat as a mechanical phenomenon of molecular motion, manifested on a statistical basis. Thus, rather unexpectedly, the mechanical view that had lost so much ground, had gained the subject of heat phenomena. At the same time, the atomistic view was firmly established in physics. History this time had moved from fluid flow to the statistical mechanics of molecules.

In the last twenty years of the 19th century, the developments of wave optics and of the theory of the electromagnetic field merged through the brilliant work of Hertz, Heaviside, and Kirchhoff, and optics became a branch of classical electromagnetic theory.

These developments, so briefly sketched above, must be kept in mind as the immediate background and fertile soil in which the beginnings of electron theory grew during the last two decennia of the last century. Maxwell's electrodynamics was available, at least in its main features, and the atomicity of matter was pretty close to an established fact.

3. THOMSON'S ELECTRON AND THE LORENTZ FORCE

The discovery of cathode rays and of the conduction of electricity through gases raised the urgent need for a better understanding of the forces that act between moving charged particles.

The model of force was, of course, Newton's great law of gravitation. And when Coulomb's torsion balance in 1785 showed quantitatively that a very similar law holds for electrically charged pith-balls the problem appeared to be settled. But then came the study of electric currents and the unexpected interplay of electric current, magnetic field, and force, and it became obvious that Coulomb's law did not tell the whole story.

The crucial step here was first the recognition that electric currents consist of moving electric charges (Fechner, 1845). It was then possible to devise generalizations of Coulomb's law which would explain (within the then extent knowledge) Ampère's law for the force between two current elements. This was attempted by Weber (1846), later improved by Riemann (1861), and tried again, in a different way by Clausius (1877).

But all these attempts at an action-at-a-distance law between charges failed and eventually had to be abandoned. The breakthrough came from those who doubted the action-at-a-distance philosophy. Their ranks had swelled since Maxwell's field theory became established, but it was not at all clear how one was to go about finding the interaction between moving charged particles via Maxwell's field.

For the static case the situation appears to be trivial. Two charges e and e' at a distance r interact by means of an electrostatic field $E = e/(4\pi r^2)$ produced by charge e. This field produces a force $F = e'E = e'e/(4\pi r^2)$ at the position of the other charge. The result is Coulomb's law.

But a moving charge also produces a magnetic field which in turn acts on the second charge and neither of these two steps were understood at the time. The pioneering work of J. J. Thomson[5] in 1881 consisted in a first attempt at such a field-mediated interaction. Maxwell's equations were not well enough understood, nor were the mathematical tools available which simplify such a calculation today. No wonder that there are several mistakes in that paper[6]. But the results are qualitatively correct for a slowly moving spherical charge of radius a and of total charge e distributed uniformly over its surface, acting on a similar charge e' of radius a'. He found that if e moves with velocity v, its kinetic energy is increased due to the presence of the magnetic field it produces, by

$$\Delta T = \tfrac{1}{2}\Delta m v^2 \tag{3.1}$$

with

$$\Delta m = f\, \frac{e^2}{4\pi a}, \tag{3.2}$$

an addition to the mass m of the particle, clearly of electromagnetic origin. The numerical factor f is of order one and quite irrelevant at the moment. Thus, a charged particle is heavier when it moves, heaving a mass $m + \Delta m$ while it would not change its mass if it were neutral.

But this was only a by-product of Thomson's work, since he was aiming at the interparticle force. Having obtained the magnetic field due to e, he then derived an expression for the action of such a field $\mathbf{B}$, say, on a charge moving with velocity $\mathbf{v}$, viz. $\tfrac{1}{2}\mathbf{v} \times \mathbf{B}$. This is only half the correct value. But he did carry through his programme and he found agreement with the force law suggested by Clausius.

His work was later corrected by FitzGerald[7], and improved especially by Heaviside who provided the missing factor 2 to the magnetic force law[8] and who found the electric and magnetic fields of a uniformly moving *point* charge of *any* speed[9]. But it took another ten years before the latter result was correctly generalized to finite size charges[10], due to the trickiness of the retardation effects in this case.

By the time of Thomson's work, a good deal of experimental work had been done on electrolysis and on the conduction of electricity through gases, so that the atomicity of electricity was also beginning to be understood. In 1894 this smallest unit of charge was given the name[11] 'electron'; but this particle was not found experimentally until three years later, in 1897. This was also achieved by J. J. Thomson[12], but independently and at the same time by Kaufmann[13]. Both were studying the properties of cathode X-rays. It must be remarked here that the knowledge of the correct force law for the action of electric and magnetic fields on moving charges was crucial in this discovery.

It was at first very surprising that the measurement yielded a ratio e/m some 2000 times larger than that which was known for a hydrogen ion. But it was everyone's guess that this was due to the smallness of m rather than due to an electron charge 2000 times larger. This belief was soon confirmed by Thomson's student, Townsend[14]. The absolute value of e, however was not measured accurately until Millikan[15].

During the ten years following Thomson's 1881 paper the understanding of the force between moving charges as mediated by electromagnetic fields had become better understood, especially because of the work of Heaviside. So that when in 1892 H. A. Lorentz set out on his ambitious project[16], he made the action of fields on moving charges one of the basic equations of his theory, parallel to Maxwell's equations for the fields produced by such charges. The force law

$$\mathbf{F} = e(\mathbf{E} + \mathbf{v} \times \mathbf{B}) \tag{3.3}$$

later became generally known as the 'Lorentz force'.

The fundamental idea of Lorentz's theory of electrons was to explain all the many diverse phenomena known at the time to be connected with charged particles, in terms of a theory of just one *fundamental particle*, the electron. If an atom is conceived to be a collection of electrons held together somehow by a positive charge very much heavier than the electrons, then the motion of the electrons, their mutual interactions and their interaction with the positive charges, must be at the basis of all electrodynamic phenomena. All macroscopic properties of matter are then to be determined by suitable averages over a large number of atoms or molecules.

Lorentz's force law can be obtained by searching for a force exerted by the electromagnetic field on a charge, such that when combined with Maxwell's equations it would yield Clausius' law of force between charges at least in some approximation. One can achieve this[17] by modifying the Lagrangian that yielded Clausius' law so that the charge *density* distribution on the electron is fully taken into account, and not only the total charge.

The fundamental equations of Lorentz thus are

$$\nabla \cdot \mathbf{E} = \sum_i \varrho_i \qquad\qquad \nabla \cdot \mathbf{B} = 0 \tag{3.4}$$

$$\nabla \times \mathbf{E} + \dot{\mathbf{B}} = 0 \qquad\qquad \nabla \times \mathbf{B} - \dot{\mathbf{E}} = \sum_i \varrho_i \mathbf{v}_i \tag{3.5}$$

together with the force law (3.3). Some years later these equations were combined into a Lagrangian formulation of the theory, not very different from the way this theory might be formulated today[18].

But we shall not pursue this theory here. What is of interest in the present context is only the description of the individual electron which is a necessary part of it.

4. THE MODELS BY ABRAHAM AND LORENTZ

After Thomson's discovery that the magnetic field produced by a moving electron results in an increase in mass, the question arose whether this additional mass Δm, (3.2), which is clearly of purely electromagnetic origin is a large fraction of the total mass of the electron. The exciting possibility presented itself that, maybe, *all* of the electron's mass is of purely electromagnetic origin.

In order to decide this question, it was thought that this mass must be studied for velocities not small compared to the velocity of light, since only then will the velocity dependence of this mass become observable and thus the electromagnetic part of the mass distinguishable from the rest. Such a velocity dependence had emerged from early calculations and when Kaufmann[19], for the first time showed experimentally that the electron mass does indeed increase when the velocity is not negligible compared to light, there was considerable excitement; at least some of the mass is indeed electromagnetic!

In the following year, Abraham computed the transverse[20] electromagnetic mass[21] and Kaufmann's results[22] were found to be in agreement with this calculation within experimental error. This was sufficient indication for Abraham and others that the electron is indeed a *purely* electromagnetic object.

A detailed study of the dynamics of the electron was then undertaken by Abraham[23] in his paper of 1903 (75 pages). The electron was assumed to be a spherical[24] rigid object with homogeneous surface or volume charge. The theory was therefore based on the Maxwell–Lorentz field equations (3.4), (3.5), for one electron, the Lorentz force equation (3.3), and the kinematic equations of a rigid (non relativistic!) body.

The most important result for future investigation was that Abraham was able to derive the equations of Newtonian dynamics also for the dynamics of an electron provided that the linear momentum due to the fields produced by the electron is

$$\mathbf{P} = \int \mathbf{S}\, d^3x, \qquad \mathbf{S} \equiv \mathbf{E} \times \mathbf{B} \tag{4.1}$$

where $\mathbf{S}$ is Poynting's vector[25], and provided that the angular momentum is

$$\mathbf{J} = \int \mathbf{r} \times \mathbf{S}\, d^3x. \tag{4.2}$$

Thus, it was apparently convincingly established that the same vector $\mathbf{S}$ which governs the flux of free electromagnetic fields also determines that part of the electron's momentum which is due to its moving Coulomb field. We shall return to this point in the following section.

Abraham also made the keen observation that $\mathbf{P}$ and $\mathbf{J}$ do not depend on the instantaneous motion only, but on the complete history of the electron. He fully realized the complication of the problem due to this fact.

For slow motion and for a surface distribution of charge on a spherical electron of radius a, he found

$$\mathbf{P} = \frac{e^2}{6\pi ac^2}\mathbf{v}, \qquad E_{mag} = \frac{e^2}{6\pi ac^2}\cdot\frac{v^2}{2}, \qquad E_{el} = \frac{e^2}{8\pi a}, \qquad (4.3)$$

for the momentum and the electric and magnetic energies due to the fields. In the general case [26], when $v^2 \ll c^2$ is *not* assumed, these expressions are multiplied by complicated functions of $(v/c)^2$. The result is that the usual relation between energy and momentum is completely destroyed, with the only exception that $E_{mag}=\frac{1}{2}Pv$ holds to all orders of $(v/c)^2$. For volume charge distributions all results are increased by a factor $6/5$.

If, as Abraham assumed, the electron is purely electromagnetic, Equation, (4.3) can be used to find its radius. An electron in slow motion then has electromagnetic mass

$$m = m_{elm} = m_{\parallel} = m_{\perp} = f\,\frac{e^2}{4\pi ac^2}, \qquad (4.4)$$

where $f = 2/3$ or $4/5$ depending on the assumption of surface or volume charge distribution. Thus, since m can be measured, the theory predicts

$$a = f\,\frac{e^2}{4\pi mc^2}. \qquad (4.5)$$

With $f = 1$ this expression later became known as 'the radius of the electron'.

In the following year, apparently in defence of his model, Abraham summarized his assumptions in detail and also criticized the new and competing model by Lorentz rather severely [27].

The most quoted article by Lorentz on the theory of electrons is his *Encyclopädie* article [28] of 1904. This article was finished in December 1903, while Abraham's long paper was finished in October, 1902. Lorentz therefore had the full benefit of this work while writing his article and he quotes Abraham extensively. In fact, as far as the structure of the electron is concerned, the article gives no new results. But it shows clearly how Abraham's work is a necessary consequence of Lorentz's electron theory once the assumptions on shape and charge distributions are made.

We cannot dwell here any longer on Lorentz's beautiful and extensive article. But one result given there must be mentioned. It is the expression for the 'resistance' which a slow electron feels under acceleration. This force (which is due to the self-inter-action just as the electromagnetic energy and momentum are) was given incorrectly in one of his earlier publications[16] and is now correctly given as

$$\frac{e^2}{6\pi c^3} \ddot{\mathbf{v}}.$$ (4.6)

It will play an important role later on.

It is interesting to note that this same model of the electron was still expounded by Lorentz as late as 1906 (in his lectures at Columbia University)[29]. Three years later, when his lectures were finally published he mentioned in the Preface: 'The publication of these lectures... has been unduly delayed, chiefly on account of my wish to give some further development to the subject... ' But he apologizes that '... neither Planck's views on radiation, nor Einstein's principle of relativity have received an adequate treatment'[29]. This was in 1909.

What happened, of course, was that after completion of his *Encyclopädie* article, Lorentz showed[30] that the Maxwell–Lorentz equations remain invariant under a certain transformation (now known as the Lorentz transformations). This transformation makes an electron which is spherical and at rest into one which is (or looks as if it were) spheroidal when moving with velocity v, and such that the axis along the direction of motion is *contracted* by a factor $1/\gamma$. Thus, Abraham's old sphere in motion which corresponded to a *prolate* spheroid at rest is replaced by a sphere at rest which becomes an *oblate* spheroid in motion. If now the forces are transformed correspondingly, then the charge distribution of the moving electron will be in equilibrium whenever that is the case for the electron at rest.

By means of his transformation of the coordinates and of the fields, the momentum of a moving electron can be calculated easily by transforming to its rest system. Using the Poynting vector (4.1) Lorentz found the electromagnetic momentum for fast electrons

$$P = m_{\text{elm}}\gamma\mathbf{v}$$ (4.7)

where the electromagnetic mass is as in (4.4). From this one immediately deduces (see ref. 20) that $m_{\parallel} = m\gamma^3$ and $m_{\perp} = m\gamma$. It is obvious that these are very much simpler results than those of Abraham. Furthermore, these results were in equally good agreement with Kaufmann's data[22].

The Lorentz transformations were independently and at about the same time derived by Einstein[31] in his first paper on special relativity. Except that for him these transformations were the description of the actual situation and not an 'as if' situation. He also derived the transformation equations for the electromagnetic field, as Lorentz did, but he added the expression for the kinetic energy of the electron as

$$W = (\gamma - 1) mc^2$$ (4.8)

showing that this is a consequence of the transformation laws and is *independent of the structure of the electron*. In the same paper, he also stated the 'principle of relativity' which was independently stated by Poincaré[32].

Thus we see that the beginnings of special relativity theory brought greater consistency to the understanding of electromagnetic momentum and energy, and much simpler expressions. But a full understanding of them was not attained until relativity was fully mastered many years later. The development of this understanding will be traced in the following section.

The full implications of Einstein's work were only beginning to dawn at that time. In Lorentz's words of that time: '...Einstein simply postulates what we have deduced, with some difficulty and not altogether satisfactorily, from the fundamental equations of the electromagnetic field. By doing so, he may certainly take credit for making us see in the negative results of experiments like those of Michelson... not a fortuitous compensation of opposing effects, but the manifestation of a general and fundamental principle.'[29]

5. THE SAGA OF 4/3 AND POINCARÉ STRESSES

We recall that J. J. Thomson first pointed out why at least part of the electron mass (and at least when the electron is moving) must be electromagnetic. We also recall that the velocity dependence of the electromagnetic mass was thought to be a characteristic not enjoyed by the non-electromagnetic mass, and that this dependence requires calculations to higher powers of $(v/c)^2$ in order to be seen.

What was not fully recognized in the early days of special relativity was that Einstein's theory requires the *same* 'velocity dependence' of all 'masses', (defined as p/v) so that in a relativistic theory this is no longer a distinguishing feature of electromagnetic mass. But incorrect application of this theory continued to obscure this fact until very recently. Several papers had been written in the intervening years, which applied the theory correctly. These, however, were either ignored or misunderstood. The result was that the velocity dependence of electromagnetic energy and momentum continued to be linked to the structure of the electron. In the present section, we shall trace the history of this long misunderstanding.

We cannot follow here the mathematical development of the theory of relativity, the recognition of the group character of the Lorentz transformations[33], the invention of Minkowski space[34] and the understanding of the vector and tensor character of various physical quantities, etc. All this is beautifully presented in Pauli's famous article.[35]

By that time, (1920), the relativistic description of the electron structure was as follows. A free uncharged particle of rest mass m moving with constant velocity $\mathbf{v}$ has a momentum $\mathbf{p} = m\gamma\mathbf{v}$ and a total energy $E = m + W = m\gamma$. Under Lorentz's transformations these quantities transform such that $p^\mu = (E, \mathbf{p})$ is a fourvector. In matter-free regions the electric field $\mathbf{E}$ and the magnetic field $\mathbf{B}$ transform such that $F^{\mu\nu}$ is a tensor under Lorentz's transformations, with $\mathbf{E} = (F^{01}, F^{02}, F^{03})$ and $\mathbf{B} = (F^{23}, F^{31}, F^{12})$. The *free* field theory (i.e. in the absence of matter) can be cast into Lagrangian form with $\mathscr{L} = -\frac{1}{4}F_{\alpha\beta}F^{\alpha\beta}$ choosing as the fundamental quantities the potentials, the scalar potential φ, and the vector potential $\mathbf{A}$. These are required to transform such that $A^\mu = (\varphi, \mathbf{A})$ is a four-vector. The fields are then determined by $F^{\mu\nu} = \partial^\mu A^\nu -$

$-\partial^{\nu}A^{\mu}$. Variation of A^{μ} yields, via the action integral $\int \mathscr{L} \, d^4x$, the free Maxwell field equations

$$\partial_{\mu}F^{\mu\nu} = 0. \tag{5.1}$$

Equivalently, if one makes the convenient choice of gauge

$$\partial_{\mu}A^{\mu} = 0 \tag{5.2}$$

(Lorentz gauge condition), one has the free Maxwell potential equations

$$\partial_{\alpha}\partial^{\alpha}A^{\mu} = 0. \tag{5.3}$$

Invariance of the action integral under translations[36] leads to the differential conservation laws

$$\partial_{\alpha}T^{\alpha\mu} = 0 \tag{5.4}$$

where

$$T^{\mu\nu} \equiv F^{\mu\alpha}F_{\alpha}{}^{\nu} + \tfrac{1}{4}\eta^{\mu\nu}F_{\alpha\beta}F^{\alpha\beta} \tag{5.5}$$

is the symmetric electromagnetic energy–momentum–stress tensor. Its components $-(T^{01}, T^{02}, T^{03})$ are seen to be just the components of the Poynting vector $\mathbf{S}$ defined in (4.1), $-T^{00}$ is just $\tfrac{1}{2}(E^2+B^2)$, i.e. is the energy density U; the nine components T^{ij} can be conveniently written as a dyadic $\mathbf{T}$ and comprise the Maxwell stress tensor. The conservation laws (5.4) thus state Poynting's theorem

$$\frac{\partial U}{\partial t} + \nabla \cdot \mathbf{S} = 0 \quad \text{and} \quad \frac{\partial \mathbf{S}}{\partial t} - \nabla \cdot \mathbf{T} = 0. \tag{5.6}$$

Integration over all space then gives the total electromagnetic energy

$$E = -\int T^{00} \, d^3x = \int U \, d^3x \tag{5.7}$$

and (by 4.1) the total electromagnetic momentum

$$\mathbf{P} = \int \mathbf{S} \, d^3x. \tag{5.8}$$

Do the four quantities $P^{\mu} = (E, \mathbf{P})$ transform as a four-vector? This is not ensured by just writing it as

$$P^{\mu} = -\int T^{0\mu} \, d^3x \tag{5.9}$$

and so clearly must depend on the properties of the $T^{0\mu}$. Let us carry out the transformation explicitly. Assume that P^{μ} is to represent the electromagnetic energy and momentum of a charged particle (an electron say), moving with velocity $\mathbf{v}$ so that the $F^{\mu\nu}$ in (5.5) are those produced by this particle. A Lorentz transformation to the rest system of the particle with coordinates $x_{(0)}^{\mu}$ gives $d^3x = d^3x_{(0)}/\gamma$ by Lorentz contraction, and one finds

$$E = - \gamma \left[\int T^{00}_{(0)} \, d^3x_{(0)} + v^2 \int T^{11}_{(0)} \, d^3x_{(0)} \right] \tag{5.10}$$

$$\mathbf{P} = - \gamma \mathbf{v} \left[T^{00}_{(0)} \, d^3x_{(0)} + \int T^{11}_{(0)} \, d^3x_{(0)} \right]. \tag{5.11}$$

Here one uses the fact that in the rest system the components of S vanish and that, because of spherical symmetry, $T_{(0)}$ is diagonal and

$$\int T^{11}_{(0)} \, d^3x_{(0)} = \int T^{22}_{(0)} \, d^3x_{(0)} = \int T^{33}_{(0)} \, d^3x_{(0)}.$$

This result shows the P^μ is a four-vector (i.e. it has exactly the same transformation law as p^μ) if and only if the 'self-stress' $\int T^{11}_{(0)} \, d^3x_{(0)}$ vanishes. Thus is established *a link between the covariance of P^μ and the stability of the electron*, because the electron is stable if and only if the self-stress vanishes. Substitution into (5.10) and (5.11) of the Coulomb field for a spherical electron with surface charge distribution (interior field $\mathbf{E}=0$) yields

$$E = m_{\text{elm}} \gamma (1 + v^2/3) \tag{5.12}$$
$$\mathbf{P} = \tfrac{4}{3} m_{\text{elm}} \gamma \mathbf{v} \tag{5.13}$$

where the electromagnetic mass is

$$m_{\text{elm}} = \frac{e^2}{8\pi a}. \tag{5.14}$$

Thus, the momentum (5.13) is exactly the result (4.7) of Lorentz, and in the slow motion limit ($v^2 \ll 1$) (5.12) and (5.13) agree exactly with the result (4.3) of Abraham when $E_{\text{mag}} \ll E_{\text{el}}$. But they are *not* the relativistic equations for E and $\mathbf{P}$ of a particle of mass m_{elm}.

These results have nevertheless been obtained by various authors since 1905, on various levels of sophistication. They are given in Pauli's article [37] and in many standard references.[38] As soon as special relativity was accepted, the strange formulae (5.12) and (5.13) were blamed on the instability of the electron, i.e., on the need for non-electromagnetic forces to keep the negative charge distribution from exploding. Thus, ironically, the velocity dependence of the 'mass' which earlier was considered proof of the electromagnetic nature of the electron (Abraham[23]), now became proof of the existence of *non*-electromagnetic forces.

The first one who spelled this out quantitatively was Poincaré[33] in 1906. He simply postulated suitable cohesive forces ('Poincaré stresses') in order to stabilize the electron. These stresses must, of course, have rather strange transformation properties in order to cancel the strange transformation law (5.12), (5.13). This fundamental paper by Poincaré is beautifully written, giving for the first time the group properties of Lorentz transformations and rotations, the correct transformation of the charge-current density, etc., but the section on the internal stresses of the electron deviates markedly from the clarity of the rest of the paper.

The first one who realized that there was something wrong with the factor 4/3 in (5.13) was Fermi [39] in 1922. But his papers used a rather sophisticated approach and were poorly written. They were either not understood or not read. In 1926, H. Mandel and in 1936, W. Wilson [40] very clearly showed how to define P^μ such that it transforms as a four-vector and independently, in 1949, Kwal [41] did the same. Finally, in 1960, not aware of these papers, I rediscovered the same for the fifth time. [42] The inclusion of this matter in my book [43] together with a detailed explanation probably helped in making it finally generally known.

The essence of these papers can be most simply presented as follows. Ordinary three-dimensional space is geometrically a three-dimensional hyperplane orthogonal to the t-axis in Minkowski space. Its covariant generalization is a three-dimensional hypersurface which is spacelike (i.e., any two points on it have a spacelike separation). Such a surface has a volume element $d^3\sigma^\mu = n^\mu \, d^3\sigma$ where $d^3\sigma$ is an invariant and n^μ the (timelike) normal to the surface. (We can here restrict ourselves to spacelike planes for which n^μ is independent of position). We can then express the electromagnetic energy and momentum of (5.9) as

$$P^\mu = -\int d^3\sigma_\alpha T^{\alpha\mu}. \tag{5.15}$$

Now a simple application of Gauss' theorem to Minkowski space shows that this integral will be independent of the choice of the spacelike surface over which one integrates if and only if $\partial_\alpha T^{\alpha\mu} = 0$, i.e. if (5.4) holds. [44]

The identification of electromagnetic energy and momentum with certain integrals over components of the energy-momentum-stress tensor, (5.7) and (5.8), were originally done for the *free* electromagnetic field, as is indicated in (5.1) through (5.8), and as is clear from the work of Poynting. [25] For such a field $T^{\mu\nu}$ is divergence free; (5.4) and the integral (5.15) is thus independent of σ. The particular choice (5.9) for σ is therefore justified.

When the electromagnetic field interacts with matter, however, as is always the case when one describes the self-interaction with the Coulomb field, the field no longer satisfies the free field equation *everywhere* in space. If $j^\mu = (\varrho, \mathbf{j})$ is the charge-current density four-vector, the Maxwell equations (5.3) are replaced by

$$\partial_\alpha \partial^\alpha A^\mu = -j^\mu \tag{5.16}$$

and $j^\mu \neq 0$ holds *somewhere* on *every* spacelike surface. The conservation law (5.4) now reads

$$\partial_\alpha T^{\alpha\mu} = F^{\mu\beta} j_\beta \tag{5.17}$$

with $T^{\mu\nu}$ as before, (5.5). Thus, the integral for P^μ, (5.15) is *not* independent of the surface σ.

The question therefore arises, which is the correct spacelike plane σ to take in the definition of the electromagnetic energy and momentum of an electron, (5.15). The choice made in (5.9) is to take for σ the plane $t = \text{const.}$, the 'now-plane' of the observer who sees the electron moving with velocity $\mathbf{v}$. If this is the definition of P^μ then, clearly, every observer will see different values for the components of P^μ, depending on his $\mathbf{v}$; but these values are *not* related by a Lorentz transformation, because P^μ is not a four-vector. Therefore, this definition has no relativistic meaning.[45]

One of the fundamental lessons of relativity is the equivalence of inertial observers. This means that all physical observables, if they are to be meaningful relativistically, must be so defined that different observers can relate them by Lorentz transformations, i.e., that all inertial observers 'see the same thing' modulo a Lorentz transformation. The definition (5.9) of P^μ does *not* satisfy this requirement.[46]

In order to arrive at a relativistic definition one must fix the plane in a Lorentz invariant way. To this end one notes that the rest mass of a particle is defined as its total energy (divided by c^2) when at rest. Consequently, P^μ must describe the electromagnetic mass m_{elm} of the electron when it is at rest, $P^\mu_{(0)} = (m_{\text{elm}}, \mathbf{0})$. For this definition σ must be the $t = \text{const}$ plane in the rest system of the electron. All other values of P^μ then follow by Lorentz transformation. Thus, P^μ defined by (5.15) will be relativistically meaningful provided σ is *always* chosen to be that plane. Mathematically, this choice of σ means that the normal n^μ is identified with the particle velocity v^μ.

This procedure ensures that P^μ will be a four-vector. Carrying out the transformation one finds [43, 47]

$$E = \gamma \int (U - \mathbf{v} \cdot \mathbf{S}) \, \mathrm{d}^3\sigma \qquad (5.17)$$

$$\mathbf{P} = \gamma \int (\mathbf{S} + \mathbf{v} \cdot \mathbf{T}) \, \mathrm{d}^3\sigma. \qquad (5.18)$$

When this is compared with (5.7) and (5.8) we see that even in the *nonrelativistic limit* the electromagnetic momentum $\mathbf{P}$ was incorrectly defined: the term $\mathbf{v} \cdot \mathbf{S}$ is of order v^2 relative to U and can therefore be neglected; but the term $\mathbf{v} \cdot \mathbf{T}$ is of the same order as $\mathbf{S}$ and can therefore *not* be neglected. Thus, the factor 4/3 goes back to an error first made by Abraham [21, 23] in using the Poynting vector as electromagnetic momentum density for moving Coulomb fields, while it is true only for *free* electromagnetic fields, i.e. radiation. Abraham was led to this erroneous notion by his 1903 paper [23] where he derived (4.1) above from the noncovariant combination of Newtonian dynamics and Maxwell's equations.[48]

One verifies easily by explicit computation that (5.17) and (5.18) yield

$$E = m_{\text{elm}}\gamma, \qquad \mathbf{P} = m_{\text{elm}}\gamma\mathbf{v} \qquad (5.19)$$

instead of (5.12) and (5.13).[49]

Thus, P^μ will be a four-vector, provided it is so defined. Its transformation properties have nothing to do with the structure of the electron, but depend only on how P^μ is defined. This realization has two important consequences:

(a) Since P^μ and the energy-momentum of an electrically neutral particle, p^μ, transform the same way, there is no way to distinguish the two parts m and m_{elm} of the total observed mass

$$m_{\text{obs}} = m + m_{\text{elm}}. \tag{5.20}$$

In particular, the old expectations that experiments on velocity dependence would permit a distinction between the electromagnetic and the non-electromagnetic mass, were completely erroneous.

(b) Whatever forces hold the electron together (Poincaré stresses in the general sense) can now be described in a covariant way and need not have rather peculiar transformation properties to match those in (5.12) and (5.13).

While the instability of the classical electron is not apparent from the transformation properties of P^μ, it is apparent from the fact that the electromagnetic energy tensor $T^{\mu\nu}$, derived from the moving Coulomb field, is not divergence-free. Something must be added to make the electron a closed system. There are, of course, many possibilities. The simplest of these is perhaps a cohesive pressure Π providing a tensor $\Pi\eta^{\mu\nu}$ so that

$$\partial_\mu(T^{\mu\nu} + \Pi\eta^{\mu\nu}) = 0. \tag{5.21}$$

This determines a covariant 'Poincaré stress' which ensures stability.[49]

But such ad hoc devices are not very attractive. The suspicion arises that other possibilities might exist for solving this problem. This was first recognized by Lorentz when he stated in his *Theory of Electrons* (ref. 29, p. 215): 'In speculating on the structure of these minute particles we must not forget that there may be many possibilities not dreamt of at present; it may very well be that other internal forces serve to ensure the stability of the system, and perhaps, after all, we are wholly on the wrong track when we apply to the parts of an electron our ordinary notion of force.'

6. DIRAC'S CLASSICAL POINT ELECTRON

The development of quantum mechanics directed interest away from a classical understanding of the electron. In fact, the expectation was that the deflated hope for an all-electromagnetic electron would now be realized by this new and promising theory which answered so many questions so well. But nothing of the sort materialized, and the old problems reappeared in a new disguise. Worse than that, new discoveries seemed to seal the hope for an electromagnetic electron: 'First, the discovery of the neutron has provided us with a form of mass which it is very hard to believe could be of electromagnetic nature. Secondly, we have the theory of the positron – a theory in agreement with experiment so far as is known – in which positive and negative values for the mass of an electron play symmetrical roles. This cannot be fitted in with the electromagnetic idea of mass, which insists on all mass being positive, even in abstract theory.' So writes Dirac[51] in 1938.

In fact, Dirac suggests to give up the idea of looking for a structure of the electron because '... the electron is too simple a thing for the question of the laws governing its structure to arise, and thus quantum mechanics should not be needed for the solution of the difficulty. Some new physical idea is now required...?[51] He then proceeds to develop a classical theory of a point charge with the aim of getting equations of motion, even at the cost of having to dodge infinite terms.

Whatever one may think today of Dirac's reasons in developing a classical theory of a *point* electron, it is by many contemporary views (and I completely concur), the correct thing to do: if one does not wish to exceed the applicability limits of classical (i.e., non-quantum) physics one cannot explore the electron down to distances so short that its structure (whatever it might be) would become apparent. Thus, for the classical physicist the electron is a point charge within his limits of observations. Can one construct a consistent point charge theory? Dirac in 1939 made the first big step in this direction.

A point charge had never before been seriously suggested because in the limit $a \rightarrow 0$ the electromagnetic mass (5.14) and with it the energy and the momentum of the electron would diverge (whether one uses (5.12) and (5.13), or (5.19) is here irrelevant). Dirac's genius made him nevertheless follow this road and, as we shall see, it led to success eventually.

Before Dirac a number of dynamical results had been known. The rate of radiation emission from an accelerated electron was first found by Larmor[52] for slow electrons, a formula well-known to all students of electrodynamics. But the corresponding results for fast electrons are less well known; they were first derived by Abraham[53] and Heaviside,[54] and later, with the benefit of relativity theory by v. Laue.[55] But the fact that this rate of energy loss due to radiation, dE/dt, is a positive definite invariant when referred to unit (ordinary) time was not fully appreciated even after Dirac's work although this is quite evident from that paper as well as from earlier work. In present day notation this invariant radiation rate is

$$\mathscr{R} = \frac{e^2}{6\pi c^3}\, \dot{v}_\alpha \dot{v}^\alpha \tag{6.1}$$

where $\dot{v}^\mu$ is the acceleration four-vector, $\dot{v}^\mu = dv^\mu/d\tau$, τ being the proper time.

Concomitant with radiation emission another effect had been known for some time. Lorentz[28] first derived the force (4.6) which he identified[56] as a 'resistance to the motion'. Its relativistic generalization, in contemporary notation

$$\frac{e^2}{6\pi c^3}\, \ddot{v}^\mu \tag{6.2}$$

was later given independently by v. Laue[55] and Abraham.[57] The zero component of (6.2) is sometimes referred to as the 'acceleration energy' or 'Schott energy',[58] the whole four-vector as 'Schott term'. It ensures energy conservation in radiation emission.

The understanding of the relativistic dynamics before Dirac's work of 1938 was the following.[55, 57] If the Lorentz electron of radius a, i.e., the Lorentz contracting charged sphere, is subject to an external force, F_{ext}^μ, and the self-interaction is taken into account one obtains the equation of motion

$$m\dot{v}^\mu = F_{ext}^\mu + F_{self}^\mu. \tag{6.3}$$

Now F_{self}^μ can only be computed by expansion in an infinite series, due to the complication of the retardation effects. This series is effectively an expansion in powers of the dimensionless operator $(a/c)(d/dt)$. Consequently, F_{self} depends on a, i.e., on the radius of the electron, as well as on its charge distribution. However, if one does not go beyond second derivatives of the velocity, the structure dependence can be completely absorbed into the electromagnetic mass m_{elm}, which appears only in the electromagnetic energy-momentum P^μ first defined by (5.9) and later corrected to (5.15). One then finds

$$m\dot{v}^\mu = F_{ext}^\mu - \frac{dP^\mu}{d\tau} + \Gamma^\mu$$

where the 'Abraham–Lorentz four-vector of radiation reaction' is denoted by Γ^μ. It is defined by

$$\Gamma^\mu \equiv \frac{e^2}{6\pi c^3} \left(\ddot{v}^\mu - \dot{v}^\alpha \dot{v}_\alpha v^\mu \right). \tag{6.4}$$

If the $dP^\mu/d\tau$ term is lumped together with the inertial term of the left side one obtains, using (5.15) and (5.20),

$$m_{obs}\dot{v}^\mu = F_{ext} + \Gamma^\mu, \tag{6.5}$$

which is a *structure independent* equation. Before Dirac this lumping together however was not carried out. It was made impossible by the '4/3 problem' discussed in the preceding section which was still prevalent. But more importantly, the result (6.5) was considered to be only *approximate*, because higher terms in $(a/c)(d/dt)$ were neglected. And the limit $a \to 0$ was prohibited by the divergence of P^μ in that limit.

　　With this background in mind we see that Dirac's elegant paper[51] solved several problems simultaneously, and put the dynamics of a point charge on a solid basis. This basis consists of the Maxwell–Lorentz equations for a point charge and the law of conservation of energy and momentum. Thus, he started with the field equations (5.16) and (5.2) and a current density

$$j^\mu(x) = e \int_{-\infty}^{\infty} \delta(x - z(\tau)) v^\mu(\tau) \, d\tau \tag{6.6}$$

where $z^\mu(\tau)$ is the world line of the point charge. He then assumed this world line to be encased in a narrow tube of radius ε and required that the energy and momentum that enter the past end should equal what emerges at the future end, together with what is lost through the casing. Eventually, the limit $\varepsilon \to 0$ was taken.

　　As a technique of great advantage he did not work with the retarded fields, but split them up as follows:

$$F_{\text{ret}} = F_+ + F_- \tag{6.7}$$

$$F_\pm \equiv \tfrac{1}{2}(F_{\text{ret}} \pm F_{\text{adv}}). \tag{6.8}$$

In the $\varepsilon \to 0$ limit, this separation yields a divergent self-interaction term

$$F^\mu_{\text{Coul}} = eF^{\mu\nu}_+ v_\nu \tag{6.8}$$

and a non-divergent self interaction [59]

$$\Gamma^\mu = eF^{\mu\nu}_- v_\nu \tag{6.9}$$

the latter being identical with (6.4). Dirac's main result was the equation

$$\dot{B}^\mu + m_{\text{elm}}\dot{v}^\mu = F^\mu_{\text{ext}} + \Gamma^\mu$$

where $\dot{B}^\mu$ must satisfy $\dot{B}^\mu v_\mu = 0$. It is at this point that Dirac must take an additional assumption, taking for B^μ the simplest possibile choice, $B^\mu = mv^\mu$. He justifies this choice by the observation that all the other possibilities 'are all much more complicated ... so that one would hardly expect them to apply to a simple thing like an electron.'[51] Apart from this ambiguity, Dirac succeeded in deriving the equations of motion (6.5) using only the field equations and the conservation laws.

Implicit in this result is of course the solution (5.15) to the '4/3 problem' since he finds the covariant result $m_{\text{elm}}\dot{v}^\mu$. The lumping together of $m\dot{v}^\mu$ and $m_{\text{elm}}\dot{v}^\mu$ into $m_{\text{obs}}\dot{v}^\mu$ is carried out here with the understanding that m must diverge with $\varepsilon \to 0$ so that m_{obs} is independent of ε and equal to the observed mass. This process became later very popular in quantum field theory under the name of 'mass renormalization'.[60]

Thus was obtained a dynamical theory of the point electron. But there were also difficulties, even after mass renormalization, and these difficulties have caused much discussion ever since. They are of three kinds; I shall discuss these here briefly and I shall also indicate how I think they can be resolved in a consistent way.

The first is the problem of *runaway solutions*. These are solutions of the Lorentz–Dirac equation (6.5) in addition to the physical solution. In the limit of vanishing external field they all reduce to solutions describing self-acceleration for a free electron. The mathematical origin of these solutions is the fact that (6.5) is of third order and there are only two physical initial conditions, position and velocity. The resulting one-parameter family contains only one solution which is physically acceptable. Dirac proposed to resolve this difficulty by an asymptotic condition on the acceleration: it should vanish for an asymptotically free particle. Different arguments leading to similar conditions on the acceleration were given by Bhabha in 1947 and by myself in 1961.[61] These arguments are based on the physical requirement that in the limit $e \to 0$ the trajectory of a charged particle should have a limit and that this limit be the trajectory of a corresponding neutral particle ('Principle of Undetectability of Small Charges'). These asymptotic conditions permit only the physical solution. We shall return to them in the following section.

The second kind of difficulty of the Lorentz–Dirac equation is that the physical solution shows *pre-acceleration*, i.e. the presence of acceleration already before the onset of a force (and correspondingly the presence of deceleration still after the cease of a force). This phenomenon is a violation of causality in its primitive meaning, or, as Dirac puts it, 'failure ... of some of the elementary properties of space-time'. In my opinion this difficulty is only apparent, because it takes place outside the validity limits of this classical theory: the causality violation involves time intervals of the order of 10^{-23} sec which are outside the domain of validity of classical physics.

It should be pointed out that the runaway problem and the causality problem are unrelated and both are unrelated to the problem of the divergent self-energy which is solved by mass renormalization. Some authors seem to blame one difficulty on the other which seems to me unjustified.

The third type of difficulty is that for certain problems no physically meaningful solution seems to exist. The example *par excellence* is the head-on collision of an electron with the attractive Coulomb field of a point source, the latter being considered infinitely heavy (ref. 43, p. 186). Here the solution becomes physically meaningless when the electron reaches a distance of the order of the classical electron radius from the point source. My answer here is the same as before, viz. that such distances are outside the domain of classical physics. A static Coulomb field must be due to a macroscopic object whose size is much larger than the classical electron radius. In that case there is no problem.

We shall return to these questions again in the context of more recent work.

7. RECENT THEORY OF THE POINT ELECTRON

The first one to argue in favour of a point electron and against its static self-interaction was apparently Frenkel. [62] Already at the time of Abraham doubt was expressed about the the mutual interaction of the various (differential) parts of the electron which leads to the self-energy. (See also Lorentz's comment quoted at the end of Section 5.) If the electron is 'elementary', how can it have parts? This among other arguments suggested to Frenkel that any question concerning such a self-interaction is meaningless and that the electron must be a point. But nobody knew how to eliminate this self-interaction and at the same time keep Maxwell's equations, keep radiation reaction (to conserve energy and momentum), and be mathematically consistent. [63] Dirac realized of course that subtracting an undesirable (infinite) term is not part of a satisfactory theory.

There were those who felt that in order to construct a theory in which an electron's field does not interact with its own source but only with other electrons, one must first of all consider a *system* of electrons. Such a system was also studied by Dirac in his 1938 paper as a direct generalization of the one-electron case. He remarked there that one can eliminate the self-interaction by a suitable asymptotic requirement on the fields. Then only the F_+ fields of all the *other* particles will act on any one electron. This is just what one obtains by means of an action principle first studied by Tetrode and by Fokker.[64] In this action principle only the electrons appear, the fields have – in a sense – all been eliminated. Thus, Fokker's method was based on an action-at-a-distance philosophy.

It is exactly for this reason that the self-interaction is easy to eliminate in Fokker's case: only particle labels occur and it is easy to omit the $i=j$ terms in a double sum. Ironically, after the great victory of field theory over action-at-a-distance theories, the latter seemed to be successful in overcoming a difficulty which field theory apparently could not cope with. In a field theory one must be aware of the fact that one measures only the *total* field at a point; there is therefore no way to exclude the component that has its origin on a particular electron.

This theory by Fokker was revived again by Wheeler and Feynman[65] and was analysed by them in great detail. Dirac's work has shown quite clearly that the Abraham–Lorentz four-vector Γ^μ comes entirely from the F_- field while the infinite self-energy comes entirely from the F_+ field (see Equations (6.8) and (6.9)). Thus one does not want to eliminate *all* self-interaction, but only the part due to F_+. This recognition overcomes the first stumbling block of previous attempts.

Let us assume that we have no external fields present so that the n electrons in interaction form a closed system. Then Fokker's action integral leads to the system of simultaneous equations

$$m\dot{v}_k^\mu = e \sum_{\substack{i=1 \\ (i \neq k)}}^n F_{i,\,+}^{\mu\alpha}(z_k)\, v_{k\alpha} \quad (k = 1, 2, ..., n) \tag{7.1}$$

where $F_{i,\,+}^{\mu\nu}.(z_k)$ is the F_+ field produced by the ith electron on the world line of the kth electron. This equation will be identical with Dirac's equation

$$m\dot{v}_k^\mu = \Gamma_k^\mu + e \sum_{\substack{i=1 \\ (i \neq k)}}^n F_{i,\,\mathrm{ret}}^{\mu\alpha}(z_k)\, v_{k\alpha} \quad (k = 1, 2, ..., n) \tag{7.2}$$

provided

$$\Gamma_k^\mu + e \sum_{i=1}^n F_{i,\,-}^{\mu\alpha}(z_k)\, v_{k\alpha} = 0.$$

But because of (6.9) this equation means

$$\sum_{i=1}^n F_{i,\,-}^{\mu\nu}(z_k) = 0 \quad \text{or} \quad \sum F_{i,\,\mathrm{ret}}^{\mu\nu} = \sum F_{i,\,\mathrm{adv}}^{\mu\nu}. \tag{7.3}$$

This is the Wheeler–Feymann 'complete absorber' condition, which requires all fields that were produced in the past to be absorbed in the future. Obviously, this condition cannot be satisfied when $n=1$ or n is small.

This theory, therefore, succeeded in eliminating the divergent self-energy problem and at the same time, it demonstrated that an action-at-a-distance theory, can be mathematically equivalent to a renormalized field-mediated interaction theory.

But we see that the Fokker–Wheeler–Feynman formulation of the point electron suffers from three conceptual difficulties:

(1) It cannot describe a single electron.

(2) It regards electromagnetic fields, including radiation fields, strictly as secondary, derived objects, the primary objects being the charges.

(3) It requires the absorber condition (7.3) as an assumption of first principle.

It should be added that there are those who do not find the absorber condition unsatisfactory. In any case, it was widely believed at the time that the absorber condition is the price one has to pay for the action principle, because it was considered impossible to derive the Lorentz–Dirac equation (6.5), which contains $\ddot{x}$, from a Lagrangian which contains only first derivatives.

The solution to these difficulties came to me in the early sixties from the realization, mentioned already earlier, that one wants to avoid only the self-interaction related to the Coulomb field and not the one related to radiation reaction, i.e., one wants to eliminate the F_+ self-interaction[66] and keep the one with F_-. Since elimination of an interaction can be done best in terms of the particle variables, as Fokker, Wheeler, and Feynman showed, *only the F_+ interaction needs to be put into this form.* Thus one is led to a theory which is of the action-at-a-distance type only for the Coulomb field (F_+-field) but which *remains a field theory with respect to the radiation field* (F_--field).

This realization agrees beautifully with the quantum mechanical understanding of electromagnetic field: only the radiation field is composed of photons (i.e. must be quantized) while the Coulomb field is not (i.e. should not be quantized). This, in turn, leads evidently to the Coulomb gauge which is, in this sense, the natural gauge. In any case, the elimination of the Coulomb field is physically easily justified, the elimination of the radiation field, however, is not, because it would mean that the photon is not as elementary a particle as the electron, a notion that I find difficult to maintain on this level of theory.

Based on these consideration it was easy for me to write down a Lagrangian for a single charge as well as for a system of point charges which interact with one another directly by means of a (covariant) Coulomb interaction and which interact also with the F_--field produced by *all* the particles.[67] For a single electron in an incident free electromagnetic field, F_{in}, this Lagrangian yields the action integral

$$I = -m \int \sqrt{-\dot{z}^\alpha \dot{z}_\alpha}\, d\lambda + e \int \bar{A}_\alpha(x)\, \delta[x - z(\lambda)]\, \dot{z}^\alpha\, d\lambda\, d^4x - \tfrac{1}{2} \int F_{\alpha\beta} F_+^{\alpha\beta}\, d^4x,$$

$$(7.4)$$

where the dot indicates differentiation with respect to the parameter λ, and $\bar{F}^{\mu\nu}$ and $F_+^{\mu\nu}$ are defined as 4-curls of the potentials $\bar{A}^\mu$ and A_+^μ. The asymptotic conditions yield the following identification of these two fields:

$$\bar{F} = F_{\text{in}} + \tfrac{1}{2}(F_{\text{ret}} - F_{\text{adv}}), \qquad F_+ = \tfrac{1}{2}(F_{\text{ret}} + F_{\text{adv}}).$$

$$(7.5)$$

This action integral yields the Maxwell–Lorentz equations and the Lorentz–Dirac equation; but the latter contains no self-energy term, so that $m = m_{\text{obs}}$.

For a system of electrons with world lines $z_i^\mu(\lambda_i)$ ($i=1, ..., n$) the above action (7.4) is easily generalized, except that now there is an additional term present which describes the *mutual* Coulomb interaction, the terms $i \gg j$ being omitted

$$\frac{e^2}{8\pi} \sum_{\substack{i=1 \\ }}^{n} \sum_{\substack{j=1 \\ (i \neq j)}}^{n} \int \dot{z}_i^\alpha \dot{z}_{j\alpha} \delta \left[(z_i - z_j)^2 \right] d\lambda_i \, d\lambda_j. \tag{7.6}$$

This is therefore an action-at-a-distance interaction.

In this theory the field F_+ emerges a s asecondary quantity, determined completely by the particle variables; this is analogous to the scalar potential in the Coulomb gauge which is also defined in terms of the charges.

Thus, I was able to give an action principle which leads to the desired field equations and equations of motion of a system of point electrons, and which contain *no* electromagnetic mass terms. This theory involves *no renormalization* and there are *no divergent expressions*.

We also note that here the single electron has *no Coulomb field*, because Coulomb interactions arise only between two different charges. And, of course, there is no longer an 'absorber condition'.

The difficulties of the Fokker–Wheeler–Feynman formulation were thus all successfully overcome; but the theory was still not trouble-free. At this point the difficulties first encountered by Dirac in 1938 are still with us. If one accepts Dirac's solution of the runaway problems by postulating no asymptotic acceleration in the distant future (e.g., $\dot{v} \to 0$ in the $t \to \infty$ limit for asymptotically free particles), this postulate must be a permanent concomitant of the Lorentz–Dirac equation.[68]

A mathematical formulation of this situation combines the third order differential equation with an asymptotic condition yielding a mathematically equivalent second order integro-differential equation. This was first done in the non-relativistic case by Ivanenko and Sokolov and by Haag[69] and then in the one-dimensional relativistic case[70] by Plass. The general equation of motion in three space-dimensions was given by me[71] in 1961:

$$m_{\text{obs}} \dot{v}^\mu = \int_0^\infty K^\mu (\tau + \alpha \tau_0) e^{-\alpha} \, d\alpha \tag{7.7}$$

where

$$K^\mu \equiv F_{\text{ext}}^\mu - \mathcal{R} v^\mu. \tag{7.8}$$

Here the invariant radiation rate, (6.1) is again denoted by $\mathcal{R}$. This fundamental equation shows that when the asymptotic condition is incorporated into the Lorentz–Dirac equation, the Schott term disappears[72] and in its place a non-local behaviour in time appears: the acceleration depends on the complete future action of the force K^μ. The latter is just the externally imposed force reduced by the energy-momentum four-vector rate of radiation loss. This non-locality, however, is strongly damped out, so that it is effectively present only over time intervals of the order of τ_0 where

$$\tau_0 \equiv \frac{e^2}{6\pi m c^3} \sim 0.6 \times 10^{-23} \text{ sec}. \tag{7.9}$$

The pre-acceleration found by Dirac is now explicitly apparent from the non-local character of the equations of motion. It is clearly a violation of causality in the classical sense. But, being governed by time intervals of the order of τ_0, it is clearly outside the applicability domain of classical physics. In fact, 10^{-23} is the life time of virtual quantum processes.

In any case, the new equation of motion has solutions which are uniquely determined by initial position and velocity data. There are no runaway solutions.

The most recent progress in this field was accomplished by Claudio Teitelboim, a young graduate student. In a series of beautiful papers he showed the following:

It was conjectured repeadly that the causality loss associated with the Schott term is related to the introduction of advanced fields into the formalism by use of F_+ and F_-. Teitelboim showed that this is not the case.[73] Using only retarded fields he replaced the separation (6.7), used by Dirac, by a different separation,

$$F_{\text{ret}}^{\mu\nu} = F_{\text{I}}^{\mu\nu} + F_{\text{II}}^{\mu\nu}. \tag{7.10}$$

This separation is Lorentz covariant; it is the very physical separation into a $1/r^2$ field and a $1/r$ field. The difference between (6.7) and (7.10) is best seen in terms of the associated energy-momentum stress tensor. Since $T^{\mu\nu}$ is bilinear in $F^{\mu\nu}$ one can define

$$T^{\mu\nu} \equiv T_{\text{I}}^{\mu\nu} + T_{\text{II}}^{\mu\nu} \quad \begin{cases} T_{\text{I}}^{\mu\nu} \equiv T_{\text{I I}}^{\mu\nu} + T_{\text{I II}}^{\mu\nu} \\ T_{\text{II}}^{\mu\nu} \equiv T_{\text{II II}}^{\mu\nu} \end{cases} \tag{7.11}$$

and one finds,[73]

$$\partial_\alpha T_{\text{I}}^{\alpha\mu} = - \int_{-\infty}^{\infty} d\tau \, \delta_4 (x - z) \left(m_{\text{elm}} \dot{v}^\mu - \frac{e^2}{6\pi} \ddot{v}^\mu \right)$$

$$\partial_\alpha T_{\text{II}}^{\alpha\mu} = - \int_{-\infty}^{\infty} d\tau \, \delta_4 (x - z) \frac{e^2}{6\pi} \dot{v}^\alpha \dot{v}_\alpha v^\mu. \tag{7.12}$$

Thus, the separation (7.10) associates the Schott term with the *inertial* energy-momentum: the Schott energy becomes part of the internal energy of the electron, characteristic of accelerated motion.[74] At the same time we see that this $\ddot{v}^\mu$ term which causes the pre-acceleration has nothing at all to do with the use of $F_+^{\mu\nu}$ and $F_-^{\mu\nu}$, i.e., with advanced fields.

The technique indicated above permitted Teitelboim to *derive* the Lorentz–Dirac equation from Maxwell's equations and the assumption of uniform motion in the distant past. Dirac's vector B^μ and the associated ambiguity no longer enters. This derivation involves only retarded fields.

Finally, this approach leads to a much better understanding of the Lorentz–Dirac equation.[75] We write the equation of a single charged particle under an external force F^μ_{ext} as

$$m\dot{v}^\mu = F^\mu_{\text{ext}} + \frac{d}{d\tau} \int F^{\mu\alpha} j_\alpha \, d^4x \tag{7.13}$$

where we used

$$\partial_\alpha T^{\alpha\mu} = F^{\mu\alpha} j_\alpha \tag{7.14}$$

which follows from Maxwell's equations. The integral which expresses the self-interaction is divergent and undefined. One can define it by the prescription that the field on the world-line be the *average*

$$F^{\mu\nu}_{\text{ret}}(z) = \lim_{\varepsilon \to 0} \frac{1}{4\pi\varepsilon^2} \int F^{\mu\nu}_{\text{ret}}(z + \varepsilon u) \, d^2\sigma \tag{7.15}$$

where u^μ is a spacelike unit vector, so that the average is done covariantly over the sphere of radius ε. The result is

$$\frac{d}{d\tau} \int F^{\mu\nu}_{\text{ret}} j_\nu \, d^4x = - m_{\text{elm}} \dot{v}^\mu + \Gamma^\mu \tag{7.16}$$

so that (7.13) yields the (unrenormalized) Lorentz–Dirac equation. This procedure is unique if reasonable assumptions are made.[75]

8. SPIN AND MAGNETIC MOMENT

The discovery of the spin of the electron is a relatively well-known chapter in the history of quantum mechanics and has been treated excellently by Jammer.[2] It is also being discussed by others at this Symposium. Therefore, there is no need for elaboration on it here. I only wish to make the following comment on this discovery:

When Pauli's exclusion principle became known to those working in the theory of atomic spectra, it was first R. Kronig early in 1925 and, later in the same year, Goudsmit and Uhlenbeck who suggested that a spinning electron would explain the two-valuedness used formally in Pauli's principle. But all these authors discarded this notion (or tried to) when they pictured a classical electron of finite size spinning with the necessary angular momentum. The peripheral velocity on the electron's surface would have to exceed by far the velocity of light, and the classical electrodynamic calculations[23] did not give quantitative agreement. Goudsmit and Uhlenbeck were luckier than Kronig because, thanks to Ehrenfest, *their* paper was published.[76]

This is an example of an often repeated struggle in our search for a correct description of nature: the acceptance of a new idea is handicapped by its contradiction with the established description which is carried too far. We do not allow for the increased sophistication, the greater degree of abstractness which accompanies progress in science. And we often do not know the limits of validity of the established description, so that we cannot allow for these limits.

The postulate that the electron spin be 1 Bohr magneton, i.e. $2 \cdot \dfrac{-e}{2mc} \cdot \dfrac{\hbar}{2}$, implied a g-factor of $g = 2$. And for over twenty years it remained inconceivable that this number could be anything else but *exactly* 2, a manifestation of the 'elementarity' of the electron as compared to the nucleons for example.

The mathematical description of the electron spin within the framework of non relativistic quantum mechanics was first given by Pauli[77] in terms of his famous σ-matrices. But non-relativistic quantum mechanics, though extremely successful in explaining the intricacies of atomic structure, was never able to account in a consistent way for various well-known spectroscopic features. From the very beginning of spin theory a completely extraneous element had to be introduced: the Thomas precession.[78] This entirely relativistic phenomenon was necessary in order to account for a mysterious 'factor' of 2 needed in the calculations of the doublet separations.

The incompleteness of the non-relativistic description of the quantum mechanical electron structure of atoms is not surprising. The interactions involved are entirely of electromagnetic origin and the combination in one theory, of the intrinsically relativistic electrodynamic field with non-relativistic mechanical motion of electrons, was bound to lead to troubles. In fact we have seen how Abraham was led to the erroneous expression for electromagnetic momentum for very similar reasons.

To be sure, a consistent quantum mechanics of spinning particles can well be constructed on the non-relativistic, i.e., Galilean invariant level,[79] but its combination with electromagnetic interactions is only of limited usefulness.

The electron, being of electromagnetic nature, is therefore consistently described only as a *relativistic* spinning particle despite the fact that a non-relativistic approximation is often satisfactory and useful. We shall now very briefly trace the development of the classical theory of such a particle.[80]

The main goal is again a set of equations which describe the dynamics of a relativistic spinning particle, electrically charged, endowed with a magnetic moment, and subject to an external electromagnetic field. The first attack on this problem was made by Frenkel[81], by Thomas[73], and by Kramers[82]. The latter's work follows Frenkel in many respects.

The idea is to generalize to special relativity the non-relativistic equation of motion for a magnetic dipole in an external magnetic field. Since the spin vector $\mathbf{s}$ and the magnetic moment $\boldsymbol{\mu}$ are related by[83]

$$\boldsymbol{\mu} = g \, \frac{e}{2m} \, \mathbf{s} \tag{8.1}$$

this equation is also the equation of motion for the spin $\mathbf{s}$.

Starting with the non-relativistic equation

$$\mathbf{s} = g \, \frac{e}{2m} \, \mathbf{s} \times \mathbf{B} \tag{8.2}$$

where **B** is the external magnetic induction, there are three possibilities to generalize s to a Lorentz covariant quantity:

(a) $s = (S^{23}, S^{31}, S^{12})$ transforms like the three space–space components of an antisymmetric tensor, $S^{\mu\nu}$. The remaining three components form a polar vector $s' \equiv (S^{01}, S^{02}, S^{03})$ which is the 'spin' associated with the electric dipole moment. It was fixed by Frenkel's assumption that it vanishes in the rest system,

$$S^{\mu\alpha}v_\alpha = 0, \qquad (8.3)$$

i.e., that the electron has no static electric dipole moment.

(b) One can use the dual tensor to $S^{\mu\nu}$, viz.

$$S^D_{\mu\nu} \equiv \tfrac{1}{2}\varepsilon_{\mu\nu\alpha\beta}S^{\alpha\beta} \quad (\varepsilon_{0123} = +1) \qquad (8.4)$$

which yields the identification $s = (S^D_{01}, S^D_{02}, S^D_{03})$ and $s' = (S^D_{23}, S^D_{31}, S^D_{12})$. This is obviously equivalent to (a) when (8.3) is again imposed.

(c) One can identify s with the space components of a four-vector $w^\mu = (w^0, \mathbf{w})$ such that in the non-relativistic limit $\mathbf{w} = \mathbf{s}$. This four-vector is often called 'polarization vector' It is uniquely determined by requiring w^0 to vanish in the rest system, i.e.,

$$w^\alpha v_\alpha = 0 \qquad (8.5)$$

or equivalently

$$\mathbf{w} \cdot \mathbf{v} = w^0. \qquad (8.5')$$

Frenkel chose to start with an action principle, but his result is identical to the alternative (a). He was led to an equation which reads in our notation,

$$\dot{S} = g\,\frac{e}{2m}\,[(1 + vv)\cdot F\cdot S - S\cdot F\cdot(1 + vv)] + v\dot{v}\cdot S + S\cdot\dot{v}v. \qquad (8.6)$$

Here we use a dot to indicate the scalar product in Minkowski space and we omit the tensor indices for simplicity. The unit tensor with components $\delta^{\mu\nu}$ is denoted by 1, the electromagnetic field tensor by F. The dot over a symbol indicates differentiation with respect to proper time.

This equation reduces in the instantaneous rest system to the two equations (8.2) and

$$\mathbf{s'} = \mathbf{s} \times \dot{\mathbf{v}}. \qquad (8.7)$$

When Kramers studied this problem eight years later[82] he generalized Equation (8.2) to the relativistic case following alternative (a), but leaving out the velocity dependent terms in (8.6). Thus, he generalized to

$$\dot{S} = g\,\frac{e}{2m}\,(F\cdot S - S\cdot F). \qquad (8.8)$$

This led him to the value of $g = 2$ as a matter of internal consistency. In his original article in *Physica* (1934) he was apparently not aware that this value was implied by his assumption, but in his article[82] of 1938 he was fully aware of the greater generality of Frenkel's equation; he preferred (8.8) because it gave the correct gyromagnetic ratio.[84]

Equation (8.8) gives in the rest system (8.2) and

$$\dot{\mathbf{s}}' = \frac{g}{2}\frac{e}{m}\mathbf{s} \times \mathbf{E}. \tag{8.9}$$

In view of the Lorentz force equation this is consistent with (8.7) only for $g=2$.

In 1941, based on earlier work by Bhabha, a more general and sophisticated paper by Bhabha and Corben[85] appeared. They used Dirac's technique of 1938 applied to the angular momentum of a particle which has both electric and magnetic dipole moments as well as different moments of inertia around the respective axes perpendicular to these moments; radiation reaction was also included. Their result when it is suitably specialized agrees exactly with Frenkel's equation. This fact is not mentioned by them nor is Frenkel listed in their references.

Twenty-five years later there appeared a short paper by Bargmann, Michel, and Telegdi,[86] in which the non-relativistic equation (8.2) was generalized by means of method (c). They derived an equation valid for any g,

$$\dot{w} = g\,\frac{e}{2m}\,(F\cdot w - w\cdot F\cdot vv) + w\cdot\dot{v}v, \tag{8.10}$$

which is usually referred to as the BMT equation. In the rest system it reduces to (8.2) with the identification $\mathbf{w}=\mathbf{s}$ and to $w^0=0$ in agreement with (8.5).

In order to relate the BMT equation to the equations by Frenkel and by Kramers one observes that the vector w^μ is just the Pauli–Lubanski[87] vector which generates the little group of the Lorentz group and which was used so successfully by Bargmann and Wigner in their discussion of relativistic wave equations.[88] It is related to $S^{\mu\nu}$ by

$$w^\mu = -\tfrac{1}{2}\varepsilon^{\mu\nu\alpha\beta}v_\nu S_{\alpha\beta} = -S_D^{\mu\alpha}v_\alpha = v_\alpha S_D^{\alpha\mu}. \tag{8.11}$$

This definition of w^μ differs from the usual one only by a trivial multiplicative constant. It is equivalent to

$$S_D^{\mu\nu} = w^\mu v^\nu - v^\mu w^\nu \tag{8.12}$$

because of Equation (8.5).

The BMT equation therefore implies an equation for $S_D^{\mu\nu}$, viz.

$$\dot{S}_D = g\,\frac{e}{2m}\,(vv\cdot S_D\cdot F - F\cdot S_D\cdot vv) - S_D\cdot v\dot{v} - \dot{v}v\cdot S_D. \tag{8.13}$$

Conversely, this equation implies the BMT equation as follows from (8.11). Thus, (8.13) and (8.10) are equivalent. But (8.13) is also equivalent to the Frenkel equation (8.6) since it is just the generalization of the non-relativistic equation (8.2) by means of method (b). Thus, the BMT and the Frenkel equations are equivalent.

In constant or slowly varying electromagnetic fields the above equation can be combined with the Lorentz force equation to yield either, from the Frenkel equation (8.6),

$$\dot{S} = g\,\frac{e}{2m}\,(F\cdot S - S\cdot F) + \left(\frac{g}{2} - 1\right)\frac{e}{m}\,(vv\cdot F\cdot S - S\cdot F\cdot vv), \tag{8.14}$$

or, from the BMT equation (8.10):

$$\dot{w} = g\,\frac{e}{2m}\,F\cdot w - \left(\frac{g}{2} - 1\right)\frac{e}{m}\,w\cdot F\cdot vv. \tag{8.15}$$

Both are useful for various applications. One notes that the Kramers equation (8.8) now follows trivially from (8.14) if and only if $g = 2$ and is equivalent to (8.15) for $g = 2$. The BMT equation (8.15) can also be written in the form

$$\dot{\omega} = \Omega\cdot\omega. \tag{8.16}$$

The antisymmetric tensor $\Omega^{\mu\nu}$ describes in a covariant way the rate of Larmor precession of the polarization vector. Parenthetically it should be mentioned that Zwanziger showed recently how polarization tensors associated with any spin s can be defined[89] as traceless symmetric matrices $\omega^{\mu\nu\cdots}$ of rank up to $2s$, and that the corresponding dynamical equation[90] is always of the form (8.16). The BMT equation is just the special case of a dipole.

An immediate consequence of the various equivalent dynamical spin equations is the conservation of spin magnitude S,

$$\dot{S} = 0;\qquad S^2 \equiv \tfrac{1}{2}S_{\mu\nu}S^{\mu\nu} = \tfrac{1}{2}S_{\mu\nu}^D S_D^{\mu\nu} = w_\mu w^\mu. \tag{8.17}$$

The discussion so far seems to indicate that the development of classical spin theory has been taking place in general agreement between the various authors. This has not been the case, however. This topic has in fact been fraught with scepticism and mutual disbelief between various schools of thought. The reason for this is the following.

As was first pointed out by Frenkel[81], and was later shown again by Bhabha and Corben[85], the presence of spin modifies the relativistic equation of motion of a charged particle. Even in the approximate absence of radiation reaction the Lorentz force equation no longer holds but is modified by terms linear and homogeneous in the spin. Such a modification might be easier to understand for an extended particle. But, as was shown explicitly by Weyssenhoff and Raabe[91], the point particle limit of an extended particle leads to the same conclusion.

In the above we have assumed that these spin dependent terms are absent. If this assumption is not made the equations become considerably more complicated. These complications have been interpreted by Corben[92] as the classical analogue of the Zitterbewegung and of pair production of relativistic quantum mechanics. Specifically, even in the absence of an external field the spinning particle does not move in a straight line but in a helical orbit or along a hyperbola in space-time.

Such classical analogues of quantum mechanical results were welcomed by those who are not satisfied with quantum mechanics and who are looking for a 'hidden variable' theory. These analogues are older than Corben's work of 1961 going back at least to Pryce[93] but were revived again in the fifties.[94] The possibility of such analogues have their origin at least partly in the difficulty of defining an *extended* relativistic particle.

The obvious Lorentz-invariant definition of a rigid body was first given by Born[95] and was quickly shown to be unsatisfactory for a rotating body.[96] Instead, one has to work with 'rigid motion' which corresponds more to a fluid or elastic body than to a rigid body in the non-relativistic sense. As a consequence, there exists a non-trivial problem of defining the centre of mass of such an object,[93] because it becomes frame dependent. All possible such mass centres form a disc of radius ϱ perpendicular to the spin angular momentum vector in the rest system,[97]

$$\varrho = S/m \tag{8.18}$$

where S is the magnitude of the spin defined in (8.17). This leads to Møller's theorem that a spinning body with positive energy density must have dimensions not less than S/m.

The controversy surrounding the classical spin theory centres around the question how seriously to take the modification of the Lorentz force equation due to the presence of spin *when applied to microscopic particles such as electrons*. If one takes the point of view that quantum mechanics is (within its limits of validity) a satisfactory theory, so that the classical theory of the electron emerges as a limit of the quantum mechanical description and that it is not a replacement of it, then the answer to this question lies in the classical limit of the Dirac equation. In the next section we shall discuss this problem and its solution as it developed during the last forty years.

But as a preliminary answer, one can argue that the validity limits of the classical theory permit only a point electron, the extended structure (of size $\hbar/m$) being the realm of quantum mechanics. In that case, the limit from an extended electron leads to $S = 0$, i.e., no spin, according to Møller's theorem. A more careful analysis, however, would involve an expansion treating S as a small quantity.[98] One then finds that the Lorentz force equation receives corrections of order S while the spin equation only of order S^2. In the limit $S \to 0$ one recovers the situation which we have assumed in the earlier discussion. But these arguments are unconvincing when it is realized that a point particle need not be defined as the limit of an extended particle.[99] The study of the classical limit of relativistic quantum mechanics is thus imperative.

In conclusion of this Section an important remark must be made on the motion of the spinning electron through inhomogeneous fields. Plahte[98] showed that the Lorentz force and BMT equations remain valid even in that case if terms of order S are neglected, or at least when $\varrho|\partial_\alpha F_{\mu\nu}| \ll |F_{\mu\nu}|$.

9. THE CLASSICAL ELECTRON AS A LIMIT OF ITS QUANTUM MECHANICAL DESCRIPTION

I believe that the generation of theoretical physicists who developed relativity theory and quantum mechanics was better educated in philosophy of science than is the present generation. They were acutely aware of the need for philosophical questioning if they were to be good theoretical physicists. In fact theoreticians from Poincaré to Philipp Frank contributed importantly to philosophy of science, not to speak of the positivist school and its 'Viennese Circle' in which they played a key role.

Thus, there was little doubt among physicists at the time that different branches of theoretical physics must be interrelated. In particular, they insisted that the special theory of relativity reduce to Newtonian mechanics in a suitable limit; that general relativity reduce to special relativity, that non-relativistic quantum mechanics reduce to non-relativistic classical mechanics, and so forth. The originators of new theories nowadays, especially in elementary particle physics, do not always pay attention to such matters, and some even doubt the existence of such relations.

In the early days of quantum mechanics this new theory was built inductively with the aid of the correspondence principle. But as quantum mechanics matured and began to stand on its own foundation, the limit to the classical theory became a non trivial problem. And when the relativistic quantum mechanics of the electron was initiated by Dirac's well-known equation[100] it took only a few years for this problem to be attacked. In 1932 Pauli wrote his famous paper about the relation of the Dirac equation to geometrical optics.[101] He started with the Dirac equation in an external electromagnetic field with minimal coupling,

$$[\gamma_\mu (\hbar \partial^\mu + ieA^\mu) + m] \psi(x) = 0 \qquad (9.1)$$

where we put $c=1$ but showed the dependence on $\hbar$ explicitly. The four-component function $\psi(x)$ is expanded as in the JWKB method,

$$\psi(x) = e^{iS(x)/\hbar} \sum_{n=0}^{\infty} (-i\hbar)^n a_n(x), \qquad (9.2)$$

each a_n being a four-component complex valued function. When the coefficients of each power of $\hbar$ are set to zero one obtains equations for S and for the a_n. While he was unable to solve the system of equations for a_n in general, he did find that the phase function S is always exactly Hamilton's principal function for a relativistic classical particle with mass m and charge e in an electromagnetic field, i.e., it satisfied the corresponding Hamilton–Jacobi equation,

$$(\partial_\mu S + eA_\mu)(\partial^\mu S + eA^\mu) + m^2 = 0. \qquad (9.3)$$

This is exactly analogous to the transition of classical wave optics to geometrical optics in an anisotropic medium.

The remarkable feature of this result is of course that the electron spin does not appear in this classical limit. This is a satisfactory result. Bohr first pointed out that there are no *classical* experiments by which the electron spin could be observed.[102] But it is also clear that, if the classical limit is understood to mean a limit to large quantum numbers, letting $\hbar$ go to zero at the same time, such that the observables remain finite but macroscopic, then the electron spin 1/2 must be absent: the fixed number 1/2 does not permit a limit to large quantum numbers.

Nevertheless, an objection was raised against Pauli's work. In 1952 de Broglie[103] argued that Dirac's equation should have as its classical limit the classical theory by Weyssenhoff, mentioned earlier,[91] where the magnetic moment of the electron affects its trajectory. This criticism went largely unnoticed until its clarification in the paper by Rubinow and Keller who continued where Pauli had left off thirty years earlier.[104] They point out that Pauli's expansion (9.2) is meant as an expansion of $\psi(x, \hbar)$ with fixed x; however, if one considers large x such that $x = x'/\hbar$ and one expands with fixed x', then a different limit is obtained and the magnetic moment would indeed affect the trajectory of the electron in the classical limit in agreement with de Broglie's argument. Thus Pauli's expansion is not valid at large distances (of order $1/\hbar$) from the trajectory due to a non-uniformity of convergence.

The main achievement of Rubinow and Keller, however, was that they succeeded in effectively solving the set of partial differential equations for the a_n by reducing them to ordinary differential equations along the trajectories. They also generalized Pauli's considerations by including an arbitrary anomalous magnetic moment term ('Pauli term'),

$$\left(\frac{g}{2} - 1\right)\frac{ie\hbar}{2m}\sigma_{\mu\nu}F^{\mu\nu}\psi \tag{9.4}$$

in the Dirac equation. When they solved the equations for a_0 they found that the vector T_μ defined by

$$T_\mu \equiv e^{iS/\hbar}\bar{a}_0 i\gamma_5\gamma_\mu a_0 e^{iS/\hbar}\exp\left[\frac{1}{m}\int_{\tau_0}^{\tau}\partial^\alpha(\partial_\alpha S + eA_\alpha)\,d\tau'\right] \tag{9.5}$$

satisfies exactly the BMT equation (8.15) and can therefore be identified with the polarization vector W_μ. They also determined the phase of a_0 which does not enter T_μ.

This proved that the approximations which entered the classical derivation of the BMT equation were consistent since they give exactly the classical limit of the Dirac equation, even when an anomalous magnetic moment is included. The heuristic arguments (see the end of the preceding section) of Plahte, given in 1966 can thus be supplemented by the rigorous proof of Rubinow and Keller given in 1963!

At this point another development must be mentioned which was quite independent of the above and which also led to the BMT equation (8.15), and in fact a couple of years earlier, although not as directly.

The difficulty of defining the centre of mass in relativistic classical mechanics which was mentioned in the last section has an analogue in relativistic quantum mechanics. It is the difficulty of defining a position operator. And as in classical mechanics this problem is closely linked to the problem of describing the polarization of the electron.[105] These quantum mechanical problems are not the subject of the present discussion, but they can be found summarized in the paper by Fradkin and Good[106] of 1961. By defining the quantum mechanical polarization vector appropriately they showed that its Heisenberg equation of motion leads in the classical limit to the BMT equation.

But this is clearly not the complete solution of the problem. The Dirac equation with an external field does not take into account the interaction of the particle with its own electromagnetic field. Consequently, the radiation reaction terms appear nowhere in the classical limit of this equation. The most interesting aspects of the classical Lorentz–Dirac equation have therefore so far not been proven to be limits of a corresponding quantum mechanical description. The latter, present quantum electrodynamics, has until now not been taken to its classical limit in an approximation which exceeds the above Pauli–Rubinow–Keller limit in an essential way, i.e. with inclusion of self-interaction.

Here, the recent work by Bialynicki–Birula must be mentioned.[107] He uses the technique of coherent states which seems essential when one goes to large numbers of field quanta as in the classical limit. But his published work has so far only derived a classical Klein–Gordon field with polynomial self-interaction and the quantum electrodynamic analogue. The derivation of classical charged point-particle equations with self-interaction has not been achieved.

10. THE CLASSICAL ELECTRON TODAY

Looking back at the history of the classical theory of the electron which began before the discovery of the electron 75 years ago, we see that it has been dominated by the self-energy and stability problem and the associated question about the electromagnetic nature of the electron mass. Three types of solutions have been proposed.

The oldest solution proposed is that of Poincaré: the electron mass is partially electromagnetic and partially non-electromagnetic, the latter compensating the former so that m_{obs} is the difference of two divergent quantities. When taken seriously this leads to the introduction of one or more compensating fields.[108] The self-energy is thus finite and the self-stress vanishes, making the electron stable.

Another very old idea (which we did not discuss) is that of modifying either the Maxwell field equations or the equations of motion to attain a stable purely electromagnetic electron.[109] Because the equations of motion are essentially derivable from Maxwell's equations, as we discussed in Section 7, modification of one implies modification of the other if the conservation laws are to remain local. At the outset such modifications seem possible in an infinity of ways, so that additional criteria are needed. The experimental criterion that these modifications have so far not been observed is easily met by choosing the modifications with small enough parameters. But this is exactly what forces these modifications into the quantum domain and makes this approach rather unattractive for those who consider the classical theory to be only an approximation of the quantum description.[110]

Finally, there is the simplest alternative, viz. to abandon any attempt at describing the mass of the electron within the framework of the classical theory. In this approach the Coulomb self-interaction simply does not exist and there is no electromagnetic self-energy. To this class belong the theories by Fokker[64] and by Wheeler and Feynman[65] and my own theory.[67] The classical theory is here recognized as being entirely phenomenological with respect to both the mass and the charge of the electron. Clearly in such a theory there is no stability problem.

There are those who feel that this approach is just 'sweeping the problem under the rug'. But there is no point forcing a theory to do what it is not able to do. The classical electron contains three universal constants, m, e, and c, corresponding to the three basic units of mass, length, and time. This is the minimum number of phenomenological constants any theory can have. One cannot expect it to explain this mass, or any part of it.

Beyond this the theory is also in fine shape if one does not push it beyond its natural validity limits. The natural length of the theory is the classical electron radius which is a factor 137 smaller than $\hbar/mc$. Thus, the electron is a point as far as classical physics can tell.

If these facts are kept in mind, the preacceleration problem is no problem. Nor is the collision of two point charges at distances of $\hbar/mc$ or less. These become problems only beyond the validity limits, as we discussed in Section 6.

There are of course still open questions. They are difficult and are primarily of a mathematical nature. I shall list some of these:

(1) The classical relativistic two-body problem remains unsolved. The few cases in which a solution has been given are of a very special nature.[111]

(2) The 'initial value problem' is unsolved. The fields at a given time depend on the history of the particles, so that no standard Cauchy problem formulation exists for the simultaneous dynamics of charges and fields.[112]

(3) The existence of solutions of the Lorentz-Dirac equation was proven.[113] But their uniqueness has not been proven. Nor is the dependence of these solutions on the charge e known. Are they analytic in e? Under what conditions do perturbation expansions converge? Can the physical requirement of uniform approach of a charged trajectory to an uncharged one in the $e \rightarrow 0$ limit be proven?

(4) The classical limit of quantum electrodynamics is still not very well understood. This is perhaps not a problem of the classical theory, but the fact that this theory is actually obtained in the limit in every detail will be the final proof of its correctness beyond experimental confirmation.

Because of the Coulomb self-energy term the theory of the point electron is not divergence-free and not mathematically meaningful until after mass renormalization, unless one accepts the formulation in which no such term arises. One can then ask for generalizations of this divergence free theory. The simplest of these is the inclusion of a magnetic monopole. The only divergence-free classical theory along this line that I know of was given in 1966.[114] The relativistic quantum theories of magnetic monopoles are at present all of the standard quantum field theoretic variety that become divergence-free at best after renormalization.

Another generalization involves the inclusion of gravitational interactions. These have been completely ignored in the present survey but there is a long and rich history of describing electrons in the presence of gravitational fields and even of attempting theories which synthesize electromagnetic and gravitational interactions into a single theory. This is clearly far beyond the present scope.

Similarly, there are attempts at combining electromagnetic and weak interactions, especially recently. But these are all on the quantum level since the latter do not seem to have a classical (nonquantum) limit.

We have seen that on the whole the classical theory of the electron is in very good shape and thus answers affirmatively the important question asked in accelerator laboratories: is there a consistent theory behind the classical equations used to compute the orbit of electrons and other high-energy particles?

But the existence of this theory also answers another question: if we were to find a mathematically consistent quantum electrodynamics, what would be its classical limit? Of course, as indicated in the above list of outstanding questions it remains for the future to verify that the theory outlined above will indeed be the classical limit of quantum electrodynamics.

Finally, there is one more use of the classical theory: if we accept that it may be the limit of quantum electrodynamics, we can expect that one can learn something from classical theory about that still not completely understood field theory. The following speculations are suggested in this connection.

(1) As is evident from the integro-differential equation of motion (7.7) classical electron theory is *not* a local theory. To be sure it is usually written as a local theory, but the asymptotic conditions are essential. When they are incorporated, the theory becomes non-local. Thus, we expect the final formulation of quantum electrodynamics to be also non-local.

(2) It is possible to formulate the theory so that the mass of the electron is *entirely* of non-electromagnetic (gravitational?) origin. In that case no mass renormalization is required. Since most people seem to agree that even quantum electrodynamics cannot account for the electron mass, we expect the existence of a formulation of quantum electrodynamics in which there is no electromagnetic self-energy.[115]

(3) If the formulation in which $m_{elm}=0$ is accepted, then Coulomb fields have meaning only as derived concepts, the primary concept being the Coulomb inter-action-at-a-distance. Coulomb effects are present only between *pairs* of electrons. This may then also hold for such things as the vacuum polarization phenomena of the quantized theory. This seems to be connected with the fact that the Coulomb gauge in some sense is a better gauge than any of the others.

In conclusion I wish to apologize for the many important contributions which I have not mentioned. This is too big a field to be summarized in one relatively short paper. But I do hope to have presented enough of the theory to show the fruits of labour by so many.

REFERENCES AND NOTES

1. E. Whittaker, *A History of the Theories of Aether and Electricity*, Vol. I, Thos. Nelson and

Sons Ltd., London 1910, revised edition, 1951; vol. II, Philosophical Library, New York 1954.

2. M. Jammer, *The Conceptual Development of Quantum Mechanics*, McGraw Hill Book Co., New York, 1966.

3. J. C. Maxwell, *Phil. Trans.* **155**, 459 (1865).

4. J. J. Waterston, *Phil. Trans.* **183**, 1 (1892). The publication of this paper was delayed by 47 years!

5. J. J. Thomson, *Phil. Mag.* **11**, 227 (1881).

6. Thomson did not apply Maxwell's equations completely consistently, nor could he have known that the charge distribution on a sphere does not remain spherically symmetric when that sphere is set in motion. Similarly, 25 years before special relativity, he combined Newtonian mechanics with Maxwell's equations.

7. G. F. FitzGerald, *Proc. Roy. Dublin Soc.* **3**, 250 (1881).

8. O. Heaviside, *Phil. Mag.* **27**, 324 (1889); reprinted in *Electrical Papers*, II, p. 504, Chelsea Publ. Co., Bronx, New York 1970.

9. O. Heaviside, *Electrician*, 23 Nov. 1888 and ref. 8 above.

10. A. Liénard, *L'Eclairage élect.* **16**, 5, 53, 106 (1898); E. Wiechert, *Arch. Néerl.* **5**, 549 (1900).

11. G. J. Stoney was the first to advocate the atomicity of electric charge at the Belfast meeting of the British Association in August, 1874 and again in February 16, 1881. (*Phil. Mag.* **11**, 381 (1881)), although it was Helmholtz's influence who advocated it independently in his Faraday lecture April 5, 1881 (*J. Chem. Soc.* **39**, 277 (1881)), which contributed most to its acceptance. The name 'electron' was proposed by Stoney in *Phil. Mag.* **38**, 418 (1894).

12. J. J. Thomson, *Phil. Mag.* **44**, 298 (1897); *Nature* **55**, 453 (1897).

13. W. Kaufmann, *Ann. Physik* **61**, 544 (1897).

14. J. S. E. Townsend, *Phil. Trans.* **143**, 129 (1899).

15. R. A. Millikan, *B. A. Rep.* 410 (1912).

16. H. A. Lorentz, *Arch. Néerland, Sci. Exact. Nat.* **25**, 363 (1892).

17. See ref. 1, vol. I, p. 393 ff.

18. K. Schwarzschild, *Gött. Nachr.*, p. 126 (1903).

19. W. Kaufmann, *Gött. Nachr.*, p. 143 (1901).

20. If the tangential velocity is $\mathbf{v}$, and the momentum $\mathbf{P}$ is parallel to $\mathbf{v}$, the component of $d\mathbf{P}/dt$ parallel to $\mathbf{v}$ is $\dot{v}\, dP/dv$; defining the longitudinal mass $m_\parallel$ by $F_\parallel = m_\parallel \dot{v}$ one finds $m_\parallel = dP/dv$. Similarly, with $\omega = v/r$ the component of $d\mathbf{P}/dt$ perpendicular to $\mathbf{v}$ is $P\omega$; defining the transverse mass $m_\perp$ by $F_\perp = m_\perp v^2/r$ one finds $m_\perp = P/v$.

21. M. Abraham, *Gött. Nachr.*, p. 20 (1902).

22. W. Kaufmann, *Gött. Nachr.*, p. 291 (1902); *Physik. Z.* **4**, 54 (1902).

23. M. Abraham, *Ann. Physik* **10**, 105 (1903).

24. Calculations for elliptic shapes, which are much more complex, were carried out too, but mainly in order to see how sensitive the results were to the assumption of a particular shape.

25. J. H. Poynting, *Phil. Trans.* **175**, 343 (1884); O. Heaviside, *Electrician* **14**, 178 and 306 (1885).

26. This calculation is carried out as follows: one observes that the equations for the potential φ of a spherical electron moving with velocity v can be transformed to the equations of a *spheroidal electron at rest*. The new potential is related to the old one by $\varphi' = \varphi/\gamma$ where $\gamma = (1 - v^2)^{-1/2}$ and the spheroid differs from the sphere in that its axis along the direction of motion is *elongated* by a factor γ. The corresponding charge distribution $\varrho' = \varrho/\gamma$ is the equilibrium distribution for this spheroid. Contrary to the case $v^2 \ll 1$ the masses are now not equal but $m_\parallel > m_\perp$.

27. M. Abraham, *Physik. Z.* **5**, 576 (1904). This was written in response to Lorentz's contraction hypothesis (see ref. 30 below). Abraham argued that this contraction would off-set the internal force balance, require non-electromagnetic forces, and thus destroy the electron theory which was based on a purely electromagnetic electron.

28. H. A. Lorentz, *Encyclopädie der Mathematischen Wissenschaften*, vol. V/2 (1904), pp. 63–144 (Maxwells Elektromagnetische Theorie) and pp. 145–280 (Weiterbildung der Maxwellschen Theorie; Elektronentheorie). A relatively recent exposition of the applied part of Lorentz's theory was given by L. Rosenfeld, *Theory of Electrons*, North-Holland, 1951; Dover 1965.

29. H. A. Lorentz, *The Theory of Electrons and its Applications to the Phenomena of Light and Radiant Heat*, Dover Publications, New York, 1952. The first edition appeared in 1909, the second in 1915; the Dover edition is a copy of the latter. It's not clear just how this publication differs from the Columbia lectures. But apart from the 'Notes' which we would call 'Appendix'

in today's scientific paper, over one third of it is devoted to 'Emission and Absorption of Heat' and to the newly developed special theory of relativity as it relates to the theory of electrons.

30. H. A. Lorentz, *Proc. Acad. Sci. Amsterdam* **6**, 809 (1904).
31. A. Einstein, *Ann. Physik* **17**, 891 (1905).
32. H. Poincaré, *Bull. des Sci. Math.* **28**, 302 (1904).
33. H. Poincaré, *Rendiconti del Circolo Mat. Palermo* **21**, 129 (1906).
34. H. Minkowski, *Gött. Nachr.*, p. 53 (1908); *Math. Ann.* **68**, 526 (1910). The latter article was published posthumously and was written up by M. Born.
35. W. Pauli, *Encycl. d. Math. Wiss.* **2**, 667 (1920); reprinted in English translation with supplementary notes by the author as *Theory of Relativity*, Pergamon Press, New York (1958).
36. E. Noether, *Gött. Nachr.* p. 235 (1918).
37. Ref. 35, English translation, p. 185.
38. C. Møller, *The Theory of Relativity*, Oxford (1952); M. Abraham and R. Becker, *Theorie der Elektrizität*, Teubner, Leipzig, all editions since the 3rd; J. D. Jackson, *Classical Electrodynamics*, John Wiley, New York (1962). R. P. Feynman, R. B. Leighton, and M. Sands, *The Feynman Lectures on Physics*, Addison-Wesley (1964); Vol. II, p. 284.
39. E. Fermi, *Physik. Z.* **23**, 340 (1922); *Atti Accad. Nazl. Lincei* **31**, 184 and 306 (1922).
40. H. Mandel, *Z. Phys.* **39**, 40 (1926); W. Wilson, *Proc. Phys. Soc. (London)* **48**, 736 (1936).
41. B. Kwal, *J. Phys. Radium* **10**, 103 (1949).
42. F. Rohrlich, *Am. J. Phys.* **28**, 639 (1960).
43. F. Rohrlich, *Classical Charged Particles*, Addison-Wesley, Reading (1965).
44. M. v. Laue, in *Ann. Physik*, *35*, 524 (1911), first proved that P^μ defined by (5.9) is a four-vector if and only if the stresses vanish. He also gave this four-dimensional generalization of Gauss' law in his book *Die Relativitätstheorie*, Braunschweig, 1st ed. 1911, 7th ed. 1961.
45. Nevertheless, this definition was used extensively. It occurs not only in the context of the structure of the electron. It was used in the relativistic explanation of the Trouton–Noble experiment, in the relativistic description of thermodynamics, and in the theory of black-body radiation in a moving cavity. References to all these are contained in Pauli's article (ref. 35). They all contain the famous factor 4/3 in the momentum.
46. It is a debated question whether quantities which are *not* geometrical objects are useful in describing observations. For example, a Lorentz contracted volume $\Delta V = \Delta V^{(0)}/\gamma$ is often used in theoretical discussions, but it has never been observed; as the Terrel effect shows (*Phys. Rev.* **116**, 1041 (1959)), attempts to observe it can lead to unexpected results. Relativistically, ΔV transforms like $\Delta\sigma^0$, as pointed out preceding Equation (5.15). See F. Rohrlich, *Nuovo Cim.* **45**, 76 (1966), for further discussion.
47. F. Rohrlich, *Am. J. Phys.* **38**, 1310 (1970).
48. I have searched for earlier usage of S as the momentum density. Apparently, it was used before Abraham, since it appears in a paper by Poincaré (*Arch. Néerland Sci. Exact Nat.* **5**, 252 (1900)) in the context of bound fields.
49. Relativistic thermodynamics and black-body radiation have received a similar reformulation in terms of covariant objects during recent years. Initiated by a paper by H. Ott, *Z. Physik*, **175**, 70 (1963), a lengthy discussion in the scientific literature followed. Earlier references can be found in N. G. van Kampen, *Phys. Rev.* **173**, 295 (1968), and V. H. Hamity, *Phys. Rev.* **187**, 1745 (1969), among many other papers. The covariant definition of P^μ was used to explain the Trouton–Noble experiment by J. W. Butler, *Amer. J. Phys.* **37**, 1258 (1969).
50. A. Staruszkiewicz, *Nuovo Cim.* **45**, A, 684 (1966); F. Rohrlich, *loc. cit.* (ref. 46).
51. P. A. M. Dirac, *Proc. Roy. Soc. (London)* A, **168**, (1938); *Ann. de l'Inst. Poincaré*, **9**, 13 (1939).
52. J. Larmor, *Phil. Mag.* **64**, 503 (1897).
53. M. Abraham, ref. 23 and *Ann. Physik* **14**, 236 (1904).
54. O. Heaviside, *Nature* **67**, 6 (1902).
55. M. v. Laue, *Ann. Physik* **28**, 436 (1909).
56. H. A. Lorentz, ref. 29, section 37.
57. M. Abraham *Theorie der Elektrizität*, Teubner, 2nd ed. (1908), vol. 2.
58. G. A. Schott, *Phil. Mag.* **29**, 49 (1915).
59. That the combination $F_{-}^{\mu\nu}$ is non-divergent on the world line was already known earlier: G. Wentzel, *Z. Physik* **86**, 479 and 635 (1933); **87**, 726 (1933). These papers were apparently known to Dirac at the time.

60. The person who knew this classical theory well and who first emphasized the importance of mass renormalization to quantum field theory was apparently H. A. Kramers at the Shelter Island Conference of 1947.

61. H. J. Bhabha, *Phys. Rev.* **70**, 759 (1946); F. Rohrlich, *Ann. Physics* **13**, 93 (1961).

62. J. Frenkel, *Z. Physik* **32**, 518 (1925), esp. pp. 526–527.

63. Attempts to modify Maxwell's equations are discussed in Section 10.

64. H. Tetrode, *Z. Physik* **10**, 317 (1922); A. D. Fokker, *Z. Physik* **58**, 386 (1929).

65. J. A. Wheeler and R. P. Feynman, *Rev. Mod. Phys.* **17**, 157 (1945) and **21**, 425 (1949).

66. This was also observed by S. N. Gupta, *Proc. Phys. Soc. (London)* A **64**, (1951).

67. F. Rohrlich, *Phys. Rev. Lett.* **12**, 375 (1964); see also ref. 43. There is also a question of time reversal invariance involved into which we cannot enter here. A recent objection to my Lagrangian, C. A. Lopez, *Lett. Nuovo Cim.*, Ser. 2, **1**, 527 (1971) was withdrawn by that author, C. Teitelboim and C. A. Lopez, *Lett. Nuovo Cim.*, Ser. 2, **2**, 225 (1971).

68. Strong arguments in favour of an asymptotic condition, based on similarity with quantum field theory were given by R. Haag, *Z. Naturforsch.* **10**α, 752 (1955). This paper tries to show that such a condition is not 'ad hoc' as it might appear to be in Dirac's 1938 paper, but goes deeply into the physical interpretation of the theory.

69. R. Haag, *loc. cit.*, ref. 68; D. Ivanenko and A. Sokolov, *Klassische Feldtheorie*, Akademie-Verlag, Berlin (1953) (Russian edition, 1949).

70. G. N. Plass, *Rev. Mod. Phys.* **33**, 37 (1961).

71. F. Rohrlich, ref. 61; see also ref. 43.

72. The Schott term has always been somewhat mysterious and has defied a clear physical understanding. In this formulation it finds its true meaning in the non-local nature of the interaction.

73. C. Teitelboim, *Phys. Rev.* D **1**, 1572 (1970) and D **3**, 297 (1971).

74. This internal energy is just the supply needed for radiation emission in uniform acceleration (hyperbolic motion); its depletion permits radiation emission by a charged particle without violation of the law of conservation of energy and momentum despite the fact that it follows exactly the same orbit as a neutral one.

75. C. Teitelboim, *Phys. Rev.* D **4**, 345 (1971); C. Teitelboim and C. A. Lopez, *Nuovo Cim. Lett.* **2**, 225 (1971).

76. G. E. Uhlenbeck and S. Goudsmit, *Naturwiss.* **13**, 953 (1925); *Nature* **117**, 264 (1926).

77. W. Pauli, *Z. Physik* **43**, 601 (1927).

78. L. H. Thomas, *Nature* **117**, 514 (1926); *Phil. Mag.* **3**, 1 (1927).

79. J. M. Levy-Leblond, *Comm. Math. Phys.* **6**, 286 (1967).

80. The classical discussion is here restricted to those equations which are of special interest in the context of the classical limit of relativistic quantum mechanics. More sophisticated theories are discussed extensively by H. C. Corben, *Classical and Quantum Theories of Spinning Particles*, Holden Day, Inc., San Francisco, 1968. This book contains many references to earlier work.

81. J. Frenkel, *Z. Physik* **37**, 243 (1926).

82. H. A. Kramers, *Physica* **1**, 825 (1934); *Quantum Mechanics*, North-Holland 1957, Dover 1964 (the latter is a copy of the corrected, second edition of 1958). This book is a translation of two articles, 'Die Grundlagen der Quantentheorie' and 'Quantentheorie der Elektronen und der Strahlung', published in vol. 1 of *Hand- und Jahrbuch der chemischen Physik*, Akad. Verlag, Leipzig, 1938.

83. In the following equations the signs are chosen for convenience to correspond to a positive electron.

84. This fact is quite clearly stated on p. 231 of the Dover edition of Kramer's book, ref. 82.

85. H. J. Bhabha and H. C. Corben, *Proc. Roy. Soc.* **178**, 273 (1941).

86. V. Bargmann, L. Michel, and V. L. Telegdi, *Phys. Rev. Lett.* **2**, 435 (1959).

87. J. K. Lubanski, *Physica* **9**, 310 and 325 (1942).

88. V. Bargmann and E. P. Wigner, *Proc. Natl. Ac. Sci.* **34**, 211 (1948).

89. D. Zwanziger, *Phys. Rev.* **137**, B, 1535 (1965).

90. D. Zwanziger, *Phys. Rev.* **139**, B, 1318 (1965).

91. J. Weyssenhoff and A. Raabe, *Acta Phys. Polon.* **9**, 7 (1947).

92. H. C. Corben, *Phys. Rev.* **121**, 1833 (1961) and *Nuovo Cim.* **20**, 529 (1961).

93. M. H. L. Pryce, *Proc. Roy. Soc.* A **195**, 62 (1948).

94. D. Bohm and J.-P. Vigier, *Phys. Rev.* **109**, 1882 (1958); D. Bohm, P. Hillion, T. Takabayasi and J.-P. Vigier, *Prog. Theor. Phys.* **23**, 496 (1960).

95. M. Born, *Ann. Physik* **30**, 1 (1909).

96. P. Ehrenfest, *Phys. Z.* **10**, 918 (1909); G. Herglotz, *Ann. Physik* **31**, 393 (1910); E. Noether, *ibid.* **31**, 919 (1910).

97. C. Møller, *Comm. Dublin Inst. Adv. Studies* A **5** (1949); see also *The Theory of Relativity*, Oxford (1952).

98. E. Plahte, *Nuovo Cim. Suppl.* **4**, 246 and 291 (1966).

99. The first to state and support explicitly that a point particle can be defined in its own right with different properties than the limit of an extended particle was probably J. L. Synge in *Relativity: The Special Theory*, North-Holland Amsterdam (1955, Second Edition 1965), p. 315 and p. 220.

100. P. A. M. Dirac, *Proc. Roy. Soc.* A **117**, 610 and A **118**, 351 (1928).

101. W. Pauli, *Helv. Phys. Acta* **5**, 179 (1932).

102. W. Pauli, Rapport du Conseil–Solvay, 1930.

103. L. de Broglie, *La Théorie des Particules de Spin* 1/2, Gauthier-Villars, Paris (1952), chapter X.

104. S. I. Rubinow and J. B. Keller, *Phys. Rev.* **131**, 2789 (1963).

105. The reason for the interrelation between position and polarization in relativistic mechanics was most clearly stated by H. Bacry, *J. Math. Phys.* **5**, 109 (1964).

106. D. M. Fradkin and R. H. Good, *Rev. Mod. Phys.* **33**, 343 (1961).

107. I. Bialynicki-Birula, *Ann. Physics* **67**, 252 (1971).

108. The quantum field theoretic form of this idea leads to one or more compensating fields. These raise the physical question of observation of such fields and the mathematical question of indefinite metric.

109. Nonlinear generalizations of Maxwell's equations were attempted by: G. Mie, *Ann. Physik*, **37**, 511 (1912); **39**, 1 (1912); **40**, 1 (1913). M. Born, *Proc. Roy. Soc.* A **143**, 410 (1934); M. Born and L. Infeld, *Proc. Roy. Soc.* A **144**, 425 (1934); **147**, 522 (1934); **150**, 141 (1935).

110. Non-local and extended theories are reviewed by Thos. Erber, *Fortschritte der Physik*, **9**, 343 (1961).

111. See for example A. Schild, *Phys. Rev.* **131**, 2762 (1963).

112. R. D. Driver, *Phys. Rev.* **178**, 2051 (1969) and references quoted there.

113. J. K. Hale and A. P. Stokes, *J. Math. Phys.* **3**, 70 (1962).

114. F. Rohrlich, *Phys. Rev.* **150**, 1104 (1966).

115. With the exception of the muon, all other charged particles are hadrons and have consequently a *finite size* due to the strong interactions. An electromagnetic contribution to their mass is then completely reasonable within a theory which is sensitive to quantum mechanical form factors.

DISCUSSION

F. Bopp: Lutzenberger and I have considered the motion of a point charge in its own electromagnetic field. In particular, we started with the Green's function $\delta(x^2)$ written down by Rohrlich. We changed this Green's function a little bit, and the particular change $\delta(x^2) \rightarrow (x^2 + l^2)$, where l is a certain constant length, yields a Lagrangian with a finite number of degrees of freedom. The strictly relativistic development according to the retardation is broken, but this breaking is an exact one. The number of degrees of freedom is just twice that of a single point particle. We then proceed to discuss this Lagrangian in a quantum mechanical way. The classical Hamiltonian turns out to be

$$H = \{ \pm \sqrt{\alpha^2 + 4r^2\,\mathbf{p}^2 + (\mathbf{p}\cdot\mathbf{r})^2} \pm \alpha \sqrt{1 + r^2} \}/r^2$$

(with $\hbar$, c and l being unity) in the case that the external momentum $\mathbf{P} = 0$; $\mathbf{r}$ and $\mathbf{p}$ are internal coordinates and momenta. One may take the first square root using Dirac matrices, and one obtains a four-component wave equation similar to that for the relativistic H-atom. It is well known how to solve such wave equations. As in the case of the relativistic H-atom, one gets the quantum numbers $k = \pm 1, \pm 2, \pm 3, \ldots$ referring to the momenta states $S_{1/2}, P_{1/2}, P_{3/2}, D_{3/2} \ldots$ as solutions of a two-component system of usual differential equations. The eigensolutions should provide us with masses of particles in different states. Unfortunately, numerical results are not yet available.

Note added in proof: Without bare masses we do not obtain stationary states. The second double sign in H is responsible for an equivalent double sign of the bare mass $M = \pm M$. If we assume $l = 4.91 \times 10^{-90} \hbar/m_p c$, $|M| = 15.3 m_p$, we obtain the following ground states for the different momenta:

$M = -|M|$ $k > 0$	State	$M = +|M|$ $k < 0$	State	$-m/M$	$|m/m_e|$
$+1$	$S_{1/2}$	–	–	3.555×10^{-5}	1
$+2$	$P_{3/2}$	-1	$P_{1/2}$	7.335×10^{-3}	206.3
$+3$	$D_{5/2}$	-2	$D_{3/2}$	10.855×10^{-3}	305.2
$+4$	$F_{7/2}$	-3	$F_{5/2}$	13.405×10^{-3}	377.1

The excited states differ from the bare mass only by some electron masses. All states are γ-unstable (life-time 10^{-17} or 10^{-26} sec) except $S_{1/2}$ and $P_{1/2}$, which may be identified with the electron and the muon. The product $l|M|$ is fitted by the ratio $m\mu/m_e$, and the length unit l by $\hbar/m_e c$.

H. C. Corben: I would like to make a comment about Professor Rohrlich's paper. It is common practice at this symposium to apply a time displacement operator backwards for 30 or 40 years. It is just about 30 years ago that Bhabha and I worked out the general classical theory of spinning particle in an electromagnetic field. This was based on the Frenkel theory, which really differs from the later development of the Bargmann–Michel–Telegdi theory in a very essential way, namely that the momentum and velocity of a spinning particle in the classical theory do not have to be proportional to each other any more than they do in a relativistic quantum theory such as the Dirac theory. In these theories the velocity is the commutator of the position with the Hamiltonian, and is quite different from the momentum. As a consequence, in the Dirac theory there is a *Zitterbewegung* as the velocity fluctuates while for a free particle the momentum stays constant. This is also true in the Frenkel theory, and therefore any of these classical theories which permit a helical motion turn out to be rather remarkably detailed classical analogues of the Dirac theory and the Majorana theory. I think that it should be recognized that the Bargmann–Michel–Telegdi theory is a kind of averaging of this helical motion, and of course it is a very accurate description of the motion of a spinning particle as seen by a macroscopic experiment. Nevertheless there is a great deal to learn, I think, from the study of the internal motions which do have their classical analogues and in fact do give a very good and most remarkable description of even such a simple theory as the motion of a spinning charged particle around a nucleus.

F. Rohrlich: I am very grateful to Professor Corben for his remarks. Within the period of one hour I have been unable to do justice to the large amount of work in this field. But because of Professor Corben's detailed book on spinning particles I felt that I could be brief on this subject. In my talk I was concerned with the classical spin equation only as far as it describes the classical limit of the Dirac equation. But I realize that this did not satisfy those who ascribe physical meaning also to classical descriptions of the electron which do not arise as classical limits of the Dirac equation; they feel that various quantum mechanical phenomena including *Zitterbewegung* and pair production, are meaningfully imitated in such a detailed classical description, while equations like the Bargmann–Michel–Telegdi equation only describe average behaviour. This is not my opinion.

R. Peierls: I would like to stress that all these considerations go beyond the classical limit. As Dr. Rohrlich has reminded us earlier, the spin is $\frac{1}{2}\hbar$, and in the classical limit that is zero. Now you can, of course, go one order of approximation beyond the classical limit and look at first order terms, but you have to be very careful because you are then running outside the range of validity of classical physics; and Corben's really very interesting model, from which one may learn something, is not the classical limit.

J. Ehlers: Since this session has been concerned to a large extent with the structure and motion of the electron – even the classical one – I wish to mention two results which may have escaped the attention of people not working in general relativity, and which show that general relativity theory, because of its flexible spacetime structure, may contribute to the construction of classical models of charged particles and to the derivation of their equations of motion. The first is: The only known exact solution of the Einstein–Maxwell equations representing the field of an object with mass, angular momentum, electric charge and magnetic moment, found by E. T. Newman and his collaborators and analyzed and physically interpreted by B. Carter and by G. C. Debney, R. P. Kerr and A. Schild [*J. Math. Phys.* **10**, 1842 (1969)], has the same gyromagnetic ratio as the Dirac electron. Unfortunately, it does not seem to be known whether other solutions exist as well. Together with a uniqueness theorem, the quoted result would be quite remarkable. Secondly, I wish to point out that E. T. Newman and R. Posadas [*Phys. Rev.* **187**, 1784 (1969); *J. Math. Phys.* **12**, 2319 (1971)] derived the classical Lorentz–Dirac equation of motion, with Abraham's radiation reaction force, from the Einstein–Maxwell theory. They did not make any assumptions about the interior of the particle, nor did they use a mass renormalization.

L. Rosenfeld: It is perhaps useful to warn against the danger of identifying symbols in an equation with physical quantities. It is not surprising that when one describes a rotating object it introduces angular momentum into an equation, and one finds something formally similar to Dirac's equation or to any equation representing an angular momentum. But the question is whether this formal appearance in an equation can be associated in an unambiguous way with the physical definition of angular momentum or magnetic moment. Of course, this is not possible, as is well-known for the terms that appear in Dirac's equation. One can indeed determine the spin, what is called the spin in Dirac's equation, but that is never done and that cannot be done by any experiment using the classical concepts of magnetic moment or angular momentum. One does it in an indirect way by Stern–Gerlach experiment or something else which essentially involves the conservation laws of energy and momentum.

17

The Development of Quantum Field Theory

Rudolf E. Peierls

As I understand my assignment today, it is to review the quantum theory of radiation up to the birth of the idea of renormalization. In doing so I may overlap slightly with what Rohrlich has talked about, but I hope not substantially.

When the principles of quantum mechanics were established it was clear that, for consistency, the electromagnetic radiation had to be described by a formalism based on the same principles. On this subject, as on so many others, the first step was taken by Dirac[1]. The idea of the quantization of radiant energy was, of course, at the starting-point of quantum theory, in the form of Planck's postulate, but consistent predictions could be expected only from a reformulation in the framework of the new quantum mechanics.

The radiation field differs from atomic systems principally by being a continuous medium, and thus having an infinite number of degrees of freedom. This may cause some difficulty in visualizing the physical problem, but is not, in itself, a difficulty for the formalism. One easy way to make the transition from quantum mechanics of particles to quantum field theory is to start from the vibrations of a crystal lattice and then think of the number of atoms increasing, and their space decreasing, indefinitely, so that a continuous medium is reached in the limit.

In fact, when Debye was discussing the quantum theory of a crystal, he went the other way about and took the crystal as approximately described by an elastic continuum because that made the mathematics simpler.

In some sense we have come from Planck's postulate for light, via de Broglie's analogy, to the wave equation for the electron, and in this sense the Maxwell fields are treated as the analogue of the wave function of an electron. In constructing a consistent quantum field theory one has to take account of the many degrees of freedom, so that the Schrödinger description would require a state function in a multi-dimensional (strictly speaking infinite-dimensional) space. The fields then are operators acting on such a function. This replacement of a wave function by a functional operator is now familiar also for particles as the 'second quantization'. Thus we may formally maintain the analogy between light and matter.

But it is useful to remember that this analogy has its limitations. Rosenfeld has already reminded us of Niels Bohr's insistence that it was not by accident that electrons first became known to us as particles and electromagnetism as a field. I would like to take a few minutes to remind you of the reasons for this.

J. Mehra (ed.), The Physicist's Conception of Nature, 370-379. All Rights Reserved
Copyright © 1973 by D. Reidel Publishing Company, Dordrecht-Holland

The first of these is that radiation has a classical limit in which the field quantities are measurable physical variables, and the uncertainty principle is of negligible importance. This situation applies in practice to the field of a radio transmitter, which may be described in terms of photons, but for which the phase, i.e. the sign of the electric or magnetic vector at any given point and given time is well known. What are the requirements for this?

The field is proportional to an amplitude, which can be written as

$$N^{1/2}e^{i\theta},$$

where N is the number of photons and θ a phase. These are non-commuting quantities, and there is an uncertainty relation which can be written (with some oversimplification) as

$$\delta N \, \delta\theta \geqslant 1.$$

So if we want to have any knowledge of the phase, the number N must be uncertain, and this is possible only if there are processes taking place which can change the number. Any experiment which can give information about the phase of the electromagnetic wave must involve the absorption or emission of photons.

For electrons the same uncertainty principle applies, but there cannot be any processes changing the electron number because of charge conservation. (There is pair creation, but this would involve the product of an electron and a positron wave function.)

Thus the field belonging to a charged (or otherwise conserved) particle can never be observable. Moreover, even if one could introduce annihilation and creation processes, and thus make the field observable, it could still never be a classical quantity. This is because the uncertainties should then be relatively small, i.e. we should have

$$\delta N \ll N \qquad \delta\theta \ll \pi.$$

From the uncertainty relation quoted above, this implies $N \gg 1$. But the number which appears in our relations is the number of particles (or photons) in one mode, which by Pauli's principle cannot exceed unity for fermions. Thus there could never be a classical field for any charged particles or for any fermions.

A third difficulty would remain even for neutral bosons. If the particle has a finite rest mass, the states of different N would of necessity have different energies (or different momenta) and a coherent superposition of such states would therefore be very hard to achieve.

On the other hand, one of the essentially particle-like properties of the electron is that its position is an observable. There is no such thing as the position of a photon. Landau and I learned this painfully by trying to formulate electrodynamics in a photon configuration space. This can be done, in spite of the existence of photon creation and annihilation, which simply means that one is dealing with a series of configuration spaces of different numbers of dimensions. However, the difficulty is that, at least in the presence of interactions, the description is not Lorentz invariant.

The content of the theory is, of course, invariant, because it has been constructed to be equivalent to the usual formalism, but this invariance is hidden, since on changing the frame of reference the function in each configuration space not only changes in a complicated way, but in part transforms into the parts in other spaces.

The basic reason for this is that, if the photon position was an observable, there should exist a probability density belonging to it, which must be the time component of a conserved four vector, and positive definite. In a field given by a tensor of integral rank such a quantity does not exist. This argument was precisely the one used by Dirac to introduce a spinor field for the electron.

One might object: If a photon produces a spot in a photographic plate, by making a grain developable, have I not observed its position? Well, within the framework of geometrical optics, I have. But if I want to locate the photon to better than the wavelength of the initial wave, I find that the chance of a grain being blackened is proportional to the square of the electric vector, since the atoms in the grain respond to the electric field. We can imagine a photographic plate in which the photosensitive reaction is caused by a magnetic dipole transition (not a very practical detector, I admit) and the most probably 'location' of the photon obtained from this would follow the magnetic vector. A moving plate responds to some linear combination of E and B. So this method cannot be regarded as a measurement.

If we work at relativistic energies the electron shows the same disease. Any experiment which can localize the electron to better than its Compton wavelength must be capable of creating additional pairs. If the observation discloses a negative charge it is not clear whether this is the electron we are looking for, or another electron created in the process. This goes together with the fact that, in a description in which negative-energy states are eliminated as possible states for the electron, and reinterpreted in terms of positrons, the electron coordinate no longer commutes with the electron number. So in this region the electron is as bad a particle as the photon. The inducement to think of it as a particle therefore comes from its behaviour in the non-relativistic region. The photon, without rest mass, has no non-relativistic state of motion.

These are old points, but they are not always as well known as they deserve. Let me return to the main story, the chronological view of field theory. I do not want to say 'historical', since this would imply a deeper analysis than I propose to present.

Having understood how to describe, and how to quantize, the free radiation field, one must turn to the description of transitions, of the emission and absorption of radiation. Dirac immediately tackled this task also, and in the course of doing so developed the formalism for time-dependent perturbation theory, which has become the standard tool for this and many other problems. No difficulty arises as long as one is concerned only with effects of first order in the interaction between matter and radiation. It was, of course, important to verify that quantum mechanics was capable of producing the expected answers. These include the expressions for the absorption of radiation by an atomic system, which, as Schrödinger showed in one of his early papers, can be calculated without quantizing the radiation field. From the absorption one can, of course, immediately write down the emission rate by using Einstein's relations, but it was reassuring to verify that quantum mechanics gave clear answers in agreement with the Einstein relations.

The difficulties start as soon as one tries to include effects of second order. After Dirac's basic paper there followed a few years of exploration, of calculations which aimed at finding out how far the predictions of the quantum theory of radiation were physically sensible, and also how far they confirmed what one expected.

An important piece of work of this kind (actually later in time than that of Heisenberg and Pauli) was the discussion of the natural line width by Weisskopf and Wigner[2]. The line width is of second order in the interaction, but since its magnitude is related to the total transition rate to all final states of the atom, which can be predicted by the simple theory, this ought to be a manageable problem. Indeed, Weisskopf and Wigner found that the width, or the imaginary part of the frequency of the emitted radiation, came out as expected. The theory was also capable of determining the line shape, which came out to be Lorentzian, as expected, and to discuss more subtle problems, as, for example, the successive emission of several photons in cascade.

At the same time, Weisskopf and Wigner also found that in addition to the imaginary part of the energy, which is the width, there appeared a shift in the real part, and this, while formally of second order in the interaction, was infinitely large.

This was one of the early examples of this kind of divergence in the interaction between electrons – and other elementary particles – and the electromagnetic field. This was a disappointment. It was not surprising that the self-energy of a point. electron should be infinite, because this was very well known from classical theory. However, the new quantum mechanics had resolved so many of the paradoxes and divergencies of the classical picture, that it was not unreasonable to hope that this one would go away also.

But it did not oblige; it stayed. And this has caused physicists very considerable headaches. It looks as if the physical cause of the infinity is the same as in the classical theory. In this respect quantum mechanics changes the magnitude of the divergent term, but not its physical nature. I would therefore like to return for a moment to the classical situation. Here my point of view will differ somewhat from that presented in Rohrlich's talk.

Consider the equation of motion for an electron in an external electric field, allowing for the radiative reaction. I shall write this down in its non-relativistic form

$$m_0 \dot{v} - eE = - m_s \dot{v} - \frac{2e^2}{3c^3} \ddot{v} + \text{const}\,\dddot{v} + \cdots .$$

Here m_0 is the 'bare' rest mass of the electron, m_s its electromagnetic self-energy, E is the external field strength. In discussing an equation of this type, Lorentz pointed out that the first term on the right-hand side is proportional to R^{-1}, where R is the electron radius; the second term is independent of R, the next one would be as R, and so on. The expression is the beginning of a power series in R, and the expansion parameter must be $R/c\tau$, where τ is a time characterizing the nature of the motion of the electron. In many practical situations the frequencies involved in the electron motion are slow compared to c/R, and the series converges rapidly. One may then neglect all but the first term on the right, finding the Newton equation of motion with mass $m_0 + m_s$, or, if one is interested in damping, include the second term as well.

Stopping after the second term leads to the equation studied by Dirac.[3] Since this contains the second derivative of the velocity, i.e. the third derivative of the coordinate, it has a three-parameter family of solutions. The new freedom consists of the famous runaway solutions. If one takes, with Lorentz, the view that R is really finite, if small, then the approximation involved in stopping after the second term is not justified, because (if m_s is of comparable magnitude to the total mass) the rate of growth of the velocity is such that the first and second terms are of the same order of magnitude, and, being an exponential function of time, successive terms will also be comparable. Thus for an electron of finite size the existence of runaway solutions is not established.

The situation is different if one proceeds to the limit $R \to 0$ in the equation, so that all terms after the second disappear. In that case m_s becomes infinitely large. One does not have to mention m_s since $m_0 + m_s$ is the observed mass, and only this combination appears. However, if we go to the limit of small R, while keeping the total mass fixed, the 'bare' mass m_0 becomes large and negative.

In this sense the runaway solutions acquire a new meaning: If the electron accelerates rapidly enough to involve frequencies comparable with c/R, the field cannot respond immediately and, because of retardation effects, the field energy will lag behind in time. This is enough for the growing field energy to be balanced by the negative kinetic energy, which in the equation is supposed to follow without delay. I submit that this picture, of the electron accelerating on 'borrowed' energy from the negative kinetic energy is a fair representation of the runaway solutions.

I do agree with Rohrlich, of course, that we should not take such calculations too seriously, because they always involve a time scale for which it is inappropriate to use classical laws. One should regard such considerations merely as illustrations of what may happen so as to ask sensible questions of quantum electrodynamics.

What we have seen here – if you can accept my interpretation – is that by 'renormalizing' the mass of a point electron, we have introduced runaway solutions.

How does the same problem look in quantum theory? We do not know. I do not want to enter into any discussion of quantum electrodynamics here because it will be the subject of Schwinger's talk tomorrow. But we have never yet succeeded in extracting any information from quantum field theory except in terms of a series in powers of the electron charge. But the problem to which we have been led in the classical case disappears as soon as one expands in a power series. It occurs only in a non-perturbative solution. The question whether the limit implied by the infinite renormalization of quantum electrodynamics would show similar difficulties for a non-perturbative solution, must remain open.

Alternatively one might prefer not to go to the limit of a point electron, and assume a finite radius. In the ordinary way this leads to the well-known difficulties which have been mentioned. They reflect the fact that the naive theory treats the electron as a rigid body and, as is well known, there can be no rigid body in relativity. Basically, a rigid body has an infinite velocity of sound, and in relativity the velocity of sound cannot exceed c.

I do not believe that one can get out of this difficulty by the prescription that the momentum, or the stress, be computed in the rest system. This prescription is easy to apply for a particle in uniform motion, but the theory must be applicable to an electron which is being accelerated under the influence of external forces. It is not at all clear how one is to extend the prescription to that case.

One way of looking at this situation is to notice that the assumption of an extended charge distribution changes the equations into integral equations. For example the electric force on the electron will now be a space integral of the charge density times the local electric field. To have equations which contain space integrations, but are local in time, is not relativistic. This argument immediately suggests a generalization in which there are integrations over space *and* time, which can easily be made relativistic.

A theory of this kind was, for the classical case, discussed many years ago by McManus[4]. As a classical theory it seems quite unobjectionable. It produces finite, unique, covariant results, and as far as one could discover, it contains no runaway solutions if the form factor is sensibly chosen. Of course one cannot be satisfied with a classical theory except as a stepping stone to a quantum treatment. To extend the idea to quantum theory one might try to quantize equations of the McManus type. But here one meets the obstacle that the occurrence of time integrations means that the equations are no longer of the canonical form, in which the time derivative of each of a number of variables is given in terms of the values of all variables at that time. We therefore do not know how to quantize a 'non-local' theory.

We once guessed a quantum form of such a theory, at least to first order in the interaction[5]. It was a fine theory, which was gauge invariant and Lorentz invariant, and had all the right limits, i.e. the McManus type theory if quantum effect were negligible, and the usual motion of a point electron if everything varied slowly enough in time. Its only trouble was that the self-energy was not finite, but more divergent than the usual theory, because of two opposite divergent terms which partly cancel, the non-local version of the theory makes one finite and leaves the other unchanged.

I have gone off at numerous tangents, and it is time I returned to the theory of quantum field theory. The next important step in the development was the work of Heisenberg and Pauli[6]. They extended the quantum treatment to the full content of Maxwell's equations, including fields in interaction with charged particles.

It is then necessary to allow for the fact that in the neighbourhood of charges the electric field is no longer purely transverse. Indeed the Coulomb field may be regarded as purely longitudinal. It is sometimes convenient for calculation to eliminate the static Coulomb field in terms of the direct inverse-square law interaction, and retain only the transverse part in the field equations, but for understanding the physical content of the equations, or for discussing their behaviour under Lorentz transformations, this is quite unsatisfactory.

Heisenberg and Pauli faced these problems and identified, and overcame, a number of technical difficulties. They brought out clearly the importance of gauge invariance for the description of the fields. One consequence of gauge invariance is that the photon rest mass is zero, and this in turn results in some of the field equations taking the form of 'subsidiary conditions' i.e. of relations not containing any time derivative, so that they must be satisfied at any given time. Such relations are then not consequences of the quantum equations of motion, which determine only the rate of change of the state function, but they restrict the choice of the state function itself. Heisenberg and Pauli studied these complications, and showed how to get around the difficulties. Another more transparent method of dealing with the subsidiary conditions was given by Fermi [7]. Heisenberg and Pauli also encountered a more formal difficulty. In the standard quantum treatment the time is treated in a different manner from the space coordinates, in that one uses a state function which represents the probability of a set of observables relating to all points of space at a given time. This description is modified in a very involved way by a Lorentz transformation, which leads to a state function concerned with constant time in the new frame of reference. They showed how to carry out such a transformation and verified that, in spite of the lack of invariance of the representation, the content of the theory was correctly Lorentz invariant.

They also noted the occurrence of 'infra-red catastrophe', which arises from the fact that in any process involving charged particles the expected number of long-wave photons emitted shows a logarithmic divergence. Since the total energy carried away by the photons is finite, and the presence of very long-wave photons of no great consequence, the 'catastrophe' is only a formal one. The first systematic way of dealing with it was developed by Bloch and Nordsieck [8].

While in this work the technical problems of handling field equations were overcome, it was confirmed that the infinities remain. I have already referred to the occurrence of an infinite self-energy in the calculations of Weisskopf and Wigner, but these were non-relativistic. Heisenberg and Pauli used a relativistic description for the electron in a one-electron picture (which is not really consistent because the electron would at once drop into states of negative energy) and still found a divergent self-energy. In fact, the magnetic part of the self-energy is more strongly divergent than the electrostatic part, and they obtained a quadratic divergence in the integration over virtual phton momenta.

Weisskopf extended this calculation to the new Dirac 'hole' theory [9]. This gives a different answer because in the virtual emission of a photon, the electron can make transitions only to states of positive energy and back, since the negative energy states are filled. The transitions of the electron from its initial state momentum p to a negative energy state of momentum p' are now forbidden. Instead, however, all the negative-energy electrons can make virtual transitions to positive-energy states and back, and this accounts for a self-energy of the vacuum, which one has to assume is compensated. In the presence of the electron in the state p, the transition from p' to p is now forbidden by the Pauli principle, so a term is missing compared to the vacuum self-energy, and this is also due to the presence of the electron. This term is numerically equal to the contribution in the one-electron theory from the transition of the p electron to p' and back, but it occurs with the opposite sign.

Actually, the argument I have given is not adequate to show the sign. Signs are always delicate matters. In fact, in Weisskopf's first paper the sign was wrong, and he had the two terms in the integral adding instead of subtracting. The error was pointed out by W. Furry, and the correct answer is that the terms involving positive and negative virtual states cancel approximately, and only a logarithmic divergence remains.

Of course, a logarithmic infinity is still an infinity, but the integral could be made finite and comparable to the real electron mass by cutting off the integration at some fantastically large momentum, or, in other words, giving the electron a fantastically small, if finite, radius; if such a cut-off really existed we would probably never be able to observe its consequences.

For this reason electrodynamics may not be the best example on which to test field theories, because the real difficulties may be so weak or may be serious only in such extreme regions, that we can easily evade them. We know – and we shall hear in more detail later – of the impressive successes of renormalized theories with infinite renormalization constants. We do not know whether this renormalization, which is done only in terms of a series of powers in the fine structure constant, is a rigorous result, i.e. whether the defining series converges. Most of us guess that it is at best asymptotic, so that it is a good approximation up to something like the 137th term, and this would give an accuracy far in excess of anything ever needed by theory or experiment.

These questions would become real challenges if we tried to set up a field theory of strongly interacting systems, for which a series expansion (even if convergent) would not be useful, and for which one would have to define renormalization (or any other device for dealing with the infinities) in a non-perturbative treatment. Of course we do not know even how to start on this problem because nobody can write a non-perturbative solution to a non-trivial field theory, even if it were free of infinities.

The same mathematical obstacles have caused the abandonment of an other attempt to change electrodynamics so as to get infinite answers, the non-linear field theory of Born and Infeld[10]. I have referred to attempts to avoid infinities by introducing the electron radius as a shortest length, or cutting off the integrals by postulating a largest momentum, which is more or less the same. The field equations would then be non-local, but still linear.

Instead of this, the idea of Born and Infeld was to introduce a greatest field strength, which would imply non-linear equations. Instead of an action density proportional to $E^2 - B^2$, one chooses a suitable function of this quantity, which causes the field to saturate. It is easy to write down a classical field theory of this kind. This replaces the Coulomb field of a point charge by one which has a finite field intensity even on the charge.

The trouble comes in quantizing this field. In the usual theory the field intensity at any point in space has infinite fluctuations, but because of the linearity these are superimposed on any impressed fields without affecting them directly. In the non-linear theory the fluctuations are bound to be finite, of the order of the maximum field strength. But because of the non-linearity this means that the response to an external source, or to a light wave, will always be strongly modified, because we are almost always and almost everywhere in the saturating regime; there is no reason that the response to a disturbance should have any similarity to that predicted by Maxwell's equations.

Born and Infeld wrote their non-linear equations in such a way that for weak fields they become the Maxwell equations. In a classical theory, where you can have weak fields, this is satisfactory. In quantum theory in which the vacuum has saturating fluctuations, it is not enough to ensure that what we see has any resemblance to the Maxwell equations. It just does not seem possible to find out what we should see, and for this reason the non-linear theory has not been studied further, although for all we know it may contain perfectly sensible answers.

This prompts a remark which goes well outside my brief, but which I cannot resist making. It seems to me that one has a very similar situation in the theories which are now fashionable, in which there are spontaneously broken symmetries. In such theories the basic equation has a certain symmetry, but its solution corresponds to a number of alternative equilibria, which are themselves not symmetric, but related symmetrically to each other. The system is always near one or other of these and therefore is not symmetric. In this case the equation for small displacements from this equilibrium will be lacking in symmetry, and that is the object of the construction. But moreover they may well differ in other ways from the equations for small displacements from the symmetric position, which is in general that of zero field. This is so because the equations must be non-linear, otherwise they could not have a displaced equilibrium. As in the case of Born and Infeld, the connection of the physical results with the original equation which motivated the whole approach may therefore become lost.

This has been a somewhat arbitrary selection of topics from the early days of field theory. I have tried to pick out the points which in retrospect still seem to hold our interest and which might even contain a lesson today, in spite of all the many important developments which have taken place since the period I was instructed to cover.

REFERENCES

1. P. A. M. Dirac, *Proc. Roy. Soc.* A.**114**, 243 (1967).
2. V. F. Weisskopf and E. Wigner, *Z. Physik* **63**, 54 (1930), **65**, 18 (1930).
3. P. A. M. Dirac, *Proc. Roy. Soc.* A. **167**, 148 (1938).
4. H. McManus, *Proc. Roy. Soc.* A .**195**, 323 (1948).
5. M. Chrétien and R. E. Peierls, *Proc. Roy. Soc.* A **223**, 468 (1954), *Nuovo Cimento* **10**, 668 (1953).
6. W. Heisenberg and W. Pauli, *Z. Physik* **56**, 1 (1929), **59**, 168 (1930).
7. E. Fermi, *Rev. Mod. Phys.* **4**, 87 (1932).
8. F. Bloch and A. Nordsieck, *Phys. Rev.* **52**, 54 (1937).
9. V. F. Weisskopf, *Z. Physik* **89**, 27 and **90**, 817 (1934).
10. M. Born and L. Infeld, *Proc. Roy. Soc.* A. **144**, 425 (1934).

DISCUSSION

C. N. Yang: I would like to make a remark concerning the question mentioned by Professor Peierls towards the end of his talk about the Born–Infeld theory. The Born–Infeld theory is derived from a desire to quench the infinite field near the origin of a source. It was, of course, a very understandable aim. However, to me, a great shortcoming of what they did was the *ad hoc* nature of the additional terms put into the Lagrangian which was used to quench the field. We know that with the gauge field, the field equations themselves are automatically non-linear and do not require the addition of a source term. In other words, even if you put the source equal to zero the equations are non-linear for the gauge field, because the gauge field itself acts as a source of its own. It seemed to me most attractive to investigate whether sourceless fields, or a gauge field without external sources would have a solution which does not have the infinities near the origin. This problem has been looked into, and one can find solutions where the singularity is much less than that of the Coulomb field. Quantization of such fields however seems to be very difficult. What I am referring to is something which has already been published by T. T. Wu and myself in the supplement to the *Progress of Theoretical Physics*, the Tomanaga Festschrift.

R. Peierls: This is certainly very interesting, but if I understand it right it is a classical calculation which is not in the spirit of the Born–Infeld theory. My main purpose in referring to the Born–Infeld theory was to point out that we have very little control over such theories because we do not know what is the nature of their quantum solutions, and I think that you would agree with that. The quantization is awfully hard, as is perhaps also the quantization of any ordinary kind of field theory because there are always non-linearities from the sources, and maybe we don't know any solution.

F. Rohrlich: I would like to make a remark which expresses my own personal opinion, not necessarily the consensus of those working in this field. In fact, I am sure that there are some who will disagree with me. But I feel that the classical theory must be viewed in perspective. Its limits of validity are clearly indicated by quantum mechanics, and especially by relativistic quantum mechanics, i.e. quantum field theory. Furthermore, I consider the old difficulties of classical electron theory fully overcome, because the theory can be formulated so that these difficulties are not encountered, without changing the fundamental equations in any way. Over the years different people have been suggesting many modifications of Maxwell's equations and of the electron's equations of motion. But I believe that these are *ad hoc* alterations, '*Verstümmelungen*' of the equations. This is true for example of the Born–Infeld theory. I consider this theory completely surpassed by quantum field theory. The latter introduces in a natural way both non-linearities and non-localities into electrodynamics. For example, the early work on vacuum polarization by Heisenberg and Euler, which was later generalized by Weisskopf and by Schwinger, shows very clearly how these non-linearities come about. One can use the exact solution of the Dirac equation in an almost constant external field to find a Lagrangian for the electromagnetic field which effectively includes vacuum polarization and describes the dielectric behaviour of the vacuum. There are no arbitrary functions or constants as in the Born–Infeld theory. After all, arbitrary non-linear terms have been suggested since Mie at the beginning of this century. Our problem is therefore to find the non-linear behaviour of quantum electrodynamics for other cases such as the Coulomb field without recourse to perturbation expansion. The latter is extremely unreliable. We know still very little about this subject of non-linearities, but a non-perturbative treatment will also be mathematically much more consistent than the present one and will obviate *ad hoc* modifications.

18

Quantum Theory of Fields (until 1947)*

Gregor Wentzel

1. EARLY QUANTUM ELECTRODYNAMICS

The basic idea of field quantization is some 60 years old. A particularly clear anticipation of later trends is found in a paper by Ehrenfest (1906). Commenting on Planck's theory of the Hohlraum radiation, Ehrenfest discusses the thermal equilibrium spectrum by means of the Rayleigh–Jeans normal modes analysis, exploiting the analogy of the amplitude of a proper vibration of the cavity with the coordinate of a material oscillator. Whereas the classical equi-partition of energies leads to the Rayleigh–Jeans spectrum, how can one modify the theory to obtain the Planck spectrum? As a possibility, suggested by Planck's quantization of material oscillators, Ehrenfest mentions the following hypothesis: the amounts of field energy residing in a normal mode of frequency v can only be integral multiples of hv. The same assumption enabled Debye (1910) actually to derive the Planck spectrum. As we know today, such a quantization of the field energy results in exhibiting the corpuscular aspects of the system; but this connection with Einstein's light quantum hypothesis (1905) remained obscure until much later. Once the wave–particle dualism was understood, after the advent of wave mechanics, Debye's theory was seen to be equivalent to Bose's photon gas statistics (1924). The characteristic features of Bose's counting of states (namely, a permutation of n identical photons does not give a new state) corresponds to the non-degeneracy of the quantum state n of a linear oscillator (Born, Heisenberg and Jordan, 1926).

If it is true that an elementary light wave is equivalent to a harmonic oscillator, with energy eigenvalues $nhv(+\frac{1}{2}hv)$, it would then seem natural to apply quantum mechanics also to the *amplitude* of the wave and interpret it as an operator or, more specifically, as a matrix with respect to the oscillator quantum number n, similar to the

---

* *Editor's note:* Professor Gregor Wentzel was the chairman of the session devoted to quantum field theory at our symposium. We had invited him to contribute a paper on the historical development of quantum field theory, but the circumstance that he now lives in retirement in Ascona, Switzerland, without easy access to reference journals, prevented him from doing so. Professor Wentzel has had a long and distinguished association with this subject, and he has exercised a profound influence on its development. For this reason, as well as to provide a detailed conceptual background of the development of this subject from its origin until 1947, we are reprinting Professor Wentzel's authoritative article. [Reprinted from: M. Fierz and V. F. Weisskopf (Editors), *Theoretical Physics in the Twentieth Century*, Interscience Publishers, New York and London (1960), pp. 48–77, by permission of John Wiley and Sons, Inc., Publishers, New York.]

coordinate matrix of the linear oscillator

$$
(n'|\, q\, |n) = \begin{cases} n^{1/2} & \text{for } n' = n - 1 \\ (n + 1)^{1/2} & \text{for } n' = n + 1 \\ 0 & \text{otherwise.} \end{cases}
$$

This idea was already used by Born, Heisenberg and Jordan (1926) to derive the mean square energy fluctuation of the Hohlraum radiation, but its major implications were discovered by Dirac (1927a) in his theory of the interaction of radiation with atomic matter. The interaction Hamiltonian (of first order) contains the matrices q as factors and will give rise, according to first-order perturbation theory, to atomic transitions in conjunction with transitions $n \to n \mp 1$ of the radiation oscillators or, in other words, with absorption or emission of photons. The energy of the combined system, atoms plus photons, is conserved within the limits of the uncertainty principle. The transition probabilities per unit time, involving the squares of the matrix elements, are proportional to the initial photon occupation numbers n in the case of absorption, and proportional to $(n+1)$ in the case of emission, where the parts $\sim n$ and ~ 1 correspond, respectively, to induced and spontaneous emission.

Today, the novelty and boldness of Dirac's approach to the radiation problem may be hard to appreciate. During the preceding decade it had become a tradition to think of Bohr's correspondence principle as the supreme guide in such questions, and, indeed, the efforts to formulate this principle in a quantitative fashion had led to the essential ideas preparing the eventual discovery of matrix mechanics by Heisenberg. A new aspect of the problem appeared when it became possible, by quantum-mechanical perturbation theory, to treat atomic transitions induced by given external wave fields, e.g. the photoelectric effect. The transitions so calculated could be interpreted as being caused by absorptive processes, but the 'reaction on the field', namely the disappearance of a photon, was not described by this theory, nor was there any possibility, in this framework, of understanding the process of spontaneous emission. Here, the correspondence principle still seemed indispensable, a rather foreign element (a 'magic wand' as Sommerfeld called it) in this otherwise very coherent theory. At this point, Dirac's explanation in terms of the q matrix came as a revelation. Known results were rederived, but in a completely unified way. The new theory stimulated further thinking about the application of quantum mechanics to electromagnetic and other fields.

In a consecutive paper, Dirac (1927b) applied second-order perturbation theory to the atom-field interaction and re-derived the Kramers–Heisenberg (1925) formula for the scattering of light by atomic systems. Finer aspects of radiative processes, like line widths due to damping, required a more elaborate mathematical treatment (Weisskopf and Wigner, 1930; for further literature see Heitler, 1954). Questions regarding the propagation velocity of light, or the coherence of scattered light, can best be answered by remembering that, for any quantum-mechanical system, the expectation values of observable quantities obey the equations of motion of the corresponding classical system. In application to the electromagnetic field operators, this leads to the classical field equations (Heisenberg, 1931), and much of classical wave optics is seen to remain valid.

In Dirac's radiation theory it is not the entire electromagnetic field that is subjected to quantization, but only its 'radiative part' consisting of plane transverse light waves. Electrostatic fields, on the other hand, are taken care of by including the Coulomb interaction energies into the Hamiltonian of the material system. This unsymmetric treatment of the two field parts not only appears contrary to the spirit of Maxwell's theory, but also raises questions from the view-point of relativity theory. A Lorentz transformation mixes the two field parts; the splitting is not invariant. The problem was now clearly this: how to formulate a general quantum theory of the Maxwell field interacting with charged particles, in a Lorentz invariant fashion. It certainly seemed a formidable programme in 1927, even though not all obstacles to be encountered could have been foreseen at that time.

One preliminary step was the Lorentz invariant formulation of the commutation rules for the charge-free field (i.e. Dirac's radiation field) by Jordan and Pauli (1928). Time-dependent field operators were introduced, and the commutators of two field components, taken at different space-time points, were expressed in terms of the, now famous, 'invariant Delta function'. The physical meaning of these results in relation to basic uncertainties in field measurements was later analysed by Bohr and Rosenfeld (1933).

A relativistic description of electrons, either force-free or subject to given external fields, was provided by Dirac's (1928) wave-mechanical theory of particles with spin 1/2. This theory, however, was beset with the difficulties coming from the 'states of negative energy', and these difficulties stayed with quantum electrodynamics until the mathematical development of the positron theory (see Section 2) brought some relief.

Heisenberg and Pauli (1929) were the first to attempt a general formulation of quantum electrodynamics by setting up a general scheme for the quantization of fields which they hoped would be applicable to the Maxwell field. Since quantum mechanics, in its usual form, starts with a Hamiltonian characterizing the system, Heisenberg and Pauli assumed that the field equations, like the equations of motion of mechanics, are derivable from an action principle which allows us to define fields P_α 'canonically conjugate' to the original field components Q_α so that 'canonical commutation relations', together with a Hamiltonian, can be constructed. The commutation relations could be proved to be invariant under infinitesimal Lorentz transformations, if the Lagrangian density is invariant (see also Rosenfeld, 1930).* This canonical formalism, which later was to become so successful when applied to scalar and other fields, turned out to be insufficient in the case it was intended for: electrodynamics. Taking the Lagrangian density for the charge-free field in the customary form

$$L = -\tfrac{1}{4} \sum_{\alpha\beta} \left(\frac{\partial Q_\alpha}{\partial x_\beta} - \frac{\partial Q_\beta}{\partial x_\alpha} \right)^2 = \tfrac{1}{2}(\mathbf{E}^2 - \mathbf{H}^2)$$

* With regard to this proof, Pauli used to say: 'Ich warne Neugierige'.

where Q_α = four-potential ($\alpha = 1, ..., 4$, $x_4 = ict$), then P_4 is found to vanish identically, and also the divergence of $\mathbf{P} = -\mathbf{E}$ vanishes. Under these circumstances, the canonical commutation relations cannot be applied. Only later was it recognized that this failure is connected with the zero rest mass of the photon and the existence of the gauge group

$$Q_\alpha \to Q_\alpha + \partial\chi/\partial x_\alpha.$$

Heisenberg and Pauli (1929, 1930) found two ways to overcome the difficulty in a formal manner. The first was to add a term $\frac{1}{2}\varepsilon(\sum_\alpha \partial Q_\alpha/\partial x_\alpha)^2$ to the Lagrangian density, prescribing the limit $\varepsilon \to 0$ to be taken in the final result. In their second paper (1930), they chose the special gauge $Q_4 = 0$ (so that a Lorentz transformation entails a gauge transformation); one has then to require that the (time-independent) space function $C = \text{div } \mathbf{E} + \varrho$ vanish identically. Both these approaches can be carried through consistently, but appear somewhat artificial and not quite appropriate to a fundamental entity like the electromagnetic field. It was a particular disappointment that the basic difficulty of the classical Lorentz theory, the infinite value for the self-energy of the point electron, seemed to survive in the quantized version (see also Waller, 1930; Oppenheimer, 1930).

Closer to the line of the later development was a new (and independent) start by Fermi (1929–30). Here, the Lagrangian is chosen such as to give the field equations in the form

$$\sum_\beta \frac{\partial^2}{\partial x_\beta^2} Q_\alpha = -s_\alpha$$

(s_α = current and charge density) which, in order to yield Maxwell's equations, have to be supplemented by the 'Lorentz condition'; i.e. the four-divergence

$$\Omega \equiv \sum_\alpha \partial Q_\alpha/\partial x_\alpha$$

should vanish. Fermi (treating all quantities in Fourier transform) observed that, in quantum theory, the Lorentz condition need not be postulated as an identity for the field operators Q_α, but it is sufficient to impose corresponding subsidiary conditions or 'constraints' on the state vector Ψ:

$$\Omega = 0, \quad \text{and} \quad \dot{\Omega}\Psi = 0 \quad \left(\dot{\Omega} = \frac{i}{\hbar}[H, \Omega]\right),$$

stating that the operators Ω and $\dot{\Omega}$ annihilate Ψ at all times. These conditions are admissible because they are equivalent to two initial conditions for Ψ (at $t = 0$, say) which are compatible (the corresponding operators commute). By solving explicitly the subsidiary conditions, one can then immediately write down how Ψ depends on the 'scalar' potential (Q_4) and the longitudinal part of the 'vector' potential (div $\mathbf{Q}$). Then, for the remaining factor in Ψ, one obtains a reduced Schrödinger equation

$$i\hbar \frac{\partial\Psi'}{\partial t} = H'\Psi'$$

where H' turns out to be the Hamiltonian of Dirac's radiation theory, including the Coulomb interactions between the charged particles (and also their infinite self-energies). (In the Heisenberg–Pauli versions, the Coulomb terms were derived in second-order perturbation theory only.) Since the primary formulation of Fermi's theory, which involves the static and wave fields in a unified fashion, can be set up in an arbitrary Lorentz frame of coordinates, it thus turns out that Dirac's quantum theory of radiation is in effect a relativistically invariant scheme provided the charged particles are also properly relativistically described, for instance by Dirac's wave mechanics for spin 1/2 electrons.

Fermi did not bother to verify explicitly that his quantization procedure (e.g. field commutation relations) remains invariant under Lorentz transformations although this might have been done by examining infinitesimal transformations and then invoking their group properties as Heisenberg and Pauli did in their versions of the theory. However, a much more transparent and elegant proof of Lorentz invariance emerged from a generalization of the theory which was discovered by Dirac, Fock and Podolsky (1932).

In all previous formulations, Lorentz invariance was far from obvious because of the non-symmetric treatment of time and space coordinates: each electron has its own position vector $\mathbf{r}_n$ (in a configuration space treatment), the fields are defined as functions of another space vector $\mathbf{r}$, but only *one* time coordinate t appears (in the state vector Ψ if the 'Schrödinger representation' is used). This lack of symmetry is removed in the 'multiple time' formulism of Dirac, Fock and Podolsky where an individual time coordinate t_n is assigned to each charged particle and the state vector Ψ, considered as a function of all four-vectors $(\mathbf{r}_n, t_n)$, is subjected to several Dirac wave equations, one for each electron. These wave equations involve the four-potential of the Maxwell field in a time-dependent representation (today, this is called 'interaction representation'):

$$Q_\alpha(\mathbf{r}, t) = \exp\left(\frac{i}{\hbar} H_F t\right) Q_\alpha(\mathbf{r}) \exp\left(-\frac{i}{\hbar} H_F t\right)$$

(H_F=field Hamiltonian without interaction); Q_α is to be taken at $(\mathbf{r}_n, t_n)$ in the wave equation for the nth electron. As a function of the field amplitudes, Ψ must again be subjected to a subsidiary condition $\Omega\Psi = 0$ which is a generalized form of Fermi's constraint. All these equations, as well as the commutation relations for the field operators, are manifestly Lorentz invariant, and also gauge invariant. If one, then, considers Ψ only in a single-time subspace (all $t_n = t$), one comes back to Fermi's theory, and finally to Dirac's radiation theory, and no further invariance proofs are necessary for these derived forms of the theory. Bloch (1934) further analysed the multiple-time theory as to internal consistency (the particles must be kept in space-like positions relative to each other:

$$c\left(|t_n - t_{n'}| < |\mathbf{r}_n - \mathbf{r}_{n'}|\right)$$

and pointed out the physical meaning of the generalized state function Ψ: it bears on measurements made on the particles at their individual times t_n.

The Dirac–Fock–Podolsky theory is the direct forerunner of later theories (Tomonaga, 1946; Schwinger, 1948) in which the charged particles are no longer described in configuration space but, instead, by a quantized field (e.g. electron field), and the set of discrete world-points $(\mathbf{r}_n, t)$ is replaced by a space-like 'surface' in Minkowski space. How this formalism served as a tool in the re-normalization programme of quantum electrodynamics will be reported elsewhere in this volume.

In the classical field theory corresponding to the multiple-time theory, the field functions depend explicitly on the particle times t_n (Wentzel, 1933). One can make use of this generalized Lorentz theory to re-formulate the classical equations of motion of charged point particles in such a way that the self-force of the particles becomes finite; in particular, the electromagnetic mass (inertia) vanishes. This is achieved by introducing a limiting process into the definition of the Lorentz force which lets the 'field point' approach the 'particle point' in time-like directions. In this way one can avoid all self-energy troubles in the *classical* theory of point electrons interacting through a Maxwell field, and it was hoped that similar limiting processes might eliminate the corresponding difficulties in the quantized theory. This hope turned out to be futile. Only the infinities of the classical, electromagnetic mass, type (diverging linearly with the cut-off momentum k_c) are affected by such a limiting process. But the quantum theory gives rise to other kinds of infinities, like Waller's (1930) term ($\sim k_c^2$) calculated for an isolated Dirac spin electron (negative energy levels empty). At a later time when the hole theory of the positron became more acceptable, one had to deal with only logarithmic singularities, but even these survived all attacks by mathematical tricks until the idea of mass re-normalization came up, with the promise of a new consistent interpretation. Meanwhile, however, the discussion about mathematical re-formulation went on for many years. Dirac (1939), after discovery of another formulation of the classical theory (Dirac, 1938), tried to modify the field commutation relations by means of a small time-like vector ('λ-limiting process') but then found that he had to introduce other artifices, like light quanta of negative energy and negative probabilities (Dirac, 1942). A review and criticism of these efforts can be found in a paper by Pauli (1943).

All this work on the self-energy problem is today superseded and of historic interest only. In spite of all failures, the general confidence in quantum electrodynamics as a supreme, though as yet imperfect, tool of atomistic theory remained alive. It is true that most special applications were based on Dirac's radiation theory, and infinities, where they appeared, were dealt with in a heuristic manner (straight subtraction or cut-off). No contradictions with experimental facts were apparent (the 'Lamb shift' was suspected but still doubtful at that time). Nevertheless, the awareness of the basic difficulties weighed heavily on our minds.

To complete the picture, it should be recalled that in the early 1930's the self-energy was not the only worry. The negative energy states of the Dirac electron have been mentioned already, and the toilsome development of the 'hole theory' will be reviewed later. There was also the infinite zero-point energy ($\sum \frac{1}{2}h\nu$) of the radiation field, and although today it may seem trivial to throw away an infinite additive constant in the energy by plain subtraction, people like Pauli (1933) went out of their way to cast the Hamiltonian density in a form which, in effect, eliminates the zero-point energy.

An infinity of a different kind is that referred to as the 'infra-red catastrophe' because the divergence occurs at small frequencies, as contrasted to the 'ultra-violet catastrophe' in the self-energies. One meets with this in problems like Bremsstrahlung if one asks for the total probability of any radiative process (photon emission); treating the field-electron interaction as small, the perturbation theory gives a result diverging as $\int d\nu/\nu$ at $\nu = 0$. Bloch and Nordsieck (1937) observed that this is really the fault of the perturbation method; the interaction is too large at low frequencies. Instead, they started by constructing the stationary states of free electrons in motion, with their proper fields attached, plus additional free photons. A weak external force (*this* is now the weak perturbation) induces transitions between such stationary states, and if initially no free photons were present there is a probability of their presence in the final state, as a consequence of the change of the proper field in the transition. It turns out that the mean number of photons emitted is infinite (concentrated at the lowest frequencies), but the total energy emitted is finite, and so is the total cross-section for the electron to be scattered into a certain solid angle. This cross-section is the same as one would calculate it from the Born approximation ignoring the field altogether. The formulae show why the expansion into powers of e (the field-electron coupling strength), though formally leading back to the original perturbation theory is inadmissible at low frequencies. Pauli and Fierz (1938) re-examined this theory, replacing the 'electron' by an extended charged body; now the cut-off (which was also needed in Bloch's and Nordsieck's work and tended to obscure a possible connection with the self-energy divergence) has a physical cause and is specified by a form factor. The size of the body is taken large enough to make the electromagnetic mass terms negligible. Non-relativistic energies are used, but otherwise the energy conservation is treated rigorously. The result of this more detailed analysis agrees with that of Bloch and Nordsieck except for the observation that the probability of very small energy losses depends on the size (form-factor) of the charged body, in such a manner as to 'render immediate application to real electrons impossible'. Although the authors thus concluded on a pessimistic note, the general expectation was that the low-frequency end of the spectrum would eventually be amenable to a consistent treatment if only one could visualize a theory of point electrons free of the more serious high-frequency divergences.

2. ELECTRON FIELD QUANTIZATION

When Jordan (1927) first sought to quantize the electron wave field, this must have seemed a rather strange idea to many of us. Contrary to photons, electrons have charge and mass (and spin $\frac{1}{2}$). Wave mechanics of many-electron systems in configuration space representation was well developed and seemed entirely satisfactory, at least as a non-relativistic theory. Pauli's exclusion principle was known to be equivalent to the postulate of antisymmetry of the state functions.

On the other hand, Dirac (1927), in the preparation of his theory of light emission and absorption, had already discussed the quantization of a (complex) Schrödinger wave function, replacing the probability amplitudes of the 'unperturbed states' by operators raising or lowering by one the occupation numbers of these unperturbed states, with commutation rules

$$b_r b_s^* - b_s^* b_r = \delta_{rs}, \qquad b_r b_s - b_s b_r = 0.$$

If the Hamiltonian is now written as

$$\sum_r W_r b_r^* b_r + \sum_{rs} V_{rs} b_r^* b_s$$

this quantized theory gives nothing new for a single particle

$$\sum_r b_r^* b_r = 1$$

the equivalent Hamiltonian matrix being

$$H_{rs} = W_r \delta_{rs} + V_{rs}$$

but N identical particles behave as they should do according to Bose–Einstein statistics, namely in accordance with configuration space wave mechanics using symmetric wave functions. In other words: for a system of identical bosons, the configuration space is not needed if the wave function in ordinary space is quantized. This procedure was called 'second quantization' or 'hyperquantization', because the use of a Schrödinger wave function amounts to a 'first' quantization. Jordan and Klein (1927) generalized this theory by admitting interaction terms $\sim b_r^* b_s^* b_k b_l$, e.g. Coulomb interactions between charged bosons. They pointed out that the Coulomb self-energy of the particles is formally eliminated by writing the non-commuting operators b, b^* in that particular order (two annihilation operators to the right).

The future prospects of this theory, in particular with regard to relativistic generalizations, were rated highly by Jordan who set out immediately to invent a similar formalism for fermions. Since the Pauli exclusion principle prohibits occupation numbers > 1, the b_r operators must be cut down to 2×2 matrices

$$\begin{pmatrix} 0 & 1 \\ 0 & 0 \end{pmatrix}_r,$$

times unit matrices ($s \neq r$). The final formulation of the theory, equivalent to configuration space wave mechanics with anti-symmetric state functions, was achieved by Jordan and Wigner (1928). Here, new operators a_r are introduced which differ from the b_r only by a judiciously chosen $\pm$ sign and obey 'anticommutation relations'

$$a_r a_s^* + a_s^* a_r = \delta_{rs}, \qquad a_r a_s + a_s a_r = 0$$

which are obviously invariant under unitary transformations ($a_r \to S^{-1} a_r S$; $S^* = S^{-1}$). In actual applications, the explicit matrix representation with the $\pm$ sign need never be written out because the anticommutation relations suffice to express all results in terms of physical quantities like expectation values of the occupation numbers $N_r =$ $= a_r^* a_r$. This fact, however, did not become common knowledge until much later (as also in the case of the Dirac matrices γ_μ: $\gamma_\mu\gamma_\nu + \gamma_\nu\gamma_\mu = 2\delta_{\mu\nu}$).

As a non-relativistic many-electron theory, the Jordan–Wigner formalism was to become a very helpful tool, for instance, for the study of the 'electron gas' in metallic conductors (a topic which cannot be followed in this report), but, was it more than a tool for special purposes? Would it be an essential element in the construction of a relativistic theory, in conjunction with quantum electrodynamics?

Dirac's relativistic wave mechanics of the spin electron, and its 'negative energy' troubles, have already been mentioned. As a possible remedy, Dirac (1930) proposed to consider as the 'vacuum' that state in which all (and only) negative energy levels are 'occupied' without, however, producing any observable charge density. A 'hole' in the negative energy 'sea' behaves like a particle of positive energy and 'positive' charge, and Dirac tried first to identify it with a proton. This interpretation soon proved untenable. Oppenheimer (1930) pointed out that, then, the hydrogen atom would lack stability, because of its rapid annihilation into two photons. A particularly lucid comment is found in H. Weyl's book *The Theory of Groups and Quantum Mechanics* (translated from the second German edition, 1931, by H. R. Robertson, New York: Dutton and Co.) on p. 263: according to Dirac's theory 'the mass of a proton should be the same as the mass of the electron; furthermore, no matter how the action is chosen (so long as it is invariant under interchange of right and left), this hypothesis leads to the essential equivalence of positive and negative electricity under all circumstances – even on taking the interaction between matter and radiation rigorously into account'.

Then, in 1932, came the experimental discovery of the positron, and the hole theory was given much credit for the prediction; but there was, as yet, no consistent mathematical formulation of this theory. Only certain perturbation calculations, pertaining, e.g. to the scattering of light by electrons, radiative pair creation and annihilation, or positron–electron scattering (with exchange effects: Bhabha, 1936), could be carried through at this stage because the difficulties appear only in higher order terms which one neglects.

The central problem was that of the 'vacuum polarization': an external electromagnetic field distorts the single electron wave functions of the negative energy sea and thereby produces a charge–current distribution which again affects the field. How is the subtraction of 'vacuum quantities', which is trivial in the field-free case, to be made now? Even if the external field is weak, it creates virtual electron–positron pairs, and their total contribution to the charge density, or polarization, appears as a divergent integral. What finite part of it, if any, will correspond to an observable vacuum polarization?

This problem, which was tackled from various points of view, proved to be very frustrating. This may be seen from the following quotations. Peierls (1934): 'One does not know whether the necessary changes in the theory will be only of a formal nature, just mathematical changes to avoid the use of infinite quantities, or whether the fundamental concepts underlying the equations will have to be modified essentially.' Furry and Oppenheimer (1934): 'The difficulties are of such a character that they are apparently not to be overcome merely by modifying the electromagnetic field of an electron within...small distances but require here a more profound change in our notions of space and time... .' It is noteworthy that Furry and Oppenheimer employed the Jordan–Wigner formalism of electron field quantization, but without much benefit.

However the pessimists were wrong this time: a workable though very complicated formulation of the hole theory was accomplished through the efforts of Dirac (1934) and Heisenberg (1934). Dirac, adopting a Hartree self-consistent field approximation, introduced time-dependent density matrices like

$$\sum \Psi_{\sigma'}(\mathbf{x}'t') \Psi_{\sigma''}(\mathbf{x}''t'')$$

(summed, e.g. over the occupied states), and examined their singularities on the light cone

$$[|\mathbf{x}' - \mathbf{x}''|^2 = c^2 (t' - t'')^2],$$

with the aim of subtracting singular functions depending explicitly on the field such that the finite remainder, for $\mathbf{x}' = \mathbf{x}''$, $t' = t''$, might serve to define observable quantities. Heisenberg used the conservation laws for charge, energy and momentum, to eliminate remaining ambiguities in the subtraction prescription. He also emphasized the shortcomings of the Hartree approximation ('anschauliche Theorie' as he called it) and proceeded to generalize the subtraction rules, now making use of the Jordan–Wigner electron field quantization which at this point starts to play its essential role. The theory is formally Lorentz invariant, and symmetric under interchange of electrons and positrons. It can be combined with quantum electrodynamics in a straightforward way and, in this form, constitutes a general theory of electrons, positrons and photons, in interaction. The smallness of the coupling constant $e^2/\hbar c = 1/137$ is essential in that expansions into powers of the electronic charge are freely made use of. (For a condensed description of this theory, see G. Wentzel, *Einführung in die Quantentheorie der Wellenfelder*, Franz Deuticke, Wien, 1943, pp. 186–191; English translation: Interscience Publishers, New York, 1949, pp. 196–202.)

However, as mentioned earlier, the self-energy difficulty survived. Compared with the single electron theory, the situation in the hole theory looked slightly less detrimental in that only logarithmic divergences remained (Weisskopf, 1934): the more strongly divergent terms, attributed to the spin energy and to the 'forced vibrations under the influence of the zero-point fluctuations of the radiation field', cancel each other (Weisskopf, 1939). Also in the terms of higher order in $e^2/\hbar c$, the divergences remain logarithmic ($\sim$powers of log k_c: Weisskopf, 1939). A suitable cut-off would then reduce the electronic self-energy to a small fraction ($\sim 1/137$) of mc^2. It was commonly taken for granted that the cut-off implied a violation of relativistic invariance unless one wanted to ignore the self-energy altogether.

More serious was the fact that, according to the positron theory, the photon also acquires an infinite self-energy, owing to its capability to create virtual electron–

positron pairs. Heisenberg's (1934) version of the theory reduces the divergence (in the term $\sim e^2$) to a logarithmic one. No remedy seemed available to save the relativistic and gauge invariance of the theory. Heisenberg expected that a solution to these problems could only come about in a theory 'which gives the Sommerfeld constant $e^2/\hbar c$ a determined value'. Many years later, 'regularization' combined with re-normalization was to provide at least a mathematically consistent framework.

While these obscure aspects of the theory had to be set aside, an extensive exploration of the vacuum polarization phenomenon came under way. Already at the Solvay Congress 1933, Dirac had reported on the case of a weak static potential, studied by first-order perturbation theory. For the charge density $\delta\varrho$, induced by the potential ϕ, he derived an expression of the form

$$\delta\varrho = \frac{e^2}{\hbar c}\left[A\Delta\phi + B\left(\frac{\hbar}{mc}\right)^2 \Delta\Delta\phi\right]$$

(Δ = Laplacian). Whereas the coefficient B has a finite numerical value, A is logarithmically infinite; however, by a suitable cut-off ($k_c \sim 137mc/\hbar$), A becomes a number of order unity. Dirac observed that $(-\Delta\phi/4\pi)$ is the 'external' charge density ϱ_0 producing the potential ϕ, hence the first term ($\sim A$) in $\delta\varrho$ means that a small fraction ($\sim 1/137$) of the external charge is 'neutralized' (at least in the static case). In Heisenberg's (1934) version of the theory, the term $\sim A\,\Delta\phi$ is automatically subtracted; this amounts to what we would call today a re-normalization of the charge. (The idea of charge re-normalization was vaguely anticipated in the papers, already quoted, by Peierls (1934), and by Furry and Oppenheimer (1934).) Then $\delta\varrho$ becomes proportional to $(-\Delta\varrho_0)$, or, more strictly speaking, to a spatial average of $(-\Delta\varrho_0)$, taken with a certain weight function over a sphere of radius $\sim \hbar/mc$ (Uehling, 1935, Serber, 1935). The potential ϕ is correspondingly modified ($\rightarrow \phi + \delta\phi$), thus altering slightly the laws of electrostatics. Observable effects of $\delta\phi$, for instance in the energies of bound electrons, were analysed by Uehling (1935); the results were disappointing, for a reason clear today: it is true that the vacuum polarization contributes to the Lamb shift, but only about $(-)2.2$ per cent of the total. Its reality is now guaranteed by the precision of the measurements.

Within the same approximation (first order in $e^2/\hbar c$), no new difficulties appeared in the more general case of time-dependent electromagnetic fields: the charge and current densities induced are given by similar formulae, with the Laplacian Δ replaced by the d'Alembertian (Heisenberg, 1934; Serber, 1935; Pauli and Rose, 1936).

If one goes to terms of higher order in $e^2/\hbar c$, one meets with the most fascinating feature of the hole theory: Maxwell's equations are to be corrected by small terms nonlinear (e.g. of the third order) in the field strengths and their derivatives. The superposition principle of electrodynamics and optics is no longer strictly valid. For instance, photons can be scattered by photons, or by electrostatic fields. This had already been suspected by Halpern (1933) and Delbrück (1933) who argued that such processes can be mediated by virtual electron–positron pairs. However, Heisenberg's subtraction procedure was needed to give reasonably low values to the cross-sections for these scattering processes, in particular in the limit of low frequencies. Actual computations of these cross-sections (without the benefit of the Feynman diagram technique!) were extremely laborious, even when leaving aside all but the simplest cases, namely, low frequencies and small scattering angles (see, for example, Euler, 1936). It was then discovered (Euler and Kockel, 1935; Kemmer, 1937) that, as far as the calculations went, all results were the same as though the vacuum field Lagrangian contained additional fourth-order terms:

$$L = \tfrac{1}{2}(\mathbf{E}^2 - \mathbf{H}^2) + \frac{1}{360\pi^2} \frac{e^4\hbar}{m^4c^7} \{(\mathbf{E}^2 - \mathbf{H}^2)^2 + 7(\mathbf{E}.\mathbf{H})^2\} + \cdots$$

(Heaviside units). A closed expression for this Lagrangian, valid to all orders of e but for long wavelengths ($\gg \hbar/mc$) only, was derived by Heisenberg and Euler (1936), and by Weisskopf (1936), through the study of special fields combined with invariance arguments, in the Hartree approximation.

As a result of all these developments, the general picture, in the later 1930's, looked as follows: apart from self-energy divergences and related difficulties ('radiative corrections', see Dancoff, 1939), the positron theory appeared to furnish well defined and plausible rules for a quantitative prediction of observable phenomena. However, experimental methods for testing the finer details of the theory (vacuum polarization) were not yet available. From the aesthetic point of view, the subtraction devices seemed too artificial to be generally appealing. Pauli, in spite of being actively interested in the theory (see Pauli and Rose, 1936), revealed his misgivings by using the deprecatory term 'subtraction physics', and another typical quotation from the same paper reads: 'The formalism of the positron theory which is accepted at present and which unfortunately is not yet substituted by a more satisfactory one... .' — As seen from our present viewpoint, the theory has proved correct in all aspects it was able to handle.

3. OTHER FIELDS, HIGHER SPINS

In the early attempts to generalize Schrödinger's wave equation so as to make it relativistically invariant, the 'scalar wave equation'

$$\sum_\alpha \partial^2 Q/\partial x_\alpha^2 = \mu^2 Q$$

with $\mu = mc/\hbar$, seemed a natural starting point (Schrödinger, 1926; Gordon, 1926; Klein, 1927), in conformity with L. de Broglie's original ideas regarding wave particle relationships. This approach was, however, criticized by Dirac (1928) who argued that the general interpretation of quantum mechanics requires linearity in $\partial/\partial t$ of the wave equation and, led by this argument (which proved to be wrong), discovered his relativistic wave equation for the spin electron. The interest in the scalar (relativistic) field subsided until Pauli and Weisskopf (1934) showed that a consistent and physically reasonable theory of the scalar field can be constructed just by applying to it the canonical rules of field quantization as set up in 1929 by Heisenberg and Pauli, in their first paper on quantum electrodynamics. In this formalism, it is the state vector Ψ of the quantized system which obeys the first-order (in $\partial/\partial t$) differential equation

$$i\hbar\,\partial\Psi/\partial t = H\Psi,$$

and the physical interpretation in terms of Ψ meets with no difficulty. The corresponding particles carry, of course, no spin, and identification with elementary particles occurring in reality was impossible in 1934. Indeed, the topic remained academic until the π meson was discovered and found to have zero spin, in the late 1940's and early 1950's.

Today, the 'charged scalar' theory is a standard example for illustrating field quantization methods, in particular the 'canonical' method used by Pauli and Weisskopf, so it will be unnecessary to describe it in detail (see G. Wentzel, sections 8 and 11 of the book quoted previously). Pauli and Weisskopf emphasized, as a gratifying feature, the fact that the field energy appears, without any *ad hoc* subtraction, as a positive-definite quantity, allowing of a corpuscular interpretation, whereas the electric charge is non-definite: the particles carry either positive or negative charges. The quantum nature of the charge (the eigenvalues are integral multiples of the elementary charge) follows from the formalism. Owing to the commutation rules supposed, the particles obey Bose–Einstein statistics, whereas anticommutation rules would lead to difficulties (see below). In other respects, the results were surprisingly similar to those derived from the hole theory of the positron. For instance, photons of energy $>2mc^2$ will, in a static Coulomb field, produce positive–negative pairs, and the cross-section for this process, as calculated by Pauli and Weisskopf, agrees in order of magnitude with that calculated by Bethe and Heitler for electron–positron pair production (*ceteris paribus*). There is also a vacuum polarization due to virtual boson pairs, with a logarithmic infinity demanding charge renormalization. A quadratic divergence ($\sim k_c^2$) appears in the electromagnetic self-energy of a charged boson (Weisskopf, 1939).

In 1937, the interest in quantized fields of a more general nature received a tremendous uplift by the discovery of the cosmic ray 'mesotrons' and the subsequent boom of the 'meson theory of nuclear forces' which will form the main topic of the last part of this report. In this present section we want to concentrate on some of the more purely mathematical questions regarding the structure of fields describing particles of a specified spin (questions which strongly attracted Pauli's personal attention).

In a systematic approach, it was natural to subdivide the question into a classical and a quantum-theoretical one: how should one generalize the classical relativistic field equations, so that after proper quantization the observable quantities can be interpreted in terms of particles of spin s, charged or uncharged, with rest mass or without? The classical (or 'c number') theories themselves could not be expected to describe single particles in a straightforward manner since even for $s=0$ and $s=1/2$ this had not been possible (for $s=0$, the probability density is not a positive-definite quantity, and for $s=1/2$ one has the negative energy states). But it could be hoped that a subsequent quantization (with hole-theoretical amendments in the case of half-odd spins) would lead to a consistent theory.

A general classical theory, mainly for force-free particles, was first set up by Dirac (1936), through a study of first-order differential equations similar in form to his wave equation for spin 1/2. The spinor notation of Van der Waerden (1929) proved a valuable tool. The structure of these theories was much clarified by Fierz (1939) (who acknowledged guidance by Pauli). In the simpler case of integral spins, the classical field is a tensor of rank s $(Q_{\alpha_1 \cdots \alpha_s}; \alpha_r = 1 \cdots 4)$, symmetric in all index pairs, with zero trace, with vanishing divergence, and obeying the Schrödinger–Klein–Gordon wave equation. For a plane wave, assuming $m \neq 0$, all these conditions leave only $(2s+1)$ amplitude components independent, as is most easily seen in the 'rest frame' $[Q_{\alpha_1} \cdots = \, = \exp(-i\mu t) \times \text{const.}]$; under spatial rotations these independent components transform according to the irreducible representation $\mathscr{D}_s$, and the corresponding single particle states are the $(2s+1)$ orientations of a spin s. Alternatively, all this can be written in spinor notation, and generalized so as to include the case of half-odd spin. (Instead, the half-odd spin can also be considered as a result of adding a spin 1/2 to an integral spin; this is the description developed by Rarita and Schwinger, 1940.) Possible expressions for energy–momentum and charge–current densities were discussed by Fierz; there is considerable ambiguity, although the total energy and the total charge are essentially uniquely defined. The energy is positive in the case of integral spin, but has either sign in the case of half-odd spin. This result is decisive for what kind of quantization is possible: commutation rules, or Bose–Einstein statistics, for integral spin; and anti-commutation rules, or Pauli's exclusion principle and Fermi–Dirac statistics, for half-odd spin. Fierz (1939 and 1950) derived a general formula for these (anti-) commutation relations.

Whereas the exclusion principle is obviously indispensable if there are negative energy states, it is much less obvious why the quantization according to the exclusion principle should be ruled out for integral spin. For Pauli, this was understandably a question of deep concern (see Pauli, 1936; Fierz, 1939; Pauli, 1940). Essential here is the postulate 'that measurements at two space points with a space-like distance can never disturb each other, since no signals can be transmitted with velocities greater than that of light'. This postulate forbids the appearance of the 'D_1 function' in the Lorentz invariant anticommutation relations, and one arrives at a mathematical contradiction if s is an integer.

The case of zero rest mass is of particular interest since it applies to the electromagnetic field $(s=1)$ and to the (linearized) gravitational field of Einstein's general relativity theory $(s=2)$. In both of these physical problems, there exists a group of gauge transformations

$$Q_\alpha \to Q_\alpha + \partial \chi / \partial x_\alpha$$

$$Q_{\alpha\beta} \to Q_{\alpha\beta} + \partial \varphi_\alpha / \partial x_\beta + \partial \varphi_\beta / \partial x_\alpha - \tfrac{1}{2} \delta_{\alpha\beta} \sum_\gamma \partial \varphi_\gamma / \partial x_\gamma$$

(possible only if $m=0$) such that all observable quantities are gauge invariant (in general relativity theory, the gauge transformation is equivalent to an infinitesimal coordinate transformation). It is then natural, for *any* spin value s, if $m=0$, to assume that any two solutions are physically equivalent if they can be transformed into each other by a gauge transformation. In application to a plane wave, this reduces (for $s \geqslant 1$) the number of independent polarizations to *two*, as is well known for light waves and gravitational waves (for $m=0$, there is no 'rest frame'). In the quantized theory, s can still be associated with the 'spin' or intrinsic angular momentum of one particle (e.g. of one photon or graviton): its component in the direction of propagation has the eigenvalues $\pm s$. Fierz (1940) has given explicit proof that the total angular momentum of a single particle state, according to this theory, is $\geqslant s$.

The existence of a gauge group in the case $m=0$, and its absence for $m \neq 0$, entails some typical differences in the formal structure of the quantized theory in the two cases. The historic example is the (real) vector field ($s=1$), with Maxwell's and Proca's (1936) field equations respectively. (Serving as a model for 'vector mesons', the quantized Proca theory was discussed in numerous papers (1938); see references quoted in Section 4.) The canonical quantization procedure, which is straightforward when applied to the three independent amplitudes of a plane wave for $m \neq 0$, does not automatically permit the limiting process $m \to 0$. In particular, the 'Lorentz condition'

$$\sum_\alpha \partial Q_\alpha / \partial x_\alpha = 0$$

which, in the canonical theory, is considered as an identity linking the four components Q_α, degenerates and becomes nonsensical for $m=0$. Indeed, this is the essence of the difficulty encountered by Heisenberg and Pauli in 1929, in their first attempts to formulate quantum electrodynamics. Fermi's proposal to replace the Lorentz condition, as an identity, by corresponding subsidiary conditions on the state vector Ψ provides the approach adequate for $m=0$. Stueckelberg (1938) invented another version of the vector meson theory (involving a redundant scalar field Q and a constraint on Ψ) which formally also covers the case $m=0$ (then $Q \equiv 0$), but this is no true union of the two theories. (For more recent studies on this question, see Belinfante, 1949; Coester, 1951.)

In all cases considered, the classical field equations are derivable from a variational principle involving a Lagrangian (density). The invariance properties of the Lagrangian (in the absence of external forces) can be used to construct conservative quantities and the corresponding densities which obey 'continuity equations' like

$$\sum_\alpha \partial T_{\alpha\beta} / \partial x_\alpha = 0$$

for the energy–momentum–stress tensor $T_{\alpha\beta} (= T_{\beta\alpha})$, or

$$\sum_\alpha \partial M_{\alpha\beta\gamma} / \partial x_\alpha = 0$$

for the 'angular momentum' tensor $M_{\alpha\beta\gamma} (= -M_{\alpha\gamma\beta})$. General expressions for $T_{\alpha\beta}$ and $M_{\alpha\beta\gamma}$ were first derived by Belinfante (1939) (with acknowledgments to Kramers and Podolanski) by exploiting the invariance of the Lagrangian under infinitesimal Lorentz transformations. Independently, Rosenfeld (1940) used the invariance under arbitrary infinitesimal coordinate transformations in general relativity theory for the same purpose, and the result agrees with Belinfante's in the limit of the special relativity metric. If *complex* fields $Q_{\alpha_1} \ldots$ are involved, the Lagrangian may be invariant under a (constant) change of phase of these fields $(Q_{\alpha_1} \ldots \to Q_{\alpha_1} \ldots e^{i\varepsilon})$, and, for an infinitesimal ε, this leads immediately to a continuity equation

$$\sum_\alpha \partial J_\alpha / \partial x_\alpha = 0$$

where J_α is now to be interpreted as the charge-current density. The physical significance attributed to these densities is, of course, that they are supposed to determine the manner in which the Q fields, or the corresponding particles, interact with gravitational and electromagnetic fields, respectively. In this context, Belinfante (1940) emphasized the fact that the Lagrangian L of a field is in general non-uniquely defined; any substitution

$$L \to L + \sum_\alpha \partial \Lambda_\alpha / \partial x_\alpha$$

leaves the field equations invariant, and whenever a four-vector Λ_α (phase invariant for complex fields) can be constructed in terms of $Q_{\alpha_1} \ldots$ (the first derivatives $\partial Q_{\alpha_1} \ldots / \partial x_\beta$ may also occur in Λ_α provided the second derivatives cancel out in $\sum_\alpha \partial \Lambda_\alpha / \partial x_\alpha$) there is then some freedom in the choice of L, and a corresponding lack of uniqueness in the definitions of $T_{\alpha\beta}$ and J_α (which, however, does not affect the total energy-momentum, $\int d^3x \, T_{4\beta}$, nor the total charge, $\int d^3x \, J_4$). For instance, in J_α, an additional term of the form $\sum_\beta \partial P_{\alpha\beta} / \partial x_\beta$, where $P_{\alpha\beta} (= -P_{\beta\alpha})$ is related to the spin polarization [e.g. for $s=1$, $P_{\alpha\beta} = i(Q_\alpha^* Q_\beta - Q_\beta^* Q_\alpha) \times \text{const.}$] may appear with an arbitrary factor; in the rest frame, this amounts to altering the magnetic moment of the particle. Hence, one has to expect that the wave equation for the charged particle in an external electromagnetic field can be written in such a way as to give its spin magnetic moment an arbitrary value (except, of course, for $s=0$). For the vector meson, an analysis by Corben and Schwinger (1940) showed this to be correct, and even earlier had it been known that Dirac's wave equation for $s=1/2$ can be modified by a 'Pauli term' so as to give the electron an anomalous magnetic moment (see equation 91 in Pauli, 1941).

The general problem of constructing wave equations and Lagrangians for charged particles of spin s in an external field (Fierz and Pauli, 1939) turned out to be very difficult for $s>1$. The usual device of substituting $\partial/\partial x_\alpha - i(e/\hbar c)\phi_\alpha$ for $\partial/\partial x_\alpha$ must be used with great caution to avoid inconsistencies. Fierz and Pauli tackled the problem by admitting, besides the leading tensor or spinor, tensors or spinors of lower rank (e.g. for $s=2$, a scalar) which, however, are made to vanish identically in the field-free case, due to a judicious choice of the Lagrangian. Roughly speaking one allows admixtures of spins $s-1$, $s-2, \ldots$, in the presence of the field, but in such a manner that, when switching off the field, one comes back to a particle of pure spin s. The theory was presented in some detail only for $s=2$ and $s=3/2$; for integral spin $\geqslant 3$, the admixtures needed were merely enumerated, with the queer result that the scheme is essentially unique only for $s \leqslant 4$.

Another question is whether particles with several spin states (e.g. one ground state and one excited) can be naturally described within the field-theoretical formalism. For instance, spins 0 and 1 can be jointly described in terms of an algebra of 16×16 matrices which, however, is reducible into representations corresponding to the spins 0 and 1 separately, without affecting the properties of these particles (Duffin, 1938; Kemmer, 1939; see also Part II, 4, of Pauli, 1941). The existence of irreducible wave equations, allowing a physical interpretation in terms of an 'elementary particle' with two different mass and spin states, has been proved by Bhabha (1952). He gives an example of a wave equation involving an irreducible set of 20×20 matrices, with eigenstates $s = 3/2$ and $s = 1/2$ having different masses. Because the charge density is positive-definite, quantization according to the exclusion principle (hole theory) is possible. The eigenfunctions representing the particle in either state are here, of course, quite different from the Dirac or Fierz–Pauli wave functions of pure spin particles.

All theories involving spins $s > 1$ are intrinsically complicated and, with the exception of the gravitational field, have not attracted widespread interest because, as it appears today, the 'simpler' cases are favoured in reality. Nevertheless, the finding that the field-theoretical formalism is far more comprehensive than reality (as we know it) is of basic importance in that it indicates that entirely new ideas will be needed to explain why the particles observed in nature have their very specific properties.

4. FIELD THEORY APPLIED TO NUCLEAR PHYSICS PROBLEMS

Once more, let us return to the late 1920's and early 1930's. Although, by then, we had learned that the emission and absorption of photons is perfectly described by interpreting the electromagnetic field amplitudes as quantum-mechanical operators, it was not commonly realized that a similar description might fittingly be applied to other creation and annihilation processes, or that even the words 'creation' or 'annihilation' of particles afforded a proper way of speaking about occurrences like the β decay. In this respect, a breakthrough came with the Pauli–Fermi theory of the β decay, and to illustrate the change in thinking we give here a translation of some of the introductory paragraphs of Fermi's (1933) paper. After having referred to Pauli's explanation of the continuous β velocity spectrum in terms of a 'neutrino' which escapes unobserved carrying away part of the energy liberated (see the article by C. S. Wu), Fermi goes on to say:

Besides the difficulty of the continuous energy distribution, a theory of the β rays faces still another essential difficulty in the fact that the present theories of the light particles do not explain in a satisfactory manner how these particles could be bound in a stable or quasi-stable manner inside a nucleus, considering the smallness of its volume.

The simplest way for the construction of a theory which permits a quantitative discussion of the phenomena involving nuclear electrons, seems then to examine the hypothesis that the electrons do not exist as such in the nucleus before the β emission occurs, but that they, so to say, acquire their existence at the very moment when they are emitted; in the same manner as a quantum of light, emitted by an atom in a quantum jump, can in no way be considered as pre-existing in the atom prior to the emission process. In this theory, then, the total number of the electrons and of the neutrinos (like the total number of light quanta in the theory of radiation) will not necessarily be constant, since there might be processes of creation or destruction of those light particles.

According to the ideas of Heisenberg, we will consider the heavy particles, neutron and proton, as two quantum states connected with two possible values of an internal coordinate ρ of the heavy particle. We assign to it the value $+1$ if the particle is a neutron, and -1 if the particle is a proton.

We will then seek an expression for the energy of interaction between the light and heavy particles which allows transitions between the values $+1$ and -1 of the coordinate ρ, that is to say, transformations of neutrons into protons or *vice-versa*; in such a way, however, that the transformation of a neutron into a proton is necessarily connected with the creation of an electron which is observed as a β particle, and of a neutrino; whereas the inverse transformation of a proton into a neutron is connected with the disappearance of an electron and a neutrino, ...

The simplest formulation for a theory in which the number of the particles (electrons and neutrinos) is not necessarily constant is available in the method of Dirac–Jordan–Klein of the 'quantized probability amplitudes'. In this formalism, the probability amplitudes ψ of the electrons and φ of the neutrinos, and their complex conjugates ψ^* and φ^*, are considered as non-commutative operators acting on functions of the occupation numbers of the quantum states of the electrons and neutrinos ....

Fermi goes on to write down in field-theoretical notation his famous β interaction (incidentally, his special choice was what is called today the 'vector' interaction). We cannot follow here the subsequent, often devious, development of β theory, except for mentioning briefly the short-lived, but historically important, 'β theory of nuclear forces'. This is the idea (Tamm, 1934; Iwanenko, 1934) that, according to second-order perturbation theory, a proton and neutron can interact by virtually emitting and re-absorbing an electron–neutrino pair. Numerically, this interaction is extremely weak, except for being strongly singular at zero distance and allowing an adaptable cut-off. Several variants of this theory (e.g. involving electron–positron pairs) were subsequently proposed, and the discussion continued even after the discovery of the 'mesotron'.

Meanwhile, Yukawa (1935) had published his ingenious hypothesis that the nuclear forces are mediated by a boson field, that is to say, by *single* particles rather than pairs, in close analogy with the electromagnetic forces, with the important difference, however, that the particles must have a non-vanishing *rest mass* such that their Compton wavelength determines the range of the nuclear forces, as indicated by the static solution ($e^{-\mu r}/r$, with $\mu = mc/\hbar$) of the 'scalar wave equation'. Yukawa estimated the rest mass of the boson as 200 electron masses. The charge of the mediating bosons determines the 'exchange character' of the resulting forces. It is interesting to note that Yukawa also suggested the boson might be β unstable so that the nuclear β decay would be explainable as a two-step process mediated by a virtual boson. (Recently, this view has become fashionable again, with the difference that a *new* hypothetical boson is needed. A different multi-step mechanism for β decay and nuclear forces, involving bosons heavier than nucleons such that the mass difference would determine the force range, was tentatively proposed by Wentzel, 1936.)

Yukawa's (1935) paper was not received, wherever it became known, with immediate consent or sympathy. However two years later it suddenly became the focus of universal attention when charged particles heavier than electrons but lighter than protons were detected, with increasing certainty, in cosmic ray experiments: Yukawa had predicted their existence! Later they were even found to decay into electrons (and presumably neutrinos). The confusion which resulted from this erroneous identification of the 'μ meson' with the 'nuclear field meson' lasted more than ten years, and it had some adverse effects, though not very serious ones, on the development of meson theory.

Meanwhile, the main guidance had come from the advancing knowledge of the properties of the nuclear forces. Their spin dependence seemed to demand a *vector* meson, rather than a scalar one, and the quantized Proca field, as well as its possible interactions with nucleons, became the subject of numerous investigations (Yukawa, Sakata and Taketani, 1938; Stueckelberg, 1938; Fröhlich, Heitler and Kemmer, 1938; Bhabha, 1938). Kemmer (1938) was the first to consider also the case of a *pseudoscalar* field which was to become so important, later, for the description of the π meson. Kemmer pointed out that, although scalar and pseudo-scalar particles behave alike when free or subject to electro-magnetic forces, their interactions with nucleons are typically different, because of the difference in parity of the boson field operators. Another important contribution of Kemmer's (1938a) was the introduction of the *isotopic spin* formalism into meson field theory; the results would then automatically be charge-symmetric, and the nuclear forces would be 'charge-independent' (in *any* approximation), as empirical data (e.g. proton–proton scattering) already then seemed to indicate.

Although the shortcomings of the perturbation method were well recognized (see, e.g. Stueckelberg and Patry, 1939–40), the second-order results were commonly used for comparison with the experimental data. Besides the two-nucleon data, the stability of heavy nuclei ('saturation') served as an important criterion. While, at first, the vector theory seemed strongly favoured, it ran into difficulties when the sign of the 'tensor force' became known from measurements (by Rabi and co-workers) of the electric quadrupole moment of the deuteron, and during the early 1940's increasing interest was bestowed upon the pseudoscalar theory. Singular ($\sim r^{-3}$) terms in the static tensor force could be cancelled out by introducing a vector field beside the pseudoscalar one (Møller and Rosenfeld, 1940; Schwinger, 1942; the struggle to make this type of theory agree with the deuteron data is reviewed in Section 3 of Wentzel, 1947).

Another, much discussed, problem was the anomaly in the magnetic moments of the proton and neutron, attributed to the 'bound meson cloud'. The idea was the same as in an earlier proposal by Wick (1935) based on the β field theory, but the stronger meson–nucleon coupling seemed to promise more satisfying results, even though one still had to put up with an arbitrary cut-off. This optimism, however, could not long be maintained; indeed, these and related questions have remained baffling up to the present day.

In view of the actual strength of the nuclear forces, it soon became an urgent desire to get away from perturbation or similar weak coupling methods. The Bloch–Nordsieck method of quantum electrodynamics, though pointing in the right direction, could not be taken over because the characteristic spin and/or isotopic spin dependence of the meson–nucleon interaction introduces essential complications (see, for instance,

Stueckelberg, 1938a). If, however, the coupling parameter g was assumed very *large*, so that an expansion into *falling* powers of g is indicated, it was then possible to make a separation of the meson field into a field bound to the (static and finite size) nucleons, and a free field which, in a higher approximation, is scattered by the compound nucleons. This 'strong coupling theory' was first carried through for the simplest non-trivial case: Yukawa's charged scalar theory (Wentzel, 1940). The most striking result was the appearance of 'nucleon isobars', i.e. excited states of the compound nucleon carrying higher charges Z, with an excitation energy proportional to $Z(Z-1)/$ $/g^2$. (The value of the coefficient, as also the 'strong coupling condition', was corrected in Wentzel, 1941; see also Oppenheimer and Schwinger, 1941.) To be applicable to pseudo-scalar and vector fields, the method had to be generalized so as to allow mesons to be bound in p, rather than s, states; there are per nucleon 3 such p states in the neutral, 9 in the charge-symmetric pseudoscalar theory (Serber and Dancoff, 1943; Pauli and Dancoff, 1942). In the charge-symmetric pseudoscalar theory (also for certain vector and mixture theories: Pauli and Kusaka, 1943; Wentzel, 1943), the isobar energy is similar to the rotational energy of a spherical top, with eigenvalues proportional to $j(j+1)$, where j (half-odd) is both the spin and the isotopic spin of the isobar state; for a given j, there are $(2j+1)$ spin orientations and $(2j+1)$ charge states $(Z=-j+\frac{1}{2},...,j+\frac{1}{2})$. The ground states $j=1/2$ are then to be identified with the ordinary neutron and proton states, while the first excited states, $j=3/2$, have precisely the same properties as the now well known resonant $p(3/2, 3/2)$ states, observed in π meson–nucleon scattering, which have an excitation energy of about 300 MeV (Brueckner, 1952). It is amusing to find this same value quoted already in a paper by Villars (1946), where the effect of isobar admixtures on the proton-neutron system (deuteron and low energy scattering) is analyzed: agreement of the strong coupling theory with the experimental data requires that the excitation energy be at least about 300 MeV (an unexpectedly high value!).

In their attempt to adapt the parameters of their mixture theory, including a cut-off k_c, to fit the experiments, Pauli and Kusaka considered it as imperative that the high energy meson–nucleon scattering cross-section be very small so as to conform to the cosmic ray observations, and this could actually be achieved by choosing k_c large enough (Oppenheimer and Schwinger, 1941). In this regard, we were all led astray by the mistaken identity of the cosmic ray mesons! Ironically, the apparent weakness of their nuclear interaction was one of the major motivations for work on the strong coupling theory, as well as for other contemporaneous speculations. Only very rarely was the suspicion expressed that there might be several kinds of mesons (Sakata and Tanikawa are quoted by Tomonaga, in a 1942 paper reprinted in the Supplement of the *Progress of Theoretical Physics*, Number 2, 1955, p. 80).

While it was thought that the isobar excitation energy might be of the order 50 MeV or even less, one had to worry whether the presence of higher isobar states in heavy nuclei might not upset the saturation. It was found (Pauli and Kusaka, 1943; Coester, 1944) that the saturation is maintained in the charge symmetric theory although the equilibrium value for the charge (more precisely: Z/A)

might tend to become too low because the Coulomb energy favours negatively charged isobars (Fierz, 1941). This defect of the theory is, of course, removed with the higher value of the excitation energy. Pauli's other objection against the strong coupling theory was its complete failure to account for the magnetic moments of the proton and the neutron; their ratio becomes -1 in the strong coupling limit, and no sufficient remedy is brought by the next correction terms (Houriet, 1945). Moreover, the neutron–proton mass difference has the wrong sign. (These difficulties of the strong coupling theory are surveyed in Section 4 of Wentzel, 1947.)

The suspicion that in reality the coupling might be neither 'strong' nor 'weak' prompted various attempts to deal with the 'intermediate coupling' case. Tomonaga (1947, and earlier work quoted there) invented an ingenious variational method which, e.g. for the charged scalar theory, allows an interpolation between the weak and strong coupling limits, for the static case (extended nucleon at rest). (Many papers by Tomonaga and his co-workers, dating back as far as 1941, which treat various versions of the theory in strong and intermediate coupling approximations, were reprinted in the Supplement of the *Progress of Theoretical Physics*, Number 2, 1955.) Another approach was made on classical or semi-classical lines, in the sense that the motions of the nuclear spin or charge coordinates were analysed in terms of classical gyrations, with emphasis on damping (reaction) effects in meson–nucleon scattering, often hopefully with regard to the alleged smallness of the cross-section (Heisenberg, 1939; Iwanenko and Sokolow, 1940; Bhabha, 1941; Fierz, 1941a; Pauli, 1946). Heitler (1941) developed a quantum theory of radiation damping which he applied to meson–nucleon scattering.

Could one be at all sure that the meson–nucleon interaction is of the type proposed by Yukawa? The idea of the older pair theories, namely, that the nucleons interact with *pairs* of particles (the interaction being quadratic in the field amplitudes), was taken up again and applied to meson pairs (Marshak, 1940). The scalar pair theory (Wentzel, 1942) has retained some mathematical interest as one of the few rigorously soluble field-theoretical problems. Spin-dependent pair interactions were treated in a strong coupling approximation by Pauli and Hu (1945) (earlier work on pair theories is quoted there), and by Blatt (1946).

Since about 1950, the rapidly accumulating experimental information has placed the development of meson theory on much firmer ground; but a major part of what is called 'theory' today is really half-empirical, and meson field theory, as a deductive scheme, has not been conspicuously successful. For some time it was believed, and many people still seem to believe, that a pseudoscalar field with a non-derivative (γ_5) coupling to nucleons accounts correctly for all facts related to π mesons and nucleons. Actually, there is very little to substantiate this claim, once one takes the crudest features, dominated by the $p(3/2, 3/2)$ resonance, for granted. This, then, leads on to the more general question which has often been raised (see, e.g. Heisenberg, 1946) whether our customary field-theoretical procedure is at all adequate; namely, to start from free fields describing 'bare' particles, and then to add an interaction term in the Lagrangian which causes the particles to become 'dressed' or 'compound', thereby degrading the original bare particles to unobservable entities. It is true that this approach, combined with re-normalization prescriptions, has resulted in a workable scheme for quantum electrodynamics, but this may be so only because, or as far as, the expansions in powers of e^2/hc converge rapidly. As to meson theory, and other more comprehensive field theories involving stronger interactions, the question is wide open.

REFERENCES

Section 1

Ehrenfest, P. (1906) *Phys. Z.* **7**, 528
Debye, P. (1910) *Ann. Phys. Lpz.* 4 **33**, 1427.
Einstein, A. (1905) *Ann. Phys. Lpz.* 4 **17**, 132
Bose, S. N. (1924) *Z. Phys.* **26**, 178.
Born, M., Heisenberg, W. and Jordan, P. (1926) *Z. Phys.* **35**, 557.
Dirac, P. A. M. (1927a) *Proc. Roy. Soc.* A **114**, 243.
Dirac, P. A. M. (1927b) *Proc. Roy. Soc.* A **114**, 710.
Kramers, H. A. and Heisenberg, W. (1925) *Z. Phys.* **31**, 681.
Weisskopf, V. and Wigner, E. (1930) *Z. Phys.* **63**, 54.
Heitler, W. (1954) *Quantum Theory of Radiation*, 3rd ed., Clarendon Press.
Heisenberg, W. (1931) *Ann. Phys. Lpz.* 5 **9**, 338.
Jordan, P. and Pauli, W. (1928) *Z. Phys.* **47**, 151.
Bohr, N. and Rosenfeld, L. (1933) *K. danske vidensk. Selsk., Math.-Fys. Medd. XII*, 8.
Dirac, P. A. M. (1928) *Proc. Roy. Soc.* A **117**, 610.
Heisenberg, W. and Pauli, W. (1929) *Z. Phys.* **56**, 1.
Rosenfeld, L. (1930) *Z. Phys.* **63**, 574.
Heisenberg, W. and Pauli, W. (1930) *Z. Phys.* **59**, 168.
Waller, I. (1930) *Z. Phys.* **62**, 673.
Oppenheimer, J. R. (1930) *Phys. Rev.* **35**, 461.
Fermi, E. (1929–30) *R. C. Accad. Lincei* **9**, 881, and **12**, 431.
Dirac, P. A. M., Fock, V. A. and Podolsky, B. (1932) *Phys. Z. Sowjet.* **2**, 468.
Bloch, F. (1934) *Phys. Z. Sowjet.* **5**, 301.
Tomonaga, S. (1946) *Progr. theor. Phys., Osaka* **1**, 27.
Schwinger, J. (1948) *Phys. Rev.* **74**, 1439.
Wentzel, G. (1933) *Z. Phys.* **86**, 479 and 635; **87**, 726.
Dirac, P. A. M. (1939) *Ann. Inst. Poincaré* **9**, 13.
Dirac, P. A. M. (1938) *Proc. Roy. Soc.* A **167**, 148.
Dirac, P. A. M. (1942) *Proc. Roy. Soc.* A **180**, 1.
Pauli, W. (1943) *Rev. mod. Phys.* **15**, 175.
Pauli, W. (1933) *Handbuch der Physik Geiger-Scheel*, vol. 24, part 1, pp. 255–6.
Bloch, F. and Nordsieck, A. (1937) *Phys. Rev.* **52**, 54.
Pauli, W. and Fierz, M. (1938) *Nuovo Cim.* **15**, 167.

Section 2

Jordan, P. (1927) *Z. Phys.* **44**, 473.
Dirac, P. A. M. (1927) *Proc. Roy. Soc.* A **114**, 243.
Jordan, P. and Klein, O. (1927) *Z. Phys.* **45**, 751.
Jordan, P. and Wigner, E. (1928) *Z. Phys.* **47**, 631.
Dirac, P. A. M. (1930) *Proc. Roy. Soc.* A **126**, 360
Oppenheimer, J. R. (1930) *Phys. Rev.* **35**, 562.
Bhabha, H. J. (1936) *Proc. Roy. Soc.* A **154**, 195.
Peierls, R. (1934) *Proc. Soc.* A **146**, 420.
Furry, W. H. and Oppenheimer, J. R. (1934) *Phys. Rev.* **45**, 260.
Dirac, P. A. M. (1934) *Proc. Camb. phil. Soc.* **30**, 150.
Heisenberg, W. (1934) *Z. Phys.* **90**, 209.
Weisskopf, V. (1934) *Z. Phys.* **89**, 27 and **90**, 817.

Weisskopf, V. F. (1939) *Phys. Rev.* **56**, 72.
Dirac, P. A. M. (1933) *Rapport du 7me Conseil Solvay de Physique*, p. 203.
Uehling, E. A. (1935) *Phys. Rev.* **48**, 55.
Serber, R. (1935) *Phys. Rev.* **48**, 49.
Pauli, W. and Rose, M. (1936) *Phys. Rev.* **49**, 462.
Halpern, O. (1933) *Phys. Rev.* **44**, 855.
Delbrück, M. (1933) *Z. Phys.* **84**, 144.
Euler, H. (1936) *Ann. Phys. Lpz.* 5 **26**, 398.
Euler, H. and Kockel, B. (1935) *Naturwissenschaften* **23**, 246.
Kemmer, N. (1937) *Helv. phys. Acta* **10**, 112.
Heisenberg, W. and Euler, H. (1936) *Z. Phys.* **98**, 714.
Weisskopf, V. (1936) *K. danske vidensk. Selsk., Math.-Fys. Medd.* **14**, 6.
Dancoff, S. M. (1939) *Phys. Rev.* **55**, 959.

Section 3

Schrödinger, E. (1926) *Ann. Phys. Lpz.* 4 **81**, 109.
Gordon, W. (1926) *Z. Phys.* **40**, 117.
Klein, O. (1927) *Z. Phys.* **41**, 407.
Dirac, P. A. M. (1928) *Proc. Roy. Soc.* A **117**, 610.
Pauli, W. and Weisskopf, V. (1934) *Helv. phys. Acta* **7**, 709.
Weisskopf, V. F. (1939) *Phys. Rev.* **56**, 72.
Dirac, P. A. M. (1936) *Proc. Roy. Soc.* A **155**, 447.
van der Waerden, B. L. (1929) *Nachr. Ges. Wiss. Göttingen*, p. 100.
Fierz, M. (1939) *Helv. phys. Acta* **12**, 3.
Rarita, W. and Schwinger, J. (1940) *Phys. Rev.* **60**, 61.
Fierz, M. (1950) *Helv. phys. Acta* **23**, 416.
Pauli, W. (1936) *Ann. Inst. Poincaré* **6**, 137.
Pauli, W. (1940) *Phys. Rev.* **58**, 716.
Fierz, M. (1940) *Helv. phys. Acta* **13**, 45.
Proca, A. (1936) *J. Phys. Radium* **7**, 347.
Stueckelberg, E. C. G. (1938) *Helv. phys. Acta* **11**, 225 and 299.
Belinfante, F. J. (1949) *Phys. Rev.* **76**, 66.
Coester, F. (1951) *Phys. Rev.* **83**, 798.
Belinfante, F. J. (1939) *Physica* **6**, 887.
Rosenfeld, L. (1940) *Mém. Acad. R. Belg.* **18**, Fascicule **6**.
Belinfante, F. J. (1940) *Physica* **7**, 449.
Corben, H. C. and Schwinger, J. (1940) *Phys. Rev.* **58**, 953.
Pauli, W. (1941) *Rev. mod. Phys.* **13**, 203.
Fierz, M. and Pauli, W. (1939) *Proc. Roy. Soc.* A **173**, 211.
Duffin, R. J. (1938) *Phys. Rev.* **54**, 1114.
Kemmer, N. (1939) *Proc. Roy. Soc.* A **173**, 91.
Bhabha, H. J. (1952) *Phil. Mag.* **43**, 33.

Section 4

Fermi, E. (1933) *Ric. sci.* **2**, No. 12.
Tamm, Ig. and Iwanenko, D. (1934) *Nature, Lond.* **133**, 981.
Yukawa, H. (1935) *Proc. phys.-math. Soc. Japan* **17**, 48.
Wentzel, G. (1936) *Z. Phys.* **104**, 34.
Yukawa, H., Sakata, S. and Taketani, M. (1938) *Proc. phys.-math. Soc. Japan* **20**, 319.
Stueckelberg, E. C. G. (1938) *Helv. phys. Acta* **11**, 299.
Fröhlich, H., Heitler, W. and Kemmer, N. (1938) *Proc. Roy. Soc.* A **166**, 154.
Bhabha, H. J. (1938) *Proc. Roy. Soc.* A **166**, 501.
Kemmer, N. (1938) *Proc. Roy. Soc.* A **166**, 127.
Kemmer, N. (1938a) *Proc. Camb. phil. Soc.* **34**, 354.
Stueckelberg, E. C. G. and Patry, J. F. C. (1939–40) *Helv. phys. Acta* **12**, 300; **13**, 167.
Møller, C. and Rosenfeld, L. (1940) *K. danske vidensk. Selsk. Math.-Fys. Medd.* **17**, 8.
Schwinger, J. (1942) *Phys. Rev.* **61**, 387.

Wentzel, G. (1947) *Rev. mod. Phys.* **19**, 1.
Wick, G. (1935) *R. C. Accad. Lincei* **21**, 170.
Stueckelberg, E. C. G. (1938a) *Phys. Rev.* **54**, 889.
Wentzel, G. (1940) *Helv. phys. Acta* **13**, 269.
Wentzel, G. (1941) *Helv. phys. Acta* **14**, 633.
Oppenheimer, J. R. and Schwinger, J. (1941) *Phys. Rev.* **60**, 150.
Serber, R. and Dancoff, S. M. (1943) *Phys. Rev.* **63**, 143.
Pauli, W. and Dancoff, S. M. (1942) *Phys. Rev.* **62**, 85.
Pauli, W. and Kusaka, S. (1943) *Phys. Rev.* **63**, 400.
Wentzel, G. (1943) *Helv. phys. Acta* **16**, 222 and 551.
Brueckner, K. A. (1952) *Phys. Rev* **86**, 106.
Villars, F. (1946) *Helv. phys. Acta* **19**, 323.
Coester, F. (1944) *Helv. phys. Acta* **17**, 35.
Fierz, M. (1941) *Helv. phys. Acta* **14**, 105.
Houriet, A. (1945) *Helv. phys. Acta* **18**, 473.
Tomonaga, S. (1947) *Prog. theor. phys. Osaka* **2**, 6.
Heisenberg, W. (1939) *Z. Phys.* **113**, 61.
Iwanenko, D. and Sokolow, A. (1940) *J. Phys. Moscow* **3**, 57 and 417.
Bhabha, H. J. (1941) *Proc. Roy. Soc.* A **178**, 324.
Fierz, M. (1941a) *Helv. phys. Acta* **14**, 257.
Pauli, W. (1946) *Meson theory of nuclear forces*, chap. III. New York; Interscience.
Heitler, W. (1941) *Proc. Camb. phil. Soc.* **37**, 291.
Marshak, R. E. (1940) *Phys. Rev.* **57**, 1101.
Wentzel, G. (1942) *Helv. phys. Acta* **15**, 111.
Pauli, W. and Hu, N. (1945) *Rev. mod. Phys.* **17**, 267.
Blatt, J. M. (1946) *Phys. Rev.* **69**, 285.
Heisenberg, W. (1946) *Z. Naturf.* **1**, 609.

19

Development of Quantum Electrodynamics*

Sin-Itiro Tomonaga

1. In 1932, when I started my research career as an assistant to Nishina, Dirac published a paper in the *Proceedings of the Royal Society*, London[1]. In this paper, he discussed the formulation of relativistic quantum mechanics, especially that of electrons interacting with the electromagnetic field. At that time a comprehensive theory of this interaction had been formally completed by Heisenberg and Pauli[2], but Dirac was not satisfied with this theory and tried to construct a new theory from a different point of view. Heisenberg and Pauli regarded the (electromagnetic) field itself as a dynamical system amenable to the Hamiltonian treatment; its interaction with particles could be described by an interaction energy, so that the usual method of Hamiltonian quantum mechanics could be applied. On the other hand, Dirac thought that the field and the particles should play essentially different roles. That is to say, according to him, 'the role of the field is to provide a means for making observations of a system of particles' and therefore 'we cannot suppose the field to be a dynamical system on the same footing as the particles and thus be something to be observed in the same way as the particles'.

Based on such a philosophy, Dirac proposed a new theory, the so-called many-time theory, which, besides being a concrete example of his philosophy, was of much more satisfactory and beautiful form than other theories presented up to then. In fact, from the relativistic point of view, these other theories had a common defect which was inherent in their Hamiltonian formalism. The Hamiltonian dynamics was developed on the basis of non-relativistic concepts which make a sharp distinction between time and space. It formulates a physical law by describing how the state of a dynamical system changes with time. Speaking quantum-mechanically, it is a formalism to describe how the probability amplitude changes with time t. Now, as an example, let us consider a system composed of N particles, and let the coordinates of each particle be $r_1, r_2, ..., r_N$. Then the probability amplitude of the system is a function of the N variables $r_1, r_2, ..., r_N$, and in addition, of the time t to which the amplitude is referred. Thus this function contains only one time variable in contrast to N space variables.

* *Editor's Note:* We had invited Professor S. Tomonaga to give a talk on this subject at our symposium, but unfortunately he could not attend. In his Nobel lecture (6 May 1966), Professor Tomonaga had given a remarkable exposition of the historical development of quantum electrodynamics. We are delighted to publish it in this volume as the statement of another distinguished physicist's view of the development of an important aspect of his subject. [Reprinted by permission of the Nobel Foundation and Elsevier, Holland.]

In the theory of relativity, however, time and space must be treated on an entirely equal footing so that the above imbalance is not satisfactory. On the other hand, in Dirac's theory which does not use the Hamiltonian formalism, it becomes possible to consider different time variables for each particle, so that the probability amplitude can be expressed as a function of $\mathbf{r}_1 t_1, \mathbf{r}_2 t_2, ..., \mathbf{r}_N t_N$. Accordingly, the theory satisfies the requirement of the principle of relativity that time and space be treated with complete equality. The reason why the theory is called the many-time theory is because N distinct time variables are used in this way.

This paper of Dirac's attracted my interest because of the novelty of its philosophy and the beauty of its form. Nishina also showed a great interest in this paper and suggested that I investigate the possibility of predicting some new phenomena by this theory. Then I started computations to see whether the Klein–Nishina formula could be derived from this theory or whether any modification of the formula might result. I found out immediately however, without performing the calculation through to the end, that it would yield the same answer as the previous theory. This new theory of Dirac's was in fact mathematically equivalent to the older Heisenberg–Pauli theory, and I realized during the calculation that one could pass from one to the other by a unitary transformation. The equivalence of these two theories was also discovered by Rosenfeld[3] and by Dirac–Fock–Podolsky[4] and was soon published in their papers.

Though Dirac's many-time formalism turned out to be equivalent to the Heisenberg–Pauli theory, it had the advantage that it gave us the possibility of generalizing the former interpretation of the probability amplitude. Namely, while one could calculate the probability of finding particles at points with coordinates $\mathbf{r}_1, \mathbf{r}_2, ..., \mathbf{r}_N$, all at the time t according to the previous theory, one could now compute more generally the probability that the first particle is at $\mathbf{r}_1$ at time t_1, the second at $\mathbf{r}_2$ at time $t_2, ...,$ and the Nth at $\mathbf{r}_N$ at time t_N. This was first discussed by Bloch[5] in 1934.

2. In this many-time theory developed by Dirac, electrons were treated according to the particle picture. Alternatively, in quantum theory, any particle should be able to be treated according to the wave picture. As a matter of fact, electrons were also treated as waves in the Heisenberg–Pauli theory, and it was well known that this wave treatment was frequently more convenient than the particle treatment. So the question arose as to whether one could reformulate the Heisenberg–Pauli theory in a way which would be more satisfactory relativistically, when electrons were treated as waves as well as the electromagnetic field.

As Dirac already pointed out, the Heisenberg–Pauli theory is built upon the Hamiltonian formalism and therefore the probability amplitude contains only one time variable. That is to say, the probability amplitude is given as a function of the field strength at different space points and of one common time variable. However, the concept of a common time at different space points does not have a relativistically covariant meaning.

Around 1942, Yukawa[6] wrote a paper emphasizing this unsatisfactory aspect of the quantum field theory. He thought it necessary to use the idea of the g.t.f. (generalized transformation function) proposed by Dirac[7] to correct this defect of the theory. Here I shall omit talking about the g.t.f., but, briefly, Yukawa's idea was to introduce as the basis of a new theory a concept which generalized the conventional conception of the probability amplitude. However, as pointed out also by Yukawa, we encounter the difficulty that, in doing this, cause and effect can not be clearly separated from each other. According to Yukawa, the inseparability of cause and effect would be an essential feature of quantum field theory, and without abandoning the causal way of thinking which strictly separates cause and effect, it would not be possible to solve various difficulties appearing in quantum field theory about which I will talk later. I thought however, that it might be possible (without introducing such a drastic change as Yukawa and Dirac tried to do) to remedy the unsatisfactory, unpleasant aspect of the Heisenberg–Pauli theory of having a common time at different space points. In other words, it should be possible, I thought, to define a relativistically meaningful probability amplitude which would be manifestly relativistically covariant, without being forced to give up the causal way of thinking. In having this expectation I was recalling Dirac's many-time theory which had enchanted me 10 years before.

When there are N particles in Dirac's many-time theory, we assign a time t_1 to the first particle, t_2 to the second, and so on, thus introducing N different times, $t_1, t_2,..., t_N$, instead of the one common time t. Similarly, I tried in quantum field theory to see whether it was possible to assign different times, instead of one common time, to each space point. And in fact I was able to show that this was possible[8].

As there are an infinite number of space points in field theory in contrast to the finite number of particles in particle theory, the number of time variables appearing in the probability amplitude became infinite. But it turned out that no essential difficulty appeared. An interpretation quite analogous to the one discussed by Bloch in connection with Dirac's many-time theory could be given to our probability amplitude containing an infinite number of time variables. Further, it was found that the theory thus formulated was completely covariant and that this covariant formulation was equivalent in its whole content to the Heisenberg–Pauli theory: it was shown, just as in the case of the many-time theory, that we could pass from one to the other by a unitary transformation. I began this work about 1942, and completed it in 1946.

3.   As I mentioned a little while ago, there are many difficulties in the quantum mechanics of fields. In particular, infinite quantities always arise which are associated with the presence of field reactions in various processes. The first phenomenon which attracted our attention as a manifestation of field reactions was the electromagnetic mass of the electron. The electron, having a charge, produces an electromagnetic field around itself. In turn, this field, the so-called self-field of the electron, interacts with the electron. We call this interaction the field reaction. Because of the field reaction the apparent mass of the electron differs from the original mass. The excess mass due to this field reaction is called the electromagnetic mass of the electron and the experimentally observed mass is the sum of the original mass and this electromagnetic mass. The concept of the electromagnetic mass had already appeared in the classical theory

of the electron by Lorentz, who computed the electromagnetic mass by applying the classical theory and obtained the result that the mass becomes infinite for the point (zero size) electron. On the other hand, the electromagnetic mass was computed in quantum theory by various people, and here I mention particularly the work of Weisskopf[9]. According to him, the quantum mechanical electromagnetic mass turned out to be infinite, and although the order of the divergence was much weaker than in the case of the Lorentz theory, the observed mass, which included this additional mass, would be infinite. This would be, of course, contrary to experiment.

In order to overcome the difficulty of an infinitely large electromagnetic mass, Lorentz considered the electron not to be point-like but to have a finite size. It is very difficult, however, to incorporate a finite sized electron into the framework of relativistic quantum theory. Many people tried various means to overcome this problem of infinite quantities, but nobody succeeded.

In connection with field reactions, the next problem which attracted the attention of physicists was determining what kind of influence the field reaction exerts in electron scattering processes. Let us consider, as a concrete example, a problem in which an electron is scattered by an external field. In the ordinary treatment, we neglect the effect of field reactions on the scattered electron, assuming that it is negligibly small. Then the behaviour of the scattering obtained by calculation (e.g. the Rutherford formula) fits very well with experiment. But what will happen if the influence of field reaction is taken into account? This theoretical problem was examined non-relativistically by Braunbeck and Weinmann[10] and Pauli and Fierz[11] and relativistically by Dancoff[12].

While Dancoff applied an approximation method, the perturbation method, in his relativistic calculation, Pauli and Fierz treated the problem in such a way that the most important part of the field reaction was first separated out exactly by employing a contact transformation method which was similar to the one which Bloch and Nordsieck[13] had published a year before. Since Pauli and Fierz adopted a non-relativistic model, and further simplified the problem by using the so-called dipole approximation, their calculation was especially transparent. At any rate, both non-relativistic and relativistic calculations exhibited several infinities in the scattering processes. *

The conclusions of these people were fatal to the theory. That is, the influence of the field reaction becomes infinite in this problem. The effect of field reaction on a quantity called the scattering cross-section, which expresses quantitatively the behaviour of the scattering, rather than becoming negligibly small, becomes infinitely large. This does not, of course, agree with experiment.

This discouraging state of affairs generated in many people a strong distrust of quantum field theory. There were even those with the extreme view that the concept of field reaction itself had nothing to do at all with the true law of nature.

---

* The main purpose of the work of the work of Bloch and Nordsieck, and Pauli and Fierz was to solve the so-called infrared catastrophe which was one of a number of divergences. Since this difficulty was resolved in their papers we confine ourselves here to a discussion of the other divergences which are of the so-called ultraviolet type.

On the other hand, there was also the view that the field reaction might not be altogether meaningless but would play an essential role in the scattering processes, though the appearance of divergences revealed a defect of the theory. Heisenberg[14], in his paper published in 1939, emphasized that the field reaction would be crucial in meson–nucleon scattering. Just at that time I was studying at Leipzig, and I still remember vividly how Heisenberg enthusiastically explained this idea to me and handed me galley proofs of his forthcoming paper. Influenced by Heisenberg, I came to believe that the problem of field reactions far from being meaningless was one which required a frontal attack.

Thus, after coming back to Japan from Leipzig, I began to examine the nature of the infinities appearing in scattering processes at the same time that I was engaged in the above-mentioned work of formulating a covariant field theory. What I wanted to know was what kind of relationship exists between the infinity associated with the scattering process and that associated with the mass. If you read the above-mentioned papers of Bloch and Nordsieck and Pauli and Fierz, you will see that one of the terms containing infinite quantities is first separated out by a contact transformation and this term turns out to be just the term modifying the mass. Besides this kind of infinity there appeared, according to Pauli and Fierz, another kind of infinity characteristic of the scattering process. I further investigated a couple of simple models which were not realistic, but could be solved exactly. What was understood from these models, was that the most strongly divergent terms in the scattering process had the same form as the expression giving the modification of the particle mass due to field reactions, and therefore both should be manifestations of the same effect. In other words, at least a portion of the infinities appearing in the scattering process could be amalgamated into the infinity associated with the particle mass, leaving infinities proper to the scattering process alone. These turned out to be more weakly divergent than the infinity associated with the mass.

Since these conclusions were derived from non-relativistic or unrealistic models, it was still doubtful whether the same thing would occur in the case of relativistic electrons interacting with the electromagnetic field. Dancoff tried to answer this question. He calculated relativistically the infinities appearing in the scattering process and determined which of them could be amalgamated into the mass and which remained as infinities proper to the scattering process alone. He found that there remained, in the latter group of infinite terms, one which was at least as divergent as the infinity of the mass, a finding which differed from the conclusion based on fictitious models.

Actually, there are two kinds of field reactions in the case of the relativistic electron and electromagnetic field. One of them ought to be called 'of mass type' and the other 'of vacuum polarization type'. The field reaction of mass type changes the apparent electronic mass from its original value by the amount of the electromagnetic mass as was calculated by Weisskopf. On the other hand, the field reaction of vacuum polarization type changes the apparent electronic charge from its original value. As was discussed in further papers by Weisskopf[15] and others, infinite terms appear in the apparent electronic charge if the effect of vacuum polarization is taken into account. However, in this talk, for simplicity, I will mention only briefly the divergence of the vacuum polarization type.

4. In the meantime, in 1946, Sakata[16] proposed a promising method of eliminating the divergence of the electron mass by introducing the idea of a field of cohesive force. It was the idea that there exists unknown field, of the type of the meson field which interacts with the electron in addition to the electromagnetic field. Sakata named this field the cohesive force field, because the apparent electronic mass due to the interaction of this field and the electron, though infinite, is negative and therefore the existence of this field could stabilize the electron in some sense. Sakata pointed out the possibility that the electromagnetic mass and the negative new mass cancel each other and that the infinity could be eliminated by suitably choosing the coupling constant between this field and the electron. Thus the difficulty which had troubled people for a long time seemed to disappear insofar as the mass was concerned. (It was found later that Pais[17] proposed the same idea in the U.S.A. independently of Sakata.) Then what concerned me most was whether the infinities appearing in the electron scattering process could also be removed by the idea of a plus-minus cancellation.

An example of a computation of how the field reaction influences the scattering process was already given by Dancoff. What we had to do was just to replace the electromagnetic field by the cohesive force field in Dancoff's calculation. I mobilized young people around me and we performed the computation together[18]. Infinities with negative sign actually appeared in the scattering cross-section as was expected. However, when we compared these with the infinities with positive sign which Dancoff calculated for the electromagnetic field, the two infinities did not cancel each other completely. That is to say, according to our result, the Sakata theory led to the cancellation of infinities for the mass but not for the scattering process. It was also known that the infinity of vacuum polarization type was not cancelled by the introduction of the cohesive force field.

Unfortunately, Dancoff did not publish the detailed calculations in his paper, and while we were engaged in the above considerations, we felt it necessary to do Dancoff's calculation over again for ourselves in parallel with the computation of the influence of the cohesive force field. At the same time I happened to discover a simpler method of calculation.

This new method of calculation was to use the technique of contact transformations based on the previously mentioned formalism of the covariant field theory and was in a sense a relativistic generalization of the Pauli–Fierz method. This method had the advantage of separating the electromagnetic mass from the beginning, just as was shown in their paper.

Our new method of calculation was not at all different in its contents from Dancoff's perturbation method, but had the advantage of making the calculation more clear. In fact, what took a few months in the Dancoff type of calculation could be done in a few weeks. And it was by this method that a mistake was discovered in Dancoff's calculation; we had also made the same mistake in the beginning. Owing to this new, more lucid method, we noticed that, among the various terms appearing in both Dancoff's and our previous calculation, one term had been overlooked. There was only one missing term, but it was crucial to the final conclusion. Indeed, if we corrected this error, the infinities appearing in the scattering process of an electron due to the electromagnetic and cohesive force fields cancelled completely, except for the divergence of vacuum polarization type.

5.   When this unfortunate error of Dancoff's was discovered, we had to reexamine his conclusions concerning the relation between the divergence of the scattering process and the divergence of the mass, in particular, the conclusion that there remained a portion of the infinities of the scattering process which could not be amalgamated into the modification of the mass. In fact, it turned out that after correcting the error, the infinity of mass type appearing in the scattering process could be reduced completely to the modification of the mass, and the remaining field reaction belonging to the scattering proper was not divergent[19]. In other words, the highest divergence of the infinities appearing in the scattering process, in the relativistic as well as in the non-relativistic case, could be attributed to the infinity of mass. The reason why the remaining part became finite in the relativistic case was due to the fact that the order of the highest divergence was only log ∞, and after amalgamating the divergence into the mass term, the remainder was convergent. The great value of this method of contact transformations was that once the infinity of the mass was separated out, we obtained a divergence-free theoretical framework.

In this way the nature of various infinities became fairly clear. Though I did not describe here the infinity of vacuum polarization type, this too appears in the scattering process, as mentioned earlier. However, Dancoff had already discovered that this infinity could be amalgamated into an apparent change in the electronic charge. To state the conclusion, therefore, all infinities appearing in the scattering process can be attributed either to the infinity of the electromagnetic mass or to the infinity appearing in the electronic charge – there are no other divergences in the theory.

It is a very pleasant thing that no divergence is involved in the theory except for the two infinities of the electronic mass and charge. We cannot say that we have no divergences in the theory, since the mass and charge are in fact infinite. It is to be noticed, however, that if we reduce the infinities appearing in the scattering process to modifications of mass and charge, the remaining terms all become finite. Further, if we examine the structure of the theory, after the infinities are amalgamated into the mass and charge terms, we see that the only mass and charge appearing in the theory are the values modified by field reactions – the original values and excess ones due to field reactions never appear separately.

This situation gives rise to the following possibility. The theory does not of course yield a resolution of the infinities. That is, since those parts of the modified mass and charge due to field reactions contain divergence, it is impossible to calculate them by the theory. However, the mass and charge observed in experiments are not the original mass and charge but the mass and charge as modified by field reactions, and they are finite. On the other hand, the mass and charge appearing in the theory are, as I mentioned above, after all the values modified by field reactions. Since this is so, and particularly since the theory is unable to calculate the modified mass and charge, we may adopt the procedure of substituting experimental values for them phenomenologically. When a theory is incompetent in part, it is a common procedure to rely on experiment for that part. This procedure is called the renormalization of mass and charge, and our method has brought the possibility that the theory will lead to finite results by the renormalization even if it contains defects.

The idea of renormalization is far from new. Many people used explicitly or implicitly this idea, and we find the word renormalization already in Dancoff's paper. In his calculation it appeared, because of an error that there still remained a divergence in the scattering even after the renormalization of the electron mass. This error was very unfortunate; if he had performed the calculation correctly, the history of renormalization theory would have been completely different.

6.   This period, around 1946–48, was soon after the second world war, and it was quite difficult in Japan to obtain information from abroad. But soon we got the news that in the U.S.A., Lewis and Epstein [20] found Dancoff's mistake and gave the same conclusions as ours, Schwinger [21] constructed a covariant field theory similar to ours, and he was probably performing various calculations making use of it. In particular, little by little news arrived that the so-called Lamb shift was discovered [22] as a manifestation of the electromagnetic field reaction and that Bethe [23] was calculating it theoretically. The first information concerning the Lamb shift was obtained not through the *Physical Review*, but through the popular science column of a weekly U.S.A. magazine. This information about the Lamb shift prompted us to begin a calculation more exactly than Bethe's tentative one.

The Lamb shift is a phenomenon in which the energy levels of a hydrogen atom show some shifts from the levels given by the Dirac theory. Bethe thought that the field reactions were primarily responsible for this shift. According to his calculation, field reactions give rise to an infinite level shift, but he thought that it should be possible to make it finite by a mass renormalization and a tentative calculation yielded a value almost in agreement with experiment.

This problem of the level shift is different from the scattering process, but it was conceivable that the renormalization which was effective in avoiding infinities in the scattering process would be workable in this case as well. In fact, the contact transformation method of Pauli and Fierz devised to solve the scattering problem could be applied to this case, clarifying Bethe's calculation and justifying his idea. Therefore the method of covariant contact transformations, by which we did Dancoff's calculation over again would also be useful for the problem of performing the relativistic calculation for the Lamb shift. This was our prediction.

The calculation of the Lamb shift was done by many people in the U.S.A. [24] Among others, Schwinger, commanding powerful mathematical techniques, and by making thorough use of the method of covariant contact transformations, very skilfully calculated not only the Lamb shift but other quantities such as the anomalous magnetic moment of the electron. After long, laborious calculations, less skilful than Schwinger's, we[25] obtained a result for the Lamb shift which was in agreement with the Americans. Furthermore, Feynman[26] devised a convenient method based on an ingenious idea which could be used to extend the approximation of Schwinger and ours to higher orders, and Dyson[27] showed that all infinities appearing in quantum electrodynamics could be treated by the renormalization procedure to an arbitrarily high order of approximation. Furthermore, this method devised by Feynman and developed by Dyson was shown by many people to be applicable not only to quantum electrodynamics, but to statistical mechanics and solid state physics as well, and provided a new, powerful method in these fields. However, these matters will probably be discussed by Schwinger and Feynman themselves and need not be explained by me. So far I have told you the story of how I played a tiny, partial role in the recent development of quantum electrodynamics, and here I would like to end my talk.

REFERENCES

1. P. A. M. Dirac, *Proc. Roy. Soc.* A **136**, 453 (1932).
2. W. Heisenberg and W. Pauli, *Z. Phys.* **56**, 1 (1929).
3. L. Rosenfeld, *Z. Phys.* **76**, 729 (1932).
4. P. A. M. Dirac, V. A. Fock and B. Podolsky, *Phys. U.S.S.R.* **2**, 468 (1932).
5. F. Bloch, *Phys. Z. U.S.S.R.* **5**, 301 (1943).
6. H. Yukawa, *Kagaku* **12**, 249 (1943).
7. P. A. M. Dirac, *Phys. Z. U.S.S.R.* **3**, 64 (1933).
8. S. Tomonaga, *Progr. Theor. Phys.* **1**, 27 (1946);
   Koba, Tati, Tomonaga, *Progr. Theor. Phys.* **2**, 101, 198 (1947);
   Kanesawa and S. Tomonaga, *Progr. Theor. Phys.* **3**, 1, 101 (1948).
9. V. F. Weisskopf, *Phys. Rev.* **56**, 72 (1939).
10. Braunbeck, Weinman, *Z. Phys.* **110**, 369 (1938).
11. W. Pauli and M. Fierz, *Nuovo Cimento*, **15** 267 (1938).
12. S. M. Dancoff, *Phys. Rev.* **55**, 959 (1939).
13. F. Bloch and A. Nordsieck, *Phys. Rev.* **52**, 54 (1937).
14. W. Heisenberg, *Z. Phys.* **113**, 61 (1939).
15. V. Weisskopf, *Kgl. Danske Vid. Sels.* **14**, No. 6 (1936).
16. S. Sakata, *Progr. Theor. Phys.* **1**, 143 (1946).
17. A. Pais, *Phys. Rev.* **73**, 173 (1946).
18. Ito, Koba and S. Tomonaga, *Progr. Theor. Phys.* **3**, 276 (1948).
   Koba, Takeda, *Progr. Theor. Phys.* **3**, 407 (1948).
19. Koba and S. Tomonaga, *Progr. Theor. Phys.* **3**, 290 (1948).
   Tati and S. Tomonaga, *Progr. Theor. Phys.* **3**, 391 (1948).
20. Lewis, *Phys. Rev.* **73**, 173 (1948);
   Epstein, *Phys. Rev.* **73**, 179 (1948).
21. J. Schwinger, *Phys. Rev.* **73**, 416 (1948); **74**, 1439 (1948); **75**, 651 (1949); **76**, 790 (1949).
22. W. E. Lamb and R. C. Retherford, *Phys. Rev.* **72**, 241 (1947).
23. Bethe, *Phys. Rev.* **72**, 339 (1947).
24. N. M. Kroll and W. E. Lamb, *Phys. Rev.* **75**, 388 (1949).
   French and V. Weisskopf, *Phys. Rev.* **75**, 1241 (1949).
25. Fukuda, Miyamoto and S. Tomonaga, *Progr. Theor. Phys.* **4**, 47, 121 (1949).
26. R. Feynman, *Phys. Rev.* **74**, 1430 (1948); **76**, 769 (1949).
27. F. J. Dyson, *Phys. Rev.* **75**, 486 (1949), 1736 (1949).

20

A Report on Quantum Electrodynamics

Julian Schwinger

My assignment is to trace the development of quantum electrodynamics. This topic is intended to be an example of the general theme of the symposium – the 'Development of the Physicist's Conception of Nature'. But, as Dirac pointed out, the phrase 'the physicist's conception' implies a degree of unanimity that rarely exists during the period of active development of a subject. Only when the material is finally frozen in the textbooks can one speak of 'the physicist's conception'. At any interesting moment during the period of development there are discordant viewpoints of individual physicists. What is the present status of quantum electrodynamics, the modern version of which is now some twenty-five years of age? Perhaps I can emphasize that the development phase is not yet terminated by telling you that I have just finished writing a book – *Particles, Sources, and Fields*, Vol. II – in which the major quantitative accomplishments of modern quantum electrodynamics have been derived, not as they were done historically, but by a conceptually and computationally simpler method. It is through this method, source theory, that I believe future generations will learn quantum electrodynamics from their textbooks.

There is nothing unusual in the displacement of a particular historical attitude toward a subject by a conceptually sounder and simpler one. For example, none of us has learned classical mechanics in the strict spirit of Newton, but much more as Euler, many years later, recast and clarified the subject. I believe that the time has come for quantum electrodynamics to be so recast and clarified.

My programme for today has two parts, of unequal length. First of all, being conscious of the historical orientation of this symposium, I shall not launch into an exposition of the new viewpoint. Rather, I shall trace the highlights in the historical development of quantum electrodynamics in order to underline, to emphasize, the conceptual difficulties in the conventional approach. In other words, I shall not be concerned today with the quantitative successes of quantum electrodynamics. They are far too familiar to merit discussion here. My attention is focused entirely on conceptual difficulties, as I see them. Perhaps I should say now that, while the difficulties of electrodynamics are on stage today, they were not the initial stimulus for the development of the different viewpoint of source theory. Rather that came from the situation in strong interaction physics.

Thanks to the continuing development of high energy accelerators, we are now acquainted with a bewildering variety of particles, which run the whole gamut of stability characteristics. And we have learned to group some of these particles into

J. Mehra (ed.), The Physicist's Conception of Nature, 413–429. All Rights Reserved
Copyright © 1973 by D. Reidel Publishing Company, Dordrecht-Holland

families, multiplets and supermultiplets, comprising particles with related dynamical characteristics, which particles can individually range from absolutely stable to long-lived to strongly unstable. We are being told that stable and unstable particles are not intrinsically different, and require a uniform kind of description. This is the situation that, to me, seemed to demand a new point of view, one that was more physically motivated, in which the practices of the experimentalist took precedence over the *a priori* definitions of the theorist. While the need for something new is most evident in strong interaction physics, electrodynamics provides the most familiar and most dynamically elementary situation in which to illustrate the ideas.

My first topic, then, is an historically oriented discussion of electrodynamics pointing toward *source theory*. The second topic, about which I shall say only a few words, is the suggestion that quantum electrodynamics, far from being an almost closed and exhausted subject, which I suppose is the prevalent opinion, may yet have a stunningly dramatic impact on the theories of strong and weak interactions. I shall list this exciting potentiality under the heading of *magnetic charge*. Perhaps I should remark here that neither of these two sets of ideas is totally new; there are several years of history behind them. Yet, relative to the time scale that has character-ized this symposium, they may pass as reasonably fresh proposals.

Essentially everything I shall discuss today has its origin in one or another work of Dirac. That will not astonish you. It is unnecessary, of course, but nevertheless I remind you that *quantum* electrodynamics began in the famous paper of 1927 [appropriately, this is the first paper in the collection, *Selected Papers on Quantum Electrodynamics*, Dover, 1958]. Here Dirac first extended the methods of quantum mechanics to the electromagnetic field. We need not be concerned now with the technical point that the quantization was applied to radiation field oscillators. The fact remains that, through the extension of quantum mechanics to electromagnetism, the electromagnetic field was promoted from a classical quantity to a quantum mechanical operator – an operator field. A cloud of formalism later obscured it all, but what Dirac did is simply this. If you think of Maxwell's equations as the equation of one photon, as Schrödinger's equation is that of one electron, the quantization to the electromagnetic field became the first example of second quantization, so-called, which is a procedure for the compact description of a multi-particle system. In this initial example, the particles obeyed Bose–Einstein statistics. Inevitably, and it is only surprising that Dirac himself did not do it, the same procedure was extended to particles obeying Fermi-Dirac statistics, with the essential difference that anti-commutation relations replace commutation relations, to express the characteristic Pauli exclusion principle. At that time, in 1928, we had a situation in which multi-particle systems obeying Bose or Fermi statistics could be described in terms of operator functions of space and time – the quantized ψ field for electrons, the quan-tized A field for photons. And these operator fields had clear and immediate physical interpretations in terms of the emission and absorption, the creation and annihilation, of the physical particles.

The year 1928 is more significant for another event – the introduction of the Dirac relativistic equation for the electron. It was first viewed as a single particle equation, which ran into the difficulty of the physically meaningless negative energy solutions. The answer proposed by Dirac in 1930 was to exploit the exclusion principle by filling up all the 'negative energy' states to form the physical vacuum state. One electron added to this infinite sea would be stable against spontaneous photon emission, as befitted a physical particle. And it was recognized that the removal of a 'negative energy' particle produced an excitation that acted physically as a particle of the same mass but opposite electric charge – the positron. Thus, the only way out of the difficulties posed by the Dirac equation was to admit from the beginning that a multi-particle system was being described. But then, why not initially adopt a second quantized description, with the operator field unifying the electron and positron as two alternative states of a single particle? With this formalism the vacuum becomes again a physically reasonable state with no particles in evidence. The picture of an infinite sea of negative energy electrons is now best regarded as an historical curiosity, and forgotten. Unfortunately, this episode, and discussions of vacuum fluctuations, seem to have left people with the impression that the vacuum, the physical state of nothingness (under controlled physical circumstances), is actually the scene of wild activity.

I shall now try to describe the resulting situation as it appeared toward the end of the third decade of our century. It was the almost inexorable outcome of a combination of ingredients. They were: the classical equations of Maxwell, the laws of quantum mechanics, the extension from the commutators of Bose statistics to the anticommutators of Fermi statistics, and, the Dirac equation with its electron–positron interpretation. The resulting mathematical formalism was clear cut. The real problem was to understand physically what had happened, and that problem would occupy a generation of physicists.

The mathematical situation is illustrated by writing down a system of field equations. They are, $(\hbar = c = 1)$,

$$[\gamma^\mu (1/i\, \partial_\mu - eqA_\mu(x)) + m]\, \psi(x) = 0,$$
$$F_{\mu\nu}(x) = \partial_\mu A_\nu(x) - \partial_\nu A_\mu(x),$$
$$\partial_\nu F^{\mu\nu}(x) = \tfrac{1}{2}\psi(x)\, \gamma^0\gamma^\mu eq\psi(x).$$

These equations only deviate from what was customary then, or indeed now, by emphasizing the symmetry between positive and negative charge – between electron and positron – through the introduction of the charge matrix q,

$$q = \begin{pmatrix} 0 & -i \\ i & 0 \end{pmatrix}.$$

This matrix acts on a two-valued index which, in addition to the four-valued spinor index, is carried by the field $\psi(x)$. The more usual formalism diagonalizes q, which has eigenvalues $+1$ and -1, and replaces the eight component ψ by the pair of four-component eigenvector fields ψ and ψ^*. In the formalism used here, the symmetry between electron and positron is expressed by the invariance of the system of field equations under the simple transformation

$$\psi(x) \rightarrow r_q \psi(x)$$
$$C: \quad A_\mu(x) \rightarrow - A_\mu(x)$$
$$F_{\mu\nu}(x) \rightarrow - F_{\mu\nu}(x),$$

where

$$r_q = \begin{pmatrix} 1 & 0 \\ 0 & -1 \end{pmatrix}$$

is the charge reflection matrix,

$$r_q^{-1} q \, r_q = - q \, .$$

The field equations are not the whole story, of course, They are supplemented by equal time commutation relations, or, to be precise, commutation relations for the fields A, F, anticommutation relations for the components of ψ. I shall not write these out explicitly.

In putting together these various equations, the physical meaning of the symbols had seemed to be clear. The charge and mass of the electron (positron) is e and m respectively, and $\psi(x)$ symbolizes the creation and annihilation of an electron (positron), while A and F represent the creation and annihilation of a photon. Is it true? Not at all. Through the innocent process of combining these equations into a non-linear coupled system, the physical meanings of all the symbols have changed. They no longer refer to the physical particles, but rather to a deeper level of description.

As a rough indication of this state of affairs, suppose we begin with an excitation of ψ, presumably referring to the presence of an electron (positron). But then there is an electric current flowing, which creates an accompanying electromagnetic field. And this field reacts on the particle field ψ to alter whatever initial excitation we assumed. In short, we have a coupled system, and the description of one physical electron is not to be written down *a priori*, but emerges only after solving a complicated dynamical problem. The situation is similar with the photon. An assumed photon field A will excite the electron-positron field ψ, and the associated electric current acts to alter the initially assumed electromagnetic field. In short, ψ and A have become abstract dynamical variables, with the aid of which one constructs the physical states, but they in themselves do not have an elementary physical interpretation. In particular, the constants appearing in the field equations, e and m, are not identifiable as the physical charge and mass of the electron (we now write them as e_0, m_0).

The question naturally arises as to what kind of direct physical interpretation can one give to the operator fields $\psi(x)$ and $A(x)$? The answer is given in terms of the vacuum state. I recall that, for us, the vacuum is the state in which no particles exist. It carries no physical properties; it is structureless and uniform. I emphasize this by saying that the vacuum is not only the state of minimum energy, it is the state of *zero* energy, zero momentum, zero angular momentum, zero charge, zero whatever.

Physical properties, structure, come into existence only when we disturb the vacuum, when we excite it. What shall we say about the excitation symbolized by $\psi(x) \mid$ vac.$\rangle$?

The reference to the specific space-time point x means that this is a localized excitation, with the complementary property that arbitrary amounts of energy and momentum are available. The spinor nature of the field implies an angular momentum of $\frac{1}{2}$, relative to the point x. And the field ψ injects a unit magnitude of charge. Evidently, one physical realization of these properties is an electron (by which we henceforth mean electron-positron). But an appropriate combination of one electron and one photon, or indeed several photons, can also qualify, to which we add one electron and one electron-positron pair, and so forth. In short, $\psi(x) \mid$vac.$\rangle$ is a superposition of all multi-particle states that can realize the particular circumstances that are symbolized by the field ψ. We indicate this situation by writing

$$\psi \mid \text{vac.}\rangle = a \mid 1 \text{ el.}\rangle + \sum a' \mid 1 \text{ el. } 1 \text{ ph.}\rangle + \sum a'' \mid 1 \text{ el. } 1 \text{ ph. } 1 \text{ ph.}\rangle + $$
$$+ \sum a''' \mid 1 \text{ el. } 1 \text{ el. } 1 \text{ el.}\rangle + \cdots,$$

where the various coefficients are probability amplitudes for the respective kinds of particle states. The analogous discussion of the vector field A gives

$$A \mid \text{vac.}\rangle = b \mid 1 \text{ ph.}\rangle + \sum b' \mid 1 \text{ el. } 1 \text{ el.}\rangle + \cdots.$$

In practice, the spectral properties of these excitations are more conveniently handled in terms of numerical functions that describe the correlations between different excitations localized at various spacetime points. Such a function is the vacuum expectation value

$$\langle \text{vac.}\mid \psi(x) \psi(x') \mid \text{vac.}\rangle \equiv \langle \psi(x) \psi(x')\rangle.$$

Through its translationally invariant dependence on the variable $x - x'$, the energy-momentum characteristics of the various excitations are displayed. Even more convenient, for technical reasons, is the time ordered product of fields, symbolized by $(\psi(x) \psi(x'))_+$, where the operator associated with the later time value stands to the left. In the particular example of the anticommuting field ψ that is now under discussion, it is advisable to counter the discontinuous appearance of a minus sign by an antisymmetrical function of the time difference, $\varepsilon(x - x')$. Then, with some additional factors that are useful, we come to the definition of the propagation function or Green's function, which is the effective mathematical instrument of quantum field theory,

$$G(x, x') = i \langle (\psi(x) \psi(x'))_+\rangle \gamma^0 \varepsilon(x - x').$$

This example refers to a Fermi–Dirac field. There is an analogous definition for the propagation function $D(x, x')$ associated with the Bose–Einstein electromagnetic field, although there are special features accompanying the gauge invariance character-istic of the Maxwell field.

These propagation functions obey infinite systems of linear inhomogeneous differential equations, obtained by adjoining analogous time ordered correlation functions involving increasing numbers of additional fields. As the first example of this system, we have the following, produced by applying the Dirac differential field equation to $G(x, x')$:

$$\left(\gamma\frac{1}{i}\partial + m_0\right) G(x, x') - e_0 q\gamma i \langle(A(x)\psi(x)\psi(x'))_+\rangle \gamma^0\varepsilon(x - x') = \delta(x - x').$$

The inhomogeneous term originates in the discontinuity of $G(x, x')$ at $x^0 = x^{0'}$, thereby introducing $\delta(x^0 - x^{0'})$, which discontinuity is measured by the equal-time anticommutator of the fields, a unit matrix multiple of $\delta(\mathbf{x} - \mathbf{x}')$. The additional factors appearing in the definition of $G(x, x')$ are such as to yield just the four-dimensional delta function. Thus, the infinite set of linear, numerical equations combines the non-linear operator field equations with the basic (anti-) commutation relations. And, through the reference to the vacuum state in the definitions, there are boundary conditions to select the appropriate solution of this equation system.

Now, imagine (!) that we have solved this infinite system and have $G(x, x')$, along with $D(x, x')$, before us. These functions will display the various multiparticle states comprising $\psi \mid \text{vac.}\rangle$ and $A \mid \text{vac.}\rangle$ in the form of additive contributions to the complete propagation function, as in

$$G(x, x') = A G_{\text{el.}}(x - x') + \cdots,$$
$$D(x, x') = B D_{\text{ph.}}(x - x') + \cdots.$$

Here $G_{\text{el.}}$ and $D_{\text{ph.}}$ are the propagation functions of the respective physical particles, the weight factors $A \sim a^2$ and $B \sim b^2$ are relative probabilities for the formation of the single particle states by the excitations $\psi(x)$ and $A(x)$, and the dots stand for the propagation functions that describe all the multi-particle excitations that are also implied by these elementary excitations.

Clearly, then, the propagation functions G and D do not describe the physical particles directly. They contain much more information, but of a kind that is rarely of immediate interest, in the nature of the background that accompanies any specific reaction that is under scrutiny. There is no difficulty, however, in isolating the parts of G and D that refer to single-particle propagation. Corresponding to the experimental requirement of following a particle for a sufficient time to measure its energy-momentum characteristics, its mass, it suffices to consider a kind of asymptotic form for these functions, such that the continuous mass spectrum of multi-particle states will lead to a suppression of those contributions. Nevertheless, the outcome of this procedure still contains a reference to the special mechanism used to create the particles, in the weight factors A and B. Accordingly, one must also remove these factors by a process of *renormalization*:

$$G_{\text{renorm.}} = \frac{1}{A} G, \qquad D_{\text{renorm.}} = \frac{1}{B} D.$$

Accompanying this change of scale is the replacement of the original parameters m_0 and e_0 by the physical mass and charge, m and e. The latter is particularly simple since the comparison of alternative and equivalent ways of expressing the coupling between two charges:

$$e_0^2 D = e^2 D_{\text{renorm.}},$$

gives

$$e^2 = e_0^2 B.$$

Renormalization, then, is the process of transferring attention from the underlying dynamical variables with which the theory begins to the physical level at which the observed particles are in evidence. It is not a concept to which there are intrinsic conceptual objections, and indeed must occur in any theory that contains structural assumptions about the constitution of the physical particles, in which abstract, underlying, dynamical variables appear.

Only now that I have described the fundamental significance of renormalization will I refer to the fact that is usually emphasized about this procedure. When one solves the coupled system of differential equations for the time-ordered field correlation functions by successive substitution, as a power series in e_0^2, the results contain divergent integrals. Very small distances or, equivalently, very high momenta make a disproportionately large contribution to the results, with the consequence that solutions to the coupled equations do not really exist – at least, not in the sense of a perturbative expansion. Whether this is merely a deficiency in the elementary mathematical approach, which could disappear in a more sophisticated scheme, or, an intrinsic failure of the physical model, I do not know. Nor am I very sure that it is particularly important to select one of the choices, since it must remain true that phenomena at very small distances play a substantial role in the whole story. What is important is that the renormalized equations, on the other hand, do have solutions – finite solutions, with numerical consequences that are in overwhelming agreement with experiment.

Here is something worth understanding. Somehow, the process of renormalization has removed the unbalanced reference to very high momenta that the unrenormalized equations display. The renormalized equations only make use of physics that is reasonably well established. In contrast, the unrenormalized equations critically involve phenomena in regions where we cannot pretend to know the physics. By what right do we, as physicists, claim that the laws of physics, even within the narrow domain of electrodynamics, will forever remain the same as we extrapolate to arbitrarily high energies? All experience suggests that new and unexpected phenomena will come into play. The resulting logical situation is very interesting. From the same system of equations emerge two very different attitudes. The unrenormalized description constitutes a model of the dynamical structure of the physical particles, which is sensitive to details at distances where we have no particular reason to believe in the correctness of the physics – an implicit speculation about inner structure – while the renormalized description removes these unwarranted speculations and concentrates on the reasonably known physics that is germane to the behaviour of the particles.

This way of putting the matter can hardly fail to raise the question whether we have to proceed in this tortuous manner of introducing physically extraneous hypotheses, only to delete these at the end in order to get physically meaningful results. Clearly there would be a great improvement, both conceptually and computationally, if we could identify and remove the speculative hypotheses that are implicit in the unrenormalized equations, thereby working much more at the phenomenological level. For electrodynamics, this may be just a *tour de force*, since it is not claimed that new numerical results will appear in this way. But, for strong interaction physics the issue is crucial. If we are caught in a formalism that has built into it implicit hypotheses, inner structural assumptions that have small chance of being correct, we shall have grave difficulties in fighting through to the correct theory. I believe that we need a more flexible kind of theory, one that can incorporate experimental results, and extrapolate them in a reasonable manner without falling into the trap of the wholesale extrapolation that infringes on unexplored areas where surprises are sure to await. I am well aware that this kind of flexible approach is anathema to the mathematically oriented [State your axioms! What are the calculational rules?], but I continue to hope that it has great appeal to the true physicist (Where are you?).

Having said all this, let us return to electrodynamics and try to identify the implicit hypothesis, the one that has introduced speculative structural assumptions. I say that it is the introduction of operator fields. Do not misunderstand me. The *field* concept is unavoidable, barring some totally new approach to the space-time continuum, which is not being advocated. But *operator* fields? That is another matter. An operator such as $\psi(x)$ is defined by the totality, or at least a sufficiently large class, of its matrix elements. And the overwhelming proportion of these refer to energies and momenta that are far outside experimental experience. Unavoidably, then, an operator field theory makes reference to phenomena in experimentally unexplored regions. It is simple enough to identify the innocent use of field operators as the culprit. But, what to replace it with?

Usually, during the development of a particular subject, several competing but equivalent formalisms are available. The example of matrix and wave mechanics naturally comes to mind. Yet one of these may be specially suited for the transition to the next levels of description. Who, in the mid-nineteenth century could have suspected the significance of the otherwise not very useful Hamiltonian formalism in the much later developments of statistical mechanics and quantum mechanics? It is important that these variant approaches exist since, in seeking a new theory, one of them may sufficiently narrow the mental gap that needs to be traversed to make this journey feasible. The human mind is not adapted to a quantum jump in ideas. A small step for mankind is all that one can reasonable expect.

I mention this in order to recall that other formalisms for quantum electrodynamics were in existence. During the 25 year period of quantum electrodynamical development, there was great formal progress in the manner of presenting the laws of quantum mechanics, all of which had its inspiration in a paper of Dirac. This paper (which is No. 26 in the collection, *Selected Papers on Quantum Electrodynamics*, Dover, 1958) discussed for the first time the significance of the Lagrangian in quantum mechanics.

I have always been puzzled that it took so long to do this, but a faint glimmering of the reason appeared when I reread this paper recently and noticed that even Dirac himself thought that the action principle required the use of coordinates and velocities rather than coordinates and momenta, despite the existence of the classical action expression

$$W = \int\limits_{t_2}^{t_1} dt \left[\sum_k p_k \frac{dq_k}{dt} - H(p, q) \right].$$

[Incidentally, this same hang-up seems to persist in recent articles claiming that the quantum action principle is inapplicable to curved spaces.] Eventually, these ideas led to Lagrangian or action formulations of quantum mechanics, appearing in two distinct but related forms, which I distinguish as differential and integral. The latter, spearheaded by Feynman, has had all the press coverage, but I continue to believe that the differential viewpoint is more general, more elegant, more useful, and more tied to the historical line of development as the quantum transcription of Hamilton's action principle.

The quantum action principle is a variational statement about the transformation function that connects states at different times, or on different space-like surfaces. Its bare expression, leaving out all the labels, is

$$\delta \langle \, | \, \rangle = i \langle \, | \, \delta W \, | \, \rangle,$$

where W is an action operator which, for particle mechanics, has just the classical form, while, for field mechanics, it is illustrated by the Maxwell–Dirac expression

$$W = \int (dx) \{ -\tfrac{1}{2} F^{\mu\nu} (\partial_\mu A_\nu - \partial_\nu A_\mu) + \tfrac{1}{4} F^{\mu\nu} F_{\mu\nu} - \tfrac{1}{2} \psi \gamma^0 [\gamma^\mu (1/i \, \partial_\mu - e_0 q A_\mu) + m_0] \psi \}.$$

The classical context of action principles is incomplete, and somewhat misleading, since the *operator* field variations have individual operator character, specifically, distinguishing F. D. fields (ψ) from B. E. fields (A, F). (A fairly detailed discussion of these two types can be found in *Quantum Kinematics and Dynamics*, Benjamin, 1970.) The quantum action principle supplies the field equations (equations of motion) as in the classical model and also the (anti-) commutation relations that are characteristically quantum mechanical. This approach emphasized the indivisible unity of the quantum principles.

The action formulation facilitates the consideration of a more general situation in which the system of interest is externally driven, perturbed. A sufficiently simple yet general form is obtained by adding to the action a linear functional of the fields, as in

$$\int (dx) [A^\mu (x) J_\mu (x) + \psi (x) \gamma^0 \eta (x)],$$

where the $J_\mu (x)$ are arbitrary commuting numbers and the components of $\eta (x)$ are arbitrary anticommuting numbers. The inhomogeneous form of the resulting field equations,

$$[\gamma^\mu(1/i\,\partial_\mu - e_0 q A_\mu(x)) + m_0]\,\psi(x) = \eta(x),$$
$$\partial_\mu F^{\mu\nu}(x) - \tfrac{1}{2}\psi(x)\,\gamma^0 e_0 q \gamma^\mu \psi(x) = J^\mu(x),$$

supplies the designation of these external quantities as sources. There are various uses of these external couplings, which are an idealized version of the interventions that constitute measurements on the system. The commutation relations of the respective fields are implicit in the source terms of the field equations, for example. And, with the possibility of exciting arbitrary states through the agency of the sources, it suffices to choose only the vacuum state in considering the over-all development of the system. Thus, the basic quantity that contains all physical information about the system is the vacuum amplitude

$$\langle \text{vac.} \mid \text{vac.} \rangle^{\eta J} \equiv \langle 0_+ \mid 0_- \rangle^{\eta J},$$

where $+$ and $-$ are causal labels designating times that follow or precede the region where the sources are manipulated.

The expansion of the vacuum amplitude functional in powers of the sources defines an infinite sequence of functions; they are the totality of Green's functions previously introduced as time-ordered field correlation functions:

$$\langle 0_+ \mid 0_- \rangle^{\eta J} = 1 + i\tfrac{1}{2}\int (dx)\,(dx')\,\eta(x)\,\gamma^0 G(x, x')\,\eta(x')$$
$$+ i\tfrac{1}{2}\int (dx)\,(dx')\,J^\mu(x)\,D(x, x')_{\mu\nu}\,J^\nu(x') + \cdots .$$

Corresponding to this functional union of all the propagation functions, the infinite sequence of dynamical equations for these functions is replaced by a small number of functional equations. The use of the action principle gives:

$$\delta \langle 0_+ \mid 0_- \rangle^{\eta J} = i \langle 0_+ \mid \int (dx)\,[A^\mu(x)\,\delta J_\mu(x) + \psi(x)\,\gamma^0 \delta \eta(x)]\,\mid 0_- \rangle^{\eta J},$$

or

$$\frac{1}{i}\frac{\delta}{\delta J_\mu(x)}\langle 0_+ \mid 0_- \rangle^{\eta J} = \langle 0_+ \mid A^\mu(x) \mid 0_- \rangle^{\eta J}$$

$$\frac{1}{i}\frac{\delta}{\delta \eta(x)\,\gamma_0}\langle 0_+ \mid 0_- \rangle^{\eta J} = \langle 0_+ \mid \psi(x) \mid 0_- \rangle^{\eta J},$$

and then the Dirac operator equation, for example, becomes

$$\left\{\left[\gamma^\mu\left(\frac{1}{i}\partial_\mu - e_0 q \frac{1}{i}\frac{\delta}{\delta J^\mu(x)}\right) + m_0\right]\frac{1}{i}\frac{\delta}{\delta \eta(x)\,\gamma^0} - \eta(x)\right\}\langle 0_+ \mid 0_- \rangle^{\eta J} = 0.$$

Successive functional differentiation of this equation now yields one sequence of the coupled Green's function equations, the other following from the analogous treatment of the Maxwell operator field equations.

The functional equations can be given a variety of other forms with different emphases. We remark, for example, that the Dirac functional equation can be expressed as

$$\left\{ i \left[\eta(x), W\left(\frac{1}{i} \frac{\delta}{\delta \eta \gamma^0}, \frac{1}{i} \frac{\delta}{\delta J} \right) \right] - \eta(x) \right\} \langle 0_+ | 0_- \rangle^{\eta J} = 0,$$

or

$$e^{iW} \eta(x) e^{-iW} \langle 0_+ | 0_- \rangle^{\eta J} = 0,$$

where W is constructed from the operator action by the functional substitutions. The formal (!) solution of this and the analogous Maxwell system is given in terms of functional delta functions by

$$\langle 0_+ | 0_- \rangle^{\eta J} = e^{iW[(1/i)(\delta/\delta \eta \gamma^0),\, (1/i)(\delta/\delta J)]} \delta[\eta]\, \delta[J].$$

Then the functional Fourier construction

$$\delta[\eta]\, \delta[J] = \int [\mathrm{d}\psi]\, [\mathrm{d}A]\, e^{i\int(\psi \gamma^0 \eta + JA)}$$

supplies the explicit integral expression (to within a constant factor),

$$\langle 0_+ | 0_- \rangle^{\eta J} = \int [\mathrm{d}\psi]\, [\mathrm{d}A]\, e^{i\{\int(\psi \gamma^0 \eta + JA) + W[\psi, A]\}}.$$

It continues to surprise me that so many people seem to accept this formal statement as a satisfactory *starting* point of a theory.

Another procedure is of more interest for our present purposes. Write

$$\langle 0_+ | 0_- \rangle^{\eta J} = e^{i\{\int(\psi \gamma^0 \eta + JA) + W[\psi A]\}},$$

where $\psi(x)$ and $A(x)$ are now *numerical* fields, numbers of the same types as $\eta(x)$ and $J(x)$, and $W[\psi A]$ is only implicitly defined through a stationary requirement:

$$\frac{\delta}{\delta \psi \gamma^0} W + \eta = 0, \qquad \frac{\delta W}{\delta A} + J = 0.$$

In consequence of the stationary property, we have

$$\frac{1}{i} \frac{\delta}{\delta \eta \gamma^0} \langle 0_+ | 0_- \rangle^{\eta J} = \psi \langle 0_+ | 0_- \rangle^{\eta J},$$

$$\frac{1}{i} \frac{\delta}{\delta J} \langle 0_+ | 0_- \rangle^{\eta J} = A \langle 0_+ | 0_- \rangle^{\eta J},$$

and the Dirac functional equation, for example, becomes

$$\left[\gamma \left(\frac{1}{i} \partial - e_0 q \left(A + \frac{1}{i} \frac{\delta}{\delta J} \right) \right) + m \right] \psi = \eta;$$

we recall again that $\psi(x)$, $A(x)$ are numerical fields. Here, then, is a formulation completely equivalent to the original one in terms of operator fields, commutation relations, and all the rest, but now expressed in the language of numerical sources, numerical fields. And it is this formulation that has the flexibility to permit a new beginning, a fresh, more physical approach to particle theory. The sources were initially tied to the operators ψ and A which describe elementary, multi-particle excitations. Why not abandon the whole operator framework and define the sources *ab initio* in terms of the excitation of single, physical particles? This is the starting point of source theory, and if any of this discussion seems moderately reasonable, I invite you to follow the systematic evolution of this phenomenological attitude in the two existing volumes of *Particles, Sources, and Fields*, Addison-Wesley, Vol. I (1970), Vol. II (1973).

The concept of magnetic charge in quantum mechanics was introduced by Dirac in 1931 and further developed in 1948. He showed that the existence of magnetic charge would provide an elementary and beautiful explanation of the empirical quantization of electric charge, which is otherwise a mysterious regularity of nature. This still remains a most compelling argument in favour of a property that, thus far, has stubbornly refused to surface experimentally. Other facets of this intriguing possibility come to light when one considers the more general situation of particles that carry both electric charge (e) and magnetic charge (g). First, let me make the following irritating statement. As far as electrodynamics is concerned, there is no observational difference between our world, in which magnetic charge does not exist, and a hypothetical world in which all atomic particles carry both charges, in a manner that is invariant under the rotation

$$e' = e \cos\theta - g \sin\theta$$
$$g' = g \sin\theta + e \cos\theta,$$

and satisfy the weak charge requirement

$$\frac{e^2 + g^2}{\hbar c} < 1.$$

This equivalence follows immediately from the charge quantization condition, referring to any two particles with charges e_1, g_1, and e_2, g_2, namely

$$\frac{e_1 g_2 - e_2 g_1}{\hbar c} = \text{integer},$$

for, under the stated charge restrictions, the integer can only be zero. Accordingly, if one were to mark, in a two-dimensional e–g space, points corresponding to various kinds of particles, these points would lie on a single straight line passing through the origin. It then suffices to specialize the arbitrary e–g coordinate system by defining this absolute line to be the axis of electric charge; all particles will now have zero magnetic charge. Note, however, that these charges need not be equally spaced, as they are empirically. For that situation to appear, there must exist other particles that violate the weak charge restriction and do not lie on the line of pure electric charge. These are magnetically (and possibly also electrically) charged particles. The charge quantization condition tells us that, in contrast with the customary electric charge magnitude,

$$\frac{e^2}{\hbar c} = \frac{1}{137},$$

such magnetic charges are measured by, essentially,

$$\frac{g^2}{\hbar c} = 137.$$

It is the very large magnitude of this unit that suggests the connection with the strongly interacting particles, the hadrons. Of course, hadrons are not magnetically charged. But neither are atoms electrically charged. In short, the emerging picture is of hadrons as magnetically neutral composites of entities that carry magnetic charge and, necessarily, also electric charge. I have christened these hypothetical particles: dyons. (An aside to any indignant member of the audience who is thinking – you warned us about introducing speculations concerning inner structure, and now you indulge in this outrageous speculation! My warning was directed against the implicit, unrecognized speculation. One must not confuse structural hypothesis and phenomenology. These are two distinct approaches which, hopefully, will eventually merge in a more fundamental theory.) And now a wholly new situation appears. The charge quantization condition is much less restrictive for dyons, permitting *fractional* charges, that is, fractions of the units of pure electric and magnetic charge. Here, unexpectedly, is contact with the empirical 'quark' model of hadrons, which uses the electric charge pattern, $2/3$, $-1/3$, $-1/3$. For a dyon this pattern extends to magnetic charge as well, and is reasonably inferred from considerations of magnetic neutrality and particle statistics. (A semi-popular account is presented in *Science* **165**, 757 (1969).)

This is the basis for thinking that, underlying the mysterious strong interactions, is another and new manifestation of electromagnetism. As for the weak interactions, let me talk only about the superweak one, that is *CP* violation, which has seemed so unrelated to the rest of physics. The point, quite simply, is that dyons have a built-in mechanism for *CP* violation. The operation of charge reflection, *C*, interchanges particles and anti-particles, producing the substitution $e, g \to -e, -g$. Space reflection, *P*, acts oppositely on electric and magnetic charges, as it does on electric and magnetic fields. Thus, the combined CP effect is $e, g \to -e, g$; for some particular fractional charge assignments, it is $2/3, 2/3 \to -2/3, 2/3$, or $-1/3, 2/3 \to 1/3, 2/3$. But the resulting charge assignments are not included in the list of particles: $(2/3, -1/3, -1/3)$, $(2/3, -1/3, -1/3)$, or of antiparticles: $(-2/3, 1/3, 1/3)$, $(-2/3, 1/3, 1/3)$, and *CP* is not an invariance transformation of the system. There is more to the story, which connects it with the ordinary weak interactions, but I refer you to the *Science* article for a short account of that.

How beautiful it would be if the logically sound concepts of magnetic charge and dyons should prove to be at the heart of the subnuclear world! But I leave you with this sobering remark. We have heard so much about the importance of beauty in physical theory. No doubt a correct theory will be beautiful (a cynic will say that our concept of beauty would evolve to make it so), but a merely beautiful theory has small chance of being correct. In short, beauty, as a criterion for validity, is necessary but not sufficient.

DISCUSSION

Question: I wonder if Professor Schwinger would comment on the possible derivation of the values of the coupling constants, in particular on Wyler's formula for the fine structure constant.

J. Schwinger: Let me interpret this question to mean what is my view about a possible physical explanation of the actual value of the fine structure constant. I have completely failed to understand the physical relevance of Wyler's contribution. It is simply a mathematical formula for a certain number. I don't know the physical basis by which to call it the fine structure constant. I'd rather talk about physics than numerology.

The real question is why it is so small compared to one. We have heard Professor Salam's suggestion of a possible connection with gravitation as an explanation of the order of magnitude of this number. I would like to point out, even though I have no idea of how to implement it, another possible connection, with the idea of magnetic charge. The reciprocal quantization of electric and magnetic charges comes from Dirac's idea, and is roughly suggested by the fact that the product of a unit of electric charge and a unit of magnetic charge is of the order of one. The question why $e^2/4\pi$ is so small, quite apart from its actual numerical value, is also the question why is $g^2/4\pi$ so large?

From this perhaps you might recognize that the real question is why e is not equal to g? After all, built into all these ideas is a notion of symmetry between electric and magnetic charge. The most naive anticipation, in the most symmetrical of all worlds, would be that e and g would be the same and of the order of one. That is *not* the situation. This way of describing things suggests its own answer: e equal to g would indeed be a more symmetrical world, but perhaps not the most stable world.

There is a possibility that the question of dynamical symmetry or spontaneous symmetry breaking is what is involved. A world with $e=g$ may perhaps have an energy of the vacuum that is vastly higher than a world in which e is not equal to g. It is their difference that is important, because obviously from the physical point of view it is the smaller of the two that we would call the unit of electric charge, and the larger of the two that we would call the unit of magnetic charge, because that is the way we would recognize them. The magnetic charge, interacting strongly, would tend to quench itself, and what remains we would recognize as electricity.

I therefore suggest the possibility that if we could indeed compare the relative stabilities of whole classes of universes, and take into account the polarization forces that would produce dynamical changes in the observed values of the coupling constants, then some future generation might be able to carry out the calculation and in that way supply a quantitative theory for the fine structure constant. However, my main point is a conceptual one. I am suggesting that what determines the actual value of the fine structure constant is primarily purely electromagnetic forces, and that gravitation has nothing to do with the answer.

C. N. Yang: I have not understood Professor Schwinger's remark about the possible connection between the existence of a magnetic monopole and *CP*-violation. I think that if one introduces the magnetic monopole, as something which interacts purely with a simple electromagnetic interaction, I do not believe that it would provide any explanation whatsoever for the decay of the K^0 into two π's. If one tampers with this, and introduces additional interactions, it is certainly true that one can explain *CP*-violation, but I don't think that it is natural to it. After all, without the magnetic monopole, we can also explain *CP*-violation by adding interactions. I would therefore think that the statement that the introduction of magnetic monopoles would render *CP*-violation natural is misleading.

J. Schwinger: What I said was that the important thing is the introduction of particles carrying both electric and magnetic charges on the same particle, in particular without the negative of that ratio also occurring. Perhaps I may add some more remarks. Suppose there are particles that carry certain fractional values of electric charges, say $1/3e$, and there are particles that also carry certain values of the magnetic charge. Let us use the same units. Let e be the unit of pure electric charge and g the unit of pure magnetic charge. Suppose that a particle like this exists. If one considers CP, the important thing here is not C; under charge reflection all charges go into their negative, producing antiparticles. The important property here is the space reflection, under which electric and magnetic charges behave oppositely. If this new particle existed, then of course one would be simply relabeling things, and there would be a CP invariance. If this particle does not exist, then there is a failure of P, and more generally CP invariance, as far as the properties of this particle are concerned. From this underlying picture to the actual phenomenological mechanism of the decay of K^0 is a long way, and I do not know how to implement it. I say only that in the structure of the underlying dynamics and the nature of the particles, there is an inherent CP-violation. If I am mistaken about this, I'd be most happy to understand it.

C. N. Yang: I do not disagree with what you said, but the question is whether you have rendered CP-violation more natural. You introduced an additional assumption of the existence of a particle, which I consider as surprising an assumption as CP-violation itself. In other words, the justification for the proposal of the existence of these additional particles could not lean on the crutch of the observation of CP-violation.

J. Schwinger: What I have just described is, of course, taken out of the context of ideas that suggest that certain patterns of particles should exist. Let me remind you of a very familiar thing, namely, that the pattern of fractional charges, the electric charges that one has known phenomenologically and which this theory of electric and magnetic charges tends to produce rather naturally, is an asymmetrical one. Let's look at the usual electric charge multiplet. With additional physical arguments I have interpreted the idea of symmetry between electric and magnetic charges to suggest that the same pattern of magnetic charges also appears: $2/3$, $-1/3$, $-1/3$. If you combine the appropriate electric and magnetic charges, $2/3$ and $2/3$ say, you find that its CP analogue is not present in this pattern. There is no $2/3$, $-2/3$; it doesn't appear here. Now I didn't struggle to do this. I simply recognized that there was a natural appearance of this triplet pattern, the same triplet pattern for magnetic charges, which has an asymmetry, but it is an asymmetry that we have become familiar with. There is nothing particularly unnatural or forced about the statement that for every electric charge you do not always have plus and minus those magnetic charges, because the pattern of charges is not symmetrical.

V. Telegdi: Assume that this extremely imaginative scheme is a cause of CP-violation, dropping Yang's objections to the scheme. Assume it is true. You are still left with a further puzzle. Why would such a mechanism display CP-violation only in the K^0 complex?

J. Schwinger: The real question is not why does it appear merely in the K^0 couplings, but why does it appear so weakly? Well, the K^0 system happens to be the only one where you can see it experimentally if it is indeed as weak as it actually turns out to be. That again has to do with the fact that I haven't elaborated the full dynamics of this scheme.

I have described, very inadequately, this idea as a purely electromagnetic one. That is not enough. As we know, there are not only purely electromagnetic forces, there are weak interactions as well. There is more to the world than electromagnetism. So there must also be more to the picture of electric and magnetic forces than merely the actions of e's and g's. Further elaboration of the theory suggests that, in analogy with the weak interactions, which I may translate as mechanisms whereby electric charges are exchanged among particles, there must also be a mechanism that exchanges magnetic charges among particles. The disparity of strength would suggest that that is a strong mechanism. Along with this is a more elaborate picture in which the particular magnetic charges that I have written here do not simply reside on particles which are the constituents of hadrons, but rather flit very rapidly back and forth. The weak exchange of charge, characteristic of weak interactions, becomes a strong exchange of magnetic charge. You will notice that the pattern is such that the average magnetic charge is zero. So there is a tendency to the quenching of magnetic charge, in such a way that the asymmetries involved here are not observable for low momenta, long distance observations, but only come into play at short distances. This kind of picture therefore automatically suggests that the *CP*-violation mechanism will be of the super weak type, occurring only at short distances and high energies. It now becomes not so incomprehensible that special systems are the only ones where it showed up. Clearly it is a much more elaborate theory than I have had the opportunity of suggesting here.

Question: How were you able to separate the problem of interaction from renormalization? Could one do it in a more familiar form that would lead to a different Hamiltonian for the interaction between electric charges and electromagnetic fields?

J. Schwinger: May I point out that when one switches a point of view and cares to deal only with the physical particles, then many of the words and phrases which one has been used to, cease to have meaning. There is no question of renormalization because there is nothing to renormalize from. One is dealing with the physical particles. The phrases bare particle and clothed particles have no meaning; these refer to the other point of view. What one has here is a phenomenologically oriented theory, which has the great advantage of being much closer to what the physicist actually does. This is not a theory the equations of which are stated in advance, but rather it involves a process of elaboration, of adding more and more phenomena, of extrapolating to higher and higher energies, of widening the field of description of the theory, in which the theory is never stated all at once, just as we never know all the physics at once. When we do know all the physics, the subject is closed.

G. Wentzel: Oh, just the same as in the normal theory.

J. Schwinger: It depends on what you are discussing. If it is electrodynamics, the predictive power is identical, with a vast simplification of the conceptual framework. In the realm of strong interactions, the theory is much more modest, much more conservative. One starts with the low energy behaviour of the known particles, makes extrapolations up to higher energies, predicts what should occur as a result of the known physics. In making the extrapolation, one recognizes of course that as one moves to higher energies, new particles come into existence. They are then incorporated into the scheme. There is no attempt at anything as all-embracing as the kind of theory that Abdus Salam has described, which attempts at the very beginning to write down the equations of the whole world. This is a much more modest approach. It is therefore hard to compare it with conventional procedures, because it is not quite the same as ordinary theories. I cannot really answer all the questions except by directing you to read enough about it, to see how it is used, because that is the only way to describe the theory.

Question: Is the theory renormalizable or not renormalizable?

J. Schwinger: To my mind, renormalizability is a statement about the high energy behaviour of the theory. Does it act reasonably or unreasonably as one goes towards high energy phenomena? The essential difference between the phenomenological point of view, that I now advocate, and the conventional one, is this. In ordinary theories, if you begin with a non-renormalizable theory, you need an infinite number of operations in order to be able to translate the formal structure into statements about physical properties. A non-renormalizable theory, from that point of view, is an absolute barrier. If I were somehow to ask for the analogue of the non-renormalizable theory in the phenomenological framework of source theory, I do not have that barrier. I begin with known phenomenology and I move up a little bit in energy, and I discover that in order to make these extrapolations I may have to introduce one or two new phenomenological constants to be taken from experiment. The theory has less predictive power, but it still has some. At no stage must I encounter an infinite number of constants, because I ask a limited number of questions about a finite domain of experience. Since one is working at the physical phenomenological level, the barrier that prevents one from making any discussions does not exist.

I would say the situation has become looser. One has more freedom to work with theories that have the non-renormalizable character in the sense of not having very good convergence properties at high energies. They require a few new phenomenological constants. As you go farther and farther, the situation becomes more serious. But nature in the strong interaction domain, to which this question really refers, has always provided us with more particles which convert the badly behaved local couplings into better behaved non-local couplings. In a procedure in which one systematically goes up in energy and incorporates this new information, the path to progress is never closed.

B. L. van der Waerden: In talks on quantum field theory, we have heard from three different authors the statement that quantum field theory begins with Dirac's paper of 1927. This statement seems to be true in the minds of most physicists, in the minds of all physicists except one. Actually, quantum field theory begins with the paper of Jordan. You cannot write the history of quantum field theory neglecting him.

There are two earlier papers on quantum field theory before Dirac's paper of 1927. Strangely enough, the papers themselves are well-known; i.e. the paper of Born and Jordan of 1925, on quantum mechanics, and the next paper of the year 1925 by Heisenberg, Born and Jordan, the famous three-men paper from which all the contemporary physicists learned quantum mechanics. At the end of these papers there is a section on quantum field theory, for which Jordan alone was responsible, not Born and not Heisenberg.

Well, as I was in Göttingen at that time, I talked with several people about it, and they said, 'Ah yes, it's true that there is some kind of quantum field theory in it, but the only thing Jordan did was to print the E and H in bold print and to say that they are matrices; as for the rest the equations are the same as those of Maxwell's.' Now I don't think this is quite true, but this was the general feeling of physicists in Göttingen at that time.

In any case, Jordan said expressly that the field forces E and H are matrices, are operators. In the second paper, the fluctuations of the electromagnetic field are discussed, first from the classical point of view and next from the quantum point of view. The field is considered as a set of harmonic oscillators, and right at the end of the paper these oscillators are quantized.

21

Progress in Renormalization Theory Since 1949

Abdus Salam

1. INTRODUCTION

Julian Schwinger has given us a brilliant exposition of the renormalization ideas associated with his, Tomonaga's and Dyson's names. My task is to take the story up from the exciting days of 1949 to the present.

As Schwinger has reminded us, even though renormalization and infinities are logically distinct, it was the persistent infinities encountered in field theories which brought the necessity of renormalization of mass and charge to the fore. The infinities arise mathematically from undefined products of distributions like

$$\frac{1}{x^2} \otimes \frac{1}{x^2}, \quad \frac{1}{x^2} \otimes \partial^2 \frac{1}{x^2} \approx \frac{1}{x^2} \otimes \delta(x) \quad \text{and} \quad \partial^2 \frac{1}{x^2} \otimes \partial^2 \frac{1}{x^2} \approx \delta(x) \otimes \delta(x),$$

when $x \to 0$. These undefined products are encountered when we solve a set of quantum field equations using a perturbation expansion. The first question which arises is this:

Are the infinities – or equivalently the undefined products – a consequence of the *bad mathematics* of a perturbation expansion? Would they appear even in *exact* solutions of field equations? In this latter event, is it the type of Lagrangian we are using which is at fault? And if the Lagrangian is to be modified in some essential manner, is there some *missing physics* which when supplied would circumvent the appearance of the infinities?

A suggestive answer to those questions has been given by Glimm and Jaffe in a series of brilliant papers published between 1969 and 1972. These authors have solved Yukawa-like field theories with the Lagrangians, ϕ^3, ϕ^4 and $\bar{\psi}(x)\psi(x)\phi(x)$ *without* using a perturbation expansion, in a space of two-space and one-time dimensions. The infinities (expected from naive perturbation theory) duly make their appearance in the *exact* solutions also. If one may extrapolate from three to four dimensions of space and time, *it is not the perturbation expansion that is at fault*. One needs to alter basically the *type* of Lagrangian one has been using in physics. One would still like the new Lagrangians to be 'local' in the technical sense, in order to preserve causality and unitarity. However, as I will show later, it is in determining the correct local modification to the types of Lagrangian we have been using that some of the physics we have been missing out – the physics of quantum gravity – which will serve as a guide. Before this, however, let me summarize what Schwinger has told us.

J. Mehra (ed.), The Physicist's Conception of Nature, 430–446. All Rights Reserved

(1) Consider the class of Lagrangians which have the simple form of polynomials in field variables. These are of the type:

$$\phi^3, \phi^4, \phi^5, ..., \bar{\psi}\psi\phi, ..., (\bar{\psi}\psi)^2, ...$$

Why we should ever restrict to the polynomial class of Lagrangians is not clear, but this is an assumption we have inherited from the history of the subject.

(2) Compute the scattering-matrix elements in a perturbation expansion, using standard Feynman rules.

(3) There is among this class of Lagrangians a small subset which we call *renormalizable*, with the property that only a limited few of the S-matrix elements for these theories are intrinsically infinite. This small subclass of Lagrangians includes ϕ^3, ϕ^4, Yukawa $\bar{\psi}\psi\phi$ and the Maxwell–Dirac Lagrangian $\bar{\psi}\gamma_\mu\psi A_\mu$. A theory is *non-renormalizable* if the class of matrix elements which are infinite is itself limitless.

(4) For the *renormalizable* class of Lagrangians we have the famous Dyson theorem on the possible types of infinities which can occur. The theorem states that for the Maxwell–Dirac Lagrangian only two matrix elements are infinite. These correspond to the self-mass δm and self-charge δe of the electron.

(5) Explicitly Dyson's theorem may be stated thus:
Write the Maxwell–Dirac Lagrangian in the form:

$$\bar{\psi}\gamma\,\partial\psi + m_0\bar{\psi}\psi + e_0\bar{\psi}\gamma_\mu\psi A_\mu.$$

Here m_0 and e_0 are the so-called 'bare' constants of mass and charge.

(6) Compute the self-mass and self-charge δm and δe.
Both turn out to be (logarithmically) infinite – i.e. have the form of undefined integrals

$$\int d^4x \frac{1}{x^2} \otimes \frac{1}{x^2}.$$

(7) Dyson's renormalization theorem states that if we replace consistently in the computation of all *other* matrix elements bare mass m_0 and bare charge e_0 by the physical mass m and physical charge e where

$$m = m_0 + \delta m$$
$$e = e_0 + \delta e$$

then all infinities in the theory are completely absorbed (renormalized) in these redefinitions. The amazing thing is that this renormalized theory agrees spectacularly with experiment. As Lamb and Telegdi will tell you, one measure of this agreement is the comparison of the anomalous magnetic moment $(g-2)/2$ for the electron with theory.

Experiment: $\frac{1}{2}(g-2)_e = 0.001\,169\,644\,(7)$,
Theory: $\frac{1}{2}(g-2)_e = 0.001\,169\,642$.

The agreement of 1 part in 10^9 is a quantitative agreement unmatched anywhere else in physics – except possibly in the Eötvös–Dicke experiment. Apart from a rather heuristic treatment of what are called overlapping infinities, Dyson's work was (almost) complete for quantum electrodynamics. For other theories it left open *two* problems, which will form the subject of my talk today.

Problem 1
Which theories are renormalizable?

Problem 2
What modification in the theory (missing physics) is needed to compute δm and δe *finitely*? I shall call this second problem the *Lorentz problem* because Lorentz was the first person to attack the problem of the computation of electron's self-mass δm in the classical theory.

One must emphasize that so long as one dealt with pure quantum electrodynamics, one could hide behind the inaccessibility of δm to experiment and therefore consider the Lorentz problem to be a pseudo-problem. It is only m – the physical mass – that experiment can determine. There is no way of measuring – within pure quantum electrodynamics – the bare mass m_0. However, going beyond electrodynamics in a higher symmetry theory where, for example, the electron and the electronic neutrino form a doublet, we *do* know the bare mass of the electron. It must equal (from symmetry considerations) the bare mass of the neutrino $(m_0)=0$. *The Lorentz problem – the finite computation of δm (and δe) – must be solved in a complete theory of particles. One half of my talk will be devoted to indicating what I consider a natural solution to the Lorentz problem. I shall suggest that Maxwell–Dirac theory is incomplete in an essential respect.* I shall suggest that if one considers the complete theory of photons, electrons and quantum gravitons, both δm and δe turn out to be finite and of the correct order of magnitude.

2. THE CLASS OF RENORMALIZABLE THEORIES

For the second development I wish to highlight, we consider an extension to the class of renormalizable theories, recently discovered. This promises to give us a lot of new physics. But before we consider this extension, let us look at the situation as it obtained before these recent developments. Throughout we shall concentrate on fermions of spin-$\frac{1}{2}$ interacting with mesons. The fermions include electrons (e), muons (μ), and the neutrino; the octet of nucleons (N), and the triplet of quarks (q). For mesons we shall consider multiplets of spin-zero (e.g. the nonet containing π, κ, η), of spin-1 (with $\varrho, K^*, \phi, \omega$ particles) and spin-2 (nonet containing $A_2, ..., f, f'$). Let these particles interact through polynomial, three-field, and wherever necessary $SU(3)$-symmetric, Yukawa-like interactions.

The problem, of which among these theories are renormalizable, was solved early during 1950–51 (Salam and Ward), following Dyson's method. This proof of renormalizability was sharpened and made mathematically more rigorous by Bogolubov, Parasuik, Hepp and Speer during the last decade. (During the last year a beautiful method of regularization has been developed by 't Hooft and Veltman at Utrecht, Bollini and Giambiagi in the Argentine, and Ashmore in Trieste. I shall, however, not speak about these somewhat technical developments.)

Table I summarizes the situation of meson–fermion Yukawa-type interactions, so far as developments up to 1967 are concerned.

TABLE I

Force	Gravity	Weak	Electro-dynamics	Strong
Coupling parameters	$G_N m_e^2 \approx 10^{-44}$	$G_F m_N^2 \approx 10^{-5}$	$\dfrac{e^2}{4\pi} = \alpha = \dfrac{1}{137}$	$g^2 \approx 1$
Meson spin $J=0$				$SU(3)$ octet π, κ, η R
$J=1$		$W^\pm$ NR	γ R	$SU(3)$ octet ϱ, K^*, ϕ R if mesons massless NR if mesons massive
$J=2$	Graviton NR			$SU(3)$ octet A_2, f, K^{**} NR

R stands for a renormalizable theory; NR is non-renormalizable.

The most interesting entry in this table is the Yang–Mills gauge theory of spin-one nonet (for example, the nonet of (ϱ, K^*, ϕ, ω) interacting with the octet of physical nucleons (N, Σ, Λ, Ξ)). Yang and Mills argued that in analogy with the case of the photon for *zero-mass* gauge particles the theory is likely to be renormalizable. Now, the nonet of the physical particles (ϱ, K^*, ϕ, ω) consists of *massive* particles. One could demonstrate that for arbitrary mass values and arbitrary coupling of these particles to nucleons, the theory is non-renormalizable.

This was the situation up to around 1967. *The new developments about which I shall speak in the second part of the talk* concern the renormalizability of the *massive* gauge-meson theories. As I said before, previous to the new developments one had considered arbitrary masses and arbitrary couplings for the gauge particles. Provided that these masses and the couplings are related in a specific manner, through a set of eigenvalue equations, *the gauge theories appear to be renormalizable.* Among the spectacular special results is the one which states that in order to achieve the *renormalizability* of weak interactions, weak and electromagnetic interactions must unite as aspects of the same internal symmetry structure. Further, either there must exist new heavy leptons or the internal symmetry group of strong interactions must be extended to $SU(4)$ or $SU(3) \times SU(3)$ rather than $SU(3)$. *And all this to secure renormalizability.* Surely this must be one of the very rare occasions when a seemingly mathematical criterion (like renormalizability) appears to be leading to new and exciting physics.

3. THE LORENTZ PROBLEM

Of these two developments, consider the solution of the Lorentz problem first. I wish to show that whereas Maxwell–Dirac theory of the electron is no more than renormalizable, the Maxwell–Dirac–Einstein theory is actually finite. The conjecture that this may indeed be the case was made long ago by Klein, Landau, Pauli, Deser, DeWitt and others. Recently Strathdee, Isham and Salam have shown that the conjecture is correct provided – and this is important – quantum gravitational effects are treated *non-perturbatively*.

Gravity is distinguished among other forces of nature by its universality, by its small coupling ($G_N m_e^2 \simeq 10^{-44}$ to be compared to $e^2/4\pi = \frac{1}{137}$) and finally by the fact that it is described by an incredibly beautiful Lagrangian.

For my purposes, the feature which distinguishes this Lagrangian from all others discussed so far, is its essential non-polynomiality. Before I demonstrate the non-polynomiality of gravity, let me say a few words about non-polynomial Lagrangians in general. A polynomial Lagrangian like $\phi^4(x)$ describes the point-interaction of four ϕ-particles at one given spacetime point

A non-polynomial Lagrangian like $\sum_{n=3}^{\infty} c_n \phi^n(x)$ describes a whole set of terms, with arbitrarily large numbers of ϕ-particles being created and annihilated at the same space-time point. With certain further restrictions which I will not state, such Lagrangians can be as *local* (and causal) in a technical sense as the polynomial Lagrangians themselves; $L = g e^{\kappa\phi}$ is an example of a *local* non-polynomial Lagrangian. The theory of such Lagrangians was developed mainly in the U.S.S.R. by Efimov, Fradkin, Volkov and others during 1963, with further developments being made by Strathdee, Delbourgo and Salam in Trieste (1969), Lee and Zumino at CERN (1969), Lehmann and Pohlmeyer in Hamburg (1970), J. G. Taylor in Southampton (1970) and recently by Honerkamp at CERN (1972).

$$\bullet \quad + \quad \diagup \quad + \quad \diagup\!\!\!\diagdown \quad + \quad \curlyvee \quad + \quad \times \quad + \cdots$$

Graphs represented by $e^{\kappa\phi} = 1 + \kappa\phi + \dfrac{(\kappa\phi)^2}{2!} + \dfrac{(\kappa\phi)^3}{3!} + \cdots$.

Now Einstein's gravitational Lagrangian, though often deceptively written in a polynomial form, is really non-polynomial – it is an infinite series in the field variable.

I shall write it in a form discussed by Isham, Strathdee and Salam, which uses a Dirac γ-basis, to exhibit its elegant structure.

$$\mathscr{L}_{\text{Einstein}} = \frac{\text{Tr. } L^{\mu}L'(\partial_{\mu}B_{\nu} - \partial_{\nu}B_{\mu} + i[B_{\mu}, B_{\nu}])}{(-\det \text{Tr. } L^{\mu}L')^{1/2}}$$

$$\mathscr{L}_{\text{Einstein-Dirac}} = \frac{\bar{\psi}L^{\mu}(\partial_{\mu} + iB_{\mu} + ie_{0}A_{\mu})\psi + m_{0}\bar{\psi}\psi}{(-\det \text{Tr. } L^{\mu}L')^{1/2}}$$

Here $L^{\mu} = \gamma_{a}L^{\mu a}$ are the sixteen vierbein fields; $B_{\mu} = B_{\mu}^{ab}\sigma_{ab}$ is the affine-connection (which, as a consequence of the equations of motion following from the Lagrangian, can be shown to be proportional to the derivative of the vierbein fields $L^{\mu a}$). The familiar Einstein metric field $g^{\mu\nu}$ is given by Tr. $L^{\mu}L^{\nu}$ ('Tr.' stands for the Dirac-matrix trace). The form above emphasizes the $SL(2, C)$ gauge-invariance of the theory.

One further remark before the theory can be used to describe quantum gravitons and their scattering. For the Feynman quantization procedure (for example) to apply it is essential that the field $L^{\mu a}$ (or equivalently the field $g^{\mu\nu}$) should exhibit asymptotic flatness, and the theory should be set up in a world with the Minkowskian background. This, together with the localizability condition on the field theory, can be shown to imply that one must parametrize $L^{\mu a}$ in the form:

$$L^{\mu a} = \exp(\kappa\gamma^{pq}\phi_{pq})^{\mu a}$$

where γ^{pq} are ten 4×4 symmetric matrices, κ is the square root of the Newtonian constant $\kappa m_{e} = \sqrt{(G_{N}m_{e}^{2})} \approx 10^{-22}$ and ϕ_{pq} is the *physical field* describing the creation and annihilation of gravitons. Note that the above expression for $L^{\mu a} (\approx \eta^{\mu a} + \kappa\phi^{\mu a} + \cdots)$ implies that $\langle L^{\mu a}\rangle_{0} = \eta^{\mu a}$, in accordance with our demand for asymptotic flatness.

We are now ready to exhibit the infinity-suppressing effects of quantum gravity on the Maxwell–Dirac Lagrangian. Suppressing the tensor indices, the non-polynomial character of Maxwell–Dirac–Einstein Lagrangian is exhibited by the factor $L = \exp(\kappa\phi)$ in the numerator of the Dirac–Maxwell term. In its essentials then, L_{Einstein} has the form $\partial_{\mu}(e^{\kappa\phi})\partial_{\mu}(e^{\kappa\phi})$ while $L_{\text{Maxwell–Dirac–Einstein}} \approx e_{0}\exp(\kappa\phi)\bar{\psi}\gamma_{\mu}\psi A_{\mu}$. The non-polynomiality of this Lagrangian implies that in addition to the direct electron–photon interaction at a space-time point x, the theory describes the emission and absorption of millions and trillions of gravitons through terms like $(\kappa\phi)^{n}/n!$ in the exponential factor $\exp(\kappa\phi)$. It is this 'atmosphere' of virtual gravitons, into which the theory immerses electrons and photons, which is the decisive element in the infinity suppression. Before we go on, one remark about the quantization procedure we have adopted. Professional relativists – some of the most illustrious of them are present here today – feel somewhat embarrassed about Feynman quantization. They suspect that, starting with the smooth boundary condition $L^{\mu a} \approx \eta^{\mu a}$, one has unduly restricted the theory and the type of space-time manifold in which one can operate. For example, they would feel that one would never, in this manner, operate in a Schwarzschild space-time manifold with its characteristic singularities.

My personal feeling is that the professional relativist ignores two important insights from the particle physicist's experience. These are:

(1) The power of the analytic continuation method.

(2) The possibility of *renormalizing* the bare Minkowskian metric $\eta^{\mu\nu}$ to the *physical* metric $g^{\mu\nu}$ – so beautifully demostrated in Thirring's classic paper through the inevitable renormalization in general relativity of *lengths* and *time intervals*.

I am labouring this point because one of the aims of this meeting is to bring closer together the general relativist's and particle-theorist's points of view. To exhibit the power of analytic continuation techniques of the particle physicist, consider the following problem. Reconstruct the *exact* Schwarzschild solution around a source-singularity, starting with quantized theory of gravitons of Feynman. In graphical terms, consider *all* (tree) graphs of the following variety:

In the static limit, the first graph gives a contribution proportional to $2MG_N/r$ where M is the mass of the static source.

In the next approximation the graphs contribute $+\cdots(2MG/r)^2$ and so on.

The challenge – and I took a bet on this last year with Carter and Penrose – is this. Can we take the exact series with *all* graphs of the above variety – with millions and trillions of gravitons exchanged – sum the series, continue the sum beyond its radius of convergence *and* recover the *exact* Scharzschild solution, manifesting all its customary space-time singularities. If we succeed, we would have recovered the singular Schwarzschild manifold, though we started our quantization procedure with a non-singular Minkowskian one. For the particle physicist, the situation is very familiar. It is similar to summing perturbation graphs for scattering to give the Bethe–Salpeter equation and then using the equation (by continuing in the energy variable) to get bound states.

The relativistic problem was considered by M. Duff for his Ph.D. thesis at Imperial College last year. *And Duff has shown that the analytic continuation technique does indeed work.* Starting with Feynman's quantization built onto a non-singular Minkowskian manifold, one can indeed recover the Schwarzschild singular manifold, *provided that the contribution of all gravitons is taken into account* in the calculation and *no approximation made.* Our quantization procedure starting with the 'bare'

Minkowskian metric ensures that all metrics with asymptotic flatness will emerge after appropriate analytic continuations.

Let us now turn back to the Lorentz problem of computing δm and δe, taking into account the *relevant* contributions of the sea of gravitons in which the theory (and physics) places electrons and photons.

As I said earlier, all infinities in field theory come about as a result of the consonance of singularities of propagators like $(-1/x^2)$ at $x=0$. In the conventional Maxwell–Dirac theory with its Lagrangian $e_0 \bar{\psi} \gamma_\mu \psi A_\mu$ the photon propagator $(A_\mu A_\nu)$ equals $\delta_{\mu\nu}/x^2$ the electron propagator $(\bar{\psi}\psi)$ equals $\gamma\partial(1/x^2)+(m/x^2)+$terms which are less singular, while the contribution of the second-order graph

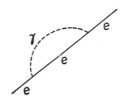

(where an electron emits and reabsorbs a photon) is essentially given by:

$$\delta m \propto e_0 \int d^4_x \frac{1}{x^2} (\gamma\partial + m) \frac{1}{x^2}.$$

Since

$$\int \frac{1}{x^2} \gamma\partial \frac{1}{x^2} d^4_x \equiv 0,$$

therefore

$$\frac{\delta m}{m} \propto e_0^2 \int \frac{d^4_x}{(x^2)^2}.$$

This integral is logarithmically infinite at the lower limit of integration (at small distances $x=0$, or from the uncertainty relation, for large energies).

Consider now what happens when the quantum ocean of gravitons is taken into account. Since

$$\langle\phi\phi\rangle = -\frac{1}{x^2}, \qquad \langle\phi^n\phi^n\rangle = n!\left(-\frac{1}{x^2}\right)^n,$$

we obtain for the propagator,

$$\langle e^{\kappa\phi} e^{\kappa\phi}\rangle = \sum_{n=0}^{\infty} \frac{1}{n!}\left(-\frac{\kappa^2}{x^2}\right)^n = \exp\left(-\frac{\kappa^2}{x^2}\right).$$

The important point about the factor $\exp(-\kappa^2/x^2)$ is that if we approach x^2 from the appropriate (time-like) direction and fill in for the other directions by an appropriate analytic continuation, the factor $\exp(-\kappa^2/x^2)$ tends to zero fantastically smoothly.

For the electron self-mass we now obtain

$$\frac{\delta m}{m} \approx e_0^2 \int \left(\frac{1}{x^2}\right)^2 e^{-\kappa^2/x^2} \, \mathrm{d}_x^4.$$

Graphically we are summing the series of graphs:

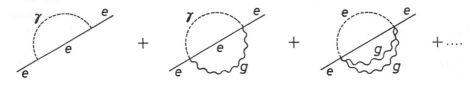

no graviton   +   one graviton   +   two gravitons + ...

This summation over one, two, three,..., gravitons, has thus given us a smoothing function $\exp(-\kappa^2/x^2)$ for the old infinity. The integral can be easily evaluated and is given in standard texts. We obtain:

$$\frac{\delta m}{m} \approx e_0^2 \left[\log(\kappa^2 m_e^2) + \kappa^2 \log(\kappa^2 m_e^2) + \cdots\right].$$

Writing,

$$R = 2G_N m_e = 2\kappa^2 m_e$$

when (R is the Schwarzschild radius of the electron) we can re-express

$$\frac{\delta m}{m} = e_0^2 \log(R m_e).$$

Note that when $R \to 0$ (i.e. when quantum gravity is ignored and $G_N \to 0$), we recover the old logarithmic infinity for δm. If gravity is not ignored, the theory has an inbuilt *cut-off at the Schwarzschild radius of the electron*. It would seem that photons of wavelength smaller than the Schwarzschild radius are simply unable to affect the electron. Their emission and reabsorption by the electron – which is what was causing the infinity at the zero end of the scale $x=0$ in $\int \mathrm{d}^4 x/x^4$ – is inhibited by the peculiar characteristics of space-time inside the Schwarzschild radius R.

Now comes an amazing numerical circumstance. Introducing all the appropriate factors of 2π, our result is

$$\frac{\delta m}{m} = \frac{3}{4\pi} \frac{e^2}{4\pi} \log \frac{1}{G_N m_e^2} + \text{terms in higher orders}.$$

The very fact that $G_N m_e^2$ is such a tiny number means that $\log(1/G_N m_e^2)$ is not small. In fact $\alpha \log(1/G_N m_e^2) \simeq \frac{105}{137}$ is a magnitude of the order of unity, so that $\delta m/m \simeq \frac{2}{11}$. The lowest-order approximation for $\delta m/m$ gives us a sensible result. The gravity correction to the electron's self-mass is not tiny, and since in the next approximation (with *two* photons and trillions of gravitons exchanged) we may expect a contribution

$\approx |\alpha \log G_N m_e^2|^2$ and so on for the third, fourth and other higher-order approximations in α – it is fully conceivable that the sum of the series for $\delta m/m$ is even nearer unity then the first approximation indicates. That is to say, gravity-modified quantum electrodynamics is likely – in a most natural manner, to fulfil Lorentz's conjecture – that *all* mass of the electron is due to its (gravity-modified) electromagnetic self-interactions ($m = \delta m$) and that the bare mass of the electron is indeed zero ($m_0 = 0$).

We can compute likewise the self-charge $\delta e/e$. This is also finite in a gravity-containing Dirac–Maxwell–Einstein theory. In this work it was important to check the electromagnetic gauge invariance of the theory. This has been done (see Isham, Strathdee and Salam, *Physical Review*, 1971). One can also show that the civilizing effect of gravity in solving the Lorentz problem is a peculiarity of tensor-gravity (spin-2^+ gravitons) and *not* scalar-gravity of Brans–Dicke variety. Since the Schwarzschild singularity of the spacetime manifold is also a peculiarity of Einstein's tensor theory, this supports the physical argument I gave – of the inter-relation of the electron's Schwarzschild radius with infinity suppression.

To conclude this part of my talk, quantum electrodynamics in its simple form is infinite because there was some missing physics – the physics of quantum gravity. Once that missing physics is supplied, the infinities are regularized in a natural manner. The amazing numerical circumstance is that the numerical value for δm, in the lowest order of the calculation accords sensibly with the conjecture of Lorentz that all inertia of the electron is due to its self-interactions.

One may in fact have started with Lorentz's conjecture in reverse, i.e., we could compute $|\alpha \log G_N m_e^2|$ from the eigenvalue equation $m_0 = 0$, i.e.,

$$\frac{\delta m}{m}(\alpha \log G_N m_e^2) = 1 .$$

One root of this equation (if the lowest order calculation is any guide) is

$$\left| \alpha \log G_N m_e^2 \right| \approx \frac{105}{137}.$$

Thus, given $\alpha = \frac{1}{137}$, we would know why the Newtonian constant $G_N m_e^2$ is so small and of the order 10^{-44} and vice versa.

4. GAUGE MESONS AND SPONTANEOUS BREAKING OF SYMMETRIES

I turn now to the second major development in the conventional polynomial theories, forgetting about the gravity modifications for the present. In 1954 Yang and Mills (and independently my pupil R. Shaw for his Ph.D. thesis at Cambridge) invented gauge meson theories corresponding to the internal isopic-spin symmetry $SU(2)$. The pattern for these theories was provided by electrodynamics – the theory of the gauge particle (the photon) corresponding to the charge-conserving internal symmetry $U(1)$.

Yang and Mills showed that for an *exact* gauge symmetry, these gauge particles must be *massless*, and their Yukawa couplings are *renormalizable*.

Now in nature, gauge particles do exist; the $(\varrho, \kappa^*, \omega, \phi)$ nonet exists corresponding to the $U(3)$ symmetry and also appears to possess Yukawa couplings with nucleons characteristic of gauge particles. However, the particles are massive and also the $U(3)$ symmetry, to which the particles are supposed to correspond, is not *exact*. Either of the two circumstances – massiveness of the gauge mesons, or the broken character of the symmetry – is enough to make the theory *non-renormalizable* in general.

I said, 'non-renormalizable in general'. This qualification 'in general' is extremely important. For *arbitrary* values of gauge meson masses and arbitrary values of the symmetry-breaking parameters the theory is indeed non-renormalizable. Could *renormalizability*, however, be restored, for *special* values of masses and a *special form of symmetry-breaking*?

The answer, developed since 1963, through the work of Higgs, Kibble, Guralnik and Hagen, and finally clearly established during 1971 by G. 't Hooft, appears to be YES. I shall state it in the form of a theorem.

THEOREM: A Yang–Mills theory of gauge mesons interacting among themselves and with spin-$\frac{1}{2}$ fermions is *renormalizable* provided:

(1) The symmetry to which the mesons correspond is broken through Heisenberg's spontaneous symmetry-breaking mechanism.

(2) Provided this symmetry breaking is introduced through a multiplet of spin-zero mesons, with non-zero expectation values (ϕ).

(3) In this event, the masses of gauge particles as well as the Fermi particles are not arbitrary but are prescribed in terms of (ϕ).

(4) If the Fermi particles are coupled through *axial*-vector gauges in addition to the vector-gauge particles, an axial-charge doubling of fermions is necessary in order to ensure renormalizability. To understand this theorem, let us introduce the concept of spontaneous symmetry breaking first.

Turn to the Dirac–Maxwell theory we considered in the first part of the lecture:

$$\mathcal{L} = \bar{\psi}\gamma\partial\psi + e_0\bar{\psi}\gamma_\mu\psi A_\mu + m_0\bar{\psi}\psi.$$

If $m_0 = 0$ (zero bare mass), the theory possesses an additional symmetry – the so-called 'neutrino' symmetry (γ_5-invariance), specifically the transformation $\psi \rightarrow (\exp i\gamma_s\theta)\psi$ leaves the Lagrangian form invariant. The bare mass term breaks this summetry.

QUESTION: Can we start with *zero* bare mass $m_0 = 0$ but compute self-consistently from within the theory a *non-zero* physical mass $m \neq 0 (m = m_0 + \delta m)$? The answer is yes. In the first part of my talk I did precisely that; I set up an eigenvalue equation for $\delta m_e/m_e$:

$$\frac{\delta m_e}{m_e} = \frac{m_e - m_0}{m_e} = \alpha \log G_N m_e^2.$$

The form of this equation is always such (from Dyson's theorem) that we can set in it $m_0 = 0$. For the special values of m_e, α and G related through $\alpha \log G_N m_e^2 = 1$, we can solve and obtain a non-zero *physical mass* m_e.

Does the γ_5-symmetry survive in the theory? In the Lagrangian, yes, since $m_0 = 0$ – but not for the physical states, since the Hilbert space of physical particles describes electrons with *mass*. The γ_5-symmetry has been '*spontaneously* and *self-consistently broken*' within the theory.

Now consider this as a Yang–Mills theory corresponding to the symmetry group $U(1)$, generated by the γ_5-transformations $\psi \to e^{i\gamma_5\theta}\psi$. The theory will now be supplemented with an (axial-vector) gauge meson Z with the interaction $ie_0\bar{\psi}\gamma_\mu\gamma_5\psi Z_\mu$. Now following Anderson, Higgs in 1963 proved a very important theorem about Z. This states that the theory *will* describe in its spectrum a spin-zero bound-state particle with an associated field ϕ, with a definite non-zero expectation value $\langle\phi\rangle$. And the coupling (f) of this field to electrons and its expectation value $\langle\phi\rangle$ *will* also be related to m thus:

$$f\langle\phi\rangle = m.$$

while the Z field will have mass $= e \langle\phi\rangle$.

To restate all the ramifications of this closely knit theorem, let me summarize it again:

Given a $U(1)$ γ_5-symmetry, one can set up a Yang–Mills gauge theory. If we introduce a spontaneous symmetry breaking, the theory *will* describe a spin-zero (bound) state with a non-zero expectation value and a definite coupling to the Fermi particles. Also the mass of the gauge particle will not be arbitrary but a definite multiple of the non-zero expectation value. *Finally* this tightness of relations between the masses and the couplings of the spin-zero, spin-$\frac{1}{2}$ and spin-1 particles has the consequence that the complete theory is *renormalizable*.

TABLE II

(a) *Situation I*
  Exact γ_5-symmetry $\to U(1)$ group
  $\quad m_e = 0$
  $\quad m_z = 0 \quad$ Theory renormalizable but not physical

(b) *Situation II*
  Spontaneously broken γ_5-symmetry:
  A. $m_e \neq 0 \to 1 = e^2/4\pi \log G_N m_e^2$ ⎱
  $\quad\; m_z \neq 0 \to m_z = e\langle\phi\rangle$ ⎰
  B. Theory must contain a scalar particle ϕ with $f\langle\phi\rangle = m_e$
  C. There must exist an *additional** heavy lepton to secure *renormalizability* with opposite γ_5-coupling to Z

---

* I have no time to go into this rather technical point, associated with the so-called Schwinger–Bell–Jackiw–Adler anomaly. It appears that the demand of *renormalizability* cannot be met whenever axial-vector gauges are present unless there is a symmetry between sets of fermions possessing equal and opposite γ_5-couplings to the axial-gauge particles. The masses of these extra fermions (they need not be leptons) are not restricted by the theory, so far as we know.

In converse, the demand of renormalizability of a gauge theory can be met only if there is a definite set of scalar particles in the theory with non-zero expectation values, and with definite relations between the masses of the gauge bosons and these non-zero expectation values.

5. A RENORMALIZABLE UNIFIED THEORY OF WEAK AND ELECTROMAGNETIC INTERACTIONS

One rather attractive instance of the application of ideas I have mentioned is that of a unified theory of weak and electromagnetic interactions. That both weak and electromagnetic interactions may in fact constitute different aspects of one symmetry scheme is a conjecture made long ago – to my knowledge, first by Julian Schwinger (1957). The conjecture was revived by Glashow (1959, 1961) and Salam and Ward (1959, 1964), who used a $U(1) \times SU(2)$ gauge group to consider four Yang–Mills-type gauge mesons, three of which would mediate weak interactions and one would be the photon. Recently Weinberg (1967) and Salam (Nobel Symposium, Gothenburg, 1968) applied the ideas of spontaneous symmetry breaking to give the three weak gauge bosons their masses. Weinberg and Salam conjectured that the resulting theory may be renormalizable. The conjecture has now been proved by 't Hooft.

Specifically, the model works with the $U(1) \times SU(2)$ gauge symmetry, with four gauge particles, $W^{\pm}$, Z^0 (mediating weak interactions) and the photon A^0. One implication of the symmetry scheme is that the weak coupling constant must equal the electromagnetic. The observed smallness of the effective weak constant G_F is then attributed to the large mass of $W^{\pm}$ particles; specifically we must have $e^2 = G_F/m_w^2$.

Now in accordance with the Higgs–Kibble theorem, the emergence of these masses from a spontaneous symmetry breaking mechanism carries with it the following implications:

(1) There must exist in the theory a scalar bound state, with an associated field ϕ, whose expectation value $\langle\phi\rangle \neq 0$.

(2) In order that the theory be renormalizable there must exist an additional heavy electron and a heavy muon with equal and opposite axial couplings to $W^{\pm}$ and Z particles.

(3) The masses of W and Z particles are given by the relations like

$$m_w = e\langle\phi\rangle,$$
$$m_z = \sqrt{2}\, e\langle\phi\rangle,$$

while $m_{\text{electron}} = f\langle\phi\rangle$.

Here f is the direct coupling of the ϕ-particle to the electron.

(4) If the theory is extended to hadrons, either the hadronic quarks should also be doubled in number (with equal and opposite axial charges) or the symmetry group of strong interaction physics should be extended from $SU(3)$ to at least $SU(4)$, if not $SU(5)$ or $SU(3) \times SU(3)$. This would of course have profound consequences for spectroscopy in hadronic physics.

Summarizing then, the renormalizability demand appears to lead to new and exciting physics. It is exciting physics in two directions. Firstly, it would appear that there is but one scheme of coupling constants and masses. There is no arbitrariness in assigning these parameters; renormalizability works for special values of these parameters and not otherwise. Secondly, there seem important restrictions on the types of internal symmetry schemes the physical particles belong to.

6. FROM RENORMALIZABLE TO FINITE THEORIES

I started this lecture by considering the prototype of all renormalizable theory – quantum electrodynamics. By considering its natural non-polynomial modification, when quantum gravitational interactions of electrons and photons are taken into account, we were able to solve the Lorentz problem – i.e. render the few remaining infinties of the theory also finite. What are the prospects of gravity rendering the renormalizable weak and strong interaction theories I have been speaking about *also finite*? After all, gravity affects protons and neutrons just as much as it affects electrons and muons.

Now gravity does affect hadrons just as much as it affects leptons but there appear to be indications that, in formulation of gravity theory for hadrons, one cannot follow Einstein blindly.

To illustrate what I mean, consider the analogous case of electrodynamics. Up to 1961, we believed that the electromagnetic interaction of protons is identical in form to the electromagnetic interaction of electrons. One believed that one would write a Dirac equation

$$(\gamma(\partial + ieA) + m_p)\psi_p = 0$$

for a proton completely similar to the corresponding equation for the electron:

$$(\gamma(\partial + ieA) + m_e)\psi_e = 0.$$

We now know this would be wrong.

Our present picture of the electromagnetic interaction of protons is a two-stage picture. Among the nonet of strongly interacting gauge particles ($\varrho, \omega \phi, \kappa^*$), there is one particle, ϱ^0, which has identical quantum numbers to the photon γ.

In quantum theory, whenever we have two particles with the same quantum numbers they *must* interconvert; there must be transitions between them. Our picture of the proton's electromagnetic interaction is therefore this. The proton emits and reabsorbs ϱ^0; the ϱ^0 then has a finite amplitude for conversion into a photon γ^0 – which in its turn can be emitted and absorbed by the electron or the muon. There are thus two distinct worlds – the world of the leptons, directly interacting with photons, and the world of the hadrons directly interacting with the gauge nonet ($\varrho, \phi, \omega, \kappa^*$) among which is contained what may be called the heavy photon ϱ^0. The two worlds communicate electromagnetically through an interconversion of ϱ^0's to γ's. The picture I have outlined above is the one borne out by numerous experiments and is generally accepted.

Now the same situation appears to hold for gravity. Among the recently discovered mesons is the nonet of spin-2^+ which includes particles like f, f', A_2, etc. The f^0 particle in particular possesses the special feature that all its quantum numbers are identical with those of the graviton. There is nothing on earth which can stop the interconversion of gravitons and f-mesons.

Isham, Strathdee and Salam have postulated the same equation as Einstein's for the f-particles. In this theory the f-meson *is* the graviton of hadronic physics – just as the ϱ^0 was its 'photon'. The f-meson possesses a universal coupling to matter, of the same non-polynomial variety as the graviton does. The gravitational force between hadrons and leptons or hadrons and hadrons proceeds through the mechanism of interconversion of f's to Einstein's gravitons, the latter in their turn coupling directly to leptons. Pictorially the respective interactions look like this:

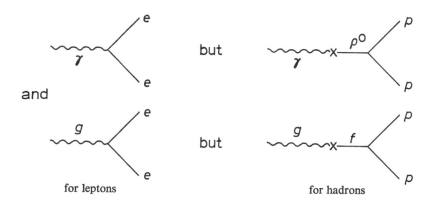

Now what is the effect of this interposition of the f-particle in hadronic physics? You may recall that the effective cut-off (mass)^2 (from the non-polynomiality) of gravity theory was proportional to the inverse of the gravitational coupling constant $1/G_N$. For f-gravity, the cut-off will come much earlier; it will come at $1/G_f$ where empirically $G_f m_N \simeq 1$.

This means that the cut-off (mass)^2 in strong interaction physics is around 1 BeV and not around 10^{19} BeV. This is precisely what is observed – the matrix elements in strong interaction physics do get damped around 1 BeV.

In summary then, the final finite theory of strong interactions may start as a re-normalizable gauge theory with a spontaneous symmetry-breaking mechanism built into it to accord with the physically observed symmetry breaking. The final finiteness would be secured by considering the intrinsic non-poly-nomiality superimposed on the theory by the universal f-gravitational interaction. We must always start with a renormalizable theory before considering the effects of f-gravity, otherwise, according to some recent work of Lehmann, we are likely to run into ambiguities in the final numbers we will compute.

I have tried to convey to you some of the excitement of this field, its liveliness, its promise – particularly in leading us on to new and physically relevant physics. There are of course hosts of unsolved problems – not the least among them being, 'does quantum gravity quench its own singularities?' And if so does this include also the singularities of spacetime manifold about which we have heard such a lot at this conference? I should like to conclude with Oppenheimer:

'If what we have learned so far … is radical and unfamiliar and a lesson that we are not likely to forget, we think that the future will be only more radical and not less, only more strange and not more familiar, and that it will have its own new insights for the inquiring human spirit.'

REFERENCES

[For the references cited in this article, see C. J. Isham, A. Salam, and J. Strathdee, *Phys. Rev.* D, **5**, 2548 (1972).]

DISCUSSION

W. Heisenberg: I would like to ask Salam concerning the relevance of gravitation to the spectrum of elementary particles. I mean it in the following sense. You mentioned at the beginning of your talk that you distinguish between two kinds of quantities: quantities like the Lamb-shift, which are already finite before you really have introduced something like gravitation, and others which are not. Now I assume a theory where the Lagrangian intends to represent the spectrum of elementary particles, of strong interaction, and assume further that this theory is renormalizable. Then I would also expect from your theory that, for instance, the ratio say between the proton mass and the gravitational constant would somehow be determined. Did you not say that the ratio between different masses is of the Lamb-shift type, that it's really not influenced by the point of the cut-off, only to say that the spectrum actually does depend on the gravitation and the cut-off? In that sense, it is not calculable like the Lamb-shift. What is your opinion on this point?

Abdus Salam: I agree with Prof. Heisenberg. However, when I deal with strong interactions I would like to talk of strong gravity, the non-polynomial theory of the *f*-mesons written with the Einstein Lagrangian. One thing which I did not mention is that when we tried to put $SU(3)$ into the Einstein Lagrangian, to our amazement we discovered that we could not just put $SU(3)$ in, but we had to buy $SL(6)$ C invariance, so that the origin of $SU(6)$ may very well lie in the Einstein-like Lagrangians of spin 2.

J. Ehlers: I would like to ask a question with respect to the importance of the non-polynomial nature of the Lagrangian. It has been very much stressed that the non-polynomial nature is important and that the gravitational Lagrangian just has this, from your point of view, desirable property. Now it is known that one can also find quite natural variables at least within the domain of classical general relativity, where the Lagrangian is polynomial. Could you comment on that?

Abdus Salam: I am glad that you asked this question, because one can easily deceive oneself. I want to give an example which is very familiar to particle physicists. Consider $g\bar{\psi}\gamma_5\gamma_\mu \, \partial_\mu\phi\psi$ which looks polynomial. It is very well known that the derivatives which occur in the ϕ field have the property of really converting it into a non-polynomial Lagrangian $\bar{\psi}(\exp\gamma_5\phi)\psi$. Although you can write down the Lagrangian of gravity in a seemingly polynomial form, when you carefully examine the graphs which I have drawn, it turns out that the property, which is important for me, of millions of particles coming out from one space-point, comes back because of the derivatives.

F. Rohrlich: You mentioned that renormalizability would play an important role in characterizing physical interactions, as to what is possible and what is not possible. Now we know that renormalizability is understood at present only within the context of perturbation expansion, so that if we had a different approximation method than perturbation expansion all theories could be renormalized without difficulty. What you are saying therefore is that there seems to be something physical in the accidental usage of perturbation expansion.

Abdus Salam: I started my discussion with the work of Jaffe and Glimm, which considers an exact solution of a polynomial field theory in two and three dimensions. The exact solution has the same infinities as the perturbation solution. So a perturbation solution is not deceiving us as to the question of infinities. I believe it is the polynomial character of the theory which is to blame for these.

R. Peierls: May I ask a related question to this? I may have misunderstood your principle, but as I understood it, you asked first of all how to make the theory renormalizable, and you prefer theories with such structures and parameters that they can become renormalizable. That, in general, still leaves you some infinite constants. Then by bringing in the presence of nonpolynomial interactions you get rid of those infinities. Question: If your theory is not renormalizable to start with, would you not also remove the infinities by virtue of the presence of the non-polynomial parts, and therefore does any argument remain for preferring the renormalizable theories?

Abdus Salam: I noticed that if I have a non-polynomial theory, I do get finite results, although there remain a number of ambiguous parameters. If I start with a renormalizable theory, I believe that those ambiguous parameters would also disappear. It is a question of what Lehmann calls the Minimality Ansatz, his minimal solution which provides for the uniqueness in the non-polynomial case. How far does it apply? Do you have to start with a seemingly renormalizable structure or not?

22

Some Concepts in Current Elementary Particle Physics

Chen Ning Yang

Many years ago, when I was still a schoolboy in China, I had the unforgettable experience of reading translations of the books by Eddington and by Jeans about the new developments in physics. They described the various conceptual revolutions in 20th century physics, starting from the special theory of relativity, leading to the general theory and to quantum mechanics. I cannot say that I understood the meaning and the necessity of the Fitzgerald contraction, the Bohr atom, or the uncertainty principle, but it was impossible for me not to catch the excitement and the enthusiasm so vividly overflowing from the pages. It was clear that here was a new gate opening onto the mystery of the structure of the universe, at once full of light and darkness, enlightenment and puzzles. How much this fascination influenced my later choice of a career I could not say, because I do not know. But even today as I recall my experience so many years ago I could still feel the mysterious excitement that had then overwhelmed me.

I tell you this because I want to say how much it was a privilege for me to hear recounted last evening the personal experiences, throughout these revolutionary periods in the development of our concept of nature, of one of the prime architects of the whole enterprise. I am grateful to him and to the organizers of this conference who have made this possible.

In his talk on the physicist's conception of nature, Dirac urged us, on more than one occasion, to abandon past prejudices. Was he urging us to abandon our traditional concepts of space-time? Was he urging us to change our views about the concept of the field? Or was he urging us to introduce some new symmetries or to abandon some old ones? He did not provide us with any definite answers to these questions. But we know that in order to search for clues for new conceptual developments of physics there are two important guides. On the one hand, we must be always rooted in new experimental findings. Detached from this root physics runs into the danger of degenerating into mathematical exercises. On the other hand, we must not be shackled all the time by the desire to fit what at each moment is accepted as experimental reality. Extrapolations based on pure logic and form are essential ingredients in many great conceptual advances in our field. Perhaps in no other physicist's work is this point more clearly exhibited than that of Dirac. In all of his work, the less important as well as the more important ones, there is that insistence on elegance of form and beauty of logic that gives his papers a unique creative flavour. Dirac himself has said that

J. Mehra (ed.), The Physicist's Conception of Nature, 447-453. All Rights Reserved
Copyright © 1973 by D. Reidel Publishing Company, Dordrecht-Holland

beauty is the only requirement. If experiments contradict a beautiful idea, let us forget about experiments.

Surely that was the feeling of Einstein when he wrote after the creation of the general theory of relativity that the theory did not need experimental proof. Surely that was the feeling of Dirac when, confronted with the necessity to explain away the negative energy states, he dreamed of the crazy idea[1] of the infinite sea with holes in it.

The dichotomy of the two not quite consistent guides for conceptual development of physics discussed above is, so to say, graphically illustrated for me in a comparison of this symposium with another meeting that I just came from: the High Energy Physics conference at Batavia, near Chicago. On the surface, there is a minimum overlap of participants in the two conferences. There is a minimum overlap in the languages used. There is a minimum overlap of those topics that seem to generate the most interest. However, if one sits back and takes a long-range view and forgets about the hustle and bustle that provides the ambient atmosphere in the new laboratory in Batavia (near Chicago), if one looks away from the corridor discussions of mysterious new high transverse momentum events, if one forgets about the new preliminary data on the decay of kaons into leptons, if one takes a longer range assessment, it will become clear that above the noise level of detailed developments there is, in fact, quite a bit of overlap in interests between the two conferences. It is in this spirit of looking at the general concepts that I should like now to discuss some developments in physics that have been under extensive discussion in the last ten years or so.

In my opinion the greatest excitement in physics during the past decade was the discovery[2] of a very weak violation of CP invariance in 1954 by Christenson, Cronin, Fitch and Turley. The precise meaning of this discovery could be understood from the following description[3] of the meaning of the operators C (charge conjugation), P (parity, or inversion), and T (time reversal operator). Consider reaction R depicted in Fig. 1 for the process

$$A + B \rightarrow C + D.$$

Given this reaction one can conceive of the reaction $(P)R$, the reaction obtained from R by reflection in the plane of the paper. If R and $(P)R$ take place at equal rates, then we say that the laws governing R and $(P)R$ obey right-left symmetry, or P invariance holds for these reactions. Similarly for C invariance and T invariance.

In 1956 it was discovered that for weak interactions P invariance is violated and C invariance is violated. It was then thought that the combined invariance CP would hold for all weak interactions. Great excitement came in 1964 with the discovery that this is not true in the decay of K^0, and this discovery led to beautiful experiments.

The symmetries C, P and T are related by a theorem[4] discovered by Lüders, Pauli and Schwinger in the early fifties. The theorem states that under very weak general assumptions, which are accepted by essentially every physicist as valid, the combined CPT invariance holds, whether or not the individual invariances C, P or T holds or not. Because of this theorem, violation of CP invariance implies violation of time reversal invariance.

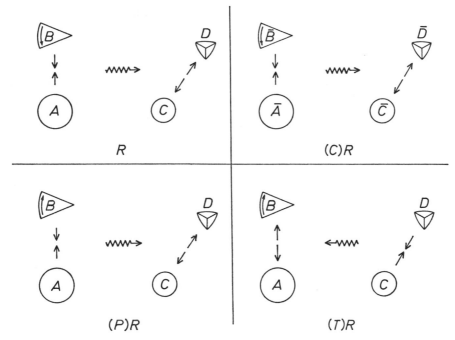

Fig. 1.   The operations C, P, and T. The reaction R describes $A + B \rightarrow C + D$ where A and C are spheres, B a spinning cone, and D a tetrahedron. The charge conjugate reaction $(C)R$ has the same kinematics as R, but with all particles replaced by antiparticles. $(P)R$ is the reaction R reflected in the plane of the drawing. Notice the reversed spin of B and the reflected tetrahedron D. $(T)R$ is the time-reversed reaction of R, describing $C + D \rightarrow A + B$. Notice the spin direction of B.

   The concepts of P invariance and T invariance were rooted originally in geometrical considerations. Such was not the case for C invariance. It was Dirac's idea of particles and holes that gave rise to C invariance. As was evident from his original paper, the hole idea is very much related to the mathematical concept of complex conjugation. For this reason, C invariance is referred to as being an algebraic concept. The introduction of complex numbers as an essential element of the algebra of physical laws is one of the new concepts brought into physics by quantum mechanics. Since that led to the invariance C, we could ask: Are there other algebras that should be brought into physics. Are there other symmetries that may be due to more complicated algebras.
   It is more than likely that the answers to both these questions are in the affirmative. On the one hand, from the side of mathematical concepts, there are exactly three associative division algebras (i.e. algebras in which addition, subtraction, multiplication and division rules exist satisfying the usual distributive and associative laws, which is commutative in addition and subtraction, and for which $ab = 0$ implies $a = 0$ or $b = 0$) that contain real numbers as a subalgebra. These are real numbers, complex numbers and quaternions. (Of these the quaternions have not yet been introduced in any essential way into the laws of physics.) We note here that the generalization of the concept of complex conjugation (i.e. $i \leftrightarrow -i$ in complex number algebra) to quaternion

algebra is the operation of a rotation between the three imaginary axes i, j and k. There are many such operations forming a group SO_3.

On the other hand, from the side of physical concepts there is the hitherto only phenomenologically based SU_2 symmetry which is very accurately observed by strong interactions. Fundamentally, this symmetry describes the striking similarity between the proton and the neutron. It has the mathematical structure of a three-dimensional rotational symmetry. That turns out to be exactly the structure of the group mentioned above which is the mathematical generalization of the concept of complex conjugation. All these suggest the possibility illustrated below of:

	mathematical operation	
algebra	leaving algebra unchanged	physical symmetry
complex numbers	complex conjugation	C
quaternions	rotation of i, j, k	SU_2

SU_2 being an algebraic symmetry resulting from the incorporation of quaternions into the laws of physics. There have been discussions of this possibility[3,5] before, but all efforts in this direction have so far remained fruitless. Nevertheless, I venture to guess that nature has indeed utilized this possibility, only physicists have not yet found the right key to the incorporation of quaternions into fundamental physics so as to explain the similarity between the proton and the neutron.

Another important symmetry, discussed extensively in the sixties, is SU_3. It was found that there are amazing systematics of particle multiplets that make SU_3 a very fruitful concept even though it is badly broken. My personal prejudice is that if we finally should have an understanding of SU_3, it will turn out to be an understanding on quite a different footing from that of SU_2.

It is interesting to ask whether SU_3 could also originate from an algebra. If we investigate the next well-structured algebra beyond the quaternions it is the octonians. (In octonian algebra one has seven square roots of -1, independent of each other by real linear transformations. It is the only division algebra besides the real numbers, complex numbers and quaternion that is an 'alternate' algebra. An alternate algebra is one in which $(xy)z - x(yz)$ is antisymmetrical in x, y and z. It follows from this definition that in an alternate algebra x^n is independent of the order of association. That is, an alternate algebra is always associative for the powers of any element. A division algebra is one in which $xy = 0$ implies $x = 0$ or $y = 0$). The group of real linear transformations that leave the octonian algebra invariant is $G2$, the first exceptional Lie group. Unfortunately $G2$ is not SU_3.

If we could turn our attention away from symmetries for a while we shall see that in the last five or six years hadron physics has been undergoing very interesting developments. The tempo of activities in this field has greatly increased, thanks to the large number of experiments on elastic and inelastic reactions studied at the high energy accelerators. While a description of a high energy hadron–hadron collision is necessarily complicated because of the large number of particles frequently produced and because of the large fluctuations in such quantities like the multiplicity of outgoing particles, the amazing thing is that the experimental results also exhibited very striking general regularities, regularities that, in my opinion, point to a direction of approach that may provide a physical description of hadronic structure. In this description hadrons go *through* each other in a high energy collision with little exchanges of quantum numbers or longitudinal momentum or energy in the centre of mass reference system. They are however excited coherently into dynamical excited states which then fragment into various outgoing particles, forming two jets. (A dynamic excited state is a superposition of states with various invariant mass values. The part consisting of the lowest invariant mass values approaches a limiting state at relatively low incoming energies. The part consisting of high invariant mass values approaches a limiting state at higher incoming energies.) In this process of going through each other, the geometrical size of the incoming particles plays an essential role, and we believe that all hadrons are approximately of the size 1.4×10^{-13} cm in diameter.

The general description outlined above is in agreement with all main features of high energy processes in both elastic and inelastic collisions. For elastic collisions one obtains thus a parameter-less relationship[6] between elastic e–p scattering and elastic p–p scattering which led, among other things, to the prediction of dips in elastic p–p angular distribution, a prediction recently confirmed experimentally at CERN. For inelastic collisions the general description led to the hypothesis of limiting fragmentation[7] which has found extensive and accurate experimental support in recent years.

What determines the basic size of the hadrons? Why are they of the same order of magnitude? Why can a hadron not be divided into geometrically smaller hadrons like droplets that we have been familiar with in physics: water droplets and nuclear matter? These are among the questions that we do not know how to answer, but it seems rather clear that we are basically dealing with a system of infinite degrees of freedom in which there is a vacuum, and above the vacuum, energy-wise, there are various excited states (of this system of infinite degrees of freedom) which we call hadrons. It would seem to me that an urgently needed task for us is to understand the physics of such a strongly interacting system of infinite degrees of freedom.

This is a difficult task but also a very concrete and challenging one. Fortunately, our experience with other branches of physics that deal with systems of infinite degrees of freedom would help us in this task.

Could it be that the above-quoted size of hadrons is related to some fundamental changes of space-time structure at such distances? I believe the answer to this question is no. Electromagnetic interactions of leptons have been studied to much smaller spatial dimensions than 10^{-13} cm and no deviations of the geometrical structure of space-time have been found.

I come lastly to a concept called gauge fields. This is a concept which is basically derived, on the side of physics, from the idea that some fundamental symmetries of the physical world should be related to invariance concepts at *every* space-time point. In particular, let us take isotopic spin invariance which states that the proton and the neutron are similar. If the electromagnetic field is 'switched off', there would be two entirely similar states of the nucleon, and which we choose to call the proton and which the neutron would be an arbitrary convention. Now if we adopt the view that this arbitrary convention should be independently chosen at every space-time point, then we would be naturally led to the concept of gauge fields. (Another way of putting this is that if I adopt a convention it should not bind my colleague in the next laboratory to adopt any specific convention whatsoever.)

On the mathematical side, the concept of gauge fields apparently is related to fibre bundles. But I do not know really what a fibre bundle is.

The electromagnetic field is a gauge field. Einstein's gravitational theory is intimately related to the concept of gauge fields, although to *identify* the gravitational field as a gauge field is not an absolutely straightforward matter.

During the last 15 years there have been repeated efforts to introduce a gauge field as a source, or the source, of strong interactions. These efforts have not been entirely successful, but the idea is fundamentally attractive. In my opinion it is likely to play important roles in the future.

In the last five years, efforts originating from the work of S. Weinberg (and earlier work of Gürsey, Schwinger, Salam and Ward and others) have led to great excitement about the possibility of amalgamating the concepts of gauge fields, electromagnetic fields and weak interactions. (And, in some versions, also strong interactions.) My colleague at Stony Brook, B. W. Lee, has just given at the Chicago Conference a rapporteur's talk[8] summarizing the intensive activities in this field during the past two years. In addition to the idea of gauge fields, an important new idea added is the concept of quantization at non-zero expectation values of field quantities. Personally, I think that these recent developments are along an important direction, but perhaps some further new ideas are still missing so that the current efforts end in highly non-unique and non-beautiful theories.

REFERENCES

1.   The more 'crazy' an idea is, the more profound it becomes when it turns out to be relevant in the description of natural phenomena. My own feeling about Dirac's introduction of the negative sea was summarized in the following sentences in a talk at the 75th anniversary celebration of Bryn Mawr College, November 6, 1959:
'The concept of charge conjugate symmetry is a purely quantum mechanical concept and is not related to any geometrical concepts such as rotational invariance. It derives its origin from the Dirac theory of the electron, which in turn is, viewed today, a logical consequence of the fusion of the quantum theory with the requirement of relativistic invariance. To first postulate the charge conjugation concept, as Dirac did about thirty years ago, was, however, a most daring and profound step, not unlike the first introduction of the negative numbers. The later experimental verification of the existence of the antiparticles constituted not only one of the most beautiful and forceful demonstrations of the practical consequences of the symmetry principles, it represented actually one of the most gratifying and far-reaching triumphs of theoretical reasoning.'
2.   J. H. Christenson, J. W. Cronin, V. L. Fitch and R. Turlay, *Phys. Rev. Letters* **13**, 138 (1964).
3.   *Vistas in Research*, Vol. 3, Gordon and Breach (1968), Lecture by C. N. Yang, October 13, 1965.
4.   J. Schwinger, *Phys. Rev.* **91**, 720, 723 (1953); G. Lüders, *Kgl. Danske Vidensk. Selsk. Mat- fys.*

Medd. **28** (1954); W. Pauli's article in *Niels Bohr and the Development of Physics*, Pergamon, London (1955).

5. See my comments in the discussion period after Tiomno's talk, Session 9, *Proceedings of the 7th Rochester Conference*, 1957 (the Interscience Publishers).

6. T. T. Chou and Chen Ning Yang, *Phys. Rev.* **170**, 1591 (1968).

7. J. Benecke, T. T. Chou, C. N. Yang, and E. Yen, *Phys. Rev.* **188**, 2159 (1969).

8. B. W. Lee in *Proceedings of the International High Energy Conference at Chicago*, September 1972, to be published.

23

Crucial Experiments on Discrete Symmetries

V. L. Telegdi

If I have accepted the invitation to speak here, it was not out of vanity alone. I thought I was invited because somebody else did not come, and the invitation to come here reminded me of a story some of you may know. It goes as follows: There was a toastmaster in either England or the United States who was to introduce some very illustrious and very witty person, and that person was prevented from coming. So the toastmaster said 'Ladies and gentlemen, tonight it was to be my signal honour to introduce to you one of the most celebrated wits of our time. Unfortunately he couldn't come, so I brought two halfwits instead.'

This is a little bit my situation today. If you know me, you will understand that my burden is much heavier than that of other speakers to the extent that I not only have to describe some exciting physics to you, but that I am supposed to amuse you as well. I hope that I shall succeed in at least one of these two tasks.

Before going into the heart of the matter, I would like to give the reason why I agreed to give this talk, since everything about its subject matter has already been said by eminent people and it has been said very well; thus, except for the jokes I do not feel that I have much to add. The reason for accepting was that I was appalled by the extraordinarily high density of theorists at this meeting. The experimentalists constituted until the very end of the planning a set of measure zero, while I feel that physics, after all, is based on experiments. I have heard the statement that a theory is true because it is beautiful, but I do not think that that is the philosophy that I would like to propagate. I think that when you do an experiment that is right and which has some result which is interesting, then ultimately the theorists will provide a framework within which that result looks beautiful. So, as seen by the experimentalist, *beauty lies in getting correct results by economical means.* Experimental physics without theory would be zoology, but theoretical physics without experiments is speculation, and not natural philosophy in its true meaning. I consider it my mission to convince you that experimentatists do useful things. Incidentally, I was a little bit disturbed last night by one of my French friends who pointed out to me, while referring to some recent incorrect experiments, that experimentalists were actually not useful. This remark is the more astonishing as the gentleman in question is a specialist in the theory of measurement.

I shall now proceed and try to enumerate the symmetries. I shall speak only of the discrete space time symmetries P, C, and T and the crucial experiments relating to them.

J. Mehra (ed.), The Physicist's Conception of Nature, 454-480. All Rights Reserved
Copyright © 1973 by D. Reidel Publishing Company, Dordrecht-Holland

Dirac has said here that great progress has always been based on abolishing prejudice, and I do not personally know of any field of physics in my time where the contribution was so much in the line of abolishing prejudice as in this one. Now let me in a very vague sense define the symmetry operations with which you are anyway all more familiar than I, since the majority of you are theoreticians: Parity, P (space reversal), time reversal T (motion reversal), and charge conjugation C (replacing the world by the antiworld, being careful to mean by charge all charges and not only the electrical one; that would be an old-fashioned point of view, so please include baryon number, lepton number, etc.).

The spirit prior to 1956 was to assume that any force law or Lagrangian that you would write down had to exhibit invariance with respect to these three symmetries. Or, in more operational terms, there was to be no way to tell right from left by means of microscopic experiments, nor was there to be a way to say be means of microscopic experiments whether we are in the world or in the antiworld, and nor was one to be able to distinguish the label of time, i.e. to decide whether it is an arbitrary choice to call the future $t>0$ and the past $t<0$ or vice versa (just as much as everybody knows that by convention Caesar died before he was born). Positive or negative time was just to be a matter of book-keeping. So this was the prejudice in the face of the $\theta-\tau$ puzzle, which consisted in the decay of a single elementary particle of unique mass into either two pions or three pions. The hypothesis of the violation of these laws was only one of the possibilities that were explored by Lee and Yang. Their celebrated paper of 1956 discussed: (A) The fact that there was no evidence for P-invariance to hold in weak interactions; (B) The experiments one could do to verify whether parity conservation was indeed violated in weak interactions. Thus there were two steps: the giant step was to challenge the prejudice, and the useful step was to suggest crucial experiments; these I shall discuss a little later.

In order to prepare the groundwork for the rest of my talk, I would like to show you a bubble chamber picture (Fig. 1), in which you can see the interaction of a π^- (coming from above) with a proton yielding a Λ and a K^0 according to the first line of the scheme reproduced in Fig. 2. This is a very beautiful picture, taken by the Louis W. Alvarez' group with LRL 72-inch chamber. In order to understand my argument, you need a 700-foot chamber, but that is no problem at all for theorists. If you had such a very large bubble chamber, and you would follow out all the reaction products until they dribble down to ordinary stable matter, you would know practically everything (or almost) that we know today about weak interactions, and certainly everything that I wish to discuss today, and that is what matters. The reaction in Fig. 1 is a 'cheap' way of producing strange particles, since two ordinary ($S=0$) particles collide to produce two strange particles whose strangenesses ($S=1$ and -1) cancel. Thus the strangeness is conserved in the (strong) production act, but can and will be violated in the weak decays, actually with the rule $\Delta S=0$ or 1. This is the essence of the Gell-Mann–Nishijima scheme. Strange particles (e.g. Λ, K^0) are produced copiously, but they decay with life-times which are of the order of 10^{-10} times too slow to be due to a strong interaction. When we look at the Λ particle in the bubble

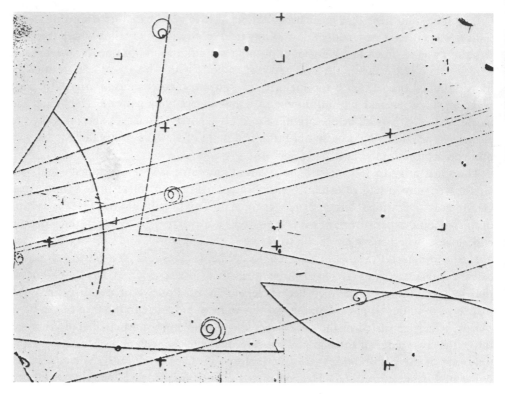

Fig. 1.   Production of a Λ and K^0 in $\pi^- - p$ collision in the Berkeley 72-inch bubble chamber (photo courtesy of Professor L. W. Alvarez).

chamber, we can see the process $\Lambda \to \pi^- + p$ (1). Next the pion could decay as $\pi^- \to \to \mu^- + v$ (3);* ultimately the μ^- will decay into the light lepton, i.e. the electron. There is a whole chain of weak interactions. The Λ itself could decay by an alternate channel, $\Lambda \to \pi^0 + n$ (1'). We come closer and closer to stable matter, and we find that the neutron decays as $n \to p + e + \bar{v}$, i.e. we re-discover nuclear beta-decay, the grand-daddy of weak interactions. On the other side (K^0) things are a little bit more complicated; the great idea of Gell-Mann and Pais was that whereas the K^0 is an eigenstate of (strong) *production*, it is not an eigenstate of (weak) *decay*. For the latter, one has linear combinations like $1\sqrt{2}(K + \bar{K})$ and $1/\sqrt{2}(K - \bar{K})$. These are eigenstates of P or of CP, but not of strangeness. Since it was assumed at that time that P was an exact symmetry of the decay mechanism, one had just these two possibilities. From the point of view of decay a K^0 is a superposition of these two eigenstates; perhaps the K_1 should be called more properly $K^{(+)}$, and the K_2, $K^{(-)}$, specifying the eigenvalues of CP. The K_1 decays into two pions, and occasionally into a pion, a lepton, and a neutrino, and the K_2 decays into three pions and most of the time into

_____

* In practice, the π^- would decay only in flight – at rest it would be captured by the hydrogen.

a pion, a lepton, and a neutrino. The 2π-decay of the K_2, and the 3π-decay of the K_1 are forbidden by CP-conservation. Again the pions decay, and the lepton, if it is a muon, decays and so forth. (There is one more branch that I should mention: the direct beta-decay, $\Lambda \rightarrow pe\bar{v}$ of the Λ, entirely analogous to neutron decay.) The great idea behind dividing the final kaon state into two orthogonal P or CP eigenstates was that the decay Hamiltonian actually does conserve these quantum numbers. This is just like a choice between linear and circular polarization, not according to the light that is emitted, but to the analyzers that one uses.

The recipe of Lee and Yang was to say that in order to find out whether parity is conserved – and I would like to underscore that in their first paper they were not really considering the other discrete symmetries at all – you have to use the principle that if you want to find out whether you have a screw you need a nut. You cannot put a screw on a scale and find out whether it is right- or left-handed. In more formal terms, you have to measure some pseudoscalar quantity, since such an observable would not have a finite expectation value in a state of definite parity. For instance, in $n \rightarrow p + {} + e + \bar{v}$ (2), the electron velocity could be correlated with the neutron polarization, with a distribution of the type $1 + A\langle\sigma_N\rangle\cdot\mathbf{v}_e/c$, where the last term is a pseudoscalar. Such a correlation corresponds to an up–down asymmetry. Similarly you would have in the μ decay $\mu \rightarrow e + v + \bar{v}$ (4), a correlation $1 + A\langle\sigma_\mu\rangle\cdot\mathbf{v}_e/c$, where $\langle\sigma_\mu\rangle$ is the polarization of the muon, and $\mathbf{v}_e$ the velocity of the decay electron. Other pseudoscalars that are useful and particularly simple are $\langle\sigma_e\rangle$ or $\langle\sigma_\mu\rangle$, each dotted into the velocity of the corresponding particle. The real statement is extremely simple: If you produce spontaneously an excess of particles of one given handedness, or *a fortiori* exclusively particles of one handedness, then you obviously do not have parity conservation. We might paraphase the law of mirror invariance of nature by saying: 'Thou shalt not erect any microscopic factories which produce screws of one handedness preferentially'. Very shortly after the suggestion of Lee and Yang the process (2) and the $\pi - \mu$-chain (3, 4) were investigated. The experiments, except for a little bit of technology for the decay (2), were extremely simple, rather straightforward and very inexpensive. Their results were most striking. Not only was parity not conserved, but it was maximally violated. In other words, correlation coefficients like A, helicities etc., did come out to be unity. So one was suddenly faced with the notion that all the β-rays that anyone had ever observed were spontaneously polarized. While people were inventing extremely clever schemes to polarize electrons, here they were spontaneously polarized, free of charge. Now, since one found a decay asymmetry of the muon, this also proved that the muon produced by pion decay (3) was polarized. That means that already in the first step (3) handed neutrinos were emitted to balance the spin zero of the pion. By the study of the $\pi - \mu$-decay chain one could thus see that parity was maximally violated in *two* successive weak interactions.

I have overdone it a little bit, because the experiment on the correlation of the electron velocity $\mathbf{v}_e$ with the neutron spin was not the first to be done. Neutron decay is clearly the most fundamental β-decay, but actually the first, the decisive measurement was done by Ambler, Wu, and their collaborators with ^{60}Co, as you no doubt know.

The muon experiments have the enormous advantage, already pointed out in our Chairman's talk, that while one can (for technical reasons) actually not compare the β-decay of the neutron and the β-decay of the antineutron, he can easily compare the β-decay of the muon (which is the μ^-) with the decay of the antimuon (which is the μ^+). I said two things that I shall illustrate in a moment: Parity was not conserved, and it was, in fact, maximally violated. This latter fact, the maximal nonconservation of parity, was not only of enormous practical advantage to the experimentalists, because it made their life much easier, but it also has a deep importance to which I shall come now. So I shall write 'maximal P violation found' on the board.

In practice, this maximal violation is equivalent to only right-handed neutrinos and left-handed antineutrinos, or left-handed neutrinos and right-handed antineutrinos, being coupled to matter (i.e. only 2 out of the available 4 states are coupled, and then these two states can be represented by 2-component spinors). This idea of maximal parity violation was anticipated on purely theoretical grounds, and there by people who loved formal beauty, namely Abdus Salam and Landau. Abdus Salam's argument was roughly the opposite of the one he presented this morning when talking about γ_5 symmetry. In those days (1956) he liked zero masses to stay zero. He said: 'The free neutrino has mass zero, and if we want to prevent it from fattening itself through (virtual) weak interactions to a finite mass, then we can accomplish this by restricting the weak interactions to those which commute with γ_5. This keeps the neutrino massless to all orders.' Now as to Landau, his recipe was similar to the well-known saying at marriages in the United States, 'I have lost a son, but I have gained a daughter,' i.e. it is not so terrible to lose the mirror symmetry of the world, because instead of requiring invariance under P, or under C, we can require invariance under the joint operation CP. The CP-mirror image of an L-neutrino is an R-antineutrino, but its ordinary mirror-image, the R-neutrino, and its charge conjugate the L-antineutrino, are postulated not to exist. Thus Landau's picture is equivalent to Salam's. The γ_5 idea of Abdus Salam leads to strictly left-handed neutrinos and right-handed antineutrinos (or vice versa), as does the Landau CP-principle. In order to impose CP invariance, one is lead to maximal parity violation!

At this point I would like to tell two anecdotes which might be illuminating. At the time of the original parity experiments, Leo Szilard, one of my most distinguished countrymen, had been completely out of physics for 10 years. I met him, at that time, in the Quadrangle Club in Chicago, and he asked me, 'Well, what's new in physics?' I said: 'There is tremendous excitement. Parity isn't conserved.' I then proceeded to describe the Ambler–Wu experiment to him. The very moment that I had very briefly explained the essential points of that experiment, he exclaimed: 'Maybe when you look in the mirror, you don't see cobalt, but you see anticobalt!' That is, he got the whole idea instantly. Wigner points out in one of his lectures that if you start with polarized ^{60}Co and everything is mirror-symmetric before the decay, while afterwards one observes an up-down decay asymmetry, then the principle of sufficient cause appears to be violated. However, this principle is *not* violated if you use a *CP*-mirror, since the initial situation is not *CP*-symmetric to begin with.

The second anecdote I can unfortunately not tell you from personal experience; it was related to me. Just as I informed Szilard in Chicago of this great discovery in physics, somebody in Cambridge appears to have taken it upon himself to inform Dirac of the fact that parity was not conserved. Dirac's reaction was terse as usual. He said: 'If you look carefully, you will see that the notion is not once used in my book.' As a collector of anecdotes, I had the great privilege to check this statement with Dirac about a year later. When asked whether this story was true, he said 'I do not recall for sure; but I always considered it [parity conservation] an unnecessary auxiliary assumption.' Of course, he was right.

There is a real point to my madness, namely that in this scheme (Fig. 2) I have put

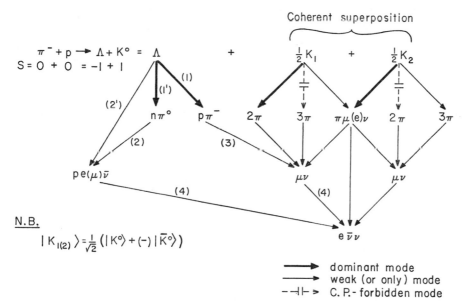

Fig. 2. Decay chains of the products in the reaction $\pi^- + p \to \Lambda + K^0$.

in not only the stable particles, but I have included quite specifically the dominant decay mode of the Λ into a pion and a proton, the one you saw in the bubble-chamber picture. Now at certain pion energies, say 1 GeV, strong dynamics is such that the Λ is produced with a spin polarization along the normal $\hat{n} = (\hat{p}_\pi \times \hat{p}_\Lambda)$ to the production plane; note that this has nothing to do with weak interactions that cause the subsequent decays. However, when you examine the Λ-decay, you can form $\langle \sigma_\Lambda \rangle \cdot \hat{p}_{rel}$ where $\hat{p}_{rel}$ is the relative momentum of the two final products π and p, and $1 + A\langle \sigma_\Lambda \rangle \cdot p_{rel}$ again represents an up–down asymmetry. You also get $\langle \sigma_p \rangle \cdot \hat{v}_p \neq 0$, i.e. observe a longitudinal polarization of the proton. Although I was involved with the study of the $\pi - \mu$-decay chain (3, 4), these facts appeared to me right away as a much greater discovery because Landau's and Salam's pictures, and the equivalent considerations of Lee and Yang on two-component neutrinos, made parity violation look very easy. The neutrino was so-to-speak personally responsible for every parity violation effect that one saw. This hypothesis was elegant, powerful, and cheap. In the Λ decays there was no such cheapness at all. I think that it is therefore completely inappropriate in the present days to start off students on weak interactions with radium β-decay. They should understand from the start that weak interactions are a very vast class of transformations, and that there are some in which no leptons at all are involved.

We thus reach the following tentative conclusion: All weak interactions are probably C and P violating and CP-invariant, and change S by 0 or 1 units. That has not been experimentally proved for all weak processes, and to this day there is no satisfactory dynamics of non-leptonic processes. But it has become a credo. As I like to say – and as I hope to demonstrate in this talk– there is a terrible and at the same time very amusing fact about experimental physics: 'Last year's sensation is this year's calibration.'' (I once said this, and Professor Feynman promptly added: 'And next year's background'.)

So now we have this very fundamental idea about C and P violation in *all* weak processes. Let us connect it with something which, as of today, I would like to call the Trieste theorem, the CTP theorem. (These are the initials of the Centre; but Trieste is also easier to pronounce.) This theorem, which was mentioned by Professor Yang in his lecture, asserts that the *product* of these three symmetries is a valid invariance in all reasonable field theories (and I don't know exactly what is meant by that) even if they were individually violated. So for maximal P violation, C is also maximally violated so that CP is still good*; and to retain CTP invariance, T remains a good symmetry. Conversely, if CP is violated, T invariance must be broken as well. I summarize these consequences of the Trieste theorem with considerable embarrassment, since I don't understand its derivation; on the other hand, I am in good company – there are many others who do not either, but it is still considered valid. Again, I do not think that this question is a matter of formal beauty, or of what class of theories we know how to write down. I believe that it is high time that we put the CTP theorem to some exacting test. That is the business of the experimentalists, or as Professor Yang would say, of fierce independence. Since he has not yet advocated any violation of CTP, it will be fierce independence if I go into it. I should say that although the first Lee–Yang paper went only into P violation, the CTP business was explored in a subsequent paper by Lee, Oehme and Yang. All the above statements and several more which I shall have the opportunity to use were made there.

We shall now discuss some of the crucial experiments. Fig. 3 shows a photographic emulsion picture of $\pi^+ \to \mu^\pm \to e^+$-decay, which I have of course kept; emulsions are a technique by now long abandoned. You can see a pion coming in, stopping, giving a muon and then see the muon getting badly scattered (this picture was taken at the

---

* Alternatively T could be violated, and PT maintained as a valid invariance.

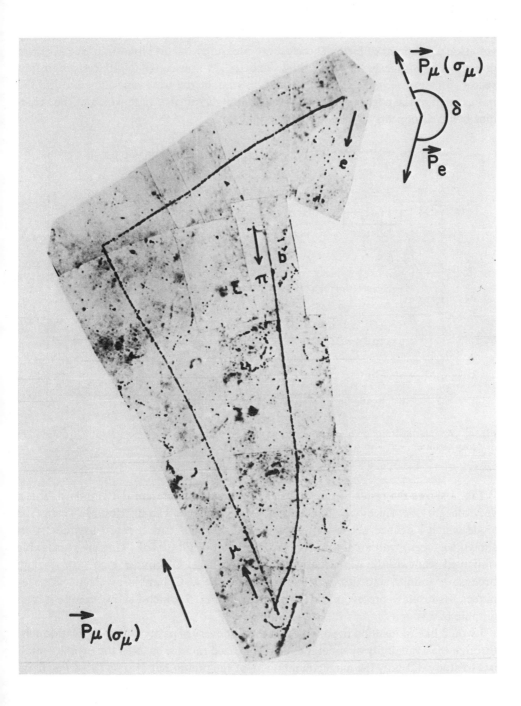

Fig. 3. Photomicrograph of $\pi^+ - \mu^+ - e^+$ decay in nuclear emulsion.

time because such 90° scatters are very unusual). The muon ran along, stopped and decayed into a positron. The muon was to acquire longitudinal polarization by parity violation at its point of birth. Therefore we measured the distribution of the positron directions $\hat{\mathbf{p}}_e$ with respect to this initial vector $\hat{\mathbf{p}}_\mu$. We measured it with respect to that vector, because we made the approximation that in the traversing the emulsion the momentum of the particle changes direction, but its spin does not, which at the velocities involved appears to be a fair approximation.

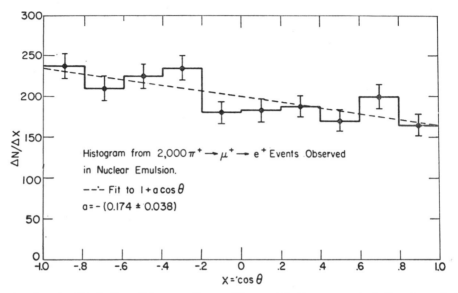

Fig. 4. Angular distribution of decay positrons with respect to the muon polarization, from studies of decay chain $\pi^+ - \mu^+ - e^+$ in nuclear emulsion (see Fig. 3).

Fig. 4 shows the result of this Chicago experiment. By present-day standards it is a very meagre one, since only 2000 decays were observed. The distribution versus the angle which I defined above was plotted, and the backward–forward asymmetry as shown was seen, with an asymmetry parameter of 0.174 ± 0.038. Thus by reasonable statistical standards it was a convincing experiment. Of course more convincing, because of infinite statistics, was its analogue done at Columbia by electronic techniques. Its result (a precession curve) is shown in Fig. 5, which I show because it is so graphic and simple.

I would like to mention in passing that the discovery of parity non-conservation has, through the availability of spontaneously polarized muons, enabled the experimentalists to study not only the magnetic moment of the muon, but also its $(g-2)$ or magnetic anomaly to an accuracy which by today is approaching that of the electron, a fact which in my opinion should inspire the theorists with awe: It is very easy to make electrons, while it is very hard to make muons. Nevertheless we know the properties of both almost comparably well, thanks to this freakish behaviour of nature.

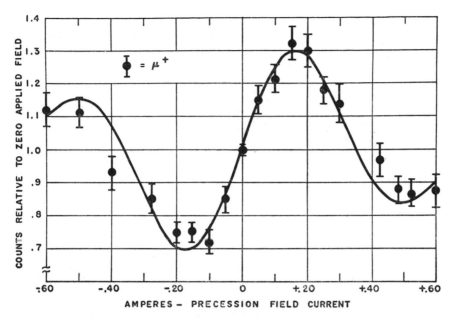

Fig. 5. Asymmetry of μ^+ decay positrons observed by electronic techniques. Muons precess in a magnetic field at a frequency ω, and the decay positrons are detected in the plane of precession. $N(\theta) = 1 + a \cos \theta$ yields $N(t) = (1 + a \cos \omega t)$.

The fact that we assume CP invariance in this scheme (Fig. 2) implies that we keep time reversal T as a valid symmetry. But this has to be checked experimentally. As I pointed out above, the check of parity conservation was the investigation of pseudo-scalar observables. A positive check of time reversal (or motion reversal) invariance correspondingly consists in the investigation of what one might call a '*time pseudo-scalar*'. With the kinematic variables of neutron decay (2) we can form the time pseudo-scalar $\langle \sigma_N \rangle \cdot (\mathbf{v}_e \times \mathbf{v}_{\bar{\nu}})$, and look for the correlation $1 + \mathscr{D} \langle \sigma_N \rangle \cdot (\mathbf{v}_e \times \mathbf{v}_{\bar{\nu}})$. Under Wigner time reversal, i.e. motion reversal, both σ_N and the velocities change sign. Thus T-invariance implies that $\mathscr{D} = 0$. In the course of investigating the β-decay of the free neutron, a group of physicists from Argonne National Laboratory and the University of Chicago indeed investigated the magnitude of this $\mathscr{D}$. This investigation consisted in a very amusing and conceptually simple experiment; the apparatus is schematically drawn in Fig. 6. A beam of polarized thermal neutrons is emerging towards you from the paper. These thermal neutrons (a small fraction of them) undergo β-decay in the evacuated volume, and the decay electrons are counted with a scintillation counter, while the decay protons are counted in coincidence – if they transverse the 'venetian blind' shutter – with a multiplier. With the shutter one can select protons emitted in a given direction, which corresponds to antineutrinos emitted in some other definite direction.

The motion reversal operation simply consists in performing the experiment with

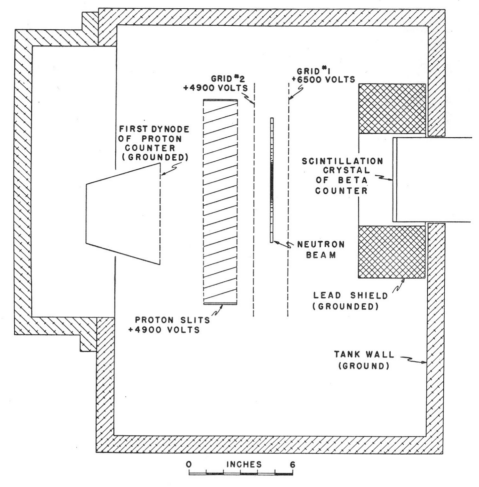

DETECTOR ARRANGEMENT FOR EXPERIMENT b) ANTINEUTRINO DISTRIBUTION WITH
RESPECT TO NEUTRON SPIN

Fig. 6.   Apparatus used to investigate time-reversal invariance in polarized neutron decay.

neutrons polarized 'up' rather than 'down', as evidenced by Fig. 7. If time reversal
invariance holds, motion reversal holds, and the probability of these two final states
should be strictly identical. They do indeed come out to be identical, i.e. one finds
$\mathscr{D}=0$.

The conclusion of such an experiment has of course to be quantified; this is done in
terms of a phase angle. In neutron decay one has two coupling constants, G_A and G_V.
If time reversal is a valid symmetry, then the relative phase of these (*a priori* complex)
constants would be 0° or 180°, or in other words they would have to be real numbers,
and the coupling would be either $V+A$ or $V-A$. The experiment just described gave
the result that this phase is $(180° \pm 8°)$. In a very recent repetition of this experiment in

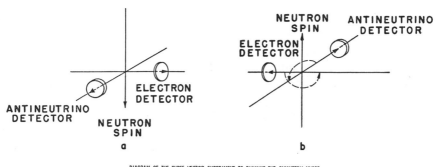

DIAGRAM OF THE THREE VECTOR EXPERIMENT TO EXAMINE THE SYMMETRY UNDER
TIME REVERSAL. a. AND b. SHOW THE "TIME REVERSED" EXPERIMENTS. THE ROTATION
ABOUT THE SPIN AXIS INDICATED BY THE DASHED LINES IN b) BRINGS THE ELEC-
TRON AND ANTINEUTRINO DETECTORS INTO POSITIONS AS SHOWN IN a).

Fig. 7.   Diagram to illustrate the effect of time (motion) reversal on the kinematical variables in the β-decay of polarized neutrons.

the Soviet Union, the accuracy of the coefficient $\mathscr{D}$ was pushed even farther down, and the phase error was reduced to $\pm 1°$.

I would like to emphasize the very positive, the very kinematic nature of this T-reversal test: You need hardly any mathematics to convince yourself that this correlation would, if it existed, indeed represent time-reversal violation. The experiments which we shall discuss next require, unfortunately, at least for me, a great deal of mathematical argumentation to get from the experimental evidence to the conclusion of time-reversal violation.

We lived in this atmosphere of (presumed) CP and T-invariance until about 1964. The paper of Lee, Oehme, and Yang had of course explored what would happen if T was also violated. Let us return to Fig. 2. The emphasis there was that K_1, K_2 were initially chosen as eigenstates of parity; but they are *also* eigenstates of CP, since $\bar{K}$ goes into K under C, with a suitable phase convention. Thus $(CP)K_2$ is equal to $-K_2$. and $(CP)K_1$ equals $+K_1$. The pre-parity violation argument that $K_2 \to 2\pi$ should not occur remains valid with the substitution of CP for P.

Before we go into the subject from an experimental point of view, let me point out a couple more things. K_1 and K_2 have masses M_1 and M_2 which are different, and decay rates Γ_1 and Γ_2 which are also different. It is not surprising that they should have different decay rates, since different final states are accessible to them. Nor is it remarkable that Γ_1 should be much (about 600 times) larger than Γ_2, since one has two-body final states for K_1, and three-body final states for K_2. It is also not astonishing that the masses M_1 and M_2 should be different, since the K^0 and the $\bar{K}^0$ can transform each other into each other virtually via two pions and via three pions. This is a bit like ^1S and ^3S positronium, where the triplet can virtually turn into one photon, while the singlet can turn virtually into two photons. Weak virtual transition with their CP selections rules produce the $K_1 - K_2$ mass difference which is very, very small. It is best expressed as $\Delta M/\Gamma_1$, which is of the order $1/2$, while Γ_1 is about 10^{-10} sec. These two predictions of the theory are both verified by experiment. One can ask with Gell-Mann and Pais why this curious super-position principle that was introduced by them for kaons is not more general; e.g. why is there no mixture of neutrons and antineutrons, leading to n_1 and n_2? The reason is that the baryon number is absolutely conserved by all interactions, and such mixtures make no practical sense. The strangeness (or hyper-charge) is conserved in the production act, but *not* in the decay interaction. Kaons are unique in exhibiting this dual property.

The next idea was to test *CP* violation simply by looking for this forbidden fruit $(K_2 \rightarrow 2\pi)$. Actually there was no great theoretical push to preform this crucial experiment. The theorists were not knocking at anyone's door saying 'please go and do it'. The story of the first experiment is not particularly well known, and since I have it from one of the participants, I shall tell it.

At one time a group at Yale led by Adair studied the interaction of K_2 mesons in hydrogen. They saw some very curious events which they interpreted as due to a *fifth* force, whatever that means. Well actually they were probably interactions with the window of the bubble chamber. At that time Cronin was planning to perform an experiment on the production of ϱ mesons, at the Cosmotron in Brookhaven, and he set up his apparatus for $\rightarrow 2\pi$. When Cronin and Fitch heard about the utterly sensational bubble chamber results of Adair, they decided to investigate jointly the interaction of K_2 mesons with hydrogen by electronic techniques, using the spectrometer intended for ϱ production. Prior to investigating the interaction in liquid hydrogen they investigated the performance of their apparatus in vacuum, and that check was the experiment that gave us *CP*-violation in 1964. The authors of this truly decisive experiment were Christenson, Cronin, Fitch and Turlay.

Fig. 8 shows the set-up of these people (if you are bored by all these technical details, then it is on purpose!). At some 30 metres from the target of the accelerator one has a neutral beam from which the charged particles have been swept out by magnets; one attempts also to remove gamma rays by first converting them into electrons. The beam emerges after the multiple collimation, the decays occur in a two-metre long He-filled region, and the two decay-products are detected symmetrically by two magnets.

By the way, by now Cronin and Fitch had transported their apparatus (Fig. 9a) built for ϱ production from the Cosmotron to the AGS – a complete change of scenery in the original plans. The decay region is actually not vacuum, but a helium bag, since that is cheaper. The decay particles are momentum analyzed by their magnetic deflection, the direction of incidence and the direction of exit being each measured with a stack of spark chambers. Now the principal mode of decay of the K_2 is into leptons, i.e. $\pi\mu\nu$ or $\pi e\nu$ these processes being about 500 times more copious than the 2π-mode one was looking for. The leptonic three-body events will not be coplanar with the beam direction, whereas the two-pion final state will be, since there is no missing neutrino. Behind each magnet there is also a Čerenkov counter to further discriminate against leptonic background.

Fig. 9b is a photograph of the set-up, for those people who really want to see it. You see also, the graduate student, I think Christenson, looking at the apparatus; a historic picture if there ever was one.

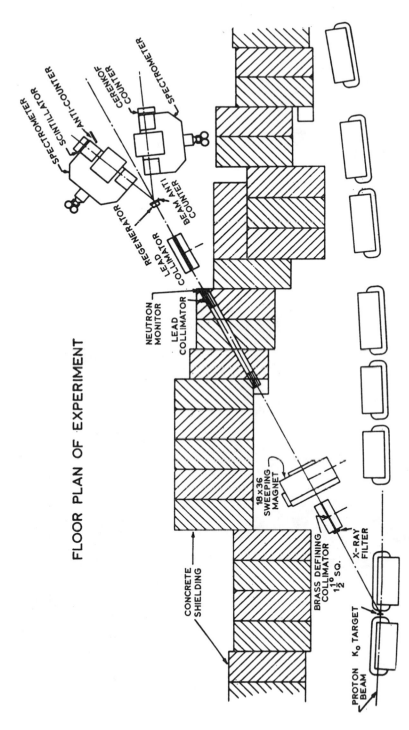

Fig. 8. Layout of the Cronin–Fitch *CP*-experiment at the AGS in Brookhaven.

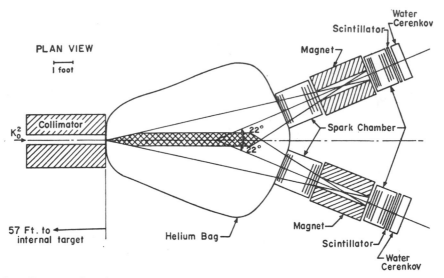

PLAN VIEW

⊢——⊣
I foot

Fig. 9a.   Set-up used to detect $K_2 \rightarrow \pi^+\pi^-$.

Fig. 9b.   Photograph of the apparatus schematically drawn in Fig. 9a (courtesy of Professor J. W. Cronin).

The analysis consisted in attributing to both particles the pion mass and to calculate the invariant mass of the decaying particle (one should, naturally, get the kaon mass). Now of course one does not believe right away that one sees 2π events. One does do a Monte Carlo calculation on the basis that one is really mislabelling $K \to \pi\mu\nu$. It is very crucial that you understand this point. From the vector momenta of the two 'pions' one also constructs the direction of the decaying kaon. One then plots the angle

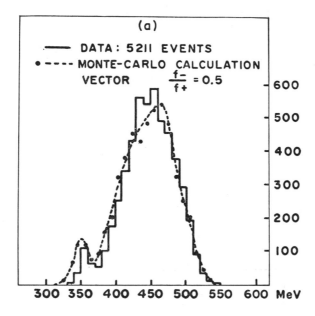

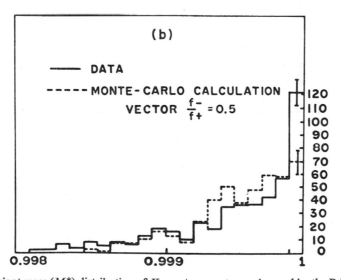

Fig. 10. Invariant mass (M^*) distribution of $K_L \to \pi^+\pi^-$ events, as observed by the Princeton group.

between the beam direction and kaon momentum for each observed event interpreted as a 2π decay. As you see (Fig. 10) the Monte Carlo calculation based on three-body decays mocking up or mimicking as 2π gives the dashed curve but there is no forward peak in that curve. Thus there is an excess of roughly 50 events that one cannot explain by assuming that they were three-body events masquerading as two-body events.

Now one can examine these events as a function of their invariant (fictitious) mass m^* (Fig. 11). The mass of the kaon is $M_K = 500$; one looks at events below M_K, above M_K and around M_K, and investigates their angular distributions. One then finds no peak whatsoever in the forward direction for events below M_K, nor above it, but a very remarkable peak for those events that have about the kaon mass. From this evidence one concludes that one is dealing with true two-body, two-pion events. The number of such events in the Cronin–Fitch experiment was about 43. Rarely has physics been changed so much by so few events. Let me point out to you that this exciting experi-

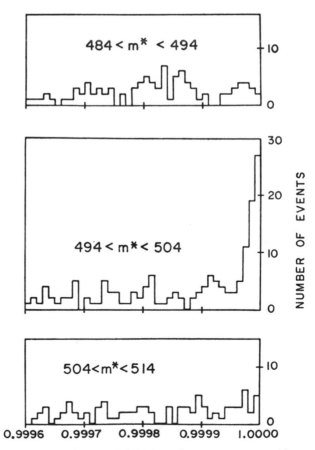

Fig. 11.   Angular distribution of momentum presumed $K_2 \to \pi^+\pi^-$ events with respect to the incident neutral beam direction, for various ranges of invariant mass m^* of the decay products (assumed to be pions).

ment could hardly have been done a couple of years earlier, because the optical spark chambers and other technologies used in it had not yet introduced. The Princeton group had developed the techniques needed to do it, and they did it. The discovery of *CP* violation is not to be decoupled from the experimental techniques of the time.

The result of this experiment which had such an enormous impact can be expressed in the following way: Assume that we start from long-lived kaon, so-far called K_2, a terminology that we can no longer use. What we used to call K_2 we shall call K_L, and what we called K_1 we shall call K_S. In other words, K_1 and K_2 are eigenstates of *CP*, i.e. mathematical objects, whereas K_L and K_S are the physical states with differing decay rates $\Gamma_{L(S)}$ and differing masses $M_{L(S)}$ which one actually observes. Let the theorists worry what the relation between K_1 and K_2, and K_L and K_S is. They know how to diagonalize a 2 by 2 matrix, and they will figure it out.

The great result can be expressed as a transition amplitude $\langle \pi^+ \pi^- | T | K_L \rangle$ which is forbidden (if $K_2 = K_L$ and *CP* holds), whereas the transition amplitude $\langle \pi^+ \pi^- | T | K_S \rangle$ is allowed. It is useful to define the number

$$\eta_{+-} \equiv \frac{\langle \pi^+ \pi^- |T| K_L \rangle}{\langle \pi^+ \pi^- |T| K_S \rangle}$$

which is complex because it is the ratio of two quantum-mechanical amplitudes. This η_{+-} can, of course, be written as $|\eta_{+-}| e^{i\phi_{+-}}$. The Princeton group measured $|\eta_{+-}|$ and found $|\eta_{+-}| = 2.3 \times 10^{-3}$. Incidentally the branching ratio $\Gamma(K_L \to \pi^+ \pi^-)/\Gamma_L$ is also about 2×10^{-3}, which is just a numerical accident. This branching ratio shows the bravura of the Princeton group in detecting an event which happens once in 2000 times, while $|\eta_{+-}|$ serves the theorists to show them how small the *CP*-violating amplitude is. It was quite clear after this experiment that similar phenomena were to be expected in the analogous decay $K_L \to \pi^0 \pi^0$, for which one can define in complete analogy an η_{00}, with a corresponding modulus and phase ϕ_{00}. Thus *CP*-violating K^0 phenomena, or at least this subset of phenomena involving pion decays, are fully defined in terms of these *four* parameters. We are now in 1972, while the Princeton experiment was done in 1964. In the intervening time something like 200 man-years of work and uncounted millions of dollars have been spent on measuring these four numbers. While the experimental situation is not yet completely settled, it is at least converging.

I shall have to make certain statements about the present experimental status; these represent my views, and you can hold me responsible for them. (When there is no unanimity about the results within a field of experimental physics, one has to use one's own judgement!) Now let us make a little graph (Fig. 12) which will be very amusing. We plot the number of $K^0 \to \pi^+ \pi^-$ decays (versus time). First we have a fast decay rate $\exp(-\Gamma_S t)$, i.e. relatively fast because $\Gamma_S \gg \Gamma_L$. At late times, we have the slow component $\exp(-\Gamma_L t)$. The level of 2π-decays at late times is what we call today the Cronin–Fitch level; these people worked very far away from the accelerator, where all possible effects due to the K_S component had died out by a factor $\exp(-100)$, i.e. they were in a pure K_L beam. It is clear that by determining a single real number like $|\eta_{+-}|$

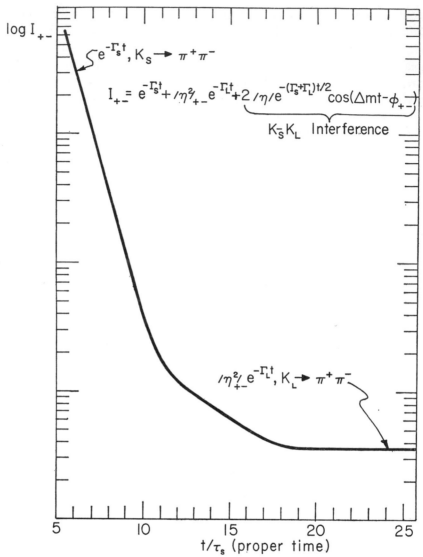

Fig. 12. Vacuum Regeneration.

one can really not make an operational statement about time reversal, since time reversal involves either motion reversal, or evolution or something like that. So we shall have to be a little careful before concluding that T is violated, while CP violation in itself is guaranteed by this observation of $K_L \to 2\pi$. Via the CTP theorem we can, on mathematical grounds, say that this experiment also implies that T is not a good symmetry, so that the product CTP may remain good. So we can write down: CP no good, T no good, CTP is O.K. and I would write underneath 'philosophy', since one would like to see something more direct.

Right after the Princeton experiment was done there came dozens of suggestions as to why there was actually no *CP* violation – like there was to be some third kind of particle or what have you – all of which have been disproved. The Princeton experiment has of course been repeated in every lab where it could be repeated and it has always been confirmed.

Let us now go back to Fig. 12, and study the $\pi^+\pi^-$ decay rate I_{+-} as a function of time. By time I of course always mean *proper* time in the rest frame of the kaons, but we don't have to write τ for that. In working out I_{+-} we shall assume, which is quite important, that we start out at $t=0$ with a pure K^0 state. In other words, we assume that we can for instance produce K^0's by $\pi^- + p \to K^0 + \Lambda$ and trigger the apparatus every time we see a Λ. So we write

$$K_0(t) = (|K_S\rangle\, e^{-i\,(M_S - i\Gamma_S/2)\,t} + |K_L\rangle\, e^{-i\,(M_L - i\Gamma_L/2)\,t})\,\frac{1}{\sqrt{2}},$$

where $|K_S\rangle$, $|K_L\rangle$ are t-independent. Then I_{+-} is proportional to the square of the decay amplitude, namely

$$I_{+-}(t) = \tfrac{1}{2}[\langle\pi^+\pi^-|T|K_S\rangle\, e^{-i\,(M_S - i\Gamma_S/2)\,t} + \langle\pi^+\pi^-|T|K_L\rangle\, e^{-i\,(M_L - i\Gamma_L t/2)}]^2 ,$$

$$I_{+-} \simeq |e^{-\Gamma_S t/2} + \eta_{+-}\, e^{-i\,(M_L - M_S) - \Gamma_L t/2}|^2 \tag{*}$$

$$\simeq [e^{-\Gamma_S t} + |\eta_{+-}|^2\, e^{-\Gamma_L t} + 2\,|\eta_{+-}|\, e^{-(\Gamma_L + \Gamma_S)\,t/2} \cos(\Delta mt - \phi_{+-})].$$

The K_L, K_S have their own 'heartbeats' in the form $M_S t$ and $M_L t$, and the relative phase shift frequency comes of course in as $\Delta mt = (M_L - M_S)t$. I believe that you really see for the first time Schrödinger rest mass frequencies beating against each other.

This experiment to measure $I_{+-}(t)$ has actually been performed, and it possesses tremendous conceptual beauty, which is what people seem to insist upon here. It is also a great pleasure to perform: since the decay length of the K_S is about 15 cm (for an energy of 3 GeV), the interference phenomenon described by (*) occurs in the laboratory over macroscopic distances of the order of metres! I at least find it very impressive to see that big wiggle due to the cross-term. Unfortunately one sees it only after two and a half years of work and then only on a piece of paper. You do not see it in space, but can you imagine that you could. This $I_{+-}(t)$ experiment has been called the 'vacuum regeneration' experiment. This is a misnomer implying that it is the vacuum that turns the K_S into the K_L, since it is very well known that this kind of conversion (called regeneration) can be induced by matter.

Fig. 13 shows the apparatus. This figure is only for those people who want to see how hard life really can be. A 12-GeV proton beam hits the target and produces a neutral beam. We would like to look at the interference between K_S and K_L, so we do not sit 20 metres from the target, but right near it. The kaons are about eight lifetimes old (for a mean momentum of 3 GeV) when they emerge from shielding wall, which consists of about thirteen tons of uranium. The V-shaped 2π-decays are measured by means of wire spark-chambers placed before and after a magnet. The Čerenkov counter rejects electron (πev) decays. The final iron wall discriminates against muons.

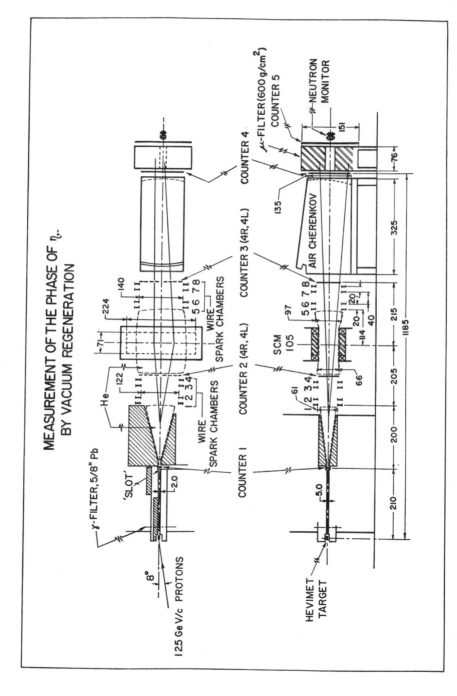

Fig. 13.   Vacuum Regeneration apparatus.

Suppressing leptonic decays is important, since $K_L \to \pi\pi$ is a very low branching ratio mode, as we emphasized earlier. Once the proper events are isolated, one simply determines the momenta of the two pions, the momentum of the parent kaon, converts the distance of the decay vertex into proper time, and plots $I_{+-}(t)$. This experiment was performed by a group from the Universities of Chicago and of Illinois at the Argonne National Laboratory, and an entirely similar one was performed at CERN, where a much refined repetition is currently in progress.

Fig. 14 shows the result. As I have said earlier, at long times, or far away from the target, we have the Cronin–Fitch level which is included in $I_{+-}(t)$ as $|\eta_{+-}|^2 e^{\Gamma_L t}$

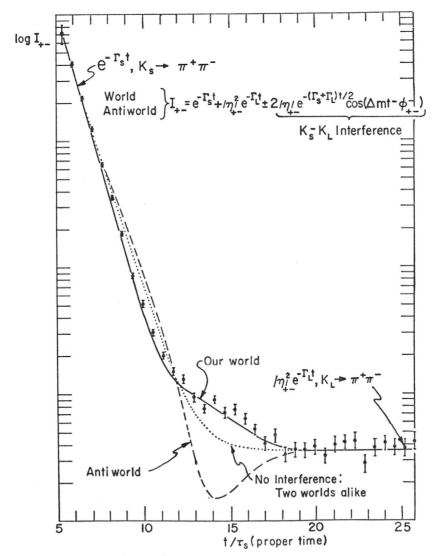

Fig. 14.   Vacuum Regeneration results.

At early times we see the well known dominant decay mode $K_S \to \pi^+\pi^-$. As, the decay mode $K_L \to \pi^+\pi^-$ is rare while the decay mode $K_S \to \pi^+\pi^-$ is abundant, the interference will become important when the K_S is old and exhausted so that its dominant decay amplitude can match the small 2π-decay amplitude of the young K_L. This happens around 12 K_S-lifetimes. If one had no interference, then we would see the time distribution indicated by the dotted line in Fig. 14; it is just the incoherent superposition of the two exponentials. The data points that we actually see are markedly different from it, and one can fit them to the expression for $I_{+-}(t)$, the more so since Δm is independently known. In fact all parameters known except Φ_{+-} are external.

Fig. 15 shows the interference term purified from its exponential dependence just by numerical manipulation, using the same data points. You see a lovely sine curve as a function of proper time (the unit, 10^{-10} sec, is about one K_S-lifetime). At a fixed distance, high momentum kaons emerge young, and low momentum kaons emerge old.

Now the curve in Fig. 14 has of course a large number of beautiful features. The most beautiful feature of it is that when you start with a pure $\overline{K}^0$ state, you witness its

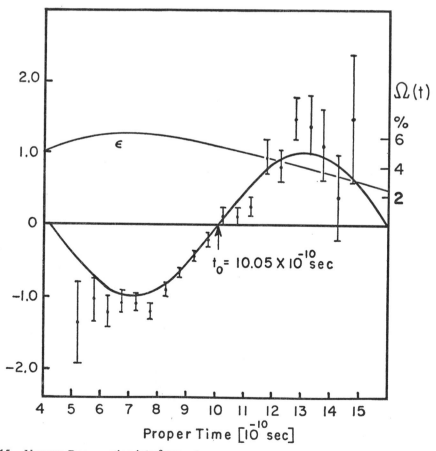

Fig. 15.   Vacuum Regeneration interference term.

non-exponential 2π-decay; asymptotically of course one has the simple $K_L \rightarrow 2\pi$ which we have discussed before. Let us assume for a moment that we would have done this experiment not with the boundary condition that $\overline{K}^0(0)=1$, $\overline{K}^0=0$, but that we would start with $K^0(0)=1$, $\overline{K}^0(0)=0$. That means that we would have performed the experiment in the antiworld, as it were. I claim that the sign of the interference term would then change to minus. The way of proving this without mathematics is to say; if I alternated two experiments, running for one minute with initial K^0's, and for one minute with initial $\overline{K}^0$'s, and the operator would not tell me when he is switching from one strangeness to the other, or if I really superposed the two beams, I would start with a 'state' of zero net strangeness (neutral as to hypercharge) to begin with. Clearly by starting with a state which has no charge, in this case no hypercharge, you cannot make any statements about charge conjugation symmetry violation. A little mathematical manipulation representing this idea shows that the interference term indeed changes sign. In an actual experiment, where both $\overline{K}^0$'s and $\overline{K}^0$'s are produced, the interference term gets multiplied by a dilution factor $[N(\overline{K}^0)-N(\overline{K}^0)]/[N(\overline{K}^0+\overline{K}^0)]$. This is just a technicality for us and we would like to have almost no $\overline{K}^0$'s, but this switch in sign for an observed time distribution is really the proof of the pudding. Fig. 14 shows also the curve predicted for the interference with the opposite sign, but otherwise with the same parameters. Now the standard trick that one uses in symmetry considerations is to try to inform a distant observer by voice communication of whether he is in the same world or not as we are. Here we instruct him to perform our experiment, and ask him to tell us whether his interference is the way we see it, i.e. whether it is constructive or destructive. This is an operational way of testing the world against the anti-world, and to establish the manifest violation of CP invariance. Then, as I said, you can from mathematical arguments further conclude that T is also violated.

This evidence for CP violation is quite compelling. There are many models to account for this violation. The most simple-minded model consists in saying the following: What is a K_L? A K_L is essentially what we thought it was, a K_2, plus a little bit of dirt which is K_1: $|K_L\rangle=|K_2+\varepsilon|K_1\rangle$, i.e. an impure state, a non-eigen state of CP. What is K_S? A K_S is a K_1 plus the same amount of dirt K_2, or $|K_S\rangle=|K_1\rangle+\varepsilon|K_2\rangle$. This is a very simple model, and by no means the only possible one. Its characteristic is that we have only one complex parameter, namely ε, and it actually turns out $\varepsilon=\eta_{+-}=\eta_{00}$. To prove this model experimentally, we have hence to prove that the moduli of η_{+-} and η_{00} and their phases are equal. The experimental situation today, after lots of confusion, is completely compatible with the identity of the moduli. A further prediction of that particular class of theories where the CP violation is in the wave function of the decaying state (rather than in the mechanism of the decay process) is that $\phi_{+-}=\phi_{00}=$ $=\tan^{-1}(2\Delta m/\Gamma_S)$, a formula I do not have time to derive here. Γ_S is the well-known K_S decay rate, and Δm the K_L-K_S mass difference, also well known. Numerically, the answer is $\phi_{+-}\simeq43°$. Present experimental results yield $|\eta_{+-}|=|\eta_{00}|$, and ϕ_{+-} in agreement with this prediction, to within a few degrees. The error on ϕ_{00}, which is much harder to measure, is so large that one cannot say. Thus this simple model seems to work. What is the meaning of this model? This model says that there are virtual transitions that are making the decay states impure states of *CP*. Among other things you shall see that $|K_L\rangle$ is no longer orthogonal to $|K_S\rangle$ as is normally supposed to quantum mechanics. Why are they not orthogonal? Because there is no discrete symmetry operation left with respect to which to orthogonalize the wave functions of the two states (all this is by the way in the paper of Lee, Oehme, and Yang).

Now the quantity ε (or at least its real part) can in turn be measured directly, and I shall conclude with the way that this is done. I'll do this mainly because a few people do not accept completely the *CP*-violation arguments given so far. Wigner, for one, didn't find them compelling, so I thought it useful to present an argument that he would buy. That is the following:

We know from other sources that there is a rule that ΔS equals ΔQ, in other words this strangeness and electrical charge change the same way in a decay. In the decay $K^0(S=1) \to \pi(S=0)$ we must hence get a π^- and *positive* charged lepton (plus ν), and for $\bar{K}^0$ correspondingly a π^+ accompanied by a *negative* lepton (plus $\bar{\nu}$). Thus, because of this extra bit of dynamics, the eletrical charge of the leptons is a monitor of the K strangeness content of the decaying state. We can define a charge asymmetry

$$\delta = (N^+ - N^-)/(N^+ + N^-)$$

in terms of the decay rates $N^\pm$. Any finite δ for a K_L beam is a most direct and violent proof of C violation. Let us rewrite K_L in terms of K^0, $\bar{K}^0$:

$$K_L = K_2 + \varepsilon K_1 = \frac{1}{\sqrt{2}} [(1 + \varepsilon)\, K^0 - (1 - \varepsilon)\, \bar{K}^0].$$
$$\qquad\qquad\qquad\qquad\quad \downarrow \qquad\qquad \downarrow$$
$$\qquad\qquad\qquad\qquad\quad e^+ \qquad\qquad e^-$$

(We ignore higher order terms in ε.)

If this extraordinarily simply Ansatz holds, then δ is also directly related to ε, and in fact one readily sees that $\delta = 2\,\mathrm{Re}\,\varepsilon$. (Note also that $\langle K_S\, K_L \rangle = 2\,\mathrm{Re}\,\varepsilon$.) These charge asymmetry experiments have by now been done by many groups. The first experiments were done at SIAC for muons and at Brookhaven for electrons. Since the experiments have been repeated and refined, their results have become consistent with each other and the formula just given. Furthermore, we know what $\mathrm{Re}\,\varepsilon$ is since $\varepsilon = |\eta_{+-}| \exp(i\phi_{+-})$. $|\eta_{+-}|$ is measured, ϕ_{+-} is measured, and we know how to take a real part. We get a σ of the order of a few parts in a 1000. Today it is known moderately reliably to 15%, which shows one knows how to measure to one part in 10000.

The simple model mentioned here is called 'superweak theory', and everything appears to agree with it. This may, of course, be satisfying to the theoreticians, because one uses a very simple Ansatz where a single parameter explains a large number of experimental numbers. However, in one sense it is a very sad result for the experimentalists, because it implies that there is no place other than in the peculiar $K^0 - \bar{K}^0$ complex to look for *CP* violations, since the superweak interaction is by definition a $\Delta S = 2$ (doubly strangeness changing) force. Thus a theory which is aesthetically satisfactory seems to be somewhat depressing from the point of view of full employment.

DISCUSSION

C. N. Yang: Thank you very much. For myself, when I first saw the interference due to the mass difference between the *K* long and the *K* short, I had the feeling that despite the complexity of the phenomena we are analyzing, it is the same beauty of the elementary interference experiments that we all do in the optics lab. I will call for questions and discussions right now. Questions or comments?

V. L. Telegdi: I cannot attribute this silence to my all-pervading lucidity.

Comment: I think we would like to hear a little more about the possible implications of a superweak interaction other than that it is no good for full employment.

V. L. Telegdi: Experimental implications or theoretical ones? Well, I think that no one has really succeeded in making a successful model of such an interaction. Even if you made one, you would have to rely on beauty it seems to me, since once you made this model it would be definition not lead to consequences outside the kaon complex. The point that I would like to make, however, is this: You cannot *prove* the superweak theory. There can be a myriad of theories which yield predictions neighboring those of the super weak. You can disprove the superweak theory by saying that the phase ϕ_{+-} is not 43°, but 95°. If you find (45 ± 2)°, some guy will come and say my theory gives almost 43°. In the same spirit you cannot check but not prove the conservation of electric charge either. You can set better and better limits, and it becomes essentially an endurance test. The tragic thing is that if you have the superweak interaction you will have propagation of *T* violation into other phenomena, like violation of the principle of detailed balancing in some complicated weak interaction, but that will creep in at such a low level that all hope is lost, at least for my generation. Hope may be lost for my generation anyhow but certainly...

E. P. Wigner: I was going to ask an experimental question. Suppose that you go into the time interval where only K_L is supposed to exist. Can you measure the ratio of 2π and 3π decays? (*Telegdi:* Yes.) How well is this ratio constant in time? Is that investigated?

V. L. Telegdi: It is investigated in the following sense, that the ratio of 2π to 3π decays has been measured in K_L beams of various momenta at various distances from the accelerators. (*Question:* Momenta of what? *Answer:* Of the K_L beam, since mean decay distance depends on momentum.) I assume that you, like the rest of us, would express the distance in K_L proper lifetimes, and that of course depends on the momentum and the geometry. These branching ratio measurements have obviously been done in different laboratories with K_L mesons of different age. I do not know of any explicit comparison of the branching ratios versus distance. As to the points which you so aptly raised in a private conversation, I think that the charge asymmetry takes care of those.

C. N. Yang: Perhaps I should ask Professor Wigner whether he could provide some background as to why he asked that question. I think that it is a very important question. Perhaps he has, as you said, been thinking about it and has only revealed his thought to us by asking that seemingly very innocent question.

E. P. Wigner: These are unstable particles and therefore they do not have definite wave functions. And the decay is usually not exponential as we all know. Now we know from the paper, now I forget who your collaborator was – the paper of Yang and Wu, that the decay is very nearly exponential but it is not... It would be interesting to see to what degree it is really exponential.

V. L. Telegdi writes 'Weisskopf and Wigner' on the board, and asks: Do you remember this paper?

E. P. Wigner: And it does point out that this is not altogether exponential, Professor Telegdi, but also gives reasonable statements about when exponential behaviour is a good approximation.

C. N. Yang: Yes, indeed, and I think that there is good reason to believe that it is good here.

V. L. Telegdi: May I summarize that? Professor Wigner, like several other people, isn't quite sure that the K_S is completely decayed by the time that we are looking at the K_L. We are generally saying that the K_S is gone, but K_L is still there. Of course he has treated this problem in his paper with Weisskopf, and Lee, Oehme, and Yang have just used the exponential approximation. But all of this is, I believe, not so terribly relevant to the charge asymmetry. I think that you would say it might be more relevant to these interference phenomena which I have shown. But then again, I had the honesty of saying that a similar experiment had been done at CERN, and I wonder why would one get a similar interference curve at a different momentum and at a different distance?

24

Superconductivity and Superfluidity

H. B. G. Casimir

I can no longer claim to be an expert on superconductivity and superfluidity: I have been unable to keep up with recent advances in theory and experiment and they are numerous and important. But I was somewhat active in this field in the thirties and it so happens that this was a rather important period in the history of the subject. And so, although I am unable to do justice to the subject in its entirety, I may at least be able to explain some features of its development.

Low temperature physics has played an important part in the early stages of quantum theory. In 1907 Einstein[1] applied the formula for the mean value of the energy of a harmonic oscillator,

$$\varepsilon = \frac{h\nu}{e^{h\nu/kT} - 1}$$

to the vibrations of atoms and molecules in a crystal. This led to an explanation of the specific heat of solids and provided a striking argument for the universal validity of the idea of a quantum of action. Einstein's simple approach in which he considers only one or a few frequencies can easily be generalized by assuming that there is a spectrum of lattice vibrations with density $\varrho(\nu)d\nu$. Then the energy content is

$$U = \int \frac{h\nu}{e^{h\nu/kT} - 1}\, \varrho(\nu)\, d\nu.$$

Of course the total number of vibrations must be equal to the number of degrees of freedom; for a simple lattice we have

$$\int \varrho(\nu)\, d\nu = 3N. \tag{1}$$

In 1912 Debye[2] made a bold and pragmatic approximation to this distribution function. He assumed that the lattice vibrations can be described as macroscopic acoustical waves. This leads to

$$\varrho(\nu) = C_D \nu^3$$

where C_D is entirely determined by the velocities of sound. Debye cuts off the spectrum at a frequency ν_m determined by Equation (1). We notice that the so-called Debye temperature Θ is defined by $h\nu_m = k\Theta$ and that the wavelength at cut-off is roughly equal to the lattice constant. Now we are here only interested in the low temperature

J. Mehra (ed.), The Physicist's Conception of Nature, 481-498. All Rights Reserved
Copyright © 1973 by D. Reidel Publishing Company, Dordrecht-Holland

limit of the specific heat and then we have

$$C_v = BT^3.$$

It is perhaps not entirely superfluous to stress that at low temperature only long waves are excited and therefore Debye's approximations are legitimate; the more we go down in temperature the better the theory becomes.* It is remarkable that in order to calculate the thermal properties of a crystal at very low temperatures we have only to know its macroscopic elastic constants.** Then, for a number of years, not much happened as far as the relations between quantum theory and low temperature physics are concerned. The Bohr theory of the atom, extremely successful in the interpretation of atomic spectra, was helpful in analyzing the behaviour of paramagnetic substances at low temperatures, but this Bohr theory did not solve the discrepancies between experiment and the classical theory of Lorentz that had found its final and most general formulation in Bohr's doctoral thesis[4]. Then quantum mechanics was created and the exclusion principle was formulated and in the course of a few years most of the essential problems were solved. I had almost said: were solved automatically but that would be doing an injustice to the brilliant work of Fermi, Sommerfeld, Bloch, Peierls and many others. Yet it is true that no entirely new concepts beyond those contained in quantum mechanics had to be introduced. Solid state physics became an important field of application of quantum mechanics; it did not contribute to the creation of quantum mechanics.

Let me recall in a very sketchy way some basic features of the theory of electrons in metals that have a special bearing on low temperature physics. Just as in the older theories the electrical current is given by

$$i = nev_d$$

where η is the number of electrons per cm^3 and v_d a drift velocity. We can write

$$v_d = \frac{e}{m} E\tau$$

where τ is the time between collisions, but whereas in Lorentz' theory we can write

$$\tau = l/v$$

where l is the free path and v the thermal velocity, we have now

$$\tau = l/v_F$$

where v_F is the velocity at the edge of a Fermi distribution. This v_F is temperature independent. Therefore, if l is temperature independent too, which is the case for

---

* At higher temperatures the situation is different. We possess today quite detailed information about the spectrum of lattice vibrations and it turns out that Debye's approximation is a very rough one indeed.

** Many years later I was able to show that also the shape-dependent thermal conductivity in ideal crystals can be calculated along similar lines.[3]

scattering by impurities, then one finds a temperature independent residual resistance. The scattering by lattice vibrations leads to a resistance proportional to T^5. At first sight this is an unexpectedly high power; it is due to the fact that at low temperatures the excited acoustical quanta are too small to change the direction of an electron by more than a small angle in one collision. At low temperatures the electrical resistance is thus given by*

$$R = R_0 + aT^5.$$

In classical theory n electrons per cm^3 would contribute to the specific heat:

$$c_v = \tfrac{3}{2}nk.$$

In Fermi–Dirac statistics, however, we have

$$c_v \approx nkT/T_D$$

where T_D is a 'degeneracy temperature' of the order of 10000 °K. The total specific heat of a metal is therefore of the form

$$c_v = AT + BT^3.$$

It so happens that in the liquid helium range for normal metals the two terms are of comparable order of magnitude. After 1930 Keesom and co-workers measured specific heats for a number of metals and alloys. If one plots C_v/T against T one obtains really a straight line and the value of A is in satisfactory agreement with theoretical predictions.

More intricate is the explanation of diamagnetism, of magneto-resistance and of thermo-electricity. But also there, a satisfactory theoretical explanation was arrived at.

Of course the successes of the theory of electrons in solids are by no means confined to low temperature physics. The theory of the structure of energy bands and the idea of positive holes, first introduced by Peierls[6] in order to explain the anomalous Hall effect, found their most spectacular application in semi-conductor physics. And we all know that semi-conducting devices have revolutionized electronics and are absolutely essential for building powerful computers.

No one can deny the successes of the theory of electrons in solids, nor the importance of its impact on technology, but there remained one obvious shortcoming: *it did not explain superconductivity.* And superconductivity had been discovered many years earlier.

Superconductivity was discovered by Kamerlingh Onnes**[7] in 1911, only three

---

* Here I am neglecting the curious phenomenon that in some impure metals the resistance increases again at very low temperatures. This effect, discovered by De Haas and Van den Berg in gold wires[5]) and studied in more detail by several investigators at Leyden, is now often referred to as the Kondo effect because Kondo was the first to propose a theoretical explanation that appears to account satisfactorily for the observed facts.

** Actually the first observations were made by Kamerlingh Onnes' assistant, G. Holst, who later became the founder and first director of the Philips Research Laboratories, but the experiments were no doubt proposed and planned by Kamerlingh Onnes.

years after the first liquefaction of Helium. It was found that the resistance of mercury became immeasurably small below a certain transition temperature. Soon afterwards it was found that also lead and tin have the same property and over the years many more metals and also alloys have been added to the list. Unfortunately it was also found that a field of no more than a few hundred gauss entirely destroys superconductivity. Kamerlingh Onnes had at once grasped the enormous potential application of superconductors for generating very high magnetic fields with practically no power. Such fields might be of technical importance and in any case they would be invaluable tools in experimental physics. These hopes were not then realized because of the low critical fields. They were realized only recently, and only to a certain extent, by means of new high critical field alloys.

More and more data on superconductors were collected, but theory made no headway. Felix Bloch's famous maxim 'Theorien der Supraleitung kann man widerlegen' (Theories of superconductivity can be refuted) stood its ground for many years. I think the main reason was that one looked at the phenomenon from the wrong angle. One had been looking for a possible phase transition, but X-ray interference experiments did not reveal any change in crystal structure. There was no change in density. There was no latent heat at the transition point. So one looked for a mechanism to explain an infinite free path in an essentially unmodified electron distribution. And that approach just did not work.

After 1930 one began to get more accurate and more subtle data on the transition phenomena. I shall describe the more important ones.

First of all it was found that for pure and perfect crystals in zero magnetic field the transition from normal resistance to zero resistance is extremely sharp – well within a few thousandth of a degree.

Next question, studied with ever greater precision since the first discovery of superconductivity: are we dealing with zero resistance or only with a very low resistance? Now I do not know any quantity in experimental physics that is as zero as the resistance of a superconductor. It is, by far, the zeroest quantity I know. I do not know exactly where the record stands today, it may be 10^{-20} or 10^{-25} times the resistance

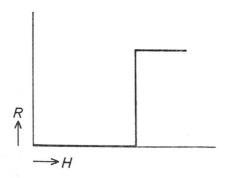

Fig. 1a. Transition curve in a longitudinal magnetic field.

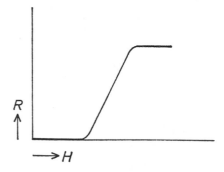

Fig. 1b. Transition curve in a transverse magnetic field.

at room temperature but, however much the methods of measurement have been refined, nobody has found a trace of resistance.

Then one studied in more detail the transition in magnetic fields.*

For a single crystal in a longitudinal field it is very sharp (Fig. 1a). In a transverse field however the situation is different (Fig. 1b).

The resistance starts to come back at a lower field, about $0.5\ H_{\text{long}}$ and reaches its normal value again at H_{long}. In the intermediate region the resistance depends strongly on the measuring current. Since H_{long} is sharply defined we can define a critical field curve (Fig. 2).

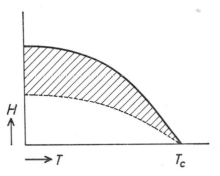

Fig. 2.   Critical field vs temperature. Full down curve: H_{long}. Dotted curve: H_{trans}. Shaded area: intermediate region.

Outside this curve superconductivity has completely disappeared. In a transverse field the transition would start at the dotted line and in the shaded region the situation is less clear-cut.

Fourthly, Keesom and co-workers, after having established the validity of the equation

$$c_v = AT + BT^3$$

for a number of metals, studied superconductors and found the behaviour indicated in Fig. 3. Above the transition point the specific heat is just as for any normal metal. At the transition point there is a sudden jump upwards – but no latent heat – and then it decreases roughly with T^3 without linear term.

The next important contribution, which had a decisive influence both on later Leyden experiments and on the work of Meissner came from Von Laue[8]. He proposed an explanation for the difference between transverse and longitudinal transition curves. His argument went as follows. Suppose a superconductor is cooled down in zero magnetic field and that then a field below threshold is switched on. What will happen? The magnetic field will not penetrate into the conductor, for if it did there would be a non-vanishing $\dot{B}$, hence a non-vanishing curl E, hence a non-vanishing E,

---

* Most of these experiments were carried out with tin. This is a very convenient material: readily available in pure form, easy to grow good single crystals of, and with a transition temperature (3.7 °K) that can be easily reached and measured.

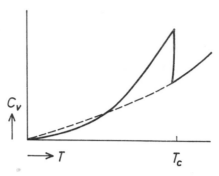

Fig. 3. Specific heat of a superconductor. The dotted line is the continuation of the 'normal curve' that would be found in fields $> H_{\text{long}}$.

hence an infinite current. You can also say this in terms of skin-effect. If the resistance gets very low the skin depth gets very small even for extremely low frequencies. Now, if we have a cylinder in a longitudinal field this eddy current effect does not appreciably modify the external field, but in a transverse field the tangential field reaches a maximum value of just twice the external field at infinity. Therefore penetration should start when the external field has reached half the longitudinal critical value.

A remarkable experiment was then carried out by De Haas and co-workers. To explain it I have first to recall the notion (an erroneous notion as was found later) of the frozen-in field. In 1919 Lippman[9] had clearly formulated the idea that the magnetic flux linked by any closed loop inside superconducting material cannot change. This rule of Lippman's explains many features of the behaviour of superconducting circuits and is also underlying Von Laue's explanation of the transition curve in a transverse magnetic field. So far, so good. But in those days it was taken for granted that when a superconductor is cooled down in a constant field no change in the magnetization will occur on passing the transition curve. But once the body is super conductive the magnetic field inside can no longer change. It will keep the value of the field in which the superconductor was cooled down, even if the external field is switched off. That is the doctrine of the frozen-in field. It is curious that nobody had taken the trouble to check it quantitatively. Fairly accurate measurements had been carried out on rings and loops, but not on solid spheres of sufficient perfection and purity. And if one uses hollow spheres or a sphere with some other imperfection there will indeed be a frozen-in field, although smaller than that predicted by the simple theory of frozen-in fields. In any case, it seems that in those days nobody had seriously questioned the doctrine. De Haas and others carried out resistance measurements on single crystals of tin that were cooled down in a constant field and that therefore were supposed to contain a frozen-in field. A year before the so-called Meissner-effect was announced, the preliminary results were communicated by De Haas during the 'Leipziger Vorträge'[10]. They can be summarized by stating that the transverse resistance curve does not depend on the magnetic history; it does not make any difference whether the wires are cooled in zero field or in a non-vanishing field. Today the conclusion seems obvious: there is no frozen-in field. But so strong was the doctrine of frozen-in fields that De Haas advocated another solution: Von Laue's theory is no good.

Yet, we of the younger generation at Leyden were not so certain that De Haas was right (and Von Laue was wrong) and experiments to test the frozen-in field doctrine had been prepared*, when Meissner and Ochsenfeld's[11] result became known, the result, you will remember, that a singly connected superconductor behaves as a body with $\mu = 0$. Suddenly the situation took an entirely different aspect. It became abundantly clear that in superconductivity we are confronted with a *new phase of the electrons*. The (longitudinal) transition curve marks a phase boundary, just as a vapour pressure curve. With due precautions one can get reversible phase transitions.** Now one can apply thermodynamics and establish relations between thermal and magnetic quantities. If f_s is the free energy of a metal in the superconducting state, in zero field, and f_n the free energy in the normal state then the critical field H_l will satisfy the equation:

$$f_s - f_n = -\frac{1}{8\pi} H_l^2 .$$

The difference in entropy is found by differentiation

$$\Delta s = \frac{1}{4\pi} H_l \frac{dH_l}{dT}$$

and for the difference in specific heat we find

$$\Delta c = \frac{T}{4\pi} \left\{ \left(\frac{dH_l}{dT} \right)^2 + H_l \frac{d^2 H_l}{dT^2} \right\} .$$

The discontinuity of the specific heat at the transition point is

$$\Delta c = \frac{T_0}{4\pi} (H_0')^2$$

where H_0' is the slope of the transition curve at the transition point T_0. This formula had already been proposed somewhat earlier by A. J. Rutgers.†[12] All these formulae were confirmed quite accurately by experiment.

The problem of finding a theoretical explanation of superconductivity appeared from then on in an entirely new light. Instead of looking for unusual conduction mechanisms one had to look first of all for a phase transition. Now one cannot learn much about the nature of a new phase by means of perturbation procedures. Starting from an ideal gas and introducing interactions one does not arrive at a workable theory

---

* In those days one had often to wait for weeks, or even months, for a run of liquid helium!

** In practice it is not at all easy to attain complete reversibility. This was already pointed out in connection with the frozen-in field.

† The thermodynamics of superconductors that was worked out by Gorter and myself[13] *after* Meissner's discovery, was thus to a certain extent anticipated by Rutgers; an early paper by Gorter[14] contains even more details.

for the elastic behaviour of crystals. One has to make an intelligent guess, pick a model and calculate the free energy. By comparing it to the free energy of the ideal gas one can derive a transition curve. Something like that is exactly what happened in superconductivity much later.

However, already in the thirties F. London[15] – partly in collaboration with his brother H. London – took a great step forward in the phenomenological theory, a step that was going to have a far-reaching influence on subsequent developments. London's *mathematics* – as distinct from London's basic ideas to which we return in a moment – were largely anticipated by a paper of Becker, Heller and Sauter,[16] which is a refinement of the argument that there can be no change of magnetic field inside a conductor. Suppose there are n electrons per cm^3 that can move without resistance. In an electric field they will be accelerated:

$$\dot{v} = \frac{e}{m} E$$

therefore*

$$\frac{di}{dt} = ne^2 \frac{E}{m}.$$

Now in a changing field we can write

$$E = -\frac{1}{c} \dot{A}$$

and therefore

$$\frac{di}{dt} = -\frac{ne^2}{mc} A.$$

If we start in a field zero with zero current then

$$i = -\frac{ne^2}{mc} A. \tag{2}$$

More generally we would have**

$$i = -\frac{ne^2}{mc} (A - A_0).$$

This equation shows that there is a penetration depth

$$l = \sqrt{\left(\frac{mc^2}{4\pi ne^2}\right)}.$$

---

* I remember that at a colloquium at Zürich in the winter of 1932–33 Pauli suggested that one should treat the electrodynamics of superconductors using this equation. At that time nobody present took up the challenge.
** This is a very simplified account of the Becker–Heller–Sauter work. In reality they take into account both the inertia of the electron and a resistance term and study in detail the transition from resistance-determined skin effect to inertia-determined skin effect.

If there is one electron per atom this is of the order

$$l \approx a_{Bohr}/\alpha$$

or, since a Bohr radius is α^2 micron about α microns. In subsequent years many experiments were devised to measure penetration depth. It was found to depend on temperature; this could be explained in terms of changing number of superconducting electrons and this again could be interpreted on the basis of a simple two fluid theory like that of Gorter and myself.[17]

London's theory, however, went a good deal further. Consider a non-relativistic Hamiltonian

$$\mathscr{H} = \frac{1}{2m}\left[p^2 - \frac{2e}{c}(A\cdot p) + \frac{e^2}{c^2}A^2\right].$$

The current now contains two terms, a kinetic term and a diamagnetic term*

$$i = \mathrm{Im}\Psi^* \frac{e}{m}p\Psi - \frac{e^2}{mc}\Psi^*A\Psi.$$

Now this diamagnetic term is exactly the same as the current in London's equation. Why then do we not find perfect diamagnetism in *all* metals? Because the kinetic term cancels the diamagnetic one. In a first order perturbation the $A\cdot p$ term will mix higher states into the original Ψ and this leads to a current that is linear in A and that almost cancels the diamagnetic term.** Almost, but not quite, there remain small quantum corrections (mathematically corresponding to the difference between a discrete sum and an integral) first calculated by Landau and then in greater detail by Peierls. These calculations led to beautiful results and stimulated important experimental work on the De Haas–Van Alphen effect, but they have nothing to do with superconductivity.

London's idea was now as follows: 'What is superconductivity?' It is a phase of the electron gas where for some reason the many particle wave function is stabilized in such a way, that the influence of perturbations by the $A\cdot p$ term is reduced. Then at least part of the diamagnetic term remains.

What about persisting currents? London's equations applied to a ring have solutions corresponding to a circulating current. This current will flow in a thin surface layer; in the interior of the material there is neither current nor magnetic field. But there is a vector potential, for the line integral of the vector potential around the ring is equal to the total flux linked. What does such a vector potential do to a wave function? If Ψ_0 is a wave function in a field free situation then

$$\Psi = \Psi_0 \exp\left[i\,(e/\hbar c)\int^{s} A\cdot ds\right]$$

* We write down the equation for one electron only, but the extension to a many electron wave function should be obvious.
** This reminds us of a well-known fallacy in the classical theory of electrons and its refutation by Lorentz and by Bohr.

is a solution in the presence of A. But if we really accept London's idea of an imperturbable many particle wave function describing a *stationary state* then this function must be a unique function. Therefore

$$\frac{e}{\hbar c} \oint A \cdot ds = n \cdot 2\pi$$

or

$$\oint A \cdot ds = n \cdot \frac{hc}{e}.$$

The magnetic flux must be quantized. This conclusion has been entirely vindicated both by theory and experiment but with one very fundamental modification. The flux quantum is not hc/e but $hc/2e$. London also thought that his argument showed that once one had explained the Meissner effect one would also have explained persisting currents. This, I think, was an error: I am coming back to this point.

I should now like to mention a paper by H. Welker.*[18] He considers a Fermi gas with a special distribution of single electron levels, viz. one that shows a gap $\Delta\varepsilon$ just above the Fermi level. To begin with the value of $\Delta\varepsilon$ is undetermined but there is assumed to exist a relation between $\Delta\varepsilon$ and a depression of the Fermi level $\Delta\varepsilon_F$. One can then write down the free energy at a temperature T as a function of $\Delta\varepsilon$ and then determine $\Delta\varepsilon$ by the conditions:

$$\frac{\partial F(T, \Delta\varepsilon)}{\partial \Delta\varepsilon} = 0 \quad F \text{ a minimum}.$$

This gives $\Delta\varepsilon$ (and hence $\Delta\varepsilon_F$) as a function of temperature. It is possible to choose the relation between $\Delta\varepsilon$ and $\Delta\varepsilon_F$ in such a way that a reasonably good fit with experimental data is obtained. $\Delta\varepsilon$ and $\Delta\varepsilon_F$ are zero above the transition temperature. Below this temperature they increase and they tend towards a constant value for $T \to 0$. That it is possible to fit the thermal data by making ad hoc assumptions about energy levels is not so surprising, but it *is* surprising that this theory comes so close to our present theories. Also it provided a 'proof' of the Meissner effect. The argument is quite simple. In a normal Fermi gas the first order perturbation of the kinetic current just compensates the diamagnetic current. Now if all matrix elements that are relevant for the first order perturbation calculation would remain unchanged but if a gap is introduced in the energy spectrum then the numerators in the expression for the current are unchanged but the denominators are increased by $\Delta\varepsilon$. Therefore the kinetic current gets smaller and part of the diamagnetic current is uncompensated. It is a beautiful argument but there is one snag: it is not gauge invariant and even a

---

* *Added in proof.* Here the author, relying too much on his personal recollections, has not given an accurate description of the historical development. Welker's papers do contain the notion of an energy gap, but not of a temperature dependent gap. (Cf) A. Sommerfeld, *Zs. Phys.* **18**, 467 (1941); see also J. C. Slater, *Phys. Rev.* **51**, 195 (1937); **52**, 214 (1937); and H. B. G. Casimir in *Niels Bohr and the Development of Physics*, pp. 118–133 (London 1955).

constant vector potential would lead to a current. Making arbitrary assumptions about energy levels without modifying the wavefunctions and hence the matrix elements is not really consistent.

Let us now look at later experimental work on superconductivity. There is a lot of it and as I said in my introductory remarks I am in no way an expert. But I think that many experiments can at least be classified – though perhaps not really explained – on the basis of what I have said so far.

There are first of all experiments dealing with superconductors that are entirely in the superconducting state. Accurate measurements of thermodynamic quantities, of thermal conductivity and of the penetration depth. Studies of the high frequency behaviour that reveal the reality of an energy gap. Studies on flux quantization. And of course always the search for new superconductors.

On the other hand there are experiments where the material is only partly in the superconducting state. It was already understood in the thirties that in the intermediate state that occurs for instance in a wire in a transverse field there must occur a domain structure of superconducting and normal matter. But then the question of surface energy enters in a critical way. In a very thin superconductor the field penetrates, therefore the magnetic free energy is reduced. So the critical field should be higher. Why can a superconductor not be riddled with very thin fibres of normal state, so that the field can penetrate whereas part of the material remains superconductive even at fields far above the value expected from the simple thermodynamics outlined above? This does not happen in the 'classical' (type 1) superconductors like Hg, Sn, Pb, and so on, but it does happen in Niobium and in high critical field alloys. In such substances (type 2) the field at which penetration begins is of the same order as in type 1 superconductors, but resistance reappears only at much higher fields. Satisfactory phenomenological theories which of necessity must involve surface energies have been developed. It turns out that one has to distinguish between the *intermediate* state where the domain structure is rather coarse and the *mixed* state which is much finer. There are still two possibilities. The fluxlines penetrating a superconductor may be free to move: this leads to a resistance which one may call a pseudo-resistance or they may be pinned to impurities or dislocations. This last situation prevails in useful high critical field alloys.*

In my lecture, as delivered at Trieste, I did not specifically mention experiments relating to the Josephson effect: I felt they might be classified as experiments on flux quantization. This may be an unjustified simplification. It is true that both flux quantization and Josephson effect provide information on the phase of a – macroscopic – wave function but the AC version of the Josephson effect tells us much more about

---

* A few historical remarks. The first (rather) high field alloys were found at Leyden in the mid-thirties and it was also found that the field strength at which penetration begins is much lower than the value at which resistance reappears[19] and so one formed a picture of a sponge-like structure. These alloys were not of much practical use because the permissible current density was low. The Russian school – Landau, Ginzburg, Abrikosov – contributed decisively to the understanding of intermediate and mixed states. The first technically useful high critical field alloys came from U.S.A.

this wavefunction than simple flux quantization, because it reveals the time dependence
of the phase of this wavefunction.

Let us now turn our attention to liquid helium itself, which in all experiments we
discussed so far, was only a faithful servant of the experimentalists. Helium becomes a
so-called superfluid below 2.2 °K. Now Kamerlingh Onnes on the very first day he
liquefied helium – 10 July 1908 – pumped away the vapour to such a low pressure
that he obtained a temperature of 1.8 °K. So on that memorable day he had not only
liquefied helium, he had also produced superfluid helium... without noticing anything
particular about it. And he continued not to notice anything particular about it until
his death in 1927. It was his successor Keesom* who started to find remarkable
properties.

As we have seen, in the case of superconductivity one had first discovered the
spectacular effect of zero electrical resistance. Much later one found the thermodynamic properties and realized that one was dealing with a new phase of the electrons.
It is an amusing twist of history that in the case of liquid helium the development
went just the other way round. There one first discovered the existence of a phase
transition. The specific heat was measured as a function of temperature (Fig. 4).

There is a transition at 2.19 °K without latent heat but with a discontinuity in
specific heat. The transition point is usually referred to as the λ-point and helium
above and below the λ-point is referred to as helium I and helium II respectively. The
first measurements were carried out at saturated vapour pressures but Keesom went
further: together with Clusius and later with his daughter, Miss A. P. Keesom he

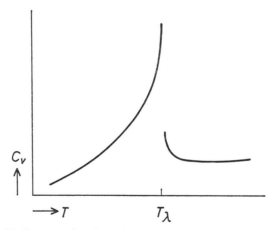

Fig. 4.   Specific heat of helium as a function of temperature.

* After the death of Kamerlingh Onnes, work at the Leyden laboratories was directed by two
colleagues, De Haas and Keesom, who were not always on the best of terms although their work
was beautifully complementary. There has even been some petty bickering about who of the two
was really successor to Kamerlingh Onnes. The obvious – though possibly not the legally correct –
answer is: both.

determined the specific heat as a function of temperature and pressure and found a transition line in a p, T diagram. These results prompted Ehrenfest[20] to introduce the idea of – and to work out the thermodynamics for – transitions of higher order. In any phase transition the free energies of the two phases are equal at the transition point. In a first order transition there is a discontinuity of the first derivative of the free energy, that is the entropy: there is a latent heat. In a second order transition there is a discontinuity in the temperature derivative of the entropy: there is a discontinuity in the specific heat.*

It was this theory of Ehrenfest that prompted A. J. Rutgers to formulate the equation mentioned before.

Still, one did not know that helium below the point is a very peculiar liquid. Then Keesom undertook to measure heat conductivity and for good reasons. In general thermal conductivity can be expressed as a product (specific heat) × (velocity of propagation) × (free path) and since the specific heat is behaving in a remarkable way one might expect a remarkable behaviour of the thermal conductivity as well. The experiments were performed by Miss A. P. Keesom for her Ph.D.[22] and the results were most remarkable indeed. There was an almost infinite thermal conductivity, which moreover did not follow classical rules; for instance heat transport was by no means proportional to the temperature gradient.

At that moment technicians in the Lab said they were not surprised. If you pass a current through a little heating spiral immersed in liquid helium then you see bubbles above 2.19 °K but you don't get bubbles at lower temperatures. But nobody had paid attention to this fact. This may be connected with the views of Kamerlingh Onnes who emphasized quantitative measurements rather than qualitative observations. 'By measurement to knowledge' was a favourite maxim of his.** It sounds better in Dutch: 'Door meten tot weten', for then it rhymes. It is my belief that in this case the laws of shift of vowels have done considerable harm to Dutch physics. But one must not be too harsh in one's judgement. After all, Keesom and his school produced a very valuable set of data, they did find a special phase and they did discover the very high thermal conductivity. Somewhat later Kapitza[23] in Russia and Allen and Misener[24] at Cambridge found an infinitely low viscosity or a very fast transport of liquid helium II through incredibly narrow capillaries or slits. Just previous to that discovery Burton

---

* Ehrenfest's ideas have been criticized by Justi and Von Laue[21] who pointed out that if two free energy curves touch each other they cannot intersect. This difficulty disappears if we assume that the free energy of one of the phases depends on an internal parameter α which because of its physical nature must be $\geqslant 0$. At a transition point $F_1(T_c) = F_2(T_c, 0)$. For $T < T_c$ we can determine α by making $F_2(T, \alpha)$ a minimum, that is from the equation $\partial F_2/\partial \alpha = 0$. It is essential that the dependence of F_2 on α is non-linear. A typical example is provided by a gas at constant volume. If it is cooled down then at a certain temperature droplets begin to form. The parameter α is here the amount of liquid present. As a matter of fact in all transitions of the second kind there is such an internal parameter: the saturation moment in ferro-magnetism, a long-range-order parameter in order–disorder transitions, a width of gap in superconductivity and so on. At Trieste it was Wigner who initiated the discussions on this point.

** At a recent summer school at Varenna I learnt that Fermi took a different point of view. He was reported to have said to a student: 'You may find agreement with theory; then you have carried out a measurement. But if you are lucky you will find disagreement. Then you have done an experiment'.

at Toronto had measured viscosity by a rotating disk method; he had found a difference in viscosity between helium I and helium II but not a very striking one. This may be explained by an – admittedly very crude – argument. We may regard liquid helium II as a kind of mixture of a superfluid state and a normal liquid, just in the same way as we say that in a superconductor we are dealing with a mixture of normal electrons. (responsible for heat transport) and superconducting electrons. In a capillary the normal part gets stuck, the superfluid part goes through. A rotating disk is not retarded by the superfluid part, but it sets the normal part into motion and that leads to damping. The experimental fact that the viscosity determined by a rotating disk method is entirely different from that derived from the flow through capillaries is in itself a most impressive deviation from classical hydrodynamics. Again technicians in the Leyden Lab were not surprised. They had noticed that you can pump against a leak in a vessel immersed in liquid helium as long as the temperature is above 2.19 °K; at lower temperatures it is hopeless. And again nobody had paid much attention to this fact.

Another striking phenomenon is the Rollin [26] film. Liquid helium II covers all walls with a thin film that can move quite fast. This film sees to it that the level of the liquid becomes equal at both sides of a dividing wall rising out of the liquid. You could not build useful dikes in liquid helium: it would rapidly crawl over the top, a disconcerting thought for a Dutchman. There is one direct and amusing consequence. Keesom had tried to obtain very low temperatures by reducing the vapour pressure and, as a good experimentalist who knew his kinetic theory of gases, he constructed a fairly wide Dewar vessel and connected it by generous tubes to a powerful pump. He did not get very far; we know today that was because the helium film was running up against the wall and there it evaporated, which led to a high rate of evaporation. Rollin pumped through a diaphragm with a small hole. This would appear to be a poor way of pumping but the evaporation is so much reduced that you get a lower pressure with a much smaller pump.

Also this phenomenon had been seen at Leyden: in 1922 Kamerlingh Onnes – notwithstanding his preference for quantitative measurements he was a keen observer himself – reported that helium can crawl rapidly over the edge of a vessel so as to equalize the level inside and outside. This phenomenon was probably thought of in terms of normal capillarity and I sometimes wonder whether people in those days may have associated it with lavender oil. Older experimentalists may still remember the use of lavender oil in a process for platinizing glass. This process was well known in the famous Leyden glassblowing shops. Now lavender oil has the unpleasant property of crawling out of bottles. The idea that liquid helium behaved somewhat like lavender oil may not have looked too far-fetched.

There are, of course, many other effects: the fountain effect, persistent vortices, vortex quantization, analogies to the Josephson effect, etc.

Again F. London [27] took a decisive step towards a theory by comparing the λ-transition to Bose–Einstein condensation. Now I do not think that theoreticians agree as to whether this type of condensation has much to do with the actual properties of liquid helium II. As a matter of fact – as was shown by Landau – a crude two fluid model can be turned into a much more satisfactory phenomenological theory by assuming that there is a gap in the spectrum of possible excitations of the liquid. In the Bose–Einstein model there is no gap of any kind. On the other hand the Bose–Einstein model has some properties that are, to say the least, suggestive. In a Bose gas without interaction a wave function for the whole system can be written as a symmetrized product of single particle wave functions and below the transition point a finite fraction of all particles is in the lowest state. Not in just a low state but in the one and only lowest state:

$$\Psi = S\left[\psi_0(x_1)\,\psi_0(x_2)\dots\,\psi_0(x_{pN})\cdot\Psi(x_{pN+1}\dots x_N)\right] \tag{3}$$

where $p<1$ but of order 1 (*not* of order $1/N$); Ψ_0 is the lowest single particle wave-function and S indicates symmetrization. Now Yang has formulated a criterion for superfluidity that makes more precise the ideas of a 'macroscopic wavefunction'. Write down the off diagonal density matrix $\varrho(x_1, x_2)$; Yang's criterion is

$$\varrho(x_1; x_2) \sim N \quad \text{even when} \quad |x_1 - x_2| \sim L$$

where N is the total number of particles and L the dimension of the vessel in which they are enclosed. It is obvious that the wavefunction of Eq (3) fulfills this condition and to my knowledge the ideal Bose–Einstein gas is the only model that has been treated completely rigorously and that shows this property.

I shall now try to say a few words about the theory of superconductivity, the real theory, the B.C.S. theory. An essential feature of the theory is that it treats the electrons as a Fermi gas with a small attraction between electrons. Now it is not so easy to arrive at such an attraction. At first sight one might expect to find a strong repulsion. Heisenberg even once tried to base a theory of superconductivity on this repulsive force and one of his main troubles was that this force is much too high to lead to phenomena of the right order of magnitude. We have first of all to get rid of that repulsion and that asks for a good deal of sophistication. Roughly one may say, the electrons in the Fermi gas are not really electrons, they are 'dressed' electrons. Two electrons do not really repel each other because of the highly polarizable medium formed by all the other electrons. The remarkable thing is that the simple picture of a non-interacting Fermi gas works so well, not only for labelling the excited states but even for calculating transport phenomena. Having got rid of the repulsion we have to find attraction. This comes about as a second order effect of the interaction with the lattice waves. The physical explanation is, as far as I can see, that one electron deforms the lattice in such a way that the other electron likes to be in the neighbourhood. The example of a double bed with a sagging mattress has occasionally been used, although a French colleague of mine criticized it by remarking that in that case there is no initial repulsion.

What happens in a Fermi gas with attraction? An exact solution is difficult but it is possible to make a very satisfactory 'Ansatz' by first combining two electrons with opposed momentum and spin vectors into so-called Cooper pairs. Then a wave function is constructed out of such Cooper pairs, which also contains states above the Fermi level, but which leads to an enhancement of the energy of attraction. The total result is a reduction of the energy and the creation of an energy gap. There exist also solutions with less depression of energy and a smaller gap, just as in Welker's model. One can then calculate free energy, specific heat, etc. The next step is the proof of the Meissner effect. The method is not too different from a Welker–London type of argument. Because of the gap the perturbation-modified kinetic current is smaller than in the ideal Fermi gas and therefore part of the diamagnetic current remains. Bardeen himself once told me that he had been profoundly impressed by this idea of London that 'a wave function in a superconductor does not change much on application of a field'. There has been some initial difficulty with gauge invariance, just as in Welker's theory. To the best of my knowledge these difficulties have been overcome.

Finally one has to explain persisting currents: as you see the theory proceeds in opposite direction to the historical development. It is not true that the Meissner effect implies the existence of persisting currents. The 'Meissner state', the state with $B=0$ is the state of lowest free energy. A ring with a circulating current is not in a state of lowest free energy, the lowest state is obviously the state with zero current. The fact that a current in a ring can be described by London's equations is not enough: we have to show that fluctuations will not lead to a dying out of the current. As a matter of fact, there is one type of fluctuations that will certainly lead to a dying out of the current: in principle temperature fluctuations might raise a whole cross-section, a whole slice of sausage, above the transition point. Of course this is extremely improbable – unless the wire is very thin and at a temperature just below the transition point – but the argument shows that a separate proof is necessary. After all, there might be other fluctuations, less clumsy than those I described.

Yang has pointed out that here flux quantization comes to the rescue. If we plot the free energy of a ring against current we obtain a parabolic curve. But because of flux quantization only discrete points on this curve correspond to reality. In between states might exist but their free energy would be higher. Therefore the state with a persisting current is a relative minimum. The current could only diminish by a jump of at least one whole flux quantum but that is a big jump, involving many electrons and therefore extremely improbable.

I should like to emphasize once more that flux quantization has really been found and that the flux quantum is $hc/2e$. This is a most beautiful experimental proof of the essential correctness of the idea of Cooper pairs.

I should like to make two final remarks. The first one is, that it seems clear that the description of the superconducting state as one large quantum state is correct but we have to remember that it is a quantum state of a many particle system.* We cannot

---

* Its mathematical treatment requires tools beyond the simple formalism of wave mechanics, viz. a further development of the methods of second quantization. Many authors and especially Bogolubov have contributed to the further refinement and justification of the B.C.S. theory.

understand it on the basis of independent single electron wave functions. Now I suppose the younger generation is better at these things than I am, but, while I have learnt to think in terms of wave functions for one particle, I am not so good at visualizing, or realizing the impossiblility of visualizing, wave functions for very large numbers of particles. Sometimes it seems to me that the very careful analysis of observability in one electron quantum mechanics that was carried out by Heisenberg, Bohr and others has not been carried out in full for many-particle wave functions, perhaps because of complete trust in the mathematical formalism. Maybe this idea of mine is just a symptom of old age, but I do feel slightly unsatisfied about this aspect.

My second remark refers to the question: 'Will we ever get superconductors at much higher temperatures?' That is, in a way, a technical question but an extremely important one. It is obvious that the technical and economic value of a superconductor at room temperature – and even of a superconductor at liquid air temperature – would be almost unlimited but at present the chances for finding it are very small. That makes it difficult for a commercial enterprise to decide whether they should invest in this type of search or not: the value of such an investment is $0 \times \infty$. However, I think the consensus of opinion is that as long as we confine ourselves to homogeneous alloys and as long as the attraction is exclusively due to coupling with lattice waves there is little chance that we shall find much higher transition points. But if it would be possible, for instance by some kind of layer structure or by imbedded organic molecules, to obtain an attraction between electrons by some additional mechanism the situation would change. I do not venture any predictions.*

REFERENCES

(*Leiden Communications* and *Communications from the Kamerlingh Onnes Laboratory* are referred to as *Comm.*)

1. A. Einstein, *Ann. Physik* **22**, 180 (1907).
2. P. Debye, *Ann. Physik* **39**, 789 (1912).
3. H. B. G. Casimir, *Physica* **5**, 495, (1938).
4. N. Bohr, Thesis, Copenhagen (1911); reprinted in collected works (ed. Rosenfeld), Amsterdam (1972).
5. W. J. de Haas, J. de Boer and G. J. van den Berg, *Physica* **1**, 1115 (1934).
   W. J. de Haas and G. J. van den Berg, *Physica* **3**, 440 (1936); *Physica* **4**, 683 (1937).
   W. J. de Haas, H. B. G. Casimr and G. J. van den Berg, *Physica* **5**, 225 (1938).
   (Also reprinted in *Comm.* 233b, 241d, 249b, 251c.)
6. R. Peierls, *Z. Phys.* **53**, 255 (1929).
7. H. Kamerlingh Onnes, *Comm.* 120b, 122b, 124c.
8. M. von Laue, *Phys. Z.* **37**, 793 (1932).
9. Lippman, *C. R. Acad. Sci. Paris*, **168**, 73 (1919).
10. *Leipziger Vortäge* (ed. P. Debye), Leipzig (1933).
11. W. Meissner and R. Ochsenfeld, *Naturwiss.* **21**, 787 (1932).
12. See note in P. Ehrenfest, in ref. 20.
13. C. J. Gorter and H. B. G. Casimir, *Physica* **1**, 306 (1934).

* During the discussion Cooper pointed out that there is at present no experimental evidence for another type of interaction. The fact that some metals do not show an isotope shift of the transition temperature is no indication of a new mechanism. It requires only a slight modification of the formalism to get interaction through lattice wave and no isotope shift.

14. C. J. Gorter, *Arch. Teyler* **7**, 378 (1933).
15. See F. London, *Superfluids I*, New York and London (1950), for further references.
16. R. Becker, G. Heller, F. Sauter, *Z. Physik* **85**, 772 (1933).
17. C. J. Gorter and H. B. G. Casmir, *Z. Techn. Physik*, **15**, 539 (1934).
18. H. Welker, *Z. Techn. Phys.* **16**, 606 (1938); *Phys. Z.* **39**, 920 (1938).
19. W. J. de Haas and J. M. Casimir-Jonker, *Proc. Amst.* **38**, 2 (1935); *Comm* 233c.
    J. M. Casimir-Jonker and W. J. de Haas, *Physica* **2**, 935 (1935); *Comm.* 237c.
20. P. Ehrenfest, *Proc. Amst.* **36**, 153 (1933); *Comm. Suppl.* 75b.
21. E. Justi and M. von Laue, *Z. Techn. Phys.* **15**, 521 (1934).
22. W. H. Keesom and A. P. Keesom, *Physica* **3**, 359 (1936); *Comm.* 242g.
    For other references on work of that period W. H. Keesom, Helium, Amsterdam (1942).
23. P. Kapitza, *Nature* **141**, 74 (1938).
24. J. F. Allen and A. D. Misener, *Nature* **141**, 75 (1938).
25. K. F. Burton, *Nature* **135**, 265 (1935).
26. B. V. Rollin, *Actes du 7e Congr. Intern. du Froid*, La Haye and Amsterdam **1**, 187 (1936).
27. See F. London, *Superfluids II*, New York and London (1954), for further references.

Part III

Statistical Description of Nature

25

Problems of Statistical Physics

George E. Uhlenbeck

1. INTRODUCTION

This morning session is devoted to problems of statistical physics, and I hope that together with my colleagues Dr Kac and Dr Cohen we will be able to show you that in this field very little is really understood, and that there remain quite general and fundamental problems which are still quite open. Since in fact even the formulation of these problems is often still controversial, let me begin with what I consider to be the basic task of statistical mechanics. It is, in my opinion, the elucidation of the *relation* between the microscopic, molecular description and the macroscopic description of the physical phenomena. This is illustrated in Fig. 1, where the microscopic description is divided according to wherther the classical or the more basic quantum theory is considered, and where only some of the macroscopic disciplines are shown. Note that I said relation and not explanation. I consider the two descriptions of nature as autonomous. It is of course true that sometimes it is possible to calculate a macroscopic property (as for instance the viscosity coefficient) from the

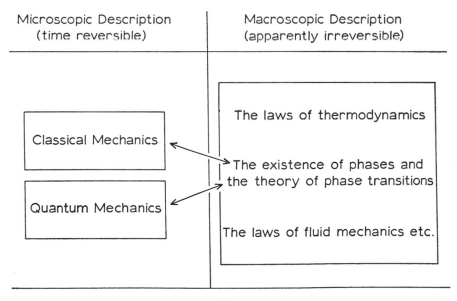

| Microscopic Description (time reversible) | Macroscopic Description (apparently irreversible) |

Classical Mechanics

Quantum Mechanics

The laws of thermodynamics

The existence of phases and the theory of phase transitions

The laws of fluid mechanics etc.

Fig. 1.

J. Mehra (ed.), The Physicist's Conception of Nature, 501–513. All Rights Reserved
Copyright © 1973 by D. Reidel Publishing Company, Dordrecht-Holland

intermolecular forces. This is very valuable as a check, but I think the more difficult and interesting questions are of a more qualitative nature. How can the macroscopic description be reconciled with the molecular dynamics? Why can one ignore in a macroscopic theory the molecular constitution of matter and still make valid predictions?

The most famous and well-established macroscopic laws are the laws of thermodynamics and statistical mechanics, in fact, started from the attempt of Boltzmann and Gibbs to reconcile these laws with classical mechanics. You know that this seemed impossible because of the apparently irreconcilable conflict between mechanics and thermodynamics with regard to the role of time. According to thermodynamics, an isolated system (that is from the molecular point of view a conservative and bounded mechanical system consisting of a large number of molecules) will, in the course of time, approach a state of thermal equilibrium in which all macroscopic variables have reached steady values. This is the so-called *zeroth law of thermodynamics*, and it is in conflict with the *Poincaré theorem*, according to which *any* bounded and conservative mechanical system will execute a quasi-periodic motion and show no trace of going to equilibrium.[1]

I believe that Boltzmann and Gibbs have found the answer to this deep dilemma and that the methods they developed reveal the general outline of *classical* statistical mechanics. There are still many unsolved problems but they are more or less 'bien posés', although very difficult![2] You will hear more about such problems from Dr Kac and Dr Cohen. I would like to discuss especially the question whether the quantum mechanics has produced an essential change with respect to the understanding of the relation between the molecular and the macroscopic description of nature.

2. THE RELATION BETWEEN QUANTUM MECHANICS AND THERMODYNAMICS

Can the basic methods of Boltzmann and Gibbs, which were developed for the classical theory, be 'grafted' in a natural way onto the laws of quantum mechanics? I think, the generally accepted opinion is that the quantum mechanics has *not* affected in any fundamental way the principles of statistical mechanics and especially not the relation with thermodynamics. I believe that this view goes back to the 1929 paper of J. von Neumann[3], which was supported and further clarified by Pauli and Fierz in 1937[4]. Let me quote from the discussion remarks of Pauli at the famous Florence conference of 1948 on the foundations of quantum statistical mechanics[5]. Pauli had first remarked that in quantum statistics not sufficient attention was given to closed systems, which is always the starting point for classical statistics. Then he goes on and I quote: 'The reason is that in quantum mechanics the observation is a process irreversible in principle accompanied by an incontrollable amount of interaction between the observed systems and the measuring instruments. It was often argued that for this reason a consideration of closed systems in statistical thermodynamics would be senseless and that the entropy increase is caused by the external influences accompanying the observations[6]. I do not believe, however, that such an argument can be maintained; on the contrary I defend the standpoint that the otherwise so important difference between classical and quantum mechanics is not relevant in principle for thermodynamical questions.'

I subscribed to this view for a long time, but in the last years I began to develop many qualms, mainly since I learned more about the superfluid properties of He II, and since I studied for the nth time in my life the curious behaviour of an ideal Bose gas. I now believe that the von Neumann–Pauli view depends *crucially* on the so-called coarse-graining procedure or on the equivalent notion of the macroscopic observer. In the classical theory one knows that such a procedure or notion is *necessary* in order to resolve the dilemma between the zeroth law of thermodynamics and the Poincaré theorem which I mentioned in the introduction. In the quantum theory, already for an ideal Bose gas, one is led to the notion that sometimes a *single quantum state* can be *macroscopically occupied*, which means that a finite fraction of the total number of particles occupies that quantum state. I think, that any two-fluid description of He II requires this notion in some form. One can even say, I believe, that the fact that the superfluid has entropy zero (as shown by the reversible thermo-mechanical effect) shows empirically the validity of this basic notion. Clearly for a macroscopically occupied single state there is no coarse-graining, and therefore one may doubt whether the approach to equilibrium or the zeroth law of thermodynamics can be explained as in the classical theory. Of course, one *knows*, that under certain circumstances there can be *persistent currents* in which the superfluid moves as if it had strictly no viscosity. The extremely slow decay of these currents takes so many years, that it is clearly qualitatively different from the classical irreversible processes. It is almost surely of a purely quantum theoretical nature.

In the following I will try to elaborate these remarks somewhat.

3. THE GENERAL ARGUMENT FOR THE APPROACH TO EQUILIBRIUM

Let me begin with a quick sketch of the classical Boltzmann–Gibbs argument[7], since it sets the pattern, I believe, for the quantum theoretical von Neumann–Pauli–Fierz argument.

Classically one says, that since thermal equilibrium is a macroscopic notion, one must introduce a small number of macroscopic variables $Y_i = f_i(x_1, ..., x_N)$, where $i = 1, 2, ..., m$ and $m \ll N$, which are functions of the phases x_k of the N particles, and whose values characterize the macroscopic state of the system. Assuming for simplicity that these values are discrete, a set of values of the $Y_i (\equiv$ macroscopic state) will correspond to a *region* in Γ-space. Assuming in addition that for large N there is one set of values of the Y_i which corresponds to a region which is overwhelmingly the largest, one arrives at the picture that a given macroscopic description will divide the given energy shell of the system into a number of *fixed* 'cells' $A_0, A_1, A_2, ...$, of which one (say A_0) is very large (see Fig. 2).

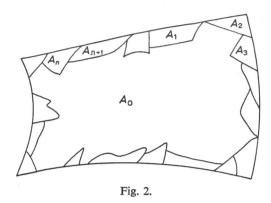

Fig. 2.

To describe the temporal evolution of the system, one should start *not* with a product of δ-functions for all the phases x_i, which may be called a pure state, since one only has macroscopic information. Instead the initial state should be represented by an appropriate probability distribution or in the language of Gibbs by an appropriate ensemble. For instance if one knows the initial macroscopic state, then one should start with a distribution $D(0)$ which is constant in the corresponding cell, say A_1, and zero everywhere else. The change $D(0) \rightarrow D(t)$ determines the evolution. Gibbs describes it about as follows. The volume of the region where $D \neq 0$ remains the same according to Liouville's theorem, but the shape will change drastically. The initial volume will be drawn out to a very long and thin ribbon, which more and more will wind through the whole energy shell, so that finally the probability that the system is in any of regions A_k, will be determined by the volume of that region. The *micro-canonical* distribution is then reached. Identifying A_0 with the thermal equilibrium state one can then say that any system starting in the macroscopic state A_1 will almost always go to the equilibrium state and once it is in that state it will almost always stay there, although fluctuations away from equilibrium will and *must* occur, because of the quasi-periodic character of the motion of the Γ-point.

Clearly this resolution of the basic dilemma depends on Gibbs' picture of the change $D(0) \rightarrow D(t)$. One calls it the *phase mixing process*, and although perhaps plausible it is of course not proved. It should be a property of the mechanical equations of notion, but it is notoriously difficult to prove. Note still: (*a*) The approach to uniformity over the energy shell will continue when t increases although each Γ-point will move quasi-periodically. (*b*) Since the distribution will always consist of a thin filament more or less uniformly wound through the energy shell, the time it takes until uniformity is reached will depend essentially on how one judges the uniformity, that is to say on the size of the finite regions A_k, which in turn are determined by the *chosen* macroscopic description. (*c*) The picture does *not* depend essentially on the total number of particles N. If N increases then the sharpness of the distribution of the areas of A_k around A_0 (and therefore the distribution of the times each macroscopic state is realized) will increase and the fluctuations around the equilibrium state A_0 will decrease.

Finally let me remind you of the so-called Gibbs H-theorem which was intended to formalize the phase mixing picture. One distinguishes between the *fine-grained* probability density $\varrho(x_1, \ldots, x_N, t)$ over the energy shell and the *coarse-grained* density $P_\alpha(t)$ obtained by averaging the fine-grained density over each of the regions A_α, so that:

$$P_\alpha = \frac{1}{|A_\alpha|} \int \ldots \int_{A_\alpha} \varrho \, dx_1 \ldots dx_N.$$

Define the H-function by:

$$H(t) = \sum_\alpha |A_\alpha| P_\alpha \ln P_\alpha.$$

Now *assume* at $t=0$ that the fine-grained and coarse-grained densities are the same. This is the preparation of the initial state. At a later time, because of the streaming of the ensemble fluid according to Liouville, one gets the phase mixing of the different values of P_α so that each region A_k will now contain a combination of the original values $P_\alpha(0)$. One proves easily:

(a) $H(0) - H(t) \geqslant 0$.

(b) H is a minimum for the completely uniform distribution. This *suggests* that any coarse-grained distribution will become the microcanonical distribution, but it is of course not a proof, because one does not show that starting from $t=0$ the streaming will continue to diminish H till the minimum value is reached.

Quantum mechanically one says that a pure state is described by a *single* wave function Ψ. One can not fix all the dynamical variables of the system. They are restricted by the uncertainty relations, which follow from the basic probability assumptions of the quantum theory. However, just as in the classical theory a macroscopic observer never knows precisely the dynamical state of a large system, similarly in the quantum theory one never knows the state of the system so precisely that it is only restricted by the uncertainty relations inherent in any *microscopic* observation. For the *macroscopic* description one must introduce a large number of wave functions Ψ_α and assume that each is realized with the probability w_α. This is called a *mixed* state and it therefore involves a statistical probability distribution in addition to the basic quantum mechanical probability assumption. They can be combined by introducing the concept of the *density matrix*. The Ψ_α can be taken in any representation. Using for instance the complete set of free particle states $u(k_1 \ldots k_N, r_1 \ldots r_N) \equiv u(K, R)$, which are products of single particle wavefunctions where one must of course (with identical) particles take into account the Bose or Fermi statistics, one can write:

$$\Psi_\alpha(R, t) = \sum_K a_\alpha(K, t) u(K, R).$$

The density matrix is then defined in this representation by

$$\varrho(K, K', t) = \sum_\alpha w_\alpha a_\alpha(K, t) a_\alpha(K', t).$$

The matrix $\varrho(K, K', t)$ is hermitean, with trace equal one, and the diagonal elements are the probabilities that the ensemble is in the state K. The $\varrho(K, K', t)$ is the analogue of the classical *fine grained* probability distribution $\varrho(\mathbf{R}, \mathbf{P}, t)$; it determines the ensemble average of any physical quantity and it evolves in time according to the quantum Liouville equation, which is again time reversible. To complete the quantum mechanical macroscopic description, one must coarse grain by averaging over 'cells' G_i which consists of a large number of quantum states. This leads to a coarse-grained density matrix (P_{ij}) and one can make plausible by a similar quasi-proof as in the classical case, using a quantum mechanical H-theorem, that for a closed system in the course of time P_{ij} will become the coarse-grained microcanonical density matrix:

$$P(E_i, E_j) = w(E_i)\delta(E_i - E_j)$$

with:

$$w(E_i) = \begin{cases} \text{const. for } E_0 < E_i < E_0 + \Delta E_0 \\ 0 \quad \text{otherwise} \end{cases}$$

and where E_i are the coarse-grained energy values.

This is only a rough sketch meant to explain the role of the coarse-graining in the general argument. For more details and also for what has actually been proved I refer to the papers of von Neumann and Pauli and also to the book of Tolman[8], where the classical and quantum theory are discussed in parallel.

4. THE IDEAL BOSE GAS AND THE PROBLEM OF THE BOSE CONDENSATION

The role of the coarse-graining is especially clear in the usual treatment of the ideal Bose gas, where one starts from the fine-grained description of the state z of the gas defined by the distribution $n_1, n_2, n_3 \ldots$ over the discrete translational energy levels $\varepsilon_1, \varepsilon_2, \varepsilon_3 \ldots$ of a molecule in the volume V. Without restrictions the energy level $E = \sum n_s \varepsilon_s$ of the whole gas would be degenerate and would correspond to $N!/n_1! \, n_2! \ldots$ different wave-functions. Restriction to the symmetric combination of wave-functions reduces the weight or a-priori probability of the state E to one, so that one can not speak of a most probable state. One now argues that this fine-grained description is much too precise and that for a macroscopic description of the state of the gas one must coarse-grain the distribution by introducing 'cells' containing groups of $G_1, G_2, \ldots$ energy levels. The macroscopic distribution Z is then defined by the numbers $N_1, N_2, N_3 \ldots$ of molecules in the successive cells, which have an average energy $E_1, E_2, E_3 \ldots$. Taking the weight of each fine-grained state to be one, one then finds for the probability of the coarse-grained distribution Z:

$$W(Z) = \prod_i \frac{(G_i + N_i - 1)!}{N_i!(G_i - 1)!}$$

by a well known combinatorial argument[9].

As a function of the N_i, $W(Z)$ has a sharp maximum for a given total number of particles N and total energy E, if N is large, and identifying the most probable state with the thermal equilibrium state, one shows, quite similar as in the classical case, that in equilibrium one gets the Einstein–Bose distribution:

$$\bar{N}_i = \frac{G_i}{\dfrac{1}{A}\, e^{\beta E_i} - 1}$$

where the two constants A and β must be determined from N and E.

Note that because of the sharp maximum of $W(Z)$ one can assume, just as in the classical case, that the time $t(Z)$ during which the state Z is realized is proportional to $W(Z)$ and one then understands the approach to equilibrium, the zeroth law of thermodynamics. Also the explanation of the other laws of thermodynamics is completely similar as in the classical theory. One finds that the entropy S is still connected with $W_{\max}$ by the Boltzmann relation:

$$S = k \ln W_{\max} = - k \sum_i G_i \{F_i \ln F_i - (1 + F_i) \ln (1 + F_i)\}$$

with $F_i = \bar{N}_i/G_i$. The constant β is still connected with the absolute temperature by $\beta = 1/kT$. The constant A is related to the chemical potential and must be determined from the given total number of particles N. It is well known that this condition leads to a paradox (I call it the Einstein paradox) because one gets (with $\lambda = h/(2\pi mkT)^{1/2}$):[10]

$$\frac{N}{V} \equiv \frac{1}{v} = \frac{1}{\lambda^3} \sum_{k=1}^{\infty} \frac{A^k}{k^{3/2}}$$

and since A can be at most one (because otherwise some $\bar{N}_i$ would be negative), there is apparently a maximum density and for given V N can not be bigger than:

$$N_c = \frac{V}{\lambda^3} \sum_k \frac{1}{k^{3/2}} = 2.61 \frac{V}{\lambda^3}.$$

But N is given! What happens if N is larger than N_c. Einstein argued that the remaining $(N - N_c)$ 'condense' in the lowest (fine-grained!) energy state $\varepsilon_1 = 0$ and do not contribute to *any* of the thermodynamic properties of the gas. Especially for $v < v_c = = V/N_c$ the pressure would remain at the constant value reached for $v = v_c$ and thus one is led to the well-known net of isotherms of the ideal Bose gas, which looks like half of the isotherm net of a real gas (because the volume of the 'condensed' phase is zero), except that there is no critical temperature.

In this way one is led to the macroscopic occupancy of a single quantum state. The argument has often been criticized, not only because of its ad-hoc character, but especially because suddenly the lowest state ε_1 is re-introduced and has thermodynamic consequences, which is in conflict with the idea that only the coarse-grained distribution has macroscopic significance. One often says [11] that for $v < v_c$, one should discard the coarse-grained description and start from the canonical partition function for the fine-grained distribution z. If one *postulates* that the Helmholtz free energy

$\psi(T, V, N)$ is given by:

$$\exp\left[-\psi(T, V, N)/kT\right] = {\sum_{(n_i)}}' \exp\left[-\frac{1}{kT}\sum_i n_i \varepsilon_i\right]$$

where the first sum goes over all values of the occupation numbers n_i with the condition (denoted by the prime) that the total number is N, then one can justify the Einstein results by a careful discussion of the thermodynamic limit $N \to \infty$, $V \to \infty$, $V/N = v$ finite and one finds indeed that in this limit the lowest energy level ε_1 is macroscopically occupied if $v < v_c$. This may seem satisfactory, but by discarding the coarse-grained description one looses the understanding of the approach to equilibrium and at least the usual justification of the canonical ensemble becomes problematic. So there is, I think, a real dilemma!

5. REMARKS ON SUPERFLUID HELIUM

Whether the Bose condensation has something to do with the He I–He II transition and especially whether the macroscopically occupied groundstate is an indication of the superfluid component, are controversial questions. I believe – and I think F. London would have agreed with me – that the theory of the ideal Bose gas gives an important hint for the understanding of the phenomena in superfluid helium. But, of course, for a real theory one must at least introduce repulsive forces between the particles and one must be able to derive the non-equilibrium and especially the hydrodynamical properties of the system from the basic postulates of quantum statistical mechanics. This is clearly a whole programme! Many interesting attempts have been made to develop a theory of non-ideal Bose systems and especially of He II, which I can not possibly review here. Let me therefore remind you only of two basic experimental results, which show the completely unclassical behaviour of He II and provide therefore a real theoretical challenge.

(a) *The thermo-mechanical effect*.[12] If two vessels of He II are connected through a very narrow constriction (a so-called superleak), then one observes that an increase ΔT of temperature in one vessel produces a flow through the superleak towards the higher temperature producing a pressure Δp. The effect can be described by saying that for slow flow the two vessels remain in equilibrium so that the chemical potential μ remains constant. This implies that:

$$\Delta\mu = -S\Delta T + \frac{\Delta p}{\varrho} = 0$$

or:

$$\frac{\Delta p}{\Delta T} = \varrho S$$

where S is the entropy and ϱ the density of the whole fluid. This relation (first derived by H. London) is well verified. In addition one finds that the effect is completely reversible for small ΔT and from the geometrical dimensions of the superleak one can conclude that the heating of the incoming fluid occurs as if it had originally the entropy zero!

(b) *The persistent current effect.*[13] Take a cylindrical vessel packed with a fine powder and filled with liquid helium. If one rotates the vessel at a temperature above T_λ, then all the liquid will rotate and contribute to the moment of inertia. If one cools the vessel, while it rotates, to a temperature below T_λ and if one then stops the rotation, at least part of the fluid continues to rotate as shown by the fact that the cylinder has internal angular momentum and behaves like a top if one tries to change its axis.

Landau[14] has incorporated both these effects in his two-fluid hydrodynamical equations which replace for He II the classical Euler equations. Landau describes the state of He II by the two velocity fields $\mathbf{v}_s$ and $\mathbf{v}_n$ of the so-called superfluid and normal fluid plus two thermodynamic variables, say the density ϱ and the entropy s. I will not write out the complete set of eight equations. They are the five conservation laws for the mass, the entropy and the components of the momentum which are quite similar to the classical equations. Then, and this is the main difference from the classical case, there is the equation of motion for $\mathbf{v}_s$. Landau assumes that the superfluid carries *no* entropy and that it moves *exactly* like an ideal fluid according to the Euler-like equation:

$$\frac{D\mathbf{v}_s}{Dt} \equiv \frac{\partial \mathbf{v}_s}{\partial t} + (\mathbf{v}_s \cdot \vec{V}) \mathbf{v}_s = - \operatorname{grad} \mu. \tag{1}$$

This remains valid (with a small addition to the chemical potential μ) if dissipative effects are taken into account as was done later in a systematic way be Khalatnikov.[15]

The assumption, that the superfluid moves *strictly* without dissipation and carries no entropy, clearly accounts for the persistent current and for the thermo-mechanical effects which I mentioned earlier. It seems to me that it is also consistent with the view that the molecular origin is a macroscopically occupied single quantum state. This would explain in a natural way why the entropy is zero and why the motion is time reversible. One may object that there is no reason yet to use the word *quantum* state, which implies that the quantum theory is essential for the description of the superfluid. But the Landau–Khalatnikov equations *do not contain Planck's constant*, although they describe the flow properties of He II apparently as well as the Navier–Stokes equations describe a classical fluid, and although they predict new phenomena (second sound!) which are experimentally verified. So where does the quantum theory come in?

We now know that the two-fluid equations are incomplete because they do *not* explain the macroscopic quantization conditions nor the macroscopic interference effects, which are the most surprising discoveries of the last decade. Let me mention only the quantization of the superfluid vorticity in He II found by Rayfield and Reif in 1963,[16] From the analysis of the motion of ions captured by vortex rings in He II these authors concluded that:

$$\oint \mathbf{v}_s \cdot \mathrm{d}s = n \frac{h}{m} \tag{2}$$

where n is an integer, h Planck's constant and m the mass of a helium atom. Note that such a quantization condition is not in conflict with the Landau equation of motion for v_s, since even with dissipation it remains of the form (1). The vorticity theorems are therefore *strictly* valid. If at $t=0$, curl $v_s=0$, it says zero, and for a contour moving with the superfluid the circulation remains constant, or:

$$\frac{D}{Dt} \oint v_s \cdot ds = 0. \tag{3}$$

One can therefore *impose* the condition (2) on the Landau–Khalatnikov equations just as one can impose the Bohr–Sommerfeld quantization conditions on the action integrals in the classical mechanics of multi-periodic systems.

This suggests that at least the superfluid should be described by a macroscopic wave equation which is time reversible and which would imply (2). Many attempts have been made in this direction, but I think we still lack a general macroscopic theory for a quantum fluid. The macroscopic occupancy of a *single* quantum state seems to me an important clue and the major difficulty is how to combine this idea with the normal, ergodic, time irreversible behaviour produced by all the other quantum states.

REFERENCES AND NOTES

1. The Poincaré theorem expresses the reversibility of the mechanical equations of motion. For a proof see for instance Chapter 1 of *Lectures in Statistical Mechanics* by G. E. Uhlenbeck and G. W. Ford, published by the American Mathematical Society, Providence, Rhode Island (1963). There one finds also a detailed discussion of this basic dilemma, which I have called the problem of Boltzmann.
2. An attempt of such an outline with a list of the unsolved problems I presented at the conference on *Fundamental Problems in Statistical Mechanics*, Vol. II, published by North-Holland Publishing Company, Amsterdam (1968), p. 1.
3. J. von Neumann, *Z. Phys.* **57**, 30 (1929) [= *Collected Works*, Pergamon Press, New York (1961), Vol. I, p. 558].
4. W. Pauli and M. Fierz, *Z. Phys.* **106**, 572 (1937) [= *Collected Papers*, Interscience Publishers, New York (1964), Vol. II, p. 797].
5. At this conference Onsager proposed that the vorticity in superfluid helium should be quantized in units h/m, which was shown to be true experimentally by Rayfield and Reif in 1963! The proceedings of the conference were published as a supplement to Vol. VI, Series IX of *Nuovo Cimento* in 1949. The discussions are still of great interest. Pauli's remarks are on p. 166 (= *Collected Papers*, Vol. II, p. 1116.)
6. This would imply that the cause of the entropy increase would be *quite* different in quantum mechanics from the Gibbs phase mixing process in classical statistics.
7. I follow the discussion given in my *Lectures in Statistical Mechanics*, see note 1.
8. R. C. Tolman, *The Principles of Statistical Mechanics*, Oxford, The Clarendon Press (1938). Compare especially Chapter VI, Section 51 and Chapter XII A.
9. Compare P. Ehrenfest, *Collected Papers*, North-Holland Publ. Comp. Amsterdam (1959), p. 353.
10. Note that *because* one has a coarse-grained distribution, the sum over the $\bar{N}_i$ is really an integral.
11. See for instance E. Schrödinger, *Statistical Thermodynamics*, Cambridge University Press, Cambridge (1967), p. 67.
12. Also called the fountain effect. See J. F. Allen and A. D. Misener, *Proc. Roy. Soc. London* A **172**, 467 (1939) and especially P. Kapitza, *J. Phys. U.S.S.R.* **5**, 59 (1941), *Phys. Rev.* **60**, 354 (1941).
13. J. D. Reppy and D. Depatie, *Phys. Rev. Letters* **12**, 187 (1964). For another version of the

experiment see H. Kojima, W. Veith, S. Putterman, E. Guyon and I. Rudnick, *Phys. Rev. Letters* **27**, 714 (1971).

14. L. D. Landau, *J. Phys. U.S.S.R.* **5**, 71 (1941) [= *Collected Papers*, Gordon and Breach, New York (1965), p. 301].
15. I. M. Khalatnikov, *An Introduction to the Theory of Superfluidity*, W. A. Benjamin, New York (1965), Ch. 9. In this book the paper of Landau quoted in 14 is also reprinted.
16. G. W. Rayfield and F. Reif, *Phys. Rev. Letters* **11**, 305 (1963); *Phys. Rev.* **136**, A 1194 (1964).

DISCUSSION

G. Ludwig: It has been said that probability comes in because of a lack of knowledge. There is, however, the physical problem of preparing the macroscopic systems that we have in nature. It is not our knowledge of the system which goes into this case. In the classical sense, the question is: How can a physicist prepare or find a macroscopic system in nature? Every system is not an isolated system for all time. It was in contact with the universe a long time ago and, therefore, it is not possible for the physicist to have such singular points in space that give rise to the diminishing of entropy. This is a physical problem, and not a problem of our knowledge.

My second point concerns the characterization given by the words 'coarse-graining' and 'macroscopic observables'. The word coarse-graining is a too specialized way of describing the passage from microphysics to macrophysics. I have the impression that the concept of the macroscopic variables has been used so as to comprise quantum mechanical phenomena as well. I believe that the difference between quantum mechanics and macrophysics is not so much that it is forbidden that Planck's constant h should come into the macroscopic description, but that there are no *complementary* observables to the macroscopic observables. Do you think that in macrophysics there are observables which are complementary in the same sense as p and q in quantum mechanics?

G. E. Uhlenbeck: Well, I have my doubts about that. I must admit that it is an open question. When I say that h doesn't occur in the macroscopic description, I just say it *practically*. When we write Landau's equations, which discard the microscopic phenomena, these equations do not contain h. That is all. There is no h. Now and then suddenly h appears, because new phenomena are discovered, such as the quantization of flux and interference in which h appears, which shows quite clearly that of course this is not a complete microscopic description of a quantum system. We are still looking for that.

R. E. Peierls: Your exposition was so clear and beautiful that it makes it very easy to say where one disagrees. I would like to take a point, and as far as I could follow it is somewhat related to what Dr Ludwig said. I agree that the coarse-graining used by von Neumann and Pauli has its limitations, and in some cases it cannot be applied. I also think that it is completely unnecessary. We do not need any coarse-graining, and the point is simply this. The naive view is that whatever state you solve from in a system, it will always necessarily reach complete statistical equilibrium in respect of every variable. That, we know, is just not true, as for example in the Poincaré cycles and recurrence. You have to make a somewhat limited state. You can limit it by saying that you don't want to look at everything in detail, that you only use coarse observations. And that is basically what von Neumann said. But you can also alternatively say that you are not willing, and it seems to me physically much more reasonable, to start the system off from a completely crazy general initial state where you take each molecule and put it into some peculiar corner and start looking in a peculiar way, because we never do that. That's not interesting.

If you restrict your initial state reasonably, then you can assert that, in fact, you will reach thermodynamic equilibrium for any observation, even a fine-grained one. Now what is this reasonable initial state? One very simple way of describing it is to say, and I think it covers all the experiments we are actually capable of doing, that you start from a system in thermal equilibrium and then alter any number of macroscopic parameters – you open a tap, or you put on a field, or you move a piston, or anything like that. A macroscopic parameter is one which comes into the Hamiltonian through the single particle or two or three particles. That covers all the experiments that we are capable of doing. That is enough. Following, for example, the method of Van Hove, one can show that from then on you really reach the fine-grained density matrix which is the correct one. Let me say that I don't agree that coarse-graining is necessary for the irreversibility, which has completely different origin.

Comment: My comment concerns the relation between macroscopic occupation of single particle states and superfluidity. My doubt is that we know that the most important manifestations of super-fluidity occur in systems which are in some way either quasi two-dimensional or quasi one-dimensional. On the other hand we have general theorems which state that the macroscopic occupation of states cannot occur in one dimension at finite temperatures, or in two dimensions either. I wonder therefore whether there there can be an essential relation between macrosopic occupation of a single quantum state and superfluidity.

G. E. Uhlenbeck: I don't know. It sounds sensible that there should be a relation, but I think that if one really would understand it then there would have to be this macroscopic quantum theory which I tried to suggest. But at present it is only a suggestion.

C. N. Yang: Maybe I could comment on the question which was just asked. I understand the question as asking that there is a theorem about the lack of magnetization in one and two dimensions, which would make it impossible to have superfluid situations. In the first place, I don't think the proof is completely relevant to the question of macroscopic occupation of some single particle states. The proof is for the lack of permanent magnetization; it's not quite the same as a proof of the lack of existence of macroscopic occupation of some single particle state. Secondly, the experimental statement that there exist one or two dimensional superfluid phenomena is not quite clear.

L. N. Cooper: The argument actually refers to fluctuations that will occur in one or two dimensional systems, and not to the macroscopic occupation of the single particle state. Now in fact these fluctuations have been seen in various cases. In a superconductor, if one goes to a two-dimensional or a one-dimensional system one actually sees fluctuations near the transition temperature, but the fact that there are fluctuations in and out of the state does not mean that one doesn't have a large occupation, or macroscopic occupation if you wish, of a single quantum state.

I have some further comments on the question Professor Uhlenbeck raised about the relation between the two-fluid model and the quantum mechanical nature of this model. It's true that h doesn't occur in the two-fluid model – h is necessary, as far as I know, for the existence of the superfluid phase, but it is just that the magnitude of h does not appear. If h were zero, the superfluid would not exist, but in dealing with the superfluid when one does not consider excitations the magnitude of h does not come in, and this is not entirely different from the hydrogen atom. The existence of the ground state requires h, but h appears explicitly only if one is considering the energy of the ground state or the separation between the first and the second levels, among other places. Actually when one deals with the excitations of the superfluid, and one of these is in a sense the state of circulation, then h begins to appear, but there are other places in more complicated excitations when one introduces perturbations and then h does appear.

There do exist the analogues of what one might call macroscopic quantum theories for the super-fluid phase. For example, the variation for the superfluid phase and in a magnetic field is given by the Ginsburg–Landau equation, which one might call a macroscopic quantum equation. And then, for example, in such things as the [A. C.] Josephson effect one can, by dealing with the wave function of the superfluid, talk about its oscillation in time across a weak link. To that extent, I believe, these equations do exist.

Professor Uhlenbeck: I quite agree with you. I think the main difficulty is to do it for finite temperature properly.

L. N. Cooper: That makes it harder, yes.

Comment: There is something deep and magical in the things which Professor Uhlenbeck has discussed. There was a remark about London in the early thirties, that superfluidity involves quantum mechanics at the macroscopic scale. This has become amply evident during the last ten years or so, and if there is quantum mechanics at the macroscopic scale then of course there is complementarity. If there is any complementarity between fields and particles, then that is there because one of the essential features of the Bardeen–Cooper–Schrieffer theory is that there are definite phase relationships and an indeterminate number of particles. In that sense statistical mechanics, or its meaning as we talk about it, looks different from that of the early thirties when one didn't have this feature.

During the last ten years or so there has been so much talk about quantum vortices and the like. There is a simple, very approximate argument that there are one-particle condensations, and that one can explain why there are quantum vortices. But on the many-body level, it is by no means simple to describe quantum vortices. One would think that something as simple as that should be very easy to grasp. It may be an indication that we look at it in a different way in quantum mechanics, and perhaps the old Bohr–Sommerfeld quantization integrals will come back again and then this will be simpler than it is now.

M. Kac: I would like to make a comment on one of the things. If you stick to the ideal Bose gas, which is perhaps unrealistic, but it is a bona-fide quantum mechanical system, then there is a theory which was somehow completely overlooked. It was given in 1957 by Blatt and Butler, and it showed that if you put the Bose gas in a cylinder and rotate it, then indeed below a certain critical velocity the condensate is stationary, and at the critical velocity the condensate develops a vortex. That is rather remarkable, because in a sense it contradicts one of the fundamental tenets of quantum mechanics. What happens is that when you look at it you find that the velocity field is that of a vortex, and because of this the phase is observable. Now you measure, or at least you believe you can measure, the velocity field of a fluid. If you have macroscopic occupancy of a state, as in this situation, you are also in the difficulty that you have to explain how it is that phases that are not supposed to be observable, all of a sudden become observable.

J. M. Jauch: Professor Uhlenbeck has explained that there are the supercurrent and the ordinary current, and the supercurrent has a quantum mechanical origin in the microscopic theory, but then the h disappears again in the phenomenological description. I wonder whether Professor Uhlenbeck could give a few more details as to why one can quantize both kinds of systems, the ordinary current and the supercurrent? What is the physical reason that normally one does not see the quantum effects in normal currents, but one can relatively easily see the quantum effects in the interference phenomena in the super currents?

G. E. Uhlenbeck: I don't know whether I quite understood the question. In my view, which is not a reductionist one, I don't think that one should say that the macroscopic phenomena can be derived from quantum mechanics. It is another description and one must see to it only that the two are in harmony with each other. It is exactly in the same sense that you can say that, if the molecules are like hard balls, as Laplace thought, then everything is known. I would say: No, still nothing is known. We still do not know how a macroscopic observer would describe an enormous assembly of them. The basic question is about the measurability of these macroscopic quanta, for which I have no answer.

26

Phase Transitions

Mark Kac

I feel a bit strange in talking about the development of the *physicist's* conception of nature because I am not a physicist by training, and I could therefore watch this development only from the sidelines. Although this position often affords a better view, it also distorts it and makes it easy to miss some of the essential plays.

Fortunately the topic which has been assigned to me is, at least today, largely mathematical, and the only experimental fact that is of any relevance to my discussion is that *all* known gases when cooled below a certain critical temperature liquefy if subjected to sufficiently high pressures.

Liquefaction of gases beginning with the discovery in 1868 of the critical point by Andrews and culminating in the liquefaction of helium by Kamerlingh Onnes in 1908 is, of course, one of the great chapters of physics. But while this development coincided in time with the development of Statistical Mechanics, the two had remarkably little influence on each other.

The reason is that the successes of the van der Waals theory were so impressive that no one thought that there was any problem left.

Well, the problem is still with us and while not in the centre of the stage with elementary particles and black holes, it stands there as a stern reminder that nature, in the words of Fourier, is indifferent toward the difficulties it causes mathematicians.

1

It is now universally accepted (at least as far as classical statistical mechanics is concerned) that the canonical ensemble is the proper way to describe a system in equilibrium with a heat bath of a given temperature. And yet even as late as 1937 doubts were raised at the van der Waals centenary Congress in Amsterdam as to whether phase transitions can be explained solely on the basis of the canonical ensemble.

Although it seems obvious that no further physical principle is needed, it is also rather difficult to see how a phenomenon so dramatic from a mathematical point of view can be deduced from a principle so innocent in appearance as that embodied in the canonical ensemble.

The problem is really twofold:

(1) Are phase transitions derivable, in principle at least, from statistical mechanics? and

(2) By what *mathematical* mechanism do phase transitions arise?

J. Mehra (ed.), The Physicist's Conception of Nature, 514-526. All Rights Reserved
Copyright © 1973 by D. Reidel Publishing Company, Dordrecht-Holland

Largely owing to the work on the two-dimensional Ising model especially to the celebrated solution of Onsager, the answer to the first question is undoubtedly yes.

The second question however, still awaits a definitive answer, and it is well to be reminded that even today there is no proof, from first principles that a gas does indeed condense.

2

The central theme of the theory of phase transitions has from the beginning been the derivation of the equation of state, and this is easy to understand in view of the interest in liquefaction of gases. This however tended to obscure a number of questions which in a way were even more basic.

Of these the question of coexistence and geometric separation of phases is probably the most fundamental, and it brings out well the subtleties of the subject.

Let me review briefly what is involved here.

First of all, a *sharp* phase transition can only occur in the limit of infinitely extended systems (although even this elementary point was rather slow in penetrating the consciousness of physicists). Since the density is non-zero, this implies that also the number of particles must also become infinite.

In other words, we should consider the so-called thermodynamic limit

$$N \to \infty, \qquad V \to \infty, \qquad \frac{N}{V} - \varrho - \frac{1}{v}, \tag{2.1}$$

where N is the number of particles, V the volume of the container v and ϱ the number density (or v the specific volume). Actually we must require that not only the volume V of the container v become infinite but that 'all dimensions' of the container become infinite.

For sufficiently low temperatures we expect the isotherm to have the well known form shown in Fig. 1, and the question is how should the fact, that for v between v_1 and v_2 we have actually two phases, be described in statistical mechanical terms.

It was, I believe, Norbert Wiener who first raised this point and who also suggested the answer.

This answer is in terms of distribution functions

$$\bar{n}_k (\mathbf{r}_1, ..., \mathbf{r}_k; v)$$

which are defined in the usual way, i.e.,

$$\bar{n}_k (\mathbf{r}_1, ..., \mathbf{r}_k; v) = \lim_{\substack{N \to \infty \\ V \to \infty \\ \frac{V}{N} = v}} \frac{N!}{(N-k)!} \frac{\int \cdots \int \exp\left\{ -\frac{U(\mathbf{r}_1, ..., \mathbf{r}_N)}{kT} \right\} d\mathbf{r}_{k+1} ... d\mathbf{r}_N}{\int \cdots \int \exp\left\{ -\frac{U(\mathbf{r}_1, ..., \mathbf{r}_N)}{kT} \right\} d\mathbf{r}_1 ... d\mathbf{r}_N} \tag{2.2}$$

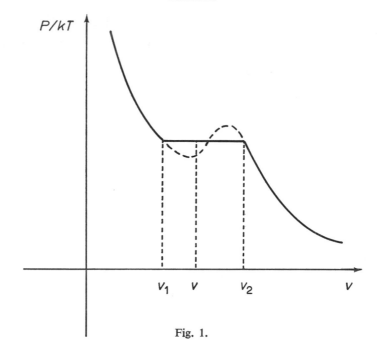

Fig. 1.

where $U(\mathbf{r}_1, \ldots, \mathbf{r}_N)$ is the interaction energy usually assumed in the form

$$U(\mathbf{r}_1, \ldots, \mathbf{r}_N) = \sum_{1 \leqslant i < j \leqslant N} \phi(\|\mathbf{r}_i - \mathbf{r}_j\|) \tag{2.3}$$

where $r_{ij} = \|\mathbf{r}_i - \mathbf{r}_j\|$ is the distance between the ith and the jth particles.

Since there are no external fields (expect the one confining the particles to the container) one should expect that for v between v_1 and v_2 we have 'blobs' of gas and liquid infinite in extent (because of the thermodynamic limit) and that therefore if one of the k particles involved in $\bar{n}_k(\mathbf{r}_1, \ldots, \mathbf{r}_k; v)$ is in a gaseous or a liquid 'blob' then so are the rest of them. It therefore stands to reason that $\bar{n}_k(\mathbf{r}_1, \ldots, \mathbf{r}_k; v)$ should be a linear combination of $\bar{n}_k(\mathbf{r}_1, \ldots, \mathbf{r}_k; v_1)$ and $\bar{n}_k(\mathbf{r}_1, \ldots, \mathbf{r}_k; v_2)$ the coefficients being the probabilities that a particle picked at random will be in the liquid blob (v_1) or in the gaseous blob (v_2). These probabilities are clearly

$$\frac{\xi_1 v_1}{v} \quad \text{and} \quad \frac{\xi_2 v_2}{v} \tag{2.4}$$

where ξ_1 and ξ_2 are the mole fractions of the two species, i.e.,

$$\xi_1 + \xi_2 = 1, \qquad \xi_1 v_1 + \xi_2 v_2 = v. \tag{2.5}$$

Thus one should expect that for $k = 2, 3, \ldots$*

$$\bar{n}_k(\mathbf{r}_1, \ldots, \mathbf{r}_k; v) = \frac{\xi_1 v_1}{v} \bar{n}_k(\mathbf{r}_1, \ldots, \mathbf{r}_k; v_1) + \frac{\xi_2 v_2}{v} \bar{n}_k(\mathbf{r}_1, \ldots, \mathbf{r}_k; v_2). \tag{2.6}$$

---

* For $k = 1$ one simply has the second of the formulas (2.5).

3

To appreciate the significance of formulas (2.6), let me consider their bearing on the van der Waals theory.

This theory as we all know was the first successful theory of the equation of state for a non-ideal gas. When it was first proposed in a doctoral dissertation in 1873, it made a justly strong impression on the physical community of the day as witness the following words of Maxwell:

The molecular theory of the continuity of the liquid and gaseous states forms the subject of an exceedingly ingenious thesis by Mr Johannes Diderick van der Waals, a graduate of Leyden. There are certain points in which I think he has fallen into mathematical errors, and his final result is certainly not a complete expression for the interaction of real molecules, but his attack on this difficult question is so able and so brave, that it cannot fail to give a notable impulse to molecular science. It has certainly directed the attention of more than one inquirer to the study of the Low-Dutch language in which it is written.

In terms which are easily understandable to us* the van der Waals theory is based on the assumption that the interaction energy of a gas can be taken to be of the form

$$U(\mathbf{r}_1, ..., \mathbf{r}_N) = U_{\text{h.s.}}(\mathbf{r}_1, ..., \mathbf{r}_N) - \binom{N}{2}\frac{\alpha}{V}. \qquad (3.1)$$

Here $U_{\text{h.s.}}$ stands for the hard sphere part of the interaction (i.e., $+\infty$ whenever two particles are closer than 2δ, the diameter of the hard sphere), and the form

$$-\binom{N}{2}\frac{\alpha}{V}, \quad (a > 0), \qquad (3.2)$$

of the contribution of attractive forces is based on the assumption that the range of the attractive force is so large that there are many molecules in the action sphere of each molecule and that therefore the potential energy of attractive forces can be replaced (for almost all configurations of molecules) by an average of the form (3.2).

One is now easily led to the equation of state

$$\frac{p}{kT} = \frac{p_{\text{h.s.}}}{kT} - \frac{a}{v^2}, \qquad a = \frac{\alpha}{2kT} \qquad (3.3)$$

with $p_{\text{h.s.}}$ being the pressure of the gas of hard spheres.

Van der Waals then also argued that

$$\frac{p_{\text{h.s.}}}{kT} = \frac{1}{v - b} \qquad (3.4)$$

* The derivation given below is due to L. S. Ornstein, Dissertation, Leiden, 1908.

with

$$b = v_0 = \tfrac{4}{3}\pi\delta^3$$

which is surely incorrect (except in one dimension).

The main difficulty with the van der Waals theory (apart from the fallacy of (3.4) and the fact that intermolecular attractive forces turned out to be short ranged) is that the interaction (3.2) is *volume dependent*. The price for this transgression against fundamental tenets of statistical mechanics is the famous 'wiggle' which was one of the errors to which Maxwell alluded in the quote above and which he proceeded to correct by introducing his equal area rule.

The flat part of the isotherm thus made its appearance by grafting on a *thermodynamic argument* to a statistical mechanical theory. Since thermodynamics is presumably contained in statistical mechanics, the need to appeal to thermodynamics appears to us now as an act of intellectual desperation. But in the days of van der Waals and Maxwell, the logical interrelations between thermodynamics and what was then called 'molecular science' were far from clear, and soon the spectacular successes of the van der Waals theory had anyway rendered objections on grounds of methodological purity a bit silly. In physics one does not argue with success and in fact, by an unwritten law, one is practically prohibited from so doing.

But gradually the difficulties and imperfections of the van der Waals theory became so evident that the pendulum swung the other way, and the famed equation of state was pushed down from the pinnacle it once occupied and demoted to the lowly role of a useful empirical approximation.

During the past few years we have understood completely how the van der Waals equation fits into the scheme of things when a series of investigations by Uhlenbeck, Hemmer, and myself have culminated in the following theorem of Lebowitz and Penrose [ref. 1]:

If the interaction potential is of the form

$$\phi(r) = \begin{cases} +\infty, & 0 < r < \delta \quad \text{(hard core)} \\ -\gamma^3 f(\gamma r), & r > \delta \end{cases} \tag{3.5}$$

then in the limit $\gamma \to 0$ the equation of state becomes that of van der Waals *augmented* by the Maxwell equal area rule provided only f satisfies certain conditions. Failure to satisfy these conditions may lead to an equation of state different from that of van der Waals, an interesting discovery of Gates and Penrose [ref. 2]. It should also be pointed out that even in one dimension with $\phi(r) = -\gamma e^{-\gamma r}, r > \delta$ one obtains in the limit $\gamma \to 0$ the van der Waals equation of state augmented by Maxwell's rule. Moreover, one can now calculate γ corrections, and one obtains van der Waals like isotherms to *all* orders in γ. And yet the actual isotherm is smooth and there is no singularity!

To make matters definite, I shall take

$$f(r) = \alpha e^{-r}$$

which is admissible. Using the Virial Theorem (which is a rigorous consequence of the canonical formalism)

$$p = \frac{kT}{v} + \tfrac{2}{3}\pi\delta^3\bar{n}_2\left(\delta^+; v\right)$$

$$- \tfrac{2}{3}\pi \int_\delta^\infty dr\, r^3 \frac{d\phi}{dr} \bar{n}_2\left(r; v\right) = p_{\text{h.s.}} - \tfrac{2}{3}\pi\alpha \int_{\gamma\delta}^\infty dx\, x^3 e^{-x}\bar{n}_2\left(\frac{x}{\gamma}; v\right),$$

(where $p_{\text{h.s.}}$ as before is the pressure of the gas of hard spheres of radius δ) we can inquire what happens in the limit $\gamma \to 0$. If we were to assume that

$$\lim_{\gamma\to 0} n_2\left(\frac{x}{\gamma}; v\right) = \frac{1}{v^2}$$

we would be led to

$$p = p_{\text{h.s.}} - \frac{4\pi\alpha}{v^2}$$

which for low enough temperatures would exhibit the 'wiggle' and be therefore wrong.
On the other hand since for

$$v = \xi_1 v_1 + \xi_2 v_2$$

we have (flatness of the isotherm)

$$p(v) = p(v_1) = p(v_2) = p_{\text{h.s.}}(v_1) - \frac{4\pi}{v_1^2} = p_{\text{h.s.}}(v_2) - \frac{4\pi}{v_2^2} = \frac{\xi_1 v_1}{v} p(v_1) + \frac{\xi_2 v_2}{v} p(v_2),$$

it follows that

$$\lim_{\gamma\to 0} n_2\left(\frac{x}{\gamma}; v\right) = \frac{\xi_1 v_1}{v}\frac{1}{v_1^2} + \frac{\xi_2 v_2}{v}\frac{1}{v_2^2}$$

which is consistent with (2.6).
Thus for the van der Waals theory at least there is an intimate relation between the flatness of the isotherm and the linear combination property (2.6), or, in other words, there are indeed two phases.

4

It may seem that to deal with the question of geometric separation of phases one is compelled to introduce correlation functions and hence go beyond thermodynamic considerations which are based only on the partition function.

This is not quite so and I shall now show how one can formulate the problem in terms of the partition function.

The question to ask is the following:

Let $z(V; v)$ be determined so that

$$\frac{N}{V} = \varrho = \frac{1}{v} = \frac{1}{V}\frac{d}{dz}\log G(z; V) \tag{4.1}$$

where

$$G(z, V) = \sum_{N=0}^{\infty} \frac{Q_N(V)}{N!} z^N \tag{4.2}$$

is the grand partition function, i.e.,

$$Q_N(V) = \int_V \cdots \int_V \exp\left[-\frac{U(\mathbf{r}_1, \ldots, \mathbf{r}_N)}{kT}\right] d\mathbf{r}_1 \ldots d\mathbf{r}_N. \tag{4.3}$$

What is the distribution of N/V in this ensemble?

We can define the probability that N/V lies between a_1 and a_2 by the obvious formula

$$\text{Prob.}\left\{a_1 < \frac{N}{V} < a_2\right\} = \frac{\displaystyle\sum_{a_1 V < N < a_2 V} \frac{Q_N(V)}{N!} z^N}{G(z; V)} \tag{4.4}$$

where $z(V, v)$ is determined from (4.1).

What is now the limit

$$\lim_{V \to \infty} \text{Prob.}\left\{a_1 < \frac{N}{V} < a_2\right\}$$

as $V \to \infty$?

The first inclination is to answer

$$\int_{a_1}^{a_2} \delta(x - \varrho)\, dx = \begin{cases} 1, & a_1 < \varrho < a_2, \\ 0, & \text{otherwise}. \end{cases} \tag{4.5}$$

But a moment's thought will make us doubt (4.5) if there are more than one phase. In fact, for two geometrically separated phases, one would expect instead of $\delta(x-\varrho)$

$$\frac{\xi_1 v_1}{v} \delta(x - \varrho_1) + \frac{\xi_2 v_2}{v} \delta(x - \varrho_2), \tag{4.6}$$

and indeed one can show (although not with complete rigour) that formulas (2.6) imply (4.6). I am not quite sure about the converse, but for the time being it is not of paramount importance.

5

Let me now turn to the quantum mechanical case. Here our ignorance is truly staggering, and one may even entertain doubts as to whether the (fine grained) canonical ensemble is the correct (and complete) way of describing equilibrium.

Taking however the canonical ensemble as a starting point, we can easily derive the analogues of (4.5) and (4.6) for the ideal Bose gas.

The result is as follows:

(*a*) For $T > T_c$ the limiting probability density of N/V given that the average of N/V is $\delta(x - \varrho)$

(*b*) For $T < T_c$ the limiting probability density if

$$
\sigma(x) = \begin{cases} 0, & (x < \varrho_c), \\[2mm] \dfrac{1}{\varrho - \varrho_c} \exp\left\{-\dfrac{x - \varrho_c}{\varrho - \varrho_c}\right\}, & (x > \varrho_c), \end{cases} \tag{5.1}
$$

where ϱ_c is the critical density.

Thus for $T < T_c$ when a condensate is present we have a completely different situation than in the case of a classical gas. In particular, there is *no* geometric separation of phases and the k particle distribution function approaches ϱ^k as the distances between the particles become infinite, even for $T < T_c$, i.e., in the presence of the condensate. There is therefore no *diagonal* long range order.

I strongly suspect that in the presence of repulsive interactions, no matter how small, the limiting probability density of N/V will be $\delta(x - \varrho)$ so that an interacting Bose gas will appear as a one phase system. In other words, the exponential result (5.1) is a fluke of the ideal Bose gas. That for low enough temperatures a macroscopic description of such systems has to be nevertheless based on a two fluid picture is therefore quite striking.

6

Returning to classical case, we might still entertain doubts about statistical mechanics being a sufficient basis for the description of phase transitions because the case of the van der Waals theory may appear not wholly convincing owing to the necessity of taking the strange limit

$$
\gamma \to 0.
$$

Let me therefore conclude my discussion by considering the only other case for which everything is more or less understood, namely, the two-dimensional Ising model with nearest neighbour interactions. To be sure, this is not the case of a real gas, but after the work of Yang and Lee we know how to interpret the Ising model as a 'lattice gas' and faute de mieux this will have to do.

Consider a $N \times M$ square lattice and let a 'scalar spin' $\mu = \pm 1$ be assigned to each of its points P. The interaction energy of the system is assumed to be of the form

$$
E = - J \sum v(P, Q) \mu_P \mu_Q, \quad (J > 0), \tag{6.1}
$$

where

$$
v(P, Q) = \begin{cases} 1 & \text{if } P \text{ and } Q \text{ are nearest neighbours} \\ 0 & \text{otherwise}. \end{cases} \tag{6.2}
$$

In the absence of an external magnetic field, the average magnetization is 0. We can inquire now what is the *distribution* of the magnetization in this case. We shall, of course, be interested in the thermodynamic limit $N \to \infty$, $M \to \infty$.

Magnetization $m(H; T)$ as a function of the external field looks like the graph in Fig. 2 for $T > T_c$ but for $T < T_c$ the graph is shown in Fig. 3. For the $T > T_c$ one would expect the distribution of the magnetization to be $\delta(m)$ but for $T < T_c$ one would expect it to be the linear combination

$$\tfrac{1}{2}\delta\big(m - m(0^+)\big) + \tfrac{1}{2}\delta\big(m - m(0^-)\big). \tag{6.3}$$

In particular we would expect for $T < T_c$ that

$$\lim_{\substack{N \to \infty \\ M \to \infty}} \frac{\sum \left(\dfrac{\sum \mu_P}{MN}\right)^2 \exp\left[-\dfrac{E}{kT}\right]}{\sum \exp\left[-\dfrac{E}{kT}\right]} = m^2(0^+). \tag{6.4}$$

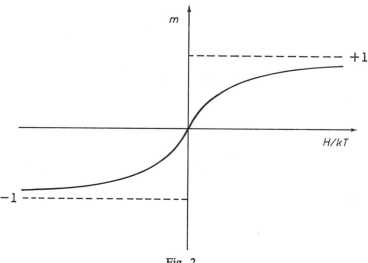

m

$+1$

H/kT

-1

Fig. 2.

It is easy to convince oneself (though a proof is needed) that the left side of (6.4) is the residual correlation at infinity, i.e.,

$$\lim_{d(P, Q) \to \infty} \varrho(P, Q) \tag{6.5}$$

where

$$\varrho(P, Q) = \lim_{\substack{N \to \infty \\ M \to \infty}} \frac{\sum \mu_P \mu_Q \exp\left[-\dfrac{E}{kT}\right]}{\sum \exp\left[-\dfrac{E}{kT}\right]} \tag{6.6}$$

and $d(P, Q)$ denotes the distance between P and Q.

Formula (6.4) states therefore that the residual correlation at infinity is equal to the square of the magnetization at field zero.

Residual correlation at infinity was calculated by Onsager and Kaufman and magnetization at field zero by C. N. Yang (leaving however unsettled a small but subtle point of rigour). Both calculations required great ingenuity, and even after many years of further work no really simple way has been found to re-derive these fundamental results.

Only very recently has A. Martin-Löf [ref. 3] succeeded in proving (6.4) directly (i.e., circumventing the explicit formulas of Onsager-Kaufman and Yang), and even

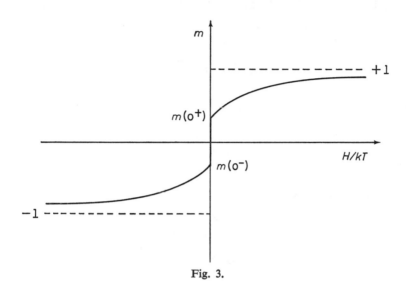

Fig. 3.

his proof holds only for *sufficiently low* temperatures and not for *all* temperatures below the critical.

It is however easy to give a heuristic justification of the result, and I shall do it in the next section.

7

If the Ising net is placed in an external magnetic field H, the interaction energy is

$$E = - J \sum v(P, Q) \mu_P \mu_Q + H \sum \mu_P \tag{7.1}$$

and magnetization is defined by the formula

$$m(H) = \lim_{\substack{N \to \infty \\ M \to \infty}} \frac{1}{MN} \frac{\partial}{\partial \omega} \log Q_{N, M}(v, \omega) \tag{7.2}$$

where

$$v = \frac{J}{kT}, \qquad \omega = \frac{H}{kT} \tag{7.3}$$

and $Q_{N, M}$ is the partition function of the model.

Introducing the $2^M \times 2^M$ transfer matrix*

$$L_{\omega, \nu}(\mu, \mu') = \exp\left[\frac{\omega}{2}\sum_1^M \mu_k\right] \times$$

$$\times \exp\left[\frac{\nu}{2}\sum_1^M \mu_k\mu_{k+1}\right]\exp\left[\nu\sum_1^M \mu_k\mu_k'\right]\exp\left[\frac{\nu}{2}\sum_1^M \mu_k\mu_{k+1}'\right]\exp\left[\frac{\omega}{2}\sum_1^M \mu_k'\right] \quad (7.4)$$

one readily verifies that

$$Q_{N, M}(\nu, \omega) = \sum_{j=1}^{2^M} \Lambda_j^N \quad (7.5)$$

where $\Lambda_1 > \Lambda_2 > \Lambda_3 \ldots$ are the eigenvalues of $L_{\omega, \nu}$.

We thus obtain

$$m(H) = \lim_{M \to \infty} \frac{1}{M}\frac{\partial \log \Lambda_1}{\partial \omega}, \quad (7.6)$$

and it is easy to see that we also have

$$m(H) = \lim_{M \to \infty} \sum_{\mu}\left(\frac{\sum_1^M \mu_k}{M}\right)\phi_1^2(\mu; \nu, \omega), \quad (7.7)$$

where $\phi_1(\mu; \nu, \omega)$ is the principal eigenvector of $L_{\omega, \nu}$. Because of the periodic boundary condition $\phi_1(\mu; \nu, \omega)$ is invariant under a cyclic permutation, and we have therefore

$$m(H) = \lim_{M \to \infty} \sum_{\mu} \mu_l\phi_1(\mu; \nu, \omega) \quad (7.8)$$

where l is an arbitrary integer between 1 and M.

Magnetization at field 0 is either

$$m(0^+) = \lim_{\omega \to 0^+}\lim_{M \to \infty} \sum_{\mu} \mu_l\phi_1^2(\mu; \nu, \omega), \quad (7.9)$$

or

$$m(0^-) = \lim_{\omega \to 0^-}\lim_{M \to \infty} \sum_{\mu} \mu_l\phi_1^2(\mu; \nu, \omega), \quad (7.10)$$

and in these formulas the order of taking limits is absolutely crucial. In fact, if we first take the limit $\omega \to 0$, we will always get the answer 0.

For $\omega = 0$ it was already observed many years ago by Ashkin and Lamb that the correlation between spins at distance r in the 'horizontal' direction (i.e., in the direction

---

* We use periodic boundary conditions so that $\mu_M = 1 = \mu_1$.

in which N first becomes infinite) is given by the formula

$$\varrho(r) = \lim_{M \to \infty} \sum_{j=1}^{2M} \left(\frac{\Lambda_j}{\Lambda_1}\right)^r (\phi_j(\mu), \mu_l \phi_1(\mu))^2, \tag{7.11}$$

where

$$(\phi_j(\mu), \mu_l \phi_1(\mu)) = \sum_\mu \mu_l \phi_j(\mu) \phi_1(\mu). \tag{7.12}$$

and $\phi_j(\mu) \equiv \phi_j(\mu; v, 0)$ are the eigenvectors of the matrix $L_{v, 0}$ (i.e., for $\omega = 0$).

Since for $\omega = 0$, $\phi_1(\mu)$ is even, the term corresponding to $j = 1$ vanishes, and in order to have residual correlation at infinity the maximum eigenvalue must become degenerate in the limit $M \to \infty$ (asymptotically degenerate). If the degeneracy is of order two, one would expect that

$$\varrho(\infty) = \lim_{M \to \infty} \left(\sum_\mu \mu_l \phi_2(\mu) \phi_1(\mu)\right)^2, \tag{7.13}$$

and this begins to look quite a bit like

$$\varrho(\infty) = m^2(0^+), \tag{7.14}$$

which is another way to write (6.4).

To understand why indeed one should expect (7.14) to be valid, let us consider the case of very low temperatures. In this case and for $\omega > 0$, $\phi_1(\mu: v, \omega)$ is approximately the vector

$$\phi^+(\mu) = \begin{cases} 1, & \mu = (1, 1, ..., 1) \\ 0, & \text{otherwise} \end{cases}$$

and for $\omega < 0$ it is approximately

$$\phi^-(\mu) = \begin{cases} 1, & \mu = (-1, -1, ..., -1) \\ 0, & \text{otherwise}. \end{cases}$$

For $\omega = 0$ we should expect approximately

$$\phi_1(\mu; v, 0) \sim \frac{1}{\sqrt{2}} (\phi^+(\mu) + \phi^-(\mu))$$

and

$$\phi_2(\mu; v, 0) \sim \frac{1}{\sqrt{2}} (\phi^+(\mu) - \phi^-(\mu)).$$

This shows that ϕ_2 is closely related to ϕ_1 being in fact what might be termed the *antisymmetric version* of ϕ_1, and if this persists for all temperatures below T_c, we have a simple explanation of why one should

$$\varrho(\infty) = m^2(0^+)$$

which as noted above is closely related to coexistence and geometric separation of two phases.

Unfortunately it seems very difficult to turn this highly appealing argument into a rigorous proof, and this is yet another indication that there are aspects of the picture which we do not understand completely.

I have connected the question of coexistence of distinct phases with the degeneracy of the maximum eigenvalue of a certain operator (transfer matrix). There are indications based on an analysis of other models that this connection is not accidental. In any case, I hope that I have convinced you that this whole area of problems is full of subtleties and that much remains to be done.

BIBLIOGRAPHY

1. J. L. Lebowitz and O. Penrose, 'Rigorous Treatment of the van der Waals Theory of Liquid-Vapor Transition', *J. Math. Phys.* **7**, 98–113 (1966).
2. D. J. Gates and O. Penrose, 'The van der Waals Limit for Classical Systems, III', *Commun. Math. Phys.* **17**, 194–209 (1970).
3. A. Martin-Löf, 'On the Spontaneous Magnetization in the Isling Model', *Commun. Math. Phys.* **24**, 253–259 (1971/72).

DISCUSSION

B. L. van der Waerden: I should like to ask Professor Kac whether he can give me a proof that these two areas are equal. The proof in the textbook is obviously nonsense. They take a Carnot circular process but this is physically impossible. Can you give me a proof?

M. Kac: In the model which started all this and in the paper of Penrose and Lebowitz, the equal area is part of the mathematical theorem, absolutely without any doubt, but you have to go through this more complicated argument. The usual proofs, I agree, leave something to be desired.

27

Approach to Thermodynamic Equilibrium
(and other Stationary States)

Willis E. Lamb, Jr.

This Symposium on the Physicist's Conception of Nature celebrates the 70th birthday of Paul Dirac. Before beginning the subject of my talk, I would like to make some remarks about ways in which Dirac has influenced me personally. The first edition of his book, *The Principles of Quantum Mechanics*, was published in 1930. In that same year I entered the University of California at Berkeley where I majored in Chemistry. I spent the summer of 1932 at home in Los Angeles, and came across Dirac's book on the shelves of the public library. I remember that it made a great impression on me, although I could not really understand very much of it. In part, due to its influence, I changed to Physics for graduate study, and during the years 1934 to 1938 worked with J. R. Oppenheimer on theoretical physics. Every spring, accompanied by most of his students, Oppenheimer went to Caltech in Pasadena, and around 1936 I had there the opportunity to hear a lecture by Dirac on magnetic poles. This was unfortunately not the occasion on which a member of the audience, addressing the speaker during the discussion period, was told of the difference between a question and a statement.

The Symposium's Sponsoring Committee had invited me to give a lecture on the history of electrodynamic level shift experiments and theory. My experiments with Retherford[1] on the fine structure of hydrogen were carried out at Columbia University in 1946/47. I first met Dirac personally at the Pocono Conference of 1948, where fine structure was one of the main topics of discussion.

During the Lindau Conference of 1959, Dirac invited me to accompany him on a long walk around the island of Mainau. He told me that he had first learned of the measurements on the hydrogen fine structure from the front page of *The New York Times* in 1947. I apologized for not sending him the news more directly, but explained that I had never thought of this work as reflecting any criticism of the Dirac equation. After a pause, he asked whether I had enjoyed making the discovery of the fine structure anomaly. I replied that I had, but that I would much rather have discovered the Dirac equation. Dirac said kindly that things like that are much harder now.

Indeed they are. On the calculational side, in the late forties Norman Kroll[2] and I were able to make a relativistic extension of Bethe's non-relativistic quantum electrodynamic level shift calculation of 1947. Although Kroll went on to more complicated calculations of this kind, this represented the limit of my own participation on the theoretical side of

J. Mehra (ed.), The Physicist's Conception of Nature, 527-547. All Rights Reserved

this subject. On the experimental side, my colleagues Retherford, Dayhoff, and Trieb-wasser at Columbia reached an accuracy of about 0.1 MHz in the hydrogen level shift of about 1059 MHz. There has been experimental work by others which slightly improved on the accuracy of our work of the early 1950's but it is still very difficult to greatly increase the accuracy. After coming to Yale in 1962 I participated in attempts to make more accurate measurements on hydrogen as well as on various states of singly ionized helium. This work turned out to be rather disappointing with respect to quantum electrodynamic level shifts, but it did lead to very interesting results on highly excited states of the helium atom.

For my contribution to this Symposium, however, I have looked for another subject related to Dirac's contributions to physics. It is: 'Approach to Thermodynamic Equilibrium (and other Stationary States)'. The announced title had an obvious connection with Dirac's contributions to physics. As a matter of fact, the subject I shall discuss also has many connections with these contributions. Dirac[3] wrote major papers on statistical mechanics even before he began his famous series on quantum mechanics in 1925. The work under discussion makes essential use of quantum mechanics and of the time-dependent perturbation theory developed by Dirac. Furthermore, the density matrix[4] of Landau, von Neumann and Dirac plays an essential role in it.

This work grew out of research on the theory of the maser and the laser. I began the former in 1955 after the development of the ammonia beam maser by Townes, and the latter in 1961 with the gas laser of Javan. There are very close connections with the quantum theory of the laser formulated by Scully and myself in 1966/67.

Dirac[3] studied electrical engineering at Bristol before turning to physics. I asked him recently in Miami whether he had published any papers on electrical engineering. His reply was negative. I didn't tell him then, and don't know whether he realizes it now that his quantum theory of radiation is of basic importance to an important branch of electrical engineering called 'Quantum Electronics'.

Since there are professional experts here among the lecturers on equilibrium and non-equilibrium statistical mechanics (Uhlenbeck, Kac, Cohen and Prigogine), I hope I may be forgiven for poaching on their field.

Newtonian mechanics developed into the theory of analytical dynamics. This subject deals primarily with the problem of an isolated system, described by a conservative Hamiltonian function. The concept of an isolated system is an abstraction, for which one must pay later on by considering more complicated situations. It is relatively easy to allow for the presence of external forces acting on an otherwise isolated system. Non-conservative forces or dissipative forces represent greater complications, but are still manageable. Quantum mechanics is most easily applied to conservative dynamical systems, but even then it is essential to consider the disturbance of the system caused by the process of measurement. Several sessions of this symposium are to be devoted to this still poorly understood subject.

On a higher level of complexity, we come to systems which are in thermodynamic equilibrium, and can be characterized by a temperature T. The system is in intimate contact with a thermal reservoir. An interplay of forces between system and reservoir occurs, with corresponding exchange of energy. In the theory of equilibrium statistical mechanics, one learns how to describe phase transitions. In some problems one finds that a system can develop spatial order, and in others there are long range quantum mechanical effects such as super-fluidity.

At the next level of complication, we find systems in non-thermodynamic stationary states. One example of this is heat conductivity where two different regions of the system of interest are in contact with thermal reservoirs, one at temperature T_1 and the other at temperature T_2. In some approximation, the thermal state of the system of interest may be described by a temperature which is a function of position. Another example of a non-thermodynamic stationary state is found in the theory of the laser oscillator. Such a device may display a well-defined temporal order. The field of biology presents us with still more complicated situations. It is doubted by some people that these are describable by present-day physical laws. Even if they are, we must clearly allow the biological system of interest to be in interaction with a more complicated environment than the thermal reservoir discussed above. It is necessary to have an interchange of matter, as well as energy. I doubt that the method of the grand canonical ensemble will suffice to describe the intake of food and processes of waste elimination.

For the discussion of equilibrium problems, the transition from thermodynamics to statistical mechanics makes it possible to treat many problems in a much more quantitative fashion. The central task is the evaluation of the partition function, or the sum-of-states

$$Z = \text{spur} \{\exp(-H/kT)\}. \tag{1}$$

A great deal of current interest centres on the theory of phase transitions. This usually involves an attempt to perform an exact evaluation of the sum-of-states. Much work is being done on various forms of the Ising model. Such problems are over idealized, but they can approximate physical reality for suitably chosen crystals.

We are concerned today with the problem of the approach to thermodynamic equilibrium. This subject[5] has been given intensive consideration for over 100 years. I mention the names of some of the great workers in this field: Maxwell, Boltzmann, Rayleigh, Gibbs, P. and T. Ehrenfest, Fokker and Planck, Pauli, Onsager, Uhlenbeck, Chandrasekhar, Bloch, van Hove, Bogolubov, and Bergmann and Lebowitz. Among the topics discussed were the H theorem, Loschmidt's and Zermelo's paradoxes, ergodic and quasi-ergodic hypotheses, Poincaré cycles and randomization of phases. A key question is: what makes the approach to thermodynamics equilibrium occur? Suppose that the system of interest is an isolated conservative dynamical system. Can this come to thermodynamic equilibrium? Clearly not, for a simple system, but perhaps so in some approximation for a very complicated system, and probably so for a small part of such a complicated system. Some authorities feel that a 'speck of dust' is needed, while others believe that the essential feature is the interaction of the system with a thermal reservoir. I strongly favour the last point of view.

One question which has received much attention is the following 'How does a causal and time reversible dynamical theory give irreversible behaviour?' I remind you that one of the earliest specific models for treating such questions is the Rayleigh problem where a heavy molecule moves into a gas of light molecules which are assumed to be maintained in thermodynamic equilibrium.

My object in this talk is to discuss the simplest possible non-trivial model for the approach to thermodynamic equilibrium. It may be regarded as playing the role of the Ising model in equilibrium statistical mechanics. The system of interest will be taken to be a single atom. It may be well to state explicitly at the outset that one is necessarily going to be considering an ensemble of such systems. This atom can be a simple harmonic oscillator, a non-linear oscillator or a two level atom. For the present discussion the system will be taken to be a quantum mechanical simple harmonic oscillator.

It is usual to think that a thermodynamic reservoir has to be a very large and complicated system which is somehow maintained in thermodynamic equilibrium. When the reservoir interacts with the system, it is changed. Some approximation or idealization has to be made in order to insure that the reservoir is not disturbed by its action on the system. The interaction of system and reservoir is not necessarily a small one, and in my view most discussions tend to sweep some difficulties under the carpet. A simpler model of a reservoir is suggested by calculations on the theory of the ammonia beam maser which were carried out at Stanford by Helmer[6] and myself in the mid-1950's. The reservoir consists of a stream of atoms characterized by a temperature T. These reservoir atoms may be simple harmonic oscillators, non-linear oscillators or two level atoms. One may use quantum mechanics for any of these models, or for the case of simple harmonic oscillators of non-linear oscillators one may use classical mechanics. For the present discussion we will use quantum mechanical two level atoms.

As stated above, the system of interest will be a quantum mechanical simple harmonic oscillator. Its cartesian coordinate will be denoted by E. This notations is carried over from the discussion of the maser in which the system is a one mode cavity electromagnetic resonator. Such a problem is dynamically equivalent to the problem of a mechanical simple harmonic oscillator. The energy eigenvalues of the system are given by the equation

$$\hbar W_n = (n + \tfrac{1}{2}) \hbar \omega, \tag{2}$$

(see Fig. 1), the eigenfunctions are denoted by $h_n(E)$ where the quantum number n can have the values 0, 1, 2, ..., ∞.

At $t = 0$, we imagine that the system is described by a wave function

$$\Psi(E) = \sum_{n=0}^{\infty} c_n h_n(E). \tag{3}$$

The probability for the state n to be found is given by

$$P_n = |c_n|^2. \tag{4}$$

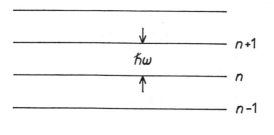

Fig. 1.   Energy levels of the simple harmonic oscillator system atom as given by Equation (2).

It is important to recall that the description by a wave function is not really of a single system but of an ensemble of similarly prepared systems. For a more general description of the system we must use the density matrix[4] ϱ. For a pure case, where there is a wave function $\Psi(E)$, $\varrho[\Psi]$ has elements

$$\varrho_{nm} = c_m^* c_n \tag{5}$$

in the n representation. The diagonal elements ϱ_{nn} are probabilities P_n for finding the nth state.

More generally, we may have a mixture of states Ψ_i with weights w_i. The density matrix for a mixture is given by

$$\varrho = \sum_i w_i \varrho[\Psi_i]. \tag{6}$$

If we are given the density matrix ϱ, we can find the weights w_i and the constituent wave functions Ψ_i by transforming the density matrix ϱ to diagonal form. The eigenvalues are the w_i and the eigenfunctions are the Ψ_i. For a physical case, we need to have all of the weights $w_i \geqslant 0$ and $\sum_i w_i = 1$.

The reservoir consist of a stream of two level atoms, with energy levels shown in Fig. 2. The transition frequency ω is defined by the equation

$$\omega = W_a - W_b. \tag{7}$$

At most, only one of these two level reservoir atoms interacts with the system at any one time. Let the duration of the coupling be denoted by T. For the present, we assume that the reservoir atom is initially in its lower state b. Its wave function is $u_b(x)$ where x is a cartesian coordinate characterizing the reservoir atom. Let the system have the starting wave function $\Psi(E)$. The perturbation energy is taken in simple dipole–

Fig. 2.   Energy levels of a two-level reservoir atom with separation $\hbar\omega$ given by Equation (7).

dipole form

$$\hbar V = - eEx. \tag{8}$$

The combined system of the reservoir atom and the atom of interest is described by two cartesian coordinates E, x. At $t = 0$ its wave functions is given by

$$\Psi(E, x, t = 0) = \Psi(E) u_b(x). \tag{9}$$

The state of the composite system is still a pure case.

During the time of interaction $0 < t < T$ the wave function for the combined system has the form

$$\Psi(E, x, t) = \sum_n a_n(t) h_n(E) u_a(x) e^{-i(W_n + W_a)t}$$
$$+ \sum_n b_n(t) h_n(E) u_b(x) e^{-i(W_n + W_b)t}. \tag{10}$$

At time T, we remove the reservoir atom from interaction with the system of interest. The joint probability distribution in E and x, given by

$$P(E, x, T) = |\Psi(E, x, T)|^2, \tag{11}$$

can describe correlations between E and x. One can learn something about the state of the system after the interaction has ceased by investigating the state of the reservoir atom, and vice versa. The 'difficulties' with the interpretation of quantum mechanics pointed out by Einstein, Podolsky and Rosen are met at this point but they do not bother one who believes that a probabilistic description of microscopic phenomena is inescapable.

We do not know whether the reservoir atom leaves in the state 'a' or in the state 'b' unless we choose to look at it. If we ignore the reservoir atom, the probability for the simple harmonic oscillator system to be in the nth energy eigenstate is

$$P_n(T) = \int |a_n(T) u_a(x) e^{-i\omega T} + b_n(T) u_b(x)|^2 \, dx =$$
$$= |a_n(T)|^2 + |b_n(T)|^2. \tag{12}$$

The system of interest is no longer describable by a wave function. The pure case has been converted into a mixture, because we don't know, or care, whether the reservoir atom departed in the state 'a' or in the state 'b'.

In order to find the time development of the wave function during the interaction we have to integrate the Schrödinger equation

$$-\frac{\hbar}{i} \frac{\partial \Psi}{\partial t} = H\Psi. \tag{13}$$

It is a convenient and valid simplification to make the resonance approximation indicated in Fig. 3. The wave equation then becomes a system of equations

$$i\dot{a}_{n-1} = V_n b_n$$
$$i\dot{b}_n = V_n a_{n-1} \tag{14}$$

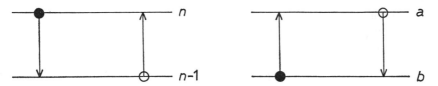

Fig. 3.   Schematic representation of resonance approximation leading to Equation (14). When the reservoir atom makes a transition from state b to state a (or a to b), the system atom may make only transitions of the type n to $n-1$ (or $n-1$ to n).

where n takes on all values $1, 2, \ldots, \infty$, and the matrix elements of the perturbation are given by

$$\hbar V_n = - e \langle n - 1| E | n \rangle \langle a | x | b \rangle. \tag{15}$$

This system of equations decouples into pairs of equations and the solution is given by

$$a_{n-1}(T) = - ib_n(0) \sin V_n T$$
$$b_n(T) = b_n(0) \cos V_n T \tag{16}$$

for all n, where $\left| b_n(0) \right|^2 = P_n(0)$. It follows that

$$a_n(T) = - ib_{n+1}(0) \sin V_{n+1} T. \tag{17}$$

The probability that the system is in the nth state at the end of the interaction is given by

$$P_n(T) = P_n(0) \cos^2 V_n T + P_{n+1}(0) \sin^2 V_{n+1} T, \tag{18}$$

so that the change in probability becomes

$$\delta P_n = P_n(T) - P_n(0) = P_{n+1}(0) \sin^2 V_{n+1} T - P_n(0) \sin^2 V_n T. \tag{19}$$

The result of the interaction of one reservoir atom with the system has been a small but finite change in the probability distribution for the states of the harmonic oscillators system. If we inject atoms at an average rate r, we can write a differential equation for the probability distribution P_n given by

$$\frac{dP_n}{dt} = r \left[P_{n+1}(t) \sin^2 V_{n+1} T - P_n(t) \sin^2 V_n T \right]. \tag{20}$$

The transition from a small change in P_n to a differential equation for its average behaviour is sometimes described as 'coarse graining' but this term is also used in statistical mechanics to describe a very different concept where a number of stationary states of a complex system are grouped together in clusters, or in cells of phase space.

Our present discussion considers that each interaction lasts a definite time T. It is very easy now to average over a distribution of possible interaction times. We will take an exponential weight function

$$W(T) = \gamma e^{-\gamma T}. \tag{21}$$

This makes it possible to consider a different model of the reservoir, appropriate for gas lasers, in which the active atoms are excited by some random process and allowed to decay in an exponential fashion by radiative decay. In this case the function $\sin^2 VT$ is replaced by its average

$$\langle \sin^2 VT \rangle = \tfrac{1}{2} V^2 / [V^2 + (\gamma/2)^2] . \tag{22}$$

Let

$$V_n^2 = nV^2 \tag{23}$$

(simple harmonic oscillator)

$$A = \tfrac{1}{2} r V^2$$
$$B = \tfrac{1}{2} r V^2 (4V^2/\gamma^2) . \tag{24}$$

The differential-difference equation for P_n becomes

$$\frac{dP_n}{dt} = \frac{(n+1)A}{1 + \dfrac{B}{A}(n+1)} P_{n+1} - \frac{nA}{1 + \dfrac{B}{A}n} P_n . \tag{25}$$

Up to this point we have injected b atoms. Let us also inject a atoms at some rate r. We use primes to denote terms corresponding to b atoms (damping atoms) and omit primes for a atoms (pumping atoms). The differential equation for P_n then becomes

$$\frac{dP_n}{dt} = - \frac{(n+1)A}{1 + \dfrac{B}{A}(n+1)} P_n + \frac{nA}{1 + \dfrac{B}{A}n} P_{n-1} +$$

$$+ \frac{(n+1)A'}{1 + \dfrac{B'}{A'}(n+1)} P_{n+1} - \frac{nA'}{1 + \dfrac{B'}{A'}n} P_n . \tag{26}$$

These are a special case of equations derived by Scully[7] and Lamb in 1966 in their quantum theory of the laser. However, as we will show, the equations can also be used to describe a realistic model for the approach to thermodynamic equilibrium for our system. We now proceed to consider a number of special cases of these equations.

Case 1. The 'a' and 'b' atoms are of the same type but their rates of injection are different so that r is not equal to r'. Then $B'/A' = B/A$ and $A/A' = r/r'$. We set $C = A'$ and $A/C = r/r'$. Equations (26) have a steady-state solution which is pictorially represented in Fig. 4. We have detailed balance if the rates represented by arrow 1 and 3 are equal and also those represented by arrow 2 and 4. These conditions are satisfied if

$$P_n = \frac{A}{C} P_{n-1} \quad \text{for} \quad n = 1, 2, 3, ..., \infty . \tag{27}$$

Our solution can be written in the form

$$P_n = (A/C)^n P_0 \tag{28}$$

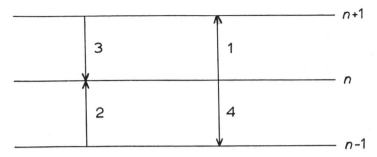

Fig. 4.   Pictorial representation of the four terms on the right hand side of Equation (26), each represented by a numbered arrow.

where

$$P_0 = [1 - (A/C)] \qquad (29)$$

if $A < C$. Write

$$P_n = P_0 e^{-n\hbar\omega/kT} \qquad (30)$$

where the 'temperature' T is given by

$$e^{-\hbar\omega/kT} = A/C = r/r'. \qquad (31)$$

We see that our reservoir of two level atoms at temperature T is able to bring the simple harmonic oscillator system to the single mode Planck distribution. This result is obtained independently of the strength of the coupling.

Case 2. $B = B' = 0$. The calculation is carried out only to first order in the coupling strength V^2. This is sometimes called the 'linear' theory. (The equations for $P_n(t)$ are linear in any case.) The equations for this case were discussed by Shimoda, Takahashi and Townes,[8] and Feller[9] devotes a chapter to similar equations.

These equations have the same steady-state solution as we had in Case 1. In both cases, there are also solutions which decay exponentially in time. They allow us to describe the approach to thermodynamic equilibrium. These solutions have the form

$$P_n(t) = \varrho_n e^{-\lambda t}. \qquad (32)$$

The possible values of the decay constant λ are called decay eigenvalues and denoted by λ_r. The corresponding eigenfunctions are $\varrho_n^{(r)}$. The steady-state solution previously found corresponds to $\lambda_0 = 0$. It can be shown that the eigenvalues are of the form

$$\lambda_r = r(C - A). \qquad (33)$$

(Note that the eigenvalue spectrum becomes highly degenerate when $A = C$, i.e., when T approaches infinity.) The general solution of the system of equations is given by

$$P_n(t) = \sum_{r=0}^{\infty} \varrho_n^r e^{-\lambda_n t}. \qquad (34)$$

Some steady-state solutions of these equations are illustrated by figures prepared using a small computer with a cathode ray oscilloscope display. Fig. 5 shows P_n as a function of n for the ratio $A/C = 0.96$. The horizontal scale has discrete spacing at intervals of 1 from 0 to 250. The points representing the P_n values should be seen as

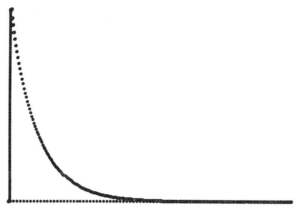

Fig. 5. Steady state (Boltzmann) distribution for the relative probability P_n for finding the n-th harmonic oscillator stationary state as a function of quantum number n, for a pumping to damping ratio $A/C = 0.96$ and saturation parameter $B = 0$. The horizontal scale runs from $n = 0$ to $n = 250$. The distribution P_n is unnormalized.

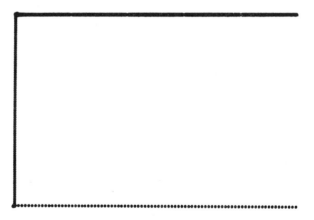

Fig. 6. Plot of relative values of $P_n = \varrho_{nn}$ for $0 \leqslant n \leqslant 250$ with $A/C = 1.0$ (corresponding to an 'infinite temperature') and $B = 0$. The distribution is unnormalizable.

discrete points, but because of the crudeness of the display this feature can only be seen on the steeper parts of the curves. The curve in this case is a simple exponential function corresponding to the Planck distribution law. In all of the figures, one will see a point slighly to the left of the lower extremity of the vertical axis. This should be ignored, as it represents a minor deficiency in the plotting routine. Fig. 6 shows the corresponding plot for $A/C = 1.0$, i.e., $T = \infty$. In this case the distribution of probabilities for finding various number of photons is uniform, and hence we have an un-normalizable distribution. As in all the other cases, the point corresponding to $n=0$ is placed at a vertical height of 1.0 for convenience. Fig. 7 shows the corresponding graph for the physically unrealistic value $A/C=1.04$, which will correspond to a 'negative temperature'. Such a concept is perfectly reasonable for a two level atom, but is absurd for a simple harmonic oscillator with an infinite number of equally spaced energy levels. The figure indicates the higher one goes in n the greater the probability P_n, and the distribution is thoroughly unnormalizable.

Let us now see computer generated plots of the first two eigenmodes of decay for $A/C=0.96$. The corresponding eigenvalues are $\lambda_1=0.04$ and $\lambda_2=0.08$. Fig. 8 corresponds to λ_1. The eigenfunction is positive for small n values and negative for larger n values. There is one node. It may well be to comment briefly on the physical significance of this curve. Suppose that the distribution of photons in the cavity resonator is given by the exponential curve of Fig. 5 which represents the stationary state. Now

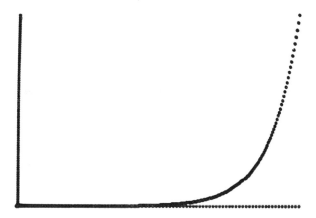

Fig. 7.   Plot of relative values $P_n-\varrho_{nn}$ for $0\leqslant n\leqslant250$ with $A/C=1.04$ (corresponding to a 'negative temperature') and $B=0$. The distribution is unnormalizable.

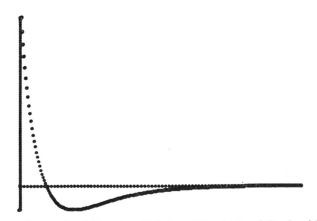

Fig. 8.   First decay eigenmode of Equation (26) for $A/C=0.96$ and $B=0$, with decay constant $\lambda_1=0.04$. The physical significance of decay eigenmodes is discussed in the text.

contemplate a distortion of this curve produced by adding to P_n the distribution shown in Fig. 8 with a small positive coefficient. The resulting distribution will be one which has more photons for small n values and fewer photons for large n values than for the equilibrium steady-state of Fig. 5. In the course of time, the resulting distribution will relax to thermodynamic equilibrium in such a way that the departure from equilibrium values of P_n decay exponentially in time with a rate constant λ_1. If, instead of adding to P_n the distribution of Fig. 8, it were subtracted with a small coefficient, the approach to equilibrium would take place in such a way that P_n for small n increase and P_n for large n decrease. Fig. 9 shows the decay eigenmode for $\lambda_2 = 0.08$.

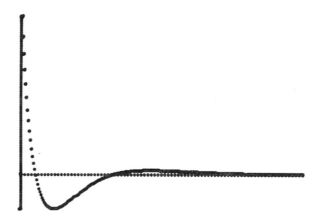

Fig. 9.   Second decay eigenmode of Equation (26) for $A/C = 0.96$ and $B = 0$, with decay constant $\lambda_2 = 0.08$. The physical significance of decay eigenmodes is discussed in the text.

This curve has two nodes and represents a more complicated way in which the distribution of photons can be deformed. There are an infinite number of higher eigenmodes of decay; that with eigenvalues λ_r has r nodes. These are needed in the expansion (34) of $P_n(t)$ to express more rapidly decaying departures from thermal equilibrium.

We now turn briefly to a discussion of the behaviour of the non-diagonal elements of the density matrix. A partial idea of their significance can be had from the observation that the ensemble average value of the coordinate E is given by the equation

$$\langle E \rangle = \text{spur}\,(\varrho E) \tag{35}$$

which is different from 0 if and only if we have non-zero elements $\varrho_{nn\pm1}$ of the density matrix. In this case we speak of an off-diagonality index of one unit. More generally, we may have elements k units off-diagonal of the form $\varrho_{nn\pm k}$, which, for a given fixed k value, will be denoted by σ_n. In Case 2, σ_n obey a differential equation

$$\frac{d\sigma_n}{dt} = -A\left(n + 1 + \tfrac{1}{2}k\right)\sigma_n + A\sqrt{n(n+k)}\,\sigma_{n-1} +$$
$$+ C\sqrt{(n+1)(n+1+k)}\,\sigma_{n+1} - C\left(n + \tfrac{1}{2}k\right)\sigma_n. \tag{36}$$

We now have more general decay eigenmodes with decay eigenvalues $\lambda(r, k)$ depending on two indices r and k. The second of these measures the off-diagonality and the first labels the possible eigenvalues in order of increasing size beginning with $r=0$. It can be shown that the decay eigenvalues of (36) are given by

$$\lambda(r, k) = (C - A)(r + \tfrac{1}{2}k) \tag{37}$$

for general values of r and k. If $k=0$, $r=0$, $\lambda(0, 0)=0$ corresponding to the steady-state distribution of photon probabilities from the diagonal elements of the density matrix in the n representation. If $k=1$, $r=0$ we get the lowest eigenvalue for unit off-diagonality $\lambda(0,1)=\tfrac{1}{2}(C-A)$. In Fig. 10 we show a plot of σ_n as a function of n for $k=1$ and $r=0$ when $A/C=0.96$. The corresponding plot for $r=1$ is shown in Fig. 11 These distributions do not represent stationary-states but decaying elements of the density matrix of off-diagonality $k=1$. As in the earlier plots the horizontal scale runs

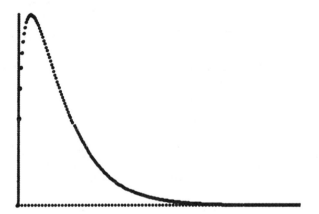

Fig. 10.   A plot of the relative values of $\sigma_n = \varrho_{nn+1}$ as a function of n ($0 \leqslant n \leqslant 250$) for the lowest decay eigenmode of Equation (36) when $A/C=0.96$ and $B=0$. The eigenvalue is $\lambda(0, 1)=0.02$.

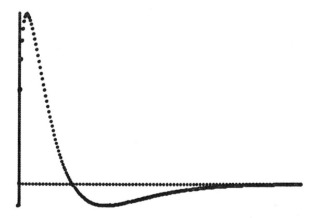

Fig. 11.   A plot of $\sigma_n = \varrho_{nn+1}$ as a function of n ($0 \leqslant n \leqslant 250$) for the second lowest eigenmode of Equation (36) for the parameters of Fig. 10. The eigenvalue $\lambda(1, 1)=0.06$.

through the integers from 0 to 250. The vertical scale is so arranged that the point for $n=0$ is taken to be at unit height.

A number of comments are in order. It is seen that even for a system with one degree of freedom there are an infinite number of decay eigenmodes. The average coordinate $\langle E \rangle$ decays exponentially to zero at a rate $\frac{1}{2}(C-A)$ in Case 2. If one thinks of analogy between the problem of the simple harmonic oscillator and the problem of electrical oscillations in a circuit containing inductance, capacitance and resistance, the decay of the amplitude at half the rate of the decay of the non-thermal energy stored in the cavity will seem reasonable. (It can be shown that terms corresponding to higher decay eigenvalues cancel out completely in Case 2.) The linear theory explodes if A is greater than or equal to C. The equation

$$e^{-\hbar\omega/kT} = A/C > 1 \tag{38}$$

means that T is less than 0. We have already mentioned that a negative temperature is a valid concept for two level atoms but absurd for simple harmonic oscillators. The trouble occurs even in the non-linear theory if we take $B/A=B'/A'$.

Case 3. Here we take $B'=0$, and again use the notation $A'=C$. This means that we are injecting non-linear pumping atoms in their upper state and linear damping atoms in their lower state. The constant B describes the 'saturation' property of the pumping atoms. In this case our differential equations takes the form

$$\frac{dP_n}{dt} = -\frac{A(n+1)}{1+\frac{B}{A}(n+1)} P_n + \frac{An}{1+\frac{B}{A}n} P_{n-1} +$$

$$+ C(n+1) P_{n+1} - CnP_n. \tag{39}$$

This case describes a simple quantum mechanical theory of a laser. The damping atoms simulate the ohmic resistance of the cavity resonator, and the pumping atoms the effect of the population inversion produced by a gas discharge or an injected beam of state selected atoms. These equations for the diagonal elements of the density matrix have a steady-state solution, and the same sort of detailed balance pairing of terms in the equations we met before. The steady-state is characterized by the equations

$$P_n = \frac{A}{C} \frac{1}{1+\frac{B}{A}n} P_{n-1} \tag{40}$$

so that

$$P_n = P_0 \prod_{j=1}^{n} \frac{A/C}{1+(B/A)j}. \tag{41}$$

This defines a 'truncated Poisson distribution'. If $A<C$, the distribution P_n falls off with increasing n, much as it did in the linear case. When $A>C$ the peak value will be for an integer close to n_p given by

$$A/C = 1 + (B/A) \, n_p$$

or

$$n_p = \left(\frac{A}{B}\right) \frac{(A-C)}{C} . \tag{42}$$

The quantity P_n as a function of n gives the probability distribution for finding n photons in a laser cavity. We now show a number of plots of P_n for $C=1.0$, and a succession of A values. Fig. 12 has $A=0.96$, $B=6 \times 10^{-4}$. It is seen that the distribution for P_n is just a little sharper than it was in the linear case. If the number of pumping atoms is now increased, keeping the number of damping atoms constant, both the quantities A and B will increase in the same proportion. Fig. 13 is shown for $A=1.0$

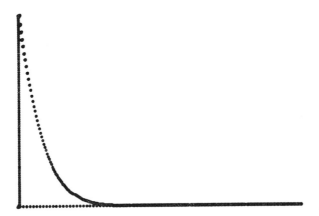

Fig. 12. Relative steady state probability distribution P_n for finding n photons in a laser below threshold, as given by Equation (40) for parameters $A/C=0.96$ and $B/A=6.25 \times 10^{-4}$. The horizontal scale runs from $n=0$ to $n=250$.

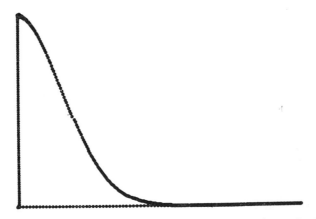

Fig. 13. Relative steady state probability distribution P_n for finding n photons in a laser at threshold as given by Equation (40) for parameters $A/C=1.0$ and $B/A=6.25 \times 10^{-4}$ with the horizontal scale used in Fig. 12.

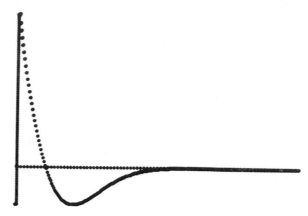

Fig. 14. First decay eigenmode of Equation (26) for parameters $A/C = 1.0$ (laser at threshold) and $B/A = 6.25 \times 10^{-4}$. The physical significance of decay eigenmodes is discussed in the text. The horizontal scale runs from $n = 0$ to $n = 250$.

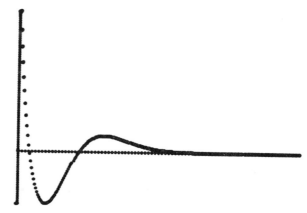

Fig. 15. Second decay eigenmode of Equation (26) for parameters $A/C = 1.0$ (laser at threshold) and $B/A = 6.25 \times 10^{-4}$. The physical significance of decay eigenmodes is discussed in the text. The horizontal scale runs from $n = 0$ to $n = 250$.

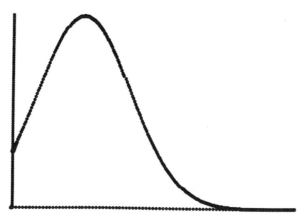

Fig. 16. Relative steady state probability distribution P_n for finding n photons in a resonator of a laser 4 % above threshold, as given by Equation (40) for $A/C = 1.04$ and the same value of the saturation parameter B/A used in Figs. 12 to 15.

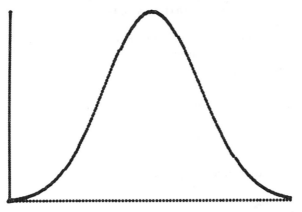

Fig. 17. Relative steady state probability distribution P_n for finding n photons in the resonator of a laser 8 % above threshold, as given by Equation (40) for $A/C = 1.08$ and the same value of the saturation parameter B/A used in Figs. 12 to 16.

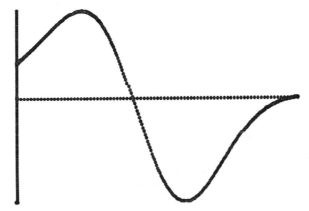

Fig. 18. First decay eigenmode of Equation (26) for $A/C = 1.08$ and $B/A = 6.25 \times 10^{-4}$ corresponding to the steady state photon distribution of Fig. 17. The physical significance of decay eigenmodes is discussed in the text.

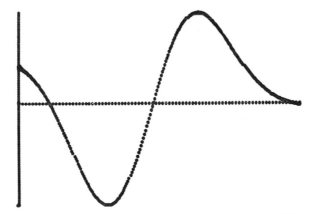

Fig. 19. Second decay eigenmode of Equation (26) for $A/C = 1.08$ and $B/A = 6.25 \times 10^{-4}$ corresponding to the steady state photon distribution of Fig. 17. The physical significance of decay eigenmodes is discussed in the text.

which corresponds to the laser threshold. This distribution has an approximately Gaussian form instead of the uniform distribution met in the linear theory. The saturation properties of the non-linear atom brings the photon probability distribution down with increasing n values. We also show in Figs. 14 and 15 the first two decay eigenmodes for this case. The corresponding eigenvalues are $\lambda_1 = 0.04983$ and $\lambda_2 = 0.12716$.

Fig. 16 corresponds to $A = 1.04$ and gives the steady-state probability distribution P_n. Notice the beginning of a well-defined peak at $n = n_p$. As before, the curve is normalized by setting the value of $P_0 = 1$.

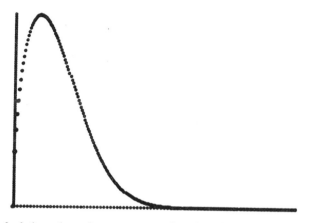

Fig. 20.   A plot of relative values of $\sigma_n = \varrho_{n\,n+1}$ as a function of n ($0 \leqslant n \leqslant 250$) for the lowest decay eigenmode of off-diagonal elements of the Scully–Lamb equations. The parameters, $A/C = 1.0$ (laser threshold), and $B/A = 6.25 \times 10^{-4}$, are the same one used in the steady state distribution shown in Fig. 13. The decay eigenvalue $\lambda(0, 1) = 0.014753$.

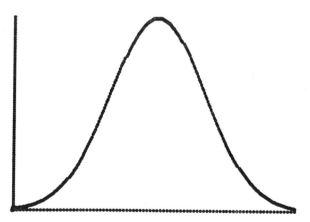

Fig. 21.   A plot of relative values of $\sigma_n = \varrho_{n\,n+1}$ as a function of n ($0 \leqslant n \leqslant 250$) for the lowest decay eigenmode of off-diagonal elements of the Scully–Lamb equations. The parameters $A/C = 1.08$ and $B/A = 6.25 \times 10^{-4}$ are the same ones used in the steady state distribution shown in Fig. 17. The decay eigenvalue $\lambda(0, 1) = 0.002483$ is much smaller than that of Fig. 20 indicating that the average electric field in a laser well above threshold decays only slowly because of phase diffusion arising from spontaneous emission.

Fig. 17 shows the steady-state distribution for a value of $A=1.08$. The peak is now very well developed. The curve is similar to, but broader than, a similar Poisson distribution. Figs. 18 and 19 show the corresponding decay eigenmodes for $\lambda_1=0.05644$ and $\lambda_2=0.09565$. As before, if the steady-state distribution is perturbed by adding a small amount of one of these decay eigenmodes, the resulting non-stationary distribution relaxes to the stationary laser distribution by the exponential decay of the admixed decay eigenmode.

We also show examples of decay eigenmodes for off-diagonal elements of the density matrix for $k=1$. Fig. 20 corresponds to $A=1.0$ and $\lambda(0, 1)=0.014753$. As before, this curve gives statistical information about the electromagnetic field in a cavity resonator which is excited just at threshold. The decay constant is relatively small. Fig. 21 gives the corresponding curve for $A=1.08$. The corresponding eigen-

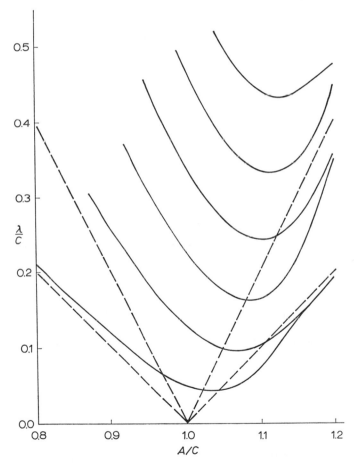

Fig. 22.   Summary of numerical calculations for the eigenvalue spectrum in Case 3 for the diagonal elements of the density matrix. The steady state solution is indicated by the axis of abscissae. The higher eigenvalues are shown by solid curves. The dashed curves correspond to the linear theory. $B/A = 6.25 \times 10^{-4}$.

value, $\lambda(0, 1)=0.002483$, becomes much smaller. Note that the shape of the curve is almost indistinguishable from that for the diagonal elements of the density matrix for the same A value.

Fig. 22 summarizes a number of results of Y. K. Wang[10] and myself in Case 3 for the eigenvalue spectrum for the diagonal elements of the density matrix. The steady-state is indicated by the axis of abscissae. The higher eigenmodes are shown by solid curves. The dashed curves correspond to the linear theory. Above the threshold, $A/C=1.0$, the absolute value of the difference $A-C$ has been used in the equation for $\lambda(r, 0)$, following a suggestion of P. Mandel.[11] Fig. 23 shows the smallest eigenvalue $\lambda(0, 1)$ for unit off-diagonality.

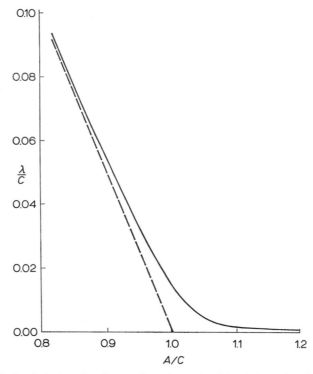

Fig. 23. Numerical calculations for the smallest eigenvalue $\lambda(0, 1)$ for unit off-diagonality of the density matrix. $B/A = 6.25 \times 10^{-4}$.

I do not have the time to discuss all of the physical ramifications of these equations. Among the topics neglected are the amplitude decay, phase diffusion, Johnson noise, and the effects of spontaneous emission from the 'a' atoms.

Only fluctuations of purely quantum mechanical nature enter our theory. They are associated with the making and breaking of contact between the system and reservoir atoms. If the system were described by a wave function before the interaction, it would be described by a mixed case density matrix after the interaction. I should mention that it would be possible to introduce equivalent random noise forces to simulate the quantum mechanical fluctuations, in a way similar to that used by Uhlenbeck and Goudsmit[12] in their treatment of the Brownian motion of a galvanometer mirror in 1929.

I have given some simple and explicit calculations, and I hope they are compatible with the more general discussion of non-equilibrium phenomena.

REFERENCES

1. W. E. Lamb, Jr., and R. C. Retherford, *Phys. Rev.* **72**, 241 (1947). For a fuller account, see W. E. Lamb, Jr., *Rep. Progr. Phys.* **14**, 19 (1951).
2. N. M. Kroll and W. E. Lamb, Jr., *Phys. Rev.* **75**, 388 (1949).
3. See J. Mehra, 'The golden age of theoretical physics: P. A. M. Dirac's scientific work from 1924 to 1933' in *Aspects of Quantum Theory* (edited by A. Salam and E. P. Wigner), Cambridge University Press, 1972.
4. See D. ter Haar, *Rep. Progr. Phys.* **24**, 304 (1961).
5. See T. Y. Wu 'On the nature of theories of irreversible processes', *Int. J. Theor. Phys.* **2**, 325 (1969).
6. J. C. Helmer, 'Maser Oscillators', Ph.D. thesis, Stanford University (1957). (Obtainable from University Microfilms, Ann Arbor, Michigan.) Also see W. E. Lamb, Jr., 'Quantum mechanical amplifiers' in *Lectures in Theoretical Physics*, vol. II, Boulder, Colorado, 1959 (edited by W. E. Britten and B. W. Downs), Interscience Publishers, New York (1960).
7. M. O. Scully and W. E. Lamb, Jr., *Phys. Rev. Letters* **16**, 853 (1966) and *Phys. Rev.* **159**, 208 (1967).
8. Shimoda, Takehashi and Townes, *J. Phys. Soc. Japan* **12**, 686 (1957).
9. W. Feller, *An Introduction to Probability Theory and its Applications*, Vol. I, 2nd ed., John Wiley and Sons, Inc. New York (1957), chapter XVII.
10. Y. K. Wang, Ph.D. thesis, Yale University (1971). (Obtainable from University Microfilms, Ann Arbor, Michigan.) Also, Y. K. Wang and W. E. Lamb, Jr., to be published in *Phys. Rev.*
11. P. Mandel, private communication.
12. G. E. Uhlenbeck and S. A. Goudsmit, *Phys. Rev.* **34**, 145 (1929).

28

Kinetic Approach to Non-Equilibrium Phenomena

E. G. D. Cohen

Abstract. The kinetic approach of Boltzmann to non-equilibrium phenomena in dilute gases is briefly discussed. Bogolubov's generalization of Boltzmann's method to dense gases is presented. The essential difficulties encountered in this attempt are exhibited and the impossibility of a straight-forward generalization of the Boltzmann equation to higher densities is concluded. Some hydrodynamical consequences of this are mentioned.

1. INTRODUCTION

One hundred years ago this year the Boltzmann equation was published.[1] This equation provided for the first time a precise mathematical basis for the discussion of the approach to equilibrium. It is an equation for the velocity distribution $f(\mathbf{r}\mathbf{v}t)$, of a gas which gives the average number of particles in the gas at the position $\mathbf{r}$ with velocity $\mathbf{v}$ at time t:

$$\frac{\partial f}{\partial t} = -\mathbf{v}\cdot\frac{\partial f}{\partial \mathbf{r}} + J(ff) \tag{1}$$

The equation expresses the rate of change of f in the form of a sum of two terms: a streaming term and a collision term. While the streaming term gives the change of f due to the fact that the particles in the gas have a finite velocity and therefore change their positions, the collision term gives the change of f due to the fact that collisions change the velocities of the particles. The collision term, $J(ff)$, has only been written down symbolically, since we do not need its detailed form here. We remark however, that: (a) in the collision term only binary collisions between particles are taken into account, as is indicated by the occurrence of two f's; this restricts the applicability of the equation to dilute gases; (b) the collision term can be written down explicitly in terms of the dynamics of 2 particles in infinite space, since the relation between the velocities of 2 particles before and after a collision are needed; (c) the collision term contains a statistical Ansatz, viz. the assumption of molecular chaos, which is the assumption of the absence of correlations between two particles that are going to collide. It is this assumption that is responsible for the irreversible approach to equilibrium.

On the basis of the Boltzmann equation all known properties of dilute gases can be understood. Thus the approach to equilibrium for any initial state can be proved on the basis of the H-theorem. The validity of the laws of thermodynamics can be established, by identifying the H-function in thermal equilibrium with $-S/k$, where

J. Mehra (ed.), The Physicist's Conception of Nature, 548-560. All Rights Reserved

S is the entropy and k Boltzmann's constant. In the years 1911–16 Chapman and Enskog[2] established the connection with the equations of hydrodynamics, which describe the last stage of the approach to equilibrium. They were able to derive from the Boltzmann equation the Euler equations for an ideal fluid and the Navier–Stokes equations for a viscous fluid with explicit expressions for the thermodynamic properties and the transport coefficients in terms of the intermolecular forces. In particular they found for the viscosity $\eta(n, T)$ and the heat conductivity $\lambda(n, T)$ as a function of density n and temperature T:

$$\lambda(n, T) = \eta_0(T)$$

$$\lambda(n, T) = \lambda_0(T)$$

where η_0 and λ_0 are well-defined functions of T only, which agree, for a realistic intermolecular potential, very well with experiment over a very wide range of temperature.

The problem I want to discuss here is the generalization of the Boltzmann equation to higher densities; or, put differently, the question: is there a generalization of the Boltzmann equation that can describe the approach to equilibrium of a dense gaseous system? I shall restrict myself to classical gases with additive, spherically summetric, repulsive, intermolecular forces of finite (short) range r_0. Many of the points I shall discuss, however, are relevant under more general conditions.

Up till the present day no one has been able to refine Boltzmann's intuitive arguments leading to the collision term $J(ff)$, in order to obtain a generalization of the Boltzmann equation that includes the effect of three and more particle collisions. As a consequence, one has had to go back to the basic equation of statistical mechanics: the Liouville equation, and somehow by a systematic expansion in the density, rederive the Boltzmann equation and then obtain correction terms to it due to three and more particle collisions.

2. BOGOLUBOV'S RESULTS

The first to achieve this was Bogolubov in 1945.[3] I shall rederive his results. However, the main point of this lecture is to exhibit certain fundamental difficulties with Bogolubov's procedure. In order to do this best, I shall present a different derivation of Bogolubov's results, than was used by himself, but which is more suitable for a critical discussion of his results.

The Liouville equation, when integrated over the coordinates and momenta of all particles but those of particle 1, yields, with appropriate boundary conditions at infinity and in the thermodynamic limit, the following equation for the first distribution function $F_1(x_1, t)$:

$$\frac{\partial F_1(x_1; t)}{\partial t} = -\frac{\mathbf{p}_1}{m} \cdot \frac{\partial F_1}{\partial \mathbf{r}_1} + n \int dx_2 \, \frac{\partial \phi(r_{12})}{\partial \mathbf{r}_1} \cdot \frac{\partial F_2(x_1 x_2; t)}{\partial \mathbf{p}_1}. \tag{2}$$

Here $F_1(x_1; t)$ is the probability density to find particle 1 in the phase x_1, i.e. at the position $\mathbf{r}_1$ with momentum $\mathbf{p}_1$ at time t. Equation (2) has the same form as the

Boltzmann equation (1). However, the collision term in (2) is exact and contains the change of F_1 due to collisions expressed in terms of the pair distribution function $F_2(x_1 x_2; t)$, the joint probability density to find the particles 1 and 2 in the phases x_1 and x_2 respectively at time t. The interparticle potential $\phi(r_{12})$ is a function of the interparticle distance $r_{12} = |\mathbf{r}_1 - \mathbf{r}_2|$ only. Similarly an equation for $\partial F_2/\partial t$ in terms of F_3 can be obtained by integrating the Liouville equation over the phases of all particles but 1 and 2 etc. This way an infinite hierarchy of coupled equations is found, expressing $\partial F_s/\partial t$ in terms of F_{s+1} $(s=1, 2, ...)$.

To obtain a generalized Boltzmann equation, F_2 has somehow to be expressed in terms of F_1, since this would lead, with (2), to a closed equation for F_1. Bogolubov gave the following argument that such an equation might well exist. First he noted that there are 3 widely separated basic lengths in the problem: r_0, the range of the interparticle forces, $\approx 10^{-8}$ cm: l, the mean free path, $\approx 10^{-5}$ cm under standard conditions (0°C, 1 atm); L, a macroscopic length, ≈ 1 cm, so that

$$r_0 \ll l \ll L.$$

Or, alternately, there are 3 widely separated basic times, which can be obtained by dividing the basic lengths by a characteristic velocity, for instance the velocity of sound: t_c, the duration of a collision, $\approx 10^{-12}$ sec; t_{mfp}, the mean free time, $\approx 10^{-9}$ sec; t_{macr}, a macroscopic time, $\approx 10^{-4}$ sec so that:

$$t_c \ll t_{mfp} \ll t_{macr}.$$

Then he assumed that the 3 basic times determined 3 basic stages in the approach to equilibrium. In particular, on the time scale of t_c, he argued that F_1 is a slowly varying quantity compared to $F_2, F_3, ...$, since F_1 is not affected in first approximation by interparticle collisions. Consequently F_1 acts as an approximate constant of the motion, to which the other distribution functions $F_2, F_3, ...$ synchronize, so that for $t \gg t_c$, F_2, F_3 will have become completely dependent on F_1. Mathematically this is expressed by saying that for $t \gg t_c$, F_2, F_3, have become time-independent functionals of F_1, with their whole time dependence determined by F_1:

$$F_2(x_1 x_2; t), F_3(x_1 x_2 x_3; t), ... \xrightarrow[t \gg tc]{} F_2(x_1 x_2 \mid F_1(; t)), F_3(x_1 x_2 x_3 \mid F_1(; t)), ...$$

To find $F_2(x_1 x_2 \mid F_1(; t))$, Bogolubov looked for a special solution of the above mentioned hierarchy of equations for the distribution functions, by means of an expansion in powers of the density n.[3]

I will not pursue Bogolubov's procedure here further but discuss now another method to obtain F_2 in terms of F_1. This method is based on a generalization to nonequilibrium of the cluster expansion for the pair distribution function F_2^{eq} of a gas in equilibrium.[4] For a gas in equilibrium, Mayer and Montroll and de Boer have derived[5] the following cluster expansion for F_2^{eq}:

$$F_2^{eq}(x_1 x_2) = F_{2,0}^{eq}(x_1 x_2) + n F_{2,1}^{eq}(x_1 x_2) + ... \tag{3}$$

where

$$F_{2,0}^{eq}(x_1 x_2) = e^{-\beta\phi(r_{12})} F_1^{eq}(p_1) F_1^{eq}(p_2) \tag{3a}$$

$$F_{2,1}^{eq}(x_1 x_2) = \int dx_3 \left[e^{-\beta \{\phi(r_{12}) + \phi(r_{13}) + \phi(r_{23})\}} - \right.$$
$$- e^{-\beta\phi(r_{12})} e^{-\beta\phi(r_{13})} + e^{-\beta\phi(r_{12})} e^{-\beta\phi(r_{23})} + \quad \text{(3b)}$$
$$\left. + e^{-\beta\phi(r_{12})} \right] F_1^{eq}(p_1) F_1^{eq}(p_2) F_1^{eq}(p_3)$$

with similar expressions for the coefficients, $F_{2,l}^{eq}$, of the higher powers of n. Here $\beta = 1/kT$ and $F_1^{eq} = 2\pi mkT)^{-3/2} \exp[-\beta p^2/2m]$, with $p = |\mathbf{p}|$.

Equation (3) expresses a many-particle property, F_2^{eq}, in terms of Boltzmann factors, i.e. in terms of (static) properties of isolated groups of 2, 3, 4, ... particles. This is the basic idea of the virial expansion. I remark that the $F_{2,l}^{eq}$ exist because of the cluster-property of the integrands. For instance, $F_{2,1}^{eq}$ exists, because the integrand on the right hand side of (3b) vanishes as soon as particle 3 is separated from the particles 1 or 2 by a distance larger than r_0. The integral over $\mathbf{r}_3$ is therefore effectively only over a region in space of $O(r_0^3)$.

A non-equilibrium generalization of the cluster expansion (3) can be made, which expresses F_2 in terms of dynamical operators associated with isolated groups of 2, 3, 4, ... particles.[4] These dynamical- or streaming-operators $S_{-t}(x_1 \ldots x_s)$ $(s=1, 2, ...)$ are defined for s-particles by:

$$S_{-t}(x_1 \ldots x_s) = \exp[-t\mathscr{H}_s(x_1 \ldots x_s)]$$

where

$$\mathscr{H}_s(x_1 \ldots x_s) = \sum_{i=1}^{s} \frac{\mathbf{p}_i}{m} \cdot \frac{\partial}{\partial \mathbf{r}_i} - \sum_{\substack{i<j \\ 1}}^{s} \left[\frac{\partial\phi(r_{ij})}{\partial \mathbf{r}_i} \cdot \frac{\partial}{\partial \mathbf{p}_i} + \frac{\partial\phi(r_{ij})}{\partial \mathbf{r}_j} \cdot \frac{\partial}{\partial \mathbf{p}_j} \right].$$

The only property of the S_{-t}-operators of interest to us here is that when acting on any function $f(x_1 \ldots x_s)$ of the phases $x_1 \ldots x_s$ of the particles $1 \ldots s$, it transforms these phases into phases $X_1 \ldots X_s$, which the s-particles had a time t earlier, if they would move backwards in time under the influence of their mutual interaction:

$$S_{-t}(x_1 \ldots x_s) f(x_1 \ldots x_s) = f(X_1 \ldots X_s).$$

The $S_{-t}(x_1 \ldots x_s)$-operators involve therefore the solution of the dynamical s-body problem in infinite space. In terms of these S-operators, the following formal expansion of the non-equilibrium pair distribution function can be obtained:

$$F_2(x_1 x_2; t) = F_{2,0}(x_1 x_2; t) + n F_{2,1}(x_1 x_2; t) + \ldots \quad \text{(4)}$$

where

$$F_{2,0}(x_1 x_2; t) = \mathscr{S}_t(x_1 x_2) F_1(x_1; t) F_1(x_2; t) \quad \text{(4a)}$$

and

$$F_{2,1}(x_1 x_2; t) = \int dx_3 \left[\mathscr{S}_t(x_1 x_2 x_3) - \mathscr{S}_t(x_1 x_2) \mathscr{S}_t(x_1 x_3) - \right.$$
$$- \mathscr{S}_t(x_1 x_2) \mathscr{S}_t(x_2 x_3) + \quad \text{(4b)}$$
$$\left. + \mathscr{S}_t(x_1 x_2) \right] F_1(x_1; t) F_1(x_2; t) F_1(x_3; t)$$

and similar expressions for the coefficients, $F_{2,l}$ of the higher power of n. Here

$\mathscr{S}_t(x_1 \dots x_s) = S_{-t}(x_1 \dots x_s) \prod_{i=1}^{s} S_t(x_i)$. We stress that the expansion (4) is purely formal. In obtaining it, one has assumed that at the initial instant of time $t=0$, the F_s factorizes, i.e. that $F_s(x_1 \dots x_s; 0) = \prod_{i=1}^{s} F_1(x_i; 0)$ $(s=2,\dots)$.[6,7] This assumption can be considered to be a generalization of the assumption of molecular chaos in the Boltzmann collision term and enables one to express $F_2(x_1 x_2; t)$ in terms of $F_1(x_1; t)$. We remark that (4) is an expansion in terms of the ratio of the first two basic relaxation: $t_c/t_{mfp} \sim nr_0^3$ since $t_{mfp} \sim l \sim 1/nr_0^2$. Like the expansion (3), one has expressed here a many-particle property, F_2, in terms of (dynamical) properties of isolated groups 2, 3, 4, ... particles.

In order to ascertain the existence of the $F_{2,1}$ and to investigate the cluster properties of the integrands one has, in view of the S-operators occurring in (4), to make a dynamical analysis of the integrands i.e. one has to determine which collision sequences of 3, 4, ... particles contribute to $F_{2,1}$, $F_{2,2}$ ... respectively.

To the three-particle term $F_{2,1}$, contributions come from[8]: (a) genuine triple collisions in which all 3 particles interact at the same time (cf. Fig. 1a); (b) sequences of 3 or more successive binary collisions (cf. Fig. 1b, c). The operator in square brackets on the right hand side of Equation (4b) is constructed in such a way that it vanishes, if only 1 or 2 binary collisions occur. Similarly, to the four-particle term $F_{2,2}$ con-

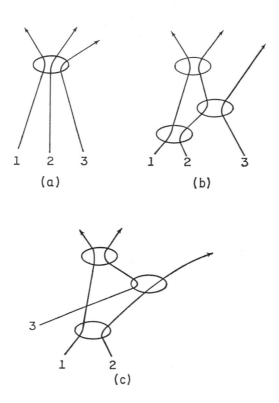

Fig. 1.   Some three-particle contributions to $F_{2,1}$.

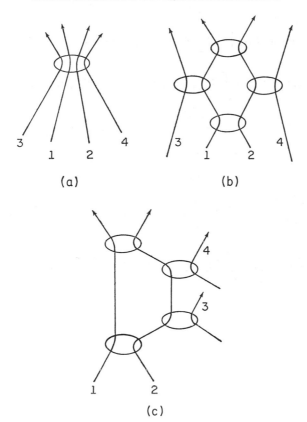

Fig. 2. Some four-particle contributions to $F_{2,2}$.

tributions come amongst others from: (a) genuine four-particle collisions (cf. Fig. 2a); (b) sequences of 4 or more successive binary collisions (cf. Fig. 2b, c).

One would expect that a cluster property of the integrands would exist, since the sequences of successive binary collisions that contribute to the $F_{2,1}$ will effectively occur in a time interval of $O(t_c)$. In $F_{2,1}$, for instance, in each of the collision sequences sketched in Fig. 1, the farther particle 3 is away from the particles 1 and 2 at $t=0$, the more difficult the recollision of the particles 1 and 2 will become. In fact for separations $\gg r_0$, the recollision seems to become virtually impossible. Thus, like in equilibrium, one expects that the integral over $\mathbf{r}_3$ in $F_{2,1}$ will extend effectively only over a region of $O(r_0^3)$, and similarly for all the $F_{2,l}(l \geqslant 1)$ in (4). If this obtains, the combination of $\mathscr{S}_t$-operators in the $F_{2,l}$ will attain their asymptotic values already for $t \approx O(t_c)$, so that for $t \gg t_c$ the $\mathscr{S}_t$-operators can be replaced by $\mathscr{S}_\infty$-operators without altering the result. If one makes this replacement, the only time dependence in F_2 is through F_1, so that then F_2 becomes a time-independent functional of F_1 and the Bogolubov result for $F_2(x_1 x_2 \mid F_1)$ is obtained:[3,4]

$$F_2(x_1 x_2 \mid F_1) = F_{2,0}(x_1 x_2 \mid F_1) + n F_{2,1}(x_1 x_2 \mid F_1) + \dots \tag{5}$$

where

$$F_{2,0}(x_1 x_2 \mid F_1) = \mathcal{S}_\infty (x_1 x_2) F_1 (x_1; t) F_1 (x_2; t) \tag{5a}$$

and

$$
\begin{aligned}
F_{2,1}(x_1 x_2 \mid F_1) = \int dx_3 [\mathcal{S}_\infty (x_1 x_2 x_3) &- \mathcal{S}_\infty (x_1 x_2) \mathcal{S}_\infty (x_1 x_3) - \\
- \mathcal{S}_\infty (x_1 x_2) \mathcal{S}_\infty (x_2 x_3) &+ \\
+ \mathcal{S}_\infty (x_1 x_2)] F_1 (x_1; t) &F_1 (x_2; t) F_1 (x_3; t).
\end{aligned} \tag{5b}
$$

Substituting this expansion for F_2 into (2), an equation for F_1 is obtained, that is of the desired form and is identical to that obtained by Bogolubov:

$$\frac{\partial F_1 (x_1; t)}{\partial t} = -\frac{\mathbf{p}_1}{m} \cdot \frac{\partial F_1}{\partial \mathbf{r}_1} + n J (F_1 F_1) + n^2 K(F_1 F_1 F_1) + \cdots \tag{6}$$

where

$$J (F_1 F_1) = \int dx_2 \theta_{12} \mathcal{S}_\infty (x_1 x_2) F_1 (x_1; t) F_1 (x_2; t) \tag{6a}$$

and

$$
\begin{aligned}
K(F_1 F_1 F_1) + \int dx_2 \theta_{12} \int dx_3 [\mathcal{S}_\infty (x_1 x_2 x_3) &- \mathcal{S}_\infty (x_1 x_2) \mathcal{S}_\infty (x_1 x_3) \\
- \mathcal{S}_\infty (x_1 x_2) \mathcal{S}_\infty (x_2 x_3) &+ \\
+ \mathcal{S}_\infty (x_1 x_2)] F_1 (x_1; t) &F_1 (x_2; t) F_1 (x_3; t)
\end{aligned} \tag{6b}
$$

etc.

Bogolubov already showed that $J(F_1 F_1)$ can be reduced to the Boltzmann collision term if one neglects the difference in the position of the two colliding molecules 1 and 2. In this connection we note that but for a change of variable, $n F_1$ is Boltzmann's f. Therefore Equation (5) represents a generalized Boltzmann equation that contains corrections to the Boltzmann equation due to three- and more particle collisions. Moveover, if one assumes that for $t \to \infty$, $F_1 (x, t)$ becomes the equilibrium distribution function $F_1^{eq} (p)$, the density expansion (5) of $F_2 (x_1 x_2 \mid F_1)$ reduces term by term to the equilibrium expansion (3).[9]

3. DIVERGENCES

In spite of this apparent success, a closer analysis of the three-body and higher order collision terms in (5) and (6) has revealed, around 1964,[10] that in 3-dimensions $(d=3)$ the four-body and all higher order collision terms diverge, while in 2-dimensions $(d=2)$ the three-body and all higher order terms diverge. This means that there is *no* cluster property of the corresponding integrands and that the above given plausibility arguments for the existence of such a property are not correct. I shall illustrate the existence of the divergences in the simplest case: the three-particle term in $d=2$. This term exhibits a logarithmic divergence due to the contributions of 3 successive binary collisions, as can easily be seen as follows.[11]

The crucial point is: what is the volume Γ_3 in the phase space of particle 3 of those

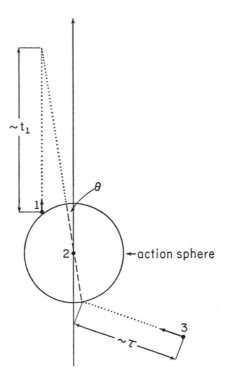

Fig. 3. Recollision event of particles 1 and 2 through an intermediate collision of particle 2 with particle 3. In the chosen coordinate system particle 2 is at rest at the origin, immediately after the first collision of the particles 1 and 2 and particle 1 moves away parallel to the z-axis. τ is the time it takes particle 3 to collide with particle 2 after the first (1,2)-collision has taken place. The (1,2)-recollision takes place at time t_1 after the (2,3)-collision. θ is the angle of the line of centres of the (2,3)-collisions, with the positive z-axis.

phases of particle 3, for which a recollision of the particles 1 and 2 occurs through an intermediate collision of either of them (say particle 2) with particle 3 (cf. Fig. 3), when the particles 1 and 2 collided before at $t=0$? In order to estimate Γ_3, it is convenient to choose a coordinate system in which particle 2 is at rest just after the first collision with particle 1 at $t=0$ and in which particle 1 moves away from particle 2 along the positive z-axis. Introducing instead of the coordinates $\mathbf{r}_3$ of particle 3, the time τ it takes particle 3 to collide with particle 2 and the angle θ of the line of centres of that collision with the positive z-axis, one has to estimate the integral:

$$\int d\mathbf{p}_3 \int d\mathbf{r}_3 \rightarrow \int d\mathbf{p}_3 \int d\tau \int d\theta \ J.$$

Here the limits of $\mathbf{p}_3$, τ and θ, are determined by the condition that a recollision of the particles 1 and 2 occurs and J is the Jacobian of the above-mentioned transformation. For $\tau \gg t_c$ a recollision of the particles 1 and 2 will occur only if $\alpha_1/t_1 < \sin \theta$

$\approx \theta \approx \alpha_2/t_1$ if t_1 is the time of the recollision. Since for $\tau \gg t_c$, $t_1 \sim \tau$, one also has $\alpha_1/\tau < \theta < \alpha_2/\tau$. Here α_1 and α_2 are functions of $\mathbf{p}_3$ and the position and momentum of particle 1 after the collision with particle 2. Since for large τ, J is independent of θ and τ, the relevant integral to consider is:

$$\int d\mathbf{p}_3 \int_{-\infty}^{\infty} d\tau \int_{\alpha_1/\tau}^{\alpha_2/\tau} d\theta \ J \sim \log \tau/t_c$$

which diverges logarithmically for $\tau \to \infty$.

Thus, when the $\mathscr{S}_t$ are replaced by $\mathscr{S}_\infty$ and no upper limit for the time of a recollision exists, the number of collision sequences of the type sketched in Fig. 3, that contribute to $F_2(x_1 x_2 \mid F_1)$ diverges logarithmically, and consequently the three-body collision term $\bar{K}(F_1 F_1 F_1)$ in Equation (6) will diverge logarithmically as well.

There exists thus an effective long range interaction of a dynamical nature between particles, due to extended sequences of successive binary collisions.

The above divergence argument for $d=2$ also shows why $F_{2,1}$ and $\bar{K}$ exist for $d=3$. For in that case, the angle θ has to be replaced by a space angle, which leads to a finite value for Γ_3.

However, a similar argument for the recollision of the particles 1 and 2 in the four particle term in $d=3$ leads to a logarithmic divergence (cf. Fig. 2). I have summarized the long time behaviour of the $F_{2,1}(l>1)$ in Table I. Although no proof exists, it is believed that for a spherically symmetric potential of the kind considered here, this behaviour also indicates the nature of the *worst* divergences that occur in the $F_{2,1}(x_1 x_2 \mid F_1)(l>1)$ and the corresponding collision terms of the generalized Boltzmann equation (6).

TABLE I

Collision Term	Behaviour for $\tau \gg t_c$	
	$d=2$	$d=3$
3-body	$\log(\tau/t_c)$	(t_c/τ)
4-body	(τ/t_c)	$\log(\tau/t_c)$
l-body	$(\tau/t_c)^{l-3}$	$(\tau/t_c)^{l-4}$

In conclusion one can say that (a) there is no straightforward generalization of the Boltzmann equation to higher densities along the lines of the virial expansion, as for instance attempted by Bogolubov. The Boltzmann equation is therefore not the first term in a virial-type of expansion and only in three dimensions is the next term meaningful. Also, (b) up till the present day not any generalization of Boltzmann's H-theorem has been given, not even for the special case of a spatially homogeneous system, where F_1 only depends on $\mathbf{p}_1$ and t and not on $\mathbf{r}_1$.

4. FINAL REMARKS

1. ORIGIN OF DIVERGENCES

The origin of the divergences of the $F_{2,1}(x_1 x_2 \,|\, F_1)$ lies in the contributions of sequences of successive collisions, with no restriction on the free paths of the particles involved. Clearly, to obtain a convergent theory, a cut-off of these free paths at the mean free path must be introduced. The mean free path is, however, a many-particle or collective property of the gas and is alien to the virial-type of expansion discussed in the previous section. Therefore, *some* collective properties of the gas *have* to be taken into account, in addition to the individual particle properties exclusively taken into account in the virial expansion procedure, if a convergent theory is to be obtained. However, the introduction of a mean free path cut-off also introduces a new length (viz. l) or time (viz. t_{mfp}) into the theory and will lead to non-local effects in space and time. As a consequence, $\partial F_1/\partial t$ will no longer depend on $F_1(t)$ alone but on all $F_1(t_i)$ with $t - t_{mfp} \lesssim t_i \lesssim t$. Although some progress in the formulation of a convergent theory along these lines has been made,[12] it is not clear at this moment whether, in general, a closed equation for F_1 – even allowing non-local effects – exists. In other words: it is not entirely certain that Boltzmann's kinetic approach *can* be generalized to higher densities.

2. COMPARISON WITH PLASMA

In the case of a plasma there are already divergence difficulties in the virial expansion in thermal equilibrium, due to the long range of the interparticle forces. One starts therefore here with an 'extreme' collective description, which leads in equilibrium to the Debye–Hückel theory instead of the virial theory and in non-equilibrium to the Vlasov equation instead of the Boltzmann equation (1).[13]

Generalization of the Vlasov equation to higher densities introduces in the next order individual particle effects and leads in the spatially homogeneous case to the Balescu–Lenard equation.[14, 15] This equation is local in time and Guernsey has shown on the basis of it the approach to equilibrium of a spatially homogeneous plasma, by proving an H-theorem.[15] Contrary to what one would expect, therefore, the neutral gas with short-range interparticle forces seems to be more difficult to treat than the plasma: a more complicated mixture of collective and individual particle effects seems to be asked for in the gas than in the plasma.

3. HYDRODYNAMICAL CONSEQUENCES

The divergence difficulties, due to the extended collision sequences discussed above, have also consequences for the hydro-dynamical equations:

No virial expansion of the transport coefficients

Applying the Chapman–Enskog procedure to Equation (6), one obtains a formal virial expansion for the transport coefficients:

$$\eta(n, T) = \eta_0(T) + n\eta_1(T) + n^2\eta_2(T) + \dots$$
$$\lambda(n, T) = \lambda_0(T) + n\lambda_1(T) + n^2\lambda_2(T) + \dots \,. \tag{7}$$

Here $\eta_1(T)$ and $\lambda_1(T)$ are finite and explicit expressions for them were first derived by Choh and Uhlenbeck in 1958.[16] However, due to the logarithmic divergence of the four-body term, one finds divergent expressions for $\eta_2(T)$ and $\lambda_2(T)$:

$$\eta_2(T) = \eta_2'(T) \lim_{\tau \to \infty} \log(\tau/t_c) + \eta_2''(T) \tag{8}$$

and similarly for $\lambda_2(T)$.

We remark that one would expect in a convergent theory a cut-off at t_{mfp} of the logarithmic divergence on the right-hand side of (8), so that one would obtain:

$$\eta_2(T) = \eta_2'(T) \log(t_{\text{mfp}}/t_c) + \eta_2''(T)$$

This would lead, with $t_{\text{mfp}}/t_c \sim 1/nr_0^3$, to an expansion of the form:

$$\eta(n, T) = \eta_0(T) + n\eta_1(T) + n^2\bar{\eta}_2(T) + n^2 \log n(T) + \dots \tag{9}$$

and similarly for $\lambda(n, T)$.

The theoretical prediction is then that a term that depends logarithmically on the density will occur in the density expansion of η as well as of λ.

There is no theoretical justification of this ad-hoc argument for the occurrence of an $n^2 \log n$-term in the transport coefficients available at the moment, except in the case of a simplified model, the Lorentz model. Here the existence of a logarithmic term has been established theoretically, restricting oneself to a certain class of dynamical events[17]; such a term seems also to have been confirmed 'experimentally' by an analysis of computer calculations of the diffusion coefficient for such a model[18]. A careful analysis of the experimentally found transport-coefficients η and λ as a function of density seems also to give some support for expansions of η and λ of the type (9) instead of (7)[19]. However, this experimental verification of (9) should be called very preliminary at the moment.

In addition, questions concerning the existence of transport coefficients as well as the precise range of validity of the hydrodynamical equations are being investigated[20].

REFERENCES

1. (a) L. Boltzmann, *Wien. Ber.* **66**, 275 (1872), or *Collected Works* **1**, 316.
   (b) For a modern discussion: (i) G. E. Uhlenbeck and G. W. Ford, 'Lectures in Statistical Mechanics' in *Lectures in Applied Mathematics*, vol. 1, American Mathematical Society, Providence, R. I. (1963), Chapter IV; (ii) S. Chapman and T. G. Cowling, *The Mathematical Theory of Non-Uniform Gases*, Cambridge University Press, England (1970).
2. (a) S. Chapman, *Phil. Trans. Roy. Soc.* A, **211**, 433 (1912); A, **216**, 279 (1916), A, **217**, 115 (1917).
   (b) D. Enskog, *Phys. Zeit.* **12**, 56 and 533 (1911); 'Kinetische Theorie der Vorgänge in Mässig Verdünnter Gasen', dissertation, Upsala (1917); *Svensk. Vet. Akad., Arkiv. Mat. Astron. Fys.* **16** (1921).
   (c) For a modern discussion, see ref. 1b(i), Chapter 6, and ref. 1b(ii).
3. (a) N. N. Bogolubov, 'Problemy Dinamicheskoi Teorii v Statisticheskoi Fisike', Moscow (1946); English translation: in *Studies in Statistical Mechanics*, Vol. I (ed. by J. de Boer and G. E. Uhlenbeck), North-Holland Publishing Company, Amsterdam (1962), p. 5.
   (b) For a modern discussion: (i), see ref. 1b (i), Chapter VII; (ii) E. G. D. Cohen in *Fundamental Problems in Statistical Mechanics* (ed. by E. G. D. Cohen), North-Holland Publishing Company, Amsterdam (1961), p. 110.

4. (a) M. S. Green, *J. Chem. Phys.* **25**, 836 (1956); M. S. Green and R. A. Piccirelli, *Phys. Rev.* **132**, 1388 (1963).
   (b) E. G. D. Cohen, *Physica* **28**, 1025 (1962); *J. Math. Phys.* **4**, 183 (1963).

5. J. E. Mayer and E. W. Montroll, *J. Chem. Phys.* **9**, 2 (1941); J. de Boer, *Rept. Prog. Physics* **12**, 305 (1949).

6. It is an open question to what extent one can prove that this assumption of molecular chaos propagates in time. The 'propagation of chaos' has been proved starting from a master equation only for special models in the low density limit[7].

7. (a) M. Kac, in *Proceedings of the Third Berkeley Symposium in Mathematical Statistics and Probability*, University of California Press, Berkeley (1955), p. 171.
   (b) F. A. Grünbaum, in *Arch. Rational Mech. Anal.* **42**, 323 (1971).

8. Only collision sequences starting with a 1,2 collision occur, because of $\partial\phi/\partial\mathbf{r}$, in Equation (2).

9. R. A. Piccirelli, *J. Math. Phys.* **7**, 922 (1966).

10. (a) J. Weinstock, *Phys. Rev.* **132**, 454 (1963); **140A**, 460 (1965).
    (b) R. Goldman and E. A. Frieman, *Bull. Amer. Phys. Soc.* **10**, 531 (1965); *J. Math. Phys.* **7**, 2153 (1966); **8**, 1410 (1967).
    (c) J. R. Dorfman and E. G. D. Cohen, *Phys. Letters* **16**, 124 (1965); *J. Math. Phys.* **8**, 282 (1967).

11. (a) E. G. D. Cohen, in *Lectures in Theoretical Physics*, Vol. VIII A (ed. by W. E. Brittin), The University of Colorado Press, Boulder (1966), Appendix III.
    (b) J. V. Sengers, in *Lectures in Theoretical Physics*, Vol. IX C (ed. by W. E. Brittin), Gordon and Breach, New York (1967), p. 335.

12. (a) J. Weinstock, *Phys. Rev. Letters* **17**, 130 (1966).
    (b) K. Kawasaki and I. Oppenheim, *Phys. Rev.* **139 A**, 1763 (1965); also in: *Statistical Mechanics* (ed. by Th. A. Bak) W. A. Benjamin Inc., New York (1967), p. 313.
    (c) Ref. 10b, last two papers.
    (d) J. R. Dorfman, 'Renormalised Kinetic Equations', University of Maryland Techn. Note BN618, University of Maryland, College Park (1969).
    (e) E. G. D. Cohen, in *Fundamental Problems in Statistical Mechanics*, Vol. II (ed. by E. G. D. Cohen), North-Holland Publishing Company, Amsterdam (1969), p. 228.

13. (a) P. Debye and E. Hückel, *Physik. Z.* **24**, 185 (1923) or collected *Papers of P. Debye*, Interscience, New York (1954), p. 217.
    (b) A. Vlasov, *J. Phys. U.S.S.R.* **9**, 25 (1945).
    (c) For a modern discussion, see, for instance, N. G. van Kampen and B. U. Felderhof, *Theoretical Methods in Plasma Physics*, North-Holland Publishing Company, Amsterdam (1967), Chapters VIII and IX.

14. (a) R. Balescu, *Phys. Fluids* **3**, 52 (1960).
    (b) A. Lenard, *Ann. Phys. (N.Y.)* **10**, 390 (1960).
    (c) Ref. 13c, Chapter XV.

15. (a) R. Guernsey, 'The Kinetic Theory of Fully Ionised Gases', Dissertation, University of Michigan (1960).
    (b) T. Y. Wu, *Kinetic Equations of Gases and Plasmas*, Addison Wesley, Reading, Mass. (1966), Ch. 6.

16. (a) S. T. Choh and G. E. Uhlenbeck, 'The Kinetic Theory of Dense Gases', dissertation, University of Michigan (1958).
    (b) Ref. 3b (ii).
    (c) E. G. D. Cohen in *Lectures in Theoretical Physics*, Vol. IXC (ed. by W. E. Brittin), Gordon and Breach, New York (1967), p. 279.

17. (a) J. M. J. van Leeuwen and A. Weijland, *Physica* **36**, 457 (1967).
    (b) A. Weijland and J. M. J. van Leeuwen, *Physica* **38**, 35 (1968).

18. C. Bruin, *Phys. Rev. Letters* **29**, 1670 (1972).

19. (a) H. M. Hanley, J. D. McCarty and J. V. Sengers, *J. Chem. Phys.* **50**, 857 (1969).
    (b) J. Kestin, E. Paykos and J. V. Sengers, *Physica* **54**, 1 (1971).

20. (a) B. J. Alder and T. E. Wainwright, *Phys. Rev. Letters* **18**, 988 (1967); *Phys. Rev. A* **1**, 18 (1970).
    (b) T. E. Wainwright, B. J. Alder and D. M. Gass, *Phys. Rev. A* **4**, 233 (1971).
    (c) Y. Pomeau, *Phys. Letters* **27 A**, 601 (1968); *Phys. Rev. A* **3**, 1174 (1971).
    (d) J. R. Dorfman and E. G. D. Cohen, *Phys. Rev. Letters* **25**, 1257 (1970); *Phys. Rev. A* **6**, 776 (1972).

(e) M. H. Ernst, E. H. Hauge and J. M. J. van Leeuwen, *Phys. Rev. Letters* **25**, 1254 (1970); *Phys. Rev.* A **4**, 2055 (1971).

(f) K. Kawasaki, *Progr. Theoret. Phys.* (*Kyoto*) **45**, 1691 (1971).

DISCUSSION

Question: The diagrams which you have drawn lead to difficulties that are effectively dealt with by the propagator and T matrix techniques of quantum mechanics. One wonders whether a non-analytic series for F_2 in terms of F_1 is an alternative proposal. Can one do anything with a non-analytical type of virial expansion, things which still might have value because they are continuous functions with respect to the parameter, but just do not permit an analytic power series expansion.

E. G. D. Cohen: It has been possible to take into account some of these extended collision sequences in a modified three-body term, which does have a form which is quite different from (5b). One suspects that there is a kind of mean free path cutoff, because there is a damping operator. In the general case which I have treated here, no real study of the properties, analytic and non-analytic, and the consequences of this operator is available. Only for special models, the Lorentz model for example, where one has a linear theory, has it been possible to go a little bit further; and there, within the restriction of the limited number of dynamic events, which one has been able to handle, one has been able to show that there is such a term. However, of all the collision processes that go on in a real gas, one has taken into account a very small number, hoping that they are perhaps the most important ones under certain conditions; but unless one goes further one really cannot prove anything for certain.

F. Bopp: What is the relation of your considerations to those of Born and Green?

E. G. D. Cohen: The starting point of Bogolubov is similar to what Born and Green wanted to do, They tried to make a simple Ansatz for F_2 in terms of F_1, an Ansatz which usually says that F_2 is some kind of binary operator acting on two F_1's. All they did was to make an extension of Boltzmann's ideas. The usual Boltzmann equation always has the fact, which is hidden in here also, that the two colliding particles are taken at the same position. If you apply this binary operator, the two particles are no longer at the same position. Thus one gets a slight extension of the Boltzmann equation, which leads to a better equation of state. That is the only tangible result which they [Born and Green] obtained. As for the triple collision and all the higher ones, Bogolubov was the one who really got a complete result first.

W. E. Lamb: In my discussion I have treated a system which approaches thermodynamic equilibrium because of its interaction with the outside world. In fact, that system would not approach thermodynamic equilibrium if it were left to itself no matter how long. Thus, to the extent that the gas you have treated is a closed Hamiltonian dynamical system, I am tempted to generalize that such a system would not approach thermodynamic equilibrium ever. Of course your discussion involves some probabilistic notions in the collision terms, and so I am sure that you are considering the possibility of the consequence that you do get thermodynamic equilibrium, but in one way or another a certain amount of dirt is being swept under the rug by all of us through the assumptions we make.

E. G. D. Cohen: Yes. It is the Boltzmann Ansatz, the statistical Ansatz of molecular chaos, which introduces the arrow of time or, in your language, the approach to equilibrium. It is the assumption of the factorization of the s-particle distribution at time $t=0$, which is a generalization of the statistical Ansatz, which introduces the irreversibility.

J. Ehlers: Even in the case where one accepts the Boltzmann equation, there is an open question whether one can deduce rigorously from Boltzmann's H-theorem that the system approaches equilibrium. This has probably been examined in the homogenous case; I wonder whether it has also been done for the inhomogenous case.

E. G. D. Cohen: I don't think so. It has been done only for the special model of hard spheres. Those proofs are of great interest to the mathematicians, but Boltzmann's proof is a physicist's proof. Otherwise, I completely agree with you.

M. Kac: I would like to make a distinction between a proof and a demonstration. A demonstration is a way to convince a reasonable man, and a proof is a way to convince a stubborn one. If it comes to the H-theorem, we have demonstrations but no proofs.

29

Time, Irreversibility and Structure

Ilya Prigogine

1. INTRODUCTION

The legacy of Newtonian mechanics is a concept of a world in which *time is essentially a parameter associated with motion*[1]. As early as 1796, Lagrange called mechanics a four-dimensional geometry[2]. The two great revolutions we have witnessed in physics during this century, relativity and quantum mechanics, have kept unmodified this character of Newton's world model.

However this description neglects essential aspects of our experience. On the level of the phenomenological, thermodynamic description as used to describe diffusion or chemical reactions, time appears in conjunction with *irreversibility*. There appears a privileged direction of time, the direction in which entropy, *time's arrow* according to Eddington's famous expression[3], increases.

There exists in addition a class of phenomena in which time is associated with coherent structures and evolution of structures. We are immediately reminded about biological structures and their evolution, but many examples of such 'dissipative' structures are also known today in physics and chemistry.

Schematically I would therefore like to distinguish three levels in the description of time:

time associated with *motion*

time associated with *entropy* (irreversibility)

time associated with *dissipative structures*.

How can we integrate these levels in the framework of the mechanical description based on the laws of classical or quantum mechanics. What are the consequences of this integration for our general conception of nature?

As this is a Symposium on the *development* of the physicist's conception of nature, I shall begin with a few historical remarks. Obviously these remarks can only be very sketchy and incomplete.

The problems we shall discuss were first formulated during the nineteenth century. The ideas of evolution and of history were central to the romantic mood of this time. As is well-known the basic importance of the second law in the context of physical evolution was clearly recognized by Clausius[4] who introduced the term 'entropy', which in Greek means 'evolution'. Moreover the first steps to integrate entropy into dynamics, to construct a microscopic model of entropy, were undertaken by Boltzmann in 1872[5], exactly one hundred years ago. Boltzmann's ideas are today still so fasci-

J. Mehra (ed.), The Physicist's Conception of Nature, 561-593. All Rights Reserved
Copyright © 1973 by D. Reidel Publishing Company, Dordrecht-Holland

nating that it is worthwhile to present them here briefly. We may even say that his ideas form the natural reference system for any new attempt to discuss the relations between dynamics and thermodynamics. He thought that the second law should result from the *replacement* of dynamics by some form of probability calculus. This idea was not new: Kronig in 1856[6], Clausius in 1857[7], Maxwell in 1859[8], Boltzmann himself in 1868[9] had already introduced probability concepts in the framework of kinetic theory. Boltzmann's scheme can briefly be summarized as follows[10]

$$dynamics \rightarrow stochastic\ process\ \text{(kinetic equation)}$$
$$\rightarrow \mathscr{H}_B\text{-}quantity\ \text{(entropy)}.$$

We see why Boltzmann's investigations are so central. They link various levels of description which had been introduced independently in the history of science: *the dynamical description expressed through the laws of mechanics, the description in terms of probabilities and the thermodynamical description.*

Boltzmann's $\mathscr{H}$-theorem has led to a very strange situation, perhaps unique in the history of science. On the one hand, Boltzmann's kinetic equation has been applied successfully to a very large range of physical phenomena, such as transport processes in dilute gases, in plasmas, shock waves, hydrodynamics and chemical reactions. Molecular dynamics experiments, such as those performed by Alder and Wainwright[11], and more recently by Bellemans and Orban[12], completely verify Boltzmann's predictions. The quantity $\mathscr{H}_B$ indeed decreases in a fluctuating fashion and the system reaches thermodynamic equilibrium after a time corresponding to a few collisions per particle. The fluctuations decrease when the number of particles is increased. There is little doubt that in the thermodynamic limit (volume $V \rightarrow \infty$, number of particles $N \rightarrow \infty$, $N/V =$ constant) we would observe for dilute gas the monotonic behaviour approach to equilibrium predicted by Boltzmann.

On the other hand, Boltzmann's ideas met with violent objections coming from theoretical physicists and mathematicians[13]. In fact, these discussions never ceased. In which sense does Boltzmann's microscopic model of entropy 'explain' irreversibility? What is its domain of validity? The classical objections are often formulated in the form of paradoxes. The most important, from our point of view, is Loschmidt's reversibility paradox: Since the laws of mechanics are symmetrical with respect to the inversion of time,

$$t \rightarrow -t,$$

to each process there belongs a corresponding time-reversed process. This seems to be in contradiction with the existence of irreversible processes.

The answer to Loschmidt's paradox is not straightforward. It is easy to make a computer experiment to test this paradox. For example, Bellemans and Orban[12] have calculated Boltzmann's quantity $\mathscr{H}_B$ for two-dimensional hard spheres (hard disks). They start with disks on lattice sites with an isotropic velocity distribution. The results are shown on Fig. 1 (white circles).

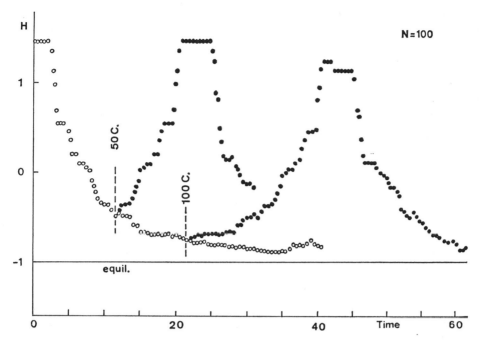

Fig. 1. Evolution of $\mathscr{H}_B$ with time for a system of 100 disks when velocities are inverted after 50 collisions, 100 collisions.

Now, if after 50 or 100 collisions, we invert the velocities we obtain a new ensemble. We may then again follow the evolution of the corresponding quantity $\mathscr{H}_B$ in time. The results are given in Fig. 1 (black circles).

We see that indeed the entropy (given by $-\mathscr{H}_B$) first decreases after the velocity inversion. The system deviates from equilibrium over a period of 50 to 60 collisions (corresponding to about 10^{-6} seconds in a dilute gas).

A similar situation exists also in spin-echo experiments or in plasma-echo experiments[14-17]. Of special interest are the spin-echo experiments in dipolar coupled spin systems as recently described by Rhim, Pines and Waugh[16], because these experiments involve an N body situation quite similar to a classical gas. There also, over limited periods of time, 'anti-Boltzmannian' behaviour corresponding to an increase of $\mathscr{H}_B$ may be observed. Rhim, Pines and Waugh[16] even introduce a 'Loschmidt demon' to deal with this situation.

Boltzmann's point of view was that his derivation of the kinetic equation depends, *in addition* to the laws of mechanics, on the initial conditions which are assumed[18]. He recognized quite clearly that, as first emphasized by Burbury, the molecules have been uncorrelated before the collision[19, 20]. He considered that this was such a 'natural' assumption that the probability of a situation in which this would not be so could be neglected.

Now both the computer experiments and the spin experiments show that we have to be careful: conditions in which correlations would exist between molecules *before* a collision (or a 'recollision') occurs between them can be realized at least temporarily. The Ehrenfests recognized quite clearly that Boltzmann's equation cannot be valid both before and after velocity inversion[13]. What is the consequence of this situation? Does the fact that Boltzmann's equation is not valid imply a set-back of Boltzmann's interpretation of entropy or of the second law itself?

Let us recall the definition of an irreversible process as introduced in thermo-dynamics[21]. According to Planck's definition it is a process which once performed leaves the world in an altered state. Planck insists: *By no experimental device, whatever the ingenuity of the experimenter, should it be possible to restore the initial state.* Now if we consider the velocity inversion experiment the positive entropy produced in the period $0 - t_0$ would be compensated by a *negative* entropy production during the period $t_0 - 2t_0$ in contradiction with the very definition of an irreversible process.

Confronted with the conceptual difficulties of Boltzmann's approach, Gibbs in 1902[22], and Einstein in 1902–03[23, 24] worked out a new much more general systematics of statistical mechanics based on ensemble theory. In a sense however it was a *return to classicism*. The complete title of Gibbs' classical memoir is: 'Elementary Principles in Statistical Mechanics, developed with special *reference to the rational foundation of thermodynamics*'. Now thermodynamics was at Gibbs' time mostly equilibrium thermodynamics. We are far from Boltzmann's ambition to derive the 'mechanical theory' of *evolution of matter*.

However once the power of the Gibbs–Einstein methods in equilibrium statistical mechanics was recognized through the pioneering work of Ursell, Mayer, Yvon and many others (see, for example, Fowler and Guggenheim[25]), it became very natural to extend these methods to non-equilibrium situations. Instead of time-independent ensembles, one has then to study the time evolution of such ensembles. This is the direction of research which led to the formulation of a new branch of statistical physics which may be appropriately called 'non-equilibrium statistical mechanics'.

A short summary of the history of this subject is presented in the introduction to my monograph on non-equilibrium statistical mechanics[26] (1962). It is impossible to summarize it here[27]. We shall see in this lecture why the progress realized in the last years leads to a much deeper understanding of the basic problems involved, such as the meaning of irreversibility and the microscopic model of entropy.

Let us now turn to thermodynamics. We may briefly formulate the second law as follows[28]. The variation of the entropy during the time dt may be written as the sum of two terms

$$dS = d_e S + d_i S \tag{1.1}$$

where $d_e S$ is the entropy supplied to the system by its surroundings, and $d_i S$ the entropy produced inside the system. The second law of thermodynamics requires that

$$d_i S \geqslant 0. \tag{1.2}$$

For isolated systems ($d_e S = 0$), equation (1.1) expresses the classical statement that entropy increases or remains constant when thermodynamic equilibrium has been reached.

Again for long time emphasis was on *equilibrium situations* for which $d_iS=0$. Fundamental contributions such as those of Maxwell or Gibbs deal uniquely with situations corresponding to equilibrium or to situations infinitesimally away from equilibrium (stability of equilibrium). Pioneering work on non-equilibrium thermodynamics was started by Duhem[29], De Donder[30] and others[31,32]. But their work remained largely unnoticed. The situation changed with the discovery of Onsager's reciprocity relations in 1931[33]. This is indeed a beautiful example of general relations independent of any detailed dynamic assumptions and therefore belonging to the 'thermodynamics' of matter. The explicit evaluation of entropy production combined with Onsager's relations lead to the formulation of non-equilibrium thermodynamics (1940–50[31,32]).

The original formulation of non-equilibrium thermodynamics corresponds to near equilibrium conditions ('linear' non-equilibrium thermodynamics). Only recently a systematic approach to the thermodynamic behaviour of matter in far from equilibrium conditions has been developed by the Brussels school[34]. The central problem here is the following: *'under which conditions may we extrapolate results obtained by equilibrium thermodynamics or by linear non-equilibrium thermodynamics to far from equilibrium conditions?'* As we shall see there exist important classes of systems in which, in far from equilibrium conditions, states of matter characterized by coherent space-time behaviour appear. We have called such states 'dissipative structures' in contrast to the 'equilibrium structures' of chemical thermodynamics.

Till now a systematic approach to such problems was only possible in the case of 'local equilibrium'. That corresponds to situations in which the variables used are the same as would appear in equilibrium. Clearly many situations such as treated in lasers[35] or in plasma physics[36] do not belong to this class. Phenomenological generalizations of thermodynamics going beyond local equilibrium have been proposed (see for example Truesdell[37]) but it is likely that the answer can come only from a microscopic model of non-equilibrium entropy.

Summarizing we may say that the problem of irreversibility had been clearly formulated during the nineteenth century. However the quantitative formulation of irreversibility both on the level of thermodynamics and on the level of statistical mechanics is still in progress. As we shall see in this lecture, we begin to have a general solution of these problems and this important progress will undoubtedly have repercussions both in physics proper and in the relation between physics and other disciplines, such as biology.

2. DYNAMIC AND THERMODYNAMICAL DESCRIPTIONS, TIME-INVERSION SYMMETRY

Let us start from classical or quantum dynamics as expressed by Hamilton's equations of motion

$$\frac{dq}{dt} = \frac{\partial H}{\partial p}, \qquad \frac{dp}{dt} = -\frac{\partial H}{\partial q} \qquad (2.1)$$

or Schrödinger's equation ($\hbar = 1$)

$$i\frac{\partial\psi}{\partial t} = H\psi. \tag{2.2}$$

Both descriptions may be unified through the Liouville–von Neumann equation for the distribution function (or density matrix [26])

$$i\frac{\partial\varrho}{\partial t} = L\varrho \tag{2.3}$$

with

$$L\varrho = \begin{cases} -i\{H, \varrho\} & \text{Poisson bracket} \\ [H, \varrho] & \text{commutator}. \end{cases} \tag{2.4}$$

In both cases, L is a hermitian operator (better *superoperator*, see [38,10,40]). Therefore

$$L = L^{+}. \tag{2.5}$$

A basic feature of equation (2.3) is its '*Lt-invariance*'. If we perform both the operations

$$\begin{cases} L \to -L \\ t \to -t \end{cases} \tag{2.6}$$

equation (2.3) remains invariant. Note that *if L is a possible Liouville operator so is* $-L^*$. On the other hand macroscopic equations involving thermodynamical quantities do not present in general such an invariance. Let us consider the heat equation

$$\frac{\partial T}{\partial t} = \kappa\frac{\partial^2 T}{\partial x^2}. \tag{2.7}$$

As the heat conductivity κ is a positive quantity, we cannot reverse its sign. Therefore the time inversion

$$t \to -t \tag{2.8}$$

now leads to the *different* equation

$$\frac{\partial T}{\partial t} = -\kappa\frac{\partial^2 T}{\partial x^2} \tag{2.9}$$

which we may call the 'anti-Fourier' equation. Both equations have a physical meaning. Equation (2.7) corresponds to the situation in which we have an *initial* value problem and we want to calculate the temperature distribution in the future. The solutions of (2.7) are 'retarded solutions'. On the contrary, in the case of (2.9) we have a *final* value problem as may arise as the result of a fluctuation in a system which was isolated for a long time. The uniform distribution corresponds then to the far distant past ($t \to -\infty$). The solution of (2.9) is an 'advanced solution'.

---

* Contrary to the operator H (the Hamiltonian), L has not a well-defined sign. For this reason L is the natural operator conjugate to time. This is also the reason why causality is expressed most naturally in superspace (and not in the Hilbert space).

We have now a pair of equations in each of which the directions of time play a different role. The time-symmetry present in the dynamical description (2.3) has been broken. There is obviously a relation between this 'symmetry breaking' and causality. This has been emphasized by many people.[26, 41, 42] But the symmetry breaking goes *beyond* causality as it leads to a description of natural processes in terms of *differential equations* in which the direction of time is specified. On the contrary, causality is a more general concept which may be applied to the *solution* of differential equations in which both directions of time play a symmetrical role.

We may say that the second law of thermodynamics for isolated systems

$$\frac{dS}{dt} \geqslant 0 \qquad (2.10)$$

summarizes in a single inequality all laws such as Fourier's law, friction phenomena, etc., which are described by phenomenological equations involving a *privileged* direction of time. This is the situation we associate with thermodynamic irreversibility.

What are the supplementary conditions which we have to require, in addition to causality, to generate thermodynamic behaviour? What is the class of systems which will manifest this behaviour? These are the questions we have to investigate now.

3. NON-EQUILIBRIUM STATISTICAL MECHANICS. THE MASTER EQUATION

Let us start with the Liouville equation (2.3). Its formal solution is

$$\varrho(t) = e^{-iLt}\varrho(0) = \frac{1}{2\pi i}\int_c dz\, e^{-izt}\, \frac{1}{L-z}\, \varrho(0). \qquad (3.1)$$

In the case of the initial value problem the contour has to be traced in the *upper* half plane corresponding to the complex variable z[26, 36] (that is for Im $z > 0$, see Fig. 2).

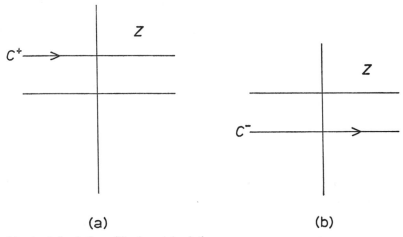

(a)   (b)

Fig. 2.   (*a*) retarded solution; (*b*) advanced solution.

With this choice of the contour we calculate the retarded solution of the Liouville equation

$$\varrho^r(t) = \frac{1}{2\pi i} \int_{c^+} dz\, e^{-izt} \frac{1}{L-z}\varrho(0). \tag{3.2}$$

Similarly the 'advanced' solution would be given by

$$\varrho^a(t) = \frac{1}{2\pi i} \int_{c^-} dz\, e^{-izt} \frac{1}{L-z}\varrho(0). \tag{3.3}$$

Both solutions satisfy the same differential equations. However causality leads to the two different integral representations (3.2), (3.3). If the evaluation of the integrals (3.2) and (3.3) leads indeed to a difference between ϱ^r and ϱ^a, we may expect 'thermodynamic behaviour'.

To proceed further it is convenient to introduce orthogonal hermitian projection operators P, Q such that

$$P + Q = 1. \tag{3.4}$$

It is then easy to verify the identities for the resolvent $R(z)\equiv(L-z)^{-1}$ [43, 44, 45]

$$R(z) = \{P + \mathscr{C}(z)\} \frac{1}{PLP + \Psi(z) - z} \{P + \mathscr{D}(z)\} - \mathscr{P}(z) \tag{3.5}$$

with

$$\Psi(z) = -PLQ\frac{1}{QLQ-z}QLP \tag{3.6}$$

$$\mathscr{D}(z) = -PLQ\frac{1}{QLQ-z} \tag{3.7}$$

$$\mathscr{C}(z) = -\frac{1}{QLQ-z}QLP \tag{3.8}$$

$$\mathscr{P}(z) = -\frac{1}{QLQ-z}Q. \tag{3.9}$$

We have here the basic operators of our formulation of non-equilibrium statistical mechanics; the choice of the projection operators fixes the 'language' in which we formulate our results. Generally P is chosen as projecting onto the diagonal elements in the representation in which some model-hamiltonian H_0 is diagonal:

$$\langle m|\, P\varrho\, |n\rangle = \langle m|\, \varrho\, |m\rangle\, \delta_{mn}. \tag{3.10}$$

For this reason $P\varrho$ is called the 'vacuum of correlations'. Of special importance is the collision operator $\Psi(z)$: reading (3.6) from right to left it corresponds to a transition from the vacuum of correlations, followed by a dynamical evolution in the correlation space, and finally followed by a return to the vacuum of correlations.

We now introduce the formal decomposition (3.5) into (3.2). We derive in this way directly the '*master equation*' for the density matrix ϱ.[26] For the diagonal elements ϱ_0 of ϱ we obtain

$$i\frac{\partial \varrho_0(t)}{\partial t} = \int_0^t d\tau G(t-\tau)\varrho_0(\tau) + \mathscr{F}(t; Q\varrho(0)) \tag{3.11}$$

where

$$G(t) = \frac{1}{2\pi i}\int_c dz\, e^{-izt}\Psi(z) \tag{3.12}$$

and

$$\mathscr{F}(t; Q\varrho(0)) = \frac{1}{2\pi i}\int_c dz\, e^{-izt}\mathscr{D}(z) Q\varrho(0). \tag{3.13}$$

It should be emphasized that the master equation (3.11) is exact. No approximations have been introduced in its derivation from the Liouville equation.

The denomination 'master equation' for (3.11) may lead to confusion. Indeed the master equation as introduced by Kac and Uhlenbeck[46] refers to a model equation incorporating features of probability theory. There are certainly supplementary assumptions to be made to permit such an interpretation of equation (3.11). Indeed let us compare it to Boltzmann's kinetic equation. We see that there are two main differences: One is the non-local character of the collision operator in (3.11), as the time change at t depends on the previous history of the system ('non-Markovian' equation). The second one is the occurrence of the memory term $\mathscr{F}$ depending on the initial correlations. The master equation (3.11) has been applied successfully to many interesting problems[47]. It has also been shown that it describes correctly the evolution of a system *both* before and *after* velocity inversion[48, 49]. Suppose that we start at $t=0$ with no correlations. During the evolution from 0 till t_0 the diagonal elements evolve according to (3.11) (with $\mathscr{F}=0$). Also correlations appear progressively as the result of the interactions. However when we inverse the velocities at $t=t_0$ we form a new ensemble. In the subsequent evolution from t_0 till $2t_0$ the memory term $\mathscr{F}$ plays an essential role as inversion of velocities leads to long-range correlations. At time $2t_0$ we recover the original state at $t=0$ with reversed velocities.

It is certainly very interesting that the master equation permits us to describe both 'Boltzmannian' and 'anti-Boltzmannian' behaviour. However the non-local form of (3.11) makes it difficult to discuss such basic questions as the *time inversion symmetry and the validity of the second law* as expressed by the 'local' inequality (2.10). We are reminded here of a most interesting controversy between Einstein and Ritz[50]. Ritz considered that it would be essential to include explicitly causality in the laws of physics such as electromagnetism. Einstein's reaction was that this would lead to a 'non-local' formulation of dynamics and therefore destroy the possibility to express simply the basic principles of physics such as the conservation laws.

Non-equilibrium statistical mechanics incorporates causality in the formulation of dynamics but the price we had to pay was precisely the non-locality. However as I shall show in the later part of this paper it is now possible to formulate non-equilibrium statistical mechanics in a new form which combines, at least in some sense, causality and locality.

Before we do so let us consider the limiting case in which (3.11) reduces to a local description. We proceed formally in the following way: we neglect the memory term $\mathscr{F}$ and assume that

$$\int_0^t d\tau G(\tau) \varrho_0 (t - \tau) \simeq \int_0^t d\tau G(\tau) \varrho_0 (t) \simeq \left[\int_0^\infty d\tau G(\tau) \right] \varrho_0 (t). \qquad (3.14)$$

Then (3.11) takes the Boltzmannian 'local' form

$$i \frac{\partial \varrho_0 (t)}{\partial t} = \Psi (+ i0) \varrho_0 (t) \qquad (3.15)$$

with (see (3.6))

$$\Psi (+ i0) = \lim_{z \to + i\varepsilon} \Psi (z) \qquad (3.16)$$

$$= \lim_{z \to + i\varepsilon} - PLQ \frac{1}{QLQ - z} QLP. \qquad (3.17)$$

The transformation (3.14) has only been made to indicate the manipulations we have to make to go from the non-local form (3.11) to the local form (3.15) and to introduce in this way the operator $\Psi(0)$.* This operator plays a central role in our whole approach and we would like to discuss it now in more detail.

4. DISSIPATIVITY CONDITION

In the limit of large systems, the sum over the intermediate states Q ('the correlations') involves an integration. Therefore a formal representation of (3.17) is given by

$$- i\Psi (0) = - \pi PLQ \delta (QLQ) QLP + i PLQ \frac{1}{QLQ} QLP \qquad (4.1)$$

where the second term is understood as a principal part. Suppose first that this expression is meaningful and that the first term does not vanish. Then we see that:

(a) $-i\Psi (0)$ may be split into an *even and an odd contribution* in respect to the transformation $L \to - L$ (see 2.6),

(b) the even part of $-i\Psi (0)$ is a hermitian operator (see 2.5), the odd is anti-hermitian,

(c) the even part of $-i\Psi (0)$ is a negative operator.

All these three properties are fundamental. The first leads to the breaking of the L, t symmetry which appeared in the original Liouville equation (2.3). The second

---

* In order to avoid misunderstanding let us stress that no claim of equivalence is made between (3.11) and (3.15).

leads to the connection between the dynamical and probabilistic descriptions, such as a description in terms of Markov chains. Finally the third, as we shall see, leads to the formulation of a general $\mathscr{H}$-theorem.

Before we go back to these important properties let us make the following remarks.

We may consider the class of dynamic systems for which the Hamiltonian is of the form

$$H = H_0 + \lambda V. \tag{4.2}$$

For example, we may consider a N-body system described by

$$H = \sum_k \varepsilon_k a_k^+ a_k + \lambda \sum_{klpq} v(klpq) a_k^+ a_l^+ a_p a_q. \tag{4.3}$$

The explicit form of the collision operator Ψ is then obtained by using perturbational techniques[28] based on formal expansion in powers of λ or c (the concentration) and on partial resummation. In the lowest order in the relevant expansion parameter one derives in this way the standard forms of the collision operator such as the Fokker–Planck operator or the Boltzmann operator. In all these cases only the even part in (4.1) is non-vanishing. The δ-function expresses simply conservation of the unperturbed energy in the collision process.

But the basic question is the existence of $i\Psi(0)$ *independently of any perturbational approach*. This question can now be answered in simple cases. Let us consider two examples. The first corresponds to a Hamiltonian of the form

$$H = \sum_n a_n^+ a_n + \lambda v \sum_k (a_k^+ a_{k+1} + a_{k+1}^+ a_k). \tag{4.4}$$

Obviously the number of particles is conserved. We may therefore consider separately the one-particle sector. It is easy to obtain from (4.4) the Liouville operator and then, using (3.6), to obtain the expression for $i\Psi(0)$. Of special interest is the infinite volume limit when the spectrum of H becomes continuous. The result obtained by Stey and Grecos for the leading terms of the matrix elements of the asymptotic expression of $i\Psi(0)$ is[51]

$$- i \langle m| \Psi(0) |n\rangle = - 4 |\lambda v| \frac{1}{2\pi} \int_{-\pi}^{\pi} d\varphi \, e^{i(m-n)\varphi} \left|\sin \frac{\varphi}{2}\right| + 0\left(\frac{1}{N}\right). \tag{4.5}$$

It is easy to verify that the general properties we have stated are satisfied (moreover the odd part of $i\Psi(0)$ vanishes in this example). Similar results may be obtained for the Friedrichs model[52, 53, 54] (which corresponds to the one particle sector of the Lee model). More precisely we consider in this case a discrete level (in the absence of interactions) which is coupled to a set of quantum states. Again the limiting process to large volumes may be performed. In the infinite volume limit we recover the Friedrichs model as originally formulated. The leading terms of the matrix elements of the asymptotic expression for $\Psi(0)$ are given by[53]

$$
\Psi(0) = \begin{bmatrix} a^{-1} & -\dfrac{2\pi}{L} a^{-1} \dfrac{\lambda^2 |v(\omega')|^2}{|\eta(\omega')|^2} \\[3ex] -\dfrac{2\pi}{L} a^{-1} \dfrac{\lambda^2 |v(\omega)|^2}{|\eta(\omega)|^2} & \left(\dfrac{2\pi}{L}\right)^2 a^{-1} \dfrac{\lambda^2 |v(\omega)|^2}{|\eta(\omega)|^2} \dfrac{\lambda^2 |v(\omega')|^2}{|\eta(\omega')|^2} \end{bmatrix}. \tag{4.6}
$$

Here ω_0 is the unperturbed energy of the discrete state, $\bar{v}_m = (1/L^{1/2})\, v(\omega_m)$ is the coupling between the discrete state and the mth lattice mode, L is the geometrical volume (length) and $\eta(z)$ is the function

$$
\eta(z) = \omega_0 - z - \lambda^2 \sum_m \frac{\bar{v}_m \bar{v}_m^*}{\omega_m - z}. \tag{4.7}
$$

In the limit of an infinite volume $\eta(z)$ has a cut along the real axis. We have then to consider the function $\eta^+(z)$ (or $\eta^-(z)$) which is analytic in the upper (lower) half plane and may be continued analytically into the lower (or upper) half plane

$$
\eta^+(z) = \omega_0 - z - \lambda^2 \int_{C^+} d\omega \, \frac{|v(\omega)|^2}{\omega - z}. \tag{4.8}
$$

Also in (4.6) a represents the integral

$$
a = -\frac{1}{2\pi i} \lim_{z \to 0} \int_{C'^+} dz' \, \frac{1}{\eta^+(z')\, \eta^-(z' - z)} \tag{4.9}
$$

where C'^+ is parallel to the real axis such that $0 < \mathrm{Im}\, z' < \mathrm{Im}\, z$.

There are two cases: either the dispersion equation

$$
\eta^+(z) = 0 \tag{4.10}
$$

admits a *real* solution, then as we have shown[52] $a^{-1} = 0$ and $i\Psi(0)$ vanishes. Or (4.10) admits no real solution then $\Psi(0)$ is given by (4.6) with

$$
a = -\frac{1}{2\pi i} \int_{-\infty}^{+\infty} d\omega \, |\eta(\omega)|^{-2}. \tag{4.9a}
$$

We may verify that $i\Psi(0)$ as given by (4.6) has again the properties we have enumerated. Here also the odd part vanishes.

We may therefore conclude that the collision operator exists in the limit of large systems.* We may therefore use this operator to classify dynamical systems (or better the limits to which dynamical systems tend when the volume and eventually the number of particles is increased):

(a) the even part of $i\Psi(0)$ vanishes;

---

* Some of the mathematical problems involved in 'dissipative systems' have been studied recently by Verbeure[82].

(b) the even part of $i\Psi(0)$ is different from zero;

(c) the even part of $i\Psi(0)$ becomes undefined.

An example of class (c) is the case of particles interacting through long-range forces such as gravitational forces[55]. Of foremost interest will be for us the second class of systems. We expect that for such systems a thermodynamic description will be possible. Indeed as $i\Psi(0)$ has an even part the kinetic equation (3.15) has lost its Lt invariance. If we reverse both time and L, the equation is no more invariant and we go over to a different 'antikinetic equation'. Because of this basic property we have called the condition

$$i\overset{e}{\Psi}(0) \neq 0 \qquad (4.11)$$

the *condition of dissipativity*. This condition when satisfied leads to the breaking of the time inversion symmetry. There are many interesting questions which arise in this connection such as the relation between dissipativity and the nature of the dynamical invariants as well as the relation between dissipativity, ergodicity and mixing. These questions are considered elsewhere.[39, 52, 53, 56]

It is also interesting to note that Uhlenbeck (see ref. 46, Appendix I by Uhlenbeck) had clearly recognized the necessity to formulate a kind of dynamical version of ergodic theory. The dissipativity condition (4.11) plays precisely this role.

In the preceding paragraph the kinetic equation (3.15) was obtained from the exact non local master equation (3.11) through drastic simplifications (see 3.14). We want therefore now to indicate a new formulation of non-equilibrium statistical mechanics which makes explicit the appearance of even terms in L, leading to Lt symmetry breaking, without going through the non-local master equation (3.11).*

5. CAUSAL DYNAMICS

Let us go back to the integral representation (3.2) for the 'retarded' solution of the Liouville equation. By a careful analysis of the singularities it can be shown that this formula may be written as[10, 40]

$$\varrho^r(t) = \Lambda(L)\,\varrho^{p,\,r}(t), \qquad (5.1)$$

where the time independent operator $\Lambda(L)$ acts on the 'physical' distribution function $\varrho^{p,\,r}(t)$ taken at the same time t. Causality is now incorporated in the form of $\Lambda(L)$. We may consider (5.1) as the transformation from one representation of the density matrix to another. The essential point is that for the advanced solution we have the *different* transformation formula

$$\varrho^a(t) = \Lambda(-L)\,\varrho^{p,\,a}(t). \qquad (5.2)$$

Indeed because of the Lt invariance, time inversion is equivalent to L inversion.

There is a condition which this transformation has to satisfy: we may perform

---

* A fuller account may be found in refs. 39 and 10.

similar transformations on observables A. Now in the Heisenberg representation A satisfies the Liouville equation (2.3) with L replaced by $-L$. Therefore for A we obtain instead of (5.1)

$$A^r(t) = \Lambda(-L) A^{p,r}(t). \tag{5.3}$$

If we want to consider equations (5.1), (5.3) as transformations to a new representation we have to require that all averages remain invariant. Therefore

$$\langle A \rangle = \text{Tr } A^+ \varrho = \text{Tr } (A^p)^+ \varrho^p. \tag{5.4}$$

Using (5.1) and (5.3) we see that this implies

$$\Lambda(L) \Lambda^+(-L) = \Lambda^+(-L) \Lambda(L) = 1. \tag{5.5}$$

We have called such transformations 'star-unitary transformations' as we have introduced the 'star' notation to denote the combined operations of taking the hermitian conjugate and reversing L[10, 38, 40]

$$\Lambda^*(L) = \Lambda^+(-L). \tag{5.6}$$

We have verified that the operator $\Lambda(L)$ we derived satisfies the basic conditions (5.5). Except in the case where $\Lambda(L)$ does not depend on the sign of L, star unitary transformations are non-unitary transformations. Still they preserve the average values of observables. They correspond therefore to equivalent (but not unitary equivalent) representations of dynamics.

Let us consider the equations of motion in the causal representation. Using (2.3), (5.1) (5.5) and (5.6) we obtain immediately for the retarded solution in the physical representation

$$i\frac{\partial \varrho^{p,r}}{\partial t} = \Phi \varrho^{p,r} \quad \text{with} \quad \Phi = \Lambda^* L \Lambda. \tag{5.7}$$

Similarly we have for the advanced solution

$$i\frac{\partial \varrho^{p,a}}{\partial t} = -\Phi' \varrho^{p,a} \quad \text{with} \quad \Phi' = -\Lambda^{*'} L \Lambda' = -\Lambda^+ L \Lambda' \tag{5.8}$$

where the prime means L inversion

$$\Lambda'(L) \equiv \Lambda(-L). \tag{5.9}$$

As we did for the collision operator in Section 4, we may always decompose Φ into an even part and an odd part with

$$\overset{e}{\Phi}(L) = \overset{e}{\Phi}(-L), \quad \overset{o}{\Phi}(L) = -\overset{o}{\Phi}(-L). \tag{5.10}$$

Then (5.7), (5.8) become

$$i\frac{\partial \varrho^{p,r}}{\partial t} = (\overset{e}{\Phi} + \overset{o}{\Phi}) \varrho^{p,r} \tag{5.11}$$

$$i\frac{\partial \varrho^{p,a}}{\partial t} = (-\overset{e}{\Phi} + \overset{o}{\Phi})\,\varrho^{p,a}. \tag{5.12}$$

The Lt symmetry of the Liouville equation is broken if and only if the even part Φ may be defined and does not vanish.

We can make some general statements about the structure of Φ. The hermiticity of L (2.5) together with the star-unitarity of Λ (see 5.5) implies that Φ has the basic invariance property:

$$[i\Phi(-L)]^+ = i\Phi(L) \tag{5.13}$$

Introducing again the 'star' notation (5.6) we may say that $i\Phi$ is a *star-hermitian operator*. Star-hermiticity may be realized in two ways:

   (a) hermiticity together with positive parity in respect to L inversion,
   (b) antihermiticity together with negative parity in respect to L inversion.

In this terminology the collision operator $i\Psi(0)$ as given in (4.1) is a star-hermitian operator. Each of its two parts corresponds to one of the two possible realizations of star-hermiticity. Also iL in the Liouville equation (2.3) is a star-hermitian operator (antihermitian and odd).

There is one more basic property of $-i\Psi(0)$ we mentioned in Section 4; it is the fact that its even part is a negative operator. In all cases where we have succeeded in constructing the transformation operator Λ, either through perturbation expansion or exactly, this is also true for $-i\Phi$. This is due ultimately to the analytical continuation introduced in the integral representation (3.2). If the operators are replaced by numbers this property is trivial.

Also it is interesting to notice that the definite sign of $-i\Phi$ guarantees the possibility of associating a *semi-group* representation to the evolution of the system. It is also a necessary condition to give a meaning to the distinction between the retarded and the advanced solutions in (5.11), (5.12). Moreover the non-vanishing of Φ is connected to the basic dissipativity condition (4.11).

For this reason we shall call *dynamic dissipative systems*, systems for which Φ can be explicitly constructed and has the forementioned basic properties.

The evolution of dynamic dissipative systems may be described in terms of a Lyapunov function. Let us consider the quadratic functional *in the causal representation* (we drop the superscripts p, r)

$$\Omega = \text{Tr}\,\varrho^+\varrho > 0. \tag{5.14}$$

It is easy to show, using (5.11), that

$$\frac{1}{2}\frac{d\Omega}{dt} = -\,\text{Tr}\,\varrho^+\,(i\overset{e}{\Phi})\,\varrho \leqslant 0. \tag{5.15}$$

Similarly, for the advanced solution Ω can only increase, it is essential to use the causal representation. In the initial representation satisfying the Liouville equation Ω would be a constant. It is the fact that the two representations are linked through a *non-*

unitary transformation which makes the introduction of the Lyapunov function Ω possible.

The existence of the Lyapunov function for dissipative systems is of course of utmost importance for the thermodynamic interpretation of the dynamical evolution. We shall discuss its link with entropy in Section 8.

It is certainly surprising that so different descriptions as given by the initial Liouville equation on one side, the causal equation (5.11), (5.12) on the other, may coexist and may even be related through the transformation Λ which guarantees the equivalence of the two descriptions (see 5.4).

Still the causal representation goes beyond the initial dynamical description through the Liouville equation, as now the retarded and the advanced solutions satisfy different differential equations. In this way, the causal representation makes *explicit* properties which appear otherwise in the *solution* of the mechanical equations of motion. This is very unexpected. Let us therefore display the two representations in a very simple example due to Grecos.

6. THE GRECOS MODEL[57]

Let us consider a model system whose time evolution is described by a Liouville type equation (2.3) with

$$L = v\delta(v - v'), \quad (-\infty < v < \infty). \tag{6.1}$$

Similarly the observables A are described by a Heisenberg type equation identical to (2.3) but with L replaced by $-L$. To this L inversion corresponds here the operation

$$v \rightarrow -v. \tag{6.2}$$

Average values are represented as in (5.4) through

$$\langle A \rangle = \text{Tr } A^+\varrho = \int_{-\infty}^{\infty} d\omega \, A^+(\omega)\, \varrho(\omega). \tag{6.3}$$

In the Schrödinger representation in which ϱ is considered as time dependent we have

$$\langle A(t) \rangle = \int_{-\infty}^{\infty} d\omega \, e^{i\omega t} A^+(\omega)\, \varrho(\omega; t = 0). \tag{6.4}$$

The main simplifying feature of this model is that L is here an operator (and *not* a superoperator).

Obviously this model is Lt invariant, exactly as the original Liouville equation.

Note also that under well-known conditions the Riemann–Lebesgue theorem leads to

$$\lim_{t \rightarrow \pm\infty} \langle A(t) \rangle = \lim_{t \rightarrow \pm\infty} \int d\omega \, e^{i\omega t} A^+(\omega)\, \varrho(\omega; 0) \rightarrow 0. \tag{6.5}$$

There is an approach to *'equilibrium' independently of the direction of time.*

Suppose that there exists a *second* description corresponding to (5.11) (we omit the superscript r)

$$i \frac{\partial \varrho^p}{\partial t} = \Phi \varrho^p \tag{6.6}$$

with

$$\Phi = \omega \delta (\omega - \omega') - iv(\omega) v^*(\omega'). \tag{6.7}$$

We impose on $v(\omega)$ the parity condition

$$v(\omega) = \pm v(-\omega). \tag{6.8}$$

The operator $-i\Phi$ has then all the properties we discussed in Sections 4 and 5:

(1) it is the sum of an even and an odd contribution in respect to the transformation $L \to -L$ (see 6.2),

(2) it is 'star-hermitian' in the sense of (5.13); the even part is represented by a hermitian operator, the odd by an anti-hermitian operator,

(3) the even part is a negative operator, as for an arbitrary vector

$$(u, \overset{e}{-} i\Phi u) = - \left| \int d\omega \, v(\omega) \, u(\omega) \right|^2. \tag{6.9}$$

Of course, equation (6.6) is no more Lt invariant. The L inversion of (6.6) would lead to equation (5.12) satisfied by the advanced solution.

The Lyapunov functional Ω as given in (5.14),

$$\Omega = \int d\omega \, |\varrho|^2 \geqslant 0 \tag{6.10}$$

has indeed the required properties. It satisfies the inequality

$$\frac{1}{2} \frac{d\Omega}{dt} = - \left| \int d\omega \, v(\omega) \, \varrho(\omega) \right|^2 \leqslant 0. \tag{6.11}$$

The retarded solution tends to 'equilibrium' in the *future* as

$$\lim_{t \to +\infty} \int d\omega \, |\varrho|^2 \to 0 ; \tag{6.12}$$

similarly the advanced solution tends to equilibrium in the *past*.

Now in this model the transformation operator $\Lambda(L)$ may be explicitly calculated[57]. The result is

$$\langle v | \Lambda | \omega \rangle = c(v) \left\{ \alpha(v) \delta(\omega - v) + i \frac{v^*(\omega) v(v)}{\omega - v} \right\} \tag{6.13}$$

with

$$\alpha(v) = 1 - i \int d\omega \frac{|v(\omega)|^2}{\omega - v} \tag{6.14}$$

and

$$c(v) = \{\alpha^2(v) - \pi^2 |v(v)|^4\}^{1/2} \tag{6.15}$$

where the second term in (6.13) and the integral in (6.14) are understood as principal parts.

In these expressions one may easily verify that the transformation satisfies (5.5) and is therefore *star-unitary*. As a result the two descriptions while deeply different from the physical point of view are 'equivalent' as average values $\langle A(t) \rangle$ are preserved. In a similar way we may calculate $\Lambda(-L)$ which would lead to the 'advanced solution'.

7. APPROACH TO EQUILIBRIUM – STATISTICAL MODEL OF ENTROPY

Causal dynamics corresponds to a *non-canonical* formulation of dynamics which takes explicitly into account the symmetry breaking introduced by causality into the equations of motion. To the time evolution as expressed in terms of the causal equations (5.11) we may associate the Lyapunov function (5.14). Let us study more closely the relation of this function with entropy.

Using the notation

$$\varrho_{ij} = \varrho_{i-j}\left(\frac{i+j}{2}\right) \equiv \varrho_v(N) \tag{7.1}$$

we may write Ω more explicitly

$$\Omega = \sum_{ij} \varrho_{ij}^+ \varrho_{ji} = \sum_N \varrho_0^2(N) + \sum_{vN} |\varrho_v(N)|^2 \tag{7.2}$$

Ω therefore includes both the diagonal elements $\varrho_0(N)$ and the off-diagonal elements $\varrho_v(N)$ (the 'correlations'). The distribution function is assumed to be normalized

$$\mathrm{Tr}\,\varrho = \sum_N \varrho_0(N) = 1. \tag{7.3}$$

It is obvious that the minimum of (7.2) subject to the constraint (7.3) is given by

$$\begin{cases} \varrho_0(N) = \text{constant independent of } N \\ \varrho_v(N) = 0. \end{cases} \tag{7.4}$$

We see that statistical equilibrium has a very simple meaning in the causal formulation of dynamics: all quantum states have the same probability and random phases. Situations which can be described in terms of the diagonal elements $\varrho_0(N)$ *alone* correspond to what George, Rosenfeld and I have called the 'macroscopic level of description' in a recent publication.[38] In such situations we have in addition to the dynamical description a second 'reduced' description. This applies to all equilibrium laws as well as to the linear range of non-equilibrium processes.[34] As we mentioned in Section 1 the corresponding thermodynamic description is the so-called 'local equilibrium description'.

We want now to go further and introduce a statistical definition of entropy which would remain valid in the whole range of the causal formulation of dynamics. The expression for the $\mathscr{H}$ quantity

$$\mathscr{H} - \mathscr{H}_{eq} = \tfrac{1}{2} \log \frac{\mathrm{Tr}\, \varrho^+ \varrho}{(\mathrm{Tr}\, \varrho^+ \varrho)_{eq}} \tag{7.5}$$

together with

$$S = -k\mathscr{H} \tag{7.6}$$

satisfies all conditions known to us. Indeed

$$\begin{cases} \mathscr{H} - \mathscr{H}_{eq} \geqslant 0 \\ \dfrac{d\mathscr{H}}{dt} \leqslant 0. \end{cases} \tag{7.7}$$

$\mathscr{H}$ is an additive function for independent systems. Moreover the factor $\tfrac{1}{2}$ in (7.5) insures that this expression gives in the neighbourhood of equilibrium identical results to the Boltzmann expression[10]. The term $(\mathrm{Tr}\, \varrho^+ \varrho)_{eq}$ has a simple meaning as at equilibrium by virtue of (7.4)

$$(\mathrm{Tr}\, \varrho^+ \varrho)_{eq} = \frac{1}{\vartheta} \tag{7.8}$$

where ϑ is the common value of the diagonal elements $\varrho_0(N)$. Because of the normalization (7.3), ϑ is equal to the number of accessible quantum states.

The most important property of $\mathscr{H}$ as given in (7.5) is that correlations are included as seen from (7.2). Let us emphasize that in agreement with (7.4) there are no *equilibrium* correlations in the causal formulation. The equilibrium correlations which would appear in other representations are *included* in the diagonal elements. The correlations which are part of the entropy (7.5) are therefore *non-equilibrium* correlations which ultimately die out. This separation between 'natural' correlations included in the diagonal elements and transient correlations introduced by initial conditions is of course very important.

It is very interesting that the Loschmidt paradox does not apply to our new definition of $\mathscr{H}$. Suppose first we start at $t=0$ with a state without correlations ($\varrho_v=0$) and isotropic velocity distribution. During time 0 till t_0, the diagonal elements tend to their equilibrium distribution. This gives rise to a decrease of $\mathscr{H}$. At time t_0 we reverse the velocities. This introduces long range non-thermodynamical correlations $\varrho_v \neq 0$ (see also the discussion in Section 3). As a consequence (see equation 7.2) of this external action on the system $\mathscr{H}$ increases. The system is ordered. During the time t_0 till $2t_0$ the abnormal correlations die out progressively. The details of the calculations are given elsewhere.[10, 59] The time variation of $\mathscr{H}$ is represented schematically in Fig. 3. At time $2t_0$ the system is in the same state as at $t=0$, but with reversed velocities. As at $t=0$ the system had an isotropic velocity distribution, this is the same state.

In contrast the Boltzmann $\mathscr{H}$-quantity has a time evolution which is represented schematically in Fig. 1. No antithermodynamic behaviour appears in Fig. 3, in contrast with the evolution of Boltzmann's $\mathscr{H}$-quantity. We can even estimate the thermodynamic price of rejuvenation or construct thermodynamic cycles in which during part of the cycle 'negative time evolution' is involved. But we cannot go into more detail here.[10]

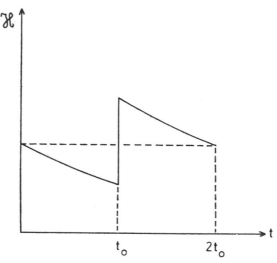

Fig. 3.   Time evolution of $\mathscr{H}$ as defined in (7.5).

We now begin to see why Boltzmann's $\mathscr{H}$ theorem was only successful in the special case of the approach to equilibrium of a dilute gas, and even then only in the case when the only source of non-equilibrium was a deviation from the Maxwellian distribution of velocities. Obviously in an N-body system there may be other types of deviations from thermodynamic equilibrium as described by correlations which were not included in Boltzmann's definition. Also it can be shown that even in the *causal* representation a trivial generalization of Boltzmann's definition, as provided by the Gibbs formula

$$\mathscr{H}_G = \mathrm{Tr}\varrho \ln \varrho, \qquad (7.9)$$

is not valid. Such a formula requires the validity of a very specific stochastic description (such as a Markov process) which is in general incompatible with the dynamical description.

Our new general definition of entropy has now been applied successfully to specific models (F. Henin[83], P. Allen[84]). It is interesting to notice that our concept of non-equilibrium entropy corresponds to a microcanonical description, while Boltzmann's concept refers to a canonical one.

Let us summarize this part of our lecture: When Boltzmann one hundred years ago proposed his kinetic equation, he had obviously in mind some scheme of successive approximations. In the case of N-body systems, we can still only proceed by approximations, but we can now study simpler dynamical systems, of the type I mentioned in this report. We can show in this way that *the physical concept introduced by Boltzmann such as the collision operator, is amenable to an exact* treatment. The basic problem is to reformulate the laws of dynamics in a way which displays explicitly in the differential equations the terms which are even in L and are responsible for the loss of the Lt invariance. Once this is done, a Lyapunov function can easily be identified and its relation with entropy discussed.

Therefore, instead of Boltzmann's scheme mentioned in Section 1,

$$\text{Dynamics} \rightarrow \text{Stochastic Theory} \rightarrow \text{Entropy}$$

we have the scheme:

$$\text{Dynamics of dissipative systems} \rightarrow \text{Lyapunov function} \rightarrow \text{Entropy.}$$

It should be emphasized that at no point concepts such as 'coarse graining' have been introduced. One could even say that the description we obtain is 'hyperfine'-grained as it displays explicitly properties of matter which in other representations appear only through the *solution* of the equations of motion.

The idea that irreversibility is due to some falsification of dynamics, to the introduction of some non-mechanical extraneous assumptions, has been, together with the mathematical complexity of dynamics of large systems, one of the main reasons for the slow progress in this field. In the remaining part of this lecture we shall adopt a thermodynamic description. Specially in the frame of the assumption of local equilibrium in which the independent variables are the same as those which appear at equilibrium, this leads to a great simplification in the description.

8. DISSIPATIVE STRUCTURES

Equilibrium structures are at least qualitatively understood in terms of classical thermodynamics and equilibrium statistical mechanics. But in addition there exists a new class of structures which may originate in conditions far from equilibrium.[34] We have called such structures '*dissipative structures*' which are formed and maintained through the exchange of energy and matter in non-equilibrium conditions. The fact that non-equilibrium may be a source of order is in fact an idea which the author had already formulated in 1945 through the theorem of minimum entropy production [60, 31], but which has been greatly clarified and extended thanks to the recent development of non-linear thermodynamics. One of the main conclusions of this theory is that there exists a class of systems showing two kinds of behaviour: a tendency to the state of maximum disorder for one type of situation and coherent behaviour for a second type. The *destruction* of order always prevails in the neighbourhood of thermodynamic equilibrium.

In contrast, *creation* of order may occur far from equilibrium and with specific non-linear kinetic laws, beyond the domain of stability of the states that have the usual thermodynamic behaviour. Traditionally, thermodynamics has dealt with the first type of behaviour, but the new extension of irreversible thermodynamics permits us to deal with the other aspects as well.

The best known examples of this duality in behaviour are *instabilities in fluid dynamics*, such as turbulence or the onset of thermal convection in a fluid layer heated from below.[61] For a critical value of the external constraints (pressure or temperature gradient, respectively), that is, beyond a critical distance from equilibrium, an instability arises that causes the convection patterns to change radically and in a discontinuous fashion. Below the instability threshold, the energy of the system is distributed in the random thermal motion of the molecules. But beyond instability it appears partly as the energy of macroscopic convection pattern.

Of great interest is the behaviour of *non-linear chemical networks*.[34] At first one would expect that those systems would always show a tendency to a disordered regime. The unexpected result was that in fact they share most of the properties of hydrodynamic instabilities with the additional important feature that the variety of the regimes beyond instability is much greater in chemical kinetics.

This is not really surprising when we realize that in chemical kinetics non-linearity may arise in a practically unlimited number of ways through autocatalysis, cross-catalysis, activation, inhibition, and so on. In contrast, the Stokes–Navier equations of fluid dynamics assume a universal form. Beyond instability, which again arises when a critical distance from equilibrium is reached, the reaction systems may become spontaneously inhomogeneous and present an ordered distribution of the chemical constituents in space. Under different conditions the concentrations of the chemicals may show sustained oscillations. Finally, other systems may exhibit a multiplicity of steady states combined with hysteresis. Let us indicate two examples of dissipative structures. The first refers to the onset of *thermal convection in a two-component fluid layer heated from below*. For one-component systems this is a well understood problem treated in detail in the comprehensive monograph by Chandrasekhar.[61] The occurrence of instability depends then on the Rayleigh number

$$R_a = \frac{g\alpha h^4}{\kappa v} \beta \tag{8.1}$$

where β is the adverse temperature gradient, α the thermal expansion coefficient, g the force (per unit mass) due to gravitation, h the depth of the layer, κ thermal diffusivity and v kinematic viscosity. Let us now consider a binary mixture.[62, 63] Because of the effect of thermal diffusion the distribution of the two components induced by the thermal gradient will not be uniform. The effect will depend on the sign of the ratio D'/D where D' is the coefficient of thermal diffusion and D the diffusion coefficient. When D'/D is positive the denser component migrates to the cold wall and this effect reinforces the effect of the adverse temperature gradient (that is lower the value of the critical value of the Rayleigh number). On the contrary if D'/D is *negative*, thermal diffusion stabilizes the system and the critical Rayleigh number should increase.

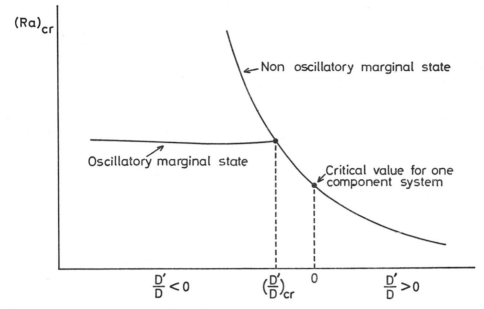

$(Ra)_{cr}$

Non oscillatory marginal state

Oscillatory marginal state

Critical value for one component system

$\frac{D'}{D} < 0$ $\left(\frac{D'}{D}\right)_{cr}$ 0 $\frac{D'}{D} > 0$

Fig. 4. Onset of thermal convection in binary mixtures.

This situation is represented schematically in Fig. 4 which gives the critical Rayleigh number as a function of D'/D. The increase of stability for negative values of D'/D was indeed observed by Legros, Rasse and Thomaes.[64a] A careful analysis has shown that this transition leads to a new oscillatory marginal state for binary mixtures for values of D'/D above a critical value.[64b] The period of oscillations has been calculated by Platten and Chavepeyer.[65] For water-methanol mixtures it is of the order of 10^2 sec. Moreover Platten and Chavepeyer have made an experimental investigation in mixtures of water-methanol and water-isopropanol. Very regular temperature oscillations of the predicted period were observed near the critical point. An example is represented in Fig. 5.

This is a beautiful physico-chemical clock. Of course, oscillatory marginal states exist in other situations which may be realized when a layer is heated from below (see Chandrasekhar [61]). But they are less surprising as they are due to the effect of rotation of the fluid or to the presence of a magnetic field. Here it is a purely dissipative process (thermal diffusion) which induces the macroscopic time organization as manifested by the oscillations.

From this point of view it is also interesting to notice that an instability may also arise if the binary mixture is heated from *above* (this corresponds to a negative Rayleigh number) and D'/D negative and sufficiently large in spite of the fact that the density is decreasing from the bottom to the top of the layer. This theoretical prediction[62, 63] has not yet been verified experimentally. If this would be so, a simple mechanical interpretation of the instability such as it exists for one-component fluids would no more be possible for mixtures.

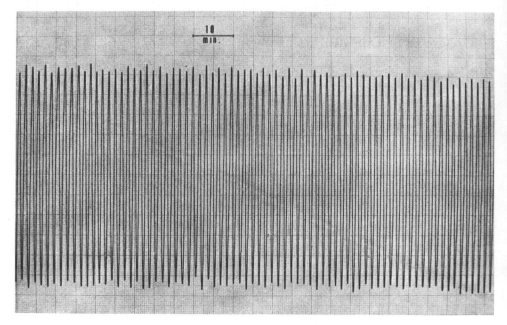

Fig. 5.   Temperature oscillations near the critical point in water-methanol mixtures. For more details see Platten and Chavepeyer.[65]

As a second example let us consider the autocatalytic chain of reactions [34, 66]

$$A \rightleftarrows X \qquad \text{(a)}$$
$$2X + Y \rightleftarrows 3X \qquad \text{(b)}$$
$$B + X \rightleftarrows Y + D \qquad \text{(c)}$$
$$X \rightleftarrows E \qquad \text{(d)}$$

$$(8.2)$$

where the initial and final products are A, B, D, E and the intermediate components X, Y; (b) is an autocatalytic step which involves Y. This component is itself produced in step (c). The concentration of A, B, D, E is prescribed and constant in time. Let us consider a model formed by two identical boxes between which X and Y may freely diffuse. As Lefer [66] has shown in certain *far from equilibrium* conditions we may have a symmetry breaking dissipative structure such that

$$x^{\mathrm{I}} \neq x^{\mathrm{II}}, \qquad y^{\mathrm{I}} \neq y^{\mathrm{II}}. \qquad (8.3)$$

On the contrary the imposed values of A, B, D, E are identical in the two boxes. Obviously we may have either

$$x^{\mathrm{I}} > x^{\mathrm{II}} \qquad \text{or} \qquad x^{\mathrm{I}} < x^{\mathrm{II}}. \qquad (8.4)$$

The situation which will be realized depends on the region in which the fluctuation

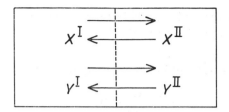

Fig. 6. Two box model.

responsible for the destruction of the initial homogeneous state arises. But once one of the states (8.4) is realized the concentration of the boxes remains different and also their subsequent evolution will in general be different.

We have a kind of *primitive memory*. On this highly simplified example we see that a '*historical*' dimension may be added when we have dissipative structures. We *need to know the succession of instabilities to understand the formation of the state of the system.*

Obviously, the occurrence of instabilities far from equilibrium is not a *universal phenomenon* in chemical kinetics. Coherent behaviour requires some very particular conditions on the reaction mechanism, whereas the equilibrium order principle is *always* valid (for short range forces). But this variety of behaviour in chemistry is welcome, if we want to account for the variety of situations observed in systems driven far from equilibrium.

9. NON-EQUILIBRIUM THERMODYNAMICS

We have insisted on the diversity of situations that may arise in non-linear systems far from equilibrium. Now we shall see that *despite this diversity, there is a general thermodynamic theory* underlying the mechanism by which a system is driven to the new regime beyond the instability of the thermodynamic branch.

Within the framework of our macroscopic description, we would like to relate instabilities to the thermodynamic properties of the systems, such as entropy, entropy production, and so on. Consider again the decomposition of dS given by Equation (1.1). Explicit calculation of d$_i S/$dt is possible in the frame of the local equilibrium assumption.[28, 34] (From the microscopic point of view, this condition implies that locally, the momentum distribution function of the system should not deviate appreciably from the Maxwellian form. Note that this restriction is compatible with large deviations from *chemical* equilibrium.) The result is

$$P = \mathrm{d}_i S/\mathrm{d}t = \sum_\varrho J_\varrho X_\varrho \geqslant 0 \qquad (9.1)$$

where the J_ϱ are the rates of irreversible processes (chemical reaction velocities, diffusion, heat flow, and so on), X_ϱ are the corresponding forces (such as differences in chemical potentials and temperature gradients). Remarkably, in this bilinear form the entropy production is expressed entirely in terms of macroscopic quantities of direct physical interest such as the flows and the forces.

Near equilibrium, Equation (9.1) becomes quadratic in the X_ϱ's. In this limit, and provided the system is subject to time-independent boundary conditions,

$$\frac{dP}{dt} \leqslant 0. \tag{9.2}$$

This is the theorem of minimum entropy production at the steady state. This theorem implies the stability of this state[60]; the proof is based on a classical analytic result known as Lyapunov's theorem, which demonstrates stability once one can construct a definite function (the Lyapunov function) whose time variation is also definite with opposite sign. This result is in agreement with the analysis of model systems discussed in the previous section.

Inequality (9.2) breaks down for states far from thermodynamic equilibrium, but we can still obtain a general inequality in this domain if we decompose dP into two terms, according to Equation (9.1):[34]

$$\frac{dP}{dt} = \sum_\varrho X_\varrho \left(\frac{dJ_\varrho}{dt}\right) + \sum_\varrho J_\varrho \left(\frac{dX_\varrho}{dt}\right)$$
$$= \frac{d_J P}{dt} + \frac{d_X P}{dt}. \tag{9.3}$$

We have introduced explicitly the variation in P due to a variation of the flows and a variation of the forces, and can now show that in the whole domain where Equation (9.1) is valid, and provided the system is subject to time-independent boundary conditions,

$$\frac{d_X P}{dt} \leqslant 0. \tag{9.4}$$

This is the extension of the minimum-entropy-production property to the non-linear domain of irreversible processes. However, in contrast to inequality (9.2) inequality (9.4) does *not* imply the stability of the steady state, primarily because $d_X P$ is not the differential of a state function in the general case. Instead, we may use inequality (9.4) to derive a stability condition for such states, and we find that stability will be ensured whenever

$$\sum_\varrho \delta J_\varrho \, \delta X_\varrho \geqslant 0. \tag{9.5}$$

Here δJ_ϱ and δX_ϱ are the excess flows and forces due to the deviation of the state of the system from the reference regime. In deriving this inequality it has been assumed that the system remains in mechanical equilibrium.

Relation (9.5) provides a universal thermodynamic stability criterion for non-equilibrium states. One can show that the inequality is always satisfied in the neighbourhood of equilibrium as well as in the absence of a feedback process of the autocatalytic type.[34] An alternative interpretation of the stability criterion in terms of a Lyapunov function is also possible.[34] Consider the entropy S as a function of the non-equilibrium state, with S_0 its value in the reference state whose stability is studied. We expand S around S_0

$$S = S_0 + \delta S + \frac{\delta^2 S}{2}. \tag{9.6}$$

Now in the domain where relation (9.1) is valid, one shows that $\delta^2 S$, which is a quadratic form, obeys the inequality

$$\tfrac{1}{2}\delta^2 S \leqslant 0. \tag{9.7}$$

On the other hand, in mechanical equilibrium, we show that

$$\frac{1}{2}\frac{d}{dt}\delta^2 S = \sum_\varrho \delta J_\varrho\, \delta X_\varrho. \tag{9.8}$$

Thus, according to Lyapunov's theorem, the reference state will be stable provided inequality (9.5) is satisfied; that is, provided the excess entropy $\delta^2 S/2$ increases in time.

So far stability has been related to deviations (arising, for example, from external perturbations) from the reference state. A much deeper insight is provided by the result of recent investigations[67] that in the whole domain of validity of relation (9.1), the excess entropy $\delta^2 S$ also determines the probability of a small fluctuation around the reference state:

$$P \sim \exp(\delta^2 S/2R). \tag{9.9}$$

This expression provides a generalization to non-equilibrium situations of the celebrated Einstein formula describing the distribution of small fluctuations around equilibrium. Now fluctuations – the spontaneous deviations from some average regime – are a universal phenomenon of molecular origin and are always present in a system with many degrees of freedom. Thus, a system in an average state close to, but below the transition threshold, will always have a non-vanishing probability of reaching the unstable region through fluctuations. When this happens, certain types of fluctuations will be amplified and subsequently drive the averages to the new (unstable) regime. In the thermal instability problem mentioned previously, these fluctuations would generate small convection currents that would be damped below the transition point but give rise to a macroscopic current beyond the instability.

In all these situations a *new order principle appears that corresponds essentially to an amplification of fluctuations and to their ultimate stabilization by the flow of matter and energy from the surroundings.* We may call this principle 'order through fluctuations'.

An additional important element should be pointed out: The formation of a fluctuation of a given type is *fundamentally a stochastic process*. The response of the system to this fluctuation is a deterministic process, obeying the macroscopic laws, as long as the system can damp the fluctuation. Now in the domain of formation of a new structure, fluctuations are amplified and drive the average values to the new regime. Thus, in this region the macroscopic description in terms of averages breaks down and the evolution acquires an essentially statistical character.

We have then a picture of a system evolving through instabilities: In the neighbourhood of a stable regime, evolution is essentially deterministic in the sense that the small fluctuations arising continuously are damped. But near the transition threshold the evolution becomes a stochastic process in the sense that the final state will depend on the probability of creating a fluctuation of a given type. Of course, once this probability is appreciable, the system will eventually reach a unique (apart from small fluctuations) stable state, once the boundary conditions are specified. This state will then be the starting point for further evolution.

10. TIME, STRUCTURE AND KNOWLEDGE

As this report shows, great progress has been realized in the integration of the levels of time in the frame of dynamics. This has of course many interesting consequences in those parts of physics where dissipative properties are of interest. We want however to devote this concluding section to problems which go beyond physics proper and touch on questions of biology and epistemology.

There seems to be little doubt that there is a close connection between dissipative structures and biological organization. Important biological processes have been shown to work beyond the stability limit of the thermodynamic branch.[34, 68, 69] This is, for example, true for important multi-enzymatic chemical cycles, such as the glycolitic cycle which is essential for the energy production in living systems.

We can also study from this point of view prebiological evolution of matter. Quantitative models are now being investigated for important steps such as prebiotic polymer formation and competition between biolymers[70, 71, 72] (see the lecture by Eigen at this symposium).

The main idea is the possibility that a pre-biological system may evolve through a whole succession of transitions leading to a hierarchy of more and more complex and organized states. As we have seen such transitions can only arise in non-linear systems that are maintained far from equilibrium; that is, beyond a certain critical threshold the steady-state regime becomes unstable and the system evolves to a new configuration. As a result, if the system is to be able to evolve through *successive* instabilities, a mechanism must be developed whereby each new transition favours further evolution by increasing the non-linearity and the distance from equilibrium. One obvious mechanism is that each transition enables the system to increase the entropy production. We may visualize this *evolutionary feedback*:[72]

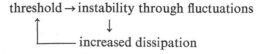

Perhaps the most interesting consequence of this approach is to propose an alternative to the belief that life originated in a single extraordinary unlikely event and that the whole evolution of life proceeded *against* the laws of physics – as a kind of giant fluctuation which would unfold over billions of years of biological evolution.

Such ideas are expressed in many authoritative books, like Monod's *Le hasard et la nécessité*.[73] He writes: 'Notre numéro est sorti au jeu de Monte Carlo. Quoi d'étonnant à ce que, tel celui qui vient d'y gagner un milliard, nous éprouvions l'étrangeté de notre condition?' (p. 160, see also pp. 139–140). Without going into the details of the argument let us observe that the idea of chance and randomness which is used in such considerations stems from *equilibrium* thermodynamics and equilibrium statistical mechanics. The occurrence of dissipative structures shows precisely that these ideas cannot be extrapolated to situations far from equilibrium. The probability of a periodic temporal process such as may occur in the two-component Bénard problem, when calculated by equilibrium statistical mechanics is vanishingly small and still it occurs with probability one, if I may say so, in conditions far from equilibrium.

Far from being the work of some army of Maxwell demons, *life appears as following the laws of physics appropriate to specific kinetic schemes and to far from equilibrium conditions.* These non-linear kinetic schemes permit the flow of energy and matter to build and maintain functional and structural order.[34, 74] But we can even go further. We have seen in Section 1 that the Newtonian world model, even in its recent versions, leads to a description in which time was associated only with displacement (the spatial time of Bergson). Of course the great successes of classical and quantum mechanics were due to the possibility of isolating simple situations, to study in detail systems with a small number of degrees of freedom. We now see that when we study, on the contrary, *large* systems *time may acquire a new meaning associated with irreversibility.* We see that the classical mechanical concept of time is the result of simplifications which are certainly no more allowed on the macroscopic level which involves a large number of interacting degrees of freedom. It is an example of what Whitehead has called rightly the *'Fallacy of Misplaced Concreteness'*.[75] It is interesting that we arrive at a model of knowledge, of intelligibility of nature which is rather near to concepts developed in a different context by psychologists such as Jean Piaget.[76]

To obtain a thermodynamic description starting with dynamics, we have to incorporate causality conditions in the dynamics. We have assumed that we can at the start *distinguish future from past*. Our scheme may be summarized as follows

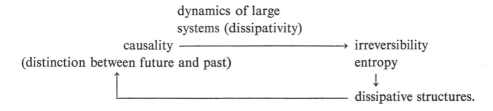

We start from the observer – a living system – who can make a distinction between future and past, and we end with dissipative structures which contain, as we have seen, a 'historical dimension'. Therefore we can now recognize ourselves as a kind of

evolved form of a dissipative structure and justify in an 'objective' way the distinction future–past we had introduced at the start.

This circularity implies no vicious circle as the distinction between future and past, once the concept of dissipative structures is recognized, is much more precise than the initial one which was assumed at the start.

It is interesting to compare this 'model of knowledge' with Jean Piaget's basic idea as formulated in his *L'Epistémologie génétique*: [76]

La connaissance ne saurait être conçue comme prédéterminée ni dans les structures internes du sujet, puisqu'elles résultent d'une construction effective et continue ni dans les caractères préexistants de l'objet, puisqu'ils ne sont connus que grâce à la médiation nécessaire de ces structures ….

Piaget goes in fact much further. In his *Biologie et connaissance* he writes: [77]

Les processus cognitifs apparaissent alors simultanément comme la résultante de l'autorégulation organique dont ils reflètent les mécanismes essentiels et comme les organes les plus différenciés de cette régulation au sein des interactions avec l'extérieur ….

In other words cognitive mechanisms would be closely related to problems of stability and autoregulation which are so well illustrated by dissipative structures. There is the double aspect to maintain such structures which are only possible in narrow limits of external conditions, and for which therefore the outside world is a potential threat, and the necessity to interact strongly with this outside world which makes possible these structures through continual exchanges of energy and matter.

Whatever the future of these ideas, it seems to me that the dialogue between physics and natural philosophy can begin on a new basis.

I don't think that I exaggerate by stating that the problem of time marks specifically *the divorce between physics on one side, psychology and epistemology on the other. Time is the starting point* of most of the investigations of nature which we may broadly associate with natural philosophy. This was true for the pre-Socratics as shown by Popper [78], and this is still true for present-day phenomenologists such as Heidegger [79], for whom *temporality is the fundamental form of all relations which makes knowing in its primordial sense possible.* [80]

I would like to reproduce in extenso the conclusions of the beautiful article 'Sens et portée de la synthèse newtonienne' by A. Koyré: [1]

Pourtant, il y a quelque chose dont Newton doit être tenu responsable – ou, pour mieux dire, pas seulement Newton, mais la science moderne en général: c'est la division de notre monde en deux. J'ai dit que la science moderne avait renversé les barrières qui séparaient les Cieux et la Terre, qu'elle unit et unifia l'Univers. Cela est vrai. Mais, je l'ai dit aussi, elle le fit en substituant à notre monde de qualités et de perceptions sensibles, monde dans lequel nous vivons, aimons et mourons, un autre monde: le monde de la quantité, de la géométrie déifiée, monde dans lequel, bien qu'il y ait place pour toute chose, il n'y en a pas pour l'homme. Ainsi le monde de la science – le monde réel – s'éloigna et se sépara entièrement du monde de la vie, que la science a été incapable d'expliquer – même par une explication dissolvante qui en ferait une apparence 'subjective'.

En vérité ces deux mondes sont tous les jours – et de plus en plus – unis par la praxis. Mais pour la theorie ils sont séparés par un abîme.

Deux mondes: ce qui veut dire deux vérités. Ou pas de vérité du tout.

C'est en cela que consiste la tragédie de l'esprit moderne qui 'résolut l'énigme de l'Univers', mais seulement pour la remplacer par une autre: l'énigme de lui-même.

In a recent article Wigner[81] expressed similar ideas when he wrote:

What is the most important gap in present science? Evidently, the separation of the physical sciences from the sciences of the mind. There is virtually nothing in common between a physicist and a psychologist – except perhaps that the physicist has furnished some tools for the study of the more superficial aspects of psychology, and the psychologist has warned the phsyicist to be alert lest his hidden desires influence his thinking and findings.

We see that physics is starting to overcome these barriers. And perhaps one of the most important contributions of the twentieth century will be to prepare the path to a more integrated view of the totality of our experience which includes both the physical world and our personal existence.[85]

REFERENCES

1. A. Koyré, *Etudes Newtoniennes*, Gallimard, Paris (1968).
2. J. L. Lagrange, *Théorie des Fonctions Analytiques*, p. 223. Imprimerie de la République, Paris (1796).
3. A. S. Eddington, *The Nature of the Physical World*, Cambridge University Press (1929).
4. R. Clausius, Read at the Philosophical Society of Zurich on 24 april 1865, published in Poggendorf's *Annalen*, 1865, p. 353; also in *Journ. de Liouville*, 2e série, t. X, p. 361.
5. L. Boltzmann, 'Weitere Studien über das Wärmegleichgewicht unter Gasmolekülen', *Wien. Ber.* **66**, 275 (1872).
6. A. Kronig, *Ann. Physik* **99**, 315 (1856).
7. R. Clausius, *Ann. Physik* **100**, 353 (1857).
8. J. C. Maxwell, *Phil. Mag.* **19**, 19 (1860).
9. L. Boltzmann, *Wien. Ber.* **58**, 517 (1869).
10. For more details see I. Prigogine, Cl. George, F. Henin and L. Rosenfeld, *Chemica Scripta* (1973), to appear.
11. B. J. Alder and T. E. Wainwright, *J. Chem. Phys.* **33**, 1434 (1960).
12. A. Bellemans and J. Orban, *Phys. Lett.* **24A**, 620 (1967).
13. P. T. Ehrenfest, 'Begriffliche Grundlagen der Statistischen Auffassung der Mechanik', *Encycl. Math. Wiss.* **4**, 4 (1911).
14. J. C. Powless and P. Mansfield, *Phys. Lett.* **2**, 58 (1962).
15. P. Mansfield, *Phys. Rev.* **137**, 1961 (1965).
16. W. K. Rhim, A. Pines and J. S. Waugh, *Phys. Rev.* B **3**, 864 (1971).
17. D. Walgraef and P. Borckmans, *Physica* **59**, 37 (1972).
18. L. Boltzmann, *Wien. Ber.* **75**, 62 (1877).
19. S. H. Burbury, *Nature* **51**, 78 (1894).
20. See specially L. Boltzmann, *Lectures on Gas Theory*, Sections 87–91. An English translation by S. G. Brush is available, University of California Press (1964).
21. M. Planck, *Thermodynamik*, Berlin and Leipzig, De Gruyter (1930), p. 83.
22. J. W. Gibbs, *Elementary Principles in Statistical Mechanics*, Yale University Press (1902), see *Collected Works*, vol. 2, Longmans, Green and Co., London and New York (1928).
23. A. Einstein, 'Kinetische Theorie des Wärmegleichgewichts und des zweiten Hauptsatzes der Thermodynamik', *Ann. Phys.* **9**, 417 (1902).
24. A. Einstein, 'Theorie der Grundlagen der Thermodynamik', *Ann. Physik* **11**, 170 (1903).
25. R. H. Fowler and E. A. Guggenheim, *Statistical Thermodynamics*, Cambridge University Press (1939).
26. 1. Prigogine, *Non-Equilibrium Statistical Mechanics*, Wiley–Interscience, New York and London (1962).
27. We hope that a more detailed history of this interesting period in statistical mechanics will appear some day; especially as the publication by R. Zwanzig, *Physica* **30**, 1109 (1964), gives an inaccurate account of the historical evolution by producing incomplete references to the papers where the ideas, on which non-equilibrium statistical mechanics is still based today, first appeared. The situation in 1956 is well summarized in the *Proceedings of the 'International Sym-*

posium on Transport Processes in Statistical Mechanics' organized by IUPAP, Brussels 1956, Wiley–Interscience, New York and London (1958).

28. I. Prigogine, *Introduction to Thermodynamics of Irreversible Processes*, 3rd ed., Wiley–Interscience, New York (1967).
29. P. Duhem, *Energétique*, Paris, Gauthier–Villars (1911), 2 vol.
30. Th. De Donder and P. Van Rysselberghe, *Affinity*, Stanford University Press (1936).
31. A historical survey may be found in: I. Prigogine, *Etude Thermodynamique des Phénomènes Irréversibles*, Paris, Dunod (1947), and in reference 32.
32. S. R. De Groot and P. Mazur, *Non-Equilibrium Thermodynamics*, North Holland Publishing Company, Amsterdam (1963).
33. L. Onsager, *Phys. Rev.* 37, 405 (1931); 38, 2265 (1931).
34. P. Glansdorff and I. Prigogine, *Stability, Structure and Fluctuations*, Wiley–Interscience (1971), French edition, Masson, Paris (1971).
35. R. Graham and H. Haken, *Phys. Lett.* 29A, 530 (1969).
36. R. Balescu, *Statistical Mechanics of Charged Particles*, Wiley–Interscience, New York (1963).
37. C. Truesdell and W. Noll, *Flugge's Handbuch der Physik*, Vol. III, Part. 3, Springer Verlag, Berlin (1965).
38. Cl. George, I. Prigogine and L. Rosenfeld, *Konigl. Dansk. Vid. Mat-Phys. Medd.* 38, 12 (1972).
39. I. Prigogine, 'Irreversibility as a symmetry-breaking process' (submitted to *Nature*).
40. Cl. George, *Physica*, to appear (1973).
41. M. Reichenbach, *The Direction of Time*, University of California Press, Berkeley and Los Angeles (1956).
42. O. Costa de Beauregard, 'Information and Irreversibility Problems', in *Time in Science and Philosophy* (ed. by J. Zeman), Czechoslovak Academy of Sciences, Prague (1971), p. 11.
43. M. Baus, *Acad. Roy. Belg. Bull. Cl. Sci.* 53, 1291, 1332, 1352 (1967).
44. L. Lanz and L. A. Lugiato, *Physica* 44, 532 (1969).
45. A. Grecos, *Physica* 51, 50 (1970).
46. See M. Kac, *Probability and Related Topics in Physical Sciences*, Interscience, New York (1959).
47. See specially the work of P. Résibois, M. De Leener and others in *Physica* and *Phys. Rev.* during the years 1966, 1969 and 1971
48. I. Prigogine and P. Résibois, *Atti del Simposio Lagrangiano*, Accademia delle Scienze di Torino (1964).
49. R. Balescu, *Physica* 36, 433 (1967).
50. A. Einstein and W. Ritz, *Phys. Z.* 10, 323 (1909).
51. G. Stey, to appear *Physica* (1973).
52. A. Grecos and I. Prigogine, *Physica* 59, 77 (1972).
53. A. Grecos and I. Prigogine, *P.N.A.S.* 69, 1629 (1972).
54. M. De Haan and F. Henin, papers to appear in *Physica* and *Acad. Roy. Belg., Bull. Cl. Sc.* (1973).
55. I. Prigogine and G. Severne, *Physica* 32, 1376 (1966); G. Severne, *Physica* 61, 307 (1972).
56. I. Prigogine and A. Grecos, Volume in Honour of H. Fröhlich (ed. by H. Haken), to appear 1973.
57. A. Grecos, private communication.
58. A. I. Khinchine, *Mathematical Foundations of Informations Theory*, Dover Publ. Inc., New York, 1957.
59. Duk in Choi, *Acad. Roy. Belg. Bull. Cl. Sci.* 57, 1054 (1971)
60. I. Prigogine, *Acad. Roy. Belg. Bull. Cl. Sci.* 31, 600 (1945).
61. S. Chandrasekhar, *Hydrodynamic and Hydromagnetic Stability*, Clarendon Press, Oxford (1961).
62. R. Schechter, I. Prigogine and J. R. Hamm, 'Thermal Diffusion and Convective Stability', *Phys. Fluids* 15, 3, 179 (1972).
63. J. C. Legros, J. K. Platten and P. Poty, *Phys. Fluids* 15, 1383 (1972).
64. (*a*) J. C. Legros, D. Rasse and G. Thomaes, *Chem. Phys. Lett.* 4, 638 (1970); (*b*) J. C. Legros, P. Poty and G. Thomaes, to appear *Physica*.
65. J. K. Platten and G. Chavepeyer, *Phys. Lett.* 40A, 287 (1972). J. K. Platten and G. Chavepeyer, submitted to *J. Fluid Mechanics*.
66. I. Prigogine and R. Lefever, *J. Chem. Phys.* 48, 1695 (1968).
67. G. Nicolis and I. Prigogine, *P.N.A.S.* 68, 9, 2101 (1971); G. Nicolis, *Advances in Chemical Physics* 19 (ed. I. Prigogine and S. Rice), Wiley and Sons, Inc. (1971).

68. I. Prigogine, 'La thermodynamique de la vie', *La Recherche* **24**, 547 (1962).
69. I. Prigogine and G. Nicolis, *Quart. Rev. Biophysics* **4** (2 and 3), 107 (1971).
70. A. Goldbeter and G. Nicolis, *Biophysik* **8**, 212 (1972).
71. A. Babloyantz, to appear *Biopolymers* (1972).
72. I. Prigogine, G. Nicolis and A. Babloyantz, *Physics Today* (1972).
73. J. Monod, *Le hasard et la nécessité*, Editions du Seuil (1970).
74. I. Prigogine, *Structure, Dissipation and Life*, First conference on Theoretical Physics and Biology 1967, North-Holland Publishing Company, Amsterdam (1969).
75. A. N. Whitehead, *Science and the Modern World*, Cambridge (1927), p. 64.
76. J. Piaget, *L'Epistémologie génétique*, Paris, Presses Universitaires de France, Collection 'Que sais-je' (1970), p. 5.
77. J. Piaget, *Biologie et connaissance*, Eds. Gallimard, Paris (1967).
78. K. R. Popper, *Conjectures and Refutations*, London, Routledge and Kegan Paul, 1963.
79. M. Heidegger, *Sein und Zeit*, 7th ed., Tübingen, Niemeyer (1953), Sections 3, 9; English Translation by J. Macquarrie and E. Robinson, *Being and Time*, New York: Harper (1962).
80. W. Bossart, 'Three Directions of Phenomenology', in *The Anatomy of Knowledge* (ed. M. Grene), Routledge and Kegan Paul (1969), p. 259.
81. E. P. Wigner, 'Epistemology of Quantum Mechanics, Its appraisal and demands', in *The Anatomy of Knowledge* (ed. M. Grene), Routledge and Kegan Paul (1969), p. 31.
82. P. J. M. Bongaarts, M. Fannes and A. Verbeure, (1973), to appear.
83. F. Henin, to be published.
84. P. M. Allen, *Physica* (1973), to appear.
85. The epistemological questions will be treated more in detail in a forthcoming paper by I. Prigogine, 'Physique et Metaphysique', to appear in the Proceedings of the Symposium *Connaissance scientifique et philosophie*, Ac. Roy. Belg. (1973).

30

The Origin of Biological Information

Manfred Eigen

1. QUESTIONS AND PSEUDO-QUESTIONS

In our days the physicist's conception of nature certainly has to include the phenomenon of life.

What is life?

Thirty years ago, Schrödinger[1], in his famous essay, was courageous enough to risk an answer to this very question. Today – after the secret of the molecular code of life has been lifted – biologists seem more cautious – to quote Crick[2]: 'It is notoriously difficult to define the word living.'

Physicists are accustomed to such difficulties, especially with respect to their most fundamental terms. In his *Principia* Newton gave a very definite answer on the question 'what is time', while Feynman[3] in his *Lectures on Physics* dares to give another definition: 'What really matters anyway is not how we define time, but how we measure it.'

Can we measure the phenomenon of life? Or better: Can we specify the criteria characteristic of life and demonstrate their validity by test-tube experiments?

We have to be aware that there are many levels of life – eventually culminating in 'nature's conception of a physicist', which is the counterpoint of the theme of our symposium.

In order to answer the above question, we certainly have to start at some lower level. It is the transition from the non-living to the living, i.e. the 'origin of biological information', on which we shall focus our interest. There are many questions referring to this scholastic theme, the most common being expressed in Fig. 1.

'Chicken and egg' in modern biology may well be substituted by 'protein and nucleic acid' (cf. Figs. 2, 3) or – in more abstract terms – by function and information. However, if 'first' is to mean a causal rather than a temporal relationship, the question 'who comes first?' appears to be meaningless. Because: if function – in order to evolve to the presently known sophisticated level – needs to be represented by information, this information, in turn, acquires all its meaning only via the function it is coding for. Indeed, in all living objects – even the most primitive ones – proteins and nucleic acids are linked up in a complex hierarchy of feedback loops (cf. Fig. 4). A causal first would have as little meaning as beginning and end of a closed cycle line.

We know from laboratory experiments, based on 'ordinary' preparative chemistry, that under prebiotic conditions both classes of biological macromolecules could have

J. Mehra (ed.), The Physicist's Conception of Nature, 594–632. All Rights Reserved
Copyright © 1973 by D. Reidel Publishing Company, Dordrecht-Holland

"Damn you, Winkle, did you have to go and ask it which came first – the chicken or the egg?"

(Drawing by Dana Fradon; © 1971 The New Yorker Magazine, Inc.)

Fig. 1.

come about independently of each other. Thus, a temporal 'first' in connection with the origin of life could be meaningful, if we rephrase the 'chicken or egg' question. For instance: Does any of the two classes of macromolecules involve sufficient executive *and* legislative power so that it was able to start autonomously the process of self-organization and evolution? Or – the alternative of this question: Was it the *inter-communication* between proteins and nucleic acids, i.e. the nucleation of a code and translation machinery, which started a converging functional self-organization – as distinguished from diverging unorganized macromolecular synthesis?

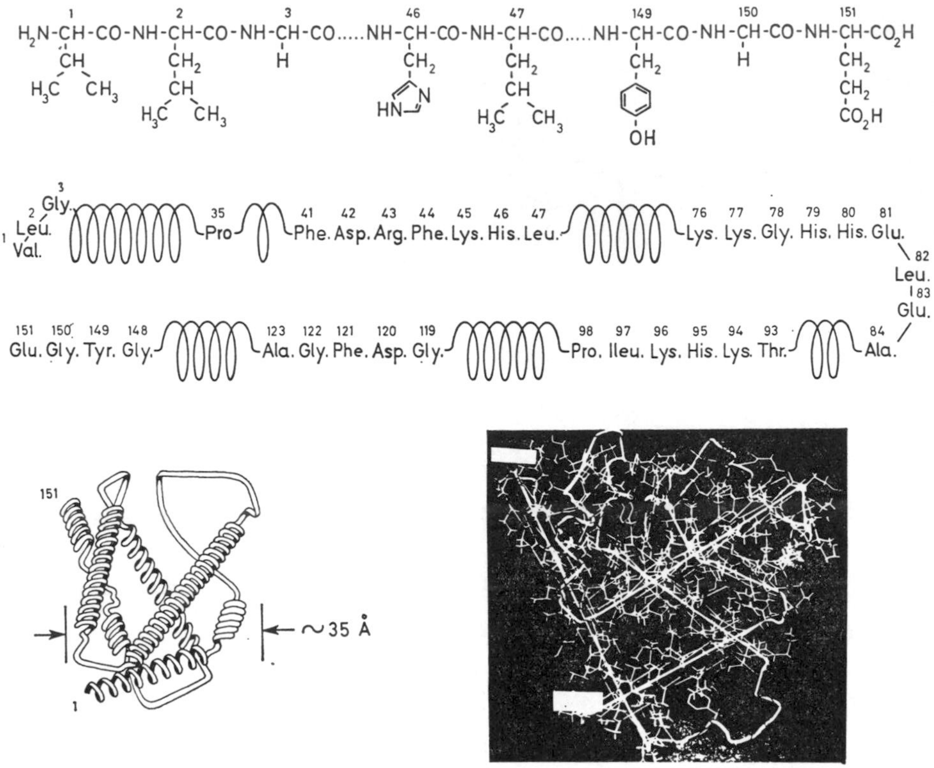

Fig. 2.   The structural levels of proteins.

(*a*) Primary: The amino-acids (ca. 20 classes) are covalently linked together to form the linear polypeptide chain.

(*b*) Secondary: The peptide bond enhances the polarity of the CO- and NH-groups in the linear chain. 'Helical' or 'pleated sheet' structures result from hydrogen bond formation among these two groups.

(*c*) Tertiary: Spatial folding, stabilized by side chain interactions, allows the formation of an 'active site' with a precisely adapted distance correlation among the functional groups. As an example, the myoglobin structure, as resolved by X-ray diffraction (Kendrew[4]), is shown.

(*d*) Quaternary: Several polypeptide chains may combine to complexes comprising regulatory and catalytic power. Control of enzyme function is induced via conformation changes upon binding of effectors at the regulatory sites. (For details cf. refs. 6, 22.)

The main problem behind all these questions is that of biological complexity. It appears already at the level of macromolecular chemistry.

2. BIOLOGICAL COMPLEXITY AND UNIQUENESS

Consider a typical small protein molecule such as myoglobin, the structure of which was shown in Fig. 2. A molecular weight of about 10^4 means that the polypeptide chain is made up of about 10^2 amino-acid residues. Given the 20 classes of natural amino-acids we may find as many as 20^{100} or 10^{130} alternative sequences of this length.

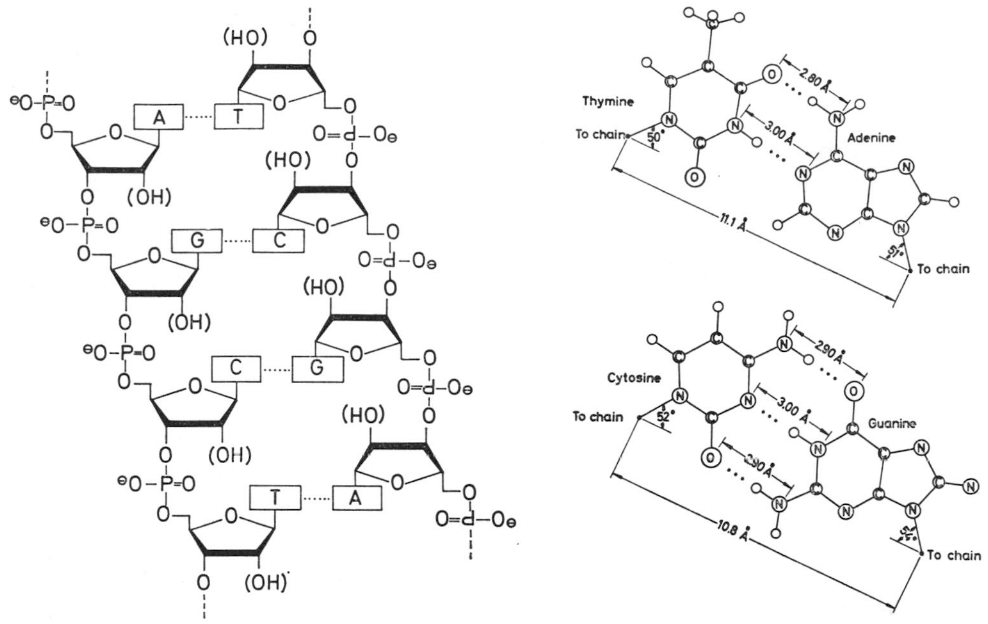

Fig. 3.   The double-helical structure of nucleic acids (according to Watson and Crick[5]).

The backbone of the poly-nucleotide strand represents an alternating sequence of phosphate and sugar (ribose or deoxyribose), covalently connected via the phosphate diester linkages. Each sugar subunit carries one of the four bases A, T, G, C (in DNA) or A, U, G, C (in RNA), which are the digits of the genetic alphabet. Complementarity is based on a preferential interaction of the purines adenine (A) and guanine (G) with the pyrimidines thymine (T, or uracil U) and cytosine (C) respectively, yielding two pairs of (almost) identical geometry (to be recognized by the polymerizing enzyme).

Actually, myoglobin is among the smaller protein molecules we find in nature. Most of the proteins are appreciably larger and may comprise up to 300 or 400 amino-acid residues in a single polypeptide chain. Moreover, such chains may aggregate to complexes of very high molecular weights, in order to combine their catalytic power with control and regulatory functions (e.g. as 'allosteric' enzymes).

These immensely large numbers have to be seen in relation to numbers we usually encounter in the physical universe, e.g. Avogadro's number, or even the total amount of matter in the universe which – expressed in proton masses – reaches 'only' the magnitude of 10^{78}. This disproportion becomes almost hopeless if we proceed to the level of whole cells, e.g. microbes – not to speak of higher organisms.

The DNA molecule that comprises the total genome of a single bacterial cell – such as *Escherichia coli* – represents one or few choices out of more than $10^{1\,000\,000}$ alternative sequences. It is obvious that only a minute fraction of all such alternatives could possibly have been tested by nature, given the barely 10^{17} seconds time of existence for a chemically relevant state of our universe. (Note that even with the help of highly adapted enzymes synthesis or degradation of a protein molecule requires a time of at least a tenth of a second up to several seconds or minutes.)

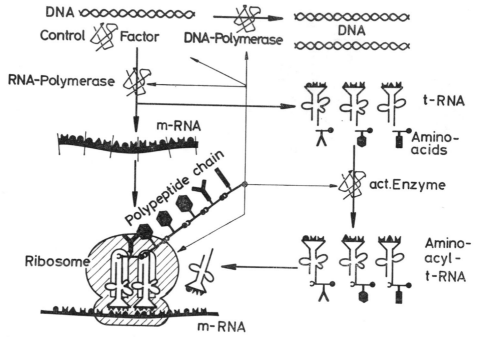

Fig. 4. Nucleic acids and proteins are intimately linked together in their reproduction cycle. The important functional links are:

(a) DNA and DNA-polymerase, to store and multiplicate the genetic information.

(b) messenger – RNA and RNA-polymerase (in interaction with protein controlfactors, i.e. promotors or repressors) to make the information available for processing.

(c) transfer – RNA, amino-acids and their specific activating enzymes (amino-acyl-synthetases), to mediate translation by linking each amino-acid to its specific (anti-condon carrying) adaptor.

(d) The ribosome, a complex consisting of RNA- and protein subunits, which binds and transports the messenger-RNA, adapts the amino-acyl-transfer-RNA and catalyzes the synthesis of the polypeptide chains.

Molecular biology has provided a deep insight in the complex hierarchy of this selfreproductive bio-synthetic cycle, utilized by every living cell (for details cf. ref. 6).

Complexity has always been recognized as a typical requisite of biological organization. Schrödinger referred to it in using the terms 'aperiodic crystal'. What really matters is an informational rather than a material quality.

Complexity on the molecular scale is characterized by an immense potential number of microstates, some of which may have unique macroscopic significance. This number is in large excess of any number of states which are or could be populated given realistic limits of time and space. After all, the classical problems of statistical mechanics also involve such large numbers of microstates, far beyond any population capacity. However, they usually refer to ergodic systems. The ensemble properties allow for suitable averaging, thereby leading to a deterministic description of the macroscopic behaviour within finite fluctuation limits. The recurrence times of such ensembles are finite – even though they may exceed by far the age of the universe.

On the other hand, the systems we shall discuss here involve non-ergodic properties.

Certain unique microstates may strongly influence the macroscopic behaviour. Upon occurrence, such 'fluctuations' may amplify and cause a breakdown of formerly stable (i.e. meta-stable) states. Evolution in fact may be considered as a sequence of such uni-directional events.

Complementary to the problem of biological *complexity* is that of *uniqueness*. Referring to the enormous complexity described above, how unique can any choice of nature be?

Information theory – as developed by Shannon[7], Wiener[8a], Kolmogoroff[8b] and others – is concerned with the probabilistic aspect of a particular choice among a set of alternatives. Consider a sequence of nucleotides encoding a genetic message. Deviations from randomness may be due to differing *a priori* probabilities of single nucleotides as well as to preferred combinations – in analogy to the occurrence of words and sentences as dictated by the structure of a language or inferred from the meaning of a given message. All these constraints can be expressed in terms of probability distributions for any k-tuple of combinations of symbols.

As Brillouin[9] in his monograph *Science and Information Theory* has clearly pointed out, the probabilistic theory does not account for 'significance' or 'value' of a message. However, it is such a *semantic* aspect of information which is behind biological uniqueness. This semantic aspect must find its ultimate explanation in functional, i.e. dynamical properties and relations, rather than in certain abundances and nearest (and further) neighbour frequencies of base pairs in nucleic acids or amino-acid residues in proteins (although the latter will certainly be important with respect to the functional aspect, as agreement on a language structure is required for communication or semantic expression).

Let us expand a little more on uniqueness of biological information as expressed by the uniqueness of function. If we analyze the dynamical performance of any biological entity – such as a single enzyme molecule, an organelle, a single cell or even a whole organism – we always encounter perfect correlation of all elementary steps involved. 'Perfect' means, for instance, that the measurable rates are at or close to the upper limits imposed by the laws of physics. The rate of any chemical turnover is limited by the collision or encounter frequency of the corresponding reaction partners. If complexes are formed, discriminative recognition is based on a certain magnitude of stability constants (i.e. free energies of interaction). An analysis shows that enzymes are optimally adapted to these requirements. Another example is the reproduction of a whole bacterial cell with its thousands of proteins and other macromolecules. The total generation time of an *E. coli* cell amounts to about 20 minutes. It includes sequential reading, translation and processing of a message which is encoded in a linear chain of about 4 million nucleotides. A message of 4 million letters corresponds to a book of more than a thousand tightly printed pages. A total of twenty minutes leaves for the processing time per letter less than a thousandth of a second. The rates of some of the elementary chemical steps of code reading and reproduction have been directly measured. A discriminative recognition of single code letters requires less than 10^{-6} seconds.[10]

Apparently, at least on the molecular level there is little if anything left for further improvement. The functional correlation among the myriads of elementary chemical processes seems to be almost perfect. Does such a perfection indicate true uniqueness of biological structures? How many alternatives have actually been tested?

A sequence containing v residues of λ classes allows for $\binom{v}{k} (\lambda-1)^k$ alternative copies having substitutions at k positions. A small protein molecule ($\lambda=20$; $v=100$, cf. above) then has only 1900 'single error' copies; but for $k=9$ – i.e. substitution of still less than 10% of positions in the original sequence – the number of alternative mutants reaches already the order of magnitude of Avogadro's number. Hence, a testing of alternatives, unless it uses some hierarchical principle, is very limited.

We may conclude this discussion of biological uniqueness by proposing three different possible explanations:

(1) The copy choice is uniquely guided by physical forces. The appearance of living entities is nothing but the manifestation of such unique interactions.
Consequence: An independent start of life, e.g. on a different planet or in the test-tube, would bring about identical structures, in particular the same code and enzymic machinery.

(2) The choice is a 'unique' incidence (or better coincidence). The expectation value for its appearance is negligibly low, but the stochastic nature of the interactions does not exclude such unexpectedly rare fluctuations.
Consequence: The probability of finding life of independent origin anywhere else in the universe, where physical and chemical conditions were similar to those on early earth, would be negligibly small.

(3) The nucleation of a self-reproducing and further evolving system occurs with a finite expectation value among any distribution of essentially random sequences of macromolecules such as proteins and nucleic acids – given suitable dimensions of space and time and favourable physical conditions. The initial copy choice then is accidental, but the subsequent evolutionary optimization to a level of unique efficiency is guided by physical principles.
Consequence: Life should be found wherever the physical and chemical conditions are favourable (i.e. – as we are told by astrophysicists – with high abundance in the universe). However, individual structures, in particular the molecular code and machinery should show only little familiarity with the systems known to us. Those structures therefore may not be directly utilized by any species on earth, although they may be built on similar principles. There should be, in particular, a 50% chance of finding a different chirality in optically active structures.

I have brought up these three hypotheses, because they lead to entirely different outcomes, which at some time, might be checked either by space or, possibly sooner, by test-tube experiments.

The difficulty of the first hypothesis especially relates to the independence of phenomena at the genotypic level with respect to those at the phenotypic level. We must admit that we do not know any simple structure–function relation which would suggest a unique guidance of evolution by inherent forces. Thus, we would have to operate with largely as yet non-understood (possibly new) physical concepts.

The second hypothesis would be beyond any theoretical treatment. We would need not just one but rather a whole series of accidents each of which having the expectation value of (almost) zero.

The attraction of the third hypothesis is that it involves a large multiplicity of copy choices to start with. For instance, the fixation of a few positions in any random globular protein structure may be sufficient for assignment of a certain catalytic advantage. It may be found in one out of as 'few' as a million, i.e. in 10^{124} out of 10^{130} possible sequences. (Any definite number, of course, would depend on what we call 'advantage'.) Again, the expectation value for the *particular* initial copy is negligibly small. Only the fact of occurrence of *some* initial copy has a sufficiently high probability. The following evolutionary optimization process has enough branching points to render any particular route completely indeterminate; but again, this needs not be true for the fact of optimization which may be the consequence of a physical law. The final outcome will be a unique structure, e.g. a k-tuple of defined positions with optimized spatial coordination. The number of alternative final destinations (and thus of alternative optimal structures) will certainly be large compared to one (i.e. 10^x with $1 < x \ll 130$). For instance, if x were 5 or 10, any such optimal structure still would be one out of an enormously large number (100^{125} or 10^{120}) of less efficient ones.

There is a certain analogy to a strategic game, such as chess. The first moves have a good chance of being correct because a large fraction of moves has the potentiality of further evolution. It is the middle phase which introduces the huge amount of branching while in the end phase the possible solutions are narrowed down to a relatively small number of equivalent alternatives with the optimum of 'mate'.

Experimental evidence obtained with random structures – as fragmentary as it is at the present time – favours some intermediate hypothesis, rather than any of the two extremes. Nevertheless, quite a number of physicists and philosophers seem to be more attracted by the first hypothesis: 'God does not play at dice', while many of the biologists are inclined to believe in the hypothesis of 'absolute chance'.

3. 'LIFE GAMES'

Von Neumann's idea of a self-reproducing automaton[11] has stirred mathematicians' interest in a particular category of games simulating proliferation and growth. Ulam[12] in his paper 'On Recursively Defined Geometrical Objects and Patterns of Growth' described a number of such games applying to cellular automata, and more recently, Conway[13] introduced an exciting game, which he called 'life' because 'of its analogies with the rise, fall and alterations of a society of living organisms'.[14] I am referring to these games here for two reasons:

(1) They demonstrate excellently how an unimaginable complexity of objects involving many functional features can be obtained from very simple construction rules.

(2) They resemble exactly the first of the three hypotheses about the origin of life: A completely deterministic course of evolution predestinated by the initial configuration, from which it follows via correct execution of the transition rules.

The games have in common a finite or infinite cellular space, in which each cell is assigned one out of a finite number of given states, e.g. 'empty' or 'filled'. Each cell furthermore has a finite number of neighbouring cells that can influence its own state. The pattern of states – by applying certain transition rules simultaneously to each cell – changes in discrete time steps. The result of a game is either extinction, growth or stationary behaviour (i.e. a stable or oscillatory pattern). Table I demonstrates the features of Conway's life games. The complexity into which patterns can grow is so enormous that – except for very few simple initial configurations – it is very hard to predict any final outcome. Especially the 'strategically' important configurations such as 'spaceships', 'glider-guns' or 'eaters' (cf. Table I) add some features to this game which make it most exciting for both the layman – who will just enjoy it – and the scientist, who might abstract new principles as to construct a universal calculator, i.e. a Turing machine.[15]

What is missing in the game is some element of randomness. Any molecular reproduction process is subject to randomly occurring 'errors'. These 'errors' or 'mutations', if properly selected, are considered to be the source of further progress, i.e. the source of new information. Conway's game does not create more information than had been put into its initial pattern by a 'fully informed creator'. There is no way of influencing the game, which will run down deterministically according to the correct execution of the transition rules.

It is, of course, trivial to devise games which exemplify the hypothesis of 'absolute chance'. Any lottery or game of dice would be suitable. Such a game may become representative of evolutionary behaviour, if appropriate selection rules are introduced.

Consider a sequence of letters, which may represent the base sequence of a nucleid acid molecule (cf. Fig. 3):

$$\text{AGUUCCGCAGGCU} \text{--------} \rightarrow \nu \text{ digits}$$
$$\lambda = 4 \text{ classes}$$

Our task will be to arrive at a specified sequence by random variation of single digits under the guidance of certain selection rules. The 'random source' may be simulated by a tetrahedral die, each face of which represents one of the four letters: A, U, G or C.

'Absolute chance' would correspond to simultaneously tossing ν dice, i.e. one for each of the ν positions, and arranging the letters accordingly. If we ask for the appearance of a specific sequence it would take on average 4^ν turns to achieve it. If ν, for instance, is about 80 – as for the smallest functional nucleic acid molecules: the amino-acid adaptors t-RNA – this game would be 'endless' (and dull, too), although, as in any lottery, the early occurrence of a 'lucky shot' cannot be completely excluded.

We can add much intellectual flavour to such a game, if we introduce criteria of evolution and define selection rules by which the system is forced to match the criteria.

TABLE I: Conway's Life Game

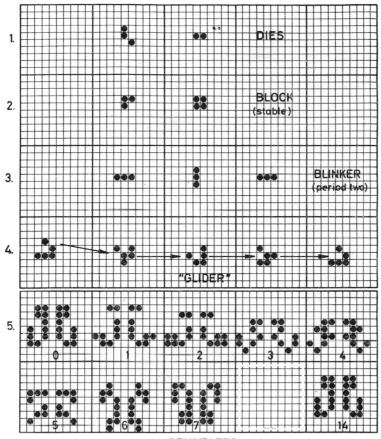

"TUMBLER"

The game is played on an 'infinite' Go-board. Each cell can assume either of two states: 'occupied' or 'empty'. A counter may be placed in a cell to denote its 'occupied' state. The state of a given cell is dependent on the states of the neighbouring cells. In Conway's space, each cell has eight neighbouring cells i.e. the four adjacent orthogonal and the four adjacent diagonal cells. Time proceeds in discrete steps, each constituting a single generation or a 'move'. The rules apply simultaneously to all cells. They are:

1. Survivals. Every counter with two or three neighbouring counters survives for the next generation.

2. Deaths. Each counter with four or more neighbours dies (i.e. is removed) from overpopulation. Every counter with one neighbour or none dies from isolation.

3. Births. Each empty cell adjacent to exactly three neighbours – no more, no fewer – is a birth cell. A counter is placed on it at the next move.

It is important to understand that all births and deaths occur simultaneously. A few typical configurations are shown. The moves are represented side by side as successive steps.

Triplets that die (1), become stable (2), or oscillate (3).

A 'glider' (4), i.e. a periodic structure that moves across the field. (Note, that the other oscillators remain at their positions.)

An oscillator (5) with a period 14. It turns upside down every seven moves. This 'tumbler' was invented by G. D. Collins Jr.[14].

(See also p. 604.)

Continuation of Table I

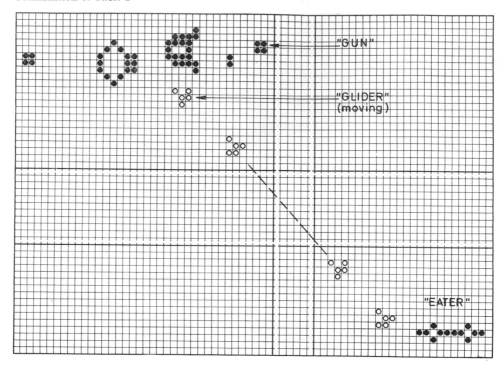

A 'glider gun' (upper left) which fires a glider, 'eaten' by the pentadecathlon (lower right). The glider gun is a stationary oscillator with a period 30, which emits gliders, one during each period. This glider is seen to proceed to the lower right where it is 'eaten up' by the pentadecathlon, an oscillator of period 15.

What is it, that we want to select for?

Information is worthless if it cannot be conserved up to the moment where it is to be read out. Any non-equilibrium structure, subject to thermal motion, will decompose. In particular, the information carrying macromolecular sequences are (and have to be) such non-equilibrium structures. Thus their 'ur-semantics' can only refer to conservation of information, i.e. survival. Advantageous are those sequences which provide further information for improving this quality: information for an enhancement of speed or fidelity of reproduction or protection against decomposition. Any self-organizing selection mechanism must operate on these properties. Actually, it is Darwin's principle which comes in at this level of evolution.

The linear polymeric chains of nucleotides, which are very easily degraded by hydrolysis, can be efficiently protected by formation of double stranded regions, utilizing the specific complementary interactions A–U and G–C, as was shown in Fig. 3. Such a polymer which manages to achieve a maximum base pairing (preferably involving the more stable GC pair) therefore will have the best chance of survival. A further clue is added by the fact that isolated base pairs are not stable. Double

TABLE II: The *t*-RNA Game.

Given: For each player a random sequence of N (e.g. 80) digits of four classes denoted by the letters A, U, G, C, and a tetrahedral die, each face of which represents one of the four letters.

Aim: By tossing dice in turn and substituting a denoted position in the sequence according to the outcome of die throw, each player should try to approach a double-stranded structure with as many as possible AU or GC pairs. The game is over as soon as the first 'complete' structure is announced. Each player counts the number of pairs, where a GC pair counts twice as much as an AU pair. In each turn, every player must toss the die only once, but he is free to announce for which position he is tossing.

Rules: (1) Complementarity rule. Two complementary digits (A and U, or G and C) in juxta position will pair, i.e. will be connected.

(2) Cooperativity rule. Those complementary pairs count only, if they occur in an uninterrupted sequence of at least 4 AU, or 2 AU+1 GC, or 2 GC pairs.

(3) Steric rule. One may form any pattern by two-dimensional folding. Folding – required in order to bring two digits into juxta position – will produce loops. Five digits of the loop region have to remain unpaired by 'steric' reasons.

A possible further constraint is the condition of end to end matching of the chain.

Note on procedure: Too many loops may reduce too severely the possibility of pairing and also may lead to restrictions with respect to the cooperativity rule. One loop – as occurring in a hair pin configuration – will limit the chance of initial pairs to the statistical expectation value. Larger fluctuations of random pairs, i.e. positive deviations from the expectation value, are to be found in special combinations of shorter regions only. To search for such regions requires formation of several loops. The advantage is due to the existence of a large multiplicity of combinations as opposed to one hairpin configuration. For 80 digits the optimum structure appears to be the clover leaf. Also the longer sequences shown in Fig. 5 are well resembled by the game.

The details of this game were worked out by Ruthild Winkler,[15] based on earlier estimates of structures by Fresco, Alberts and Doty (cf. ref. 27).

---

stranded regions can only form via a nucleation process involving the cooperative interactions between either 2 GC, or 1 GC and 2 AU, or 4 AU pairs. Furthermore, there are steric constraints associated with the folding which restrict the maximum number of pairs that can be formed in a particular structure type. All these constraints will be taken into account in our game by special rules.

In Table II such an evolution game is described. The quantitative rules are based on experimental data which were obtained with isolated oligo-nucleotides of known sequences.[16] The goal is to achieve, within the shortest possible time (i.e. with a minimum number of tossing dice), a structure which is characterized by a maximum number of AU and GC pairs (where the more 'valuable' GC counts twice as much as the less stable AU pair). The resulting types of (secondary) structures (e.g. 'clover leaves') which appear quite reproducibly as winners in this game, agree remarkably well with those 'natural' structures which so far have been checked by sequence analysis[19, 20, 21] (cf. Fig. 5). Most of these RNA molecules have been selected for their functional properties (e.g. as amino-acid adaptors) rather than for their ability to encode protein structures. For these molecules, stability indeed is of essential evolutionary value. This can be directly demonstrated in test-tube experiments where a certain functional property is subjected to a particular selection strain (cf. Section 5).

The game described in Table II simulates almost the extreme of a natural selection process. A true extreme is accomplished, if selective advantage refers to the correct occupation of any *single* position, independent of the status of other positions. Pairing

– in absence of any additional constraint such as cooperativity or steric hindrance –
would resemble such a case. In order to achieve a given occupation at v positions we
would require only the minimum of $v(\lambda-1)$ trials. (Each of the v positions is tried λ
times on average, but one may start from an initial pattern in which v/λ positions are
expected to be correctly occupied. If 'pairing' is the only constraint, v involves only
half of the pairable positions, which have just to match the other half.) The number
$v(\lambda-1)$ has to be seen in comparison to λ^v, the number of unconditioned trials in

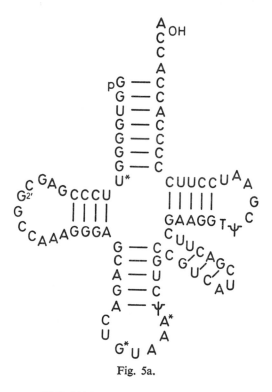

Fig. 5a.

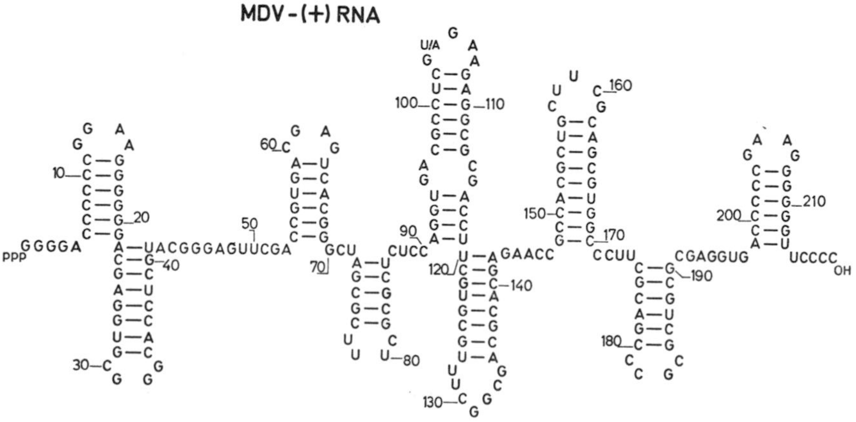

MDV-(+) RNA

Fig. 5b.

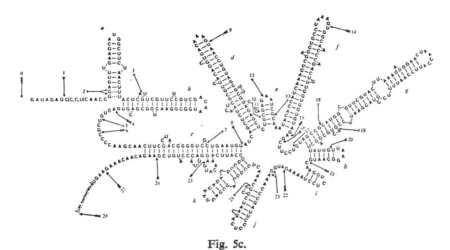

Fig. 5c.

Fig. 5. 'Secondary' structures of functional RNA-molecules.
  (*a*) transfer-RNA (adaptor of phenylalanine).
  (*b*) 'midi-variant' obtained in evolution experiments using the specific replicase of phage $Q\beta$ (Spiegelman[20]).
  (*c*) 'coat' gene obtained by nuclease digestion of phage MS2-RNA (Fiers et al.[21]).
  This gene codes for a sequence of 129 amino acids.
  All structures are further spatially folded.

absence of any selection rule (e.g. for $\lambda = 80$ and $\nu = 80$: 240 as compared to $\sim 10^{48}$). This extreme of a non-cooperative selection process can hardly be found in nature. It would appear to be deterministic since it is bound to reach a true maximum, independent of the route. The requirement of cooperativity, which is much more pronounced at the phenotypic level (e.g. in the three-dimensional structures of proteins and their interplay in control functions), will increase the number of trials tremendously. It will cause the route to branch considerably so that the final destination will depend on the historical sequence of the indeterminate stochastic events. This procedure of optimization differs appreciably from a thermodynamic equilibration process, where a true extremum (minimum of free energy) is reached independent of the route and is only characteristic of the state properties.

If we return now to events of 'natural selection' we have to find out how the player who in the game tosses dice and selects according to the rules, is to be replaced by a mechanism which is inherent to the molecular machinery. If Darwin's principle extends to the molecular level, we have to associate it with certain molecular properties and derive it from more fundamental physical principles.

4. DARWIN AND THE MOLECULES

4.1. *Stability and Selection*

The ability of selective self-organization must be associated with peculiar dynamical properties of the material and will be found only under special circumstances.

Let us consider some chemical entity i, which may consist of a single compound, a macromolecular sequence or even a whole ensemble of interacting species with cooperative reaction behaviour. This entity is steadily formed and decomposed, possibly in a complex multimolecular process involving any parallel or consecutive reaction steps of arbitrary order:

$$\Rightarrow i \rightarrow .$$

The population variable of i is denoted by c_i and its variation δc_i by x_i. The concentrations c_i assume certain stationary values if the corresponding net rates of production and removal (involving decomposition or transport) match each other. In particular, the stationarity of c_i corresponds to an equilibrium distribution if the two arrows connect identical states, i.e. if one represents the reverse of the other.

We focus our attention on the fluctuations of i, occurring in the population variable and their effects on both the net formation and net decomposition rate. If the variation of the rate has the same sign as that of the population variable, we may denote this fact by a ' $+$', if the signs are opposite by a ' $-$', and if the rate is independent of x_i by a '0'. For coupled reactions a complete system analysis is required in order to decide about the sign of the fluctuation response of the rate terms. The nine possible situations then can be represented in form of a 'pay off' matrix.

		Formation		
		$-$	0	$+$
Decomposition	$+$	S_{11}	S_{12}	D_{13}
	0	S_{21}	D_{22}	I_{23}
	$-$	D_{31}	I_{32}	I_{33}

The main diagonal of this matrix includes the three principal outcomes: stable state (S), random drift (D) and instability (I). The indices may help to identify the different cases.

Any process near true equilibrium will fall into the category S_{11}; it will be discussed in more detail below in connection with Ehrenfest's urn model. For open non-equilibrium systems the first row and the second column represent the more common reaction behaviour: first order decomposition processes and absence of any feedback from reaction products to reactants. More specialized properties are required for the two other rows and columns respectively. They can be found especially in biochemical systems and may include auto-, cross- or cyclic catalysis (third column), product inhibition (first column), substrate saturation of the catalytic site (second row) or more sophisticated control as exhibited by allosteric regulation[22] (third row and column). The two latter cases generally work only within a limited range of substrate concentration.

The 'pay off matrix' reflects a symmetry with respect to stable and unstable situations, i.e. S_{11}, S_{12}, S_{21} as compared to I_{33}, I_{32}, I_{23}. According to Lyapunov's criteria [23] these six outcomes resemble the 'easily decidable' cases. If – in the neighbourhood of a steady state $(dc_s/dt=0)$ – we expand the system of rate equations in a power series of $x_1, ..., x_n$, it can be written in the form:

$$\frac{dx_s}{dt} = \sum_{i=1}^{n} p_{si}x_i + F_s(x_1, ..., x_n). \tag{1}$$

The function F_s comprises all non-linear (but none of the linear) terms and hence can be assumed to converge for sufficiently small x_i. Let λ_i be the eigenvalues of the linear system, i.e. the roots of the equation

$$|P - \lambda E| = 0; \quad P = \|p_{ik}\|; \quad E = \text{unity matrix}. \tag{2}$$

Two cases then can be immediately distinguished for which the trivial solutions $(x_s=0)$, regardless of the form of F_s, are either 'asymptotically stable' or 'unstable'. It was shown that for both alternatives the system (1) has families of solutions in the form of series (including products of exponentials and polynomials in t) which converge either for $t>0$ (stable) or for $t<0$ (unstable). The stable case results if all Re $\lambda_i<0$, whereas instabilities occur if among the λ_i, there is at least one λ_j for which Re $\lambda_j>0$. A decision is not so easily possible if none of the λ_i are positive, but there exists at least one λ_k for which Re $\lambda_k=0$ (i.e. a zero or purely imaginary root). Only the anti-diagonal $(D_{13}D_{22}D_{31})$ of the 'pay off matrix' could include those 'doubtful' cases, although it is not exclusively represented by them. Depending on the particular form of the original rate equation, especially the non-linear terms, stable or unstable situations may show up if the expanded equations are inspected according to the method of Lyapunov. Prigogine [24] in his lecture has given a very general formulation of the stability criterion applying to steady states far from equilibrium.* He referred to the effect of fluctuations on the dissipation term

$$\delta_X \sigma = \sum_k \delta J_k \, \delta X_k$$

where σ is the entropy production dS_i/dt from irreversible processes inside the system, J_k a generalized flow and X_k its conjugate generalized force. For chemical reaction systems, flows and forces are represented by the (non-vectorial) reaction rates V_k and the corresponding affinities A_k. Affinities, as derived from chemical potentials, contain logarithmic concentration terms, leading to a positive response on X_i for the decomposition of i and to a corresponding negative response for the formation of i. (Note that there are at least two affinities which have to represent formation and decom-

---

* Notice his definition of local equilibrium [25] which allows the variables temperature and pressure as well as entropy, chemical potentials etc. to have the same meaning as in equilibrium thermodynamics. This assumption holds for any application to chemical processes far from equilibrium in the context of this paper.

position of a compound i at a non-equilibrium steady state.) If we analyze now the contributions $\delta V_k \, \delta A_k$ for each of the two processes and compare with the 'pay off matrix' introduced above, we find in agreement with the Glansdorff–Prigogine [25] criterion: $\delta_X \sigma > 0$ for all S-cases, $\delta_X \sigma < 0$ for all I-cases and $\delta_X \sigma = 0$ for D_{22}. For D_{13} and D_{31} the contributions from formation and decomposition have opposite signs, so that any of the three outcomes is possible, depending on the particular form and magnitude of the rate terms.

4.2. Stochastic Models

The versatility, either to assume a stable distribution or to scan randomly through different states or to change irreversibly as a consequence of an instability, will turn out to be a most important prerequisite for evolutionary self-organization. A confrontation of these prototypes of behaviour and a discussion of their stochastic nature may further illustrate and emphasize this point.

Stable behaviour is best exemplified by Ehrenfest's urn model – some physicists also refer to it as the dog-flea model. My description of this classical problem of statistical mechanics can be brief, though I can't omit it completely in view of the subsequent discussion of other models. Think of two urns – or boxes – among which we distribute arbitrarily $2N$ spheres, each carrying one of the numbers $1, \ldots, 2N$. In addition we have a box containing lots, again one and only one for each number. This set of lots has to be kept complete during the game in order to guarantee equal *a priori* probabilities of all spheres. The procedure of the game is to draw lots and transfer the corresponding sphere from the box in which it happens to be to the other. The well known outcome is an equipartition of spheres among the two boxes, independent of the initial distribution. There will be on average N spheres in each box with mean fluctuations $n = \pm \sqrt{N}$.

The game simulates (chemical) equilibration between two states of equal free energy and could, of course, easily be generalized to distributions among any k states having different weights of being populated. What is important, is the stability of the distribution. The probability P_n for the occurrence of a fluctuation n is [26]

$$P_n = 2^{-2N} \frac{(2N)!}{(N+n)!\,(N-n)!} \approx \frac{e^{-(n/\sqrt{N})^2}}{\sqrt{(\pi N,}}.$$

It follows, in particular, that $P_{n=0} = (\pi N)^{-1/2}$ while $P_{n=\pm N} = 2^{-2N}$, which is also the probability for *any* single microstate. For large numbers of N (realistic for most problems of statistical mechanics) this probability is so small that extreme fluctuations (almost) never occur, while 'small' fluctuations (i.e. within the mean limits of $\pm\sqrt{N}$) are very frequent. This behaviour is brought about by the self-regulatory properties of the fluctuations. The larger they are, the higher is the probability of their reversal.

This property, for instance, is missing in the game of coin tossing, representative of case D_{22}. Here, previously occurring fluctuations have no influence on the probability of the next outcome. If in such a game I would start to count after an accidental series of success and include these counts, I would have favoured my position in an unfair manner, because for any further throw I still have a probability of one half for being successful. Unlike in Ehrenfest's game, the fluctuations drift without self-control so that the distribution becomes steadily broader with increasing time.

A fluctuation behaviour as associated with Ehrenfest's model is just 'counter-indicated' for a Darwinian selection mechanism. In view of the enormous complexity described in Section 3, it is absolutely prohibitive to use stable equilibrium distributions. Such a system could never make a unique choice, since it would populate all alternative states according to their relative free energy contents. What rather is required, is a mechanism involving instabilities, which – under the constraint of growth limitation – allows for the complete removal of the less advantageous copies.

Such a behaviour can be simulated by another game: Given a supply of N^2 spheres, in which each of the numbers 1 to N is represented with N copies: Out of this supply we take one representative of each number – i.e. a total of N non-degenerate copies – and enclose them in a box. The procedure of the game is, to draw arbitrarily spheres and subject them alternatingly to one of the two fates: removal or duplication. A strict alternation of these fates guarantees a constant population of N spheres in the box during the whole course of the game. The unavoidable result will be: Selection of one number appearing with N-fold degeneracy.

Actually, there are two ways to play this game. Instead of alternating strictly removal and duplication one might do it at random. Then the number of spheres in the box will not remain constant but fluctuate around the mean value N. These fluctuations are not self-controlling as they were in Ehrenfest's model. Thus, the number of spheres will steadily depart from N (in both directions) with increasing time, until a 'fluctuation catastrophe' extinguishes the whole population. This fate is as unavoidable as is unique selection in the first case. Before the fluctuation catastrophe happens the variety of initially N non-degenerate number will narrow down to a few (or even one), but these will be present with many (on the average with a total of N) copies.

The game qualifies as a stationary Markov process. A stochastic treatment[27] using a method introduced by Doob[28] and applied to similar problems of population genetics by Bartholomay[29] reveals quantitatively the described results. The probability for the appearance of a fluctuation catastrophe – and similarly for an amplification of one sphere to N copies, i.e. unique selection at constant population – reaches e^{-1} after N^2 removals or duplications of spheres. The average lifetime τ of a sphere which is assumed to be equal to the average duplication time, is equivalent to N trials in the game, because each of the N spheres has a chance during the time τ either to duplicate or to be removed.

In this game there is no property which distinguishes one sphere from any other, hence there is no way to predict which sphere is to be selected. We can only say that there is one number – and only one for constant N – which will be left over. This property is inherent to the kind of system, due to the peculiar type of fluctuation behaviour. The fluctuations drift, but a dissymmetry favouring the persistence of highly degenerate copies is introduced by the singularity of irreversible extinction occurring at the population number zero. It has a similar effect as 'true' amplification of fluctuations, which would require an excess of duplication over removal rate and therefore cannot occur in the particular game described.

Does the game in this form simulate Darwinian selection?

Darwin adopted for his principle of natural selection Spencer's formulation 'survival of the fittest'. It has been argued[30] that – if fittest is to mean nothing but the fact of survival, i.e. if there is no other way to determine 'fitness' than by the fact of survival – the principle in this form plainly expresses a tautology, namely 'survival of the survivor'.

This, indeed, is true for the game I have just described. The initial indeterminacy is not due to a lack of knowledge but is inherent to the fact of equal *a priori* probabilities of removal and duplication for all competitors.

Such an assumption applied to molecular decomposition and reproduction processes is not only very unrealistic, it would even exclude selection under the constraint of constant population. All molecular processes are subject to thermal 'noise'. A correct (error free) duplication of any sequence of digits representing a message would then require either (unrealistically) extreme differences of interaction energies for a precise discrimination of digits or – supposing the known relatively small interaction energies – the absence of thermal 'noise'. In fact, it is always only a fraction $Q < 1$ of the reproduced sequences which is correct, regardless of the level of evolution, during which the error rate of single digit recognition is reduced to less than one in 10^{10}.

We may account for imperfect replication and modify our game accordingly. Every kth sphere drawn from the box in the duplication phase may be supplemented by a sphere with a 'new' number $> N$. (E.g. if $Q = 0.9$; $k = 10$). If the total rates of spheres leaving and entering the game match each other, then for any sphere present in the box the probability for removal will exceed that for duplication. This is obvious since steadily new numbers come in. Hence any population of a given number can have only a finite life time, which means that no stable selection is possible – at least as long as the *a priori* probabilities for removal and duplication for all spheres are degenerate.

Reproducible selection in presence of noise can only occur if the individual species differ in their reaction behaviour. Then 'fittest' really refers to an individual property, which now can be checked by independent experiments. It refers to an optimal combination of reduplication rate, accuracy, and life time. Fluctuations which bring about those optimal species, will truly amplify, while those involving less advantageous copies will decay. The result is a selection process which is strongly biassed by individual behaviour. The t-RNA game discussed in Section 3 provides an instructive example. The 'player' now is substituted by a mechanism of self-organization which is inherent to the system.

4.3. *Phenomenological Representation*

Selection and evolution at the molecular level are processes of fundamentally stochastic nature. They usually start out from a single copy produced in an indeterminate elementary process.* If an advantageous copy appears, it has to reach a certain abundance – depending on the magnitude of its selective advantage – before a deterministic selection is ensured. On the other hand, this indeterminacy then refers essentially to the copy choice, rather than to the process of amplification.

The mechanism of selection can be described by a deterministic theory,[27] which essentially is based on three phenomenological parameters:

A_i a rate parameter characteristic of the template instructed *reproduction* process. This process is of inherent autocatalytic nature and therefore leads to an amplification of copies.

Q_i a *quality* factor describing the fidelity of reproduction. Exact reduplication then is characterized by the term A_iQ_i, while $A_i(1-Q_i)$ describes the production of error copies k which are relatives of i (belonging to classes of different 'degrees of affinity'). The contribution of the copy i to the production of any copy k may be described by a probability distribution φ_{ik} with $\sum_{k \neq i} \varphi_{ik} = 1$.

D_i again a rate parameter responsible for the *decomposition* of the copy i.

These parameters represent *specific* rates, hence the rate terms occurring in the phenomenological equations are products of A_i, A_iQ_i or D_i with the population variable c_i. The reaction types under consideration are such that they belong to the category D_{13} introduced above, otherwise the three parameters may be arbitrary functions of c_i or any other population variable c_k, as well as of certain environmental factors. The quantity A_i, in particular, always contains some stoichiometric function of the concentrations of the energy rich (monomeric) reactants – which may or may not be buffered to constant levels.

In the rate equations the three parameters occur as a combined term

$$W_i = A_iQ_i - D_i$$

which may be called *selective value*. It is essentially this quantity which is decisive in the contest of natural selection.

The selective value exhibits threshold properties ($A_iQ_i \gtrless D_i$) but true selection – as to be distinguished from mere segregation – requires a steadily adjustable threshold. This can be achieved by introduction of control: constant flow, i.e. limitation of matter and free energy supply, or constant overall force, i.e. buffering of reactant and of overall concentration of reaction products (constant overall affinity).

No selection process occurring in nature has to adhere to either of these defined constraints – as no steam engine would work under the idealized conditions of the Carnot cycle, which nevertheless serves to determine its maximum efficiency.

In the game, constant population was achieved by strict alternation of duplication and removal. In chemical systems we have to adjust fluxes in order to compensate steadily for the excess production $\sum_{k=1}^{N}(A_k - D_k)x_k$. The simplest systems exhibiting

---

* The possibility of macroscopic indeterminacy resulting via amplification of indeterminate quantum phenomena has been considered long ago by Jordan.[31]

selective behaviour are those for which the selective values W_i are constants.* At constant population all species whose selective value exceeds the average productivity $\bar{E} = \sum_{k=1}^{N} (A_k - D_k) \, x_k / \sum_{k=1}^{N} x_k$ will grow, all others will decay. Thereby they shift $\bar{E}$ to a value determined by the ensemble maximum of W_i:

$$\bar{E} \to W_{\max}.$$

This 'selected' species will dominate the population. It is accompanied by a whole family of reproducibly occurring mutants. The rate of appearance of a mutant will be the lower, the more distant its 'degree of affinity' with respect to the selected copy is. If among the non-populated, more distant relatives, and advantageous copy $(W_i > \bar{E})$, eventually is brought about by a stochastically significant fluctuation, it will grow and cause a breakdown of the formerly (meta-) stable population. Evolution in such a system is characterized by a monotonic sequence:

$$W_{1m} < W_{2m} < \cdots < W_{\mathrm{opt}}.$$

One can show that at constant flows the occurrence of an advantageous mutant corresponds to a negative fluctuation in the dissipation term

$$\delta_X \sigma < 0.$$

The resulting reorganization at constant reaction flows will yield a higher degree of order reflected by a decrease of the internal entropy (if we assume the competitors to be degenerate with respect to their free energy content). The internal entropy production, of course, remains always positive and the decrease of internal entropy at constant turnover occurs at the expense of external fluxes.

The simple monotonic extremal behaviour is met generally with 'linear' systems.** In the case of coupled – still linear – reaction systems, the same criteria apply to the normal modes which can be obtained by an affine transformation of the population variables. Cyclic catalysis then is characterized by a positive eigenvalue attributed to one of the k normal modes of the cycle. The rest of $k - 1$ eigenvalues is negative and describes some kind of equilibration among the relative population levels of the cyclic ensemble.

There are also important classes of non-linear reaction systems to which the straightforward monotonic optimization principle still applies. Such systems usually include ensembles of species which are linked together via non-linear reaction couplings causing the whole ensemble – after nucleation of the cyclic connection – to behave like a single individual. A model system relevant with respect to the origin of translation is described in more detail in Section 5.

From a more general point of view, optimization does not necessarily involve monotonic temporal behaviour, nor does 'fittest' necessarily refer to an ensemble maximum of the respective population variable.

---

* Those systems always have to be far from equilibrium. Close to equilibrium all rate equations – independent of their form – can be linearized yielding negative eigenvalues which are always real (as a consequence of Onsager's relations expressing microscopic reversibility).
** The term 'linear' refers to the W_i-term. Non-linear terms are introduced only via external control and therefore appear in the E-term with a negative sign.

The phenomenological parameters, regardless of their particular form, qualify for selective and evolutionary behaviour if they describe a system which is distinguished by the following characteristics:

(1) Metabolism, as represented by at least two independent rate terms with positive affinities. They refer to the formation of an intermediate via the consecutive turnover of energy-rich into energy-deficient material.

(2) Self-reproduction, as an inherent property of the intermediates. This quality is expressed by the particular form of the amplification rate term, requiring $A_i Q_i > D_i$

(3) Mutability of the intermediate as expressed by a quality factor $Q < 1$. This condition requires the existence of alternative intermediates, some of which have to be self-reproductive.

These prerequisites are *necessary* for selection, but – as will be shown – not *sufficient*, at least as far as an unlimited evolution to the known level of cellular life is concerned.

Of particular importance in this connection is the quality factor Q. The threshold property of selection requires a minimum value of Q, dependent on the relative variance of the rate terms, i.e.

$$[Q_m]_{min} = \frac{\bar{A}_{k \neq m} + D_m - \bar{D}_{k \neq m}}{A_m}$$

where the index m refers to the selected species and the averages are taken over all competitors of m.

On the other hand, Q refers to an information carrier representing a message of v digits. We attribute to each digit class an elementary fidelity factor q_i which can be expressed in terms of digit abundances and free energies of interaction (Section 5). For λ distinguishable classes (possibly including pair combinations if recognition is based on a cooperative interaction) we obtain

$$Q_i = \prod_{k=1}^{\lambda} q_k^{v_{ik}} \qquad \sum_{k=1}^{\lambda} v_{ik} = v_i$$

or in the simplest possible case of uniform (or average) fidelity $q : Q_i = q^{v_i}$. This relation imposes an upper limit v_{max} of digits which can be safely reproduced, given a certain variance of rate parameters supposing $(1 - q_k) \ll 1$ for any digit k:

$$v_{max} = \frac{|\ln Q_{min}|}{1 - \bar{q}}; \qquad \bar{q} = \frac{\sum_k v_k q_k}{\sum_k v_k}; \qquad \sum_k v_k = v.$$

Unless Q_{min} approaches very closely one (i.e. for every small selective advantages) the term $|\ln Q_{min}|$ will be of the order of magnitude of one (otherwise it will be smaller).

In other words, v_{max} can not be appreciably influenced by selection for larger variances of the rate terms. Even if the rate parameters of the selected species differ by orders of magnitude from those of their competitors, the effect on v_{max}, due to the logarithmic relationship, would be insignificant. On the other hand, any improvement in $\bar{q}$ could have a tremendous influence on v_{max}. During the course of evolution the single digit quality factor for reproduction of genetic messages, i.e. the $\bar{q}$-value for nucleotide recognition, increased from about 0.99 for non-enzymic cooperative nucleotide interaction, to $(1-10^{-5})$ for enzymic recognition *in vitro*, to better than $(1-10^{-10})$ for higher organisms *in vivo* (involving code checking by 'repair-enzymes' as well as redundancy due to multiple gene representation). The length of genetic messages could increase accordingly from about one hundred to well over ten billion nucleotides in higher organisms. Evolutionary optimization was concerned with rates – as expressed by the absolute value of W_i or $\bar{E}$ – mainly in an initial phase, where the enzyme machinery had to be adapted. Later stages are more dominated by the increase of information capacity (ν_{max}) requiring a concomitant increase of fidelity of reproduction. This allowed the species to become less and less dependent on the special environmental prerequisites of their origin. The selective value in relation to the average production reflects only the 'fitness' with respect to a given environment (including external factors as well as potential internal couplings).

4.4. *Evolution Experiments*

The fact that selection is inherent to certain well defined conditions of mattter qualifies Darwin's principle as deducible from more fundamental physical laws. The principle then does not just reflect the peculiar historical event of evolution, nor does it require for its application the pre-existence of any form of 'life' (as often has been claimed). Given adequate boundary conditions, selection and evolution appears to be an inevitable process. Therefore, it should be possible to reproduce this process in the test-tube, at least as far as principal steps are concerned. Indeterminacy of copy choice and route dependence of the optimization procedure may exclude a too close reproduction of any historical course.

A typical test-tube experiment resembling Darwinian selection has been carried out by Spiegelman and his co-workers.[32] It is schematicaly represented in Fig. 6. Flux control allows for a regulation of constant levels of energy-rich reactants (i.e. the nucleotide-triphosphates ATP, UTP, GTP and CTP or activated amino-acids if protein synthesis is involved) as well as of a constant overall-population of polymeric reaction products. Semi-permeable filters are used for a separate supply and removal of solvent, monomeric reactants and polymeric products. The experiment in this form is almost exactly simulated by the selection game with alternating removal and noisy duplication, supposing a non-uniform distribution of *a priori* probabilities.

A systematic experimental investigation of molecular self-organization is still in its initial phase.[27] So far it was concerned with evolutionary processes at the level of nucleic acids using enzymes as environmental factors only. This large diversity of structures and their role in phenotypic expression, as revealed by such studies involving phage and artificial ribo-nucleic acids, was quite unexpected.[20] The results modified considerably our view about reciprocity of recognition among individual structures of proteins and nucleic acids. They stressed the requirement of certain symmetries in the distribution of base pairing regions in a selected RNA molecule (cf. Fig. 5). Such a

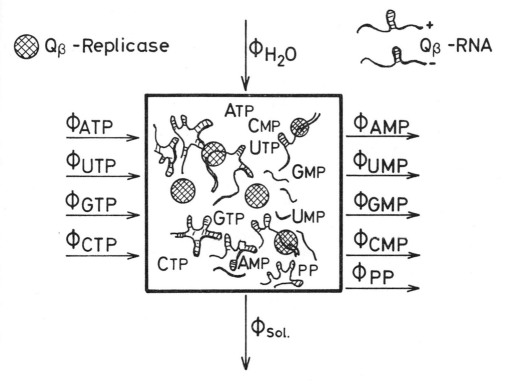

Fig. 6. The selection game described in Section 4.2 resembles an evolution experiment, which – in principle – was first carried out by Spiegelman, using $Q\beta$-phage-RNA and replicase.

The 'box' contains a reaction mixture providing optimal conditions (pH, ionic strength, etc.) for template-instructed RNA-synthesis. The system is provided with a constant level of replicating enzyme, i.e. the phage specific $Q\beta$-replicase. All fluxes, such as the influx of energy-rich reactants i.e. the four nucleoside triphosphates) and of solvent, as well as the outflux of energy deficient decomposition products and of solution (including the polymeric reaction products) are controlled. Polymerization is initiated by $Q\beta$-RNA primers acting as templates. Experiments can be carried out at constant fluxes or at constant levels of concentrations (i.e. constant overall affinities). The latter condition was approximated in Spiegelman's experiments by 'serial transfers'.[32]

symmetry allows both complementary strands to be recognized by the same enzyme. The preference for certain non-linear models of protein-nucleic acid interaction is strongly biassed by these findings.

Those experiments have still to be supplemented by a study of the evolution of enzyme function. Experiments which include protein synthesis are much more difficult to carry out. They require reliable – if possible automatized – in vitro methods for a sequential polypeptide synthesis. An evolutionary programme could be largely played through by a computer. It requires only an occasional, but systematic, checking of the random variations brought about by a programme representing the genotypic level. The strategy would be, not to try to synthetize 'whole' living objects de novo, but rather to substitute successively single components of an otherwise complete bio-synthetical ensemble – analogous to the principle of a 'music minus one' record, where only one instrument is to be played by a dilettante while the rest of the ensemble consists of professional musicians. Those experiments are not designed to produce a 'homunculus' in the test-tube, but rather to check the assumptions underlying the various models of self-organizing evolutionary systems.

5. REALITY AND MODELS

A theory has only the alternative of being right or wrong. A model has a third possibility: it may be right, but irrelevant.

The existing theory accounts for the principles of evolution and their foundation in the laws of physics. It includes the origin of self-organization and its unique outcome 'life'. Moreover, it provides a quantitative basis for the evaluation of laboratory experiments and may thus guide us to relevant models via exclusion of inadequate alternatives. However, what such a theory will never reproduce is the precise historical route of evolution, the 'never' being a consequence of the tremendously large multiplicity of possible choices.

Models have to fulfil two requirements:

(1) They have to start out from realistic prerequisites.

(2) They – or their consequences – have to be in agreement with presently known facts.

The prerequisites are mainly of chemical nature. The large variety of compounds brought about during the prebiotic phase includes all essential precursors of biological macromolecules, in particular self-instructive and catalytic structures as associated with more or less random poly-nucleotides and poly-peptides. This can be demonstrated by laboratory experiments in which appropriate conditions are simulated.[33–36] Not all questions concerning this phase of evolution have been answered so far. However, the problems are of a typical chemical nature, and quantum mechanics – in principle – can account for all the diversification of matter upon which biological complexity rests. The models of self-organizing chemical networks we shall discuss, presuppose the presence of – yet unorganized and largely irreproducible – polymeric structures of the nucleic acid and protein type. An important question, related to reality, is that of abundance or concentration.

It is true that the oceans could have provided a vast supply of 'soup', but it is very unlikely – by the very reason of 'supply' – that life started in the bulk of the oceans. The synthesis of macromolecules requires a certain concentration level of energy-rich building material. To saturate the oceans with monomeric reactants would be quite an uneconomical utilization of the energy resources. Self-organization, furthermore, is a cooperative process involving the interaction among different specified partners and therefore depending on some kind of 'nucleation'. Tidepools, ponds or even puddles, rather than oceans would be predestined to hold any ensemble together and protect it against dilution. Moreover, as the work of Fox[34] and others has shown, they could provide more favourable conditions for macromolecular synthesis, e.g. large temperature fluctuations, surface catalysis etc. Actually, those were the conditions which in the laboratory yielded functional, high molecular weight proteinoid compounds.

5.1. *True Self-Copying*

The selection game described in Section 4 represents the simplest model of a self-organizing Darwinian system. We may attribute to each class 'i' of spheres an individual selective value W_i, but there is no further internal coupling among spheres, and hence W_i is independent of any population variable. Invariance of overall population is effected by external growth control.

Asexual multiplication of micro-organisms in a constant medium – requiring regulation of environmental factors and of overall population density – could be described by such a model.

Do we know of any counter-example at the molecular level?

It is easily possible to quote chemical structures suitable for such a *self-* reproductive mechanism. However, nature did not make use of them, for an obvious reason. The evolutionary potential of macromolecular sequences is based on their capability of encoding information. This requires at least two classes of digits, i.e. two classes of monomeric reactants. It would correspond to a very unique accident, if the natural abundance of these two (or more) digit classes would be such, that both are included in the polymeric sequence with comparable *a priori* probabilities. Otherwise, a quite *uniform* sequence of the most frequent digit would possess the highest selective value and it would outgrow any other non-uniform sequence.

In fact, we may expect quite extreme differences for the primordial appearance of the four nucleotides. In a reducing atmosphere the base adenine (A) could form more easily than its oxygen-containing analogues. As was recently shown[37], it can be obtained even by straight-forward condensation of HCN. In the presence of oxygen-containing compounds, the abundances may shift. As the simplest representative of oxypurines, hypoxanthine, the base constituent of inosine (I), must have occurred. Inosine resembles guanine (G) in its complementary interaction with cytosine (C). It may well be, that only at a later stage of chemical evolution 'I' was substituted by 'G', since the GC pair is more stable than IC. 'I', furthermore, may have been favourable at early stages, since it resembles some kind of a 'joker' in its interactions with other bases.[38] The anticodon region in some transfer-ribonucleic acids still utilizes this capability. There are two reasons for a wide spread of abundance: (*a*) differences in the stabilities of the purines (A, I, G) and pyrimidines (U, T, C) and (b) individual stabilizing effects via nearest neighbour interactions in the polymeric chain. Such a 'stacking' of nucleotides is the main cause of the cooperativity of base pairing. All these influences would strongly bias the appearance of uniform sequences in a linear self-copying mechanism.

5.2. *Complementary Instruction*

The simplest trick to ensure the build up of a large macromolecular information storage capacity is to utilize complementary, rather than self-instructive interactions. The abundant digit then cannot reproduce unless it introduces a corresponding amount of the less frequent complementary digit. In this way the system can develop at least a binary alphabet.

How far could nucleic acids evolve on the basis of complementary interactions?

A model for self-organizing reproduction of nucleic acids utilizing the cross-catalytic effect of complementary instruction has been developed on the basis of the theory described in Section 4. It applies to non-enzymic template instruction as well as to enzyme-catalyzed replication, inasmuch the enzyme can be regarded as an environmental factor rather than an evolving reaction product (cf. the $Q\beta$-experiment described in Fig. 6). The dynamics of selection, e.g. at constant overall population, is described by a system of coupled differential equations.[27] Each species is represented by a pair of equations referring to the two complementary strands. Non-linear terms, as introduced by the constraint of constant overall population, are always negative. The selective value: $W_i = A_i \cdot Q_i - D_i$, applying to a species with linear self-copying, is now replaced by the two eigenvalues of the matrix

$$\begin{pmatrix} -D_{+i} & A_{+i}Q_{+i} \\ A_{-i}Q_{-i} & -D_{-i} \end{pmatrix}$$

One of these eigenvalues – we may call it $W_{\pm i}$ – can be positive, namely if $A_{+i}Q_{+i} \cdot A_{-i}Q_{-i} > D_{+i}D_{-i}$, characterizing the growth mode of the complementary pair. The other eigenvalue is always negative, describing an 'equilibration' mode, i.e. approach of a fixed ratio of population variables of the positive and negative strand. The final outcome again is selection of one species, including both complements and their families of mutants. This species is distinguished by a maximum selective value

$$W_{\pm m} > \bar{E}_{k \neq m}$$

where $\bar{E}_{k \neq m}$ is the residual ensemble average of excess production, the index k referring to all species except m, present at any time interval during the selection procedure. It applies finally to all mutants of m present with a population number > 0.

Two conclusions are of particular importance with respect to our question about the 'origin of biological information'.

(1) Selective value now refers to a term containing the geometric mean of the rate parameters AQ and D. Therefore it is possible to select with equal weight for advantages occurring in either of the two strands. Such a simultaneous optimization of both strands were not possible if arithmetical rather than geometric means of rate or time constants were decisive. Wherever the rate parameters are effected by structural properties (stability, as in t-RNA, or specific recognition by the replicating enzyme, as in $Q\beta$-RNA), the requirement for optimization of the geometric means will bring about a certain symmetry of both strands, as is reconfirmed by experimental facts (cf. structures in Fig. 5).

(2) The information capacity resulting from the purely competitive linear selection procedure is limited to the digit content of one single species (including both strands). Further evolutionary progress could come about only via a successive occurrence of rare mutants of this particular species.

The maximum information capacity which could originate in this way corresponds to the quantity v_{max} (cf. Section 4) and hence depends critically on the average of the single digit quality factors q_i. For the simple model of complementary instruction, q can be easily correlated with measurable interaction parameters such as the free energy ΔG_{ik} or the stability constant $K_{ik} = \exp(-\Delta G_{ik}/RT)$ for complex formation among any digit i and k. Let λ be the number of alternative digits in a sequence and $m_1 \ldots m_\lambda$ the molarities of their monomeric precursors. The quality factor for recognition of a template digit i by its complement j then can be expressed as

$$q_{ij} = \frac{m_j K_{ij}}{\sum\limits_{k=1}^{\lambda} m_k K_{ik}}.$$

If recognition includes cooperative interactions the factors may differ for the various nearest neighbour combinations. Experimental data obtained with oligo-nucleotides of varying lengths – comprising the laboratory work of several years[16-18] – suggest mean q-values around 0.99, depending on the relative AU- and GC-content. These values refer to optimal conditions of enzyme-free recognition. They could not be enhanced appreciably by alterations of the physical environment, which simultaneously would affect the complementary as well as the competitive interactions. Only exclusively specific recognition adapted to the complementary pair could bring about a profound improvement. For instance, the error rate in the presence of $Q\beta$-replicase as compared to enzyme-free recognition, is reduced by several orders of magnitude.

There are three types of mis-copying: substitution, deletion and insertion of digits, all of which have been demonstrated in test-tube experiments.[32] The latter two, i.e. insertion and deletion can also be artificially provoked by certain mutagens. Their consequences usually are more serious, especially if – at a later stage of evolution – translation of encoded massages is involved. They lead to frame shifts and thus may destroy the whole subsequent message. The quality factor Q can formally account for all those effects.

Our conclusion about the linear model of complementary reproduction is as follows: Given realistic values of digit quality factors (q_{ij}), the information capacity that could be accumulated reproducibly in a selected polymeric chain turns out to be much too low to allow for any encoding of an enzyme machinery as would be required for translation and further enhancement of reproduction. A sufficiently large capacity could be achieved only via cooperative reproduction of different chains requiring specific coupling factors, most likely to be materialized by specific protein structures.

5.3. Cyclic Catalysis

The large functional capacity associated with the three-dimensional structures of proteins justifies the question, whether these compounds were not able to organize themselves to self-reproducing functional entities.

There is certainly no property inherent to the protein structure which could guaran-

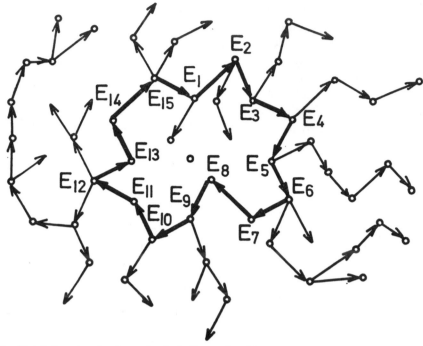

Fig. 7. Catalytic network of proteins including a closed loop.
Any enzyme E_i in this loop is produced from precursors with the help of the enzyme E_{i-1}.

tee anything like self- or complementary instruction. On the other hand certain enzyme complexes – admittedly of highly sophisticated structure – have been found to link up amino-acids into well defined sequences, the longest of which comprise as many as 15 amino-acids residues in a single chain. Other enzymes may weld those oligopeptides together into functional protein structures. Such a – quite specific and unique – ensemble of proteins may become self-reproductive if it forms a catalytic cycle, as schematically represented in Fig. 7.

The rate equations describing a cyclic network may be linearized if the reactants (i.e. amino-acids and oligo-peptides) are present in sufficient excess over the synthetizing enzymes, the representatives of the catalytic cycle. In analogy to the model of complementary instruction the competitive dynamical properties of each individual cycle can be represented by a matrix of the form:

$$
\begin{pmatrix}
-D_1 & 0 & \cdot & \cdot & \cdot & 0 & +F_1 \\
+F_2 & -D_2 & 0 & \cdot & \cdot & & 0 \\
0 & +F_3 & -D_3 & \cdot & \cdot & & \cdot \\
\cdot & \cdot & \cdot & \cdot & \cdot & \cdot & \cdot \\
\cdot & \cdot & \cdot & \cdot & \cdot & \cdot & \cdot \\
\cdot & \cdot & \cdot & \cdot & \cdot & \cdot & 0 \\
0 & \cdot & \cdot & \cdot & 0 & +F_n & -D_n
\end{pmatrix}
$$

(F_i corresponds to $A_i Q_i$ in the second model.)

This matrix has $k-1$ negative and – possibly – one positive eigenvalue, i.e. if $\prod_k F_k > \prod_k R_k$. Again, the positive eigenvalue containing the geometric mean of the catalytic coupling constants characterizes the selective value of a single cycle, which may compete with other cyclic ensembles.

At first glance, such a system promises to be quite interesting with respect to our question, because it seems to be able to collect a considerable amount of information. There are, however, two shortcomings, which seriously would limit any progress of evolutionary optimization.

(1) The cycle can never free itself from parasitic branchings which accumulate and finally link up all competing structures (cf. Fig. 7).

(2) Any error in order to be reproduced has to propagate through the whole cycle. In other words, it has to produce a defined sequence of errors, all having the unique property of leading to their own reproduction. Each step of adaptation to higher catalytic efficiency then would involve the origination of a completely new cycle, which is revolution rather than evolution. Progress in such a system would be extremely slow, since the required coincidence of errors must be considered a very rare event. Evolution requires instructive properties, which are inherent to the system and not the unique outcome of some complex interactions.

5.4. *Hypercycles*

The models considered so far are prototypes of linear reaction systems differing only in their degree of complexity. The requirements of

(1) inherent self- or complementary instruction, and

(2) adaptable catalytic power of practically unrestricted capacity can be fulfilled only by a cooperation among several reaction partners, most likely belonging to different classes of macromolecules, i.e. nucleic acids *and* proteins. Cooperative behaviour is reflected by intrinsically non-linear reaction mechanisms.

A simple example of a non-linear model, involving only one class of macro-molecules: the nucleic acids, is schematically represented in Fig. 8. The primary cyclic ensemble of mutually reproducing complementary chains of limited length may specifically reinforce each other via a cyclic catalytic coupling. Such a cyclic hierarchy of catalytic processes may be called a 'hypercycle'. The particular system shown in Fig.8 is hypothetical. There is no clear experimental evidence yet, that nucleic acids could exert such catalytic influences, as specific enhancement of speed and accuracy of replication or protection against decomposition.

The evolutionary significance of any system depends on its capability to bring about access to the next higher level. For the system shown in Fig. 8 the next higher level would be the utilization of the larger functional capacity of proteins. This requires some intrinsic assignment of nucleotide codons to amino-acids, i.e. the origin of a code and translation. Again, there is no experimental evidence so far, that nucleic acids – without help of proteins – could facilitate such assignments. This does not exclude that there exists some sophisticated – yet unknown – mechanism, but it certainly

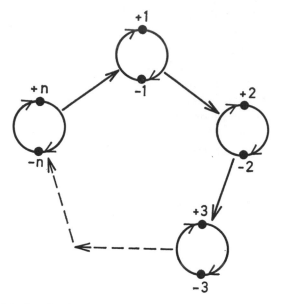

Fig. 8.   A 'hypercycle' of replicative RNA-chains.
  The complementary reproduction cycle of each individual chain ensemble is enhanced by specific interaction with one strand (or both) of the preceding ensemble. The linkage between the different ensembles is cyclically closed.

strongly disfavours the presence of any straightforward amino-acid–codon relationship.

Suppose there would be no intrinsic relation among codon, anticodon or adaptor and amino-acid. Is there any chance of a finite probability that translation could have originated out of a fluctuation of random assignments – mediated by previously uncommitted protein structures? This is a fair question, because we have the evidence that proteins in principle can exert such a function, and that any possible assignment may occur with finite chance in a random set.

Let us assume we want to translate λ digits

$$a, b, c, \dots,$$

with the help of a corresponding number of adaptors

$$A, B, C \dots.$$

There are λ^2 possible assignments, i. e. for any digit with any adaptor. Each assignment may be brought about by a specific catalyst, and each such catalyst may be found with the same *a priori* probability among a random set of proteins. What we have to select for is a set of λ catalysts, which facilitate a unique assignment, where each type of digit is associated with one and only one type of adaptor and vice versa. There are $\lambda!$ such unique sets occurring among the total of $C_\lambda = \binom{\lambda^2 + \lambda - 1}{\lambda}$ sets. C_λ is obtained by combining all θ^2 possible assignments (i.e. 'combinations with repetition') in classes of λ elements.

Now let us assume compartments, large enough to enclose just λ of those catalysts which facilitate some kind of assignment. At the same time the compartments should contain an equivalent amount of random sequences of digits which according to any kind of translation yield functional structures among which we find again λ catalysts for assignment. How often can we expect a compartment, in which

(a) unequivocal translation is initiated by a unique set of catalysts, and

(b) the products of such a translation reinforce the same type of assignment?

It would be one out of $\dfrac{C_\lambda^2}{\lambda!} = \dfrac{([\lambda^2+\lambda-1)!]^2}{[\lambda!]^3 \, [(\lambda^2-1)!]^2}$ compartments (cf. Table III).

TABLE III

λ	2	4	8	20
$\dfrac{C_\lambda^2}{\lambda!}$	~ 50	$\sim 7 \times 10^5$	$\sim 3 \times 10^{15}$	$\sim 10^{50}$

Note that only a coincidence of functions rather than of sequences is required, otherwise these numbers would be incomparably larger. It is this coincidence which originates, out of a transient event, the semantics of genotypic information. How often such a fluctuation really would occur, depends on the 'size' of the hypothetical compartment, i.e. on the chance to find assignment functions among any set of random protein structures. Any bias yielding deviations from uniform *a priori* probabilities will increase these chances.

The numbers, some examples of which are given in Table III, indicate that a random start becomes very unlikely for any λ-value above eight or ten. A number of 10^{50} alternatives as found for $\lambda=20$, of course, could never be accommodated on earth within any realistic span of time. A precursor code may have started with a classification of amino-acids rather than with individual assignments. A distinction of four classes, such as non-polar, polar (but neutral), positively and negatively charged, may correspond to a minimum requirement, as far as the structure of primitive proteins is concerned. Functional needs may increase the number of classes, to include catalytically active groups, e.g. with particular acidic or basic properties. Eight classes would resemble the maximum capacity of a triplet code using binary digits. The table of the genetic code indeed strongly suggests the pre-existence of some type of classification (cf. ref. 27).

The appearance of translation opens a new dimension of phenotypic expression. However, this can be utilized for further evolutionary progress only, if the representative genotypic ensemble organizes itself in such a way that it can be selected as a whole and thereby preserve its information content.

The model of a self-replicative hypercycle involving cooperation among nucleic acids and proteins and capable of selective stabilization of a biosynthetic translation apparatus is represented schematically in Fig. 9. It is distinguished by the following properties (for a detailed treatment cf. ref. 39):

(1) Each cycle as a whole has self-enhancing growth properties. Due to the non-linear amplification terms, singularities in the population variable may occur within finite times. (The types of non-linear rate equations for the I- and E-systems are described in the legend of Fig. 9.)

(2) Under growth control (e.g. constant fluxes or overall reaction forces) indepen-

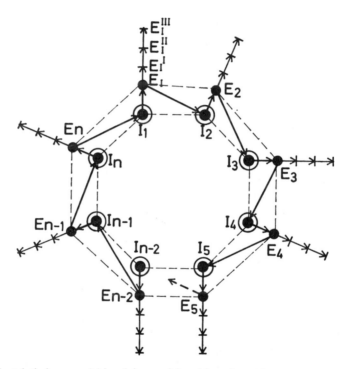

Fig. 9. The 'catalytic hypercycle' involving nucleic acids and proteins.

I_i represents a complementary chain ensemble of RNA-strands. E_i – encoded by the 'positive' strand of I_i – is a protein which provides the functional linkage to a subsequent information carrier I_{i+1}. These linkages must close up, i.e. there is finally a protein E_n which enhances the formation of I_1 Enhancement may include specific replication, repression, or protection against decomposition. Only part of the information of I_i may be utilized for the factor providing the coupling $E_i \rightarrow I_{i+1}$, whereas other parts may include functions of general utility (polymerization, translation, metabolic functions etc.). However, the readout of the total information in I_i must be turned on by one common coupling factor in E_{i-1}.

The hypercycle is described by two sets of non-linear differential equations, one for the information carriers I_i, and the others for the carriers of function E_i. The production terms for I_i contain expressions including the population variables of both I_i and E_{i-1}, whereas the production terms for E_i are determined by the population variables of I_i. The expressions are based on the results of the stochastic theory of template-instructed enzymic polymerization processes. The calculations for various cycles were carried out with the help of a computer programme developed by P. Schuster (cf. ref. 39).

dent cycles will compete for selection. As a consequence of the non-linearity selection is extremely sharp, resembling 'all or none' behaviour.

(3) If one class of the macromolecular components (I or E) is present in large excess the system may assume quasi-linear behaviour. In the non-linear range the steady-state solutions involve undamped oscillations. Normal mode analysis (as carried out by Schuster[39]) following the procedure of Chandrasekhar[40]) reveals stable stationary states as long as the hypercycle does not include more than three constituents of each class. Undamped harmonic oscillations (purely imaginary roots) occur with four, while for more than four constituents of each class unstable solutions are found. Any deviation from the steady state then will amplify, but – due to the growth limitation – finally reverse again. The concentration changes of the different constituents run like waves around the cycle, as shown for an example in Fig. 10.

(4) Elimination of the time variable and numerical integration leads to phase diagrams in a k-dimensional space (for a cycle with k constituents of each class). A projection to a two-dimensional subspace as spanned by the population variables of two subsequent constituents I_i and I_{i+1} yields integral curves which approach a limit cycle for $k>4$. The limit cycle degenerates into a triangle for large values of k, resembling concentration waves of almost rectangular shape. This kind of behaviour is important for the speed and sharpness of selection.

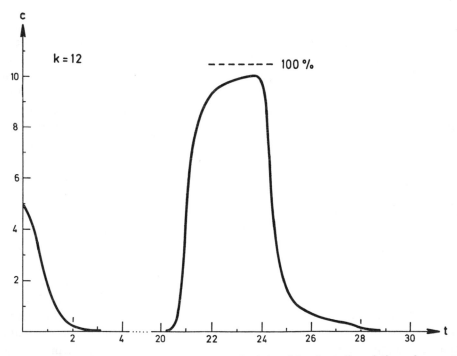

Fig. 10. Example of a solution (concentrations vs. time) describing the cyclic variations of concentrations in a 12-membered hypercycle (i.e. involving $I_1...I_{12}$; $E_1...E_{12}$). The rate constants for all steps were assumed to be equal. Time is given in units of reciprocal rate-constants (cf. ref. 39).

(5) The particular non-linear properties enable the system (*a*) to utilize very small selective advantages occurring to a stochastically significant extent, (*b*) to evolve very quickly, and (*c*) not to tolerate any new competitor, after a given hypercycle has amplified to a finite population level. Hence, even with the random start of arbitrary codon assignments, universality of the code, and also uniform chirality of all polymeric structures, will be guaranteed by the high threshold value for nucleation of new hypercycles. The non-linear competition will furthermore strongly disfavour any 'newcomer', starting at a low population level.

(6) The cyclic coupling provides a sufficiently large information capacity which never could be reached by single chains (cf. v_{max}), even in the presence of (yet weakly adapted) enzymes. Calculations show that the cycle will automatically economize its size to the functional needs which guarantee its existence. Genotypic changes, occurring in a single chain I_i, can be utilized immediately, if they strengthen the interaction with E_{i-1}. They do not have to propagate further changes around the cycle, as was necessary for the straightforward coupling in linear systems.

(7) Parasitic branches (i.e. couplings to other I-cycles) outside the closed loop defining the hypercycle, are not tolerated by the system. Those branches either interfere with the origin of the hypercycle, or the system selects against them. If they appear after the cycle has reached a finite population level, they have no chance of growing as a consequence of the non-linear selection behaviour.

(8) There is only one way to evolve branches. They must be parts of the I_i-chains inside the loop. However, they may contain more information than needed in order to code for the coupling factor E_i connecting I_i with I_{i+1}. This automatically will lead to an operon-structure of the genome, whenever the single I-units are linked together. The chromosomes of presently encountered cells exhibit such an operon structure. They consist of genetic units, usually involving several genes that function cooperatively under the control of inducers or repressors (= protein factors) which interact with the control site at the DNA. The evolution of such branches requires an adaptation of $(1-\bar{q})$ to $<10^{-3}$. All functions of general utility, expressed by polymerization, translation and metabolic enzymes, will finally occur in these branches, while the E_i are the specific coupling factors, such as 'coat proteins' which act via protection, repression–derepression, or any other type of induction.

(9) The evolutionary utilization of selective advantages occurring in operon branches requires finally some kind of compartmentalization. The enclosure into compartments will also protect the population against deleterious mutations occurring in single species. It could have been quite important already at very early stages because of problems of dilution. The nucleation of the code and translation apparatus may actually have occurred in a compartmentalized system. These early compartments should be quite open structures, e.g. complexes like the ribosomes or coacervates as proposed by Oparin.[41]. Microspheres as frequently found in laboratory experiments simulating prebiotic conditions (cf. Fox[34]) may later offer further advantages in protecting a given ensemble.

(10) A compartmentalized system will finally also have to 'individualize' by linking

together all single I-units into one stable chain (possibly a double-stranded DNA). The function to link up 'sticky' ends of nucleic acid chains may have evolved in one of the branches. Those enzymes, i.e. ligases, are wellknown today. The cyclic gene maps (cf. Fig. 11) as well established for the bacterial chromosomes[6] could have evolved in this way. Another consequence is the random arrangement of the different operons. The coupling factor (resulting from gene I_i) may be quite distant from its target gene (I_{i+1}), while proximity is met only for coordinated functions enclosed in a given operon.

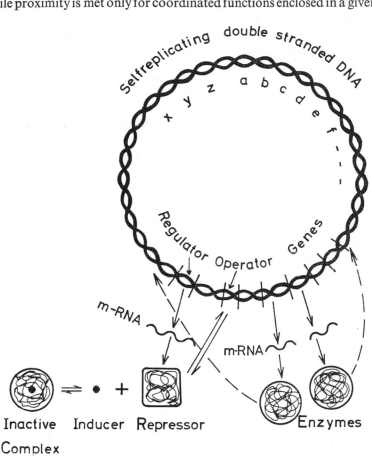

Fig. 11. The final evolution product of a hypercycle resembles the gene and operon structure of the cyclic chromosome of a bacterial cell. The genes derived from the single information carriers I_i, which are linked into a cycle by a ligase (which closes up open ends of polynucleotide chains). In such a cyclic genome the genes I_i and I_{i+1} (coupled via a factor of E_i) need not be neighboured (cf. ref. 6).

The two last mentioned effects are not immediate consequences which could be derived from the mathematical structure of the particular system of non-linear differential equations. However any such system, offered the possibility of compartmentalization and individualization, will make use of it, because it is of advantage for survival. They show that the hypercycle is not a dead end, but rather leads to a new level of evolution, which may be identified with a primitive precursor of a bacterial cell.

6. SOME ANSWERS

This paper started from a number of questions:

Can we define the criteria of life?

Are they based on known physical laws?

Was the phenomenon 'life' determined to come about?

Do we know the (macro-)molecular precursors and the sequence of their appearance?

Although each of these questions is posed in a form that it could be answered by a simple 'yes' or 'no', my answer, in conclusion, will be a little more detailed.

Let us start with the three 'classical' criteria of life: [41]

(1) self-reproduction – to preserve biological information despite a steady destruction.

(2) variability and selection – to enlarge the information content, biassed by certain value criteria.

(3) metabolism = free energy flow – to compensate for the steady entropy production, connected with the fulfilment of the criteria 1 and 2, and thereby to prevent global equilibration.

These criteria are to be met by particular molecular properties as well as by maintenance of certain physical conditions. They are necessary prerequisites of life. They lead to self-organization of material systems. Instabilities steadily narrow down the number of possible alternatives of states, sometimes reaching uniquely defined optima. This narrowing down of the number of alternatives in a given set corresponds to a 'gain of information'.* Although the copy choice is a fundamentally stochastic event, the process of optimization is deterministic within defined fluctuation limits. A value principle can be formulated, which governs the selective optimization procedure. It is based on general stability criteria and in agreement with the principles of non-equilibrium thermodynamics, as detailed in the lecture of Prigogine.[24]

These criteria certainly are *necessary*; but are they also *sufficient* to describe the phenomenon 'life'? If we refer to the term life, we must understand it as the phenomenon we actually encounter nowadays, e.g. in the form of bacteria, plants or animals. As the preceding discussion shows, these criteria do not even suffice to describe the various models adequately. Additional more specific requirements have to be fulfilled to guarantee an open-ended evolution from any given to the next higher level, distinguished by new features. Models, of course, can only reveal the principal potentialities at a given level and uncover their limitations. It should be clearly emphasized, that the above models were not intended to resemble the historical course of evolution (which actually may have passed through many intermediates before anything as sophisticated as a living cell came about). However, they tell us which possible alternatives are to be excluded. Moreover, they specify which properties a system must bring along in order to serve as a true precursor of the next level. On the other hand, evolution need not be

---

* According to a definition given by Renyi.[42]

an entirely monotonic process, but rather may include discontinuities, i.e. global instabilities, to which we then may refer using the term 'origin'. The information on which the hypercycle is built largely 'nucleates' from previously uncommitted structures of proteins and nucleic acids and thus originates at this very level.

Do the known laws of physics provide a complete basis for an understanding of the phenomenon life?

All we can do at the moment is, to disprove the negation of such a claim by quoting counter-examples. In other words we can use physical models to disprove any claim that the known laws of physics are not sufficient to describe the phenomena which specify a living object. The hypercycle is one example.

Such a statement is at least true for the more primitive forms of life, to which our discussion – referring to 'origin' – is restricted. There is no reason, why such a statement should not hold also for any of the subsequent stages of evolution:

(a) Sexual reproduction – characterized by exchange of information among different species providing a faster utilization of advantages. As Jacob[43] emphasized, death becomes an evolutionary prerequisite at this level.

(b) Differentiation and morphogenesis – a self-organization process at the cellular level, but under the control of molecular mechanisms distinguished by similar features[44] as discussed in this paper.

(c) Communication with the environment via nervous systems, again involving several stages ranging from : (i) simple control, (ii) a new kind of self-organization of 'experienced' information,[45] under the guidance of a 'secondary' value system, to finally (iii) an abstract replay of information and selective focussing via valuated responses, which we associate with the term consciousness.

The new basic ideas of physics in the 20th century did not result from a lack of knowledge, but rather from a discrepancy between experimental facts and logical conclusions based on previously accepted concepts. There is still much lack of knowledge about this advanced level of organization of matter, we call 'life', and its novel non-material consequences. However, at this stage it is simply 'lack of knowledge', and not 'discrepancy' with the present concepts of physics.

ACKNOWLEDGEMENTS

I should like to thank Dr Ruthild Winkler for many suggestions in the preparation of the manuscript.

REFERENCES

1. E. Schroedinger, *What is Life?*, Cambridge University Press (1944).
2. F. H. C. Crick, *Of Molecules and Men*, University of Washington Press, Seattle and London (1966).
3. R. P. Feynman, R. B. Leighton, and M. Sands, *The Feynman Lectures on Physics*, Addison Wesley Publ. Comp. Inc., Reading, Mass. (1965).
4. J. C. Kendrew, *Sci. Amer.* **205**, 100 (1961).
5. J. D. Watson and F. H. C. Crick, *Nature* **177**, 964 (1954).
6. J. D. Watson, *Molecular Biology of the Gene*, W. A. Benjamin Inc., New York, 2nd ed. (1970).

7. C. E. Shannon and W. Weaver, *The Mathematical Theory of Communication*, University of Illinois Press, Urbana (1949).

8a. N. Wiener, *Cybernetics or Control and Communication in the Animal and Machine*, Boston (1948).

8b. A. N. Kolmogoroff, *C. R. Acad. Sci.* **208**, 2043 (1939).

9. L. Brillouin, *Science and Information Theory*, Academic Press, New York (1963).

10. M. Eigen and D. Pörschke, *J. Mol. Biol.* **53**, 123 (1970).

11. J. v. Neumann and O. Morgenstern, *Theory of Games and Economic Behaviour*, University Press, Princeton (1963).

12. S. Ulam, in *Essays on Cellular Automata* (ed. A. W. Burks), University of Illinois Press, Urbana (1970).

13. J. H. Conway, in *Mathematical Games* (by M. Gardner), *Sci. Amer.*, Oct. 1970, pp. 120 ff.

14. M. Gardner, *Sci. Amer.*, Feb. 1971, pp. 112 ff.

15. M. Eigen and R. Winkler, *Mannheimer Forum*, 1973 (to be published).

16. D. Pörschke, *Biopolymers* **10**, 1989 (1971).

17. S. K. Podder, *Biochemistry* **10**, 2415 (1971).

18. O. C. Uhlenbeck, J. Baller, and P. Doty, *Nature* **225**, 508 (1970).

19. P. Philippsen, R. Thiebe, W. Wintermayer, and H. G. Zachau, *Biophys. Biochem. Res. Comm.* **33**, 922 (1968).

20. D. R. Mills, F. R. Kramer and S. Spiegelman, *Science* **180**, 916 (1973).

21. W. Min Jou, G. Haegeman, M. Ysebaert, and W. Fiers, *Nature* **237**, 82 (1972).

22. M. Eigen, Nobel Symposium 5 on *Fast Reactions and Primary Processes in Chemical Kinetics*, Almquist and Wiksell, Intersc. Publ., Stockholm (1967).

23. V. I. Zubov, *Methods of A. M. Lyapunov and their Application*, Noordhoff Ltd., Groningen (1964).

24. I. Prigogine, this volume, pp. 561–593.

25. P. Glansdorff and I. Prigogine, *Thermodynamic Theory of Structure, Stability and Fluctuations*, Wiley-Interscience, New York (1971).

26. M. Kac, *Amer. Math. Monthly* **54**, 369 (1947).

27. M. Eigen, *Naturwissenschaften* **58**, 465 (1971).

28. J. L. Doob, *Stochastic Processes*, J. Wiley, New York (1953).

29. A. F. Bartholomay, *Bull. Math. Biophys.* **20**, 97 (1958); **20**, 175 (1958); **21**, 363 (1959).

30. C. H. Waddington, *The Nature of Life*, Unwin Books, London (1961).

31. P. Jordan, *Die Physik und das Geheimnis des organischen Lebens*, F. Vieweg und Sohn (1945).

32. S. Spiegelman, in *The Neurosciences*, 2nd Study Program (ed. F. O. Schmitt), The Rockefeller University Press, New York (1970).

33. S. Miller, *Science* **117**, 528 (1953).

34. S. W. Fox, *Naturwissenschaften* **56**, 1 (1969).

35. C. Ponamperuma, C. Sagan, and R. Mariner, *Nature* **199**, 222 (1969).

36. M. Calvin, *Chemical Evolution*, Clarendon Press, Oxford (1969).

37. L. Orgel, private communication.

38. G. Felsenfeld and H. T. Miles, *Ann. Rev. Biochem.* **36**, 407 (1967).

39. M. Eigen and P. Schuster, to be published.

40. S. Chandrasekhar, *Hydrodynamic and Hydromagnetic Stability*, Clarendon Press, Oxford (1961).

41. A. I. Oparin, *Genesis and Evolutionary Development of Life*, Academic Press, New York (1968).

42. A. Renyi, in *Berkeley Symposium on Mathematical Statistics and Probability* (ed. J. Neyman), University of California Press, Berkeley (1961), p. 547.

43. F. Jacob, *La Logique du Vivant – Une histoire de l'hérédité*, Editions Gallimard (1970).

44. A. Gierer and H. Meinhardt, *Kybernetik* **12**, 30 (1972).

45. H. R. Wilson and J. D. Cowan, *Biophys. J.* **12**, 1 (1972).

Part IV

Physical Description, Epistemology, and Philosophy

31

Classical and Quantum Descriptions

C. F. von Weizsäcker

CONTENTS

1. THE PROBLEM

1.1 INTRODUCTION

The title[1] 'Classical and quantum descriptions' is a technical version of the problem of a semantical interpretation of quantum theory. The question is: what does quantum theory mean? Hence the task is not to improve quantum theory by new additions, but to understand it. Its mathematical structure is not disputed. On the existing interpretation – usually called the Copenhagen interpretation – there have been decades of discussion. They have their origin in the fact that we can only accept this interpretation if we take some rather fundamental philosophical decisions. In this paper I do not try to argue for these decisions except for an attempt to contribute to a clarification of their meaning. I take a particular approach in starting from the idea of a quantum logic. This is in my view not a peculiar 'empirical' logic, but a specification of a general logic of temporal statements, that is of statements on facts and possibilities. Statements on facts are 'classical descriptions', statements on possibilities are embodied in the quantum state vectors.

The problem is described in Section 1. Section 2 gives the general logical background. Section 3 interprets quantum theory as a refined theory of probability. Section 4 draws the consequences concerning the meaning of the concepts of reality, of fact and event in quantum theory. Thus in its main parts the paper contains no more than an analysis of the meaning of the given theory. I think that such an analysis may lead to rather far-reaching consequences both in philosophy and in the further development of physics. The next steps in physics may depend on our overcoming the apparently slight remaining semantical inconsistencies in present-day quantum theory. These problems are mentioned in the concluding remarks.

J. Mehra (ed.), The Physicist's Conception of Nature, 635-667. All Rights Reserved
Copyright © 1973 by D. Reidel Publishing Company, Dordrecht-Holland

1.2   BOHR'S VIEW

Niels Bohr held the view that every event about which we can meaningfully speak in physics, i.e., every actual or possible phenomenon or measurement, must be described in classical terms. If that is correct, and if quantum theory is correct, and if a 'quantum description' means a correct description of an event according to quantum theory, then a quantum description is a classical description. 'Description' according to Bohr means 'classical description'.

Bohr's statement seems to lead to a paradox. Classical terms mean terms definable in classical physics. Bohr himself stressed that classical physics has been replaced by quantum theory for atomic events, and he admitted that the classical description of macroscopic bodies is probably to be explained as an approximation or a limiting case of an application of quantum theory for large quantum numbers. Are we hence not permitted to speak of events in the atom, and is the description of events or of phenomena confined to an essentially approximative language? Bohr himself was convinced that his statement was a necessary condition for understanding that quantum theory does not lead to paradoxes, and that the so-called paradoxes of quantum theory all arise from a lack of understanding of the truth expressed in his statement.

Historically there were two reasons in favour of Bohr's statement that did not depend on its use for solving paradoxes.

First, quantum mechanics and quantum field theory were developed under the guidance of the correspondence principle. The quantities of classical particle mechanics and classical field theory were known, and certain operators in the Hilbert space of abstract quantum theory were given a physical meaning by being identified with those known and measurable quantities. In modern logic it is usual to distinguish between syntax and semantics. A formal system which is established by giving elementary symbols and syntactic rules for their combination is made a branch of logic or of mathematics by giving an interpretation to its formulae; the science formulating this interpretation is called semantics. In physics we can speak of an iteration of this process. The mathematics of Hilbert space presupposes a mathematical semantics of its symbols: the letter ψ means a complex vector, the letter H means a self-adjoined linear operator, etc. Yet for the physicist this is still an uninterpreted 'formalism'. He will have to say that H is the Hamiltonian of a certain physical system, and that its eigenvalues can be measured by an instrument able to measure energies, etc. In this 'physical semantics' classical physics still plays an unchallenged role. It is true that non-classical observables like spin, iso spin, parity, strangeness have been added, but they cannot be measured except indirectly, by directly measuring classical quantities.

Secondly, when Bohr was exposed to the question whether this role of classical physics was more than an historical fact, he turned to epistemological arguments. Physics is about what we can know, or what we can speak about. In order to be able to know an event in nature, we must be able either to experience it directly or to connect it with our experience in an unambiguous manner. Direct experience is always in space and time. Unambiguous conclusions from direct experience to events not directly observed can only be drawn if a strictly causal chain connects them. Space-time-description and causality must go together in experiments. But they go together only in classical physics. Hence classical concepts are needed for all descriptions of phenomena in physics.

Even if we accept the basic epistemology of the second argument, it is neither quite clear why an experiment should need the precise connection of space-time-description and causality nor why this connection should be available only in the historical form of classical physics. The second argument is essentially an attempt to give a structural rather than a historical description of the term 'classical'. Yet this structure seems to need further clarification. This is the aim of the present paper.

1.3 EINSTEIN'S PRINCIPLE

Heisenberg founded matrix mechanics in 1925 on the idea that the theory should only contain observable quantities. Einstein in a talk with Heisenberg called this philosophy nonsense, even if he might have held it himself in his younger days. He said: 'Only the theory itself determines what is observable.' (See Heisenberg, *Der Teil und das Ganze*, p. 92.) Bohr said: 'Only classical quantities are observable.' How to resolve this clash between these apparently mutually contradictory views?

Heisenberg's principle always remains a useful heuristic principle. What we already have observed, is certainly observable and must be included in the theory. For those quantities which are not actually observed we are so far free to assume their existence or their non-existence. Einstein's principle says that only the final theory itself will decide this question. The theory may even change our description of what we had observed before we knew the theory.

Heisenberg's uncertainty principle of 1927 was no longer positivistic. It was rather a consequence of what Heisenberg had learnt from Einstein. It is a consideration of consistency. Quantum mechanics does not offer a description of states in which both position and momentum would be simultaneously defined. Hence in the frame of this theory such states cannot possibly be observed; you cannot observe the non existent. If somebody raises the objection that you may observe such states in, say, a gamma-ray microscope then Heisenberg's uncertainty principle tells him why this is not exactly possible if quantum theory is correct.

Bohr's view is epistemological and a priori. Any observation is to be made in space and time and to be described causally. This seems undeniable. Our present question is how quantum theory accounts for this nature of all possible observations, that is how Einstein's principle satisfies Bohr's condition.

1.4 THE LOGIC OF QUANTUM THEORY

What is the essential difference between classical and quantum theory? We can leave aside non-classical observables for this step in our analysis, assuming that the difference must admit of an expression in the case of the different treatment of the 'same' observable by the two theories. Hence in comparing the theories we will semantically identify the classical quantities with those observables of quantum theory that are understood to correspond to them. Semantical identification means that the same measurement, described in both theories, will yield the same values of the identified observables. We know that this can be assumed without implying a contradiction; all eigenvalues of quantum observables are real numbers, and they can be identified with the real values of classical quantities. The difference of the theories then lies in the predictions on the outcome of measurements. Some classical values are impossible in quantum theory, further there is no state in which the outcome of all possible measurements is predictable, and the probabilities in quantum theory differ from the classical ones. We can subsume all these differences under the statement that the probability predictions differ in the two theories.

This difference turns out to be a fundamental one. The very calculus of probabilities is changed in quantum theory due to the superposition principle. Kolmogorov's axioms of probability begin by defining a Boolean lattice of possible events to which real numbers are assigned as their probabilities. In quantum theory the axioms themselves can be retained, if carefully interpreted, but the mathematical structure of the lattice is changed into a projective geometry whose elements are the subspaces of the Hilbert space. Now the Boolean structure of the lattice in classical probability is founded on the propositional calculus of classical logic; an 'event' is the logical sum of a certain number of elementary events: $A = a \vee b \vee c \dots$. Hence it seems that classical logic breaks down in quantum theory. J. v. Neumann has introduced the term 'quantum logic' for the propositional calculus of the subspaces of the Hilbert space. It even seems that this change in logic is the only independent change when we replace classical by quantum theory. All other changes can be understood as its consequences.

Yet there are grave objections against the idea of a quantum logic. First, a lattice of propositions is not yet a logic. It corresponds to the lowest type in Russell's hierarchy of logical types, it contains no predicates of predicates, no propositions about propositions. P. Mittelstaedt[2] has done some steps towards introducing higher types, but more is left to be done. Secondly, however, it seems inherently paradoxical that quantum theory, being a mathematical theory which presupposes classical or at least intuitionistic logic, should result in a different logic. More specifically, can we expect to change logic under the impact of special experience?

In Section 2 I shall give a detailed answer to this objection. In outline it says that quantum logic reminds us of a gap that has always existed in classical logic, namely its insufficient description of temporal statements, that is of statements on events which happen in time. The propositions of which quantum theory consists as a theory are of the nature of laws or 'timeless' propositions. But the statements which form a non-Boolean lattice are not general laws but singular statements of the structure 'the physical object x has now the property p' or 'the observable L has now the value λ'. We will call them temporal statements or temporal propositions. We will see that they obey different logical laws, dependent on whether they are applied to facts or to possibilities, or, as we can say less precisely, whether applied to the past or to the future. If applied to facts they obey the classical logic, if applied to possibilities they obey a modal logic which under certain additional postulates turns out to be quantum logic.

The logical analysis will serve to interpret Bohr's statement. What Bohr calls a phenomenon or an event is a fact in the terminology of temporal logic. Its description will obey classical logic. What a mathematical physicist of the von Neumann school calls a state is called a list of possibilities in temporal logic. Its description obeys modal logic. Where quantum logic can be approximately replaced by classical logic, quantum physics is approximately replaced by classical physics.

2. CLASSICAL AND TEMPORAL LOGIC[3]

2.1 WHAT IS LOGIC?

Logic has historically developed as the science of correct inference. In the syllogism 'All Greeks are human. All humans are mortal. Hence all Greeks are mortal' the correctness of the inference does not depend on its being drawn about Greeks and human and mortal beings. The general inference form 'All A are B. All B are C. Hence all A are C' will yield a correct syllogism, whatever terms are inserted for the variables A, B, C. The inference will be correct even if the concluding proposition will be wrong. 'All members of party X are honourable men. All honourable men are truth-speaking. Hence all members of party X are truth-speaking' will yield a false conclusion for practically all political parties. Since the inference is correct, we must conclude that at least one of the premises must be wrong (reductio ad absurdum). Besides the universally correct inference forms, logic knows of propositional forms which are universally true, or, as we can say, propositional functions which have the value 'true' for every value of their arguments. 'If all A are B and all B are C, then all A are C' is such a form; the law of contradiction $\overline{a \wedge \bar{a}}$ ('not: a and not a') is another one. Such proposition forms are called analytically true. An analytically true propositional form can be known to be true in advance, or, to say it in Latin, a priori, whatever proposition we insert in it.

How can human beings possess a priori knowledge, even if it seems to be of so trivial a nature? Is the law of contradiction trivial at all? A preliminary answer is this. We are not denoting all facts, things or truths by peculier names, but we speak of them by means of universal concepts that apply to more than one case. A language consisting of a finite number of words will not be able to form as many expressions as there are objects to be named. And a being like man who learns to think by means of language will not be able even to think of the objects otherwise than by subsuming them under universal names. Hence all our utterances (and, probably, thoughts) have two kinds of 'meaning' at the same time: they mean the object to which they refer, and they mean the way in which they mention that object. Frege[4] called the object denoted by a name the 'Bedeutung 'of that name, and the way in which the name mentions the objects the 'Sinn' of the name; a 'name' here may consist of one word or of a group of words. Following Church[5] I shall call the object of a name its denotation and its way of mentioning the object its sense. Frege's standard example of different senses for the same denotation is 'the morning star' and 'the evening star' for the planet Venus; Russell's standard example is 'Sir Walter Scott' and 'the author of Waverley'. Both examples show that the difference of senses is objective in the sense that one may know both senses without knowing the identity of their denotation. Probably a mind would have to be omniscient in order not to need sense as distinct form denotation.

The identity of a denotation of two senses can be an empirical fact; so it is in the two examples just mentioned. It can, however, also follow from an understanding of the senses directly, 'analytically'. This seems to be the case in mathematics: '$3+1$', '$1+3$', 2×2, 'the square of the smallest prime number' are expressions with different senses, all denoting the same number which we also describe by the word 'four' and the symbol '4'. Similarly in logic. We shall now study an example which at the same time will introduce us into temporal logic.

2.2   TEMPORAL STATEMENTS

Russell's theory of logical types takes certain judgements to be elementary, ascribing elementary predicates to elementary objects. For the purpose of his logical analysis it does not matter what those objects and predicates are, provided they obey the rules of classical logic. Examples used by logicians generally take rather complicated physical objects as elementary: 'Venus is a planet', 'Trieste is on the Adriatic shore', 'this rose is red'. The physicist will observe that all these are objects of classical physics. Under the viewpoint of physics they are only approximately enduring objects to which changing predicates can be ascribed. They consist of simpler objects like atoms or elementary particles whose interaction obeys the laws of quantum theory. The physicist, in relativity as well as in quantum theory, will take a statement about an event as his example of an elementary judgement: 'the counter coincides with figure 5 on the scale' 'the electron is at position X'. An event will in general last a while; 'it is raining', 'the moon is shining' are permissible examples.

We are quoting these examples in the form in which they appear in everyday life. From the point of view of logic they do not yet seem to be full propositions. 'It is raining' may have been true yesterday but false today. The same phrase refers to two different events in the two cases: to the factual rain of yesterday, and to the rain that might have fallen today. Even if it is actually raining today, today's rain is a different event from yesterday's rain. Temporal statements are given in conceptual form. The phrase 'it is raining' uses an empirical concept 'rain' or 'raining' under which meteorological events at many times and in many places may fall. Logically 'rain' is a predicate of events, and it defines a class of events, precisely of those events which we describe by saying 'it is raining'. Grammatically, our examples were formulated in the present tense. This grammatical form can carry one of two different meanings (besides others that are not relevant for our purpose). 'It is raining', 'I see the rain' may express an event that happens just now (in general now and here). This is the actual present tense; it expresses the mode of time called the present. On the other hand the same phrases can be used in order to quote the statements; they are quoted in the form in which they would be made at any other time, in the past or in the future, at any place however distant from here, by any English-speaking person, if he is speaking about what is present to him. This is a way of mentioning the statements, as distinct from asserting them, and the present tense is commonly used for it, too. This case I shall grammatically call the neutral present tense.

In logic, generality is usually expressed by variables. In our initial example of a generally correct inference, the variables A, B, C indicated that any term could be inserted in their places. We called the expression containing the free variables an inference form. Similarly, if f is a fixed predicate that can be attributed to a class of objects, the letter t may denote the free space in a proposition where the name of any member of the class can be inserted. Then $f(t)$ is a propositional form which yields a proposition if we replace t by any permitted object-name. $f(t)$ is also called a propositional function where the 'variable object' t serves as an argument, the fixed predicate defines the function, and the values of the function are propositions on the objects. In our example f may stand for 'it is raining', and t would represent the situation (place and time). In logic generally a propositional function, a predicate, a concept, and a class mean different aspects of the same structure. In temporal logic, a temporal statement in neutral present tense expresses a propositional function whose possible arguments are situations and whose values are events (logically: names of situations and statements of events).

Our next question is how to name the situations, especially the times. A time can be expressed in two different ways: relative to the present, or objectively by calendar and clock. In describing phenomena, that means in the descriptions which are the subject matter of this paper, expression relative to the present (tense expression) is the natural way. 'It is raining now, look at the rain!' 'It was raining yesterday, my coat is still wet.' 'It may be raining tomorrow, take your umbrella with you for the trip!' Physics in the making refers to experience in the same fashion. 'Anti-protons have actually been observed.' 'So far nobody has seen a quark, but evidence may be forthcoming in the future.' Experience means to learn from the past for the future. Theories in physics rely on past experience and are tested by their ability to predict future experience (from which we will have learnt when it will have become past experience). In Subsection 2.5 we will see that statistical thermodynamics presupposes an explicit use of tenses. Hence we will mainly study the tense expression of time.

We further distinguish between facts and possibilities. 'Napoleon died in St Helena' is a fact. 'Mr Fritz Berg may die in St Helena' is a possibility. Again, we distinguish two kinds of possibility, formal and actual. A formal possibility is whatever may be conceptually described as a possible event. 'To die in St Helena' is a formally possible event for a human being. Temporal statements in neutral present tense express formal possibilities. In physics a theory must express what formally possible events can happen to its objects. The mechanics of point-masses says that a point-mass has a state defined by its position and its momentum; hence, e.g., a particular position x is a formally possible predicate of a point-mass. An actual possibility, on the other hand, is what might actually happen, given a situation. When Napoleon had arrived in St Helena, it turned out to be actually possible that he would die there; at some time in in his last illness it had become actually impossible that he would die in Paris. Deterministic physics considers only one future state for a given object at any time as actually possible. Quantum theory describes all one-dimensional subspaces of the Hilbert space of a given object as formally possible states, and all the eigenstates of the observable which is going to be measured now as actually possible results of the measurement, provided they are not orthogonal on the actual state. When I simply speak of possibilities, I shall generally mean actual possibilities. In the direct sense of the words, a fact is a present or past event, a possibility is a future event, and both must be formal possibilities. In an oblique sense we also speak of future facts and of possibilities of the past. A future fact is a possible event, considered in the way it will appear after it will have happened. We may think it to be not only possible but even necessary; yet we cannot be really sure as long as it has not actually happened, that is as long as it is not a fact. A past possibility similarly is what was possible when it had not yet happened. In another sense we speak about unknown facts of the past as possible facts; in science this subjective possibility is usually meant to imply that at least in principle the unknown might become known in the future.

2.3   FORMALLY POSSIBLE STATEMENTS ON FACTS

When we speak of a fact of the past we mean to say that an event has happened in the past, and its having happened is now a fact. It is now a fact that the first word of this sentence (the word 'it') has been written down, but when I state this fact the event of writing the word is already in the past. Whenever I speak it is true to say that 'I am speaking now'. Hence whenever I speak sincerely about a fact known to me I speak about a present fact which consists in an event having happened in the past or perhaps happening just now. We now leave aside the case of present events. In order not to forget that we speak about past events we shall quote the statements about them in the past tense, even if they are meant not to refer to a particular time and event but to express formally possible facts.

We assume that a statement about a particular fact is either true or false, even if we do not know it. This is what we mean by a fact. Since we are concerned about the operational meaning of terms we assume that a fact is an event about which it is at least in principle possible to decide whether it has happened or not. Thus the logic of statements on facts is two-valued. We can construct a Boolean lattice of formally possible facts referring to the same past situation. Let x_λ ($\lambda = 1, ..., l$) denote l different statements of formally possible facts. They are taken to be independently possible. We call $\bar{x}_\lambda$ the statement that x_λ is false. In two-valued logic $\bar{x}_\lambda$ is taken to be equivalent to x_λ. $\bar{x}_\lambda$ is a truth-function of x_λ.

The introduction of the concept of a truth-function deserves some care. Two-valued logic ascribes one of two formally possible truth-values 'true' and 'false' to any proposition. A truth-function $f(x_1, x_2, ..., x_l)$ is usually defined as a propositional function of the variables x_λ such that the truth-value of that proposition which is the value of the function for given values of the arguments depends exclusively on the truth-values of the propositions which are the values of the arguments. Frege who introduced this idea was justified in so doing by his theory that the denotation of a proposition is its truth-value. In Frege's sense a truth-function is originally a function whose arguments and function-values are not the senses but the denotations of propositions. I prefer to say that a temporal proposition denotes a formally possible fact, and that different propositions denote different facts. This denotation is still to be distinguished from the sense of the proposition. Two propositions of different sense can denote the same fact, like 'the Russians have landed a spaceship on the morning-star' and 'the Russians have landed a spaceship on the evening-star'; we shall see better examples presently. Now what is the formally possible fact denoted by a proposition which is defined as a truth-function, say by a negation? Let x denote 'it was raining', which will come true or false depending on the situation. $\bar{x}$ denotes 'it is false that it was raining', or generally: $\bar{x}$ denotes 'x is false'. This is, literally taken, not the form of a statement on an event, but the form of a statement on a statement. There is the corresponding function that may be called assertion: $\underline{x}$, denoting 'x is true'. $\bar{x}$ and $\underline{x}$ may be called reflections on x. Frege[6] says, however, that 'x' and 'x is true' are asserting the same sense (Gedanke) and have the same denotation. This is essentially the assumption of two-valued logic. We follow him in this point for the present subsection.

If we formally consider all two-valued functions of l two-valued arguments we find $n=2^{2^l}$ different functions; for $l=1$, $n=4$, for $l=2$, $n=16$, as is well-known from propositional calculus. For $l=1$ we get the four functions of Table I, where t and f denote 'true' and 'false'. T is the function which is always true and is called Truth

TABLE I

x	$\underline{x}$	$\bar{x}$	T	F
t	t	f	t	f
f	f	t	t	f

(Frege: 'das Wahre'); F is always false and is called Falsehood ('das Falsche'). T and F exist for any l. They are the only truth-functions which assert nothing about the facts under consideration; there is no particular fact asserted by them. We shall denote the n truth-functions of l variables x_λ by the symbols Y_ν ($\nu=1,...,n$). They form a Boolean lattice. Its atoms (lowest elements) are $m=2^l$ functions $X_\mu=Z_1...\wedge Z_l(\mu=1,...,m)$, where $x\wedge y$ means the conjunction or the 'logical product' of x and y, i.e., the truth-function which is true if and only if x and y are both true, and where z_λ is a variable for which x_λ or $\bar{x}_\lambda$ can be inserted; there are m different combinations of this sort. T is the one-element, F the zero-element of the lattice.

Propositional logic makes explicit use of the truth-functions of two variables, especially those given in Table II. The Y_ν can be built up from the X_μ in the form $Y_\nu=X_{\mu_2}\cdot\vee$ $\vee X_{\mu_2}\vee...$. The number of different summands X_{μ_i} is arbitrary but $\leq m$; F consists of zero, T of m different summands. The expression of the truth-functions $Y_\nu(x_1,...,x_l)$ by means of the classical propositional functions $\bar{x}$, $x\wedge y$, $x\vee y$,..., is redundant. The same truth-function is described by infinitely many expressions using these proposi-

TABLE II

x y	$x \wedge y$	$x \vee y$	$x \rightarrow y$	$x \leftrightarrow y$
t t	t	t	t	t
t f	f	t	f	f
f t	f	t	t	f
f f	f	f	t	t

tional functions. Two such expressions A and B are called equivalent if they denote the same truth-function; then the expression $A \leftrightarrow B$ is equivalent to T. We can call a truth-function $Y_v(x_1, ..., x_l)$ a concept or a predicate which is extensionally defined. We can further call any expression for Y_v in the classical propositional functions an intensional definition of that concept; we can call its way of composition from these functions its sense, and the concept itself its denotation. With these definitions every propositional function which is always true, like e.g. $x \wedge \bar{x}$ (non-contradiction) or $x \vee \bar{x}$ (tertium non datur), denotes Truth, every one that is always false, like e.g. $x \wedge \bar{x}$ (contradiction), denotes Falsehood. Thus the two functions which do not denote any particular fact denote analytical truth and analytical falsehood, and precisely for these cases Frege's view that the denotation of a proposition is its truth-value seems natural.

However, T and F can also be taken to be statements on facts of a higher level of reflection. T then means that there is a situation in which $x_1, ..., x_l$ are meaningful formally possible statements, and F means that there is no such situation. The situation existing if T is true can also be described by saying that there is a complete m-fold alternative consisting of the m formally possible statements X_μ which are mutually exclusive (alternative) and one of which must be true (complete).

2.4 Mathematical Logic

The intention of this paper is not to discuss mathematical logic, but only to indicate its connection with temporal logic. The connection is two-fold, and in a way circular.

As described in Russell's theory of logical types the propositions of mathematical logic do not belong to the lowest type of propositions about facts but to the higher types of propositions about propositions. They formulate laws about propositions. In classical logicism number is defined as a class of equivalent classes; thus any class of numbers about which elementary arithmetic may prove a theorem is a class of classes. Here the lowest class may be a class of anything distinguishable, for instance of events. We have not yet treated possibilities, and hence we have not yet formulated laws about events that would differ from classical logic. But we can already guess that it will not be contradictory to formulate non-classical laws of events while the classical laws of mathematics and of mathematical logic hold for the statements on those laws. A law about events will admit of a formulation as a propositional function of statements on events which is always true – either analytically true or empirically true as a law of nature which excludes certain formally possible events. Let f be a propositional function on facts, and let Tf be the statement that this function is always true in one of the two senses just mentioned. Then we can say that Tf expresses a formally possible fact of the second logical type. If further we can be sure that Tf must necessarily be either true or false, we can apply classical two-valued logic to it, even if f should happen to depend on temporal propositions which are not treated according to two-valued logic. This settles the a priori objection against the possibility of quantum logic.

But there is a difficulty for classical mathematical logic itself. How can we be certain that a statement like Tf must necessarily be true or false? If f is to be universally true on empirical grounds we run into the problems of the foundation of empirical science (see Subsection 5.1). If f is to be analytically true there are the well-known questions that have arisen out of the logical paradoxes. In our presentation we kept the numbers l, m, n finite. Then there is no difficulty. But for infinite alternatives modern mathematical logic requires proofs of consistency which must themselves be carried out constructively. In constructive mathematics concepts of possibility cannot be eliminated. 'It is possible to prove a theorem A' is not logically equivalent to the statement of fact 'there is a proof for A'. If temporal logic is the logic of possibility and fact, the presuppositions of modern mathematical logic would have to be described by temporal logic.

I cannot discuss this problem here, and I conclude the subsection by a methodological remark. It is not a vicious circle to assume that mathematical logic should at the same time apply to the laws of temporal logic and be subject to those laws. Self-application is inherent in the idea of a logic. Logic is a science about concepts, judgements and inferences. Being a science, it consists itself of concepts, judgements and inferences. Hence, if it is a true science it must obey its own laws. Our distinction of mathematical and temporal logic would finally disappear again if constructionism were right in thinking that temporal logic applies to the higher Russell types. This, by the way, would not imply an identity of intuitionistic logic and quantum logic, since their special problems are different. The problem of intuitionism is infinity, the problem of quantum theory is complementarity. These two problems are different aspects of the central problem of possibility.

2.5 TEMPORAL LOGIC AND THERMODYNAMICS

The first indication that fact and possibility must be distinguished in physics comes from statistical thermodynamics.[7] Mechanics is reversible, and statistical mechanics must be so, too, unless it introduces irreversibility explicitly or implicitly. From Boltzmann's H-theorem it follows correctly that in a state s with non-maximal entropy an increase of entropy during the further development of the system is overwhelmingly probable. This means that a physicist observing s at a time t_0 will predict with confidence that the original entropy, say H_0, will increase in the course of time towards values $H > H_0$. We predict: $H > H_0$ for $t > t_0$. But mathematically there is no difference between positive and negative time-intervals. The same calculation would lead to the 'retrodiction': $H > H_0$ for $t < t_0$. This is nearly always empirically wrong. We know the second law by experience, that means only for times now past, while the present argument would force us to apply it with the wrong sign of $\dot{H}$ to the whole time which is now past, and with the correct sign only to the future which we have not yet seen.

The answer to the apparent paradox is that past events are facts, and that the concept of probability applies only to possibilities and not to facts, hence only to the future and not to the past. Probability is a quantitative version of possibility. Future events are possible, past events are facts. Every past event was once a future event. At that time it was only a possibility, and a probability could meaningfully be assigned to it. Thus an observer at any past time would have predicted an increase of entropy as highly probable for what was future to him. Hence, if the concept of probability is empirically justified at all, the second law must have held in the past and can be expected to hold in the future.

A consideration of consistency must be added. We do not ask how probable is a known fact once it is known. Indeed, a particular fact is always one of a large class of formally possible facts and would have to be assigned a very small a priori probability, but knowing it has happened we are not asking that way. Yet, why do we know facts of the past and not of the future? In a phenomenological description of our own experience (called 'subjective' because it is the experience of real human beings whom we have agreed to call 'subjects') we can only say that the past has actually happened and we know part of it; this direct knowledge we call memory. The future has not yet happened and cannot therefore be remembered. These statements are as simple as any descriptions of experience, and if physics is to be an empirical science we may well rest satisfied with them. However, we wish to apply physics to those parts of the history of nature in which we have not lived, and perhaps even to explain the behaviour of organisms including human beings, from the laws of physics, including the second law. Does anything exist in the 'objective' world that would correspond to the remembrance of things past and to the lack of a 'memory of the future'? It exists indeed. It is the fact that there are objective documents and traces of the past (fossils, minerals, light coming from distant stars which we describe by retarded potentials) but none of the future (no advanced potentials, no fossils of future organisms). If we ask why this should be so, the only tenable answer is that it is because time has objectively been streaming and ergo the difference between fact and possibility has prevailed all through the history of nature. Retarded potentials mean relativistic causality, i.e., they mean that the future and not the past is influenced by the present. [8] Documents are improbable states of matter. If the second law holds, they must be produced by even less probable events while they will in the end produce more probable ones. That means they contain much information about the past and little information about the future.

This is nothing but an argument of semantical consistency. We already, though inexplicitly, understand time when we have any experience. This understanding is explicitly used in interpreting the formalism of statistical thermodynamics. Applying thermodynamics to our everyday experience we partly explain those traits of experience which were used for its interpretation. But there is no full symmetry or circularity, there is just consistency. The laws of physics presuppose an understanding of time but they cannot fully explain time. Grünbaum has rightly pointed out that the laws of physics contain no 'objective quality' by which the present state might be distinguished from states at a time which is not now. This is what we must expect. The laws of nature describe nothing but formal possibilities since they are expressed by conceptual language. They are made in such a way as to apply to all times.

This analysis does not intend to make dogmatic a priori statements on time. The analysis is strictly conditional: If we accept the phenomenal structure of time we can explain the second law; if I see it rightly no other explanation has withstood the criticism of being viciously circular. But both the structure of time and the second law might turn out to be only preliminary or regional aspects of something more profound.

2.6 Statements on Possibilities

We avoid the ascription of the traditional truth-values to statements on future events, like 'it will rain tomorrow'. The operational justification for a two-valued logic would be very weak in their case. In everyday life a man may predict rain and rain may come at the predicted time, and still we might not consider this a strong test for his prediction coming from true understanding. In order to impress us he ought to be right many times. What is tested then is not the particular statement on one future event but a law, perhaps a law of probability. Hence we prefer to say: a formally possible future event can be necessary, impossible or contingent (i.e., neither necessary nor impossible). We write Nx, Ix and Cx for the three cases. No full empirical decision process exists for these three modalities. If we observe the event we falsify Nx when $\bar{x}$ is found, and Ix when x is found. We cannot verify Nx by finding x, since Cx may as well have been the case. Analogously we cannot verify Ix by finding $\bar{x}$. And we cannot test Cx by any single observation. Empirical testing of futuric modalities is generally done by replacing them by probabilities. Thus we use the futuric modalities only as conceptual material for the theory of probabilities.

Theoretically futuric modalities are deduced as conditional modalities: 'if x, then y will be necessary, impossible, contingent'. We write this: xNy, xIy, xCy. These modalities will transform into conditional probabilities. Conditional modalities or probabilities can also be applied to the past. This is what we did in deducing the second law for the past. They seem to be the natural expression for counter-to-fact-conditionals.

These futuric modalities do not permit some of the operations that are usual in most forms of modal logic. In general modal logic, besides the modalities like the 'necessary', etc., a value 'true' or 'real' and similarly 'false' or 'unreal' can be ascribed to a proposition, and there are implications like: 'what is necessary, is real' and 'what is real, is possible'. We strictly avoid this contamination. Facts are real, statements on facts are true or false, possibilities are qualified by the modalities but they are not real (they are 'real possibilities').

It is also a question whether the modalities can be iterated. We think that Nx, Ix, Cx can be true or false. But we also think that they can express formally possible facts that may arise in the future, and then an iteration may be meaningful: 'it is impossible that there will be weather tomorrow that would make rain necessary for the day after tomorrow'. This is logically connected with another iteration to be discussed in Section 3. It is meaningful to ascribe a probability to a probability; it refers to ensembles of ensembles. If the ensemble to be tested consists of future events, an ensemble of ensembles consists of more distant future events.

One remark is important. In quantum theory the modalities turn out to be doubly conditional: 'If the situation x prevails, *and* if we look for y, then y will be necessarily found.' This is very deeply connected with the quantum-theoretical concept of an event and will be discussed there (Subsection 3.2).

3. QUANTUM THEORY OF PROBABILITY

3.1 CLASSICAL THEORY OF PROBABILITY

The mathematical foundations of the classical theory of probability are sufficiently clear, e.g., in the form of Kolmogorov's axioms. A real number $0 \leqslant p \leqslant 1$ is attributed to any element of a Boolean lattice whose elements are called events. The semantical problem is what we mean by an event and by its probability in physical reality. In this paper I shall use in principle the traditional 'objective' interpretation connecting probability with relative frequency. It fits easily into temporal logic. Only one point deserves consideration. A careful introduction of the concept of probability will have to proceed in steps or levels of reflection which correspond to Russell's hierarchy of logical types. This is of no practical relevance in classical probability but it is meaningful in the quantum theory of probability where it gives an interpretation to the idea of second or multiple quantization (Subsection 3.3).

We take a probability to be a prediction. The 'event' to which it refers is a formally possible event. In the direct meaning of 'probability' no probability can be assigned to a fact. The 'direct' probability of an event is the probability that the event will happen. A fact has happened. Analogously to possibility there are two oblique senses in which a probability can be ascribed to a fact: its probability before it happened, and the probability for an unknown fact of turning out to be a real fact. These oblique senses we leave aside here.

Is a probability to be assigned to a single event or to an ensemble? When we assign it to a formally possible event we assign it to a concept or a class. The possible event of finding 5 spots on the upper side of this die when I will cast it now is a member of the class of formally possible events of finding 5 spots on a die whenever it may be cast. To this class we assign a relative frequency $f = 1/6$, and this is what I mean when I say that the particular event of which I have been speaking has the probability $p = 1/6$. I cannot measure a probability in observing a single event, but I can measure the relative frequency of a formally possible event in a large ensemble. Thus I interpret a probability as a prediction of a relative frequency.

Yet there is a problem left. Taking a particular empirical ensemble we will mostly find empirically that the relative number of occurrences of 5 spots is not 1/6. We exclude the reason that the die may not have been symmetric; to this problem we shall return very briefly at the end. Even for a 'good' die the theory of probability predicts a fluctuation of the actual values of f around the expectation value $\bar{f}$ which we take to be $1/6$. It seems that the probability $p = 1/6$ was not meant to be the prediction of f but of its expectation value. But it is only f that can be actually measured in an ensemble. Only if we do not strictly predict f but if we assign a probability to every formally possible value of f can we define $\bar{f}$. But now we have iterated the introduction of the concept of a prediction of a relative frequency. The probability of f is the prediction of the relative frequency of a relative frequency, referring to an ensemble of ensembles. This same iteration is inherent in the definition that a probability is the expectation value of a relative frequency.

The iteration cannot be avoided. It leads to an infinite regress in definition which can practically be ended by identifying probabilities very close to 1 and 0 with certainty, or, if we use a modal logic of possibilities, with necessity and impossibility. Due to the mathematical law of large numbers all the relevant higher probabilities tend towards 1 or 0. For practical calculations the hierarchy of ensembles can be replaced by one large ensemble. But in principle the infinite regress belongs to the meaning of objective probability. Logically it is closely akin to the distinction between 'x' and 'x is true' (Subsection 2.3) which also formally leads to an infinite regress. Two-valued logic avoids the regress by taking the two statements as equivalent. This is no longer possible if we introduce modalities. 'x' is not equivalent to 'x is possible'; we may doubt whether we should consider 'x' and 'x is necessary' as equivalent. 'x', 'x has the probability $p(x)$', '$p(x)$ has the probability $p'(p(x))$' ... are clearly different statements, each of which has a definite and different meaning. In the empirical determination of probabilities according to Bayes it is usual to speak of the probability of a certain probability distribution. Hume's problem (see Subsection 5.1) may make us understand why we cannot expect more.

By ascribing probabilities to classes of events only, we introduce conditional probabilities from the outset. A class of events will only have a well-defined expectation value of relative frequencies if the conditions are well-defined under which the events belonging to the class are supposed to happen. Two different observers with different knowledge may subsume the same particular event under two different classes of events, having different probabilities. Both may be able empirically to justify their probability assumptions, since each one will test 'his' probability under the conditions defined by his knowledge. Let e.g. observer A predict the relative frequency of an event E which is defined by a total of 12 spots appearing in two successive casts with a die. He will assign it the expectation value $P_A(E) = 1/36$. Another observer B will be permitted to make his prediction after having seen the result of the first cast. His prediction P_B will depend on what he has seen. If the first cast yielded a 6, he will predict $P_B(E) = 1/6$; if the first cast gave a figure different from 6, he will predict $P_B(E) = 0$. Both observers will be able to test their predictions in large samples of cases selected according to their respective knowledge. This is the classical analogue to the 'reduction of wave packets'.

The determination of the conditional probability for a class of events can be done theoretically, e.g. by symmetry-considerations according to Laplace, or empirically according to Bayes. There is no contradiction between the two methods, since the Bayes method must start with an assumption on 'a priori-probabilities'. Thus the theoretical assumptions on probabilities can be empirically tested. Such an empirical test of a probability always contains an unknown error. Like all other empirical quantities a probability can only be assigned an empirical value with some degree of probability. The infinite regress in definition follows from the fact that objective probability is meant to be an empirical quantity.

3.2 THE PROPOSITION LATTICE OF QUANTUM LOGIC

Axiomatic quantum theory aims at building up abstract quantum theory from axioms that carry some a priori likelihood. It will strive not just to assume but to deduce the use of a Hilbert space. The most sucessful attempts in this direction have been made by a direct approach to the lattice of formally possible events (or of propositions on these events). Jauch, Piron and others have done most in this field.[9] I shall rather briefly review here another attempt by Drieschner.[10] He uses eight axioms and some verbal comment which I have tried, in addition to the axioms to condense into seven postulates.

A. *Postulate of alternatives*

Physics formulates probability predictions on the outcome of future decisions of empirically decidable alternatives. The list of the $x_\mu (\mu = 1, ..., m)$ in Subsection 2.3 may serve as an example of an alternative. The salient point here is the concept of empirical decidability. It results in the two-valued logic for the permitted answers. Yet the axiom of indeterminism (axiom 5, below) will state that, given any decidable alternative, there are formally possible events in the same situation (states of the same object) which are not contained in the Boolean lattice of events erected on the alternative. They can only become actual events if an alternative different from the one under consideration is decided. It is characteristic of quantum theory that the question, which formally possible events become actually possible depends on the alternative imposed upon the situation by the observer. This is perhaps the most profound riddle of quantum theory: since there are so many formally possible events (Hilbert vectors) in a situation, why should the events that become simultaneously actually possible be always a very small choice from them (an orthonormal set), defined as the answers to a decidable alternative? From this point of view the present postulate is very far from formulating a triviality. We may venture to say that it expresses a consequence of the possibility meaningfully to use the concept of a fact. Quantum states 'coexist' in a manner described by the superposition principle. Facts do not coexist in this sense. Two facts are either identical or distinct, and it can be decided in principle which of the two they are. 'Coexistence' is a property of possibilities. If possibilities are possibilities of facts, then there must be decidable alternatives. What Bohr calls an unambiguous description is the description of a fact; it must obey classical logic.

B. *Postulate of objects*

The answers of an alternative ascribe formally possible properties to an object. The concept of an identical physical 'object' (or 'system') lasting through time is commonly used in expressing quantum theory. It is no less a 'classical' concept than the concept of a decidable alternative, even more so. If there were no lasting objects there might be no lasting facts and no fixed meanings of words (probably no words at all). But it is another question to what degree or in what approximation nature actually offers these prerequisites of classical thought. Given an object, quantum theory offers decidable alternatives, represented by linear self-adjoint operators in the Hilbert space of the state-vectors of that object; complementarity then means that not all these alternatives can be decided together. But the quantum mechanical composition law (see postulate E) shows that a formally possible object does not in general actually exist. And elementary particle physics shows that all objects can be transformed into other objects. There seems to be no ultimate object; perhaps, at the utmost, there are ultimate single alternatives from which objects can be composed. (See Subsection 5.3.)

C. *Postulate of atomic propositions*

To every object there are atomic propositions, and ultimate alternatives whose answers are atomic propositions. A proposition expressing a formally possible event is called atomic according to lattice theory if it is not implied by a proposition different from itself (except by Falsehood Γ). This postulate of the atomicity of the lattice of quantum propositions will follow from a consistent application of the postulate of finitism to ensembles.

D. *Postulate of finitism*

The number of the answers to any decidable alternative of a given object is not greater than a given positive integer m, characteristic of the object. This postulate is not fulfilled in ordinary quantum theory. It implies a finite number of dimensions for the Hilbert space of any object. Yet we take it to follow from the meaning of the concepts of fact and decision. Only a finite number of decisions can have been made at any time by any observer or by any measuring instrument. Of course we may assume the number m to increase with time. This we will consider in Subsection 5.3. Reasons may be given for assuming that such an increase in m means a change in the nature of the objects, and that the postulate as formulated here is appropriate for an object of constant nature. In the present section we can think of m being very large. We then get a simplification of mathematical procedure which will not essentially change the unrelativistic quantum theory which we build up here. On how finitism is to be applied to be continuum of Hilbert vectors see Subsection 3.3.

E. *Postulate of composition*

Any two objects define a composite object whose parts they are. The direct product of any two ultimate alternatives of the two parts is an ultimate alternative of the composite object. In classical logic this may seem to be a triviality. We know than in quantum theory the space of possible states of the composite object is vastly larger that the set of those states within it in which we can say that the two parts actually exist. These are the product states; only for them will the 'analytic' propositions T_1 and T_2 of the two parts be actually true (see Subsection 2.3). In full rigour this cannot be the case at all if the two parts are of identical nature and only a symmetrized part of the ultimate alternatives of the composite object is used; on the other hand it is always sufficiently true if the two objects are clearly separated in space and/or in time.

F.   *Postulate of the probability function*

Between any two states of the same object a probability function $p(a, b)$ is defined, giving the probability of finding b when a is necessary. 'a is necessary' means 'a will be found if looked for'. We assume that two states are not well-defined if there is not a conditional probability connecting them. If this probability depends on the state of the environment, then the definition of the two states a and b also depends on it.

G.   *Postulate of objectivity*

If a certain formally possible object exists actually, there is at any time an atomic proposition about it which is necessary. This is a strong postulate. It says that if there is an object there is always an ultimate fact about it. The object exists if the proposition T of its lattice is true (or 'necessary'). The postulate says that T can only be necessary if implied by a necessary atomic proposition of its lattice. For a detailed discussion see the papers quoted in note 10. I expect a full elucidation only from a theory of ultimate alternatives (Subsection 5.3).

We can now erect the lattice of formally possible propositions on an object X. It will contain the Boolean sublattices erected on the atomic propositions belonging to its decidable alternatives. But as a whole it will not be Boolean since axiom 5 will postulate that all its atomic propositions are not simultaneously decidable. The lattice consists of a choice from all the subsets of the set of all its atomic propositions. Such a subset A is supposed to belong to the lattice, i.e., to correspond to a decidable proposition, if it can be built up from atomic propositions by the two elementary proposition functions $\bar{A}$ (negation) and $A \wedge B$ (conjunction) from which the logical sum is explicitly defined by de Morgan's law $A \vee B = \overline{\bar{A} \wedge \bar{B}}$. $\bar{A}$ consists of all atomic propositions that are mutually exclusive with all propositions of A. Two propositions a and b are called mutually exclusive if $p(a, b) = p(b, a) = 0$. $A \wedge B$ is the intersection of A and B. The proposition A can be read 'the necessary atomic proposition lies in the subset A'; A's being necessary, contingent or impossible will depend on which atomic proposition happens to be necessary. The structure of the lattice is defined by the axioms.

1.   *Axiom of equivalence*
If a and b are atomic propositions, then $p(a, b) = 1$ is equivalent with $a = b$.

2. Axiom of finite alternatives
If m mutually exclusive atomic propositions ai $(i=1, ..., m)$ are given, for any atomic proposition b:

$$\sum_{i=1}^{m} p(b, a_i) = 1.$$

Here m denotes the upper limit mentioned in postulate D. The axiom expresses what we mean by calling an alternative complete.

3. Axiom of decision
For any A there is an alternative $a_1, ... a_m$ thus that $a_1, ..., a_r$ are elements of A, and $a_{r+1}, ..., a_m$ are elements of $\bar{A}$. This axiom is equivalent to Jauch's axiom P of weak modularity (Jauch, p. 86).

4. First axiom of completeness
To any set of $k < m$ mutually exclusive atomic propositions $a_1, ..., a_k$ there is an atomic proposition a such that $p(a, a_i) = 0$ $(i=1, ... k)$.

4b. Second axiom of completeness
To any set of $m-2$ mutually exclusive atomic propositions $a_3, ..., a_m$ and any atomic proposition b there is an atomic proposition a_2 with $p(a_2, a_i) = 0$ $(i=3, ..., m)$ and $p(a_2, b) = 0$. The two axioms 4 correspond to Jauch's A_2 (p. 87), the 'covering law'.

5. Axiom of indeterminism
For any pair of mutually exclusive atomic propositions a_1 and a_2 there is an atomic proposition which excludes none of them. This axiom excludes classical physics, i.e., a Boolean lattice. It corresponds to Jauch's principle of superposition (p. 106). In the form given here it excludes super-selection rules; it may be weakened so as to admit them.

6. Axiom of exclusion
For atomic propositions: if $p(x, y) = 0$, then $p(y, x) = 0$. This axiom were better placed after 2.

Axioms 1, 2, and 6 seem to express what we mean by alternatives. Axioms 3 and 4 are not implausible but not evident. Axiom 5 formulates that possibilities cannot be reduced to unknown facts. These 6 axioms suffice to show that the class of admitted propositions is a projective geometry of $m-1$ dimensions, isomorphic to the lattice of subspaces of an m-dimensional vector-space in which the probability function defines a metric. For the further arguments implying that the vector space is erected over the complex numbers see the papers by Jauch, Finkelstein, Stuckelberg et al. (Jauch, p. 131). The equation of motion in general form follows if we assume that motion is a continuous function of time mapping the Hilbert space on itself and preserving the metric. Apart from motion, this general scheme of quantum theory can indeed be called a quantum theory of probability. It does not use any particular concept of physics, and it is non-classical due to indeterminism.

3.3 MULTIPLE QUANTIZATION

For a proof of semantical consistency we must show that probabilities obeying the laws of quantum theory will also correspond to the definition that they are expectation values of relative frequencies. We must reproduce the sequence of levels discussed in Subsection 3.1. This is in fact possible. It is done by the well-known procedure of second quantization, and it reveals the meaning of that procedure. We discuss it for the case of Bose statistics.

An alternative a_k ($k=1, ..., m$) gives rise to a quantum state space of m dimensions. Let $\psi(k)$ be the state vector. Now replace $\psi(k)$ by an operator ψ_k with the commutation relations

$$\psi_k \psi_l^* - \psi_l^* \psi_k = \delta_{kl}; \quad \psi_k \psi_l - \psi_l \psi_k = \psi_k^* \psi_l^* - \psi_l^* \psi_k^* = 0$$

Then the operator $N_k = \psi_k^* \psi_k$ must have the eigenvalues $N_k = 0, 1, 2, ...$. ψ_k and N_k operate on an Hilbert space whose vectors can be written as functions $\phi(N_k)$. They represent possible states of a real ensemble of equal objects in each of which the alternative a_k can be decided. N_k is the number of objects for which a_k is true (or necessary). $N = \sum_k N_k$ is the total number of objects in the ensemble. If all the objects of one ensemble are in the same state x whose normalized state vector is $\psi_x(k)$ it is easily shown that

$$p(xk) = |\psi_x^*(k) \psi_x(k)| = \exp_x (N_k/N)$$

That means that the probability of a_k under the condition x is indeed the expectation value of the relative frequency of objects with a_k in an ensemble of objects with x.

We see that the complex state vector $\psi(k)$ is the classical limiting case of the operator ψ_k and that the measurable values of the wave intensities $\psi_k^* \psi_k$ are always non-negative integers. In this sense quantum theory justifies the postulate of finitism also for the description of the Hilbert space, if we ask for a truly operational meaning of ψ. The absolute square of any component of ψ is now not a real number but an operator with discrete eigenvalues. This means exactly that the occurrences of the state denoted by k can be counted. This simple consideration does not, however, give an upper limit of N_k (see Subsection 5.3). We may call this formalism a quantum arithmetic, in analogy to quantum logic. Natural numbers if taken as results of the real operation of counting become operators with the classical natural numbers as their eigenvalues; if not actually counted their values do not necessarily exist. Complex numbers $\psi(k)$ and real numbers $|\psi(k)|^2$ become classical limiting cases of operators. The process can be iterated, and only the highest level under consideration is described by classical logic. In both cases it is open to doubt whether this is the final truth.

Multiple quantization is not only a formalism; it produces a wealth of new structures which carry physical meaning. Thus many-particle physics can be deduced by 'second quantization' from a classical field theory which in some cases can be read as a one-particle quantum theory. We shall return to this point in Subsection 5.3. Here we just mention one application to an analysis of dynamics. Classical point-mechanics can be deduced from a variational principle. According to Dirac[11] the validity of this principle can be deduced from wave mechanics in a manner analogous to the Huygens principle in optics. Now wave mechanics, too, can be deduced from a variational principle. It is to be expected that this admits of a corresponding explanation by second quantization.

4. REALITY IN QUANTUM THEORY

There seems to be agreement that quantum theory (at least non-relativistic quantum theory) is mathematically consistent and even basically simple. Yet there is a widespread view that its interpretation is difficult and perhaps paradoxical. This makes us suspect that it is not the truth expressed in the formalism but the concepts used for its interpretation that lack clarity. The main difficulty seems to be presented by the question in which way quantum theory refers to reality. Thus we suspect the concept of reality of not being clarified. We begin by a review of some of its meanings.

4.1 DIFFERENT CONCEPTS OF REALITY IN CLASSICAL PHYSICS

Classical physics describes things (bodies or fields) in space and time in the manner in which we can deduce laws about them from human experience. It takes things, space, time and observers to be real. But there were at best different concepts of reality and some unsolved problems in the ontology of classical physics.

(a) Things
Reality for a thing in classical physics meant to be somewhere, that means somewhere in space. It had to be in some place at any time during its existence. Things were understood to be objectively in space and time. The origin of quantum theory (Planck) makes us suspect, however, that continuous bodies and fields obeying classical continuum dynamics are thermodynamically impossible. Thus 'being in space and time' is perhaps not a clear, self-consistent concept at all.

(b) Space
Space is taken to be real, but its reality cannot consist in being somewhere in space. While classical physics worked well, there was never philosophical agreement whether space was an entity by itself (Newton), a class of relations between bodies (Leibniz, Mach) or a form of intuition (Kant). Its epistemology was equally difficult: is it an object of experience? Recently the meaning of its continuity has become doubtful.

(c) Time
For a thing to be in the strict sense is to be now. The city of Nineveh is no more, the city on the moon is not yet built. Aristotle (*Physics*, Book IV, 217b, 33–34) wonders whether past and future can be said to be at all. Augustine (*Confessions*, Books X and

XI) places the past into the memory, the future into the anticipation, hence both into the human mind. Thus time came under the suspicion of being 'only subjective'. But the physicists very sensibly went on to believe in events that are not now, and they invented an 'objective time' in order to express clearly what everybody understands by time. They describe objective time like a space coordinate. But then the question arises, how subjective time is connected with it. How does the mind 'move its "now" through objective time'? This question is probably absurd, since it seems to presuppose what it denies, but it is a necessary question once we have distinguished subjective from objective time.

(d) Observers

Human beings observe nature by their bodies being causally connected with the things of nature of which they are part. Man knows nature. But his knowledge appears to be different from his body. It seems to be in a thing that we have accustomed ourselves to call the mind. If all knowledge is in the mind, we seem to know nothing outside the mind. Then the real existence of the whole world from which we started would seem to be a mere guess, an unprovable hypothesis. We feel that this 'problem' must be the result of some confusion of thought, but we find it difficult to say how and why. For this reason most physicists prefer not to speak of the mind at all. Classical physics speaks of the things known but not of the way we know them.

4.2 PROBLEMS OF REALITY IN QUANTUM THEORY

That paradise of ignorance has been lost for quantum theory. We must reflect now upon our way of knowing the things we know. This is forced upon us by the appearance of a fifth concept, that of the state vector. What is its relation to reality? The problem can most easily be formulated when we remember the two ways of change of the state vector,

  (α) continuously, according to the law of motion
  (β) discontinuously, according to the change of knowledge.

If there were only (α) we would say the state vector is a property of things, it is their objective state. If there were only (β) we would say it is an expression of our knowledge. All the so-called paradoxes arise from unsuccessful attempts to say both things together. Yet we cannot get rid of any one of the two interpretations. We cannot get rid of (β) since ψ is empirically defined by the probabilities it yields, and probabilities change with changing knowledge (see the end of Subsection 3.1.) (I shall discuss the theory of Everett which apparently avoids the reduction (β) in Subsection 4.3.) We cannot either get rid of the description of the state of things, because, to put it abstractly, this is what we mean by knowledge. 'I see an apple-tree behind my window' does not mean 'there is an apple-tree behind a window in my mind' but 'there is an apple-tree behind my window, and I see it'. That would not create a difficulty if the change (β) would only mean that we become aware of facts which existed but which we did not know before the measurement. This is the solution aimed at by all theories of hidden parameters. It is not my intention to prove that such theories are impossible. But they are certainly not part of quantum theory as it stands. The mathematical consistency and simplicity of the existing theory must admit of a verbal interpretation that does not need hypotheses on hidden parameters. I try to analyze quantum theory, not to change or supply it.

The answer is of course given by the Copenhagen interpretation. We must ascribe the observed properties to the things but we cannot suppose all the decidable alternatives on things to be objectively decided even when not observed. This is what I have tried to describe by the distinction between fact and possibility. We will not expect contradictions if we stick to this rule: *Quantum theory is a theory on the probabilistic connection of facts. Facts themselves are to be described classically. Where a classical description is not possible there is no fact.* Subsection 4.3 will elaborate on this rule in slightly more detail.

Still there are three justifications for asking questions that go beyond this 'Golden Copenhagen Rule'.

First the avowedly unclear classical concepts of reality induce us, at least psychologically, to ask such questions as: What happens to the things between the events called facts? Is it the measuring interaction or the mental act of taking cognizance that reduces the state vector? Can we hope to describe acts of consciousness by quantum theory? Even if we suspect these questions to be partially meaningless we wish to be allowed to consider them in order to clarify the concepts that produce them and thus to draw a line between sense and nonsense in them.

Second, there is an intrinsic problem of quantum theory in the question under what conditions the classical description of facts is in agreement with quantum theory. We consider classical physics to be a limiting case of quantum theory. The limit is not contained in the infinite sequence of quantum states converging towards it. A classical description cannot be strictly correct. (I am not discussing Ludwig's idea that classical and quantum physics are rather two different limiting cases of an unknown theory. Here, too, I stick to an analysis of existing quantum theory.) Hence, in full strictness, there are no facts at all. How are we to describe semantically the convergence of situations that are not strictly facts toward the ideal case of facts?

Thirdly the unsolved problems of quantum field theory (see Subsection 5.3) may well be connected with a lack of clarification of some intrinsic problems of quantum theory. In all quantum theory, time is treated like a classical variable; in quantum field theory this is the case for space and time. We use the mathematical point-continuum for a decription of space-time as though its points were facts. If there are no strict facts about things, why should there be strict facts about space-time? Multiple quantization (Subsection 3.3) makes us suspect that space-time coordinates ought to be operators. The meaningful introduction of this idea will probably rest on the answer to our second question.

4.3 QUANTUM THEORY AS A PROBABILISTIC CONNECTION BETWEEN CLASSICAL FACTS

Quantum theory formulates the probabilities for all formally possible results of all formally possible observations. The state vector of an object X contains all the information about the future that can be deduced from all observations so far made on the object. If the last observation was a maximal observation, its result is sufficient to determine this information. A new observation gives new information and thus produces a new state vector. The semantics of the theory will be consistent if the information possessed by two observers (or one observer at different times) cannot contradict each other.

Consider the Einstein–Podolsky–Rosen[12] experiment in a simplified version essentially due to Jauch (pp. 185–187 of his book). Let a spin 0-particle decay into two spin 1/2-particles without an exchange of orbital and spin angular momentum. The two particles are observed at later times at two points x_1 and x_2 far apart. Both observers have a choice of measuring the spin component of their respective particle either in the y-direction or in the z-direction. We speak of three observers, A, B_1, B_2. A possesses the initial information as given in our last sentences. B_1 measures at x_1, B_2 at x_2. After the measurements they communicate their results and test their predictions. A and B_1 do not know initially which one of the two available experiments B_2 will choose to perform. We call these two experiments y_2 and z_2, and their possible results y_2^+, y_2^-, and z_2^+, z_2^-. Neither do A and B_2 know which one of the experiments y_1 and z_1, with possible results y_1^+, y_1^- and z_1^+, z_1^- will be performed by B_1. A has a list of four conditional probabilities p depending on the choices of the B's.

	p		p		p		p
$y_1^+ y_2^+$	0	$z_1^+ z_2^+$	0	$y_1^+ z_2^+$	1/4	$z_1^+ y_2^+$	1/4
$y_1^+ y_1^-$	1/2	$z_1^+ z_2^-$	1/2	$y_1^+ z_2^-$	1/4	$z_1^+ y_2^-$	1/4
$y_1^- y_2^+$	1/2	$z_1^- z_2^+$	1/2	$y_1^- z_2^+$	1/4	$z_1^- y_2^+$	1/4
$y_1^- y_2^-$	0	$z_1^- z_2^-$	0	$y_1^- z_2^-$	1/4	$z_1^- y_2^-$	1/4

Now assume that B_1 measures y_1 and finds y_1^+. He will then have two conditional probabilities for the result which B_2 is going to report to him:

	p		p
y_2^+	0	z_2^+	1/2
y_2^-	1	z_2^-	1/2

Assume that B_2 measures y_2. A, before learning from B_1's result, will predict $p=1/2$ for both possible results y_2^+ and y_2^-. B_1 after his measurement knows for certain that B_2, if he has measured y_2 will report the result y_2^-. If quantum theory is correct, B_2 will actually find y_2^-, and he will now predict that B_1, if he has measured y_1, will report the result y_1^+.

We can go through all the listed possibilities and we will find no contradiction, and as in the dice case of classical probability (Subsection 3.1, near the end) all observers will be able to test their probabilities by repeated measurements in the ensembles corresponding to their respective knowledge. Especially A can go on predicting the later future with his original state vector if he does not know the results of B_1 and B_2; only he must be careful not to forget the unknown phase-shift effected by the fact that B_1 and B_2 have made some measurements. This, I feel, is all that is behind Everett's[13] contention that we do not need to reduce the wave function. The unreduced wave function is the prediction possible to an observer who does not know the outcome of later observations. Everett's way of expression is that in any observation all the formally possible events actually happen and that the branches into which the wave function splits in this process will no longer interact due to the irreversibility of the process. Hence they will not know of each other, and in each branch the observer will say that what he observed is what has actually happened. If irreversibility is absolute, this way of expression amounts to a re-definition of words, calling 'actual' what is usually called 'possible' and calling 'known' what is usually called a 'fact', without changing anything in the structure described by the words. I prefer the usual words. The really important fact is that irreversibility is only an approximation; the usual words are perhaps best adapted for discussing this fact. The rest is the philosophical problem of linguistic convention.

Why, then, should facts be described classically? A fact is a fact whether somebody knows it or not. Probabilities of unknown facts obey classical statistics, without superposition. If a self-registering instrument documents the outcome of an experiment on Sunday morning by punching a hole into a constantly moving tape, the observer looking at it on Monday morning knows from the place on the tape that this hole has existed for 24 hours. If quantum theory is to respect facts in describing them the way they are known to observers, it should not say that this fact has arisen by the mental process of the observer. We are permitted to accept the language of the observer without having to fear a contradiction precisely if the event has been irreversible. *Irreversibility of facts is what we mean by a classical description.* The historical theories of classical physics are limiting cases of the quantum theories of their respective objects which connect the irreversible limiting cases of quantum events.

This reduction of fact to irreversibility is, however, a problem rather than a solution. We started with an 'external semantic', supposing we knew what we mean by a fact, and ascribing to quantum theory only the task of formulating probabilities for possible facts. Then we admitted that quantum theory itself defines what are formally possible events. They are measurements of the eigenvalues of some operators of the Hilbert space of the objects. Quantum theory of measurement sets out to describe the interaction between the object and the instrument; the operator to be measured is then the object part of the Hamiltonian of the interaction. But all this does not eliminate the truth that the state vector only defines probabilities of events, and that it must be reduced when an observer knows that an event has happened. The time-development of the Schrödinger function of the instrument only leads to an irreversible distribution of probabilities of results from which the choice is made by any observer when he looks at it. Due to the irreversibility he is only uncontradictedly permitted to state that the event has not taken place when he looks at the punch-hole but 24 hours earlier. But how can he refute an adversary who may say: 'The event your instrument measured was, say, a radioactive decay. It produced in the instrument a probability of finding a punch hole on the tape, depending continuously on the location of the hole on the tape which you take to indicate the objective time of its happening. By your looking at the tape on Monday your consciousness produced the situation that you call a document of an event on Sunday morning.' He is sure that the adversary is somehow talking nonsense. But how is he to prove that?

From his knowledge our observer can quantum-mechanically deduce that if somebody had looked at the tape on Sunday at noon, he would with certainty already have found the hole in the tape. He and the earlier observer will as little get into contradicting statements as the two Einstein–Podolsky–Rosen observers B_1 and B_2. Neither is the situation changed by endowing the instrument itself with consciousness, or, what amounts to the same, by replacing the tape by an observer and, on Monday morning, instead of asking him meaningful questions looking for a document of his observation in his brain. I take this to be the problem of 'Wigner's friend'. If the observation on Sunday morning, including the effects in the brain and the mental act which constitute it together (whatever their relationship or possible identity), has been irreversible, there is no contradiction in the assumption that what the later observer observes on Monday has been a fact since Sunday morning. It is not a contradiction or a paradox that if we *know* that no conscious observation was made on Sunday we know that there will be no witness to testify for the fact, and this problem is not peculiar to quantum theory.

The remaining question is how we describe irreversible facts in quantum theory. There is a continuous series of possible changes, from the fully reversible through a number of cases with decreasing probability of return down to practical irreversibility. We must try to understand this 'limiting process' not only mathematically but semantically.

4.4 A DIGRESSION: A MIND KNOWING ITSELF IN TIME

Since we cannot avoid speaking of the mental act of knowledge we may learn from the apparently simpler case of a mind knowing itself in time. The statement that past events are facts, relies on memory in the first place. Memory is, according to Augustine, the presence of the past. Thus it is worth a study, *how* the past is present in memory.

The philosopher Husserl[14] has tried to give a phenomenological description of retention, i.e., of the mental presence of the past part of a longer-lasting coherent process or sequence of events while the process is still going on. He illustrates it by a two-dimensional diagram (see Fig. 1). Here *APE* means the series of moments in time (of 'nows'). What was a present event at *A* is 'sinking down' in the way in which it is mentally present into the 'retained past' *A'*. *EP'A'* is the presence of past events to the mind in the moment *E*.

This simple representation poses a problem. On the one hand it reminds us that a remembrance of a past event is a present mental fact. Husserl tries to describe this presence of the time flow in saying that the flow of 'time-constituting consciousness' is such as to 'necessarily contain a self-representation of the flow' (*loc. cit.*, p. 83). It must be remembered that now we can only speak meaningfully of such past events that have left a document existing now (Subsection 2.5). On the other hand it seems impossible that every past event should be represented by a separate present fact. Husserl's diagram would then have to be not two-dimensional, but infinite-dimensional. A fact in the past representing an earlier past event is itself a past event that would have to be remembered now. We can indeed, if we desire to do so, remember our past remembrance of an even earlier event. In the diagram the point P contains a representation of the full line-segment AP, designated by PA_p. This whole line-segment PA_p must be represented in P', etc. Even if we cut the flow of consciousness into discrete steps there would be an impossible multiplication. n steps would produce 2^n distinct mental facts. If a step lasts one tenth of a second, one minute would produce $2^{600} \approx$ $\approx 10^{180}$ mental facts, to be carried by our 10^{10} ganglions!

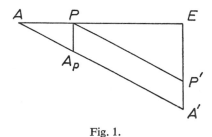

Fig. 1.

The solution (given by Husserl himself, pp. 118–19) must be that retention is only potential; it is only the possibility for an explicit act of remembrance. This can perhaps more easily be understood in the example of an external object which serves as a document of past events. Take a fossil, say a petrified Jurassic ammonite. If we understand to ask the right questions, the same snail-formed piece of limestone on my desk serves as a document of many events: of the life of a particular animal 120 million years ago, of its becoming embedded in a lime sediment, of its petrifaction, of the persistence of the sediment through 120 millions of years, of its getting to the surface by erosion in the 20th century of our era, of my walk in the Göttingen forest in 1955 on which I found it, of the good order kept in my home by my wife ever since.... All these are events or sequences of events in which the stone interacted with the environment in a manner which is documented in the present state of the world of which I can be sufficiently aware to draw all the conclusions mentioned here.

We learn from this that facts are themselves present as possibilities of new events. A mental fact is the possibility of a mental event, say of a remembrance to be produced by reflection or by some other interaction. A mental event, on the other hand, is not fully defined while it happens. I can ask: 'What is it I experience just now?' 'What am I actually thinking?' Reflection is needed in order to define the event. But reflection already uses the retention of the past part of the event. Consciousness is, so to speak, an unconscious act[15]; it is a possibility of more consciousness.

We shall now apply the result to facts in physics.

4.5   A Theory of Events

In this subsection I shall state the results of the previous analysis in a rather thetic form.

An event in the strict sense is a present event, something that is actually happening. Quantum theory as a general, conceptual theory can only describe formally possible events. They are in principle always described by state vectors (see Subsection 3.2, G, the postualte of objectivity), even if we are confined to knowing linear sets of state vectors or statistical mixtures of such linear sets. The probability function $p(a, b)$ defines the probability for b to happen now under the double condition that a has just happened and that b is made possible by the interaction Hamiltonian with the environment. What that precisely means is to be discussed presently. An event that has happened is a fact. A fact is the possibility of a class of events documenting that fact.

The second condition for $p(a, b)$ indicates that events can only happen if there is interaction. A strictly isolated object does not belong to the world, it is devoid of events. What we idealize as an isolated object is operationally defined by an interaction in the past and the possibility of measurements on it in the future. Thus it is a fact, i.e., a state vector.

An event that is happening in the interaction of a small number of objects is apt to disappear again, i.e., not to become a fact. It is a still undefined event. We can speak of it theoretically but we cannot know it. It can easily be described by combining its participant objects into one composite isolated object, that is as a non-event.

If an event distributes an amount of energy on many objects, its probability of being undone becomes extremely small. Then it can be said to be a fact.

Even in the case of a fact emerging, the development of the state vector according to the Schrödinger equation will not lead to a well-defined fact but to a probability distribution for facts. Take an idealized example. A light object, say a match-box, is located at a high place[16]. If a silver atom in a Stern–Gerlach experiment shows spin $+1/2$, the match-box is touched so as to fall down to its right side, if $-1/2$, to its left side. Let the probabilities of both cases be $1/2$. But if one atom of the match-box falls to the right, the probability for any of its other atoms falling to the left is practically zero. The event really happens that the silver atom takes on either spin $+1/2$ of spin $-1/2$. It is made irreversible by the match-box falling down this side or the other. By the interaction with the match-box the phase of the silver atom is lost for further knowledge. If the two rays are recombined after the interaction with the match-box they will be incoherently superposed. But if there is no match-box to interact with the two parts of the wave function of the silver atom may be coherently recombined and we can say that the event of going through one of the rays has 'disappeared'. In this case the lack of interaction with the environment is mathematically described by saying that there is a wave function of the single object 'silver atom' which is smoothly developing.

Thus we would say that events happen when things interact. They have probabilities which are described by the state vector. In principle all events are reversible (according to quantum theory as analyzed here). The simplest way of reversal of an event is the absence of an interacting partner, that means the not-happening of the event. Quantum theorists have a jargon well adapted to this situation. They speak of a state virtually being all the states from which it can be made by superposition. These states are representing the events that might happen to the object. The expression of an event 'happening virtually' precisely describes the situation. An event in which many objects have participated has a very small probability of reversal. If a human person becomes a participant in it he or she will say this or that happened. But this mental event is itself not finally defined (see end of Subsection 4.4). We cannot deny the possibility (even if extremely improbable) of its being reversed and undone. Only to a Cartesian ontology of the mind will this seem impossible.

Is this description justified by quantum theory or is it a fairy tale? The distinction of possibility and fact offers an answer if applied to the description itself. It is precisely a description of the sequence of the possible results of the possible observations on the developing system of the objects described. One might interrupt the course of events at any moment, and one would find the objects in their respective states with the corresponding probabilities. Of course, after the observation the further development would differ from what it would have been without the observation. If we took a large ensemble of objects initially in the same states, and if we made all possible measurements in the ensemble often enough to measure the probabilities of their results, using every member of the ensemble just once, the sequence of our measurements would follow the story told by the development of the state vector. This is what we – the theorists – can know. If, however, we renounce any measurement, then we know that the system has not been disturbed and that we can know no more. That is not the fault of quantum theory.

Thus we see that our description stays within the Copenhagen Golden Rule as applied to the description itself. The rest is either philosophical interpretation or application.

5. CONCLUDING REMARKS

5.1 EPISTEMOLOGY: CONDITIONS OF EXPERIENCE AND SEMANTICAL CONSISTENCY

The concepts of fact and possibility are related to the basic epistemological problem of empirical science. Facts refer to the past; they can be known by experience. Possibilities refer to the future; how can they be known? We deduce them from general laws. How can we know the general laws to hold for cases which we did not know when we stated the laws? We rely on a regularity of all nature. If we know this regularity from the past, what entitles us to expect it for the future? This is Hume's problem. It has not been answered by the modern philosophy of science.

The problem can be sharpened. Popper[17] points out that for logical reasons a law of nature, if stated as a universal judgement, cannot be empirically verified in a strict sense since we do not know all the cases to which it refers. It can be empirically falsified by a counter-example. But an observation is only a counter-example if interpreted by a theory in which we can trust. With 'robust realism'[18] Popper sees the development of science as a Darwinistic survival of the fittest, that is of those theories which have so far most successfully withstood empirical falsification. Kuhn[19] describes this evolution as a sequence of successful paradigms. The greatest paradigms are rather 'closed theories' in the sense of Heisenberg[20].

Quantum theory is probably the most successful of closed theories. It has withstood millions of formally possible falsifications, and it is basically simple. How is such a theory possible? The questions is not, how it could have been discovered if true, but what must be the case in order that it should be true?

An analysis in the Copenhagen spirit will ask what must be the case in order that such a question can be meaningfully asked. What, e.g., is presupposed by the assumption that there are facts which can possibly falsify a theory? A theory must describe formal possibilities, it must predict events, that is at least actual possibilities and probably probabilities, and there must be facts which fit sufficiently well into the scheme of formal possibilities for being compared with the predictions of the theory. That means: the logical scheme of fact and possibility is a precondition of what we call an empirical theory.

It was Kant's idea that the fundamentals of science, while discovered in the historical process of experience, are not justified by special experience, but by being preconditions of all experience. We do not therefore know that they cannot possibly fail, but we understand that when or where they would fail there would be no experience. More than that cannot be answered on Hume's problem.

Now physics is approaching a very high degree of unity. It is an interesting question how much of the basic assumptions of quantum theory or of the coming theory of elementary particles is already determined by the conditions of experience. Time, fact and possibility certainly belong to these conditions. This is why I stressed their importance; they are preconditions of all physics. I propose the working hypothesis that the evolution of physics converges towards a general theory of preconditions for decidable alternatives. In such a theory all empirical laws, in as far as they state more than the general principles of the theory, will only express the historical presence of some special facts out of the list of formally possible facts given by the theory.

Semantical consistency is a guiding principle for such a theory. Its striking examples are the theories of measurement in relativity and in quantum theory. Historically the concepts which we use to interpret our formalisms are older than the respective formalisms. If the formalism is universally valid, it will also apply to the experience in which those concepts were originally framed. This gives a condition of consistency that may lead to a re-interpretation of the original concepts. The hierarchy of levels in logic, probability and quantum theory (Subsections 2.4, 3.1, and 3.3) is an example of such a consideration, the discussion of documents and irreversibility (Subsections 2.5 and 4.5) is another one.

5.2 ONTOLOGY: THE UNITY OF NATURE

If physics will end up as a general theory of empirically decidable alternatives we would expect that everything which admits of empirically decidable alternatives were subject to physics. This is mainly a matter of the further development of physics (see last subsection). Here we review the classical 'realities' (4.1) in the light of this idea.

We expect things to be reduced to elementary quantum fields. These fields are described in space and time. Their laws seem to be determined by fundamental symmetry groups. The idea that living organisms can be described by the laws of general physics seems natural to most modern biologists. The central remaining question is whether the mind, too, can be subject to these laws. If our analysis was correct, it showed that time and knowledge with their characteristic structure must be presupposed if we wish to understand physics. This is not surprising since physics *is* knowledge acquired in time and referring to events in time. Thus the question is rather: can we assume a monism in which all events are supposed in principle to be of the same nature as those which we know under the name of mental events (memory, knowledge, volition, action, feeling, emotion), or must we admit a dualism in which matter is different in nature from mind? Monism of course is not the idea that matter consists in our ideas of matter, but that there is a continuous sequence of events connecting human consciousness, animal behaviour, organic self-reproduction and growth, and matter in motion. The general description of events in Subsection 4.5 would not contradict such a view. There is no contradiction in assuming that a mind can state empirical facts and make testable predictions on the behaviour of a mind exactly in the way described by quantum theory. The object-mind can also happen to be himself. The non-contradiction between the contents of the knowledge of two observers will persist if the two observers are understood to be one observer at different times, observing himself. There is no contradiction either in assuming that all events 'have the essential nature of events in minds'; only the meaning of the words in quotation marks will depend on a semantical interpretation of the continuity between human consciousness and matter in motion.

In this case a mental event or an 'I' would be no more an absolute than a material event or a thing. No concept will be better confirmed than the irreversibility of the facts on which it rests. For a pragmatic approach facts are sufficiently reliable. Philosophically it is important that a fact is not an ultimate reality. All we can deduce from the supposed semantical consistency of physics is that a method of thought that takes fact and possibility as its starting points will confirm their validity as good, but limited, approximations. That method of thought has developed in human history; it is characteristic of the age of science. What other aspects will reality show if differently questioned?

5.3   FURTHER DEVELOPMENTS IN PHYSICS

We return to physics. Physics tends towards even greater unity than it possesses today. Three great groups of unsolved problems are seen: elementary fields, cosmology, and organic life.

The first problem of quantum field theory is the fusion of quantum theory with special relativity. It is solved for free fields. But in the language used in Subsection 4.5 free fields are fields without events. The apparent singularities in interaction theory have made some physicists wonder whether space and time ought to be treated as mathematical continua at all. The method of multiple quantization (Subsection 3.3) makes us doubt the use of any c-number quantities.

A strictly finitistic quantum theory, with a finite-dimensional Hilbert space for any object, will not possess position operators with a continuous spectrum. The decision on the relationship between finitism and continuity will, however, depend on a better understanding of time. Facts are discrete. This is not only a phenomenological description: we never know more than finite numbers of facts. It also corresponds to the objective description of facts as irreversible; every fact needs a particular cascade of events for being irreversibly established (Subsection 4.5). Hence the discrete structure of facts seems definitely to belong to nature and not only to our limited brain (or, to say it in an epistemologically guarded manner, it belongs to our description of nature and not only to our description of ourselves). Infinity (and continuity in particular) expresses possibility as going beyond any established fact; it expresses the open future. One would expect a measurable time not to be a real-number parameter but a counting operator (Subsection 3.3), perhaps counting the facts, that is the events which have happened up till now. Finitism in quantum theory (Subsection 3.2, postulate D) then would mean an operator m limiting the number of decisions about any object and about the world as we can describe it in the object-language; the expectation value of m will be finite at any time but increasing with time; it may be taken as measuring time. This idea can be formulated relativistically. For two observers moving in different frames different events will be simultaneous and hence, even when they meet in the same place at the same time they will consider a different number of facts as having happened in the world so far. Their different descriptions will no more entail a contradiction than those of the Einstein–Podolsky–Rosen observers (Subsection 3.3).

In such a theory multiple quantization would not have to start with a particle in an infinite continuous space. I would not have to accept space as given but to construct it. The simplest and most abstract construction will begin with a two-fold alternative, i.e. with one yes–no decision. Multiple quantization starting there would mean to build up the objects in the world as ensembles of such decisions. The number corresponding to one object would mean the number of decisions needed for fully defining that object. Now the quantum-mechanics of an ensemble of two-fold alternatives nearly automatically leads to a description in a three-dimensional real space and hence in the higher steps to the usual field theories in three-dimensional space[21]. Mathematically this introduction of space corresponds to Schwinger's[22] introduction of angular momentum by a Bose-ensemble of spins1/2. It has been studied by Penrose[23] and especially by Finkelstein[24] who introduced the discrete counting of time ('chronons') as the basic idea. A quantum-theoretical construction of space would in the same step lay the foundation for field theory and cosmology. It would thus establish the unity towards which physics is moving. The essential problem is how to define the interaction of the elementary alternatives.

The idea of semantical consistency would imply that elementary alternatives are not to be dogmatically introduced like the atoms of earlier atomism. They would only formulate the fact that all decisions which can be made in physics are reducible to finite numbers of yes–no decisions. Semantical consistency would mean not to introduce any structures that cannot be analyzed into the simple decisions defining them. In this way the envisaged unity of physics would come closer to the ideal that physics should only formulate the preconditions of experience.

In the theory of organic life the evolution of new forms is one of the basic problems. It takes place in time, and it seems to correspond to the emergence of ever new possibilities out of an accumulation of more and more facts, that is of more and more established structures. Hence the theory of time envisaged in the present paper may make its fusion with physics easier.

REFERENCES AND NOTES

1. This title was proposed to me by Jagdish Mehra, and I gladly accept it. I also accept his proposal to give a somewhat more extended presentation in the written paper than can be given orally.
2. P. Mittelstaedt, *Philosophische Probleme der modernen Physik*, Mannheim 1963, and a paper presented in München 1972.
3. This and the following sections are shortened versions of parts of a book which M. Drieschner and I are preparing. Some of the philosophical ideas are more broadly developed in my book *Die Einheit der Natur,* Munchen 1971.
4. G. Frege, Über Sinn und Bedeutung, 1892, reprinted in "Funktion und Begriff", ed. G. Patzig, Göttingen 1963, and in "Kleine Schriften", ed. I. Angelelli, Darmstadt 1967. My use of the two concepts does not fully agree with Frege's.
5. A. Church, Introduction to Mathematical Logic I, Princeton 1956, p. 4.
6. Über Sinn und Bedeutung, p. 34.
7. See my paper in *Ann. Physik* **36**, 275 (1939), reprinted in *Einheit der Natur*, pp. 172–82, and the *History of Nature*, Chicago 1950, 4th lecture.
8. The Feynman–Wheeler theory is forced to introduce the same so-called causal asymmetry (that is the phenomenal meaning of causality and time) by an absorption to which there is no corresponding emission or by an expanding rather than a contracting universe.
9. See J. M. Jauch, *Foundations of Quantum Mechanics*, Addison Wesley (1968).
10. M. Drieschner, *Quantum Mechanics as a General Theory of Prediction* (1968). It is described in my contribution to *Quantum Theory and Beyond* (ed. T. Bastin), Cambridge (1971), reprinted in German in *Einheit der Natur*, pp. 249–63.
11. P. A. M. Dirac, *Physik. Zeitschr. der Sowjetunion* **3**, 64 (1933).
12. A. Einstein, B. Podolsky, N. Rosen, *Phys. Rev.* **47**, 777 (1935).
13. H. Everett III, *Rev. Mod. Phys.* **29**, 454 (1957).
14. E. Husserl, *Zur Phänomenologie des innereen Zitbewusstseins (1893–1917)*, (ed. R. Boehm), den Haag 1966 (Husserliana X), p. 28.
15. This formula came to my mind like an answer after months of reflection on what Bohr had said when I first met him (1932). Later on I learned from K. Meyer-Abich that it is a sentence written by William James whom Bohr had read just then.
16. G. Süssmann used this example in our discussions 15 years ago.
17. K. Popper, *Logic of Research*. See my remarks on it in *Quanten und Felder* (ed. H. P. Dürr), 1971, p. 20–24.
18. Addition (1968) to chapter V of *Logic of Research*.
19. Th. S. Kuhn, *The Structure of Scientific Revolutions*.
20. W. Heisenberg, *Der Teil und das Ganze*, München (1969).
21. C. F. v. Weizsäcker, E. Scheibe, G. Süssmann, *Naturforsch.* **13a**, 705, (1958). See also my contribution to *Quantum Theory and Beyond*, note 10 above.
22. J. Schwinger, reprinted in *Quantum Theory of Angular Momentum*, Dover.
23. R. Penrose, in *Quantum Theory and Beyond*.
24. D. Finkelstein, *Phys. Rev.* **184**, 1261 (1969) and **D5**, 320 (1972).

32

Wavefunction and Observer in the Quantum Theory

Leon N Cooper

INTRODUCTION

It is remarkable that after almost a half century of development and continuing success, there still is no general agreement on how the quantum theory is to be interpreted, or indeed whether a consistent interpretation, using no external concepts is possible. We are aware that the meaning (or the interpretation) of the theories we invent can change. Consider, for example, the equation Paul Dirac proposed for the electron – surely one of the most beautiful creations of the twentieth century. Conceived originally as a wave equation for the electron consistent with the principle of relativity, it is now interpreted as the field equation for a particle of spin $\frac{1}{2}$, while almost every structural relation as originally written down by Dirac is maintained. We are also aware that we can proceed even when it is likely that our theories are inconsistent. This could not be more clearly demonstrated than by the post World War II success of the renormalization programme in quantum electrodynamics – the union in the quantum theory of the fields of Dirac and Maxwell. There the incredible numerical agreement between theory and experiment is balanced by the strong possibility that at their roots the axioms of the theory as usually stated are inconsistent.

We may learn from this that neither inconsistency nor lack of understanding should deter the physicist from his task of forging the struts and links that make up the structure of his science. For we proceed on a day-by-day basis as the craftsman who senses with his fingers what is possible with the material he is moulding: how far to bend without breaking, how deeply to chip without shattering. As in pure mathematics the structure of relations that is a scientific system has a life almost independent of its meaning; it possesses an internal sense even in the absence of a completely worked out connection between the elements of the theory and the 'real world'.

But just as the craftsman, absorbed in the sculpture of a gargoyle or the fabrication of a column must occasionally, in the decades of construction, have wondered, viewing the slowly growing cathedral in the setting sun, what this enterprise would mean, we also perhaps can question, when the day's work is done, the consistency and/or the meaning of the enterprise that consumes our energy. We may even hope that it is not impossible (though one hardly palpitates with anticipation) that, if there is a deep inconsistency in the quantum theory, a clear enough understanding could suggest a fruitful new departure which would be relevant (if one is permitted to use

J. Mehra (ed.), The Physicist's Conception of Nature, 668–683. All Rights Reserved
Copyright © 1973 by D. Reidel Publishing Company, Dordrecht-Holland

such a word) to the more immediate and day-by-day problems of understanding experiments.

There have been numerous attempts, dating almost from the origin of the quantum theory to introduce modifications that would make the concepts of the theory more akin to those of the classical theories that preceded. The interpretation of von Neumann[1], which accepts the probability interpretation of Born, requires a process (often called wavefunction reduction) which does not develop according to the Schrödinger equation. We will present here an attempt to interpret the quantum theory in which no device inconsistent with the time development of the wavefunction according to the Schrödinger equation such as the type I process of von Neumann is employed. Nor is any special assumption made concerning the interaction of living or conscious beings with the physical system being observed. If this approach is correct we will have shown that there is no new direction to be found by examining the meaning or the interpretation of the theory. While this may be discouraging to some, this does not mean that some special relation between the mind and the quantum theory does not exist. Rather what we attempt to demonstrate is that no such relationship must be assumed at present to preserve the consistency of the theory.

Some of the argument presented here has recently been put forth by D. Van Vechten and myself[2]. In addition, many authors have presented points of view similar in various ways. In particular, let me mention A. Wheeler[3], H. Everett III[4], K. Gottfried[5] and B. DeWitt[6].

MEASUREMENT AND IRREVERSIBILITY

We attempt first to show that even in the traditional definitions the process of measurement, or the preparation of a state, involves an irreversible interaction of the 'system to be measured' with another system 'the measuring apparatus'. We mean irreversible in the sense of statistical mechanics. Since we do not postulate any irreversible processes at the level of microscopic interactions this definition appears to lack a fundamental quality. It is our point of view, however, that there is nothing 'funda-

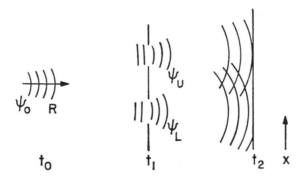

Fig. 1.   Traditional arrangement resulting in an interference pattern.

mental' about measurement. In the absence of irreversible interactions the idea would not have been introduced.

Consider a traditional arrangement (Fig. 1) which results in the interference pattern due to the passage of an electron through a barrier with two slits. The wave function of a single electron localized in the region R, at the time t_0, moving to the right in the horizontal direction may be written

$$\Psi(t_0) = \psi_0(t_0). \tag{1}$$

This develops according to the Schrödinger equation so that at time t_1 the packet is passing through the slits and we might write for the wave function

$$\alpha\psi_U(t_1) + \beta\psi_L(t_1), \tag{2}$$

where $\psi_U(\psi_L)$ is the part of the packet that passes through the upper (lower) slit and is normalized. At time t_2 when the packets have reached the screen, the wave function can be written

$$\Psi(t_2) = \alpha\psi_U(t_2) + \beta\psi_L(t_2). \tag{3}$$

The probability amplitude that the electron arrives at the point x_0 on the screen is given by

$$\langle \Psi(t_2, x) \,|\, x_0 \rangle = \alpha \langle \psi_U(t_2, x) \,|\, x_0 \rangle + \beta \langle \psi_L(t_2, x) \,|\, x_0 \rangle. \tag{4}$$

We cannot say whether the electron has 'gone' through the upper or lower slit since the amplitude is a coherent sum of ψ_U and ψ_L.

In the double Stern–Gerlach experiment, discussed by both Bohm[7] and Wigner[8], an electron prepared in a state which is an eigenfunction of σ_z is acted upon by an inhomogeneous magnetic field in the x direction so that

$$\psi_U \text{ is correlated with spin } \uparrow$$

and

$$\psi_L \text{ is correlated with spin } \downarrow.$$

We thus have at t_1

$$\Psi(t_1) = \alpha u_\uparrow \times \psi_U(t_1) + \beta u_\downarrow \times \psi_L(t_1). \tag{5}$$

Is this to be regarded as a measurement of the spin state of the electron?

In the usual sense the answer must be no, since at the screen at time t_2 the amplitude that the electron arrive at x_0 is

$$\alpha u_\uparrow \langle \psi_U(x, t_2) \,|\, x_0 \rangle + \beta u_\downarrow \langle \psi_L(x, t_2) \,|\, x_0 \rangle. \tag{6}$$

Its wave function is thus (as expected) a superposition of the two spin states $u_\uparrow$ and $u_\downarrow$. This is made more graphic if we recombine the spatial wave packets (by the introduction of a current which reverses the effect of the inhomogeneous magnetic field) in such a way that

$$\alpha\psi_U(t_1) + \beta\psi_L(t_1) \rightarrow \psi_0(t_2, x), \tag{7}$$

where $\psi_0(t_2, x)$ is the original wave packet displaced along the direction of motion.

The wave function would then be

$$\Psi(t_2, x) = \psi_0(t_2, x)(u_\uparrow + u_\downarrow). \tag{8}$$

and the electron is again in an eigenstate of σ_z.

What then constitutes a measurement? To discover whether the electron has passed by the upper or lower path, we might place a detector at each slit. Call a detector system $A \equiv (a_U \times a_L)$. The wave function of the entire system (electron+detector) is then written

$$\Psi = \psi \times A. \tag{9}$$

At t_1 the electron can interact with the detector and this interaction puts the detector system into the states

$$A_U \equiv (a_U^+ \times a_L)$$

and

$$A_L \equiv (a_U \times a_L^+). \tag{10}$$

The meaning of the latter two equations is that the electron interacts with the upper detector and not with the lower, or with the lower detector and not with the upper.*

We assume in what follows that the state represented by A_U is orthogonal to that represented by A_L so that for all time

$$\langle A_U \mid A_L \rangle = 0. \tag{11}$$

The wave function

$$\Psi(t_0) = \psi_0(t_0) \times A(t_0) \tag{12}$$

can now develop in time to give at t_1

$$\Psi(t_1) = \alpha \psi_U \times A + \beta \psi_L \times A. \tag{13}$$

If we assume that the detectors are so designed that there must be an interaction in order for the wave function to pass through the slits, we obtain at $t_1 + \delta$ the correlations

$$\psi_U \times A \rightarrow \psi_U \times A_U,$$
$$\psi_L \times A \rightarrow \psi_L \times A_L. \tag{14}$$

Thus all together we have

$$\Psi(t_1 + \delta) = \alpha \psi_U \times A_U + \beta \psi_L \times A_L. \tag{15}$$

This wave function develops further so that at t_2 we have

$$\Psi(t_2) = \alpha \psi_U(t_2) \times A_U(t_2) + \beta \psi_L(t_2) \times A_L(t_2). \tag{16}$$

---

* This of course is an hypothesis in addition to the Schrödinger equation. The electron is indivisible. In quantum field theory we would write

$$[(\psi_U^* \psi_U)(\psi_L^* \psi_L), \psi^*(x)] = (\psi_U^* \psi_U)\psi_L^* \delta(x - x_L) + \psi_U^*(\psi_L^* \psi_L)\delta(x - x_U).$$

However, even if the electron did not possess this property we might design a system which were triggered only if the amount of charge interacting exceeded for example $e/2$ and accomplish the same result.

Since $\langle A_U \mid A_L \rangle = 0$ it follows that the probability amplitude that the electron arrive at x_0 and that the detector read A_U (the electron has gone through the upper slit) is

$$\langle \Psi(t_2) \mid x_0 A_U \rangle = \alpha \langle \psi_U(t_2) \mid x_0 \rangle \tag{17}$$

while the amplitude that the electron arrive at x_0 and that the detector read A_L (the electron has gone through the lower slit) is

$$\langle \Psi(t_2) \mid x_0 A_L \rangle = \beta \langle \psi_L(t_2) \mid x_0 \rangle \tag{18}$$

and of course there is no interference.

Therefore unless A_U and A_L can develop into states which are not orthogonal to each other, or if one requests a matrix element which is a superposition of A_U and A_L there can be no further interference between ψ_U and ψ_L and the wave function (Equation (16)) is not distinguishable from a mixture. We postulate no reason in principle that a superposition of A_U and A_L cannot be constructed. (This will be discussed below.) If it can be, however, we would not regard the interaction with A as a measurement.

Whether A_U and A_L will develop into states which are not orthogonal to each other depends upon whether the states A_U and A_L are reversible. In a microscopic sense, of course, all states are reversible. But for large systems in the thermodynamic sense certain states are not.

It is this that distinguishes the double Stern–Gerlach experiment from what we usually understand as a measurement. For the former there exists an easily realizable interaction (that produced by a current loop for example) which will reverse the effect of the original inhomogeneous magnetic field and thus bring ψ_U and ψ_L back to ψ_0. The interaction in what is usually understood as a detector (e.g., an interaction in a photographic plate) cannot usually be made to reverse itself. Therefore, the states A_U and A_L do not, in the normal course of events, develop into states which are not orthogonal to each other. We take the point of view that the existence of interactions which are macroscopically or thermodynamically irreversible is what removes the possibility of future interference and makes a coherent wave function indistinguishable from a mixture. Therefore, a measurement or the preparation of a state from this point of view consists of the interaction of a system to be observed or prepared with another system which can be put into a state that is irreversible for reasons of entropy, thus eliminating the possibility of future interference. This second system (often called the apparatus or even the observer) is often thought of as being classical. But this now is not necessary. Any quantum system (a nucleus which can undergo fission for example) which can suffer a large increase in entropy due to the interaction will do as well.

Whether or not a matrix elements such as

$$\langle \psi(t_2) \mid x_0, (A_U + A_L)\sqrt{2} \rangle \tag{19}$$

can be observed in practice depends upon whether one can in fact prepare the coherent sum $A_U + A_L$. If this can be done we would no longer regard the interaction with A as a measurement.

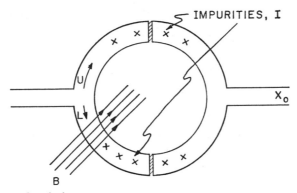

Fig. 2. A phase measuring device.

This situation can be illustrated with the following example. Suppose instead of a double slit the electron moves in a magnetic field through a circuit made of conductors with impurities as in Fig. 2. As is well known the relative phase of the electron wave function as it traverses the upper or lower path depends upon the enclosed flux

$$\Phi = \oint A \cdot dl = \int B \cdot dA. \tag{20}$$

If there are impurities however (which we represent by I) then in a manner entirely analogous to the argument leading to (15) we obtain

$$\Psi(t_2) = \alpha \psi_U \times I_U + \beta \psi_L \times I_L. \tag{21}$$

If we now ask for the probability amplitude that the electron be at x_0 (which can be related to the current) even though we have never looked we assume that what we are really requesting is either

$$\langle \Psi \mid x_0, I_U \rangle$$

or

$$\langle \Psi \mid x_0, I_L \rangle \tag{22}$$

and the probability is written as the sum

$$(|\langle \Psi \mid x_0, I_U \rangle|^2 + |\langle \Psi \mid x_0, I_L \rangle|^2) \tag{23}$$

so that there is no interference and this probability is independent of the enclosed magnetic field.

There is no principle that prevents us from asking for

$$\langle \Psi \mid x_0, (I_U + I_L)/\sqrt{2} \rangle \tag{24}$$

but we cannot construct an apparatus that corresponds to this question.

The situation is dramatically changed if the conductor becomes a superconductor and the appropriate Josephson Junctions are inserted. Now in effect (the wave function now refers to an electron-pair rather than a single electron but there is no change of the principle) there is no excitation of the impurity states, I remains I and we can observe the matrix element $\langle \Psi \mid x_0 I \rangle$ which shows all of the expected dependence on the enclosed flux[9].

INTERACTION WITH CONSCIOUS SYSTEMS

From the point of view developed above a measurement is a completely objective event. In no sense is the presence of a living or conscious observer required. If there is a conscious observer she or he can be made a part of the development of the wave function in the following manner.*

We propose that a mind is a system which, though large and capable of essentially irreversible changes, can be described within the Schrödinger equation. Thus a mind coupled (via the usual sensory organs) to the apparatus $A \equiv (a_U \times a_L)$ of Equation (10) is to be denoted by a wave function of the form

$$\Psi = \psi \times A \times M. \tag{25}$$

As the wave function develops in time we have eventually

$$\Psi = \alpha \psi_U \times A_U \times M_U + \beta \psi_L \times A_L \times M_L. \tag{26}$$

It is important to recall that just as

$$A_U \equiv (a_U^+ \times a_L)$$

and

$$A_L \equiv (a_U \times a_L^+), \tag{27}$$

so also M_U means cognition of a registration in the upper detector and no registration in the lower while M_L means the reverse.

Thus finally the wave function is a linear superposition of the two states

$$\psi_U \times A_U \times M_U$$

and

$$\psi_L \times A_L \times M_L. \tag{28}$$

If we now agree that one of the instruments (say A or M above) interacts in an irreversible manner in the sense discussed, then once this interaction occurs every instrument coupled to U or L in the future registers invariably U or L guaranteeing a complete correlation between the state of the interacting instrument and the branch of the wave function with which it is associated and the two branches of the wave function can no longer interfere.

If the wave function of a quantum system coupled to instruments, minds, etc. at t_1 is

$$\Psi(t_1) = \alpha \psi_U \times A \times M \times \cdots M^e + \beta \psi_L \times A \times M \times \cdots M^e, \tag{29}$$

* We don't insist that a conscious creature can in fact be so described. (It could turn out that this is not possible.) Rather the argument is that such an assumption poses no problems of principle for the interpretation of the quantum theory.

and develops at some later time into

$$\Psi = \alpha \psi_U \times A_U \times M_U \times \cdots M_U^e + \beta \psi_L \times A_L \times M_L \times \cdots M_L^e, \tag{30}$$

and if we, M^e, then regard this system ourselves, we 'discover' for example that it is in the state U. This according to tradition 'reduces' the wave function (the discontinuous process of von Neumann not describable by a unitary transformation) so that upon cognition,

$$\Psi = \alpha \psi_U \times A_U \times \cdots M^e + \beta \psi_L \times A_L \times \cdots M^e \tag{31}$$

suddenly becomes

$$\Psi = \psi_U \times A_U \times M_U \times \cdots M_U^e, \tag{32}$$

and is renormalized. This reduction, however, is not incorrect only in the case in which at least one of the devices (mind included) within the correlated system cannot reverse itself. Were such total reversal to occur, the upper and lower correlated branches would again interfere, and by reducing the wave function we would be throwing away a branch needed to produce the resulting interference pattern. Thus the practice of reducing the wave function upon coming to 'know' the state of the system is either a manner of speaking or incorrect.

We can perhaps make this more graphic by considering what might be called a reversible mind. For simplicity, we assume that the mind may register only two states, U and L. (As a concrete example, consider a 2-level atom and a photon of the right energy, Fig. 3.) Transitions between the two states are possible so that if there are no

L   U

Fig. 3.   A reversible simple mind.

irreversible devices coupled to the system, it oscillates between U and L. Thus the state $M_U(t_1)$ does not preclude the possibility that this could develop into $M_L(t_1 + \delta)$. In this process, of course, the mind would retain no memory of its previous state. In this kind of 'mind reversal' the two branches U and L would be expected to interfere just as in the Stern–Gerlach experiment with the introduction of the current loop. Thus a 'reduction' of the wave function which discards one of the branches would be incorrect.

From this point of view the concept of 'knowing something' has been introduced to order a world in which 'mind reversals' do not normally occur. If 'mind reversal' were a relatively frequent event, 'knowledge' or 'memory' would not exist in the usual sense and the concept of 'knowing something' would not be likely to have been introduced in the same way.

WAVE FUNCTION AND EXPERIENCE

A wave function which is a superposition of various amplitudes however does not contain any information which indicates in which amplitude the system 'really' is. (The type I process of von Neumann (wave function reduction) is designed precisely to put this information into the wave function.) We therefore must find some entity in the theory which corresponds to the seemingly evident fact that knowledge is possible.

When we are aware at the time t_1 that we (M^e) are in the state M_U^e, we no longer can require that the wave function have the form

$$\Psi(t_1) = \psi_U \times A_U \times B_U \times \cdots M_U^e(t_1), \tag{33}$$

although if it has the above form we will find ourselves in the state M_U^e. Rather we propose that what is required when we say that we 'know' a system is in the state U is that there will be agreement between ourselves, other observers and detectors (all systems coupled to us) that the system is in U. At any one time this agreement is contained in the matrix element

$$\langle \Psi(t_1) \mid \psi_U \times A_U \times B_U \times \cdots M_U^e(t_1) \rangle = \alpha, \tag{34}$$

which is the probability amplitude for a simultaneous observation of ψ_U, $A_U \ldots M_U^e$ at t_1. The probability amplitude for disagreement among systems or observers – the probability amplitude for a simultaneous observation of say ψ_U, A_U, $B_L \ldots M_U^e$ at t_1 is

$$\langle \Psi(t_1) \mid \psi_U \times A_U \times B_L \times \cdots M_U^e(t_1) \rangle = 0. \tag{35}$$

The correlation among systems or observers makes a matrix element such as Equation (35) zero. Because of this, our knowledge that our own mind is in some state need not be reflected in the wave function. Rather, it is expressed in the way we pose the question. That a system is in the state U is equivalent to the statement that all coupled 'good' systems and sane minds will agree that the system is in the state U and this agreement is contained in Equation (34). The probability amplitude for disagreement (Equation (35)) is zero.

If we have a system of coupled detectors and minds which develop in time, we must construct some entity that corresponds to our experience that things happen at particular times and that we can know and agree with one another that they happened at those times.

The entity we propose is the time-dependent correlation function which gives the probability amplitude for the measurements Z_n, $Y_m \ldots A_i$ at the different times t_z', t_y', $\ldots t_a'$:

$$Z_n(t_z'), Y_m(t_y') \ldots C_k(t_c'), B_j(t_b'), A_i(t_a'). \tag{36}$$

This is

$$\langle \Psi(t_z') \mid A_i^*(t_a') B_j^*(t_b') \ldots Z_n^*(t_z') \mid 0 \rangle. \tag{37}$$

In the expression above, $|0\rangle$ is the vacuum, $t_a' \leqslant t_b' \leqslant t_c' \cdots \leqslant t_z'$, $\Psi(t_z')$ is the wave function of the system (as obtained from the Schrödinger equation and the initial conditions) at the time t'_z, and $A^*_i(t'_a)...Z^*_n(t'_z)$ are the Heisenberg operators that create the state $|A_i(t'_a)\rangle...|Z_n(t'_z)\rangle$ from the vacuum.

We denote the correlation function (37) by

$$\mathfrak{A}(A_i(t'_a)...Z_n(t'_z)). \qquad (38)$$

and we postulate that this gives the probability amplitude that we *could* experience Z in the state Z_n at $t'_z...A$ in the state A_i at t'_a. Whether or not we actually have this experience depends of course on whether or not we look. Since however the events are there to be experienced when we wish, we may, in the classical sense, assign to them an objective existence.

Now assume that the various detectors (which now may include minds) are sufficiently decoupled so that the wave function can be written as a product, and that the Hamiltonian (other perhaps than for interactions over a very brief period) can be written as a direct sum. Excluding the times $t_a, t_b, t_c, ..., t_z$ the Hamiltonian for the entire system is

$$H = H_A + \quad_B + \cdots H_Z. \qquad (39)$$

At t_a there is a possible interaction H_{0A}; at t_b there is a possible interaction H_{AB} etc. The interaction H_{AB}, for example, acts at the time t_b in such a way that

$$A_U(t)B(t) \rightarrow A_U(t_b)B_U(t_b)$$
$$A_L(t)B(t) \rightarrow A_L(t_b)B_L(t_b). \qquad (40)$$

At t_c a similar possible interaction H_{BC} takes $B_{U(L)}C \rightarrow B_{U(L)}C_{U(L)}$, and so on. A time t'_i is defined to be any time other than the instants at which the detectors are changing their state (a process thought of as being short compared with the times over which the entire system develops). Thus if for example

$$t_i < t'_i < t_{i+1}$$

at the time t'_i all of the systems $AB...I$ have registered U or L while the systems $J...Z$ have not yet registered.

With these simplifications, the correlation function (37) takes the form given in Equation (38) of Ref. 2:

$$\mathfrak{A}(A_i(t'_a)...Z_n(t'_z)) = \langle\Psi(t'_z)|\exp\{-i[(H_A + H_{0A})(t'_z - t'_a)$$
$$+ (H_B + H_{AB})(t'_z - t'_b) + \cdots$$
$$+ (H_Y + H_{XY})(t'_z - t'_y)]\}A_i(t'_a) \times B_j(t'_b) \times \cdots Y_m(t'_y) \times Z_n(t'_z)\rangle. \qquad (41)$$

If the times $t'_a, t'_b...$ all occur after the detectors $A, B...$ have registered (i.e., $t'_a > t_a$, $t'_b > t_b...t'_z > t_z$), this becomes

$$\mathfrak{A}(A_i(t'_a)B_j(t'_b)...Z_n(t'_z)) = \langle\Psi(t'_z)|\exp[-iH_A(t'_z - t'_a)]A_i(t'_a)$$
$$\times \exp[-iH_B(t'_z - t'_b)]$$
$$\times B_j(t'_b)...\exp[-iH_y(t'_z - t'_y)]Y_m(t'_y)Z_n(t'_z)\rangle. \qquad (42)$$

As an example consider

$$\mathfrak{A}\left[A_U(t_a')\,B_j(t_b')\,C_k(t_c')\right] = \langle \Psi(t_c') \,|\, \exp\{-i\left[(H_A + H_{0A})(t_c' - t_a')\right.$$
$$\left. + (H_B + H_{AB})(t_c' - t_b')\right]\}\, A_U(t_a')\,B_j(t_b')\,C_k(t_c')\rangle . \quad (43)$$

Let us assume that $t_a < t_a' < t_b$ and $t_c' > t_c$ so that the detectors A and C have registered.
(1) If $t_b' > t_b$, the inner product becomes

$$\langle \Psi(t_c') \,|\, \exp\{-i\left[H_A(t_c' - t_a')\right]\}\, A_U(t_a') \times \exp\{-i\left[H_B(t_c' - t_b')\right]\}\, B_j(t_b')\,C_k(t_c')\rangle =$$
$$= \langle \Psi(t_c') \,|\, A_U(t_c')\,B_j(t_c')\,C_k(t_c')\rangle . \quad (44)$$

An essential point in the argument is just that the state recognizable as A_U at t_a' develops into a state still recognizable as A_U at t_c'. (The blackened grain on a photographic plate remains recognizable as a blackened grain.)
The wave function at t_c' is

$$\Psi(t_c') = (\alpha\psi_U \times A_U \times B_U \times C_U + \beta\psi_L \times A_L \times B_L \times C_L) \times D \times E \dots .$$

Therefore the correlation function is zero unless j and k are both U. Thus given that A has registered A_U at t_a', we can conclude that B will register B_U at t_b' and C will register C_U at t_c'.
(2) If $t_b' < t_b$ the detector B has not yet registered $(B_j = B)$ and the inner product becomes

$$\langle \Psi(t_c') \,|\, \exp\{-i\left[H_A(t_c' - t_a') + (H_B + H_{AB})(t_c' - t_b')\right]\}\, A_U(t_a')\,B(t_b')\,C_k(t_c')\rangle; \quad (45)$$

at t_b, $A_U B \rightarrow A_U B_U$ due to the interaction H_{AB}; at the later time t_c' the scalar product becomes

$$\langle \psi(t_c') \,|\, A_U(t_c')\,B_U(t_c')\,C_k(t_c')\rangle , \quad (46)$$

which equals zero unless $k = U$.
Thus for $t_a' > t_a$, $t_b' < t_b$, $t_c' > t_c$ (and as before $t_a' < t_b' < t_c'$) we find a non-zero amplitude for A_U, B unregistered and C_U. The interaction H_{AB} converts B unregistered at t_b' into B_U for times later than t_b.
We can now say that something occurred (e.g., a measurement) after the first irreversible interaction. For after t_a (when A has registered) we can say

$$\text{given} \quad A_U(t_a') \rightarrow B_U(t_b') \rightarrow C_U(t_c') \quad \text{etc.,}$$
$$\text{if} \quad t_b' > t_b \quad \text{and} \quad t_c' > t_c$$

(if the detectors have registered). Or

$$\text{given} \quad A_L(t_a') \rightarrow B_L(t_b') \rightarrow C_L(t_c') \quad \text{etc.}$$

Before the irreversible interaction we could not say this because $\psi_U(t_1)$ might develop into $\psi_L(t_2)$. It is only with an irreversible interaction that we can be sure that the entire sequence of correlations will follow. We can therefore say that something happens when no interaction (in the usual sense) can reverse the situation: when given $A_U(t_a')$ implies that we will not find $A_L(t_b')$.

Our mind in this respect is like any other irreversible system. We can say that given $A_U(t_a')$ our mind will register $M_U^e(t_m')$ as

$$\mathfrak{A}[A_U(t_a')\ldots M_L^e(t_m')]=0.\tag{47}$$

Therefore we can say that if U occurred I will see it as U or if I see it as U it was U. There is nothing special about cognition. It is one act in a correlated chain.

If the mind were not thought to be irreversible, $M_U^e(t_0)$ could develop into $M_L^e(t)$, or to some other state not orthogonal to M_L^e. The essence of the assumption of irreversibility is that $M_U^e(t_0)$ develops into a state always recognizable as M_U^e and always orthogonal to M_L^e. Thus the amplitude that the mind be in the state U, given that we found ourselves in U at some previous time, is always the same and always comes from the same branch of the wave function. The amplitude that we find L, given that we found U previously, is zero since M_L cannot develop from M_U.

WIGNER'S FRIEND IN THE OTHER WORLD

Since Max Born's probability interpretation of the quantum theory it has become textbook wisdom that the wave function does not contain complete information of what will happen in the future. This is unusual from a classical point of view, but still comprehensible if we accept a quantum theory that is a-causal (not deterministic): 'Equal causes do *not* produce equal effects.'

If we add to this the additional limitation that *the wave function does not contain complete information of where we are now*, we extend to the present what has since Born been accepted for the future. Here we depart from a traditional view of the relation between the wave function and the observer as expressed for example by Dirac[10]:

If the system is in a state such that a measurement of a real dynamical variable ξ is certain to give one particular result...then the state is an eigenstate of ξ and the result of the measurement is the eigenvalue of ξ to which this eigenstate belongs.

For we entertain the possibility that even though, in actual measurements, we may obtain the result ξ' for the real dynamical variable ξ and that this is as certain as anything we encounter in our experience, the wave function may not be an eigenstate, $|\xi'\rangle$, of ξ. [It follows that given the Hamiltonian for the universe, unless we know the entire wave function at some time (say $\Psi(t_0)$), the range of our present possible observation might not enable us to reconstruct our past.[11]] This perhaps makes the theory incomplete, but no more so that it already was when the possibility of a determined future was abandoned.

If we accept such an approach a number of questions are posed. Perhaps the most puzzling arise from the attempt to reconcile our certain feeling that something has occurred, that we have a definite present, with the absence of any such indication in the wave function itself. [It is just the attempt to put our knowledge of the present into the wave function that leads to the concept of wave function reduction.] We

have no difficulty for example describing a system of instruments, other minds (or
even our own mind if we are speaking of the future) and stating that the system (instruments, other minds) is, will be, or was, either in state U or L, each with the probability amplitude given by the proper inner product.

The process of discovery or cognition however is difficult to describe. We can say
of our own mind that it will be the state U or L with probability $|\alpha|^2$ or $|\beta|^2$ at a
future time. But to say that our own mind was in either of the states U or L in the
past means that we have forgotten; to say that right now our mind is either in the
state U or L means that we do not know the state of our own mind. When one speaks
of cognition it is tautologically implied that one knows the state of one's own mind –
that one has discovered which of the various possibilities in actual fact has come to
pass.

It is here that we sense the discomfort that Wigner and/or his friend have felt
and described so graphically.[12] For if the wave function is a linear superposition
of ourselves in the state M_U^e and M_L^e, (30) how can we account for our unshakable
conviction (as firm as any conviction we have) that we are in fact in one or the other
but not in both. Or if the wave function, (30), contains two branches (U and L) which
way has the world gone? Does one have to say with De Witt[6] that there are now two
worlds U and L (soon to become a non-denumerable infinity of worlds) miraculously
separated from each other so that they do not interfere? This seems so intrinsically
paradoxical and uneconomical that one is tempted to reject any such approach without further consideration. But need matters be so bad? Let us at least attempt to
develop the point of view that they are not.

The theory as proposed above contains the following entities:

(1) the Hamiltonian of the entire system, H
(2) the wave function at some initial time t_0, $\Psi(t_0)$
(3) the correlation function $\mathfrak{A}(A_i(t'_a)...Z_n(t'_z)...)$.

The wave function develops according to the Schrödinger equation

$$\Psi(t) = e^{-i/\hbar H(t - t_0)} \Psi(t_0)$$

and this determines the correlation function (37). It is this last which is put in correspondence with our experience. In one sense, of course, the possible forms of (37) are
dictated by the Hamiltonian of the complete system. (In such a strict sense it is also
true in classical physics that the possible observations an observer could make are
dictated by the Hamiltonian and the initial conditions.) If for convenience (and only
for convenience) we allow the observer to look where he wishes, then all of his experience can be put into correspondence with (37), but we must have the freedom to
choose this to correspond to the experimental arrangement he has constructed.

We may now, perhaps, provide some comfort for Professor Wigner and his friend.
If the wave function (even after an 'observation') has the form (30), it is a linear
superposition of the mind in the state M_U^e and M_L^e. How is it that M^e does not become
schizophrenic? We propose that M^e feels no discomfort because discomfort is equivalent to a sense of the immediate presence of both U and L; but the probability

amplitude for any observation of the form

$$\langle \Psi(t) \mid \ldots_U \ldots_L \ldots \rangle = 0.$$

Our experience of the consistency of our world is contained in the agreement between the various events described in the correlation function. Our certain sense that we are in the state U (or L) is given to us by the nature of the physiological system called mind which with the detector (10) is postulated to have the states

$$M_U = (m_U^+ \times m_L \ldots) \quad U \text{ and not } L$$

and

$$M_L = (m_U \times m_L^+ \ldots) \quad L \text{ and not } U. \tag{48}$$

These states of mind can, via the time-dependent correlation amplitude, be consistent with all the experience M has of its environment.

We propose that even though the wave function is a linear superposition of state U and L, this need not be disturbing if the wave function no longer is considered to contain complete information of our present state. This information – just as certain information concerning future is not contained in the theory. We seem to be able to accept the statement that the wave function does not contain complete information of what will happen in the future (even though we are completely aware that it is the nature of the world that something will happen). But when we attempt to extend this to the present we draw back because we know where we are at present.

This is closely related to a second difficulty which is associated with the view expressed by various authors[13], that since the wave function is a superposition of several amplitudes, each of which in a certain sense corresponds to a possible world, we must accept the idea that there are in fact many worlds developing simultaneously but separated from each other because of the impossibility (or improbability) of interference between them. Besides its intrinsic paradoxical character this view suffers from the difficulty of not really being able to assign an instant when the worlds 'split'. Since interactions range from the easily reversible to those which never reverse there is no point in the scale of lifetimes at which we can say: 'There the world has split and we are on one branch.' Further, though it is the case that the wave function may be regarded as a sum of several amplitudes (say U and L) it may just as well be regarded as a linear superposition of

$$\frac{(U + L)}{\sqrt{2}} \text{ and } \frac{(U - L)}{\sqrt{2}}.$$

When then does the world split?

These problems arise from the attempt to associate the wave function itself with experience. If however we associate the correlation amplitude (37) with experience no difficulty of this form arises. We assert that there exists a universe that is or can be experienced. The probability amplitude for any given experience or sequence of experiences is given by (37). How then do we interpret the non-zero amplitude

$$\langle \psi \mid \psi_L \times A_L \times \cdots M_L^e \rangle = \beta, \tag{49}$$

when we 'know' that the system is in the state U? This amplitude we assert is in fact non-zero and corresponds to the amplitude that the system and all of the minds (including our own) are in the state L. (We don't usually ask for the amplitude that our mind be in the state L if we know it to be in U.) That this amplitude is non-zero has no effect on the other amplitude (U) if the two can never interfere. If we arrange another double slit so that the upper branch is split into two parts, one is concerned only with the conditional probability (given M_U^e for the first set of slits) for the various possible outcomes of the second process. As long as there is no possibility of interference, and our questions are prefaced by 'given M_U^e,' we can discard the amplitude $\psi_L \times \dots$ and renormalize $\psi_U \times \dots$. Thus the so-called reduction is really a renormalization. If, however, the systems A, M, etc., can be reversed (brought back to a state in which the two amplitudes can interfere) then it is essential that the amplitude $\psi_L \times \dots$ be retained so that the possibility of interference be preserved. We know something, therefore, because of the possibility of an infallible correlation between the state of our mind and other minds and systems, and not because the wave function of the world has no amplitudes for other possibilities. The other amplitudes reveal themselves only in mind reversals and reversals of macroscopic systems that do not ordinarily occur. *The wave function is thus a dynamical quantity never converted to 'knowledge' in the sense of a classical probability. Conversion to knowledge occurs entirely in the mind and is described by*

$$M^e \to M_U^e \quad \text{or} \quad M^e \to M_L^e. \tag{50}$$

In those cases in which a measurement at an earlier time does not determine the result of a later measurement (visualize a sequence of double slits with detectors A^1, $A^2 \dots A^k \dots$) and the correlation amplitude

$$\mathfrak{A}\left(A_{i_1}^1\left(t_1'\right) A_{i_2}^2\left(t_2'\right) A_{i_3}^3\left(t_3'\right) \dots A_{i_k}^k\left(t_k'\right) \dots\right) \tag{51}$$

where $i = U$ or L for the case of a double slit, a sequence of 'measurements' (values of A) are required if we want to distinguish the world we are experiencing from those we might have, or might be experiencing. The world as it is presently was not foretold by the wave function in the past and cannot be contained in the wave function in the present. For such an uncorrelated sequence one cannot say what the registration in A^{i+1} will be, given a particular registration in A^i. One can only give probability amplitudes for the various different possible outcomes. Our desire to somehow have a dynamical connection between the various outcomes would in effect, if carried out, mean a return in one form or another to a deterministic theory.

I have attempted to show that there is a consistent interpretation of the quantum theory which requires no new external concepts, retains the probabilistic character of the theory and is in agreement with experience. This does not demonstrate that no new concepts will be necessary in order to produce a theory in agreement with all future or even present experience. Rather it is proposed that no such new concepts are required for the interpretation of the quantum theory.

ACKNOWLEDGEMENTS

This work was supported in part by the Advanced Research Projects Agency and the U.S. National Science Foundation.

REFERENCES AND NOTES

1. J. von Neumann, *Mathematical Foundations of Quantum Mechanics* (translated by Robert T. Beyer), Princeton University Press, Princeton, New Jersey (1965).
2. L. N. Cooper and D. van Vechten, *Amer. J. Phys.* **37**, 1212–1220 (1969).
3. J. A. Wheeler, *Rev. Mod. Phys.* **29**, 463 (1957).
4. H. Everett III, *Rev. Mod. Phys.* **29**, 454 (1957).
5. K. Gottfried, *Quantum Mechanics*, W. A. Benjamin, Inc., New York (1966), vol. I, pp. 165–189.
6. B. DeWitt, *Physics Today* (1970).
7. D. Bohm, *Quantum Theory*, Prentice-Hall, Inc., Englewood Cliffs, New Jersey (1951), Chap. 22.
8. E. P. Wigner, *Amer. J. Phys.* **31**, 12 (1963).
9. R. C. Jaklevic, J. Lambe, A. H. Silver and J. E. Mercereau, *Phys. Rev. Letters* **12**, 159 (1964); R. C. Jaklevic, J. Lambe, A. H. Silver and J. E. Mercereau, *Phys. Rev. Letters* **12**, 274 (1964).
10. P. A. M. Dirac, *The Principles of Quantum Mechanics*, Oxford, third edition (1947), p. 35.
11. J. S. Bell, *International Colloquium on Issues in Contemporary Physics and Philosophy of Science, and Their Relevance for our Society*, Penn State University, September 1971.
12. E. P. Wigner, *The Scientist Speculates* (edited by I. J. Good), London (1961), reprinted in *Symmetries and Reflections*, Bloomington (1967), p. 171.
13. B. DeWitt, ref. 6 and B. d'Espagnat, *Conceptual Foundations of Quantum Mechanics*, W. A. Benjamin Inc., Menlo Park, California (1971). See especially chapter 20.

33

The Problem of Measurement in Quantum Mechanics

Josef M. Jauch

The problem of measurement in quantum mechanics concerns the question whether the laws of quantum mechanics are consistent with the acquisition of data concerning the properties of quantal systems. This consistency problem arises because the system to be measured as well as the apparatus used for the measurement are themselves systems which are presumed to obey the laws of quantum mechanics. Therefore the evolution of the state of such systems is governed by a Schrödinger-equation.

One can see without any calculation, reasoning from very general properties of such evolutions, that the measuring process exhibits features which are apparently inconsistent with the Schrödinger type evolutions.

(*a*) The typical process ends with the establishment of a permanent and irreversible record. This contradicts the time-reversible Schrödinger equation.

(*b*) In a given individual experiment the result of the measurement is one of several alternatives. A repetition of the experiment under identical initial conditions may lead to another of these possible alternatives. This is incompatible with the unitary evolution of Schrödinger which always transforms pure states into pure states.

Several 'solutions' have been proposed for the explanation of this apparent inconsistency. Let me mention some:

1. The Schrödinger evolution is *not generally valid*. It somehow breaks down in the presence of sufficiently large systems.

2. Quantum Mechanics is an *incomplete* theory which does not describe all the real physical properties of individual systems but only certain statistical properties of ensembles of such systems. A complete theory would involve the presence of *hidden variables*. The values of such hidden variables would be responsible for the outcome of the different alternatives in the measuring process.

3. Quantum Mechanics involves an interaction between consciousness and the physical world. Just as the physical world acts on the state of consciousness, so in a kind of generalized law of action and reaction, consciousness should react back on the physical states of microsystems in just such a manner as to produce the observed behaviour of quantal systems under measurements.

In my opinion none of these explanations of the enigma of the measuring process is promising. Let me give you some reasons why I think so.

(ad 1) If one disregards the measuring problem there is absolutely no evidence that the Schrödinger equation for closed systems should not be generally valid. In fact it is precisely the quantal behaviour of macroscopic systems which continue to be discovered

J. Mehra (ed.), The Physicist's Conception of Nature, 684-686. All Rights Reserved
Copyright © 1973 by D. Reidel Publishing Company, Dordrecht-Holland

that furnish startling extensions of the validity of quantal laws. Among such systems are the superconducting metals, the superfluids, the lasers. Hence it would be contradictory to all experiences with such systems if one were to adopt an ad hoc hypothesis of a break-down of quantum theory for large systems in order to explain the measuring process.

The situation is different for open systems. In fact the subsystem of a composite system, consisting of two quantal systems cannot evolve in accord with a Schrödinger equation since for such a system pure states are generally changed into statistical mixtures. This fact is in accord with the principles of quantum mechanics. But it is irrelevant for the measuring process since we can always consider the evolution of the closed system to be measured together with the apparatus of measurement. This apparatus may be in the form of a cat enclosed in a box as Schrödinger assumed with his 'hellish contraption' and the alternatives in this case would be the cat alive or dead.

(ad 2) Hidden variables have been considered by many people, already before von Neumann made his famous proof that they are inconsistent with the rest of quantum mechanics. As is well known this proof was criticized and hidden variables continued to sprout like weeds in the garden. There are essentially two possibilities: hidden variable theories may be so constructed that they reproduce all of the measurable consequences of quantum mechanics. This can and has been done, although these variables have some very strange properties. The other possibilities are hidden variable theories which lead to physically observable differences in certain crucial experiments. Then the theories can be experimentally tested. Such tests have been made and the results have been negative and thus give an overwhelming confirmation of unadulterated quantum mechanics.

(ad 3) The involvement of consciousness at this stage seems hard to reconcile with the fact that the experimental situation can be so arranged that consciousness plays actually no part whatsoever during the act of measurement for instance by storing the permanent record in the memory of a large computer. This information can be recalled at a much later time involving only the observation of macroscopic and classical systems. It is hard to see why this last step of becoming aware of these macroscopic and classical data in some consciousness should have a decisive influence on the actual measuring process which was completed a long time ago.

There are other, more fantastic schemes, which have been invented to explain the measuring process, but it is not my intention to give a complete inventory since I do not believe that any of them solve the problem of measurements in a satisfactory manner.

It seems to me no such solution is possible without taking into consideration Bohr's essential contribution to this problem by insisting on the essentially classical feature of the result of any measurement. The difficulty with this remark comes from the fact that the notion of a classical system in the context of quantum theory is difficult to describe with precision. It is easy to give a mathematical definition of a classical system: It is one for which all observables commute with one another. But it is difficult to show that such systems exist. It would seem to be almost certain that they can only exist in some approximative sense.

At this point it is perhaps helpful to compare the quantal measuring process with the classical one. Of the two major problems which I have formulated, the first one appears already in certain classical types of measurements. For example measuring the temperature of a body with a thermometer involves an irreversible flow of heat from the system to the apparatus (or vice versa) and this is irreversible in spite of the fact that the equations of motion for classical systems are invariant under time-reversal. But even the second one has some analogy with the fact that classical quantities, which are usually averages over a large number of molecules, invariably are subject to fluctuations and it is precisely the occurrence of these fluctuations which enable us in principle to reconcile the arrow of time with the time-reversibility of the equations of motion.

This analogy strongly suggests that the solution of the measuring problem in quantum theory is closely related to the problem of deriving the laws of irreversible evolutions of thermodynamic variables from the reversible equations of motion for the constituent molecules of a large system. This is indeed the case. The key concept in both situations is *coarse-graining*.

I have shown in 1964 that the mathematical definition of a classical system implies that the states of the system fall into equivalence classes which by no measurement with this classical apparatus can be distinguished from each other. These equivalence classes are therefore the quantal analogue of the classical coarse-graining. These equivalence classes I have called *macro states*. They evolve not in a deterministic but only in a stochastic manner in complete accord with the prediction calculated on the basis of the usual rules of quantum mechanics. [J. M. Jauch, *Helv. Phys. Acta* **37**, 293 (1964).]

With this observation the problem of measurement is not completely solved, however, because it should be possible to derive the stochastic master equation for the evolution of macrostates with some realistic model. Much work has been done on this part of the problem during recent years. I mention here especially the attempts of Van Hove, Prosperi, Prigogine, Rosenfeld and others. Although I believe that the ideas developed in these papers are essentially correct, I have some reservations from the mathematical point of view concerning this phase of the problem. I shall not dwell on this point, but merely express my conviction that non-relativistic quantum mechanics is consistent with the measuring process and a complete proof needs merely the solution of certain technical but unfortunately rather difficult mathematical problems.

34

Subject and Object

J. S. Bell

The subject–object distinction is indeed at the very root of the unease that many people still feel in connection with quantum mechanics. *Some* such distinction is dictated by the postulates of the theory, but exactly *where* or *when* to make it is not prescribed. Thus in the classic treatise[1] of Dirac we learn the fundamental propositions:

'... any result of a measurement of a real dynamical variable is one of its eigen-values ...',

'... if the measurement of the observable ξ for the system in the state corresponding to $|x\rangle$ is made a large number of times, the average of all the results obtained will be $\langle x | \xi | x \rangle$ ...',

'... a measurement always causes the system to jump into an eigenstate of the dynamical variable that is being measured ...'.

So the theory is fundamentally about the results of 'measurements', and therefore presupposes in addition to the 'system' (or object) a 'measurer' (or subject). Now must this subject include a person? Or was there already some such subject–object distinction before the appearance of life in the universe? Were some of the natural processes then occurring, or occurring now in distant places, to be identified as 'measurements' and subjected to jumps rather than to the Schrödinger equation? Is 'measurement' something that occurs all at once? Are the jumps instantaneous? And so on.

The pioneers of quantum mechanics were not unaware of these questions, but quite rightly did not wait for agreed answers before developing the theory. They were entirely justified by results. The vagueness of the postulates in no way interferes with the miraculous accuracy of the calculations. Whenever necessary a little more of the world can be incorporated into the object. In extremis the subject–object division can be put somewhere at the 'macroscopic' level, where the practical adequacy of classical notions makes the precise location quantitatively unimportant. But although quantum mechanics can account for these classical features of the macroscopic world as very (very) good approximations, it cannot do more than that.[2] The snake cannot completely swallow itself by the tail. This awkward fact remains: the theory is only *approximately* unambiguous, only *approximately* self-consistent.

It would be foolish to expect that the next basic development in theoretical physics will yield an accurate and final theory. But it is interesting to speculate on the possibility that a future theory will not be *intrinsically* ambiguous and approximate. Such a theory could not be fundamentally about 'measurements', for that would again imply

J. Mehra (ed.), The Physicist's Conception of Nature, 687–690. All Rights Reserved

incompleteness of the system and unanalyzed interventions from outside. Rather it should again become possible to say of a system not that such and such may be *observed* to be so but that such and such *be* so. The theory would not be about '*observables*' but about '*beables*'. These beables need not of course resemble those of, say, classical electron theory; but at least they should, on the macroscopic level, yield an image of the everyday classical world[4], for 'it is decisive to recognize that, however far the phenomena transcend the scope of classical physical explanation, the account of all evidence must be expressed in classical terms'.[5]

By 'classical terms' here Bohr is not of course invoking particular nineteenth-century theories, but refers simply to the familiar language of everyday affairs, including laboratory procedures, in which objective properties – *beables* – are assigned to objects. The idea that quantum mechanics is primarily about 'observables' is only tenable when such beables are taken for granted. Observables are *made* out of beables. We raise the question as to whether the beables can be incorporated into the theory with more precision than has been customary.

Many people must have thought along the following lines. Could one not just promote *some* of the 'observables' of the present quantum theory to the status of beables? The beables would then be represented by linear operators in the state space.[6] The values which they are allowed to *be* would be the eigenvalues of those operators. For the general state the probability of a beable *being* a particular value would be calculated just as was formerly calculated the probability of *observing* that value. The proposition about the jump of state consequent on measurement could be replaced by: when a particular value is attributed to a beable, the state of the system reduces to a corresponding eigenstate. It is the main object of this note to set down some remarks on this programme. Perhaps it is only because they are quite trivial that I have not seen them set down already.

The state vector (or density matrix) in what follows will always be that of the Heisenberg picture: all time dependence is in the operators and the state refers not to a single time but to a whole history. This permits us, if we wish, to define the 'system' under study simply as a limited space-time region. This seems a less intrinsically ambiguous and unrealistic way than any other I can think of to separate off a part of the world from the rest. Of course, one could try to think of the world as a whole, but it is less intimidating to think of only a part. In the approach[8] known as the 'theory of local observables' a Heisenberg state (pure or mixed) can indeed be attributed to any limited region of space-time. It gives, roughly speaking, the expectation value of all functions of the Heisenberg field operators with space-time arguments in that region. If something like a Lorentz-invariant causal connection between field operators is postulated then the region of relevance of the state vector can be extended by including all points whose forward or backward light cones pass entirely through the original region, as in Fig. 1. It is then the Heisenberg state of the extended region which reduces, whenever a 'local beable' in that region is attributed a particular value, to its projection in the subspace with the given eigenvalue. Whatever the particular space-time location of the beable considered, there is no question of any particular space-time location of the associated state reduction, which is coextensive with the whole history of the system under study.

Whereas 'measurement' was a dynamical intervention, from somewhere outside, with dynamical consequences, it is clear that 'attribution' must be regarded as a purely conceptual intervention. It is made, say, by a theorist rather than an experimenter; he is quite remote in space and time from the action, and simply shifts his attention from the whole of a statistical ensemble to a sub-ensemble. It follows that attributing a particular value to some beable cannot change particular values already attributed to some other beables. It follows that only those states can be allowed which are simultaneously eigenstates of all beables, or superpositions of such states. Moreover, we

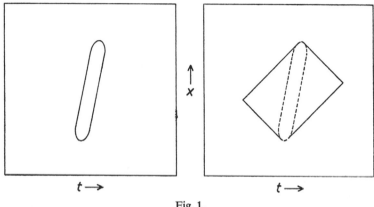

Fig. 1.

need only consider incoherent superpositions, for the beables, unable to induce transitions between different eigenstates, are insensitive to any coherence. Now the beables may not be a complete set, and a list of their eigenvalues may not characterize a state completely. However, the converse is true: when a particular member state of the incoherent superposition is specified, definite values are specified for all beables. Thus the theory is of deterministic hidden-variable type, with the Heisenberg state playing the role of hidden variable. When this state, which may originally refer only to the limited region in the figure, is specified, all beables in the extended region are determined.

I suspect that a stronger conclusion would be possible, that one cannot in fact find interesting candidates for beables in interesting quantum mechanical systems. But my own indications in this direction seem to me unnecessarily elaborate and I will not attempt to present them here. The preliminary conclusion is in a way more striking. In the basic propositions quoted from Dirac there was in fact another element, in addition to the vague subjectivity, which could have disturbed a nineteenth-century theorist. That is the *statistical undeterministic* character of the basic notions. In following what seemed to be a minimal programme for restoring objectivity, we were obliged to restore determinism also.

REFERENCES AND NOTES

1. P. A. M. Dirac, *The Principles of Quantum Mechanics*.
2. In this connection there are many very relevant investigations involving considerations which may be roughly identified by the words 'ergodicity' or 'irreversibility'. They tend to show that the effect of wave packet reduction associated with macroscopic observation is macroscopically negligible. (Or it may even be shown that the effect is accurately zero in some hypothetical limit: e.g., K. Hepp[3]) takes infinite time.) The relevance of these investigations is of course to the question of the sufficient unambiguity of the theory for practical purposes, and not at all to the question of principle considered here.
3. K. Hepp, *Helv. Phys. Acta* **45**, 237 (1972).
4. A more extreme position would be that the beables need refer only to mental events.
5. N. Bohr.
6. Such beables would be related to the 'classical observables' of Jauch and Piron (see for example the contributions of these authors in *Foundations of Quantum Mechanics*, Proceedings of the International School of Physics 'Enrico Fermi', Course IL, Academic Press, New York, 1971; also H. Primas[7]). However, these authors (*loc.cit.* and private communications) intended their 'classical observables' to refer only to 'apparatus' while not in interaction with 'quantum systems' and perhaps to be only approximately 'classical'. Here we wish to avoid any arbitrary division of the world into 'systems' and 'apparatus', and any arbitrary limitation on the range and duration of interactions, and are concerned with the question of principle and not with that of practical approximation.
7. H. Primas, *Advanced Quantum Chemistry of Large Molecules*, Vol. 1: 'Concepts and Kinematics of Quantum Mechanics of Large Molecular Systems', Academic Press, New York (1973), and preprint (July 1972).
8. See, for example: R. Haag, in *Lectures on Elementary Particles and Quantum Field Theory*, 1970 Brandeis Lectures (Editors S. Deser, M. Grisaru and H. Pendleton), M.I.T. Press (1970). In this theory the over-all system need not be finite. The idea that the measurement problem might be significantly different in such a context has sometimes been expressed.[3, 7, 9].
9. See, for example, the preface to B. d'Espagnat's *Conceptual Foundations of Quantum Mechanics*, Benjamin, New York (1971).

35

Subject, Object, and Measurement

R. Haag

The laws of quantum physics are formulated in terms of predictions for the outcome of experiments. Traditionally in physics the motivation for the performance of experiments was the desire to learn about the 'outside world', i.e. the world abstracted from the presence of conscious, thinking and planning beings. One may ask first whether quantum physics teaches us that such an abstraction is grossly illegitimate. If the answer is no[1,2] then we may ask further: what can we learn from the laws of quantum physics about the properties of this outside world? Specifically we may ask which concepts can be used, which attributes can be assigned to describe the outside world? We shall call such attributes 'real' or 'objective' quantities.

In quantum physics just as in classical physics we talk about systems and processes. This means that we divide the world into small pieces to which we assign an individuality. It is quite clear that this division procedure cannot be carried through arbitrarily and it may have its limitations. Niels Bohr emphasized the indivisibility of certain processes as one of the essential lessons of quantum physics. If we go to the extreme and paraphrase this feature saying 'the whole is more than the sum of its parts' then we have to admit that no system (apart from the whole universe) and no event can be completely 'objective' because in isolating it, in assigning an individuality to it, we (subjectively) introduce some falsification of the real world. On the other hand it is evident that under suitable circumstances we can consider systems and events as real (objectively existing) individuals within a sufficient degree of approximation. We are thus led to the approximate (or asymptotic) concept of 'irreducible system' and 'irreducible process'. Let us make this more concrete in the example of quantum *mechanics*. There a system will be a collection of particles such that at the time under consideration the ties of this set of particles to the remainder of the world may be (practically) ignored. It may be called irreducible if, at this time, no subset of these particles may be considered as isolated from the others to the desired degree of approximation.

Thus, the notion of 'system' (to which we can assign an individuality in an objective sense) will vary with time. We have so far used as the criterion for application of this notion large spatial separation from other bodies (relative to the range of forces) at a given time. Let us see whether this is sufficient to arrive at a reasonably complete, objective description of the outside world. According to this picture we have a decomposition of the outside world into 'irreducible systems' existing at a particular time and 'events' which correspond to collision processes between systems and lead in

J. Mehra (ed.), The Physicist's Conception of Nature, 691-696. All Rights Reserved
Copyright © 1973 by D. Reidel Publishing Company, Dordrecht-Holland

general to a change of the systems. The schematic picture of a (macroscopic) space-time region of the outside world is then a network in which the lines correspond to irreducible systems, the vertices to events. The direction arrows on the lines indicate the temporal sequence (see Fig. 1).

In order to come from this picture to a reasonably complete description which satisfies our desire for (at least statistical) causality, the simplest assumptions would be

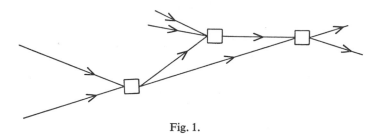

Fig. 1.

(a) each system has attributes summarily denoted by ξ (its individual state).[3]

(b) for a macroscopic system the well-known macroscopic attributes (classical or thermodynamic state quantities) belong to ξ.

(c) An individual event is characterized by a transition from an initial state $\xi_1,...,$ ξ_n (ξ_i being the individual state of the ith incident system) to a final state $\xi'_1,...,\xi'_m$. Given the initial state the laws of nature determine the probability $W(\xi'_1,...,\xi'_m \mid \xi_1, ...,\xi_m)$ for this event.

The theory would then have to classify possible types of systems and events, provide information about the set through which ξ may range for each type of system and determine the probabilities W for each type of event.

Let us see now, whether the above picture with the simple assumptions (a), (b), (c) is in accordance with the laws of quantum physics. The answer will be: not entirely, but in important special circumstances yes. We discuss some characteristic examples first and shall then consider the necessary modifications of the above assumptions.

EXAMPLE 1: *Passage of cosmic radiation through the atmosphere*

Here the simple scheme is adequate. The 'systems' are single particles such as molecules in the air, protons, mesons etc. The events are binary collisions. The individual state of each particle is characterized by a *momentum p and a spin wave function u* (direction in a $(2S+1)$ dimensional complex space). The probability $W(p'_i u'_i \mid p_1 u_1, p_2 u_2)$ is replaced by the differential cross-section which is computed according to quantum physical laws for a *single* binary collision process. With this input and an ordinary stochastic treatment of our ignorance (including the averaging over all spin wave functions) we obtain a good description of the observed phenomena. Why do we know that each particle has a definite momentum though we may not know its value and we have not attempted to measure it? This comes from the circumstance that the spatial distance between subsequent events is large compared to the range of the interaction forces. Therefore, on the one hand (using now customary language of wave mechanics) a particle emanating from one event will have a wave function which is practically a plane wave within the subsequent interaction region. On the other hand the network will be practically always a 'tree diagram' (no loops occur) so that 'interference phenomena' between different branches of complicated multiparticle wave functions are excluded.

Comments: In spite of the special circumstances and in spite of the approximations used in the treatment, this example illustrates one very important and general aspect. The concept of an individual state for a (reasonably isolated) system is introduced for the purpose of separating the past history from the future fate of this system. We may consider each as a part of the network, the former consisting of those events which are reached from the line of the system under consideration by proceeding exclusively along lines in the opposite sense of the time arrow, the latter of those reached by following the lines in the direction of the arrow. The union of the two parts, of course, still does not give the whole connected part of the network. The method of theoretical description to which the assumptions (*a*), (*c*) above are geared is such that we envisage a class of networks in which the past history of one or several systems is fixed whereas the future parts vary. The essential simplifying feature in our Example 1 lies in the fact that we do not have to envisage all possible future types of events but only a limited class of them. The limitation is not a subjective one, but arises from the (real) limitation in the types of systems which are available as collision partners. Corresponding to this limitation the parts of quantum mechanical wave functions belonging to far separated regions need no longer be regarded as members of one inseparable unit but may each be considered as individual entities describing mutually exclusive possibilities. In the statistical description of an ensemble of such processes this is the step of replacing the wave function by an incoherent mixture. It arises here not as the result of a planned measuring act but as a consequence of qualitative limitations in the possible types of reaction processes under the prevailing circumstances of the part of the outside world under consideration.

One may object that this argument – and hence the assignment of the individual state to each system – depends crucially on an approximation and that the realistic description is untenable if the process is treated really rigorously. However, one has to bear in mind that in all cases in which we carve out an individual element from a complex structure there is an approximation involved; very large quantitative differences become qualitative differences and in every area of science we are concerned with idealizations based on the disregard of absolute precision. If we wanted absolute precision we would also have to negate the existence of individual human beings and there would be no sense to our discussion.

EXAMPLE 2: *Einstein, Podolsky and Rosen*

This is the most transparent illustration of the fact that the simple assumptions (*a*) and (*c*) above have to be modified, i.e. that the spatial isolation at one time is not always enough to guarantee that a system has an objective individual state in the sense used in assumption (c).

Let us consider the decay of a spin zero particle A into two charged spin $\frac{1}{2}$ particles B, C, which subsequently may interact with Stern–Gerlach magnetic fields S_i and ultimately with photographic plates P_i. The network corresponding to such a process is shown in Fig. 2.

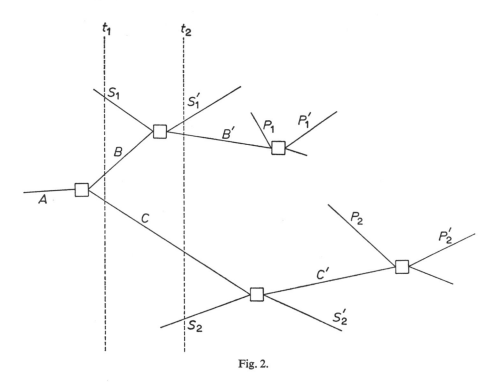

Fig. 2.

The magnets S_i are characterized by unit vectors e_i which give the direction of the inhomogeneous magnetic field; the plate P_i is so placed that it detects a particle if it is deflected from its original path in the directions e_i and it is not hit by a particle deflected in the $-e_i$ direction. We have also indicated on Fig. 2 two time cuts, the first (time t_1) in which we have the practically isolated systems B and C before any interaction with a magnet; the second at t_2 where we have B' and C and B' has undergone a scattering process by the magnet S_1 whereas C has not interacted yet.

According to the quantum mechanical calculation the probability of obtaining a speck on both plates is given by

$$P(e_2, e_1) = \tfrac{1}{2} \sin^2 \frac{\theta}{2}$$

where θ is the angle between e_1 and e_2.

If the assumptions (a) and (c) were good for the first time cut then we should be able to write $P(e_2, e_1)$ in the form

$$P(e_2, e_1) = \sum_{\xi_1, \xi_2} F(\xi_1) F(\xi_2) W(e_1, \xi_1) W(e_2, \xi_2)$$

where F, G are probability distributions and W is a transition probability matrix. Such a decomposition of P is, however, impossible.[4]

Therefore, in this example it is not possible to attribute individual states to the systems B and C separately at time t_1 although the criterion of large spatial separation is satisfied. The attribute of individual state at time t_1 can only be given to the combined system $(B\ C)$.

The assumptions (a), (b), (c) have to be modified by replacing the word 'system' everywhere by another concept which we might call an 'irreducible complex'. Such a complex may possibly consist of several irreducible systems, coherent in spite of large spatial separation due to common past history.

In the course of time as the network grows there is a tendency of breaking up complexes (as well, of course, as the possibility of forming new ones). Thus in Example 2 we have at t_1 the irreducible complex (B, C) but already at t_2 the irreducible complexes are the systems B and C separately. The coherence is broken by the same mechanism as described at the end of Example 1. The magnet S_1 decomposes the total wave function into two disjoint parts with ultimately far separated support regions in configuration space so that no interference between them is possible in any practically attainable future event. Each of these two disjoint parts describes a definite spin state for B and for C.

The Einstein–Podolsky–Rosen example demonstrates in the extreme the fact that the concept of an objective individual state is not always a causal one. Here the existence of a specific state for the system B emerges somewhere in the middle of the B-line as a consequence of an event which can have no causal influence on B. This disconcerting feature was encountered in a different and milder form already in the discussion at the end of Example 1. Namely, the real, individual state of a system (or complex) is not strictly a function of the past history. It becomes independent of the future only if some qualitative restrictions concerning future events are added. We must now say also, that it is not always independent of events at space-like distances. One may conclude from this that the concept of a real, individual state should better be avoided altogether. I do not, at this moment, hold this opinion. However, be this as it may, the assumption of a real outside world and the possibility of describing it are not contradicted by the laws of quantum physics. The question is only how the outside world may be carved up into pieces to which one may still attribute an objective individuality and real attributes. Apart from its epistomological aspect this statement may even be useful to bear in mind for physics itself since it may be that at one time the selfconsistency between the available types of measuring apparatus and the dynamical laws becomes important.

Let me close this discussion with a very brief remark concerning the measuring process itself. Essentially it is assumption (b) above which is relevant here. It is not an independent assumption but should be derived from the microscopic laws. This derivation is indeed a well-known programme which may be paraphrased as the 'fundamental task of quantum statistical mechanics'. While at present no complete and satisfactory solution of this task has been accomplished, there do exist partial results[5] which strongly indicate that indeed the assumption (*b*) is a consequence of the microscopic quantum physical laws.

NOTES

1. I believe that the answer is no for the following reason: The relevant experiments and observations may be automatized to the extent that the role of the human observer is reduced to sheer acts of cognition which do not alter the phenomena any more. Against this one may raise the following objection. It is sometimes argued that a rigorous application of the principles of quantum physics would lead to the conclusion that the speck on a photographic plate or even the number printed out by a computer can never be a fact but only a potentiality as long as nobody looks at them and that these things become facts only after the instruments have been read by an intelligent observer; in other words, the termination of the measurement is only in the mind. According to this claim quantum theory tells us that an unobserved photograph stored in my files has no picture on it; it has the potentiality of having one of various pictures, and only when I look at the photograph then one of the potential pictures is created as factual. If this statement could be claimed to be an inescapable consequence of the laws of quantum physics, then I personally would conclude that there must be something fundamentally wrong in our understanding and application of these laws. However, I do not believe that this claim can be maintained and we shall return to this point below.

2. The assumption of an 'outside world' does not touch the dispute between idealistic and realistic philosophies since we talk, after all, only about a limited range of phenomena, the 'physical universe' in the sense of the above abstraction.

3. To indicate the scope of possible candidates for ξ: it might be a collection of 'hidden variables', it might be a wave function.

4. This is easily seen since $P(e, e) = 0$ and since the three functions F, G and W should all be non-negative.

5. See for instance the book by D. Ruelle, *Statistical Mechanics, Rigorous Results*, and the article by K. Hepp, 'Quantum Theory of Measurement and Macroscopic Observables', Zurich preprint (1972).

36

Measurement Process and the Macroscopic Level
of Quantum Mechanics

Ilya Prigogine

The measurement process remains an exciting epistemological problem in quantum mechanics as witnessed by the interest and even the passion this problem still arouses.

In my lecture[1] I have indicated that it is now possible to present a unified formulation of dynamics and thermodynamics for a large class of dynamic systems. Indeed, in the 'causal representation' of dynamics we have introduced, there exists a Lyapunov function which decreases monotonically and which may therefore be related to entropy.

The class of systems, for which such a Lyapunov function exists, is formed by large dynamical systems which satisfy the *condition of dissipativity* which I have introduced and briefly discussed in my lecture. When satisfied this condition leads to a *symmetry-breaking* of the equations of motion. The invariance of the equations with respect to the inversion of both the Liouville operator L and time t (the so-called L, t invariance) is lost. We then obtain the *causal formulation* of dynamics which replaces the usual dynamical group by *two* semi-groups, applicable respectively to the retarded and the advanced solutions.

It is often asked how large such systems have to be. It has been proved that a causal dynamics exists for dissipative systems when, briefly speaking, the number of degrees of freedom N tends to infinity. Nothing prevents us in the asymptotic expansions to retain 'correction' terms which may be of the form N^{-1}, N^{-2},.... This can be done exactly in simple situations. Such terms give us for large N very small corrections without destroying the asymptotic results for $N=\infty$.

A comparison with the phase transition problems is in order. The very existence of phase transitions is an asymptotic result and the correction terms are negligible in macroscopic systems.

Two points should be noted. First, the causal formulation of dynamics refers to *superspace* using the density matrix and *not* to the Hilbert space. The reason was indicated in my lecture 'Time, irreversibility and structure' in Part III of this volume: the dissipativity condition leads to the appearance of contributions in the equation of motion which are *even* in the Liouville–von Neumann operator L. The even operators describe, in the causal dynamics, the difference between the retarded and the advanced solutions. It seems impossible to introduce this description in the frame of the Hilbert space description (see [2,3]).

The incorporation of causality introduces therefore into quantum mechanics a kind

J. Mehra (ed.), The Physicist's Conception of Nature, 697–701. All Rights Reserved
Copyright © 1973 by D. Reidel Publishing Company, Dordrecht-Holland

of 'non-linear' element when the probability amplitudes are concerned, as it forces us to consider the equations of motion for the products of amplitudes in the superspace.

The second point is that the entropy defined through the Lyapunov function refers to a more general situation than considered usually in macroscopic physics as it includes all types of correlations which may be introduced through initial conditions or external actions on the system.

Therefore we have still in the frame of the causal formulation of dynamics to identify the 'macroscopic level' which contains the laws of continuum mechanics and of equilibrium thermodynamics. Also non-equilibrium thermodynamics should be included in the macroscopic level at least when limited to situations which may be expressed in terms of variables which have a meaning at equilibrium (this corresponds to the so-called *local* equilibrium assumption).

From the experimental point of view, the existence of such a level is a direct consequence of classical thermodynamics and of the validity of phenomenological equations such as the Fourier equations.

Fortunately, starting from the causal formulation of dynamics, it is straightforward to identify this 'macroscopic level of quantum mechanics'[1, 2].

In this formulation the diagonal elements satisfy a *separate* equation of motion (this situation corresponds to a '*subdynamics*' as we have defined it[1, 2]), and moreover all off-diagonal elements vanish asymptotically[1]. Therefore the macroscopic level is described by the ensemble of *diagonal elements* of the density matrix in the causal representation.

We may also reach the macroscopic level starting from the Liouville–von Neumann equation and applying a suitable projection operator (we call it Π) to the dynamic operator $\exp(-iLt)$[2]. This operator Π commutes with L and as a result we again obtain a closed subdynamics. In fact this is the method used in our joint paper with George and Rosenfeld[2]. The causal representation has been obtained in full generality only later[3]. The essential point is that the evolution in the macroscopic level is determined by the *diagonal* elements of the density matrix. (If we work in an arbitrary representation off-diagonal elements contribute to the macroscopic level but they are *functionals* of the diagonal ones.)

The macroscopic level corresponds to a well-defined and autonomous description in the *superspace* which exists only for dissipative systems. Still this description is incomplete. Therefore we cannot associate with situations described in terms of the macroscopic level alone any wave function (we have not enough information to take the 'square root' of the density matrix). *A fortiori*, the quantum superposition principle is not applicable to macroscopic states.

We have here a very clear example of complementarity as formulated by Heisenberg in the following way[4]. 'The characterization of a system by its Hilbert vector is complementary to its description in terms of classical concepts.' However, it should be emphasized that the macroscopic level still contains Planck's constant whenever quantal effects occur on a macroscopic scale. An obvious example would be the equation of state of liquid helium.

The transition to the macroscopic level corresponds to a projection onto a subspace of the motion. We could here, if we want to do so, speak of 'coarse graining' as Professor Jauch does[5]. However, it should be emphasized that in our approach this introduces no foreign element into dynamics (as the projection operator Π is itself constructed in terms of dynamical quantities). Expressions such as 'coarse graining' also give a wrong impression of generality. In fact the macroscopic level exists only for systems with broken symmetry. This is really the basic feature which connects concepts such as irreversibility, the macroscopic level and the measurement process.

Concerning this last point we follow exactly Bohr's argument[6] who pointed out that any measurement is essentially a codified registration of some characteristic signal. Therefore, an element of macroscopic irreversible thermodynamics is involved. In a more precise way we make the assumption that a measurement apparatus admits a description at the macroscopic level. Combining this assumption with the description in terms of probabilities valid on the macroscopic level, we come to the following scheme.

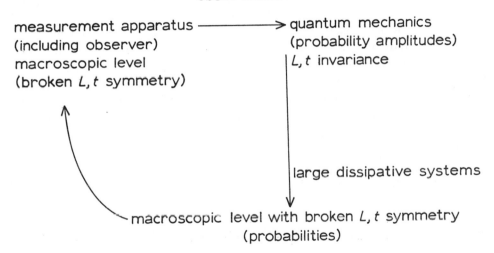

This scheme is very similar to the one I used to introduce irreversibility in my lecture[1]. It starts from the measurement apparatus including the observer. The properties ascribed to the measurement apparatus such as the irreversible recording of the results *presuppose* the broken L, t symmetry which leads then to a thermodynamic description of macroscopic systems. This presupposition is justified by formulating *first* the laws of quantum mechanics for simple systems for which the L, t invariance holds and *then* by showing how for large dissipative systems this invariance is broken.

The scheme I have indicated gives a solution to the so-called 'closure problem of quantum mechanics' (see ref. 7, for example), which corresponds to the transition from probability amplitudes to probabilities in the measuring process. This problem seemed difficult and even impossible to solve in the frame of quantum mechanics because it was not recognized that asymptotic elements may lead to radically new features. Till now at least these new features can only be incorporated in a formulation of quantum dynamics in the *superspace*. However, at no point the principles of quantum mechanics are violated. *Our method solves the closure problem strictly in the frame of non-relativistic quantum mechanics.*

There is no necessity to go beyond quantum mechanics through the introduction of hidden variables or the addition of new, non-linear terms.

The importance of the macroscopic level to elucidate the closure problem was already emphasized by many people (see e.g. Feyerabend[8], Rosenfeld[9], Ludwig[10], Lanz, Prosperi and Sabbadini[11]). The important new result is that we have now identified the dynamic feature (the dissipativity condition) which permits to characterize in a precise fashion this macroscopic level.

The scheme we have described involves both the macroscopic level *and* the microscopic world on which we make observations. Neither of them can be considered in isolation. Obviously this scheme has the property beautifully expressed by d'Espagnat in his communication[12] as 'notre connaissance de la non-séparabilité de quoi que ce soit que l'on puisse nommer réalité extérieure à nous'.

There are obviously many ways to improve our understanding of quantum mechanics. The study of large systems is one, the study of quantum logic another. However, it seems to me that we shall not avoid the need to consider carefully the many body aspects involved. This may be a bitter pill to swallow for those who remain attached to highly idealized models involving only a few degrees of freedom. But one of the main changes in our outlook in the last decades has come from the gradual understanding that such models are severely restricted both in the direction of macroscopic and microscopic situations. Even the symmetry properties of such simplified models have to be 'broken' to include the macroscopic world in terms of which the measurement process has to be defined.

The point of view summarized here has been developed in greater detail in two papers in collaboration with Claude George, Françoise Henin and Léon Rosenfeld[2, 3].

REFERENCES

1. I. Prigogine, Time, Irreversibility and Structure, lecture presented at this symposium.
2. C. George, I. Prigogine and L. Rosenfeld, 'The Macroscopic Level of Quantum Mechanics', *Det Kong. Danske Vidensk. Selsk. Mat. Phys., Medd.* **38**, 12 (1972).
3. I. Prigogine, C. George, F. Henin and L. Rosenfeld, *Chemica Scripta, Stockholm*, to appear (1972).
4. W. Heisenberg, in *Niels Bohr and the Development of Physics* (ed. by W. Pauli), London, Pergamon Press (1955).
5. J. M. Jauch, communication at this symposium.
6. See N. Bohr, *Dialectica* **2**, 312 (1948); other references may be found in ref. 2.
7. A. Petersen, *Quantum Mechanics and the Philosophical Tradition*, M.I.T. Press (1968).
8. P. K. Feyerabend, in *Observation and Interpretation* (ed. by S. Körner), Butterworth's Scientific Publications, London (1957), reprinted by Dover, New York.

9. L. Rosenfeld, *Nature* **190**, 384 (1961).
10. C. Ludwig, in *W. Heisenberg und die Physik unserer Zeit* (ed. by F. Bopp), Fr. Vieweg und Sohn, Braunschweig (1966).
11. L. Lanz, G. M. Prosperi and A. Sabbadini, *Nuovo Cimento* **2B**, 184 (1971).
12. B. d'Espagnat, communication at this symposium.

37

Why a New Approach to Found Quantum Theory?

G. Ludwig

The mathematical elegance of the theory of quantum mechanics and the crucial conceptual problems arising from it are most fascinating. The well-known books of Dirac[1] and of von Neumann[2] are most useful if one tries to 'understand' what is meant by the 'physical interpretation' of quantum mechanics. Concerning the problem of the measuring process, Heisenberg's book *Die physikalischen Prinzipien der Quantentheorie*[3] is very fundamental. The beauty of the mathematical structure of quantum mechanics can be felt by studying the books of Wigner[4], van der Waerden[5] and Weyl[6], which also describe the connection between group theory and quantum mechanics. Although group-theoretical methods are essential for a logical construction of quantum mechanics (e.g. the representations of the Galilei-group in non-relativistic quantum mechanics), I would like to discuss here the problems of the foundation of quantum mechanics only.

As a student I was fascinated by the clear and precise way in which von Neumann[3] elaborated the problems of the measuring process, especially by his manner of dealing with the problems of the measuring chains, including the position of the cut between the observed object and the observing subject. The so-called subjectivistic interpretation of quantum mechanics – which says that every measuring process can be completed in the consciousness of the observer only – has seemed to me not only unavoidable with respect to the arguments of von Neumann, but I saw in it a deep change in man's thinking about nature. The very interesting example, discussed by von Weizsäcker, that an electron is observed by means of a microscope and photographic plates which are placed in the focal plane or the image plane of that microscope, convinced me of the importance of this change in man's thinking about nature.

The question, as to what is really meant by the concept of 'probability', has become more and more misty, the deeper I have thought about the problems of the interpretation of quantum mechanics.

Somehow I became convinced by considering real and 'Gedanken' experiments that a measurement is actually closed in the macroscopic measuring devices, i.e. there is no necessity of a subjective observer and his consciousness. I felt certain that the 'objectivistic' description of the measuring devices was an *unavoidable* assumption of the interpretation of quantum mechanics, i.e. the consciousness of the observer had disappeared out of physics (see refs. 8 to 11).

It would take me too far afield if I were to give detailed arguments to account for this change in my beliefs concerning the interpretation of quantum mechanics.

J. Mehra (ed.), The Physicist's Conception of Nature, 702-708. All Rights Reserved
Copyright © 1973 by D. Reidel Publishing Company, Dordrecht-Holland

You know the arguments which Wigner gives to show that his famous 'friend' cannot be in a superposition state (i.e. a superposition state which consists of the observed micro-object, the measuring device and the friend) if this friend had made an observation. Wigner's arguments are my arguments also, but I do not see why a computer is not as suitable as a friend to tell me the result of an observation. (By what I naturally do not want to say that a computer is able to replace a friend in every respect.) But if the arguments are valid with respect to a computer, then the results of an observation, which are evaluated and recorded by the computer, *can be taken as objective facts* to interpret quantum mechanics.

I subscribe to the notion of 'complementarity' as given by Bohr, namely a complementarity between the classically describable measuring devices and the measured results (which one gets by the help of these classically described devices) being no more describable by classical physics. With this remark I do not wish to arouse a discussion about the 'correct' interpretation of Bohr's ideas, I only wish to point out that already in the first few years following the formulation of quantum mechanics, it was known that quantum mechanics was to be interpreted by means of classically describable results of measurement.

Before discussing whether the objective description (often called 'classical description') of the measuring devices is the necessary assumption to interpret quantum mechanics, I want to stress a second problem that forces us to try a new way of founding quantum mechanics. This second problem is very closely connected with the problem of a new logic, i.e. a probability logic; it is the notion of the 'probability of *one* event'.

I must say that I have never succeeded in understanding as to what is meant by the 'probability of one event' in *physics*. Moreover, I have never succeeded in understanding in which respect my more or less certain *knowledge*, whether an event has taken place or not, has anything to do with physics.

There are many people who think that it is impossible to work in quantum mechanics without using the basic notion of the probability of *one* event. They say that what one is doing is impossible, and if you cannot understand the basic conception of the 'probability of one event' one will never succeed in understanding quantum mechanics. It is my belief, however, that quantum mechanics can be explained to all those who cannot (as I cannot) consider the probability of *one* event as a *basic conception of physics*.

There is the opinion that such an attempt is bound to fail, because it is *proved* that it is impossible to introduce any notion of probability without understanding the basic notion of the probability of *one* event. But if one examines these so-called *proofs*, one only sees arguments but no proofs. As mentioned above, I also cannot *prove* it to be unavoidable to start an interpretation of quantum mechanics by using objectively described measuring devices only. I can only give arguments which convinced me to try just this way. Corresponding to that there exist no *proofs*, but arguments only, for starting an interpretation using a probability logic – and, of course, one can try other ways than I do.

All arguments by means of which one thinks of proving that a theory of probability cannot be founded without using the notion of a 'probability of *one* event', include the assumption that a notion of a probability of *one* event is already given. That is why one finds it difficult to understand these arguments.

Conclusion: I am forced to try founding quantum mechanics in a way that assures, on the one hand, that the measuring devices are described objectively; on the other hand, we must be sure that no notion of a 'probability of one event' needs to be used in *founding* quantum mechanics.

If we start with these two preliminary decisions it is necessary to give up another belief, namely that quantum mechanics is a theory completely describing the macroscopic measuring devices too, i.e. describing these macroscopic devices in such a way that there remain no unsolved problems of macrophysics. The typical non-classical structure (i.e. non-Boolean structure) of quantum mechanics cannot account for an objective classical (i.e. Boolean) description of the measuring devices. According to this idea, quantum mechanics does not completely describe either biological objects or the inanimate macroscopic measuring devices[26]. If I understand it correctly, it is also Wigner's standpoint that quantum mechanics does not completely describe biological objects.

The main problem concerning Bohr's 'complementarity principle' was the following: How is it possible to explain that the microscopic region can only be described by quantum mechanics whereas the macroscopic measuring devices are describable by means of classical physics? My own conclusion is not to regard quantum theory as the correct theory describing macroscopic systems too. (Compare with the conclusion of von Weizsäcker, as expressed in his contribution to this symposium and in ref. 7.)

Many people think that there exists some kind of a 'proof' according to which my notion may be impossible. Therefore it is to be shown that it is possible to found quantum mechanics in my way, i.e. by building up a foundation of quantum mechanics by starting only from the objectively described devices and neither assuming anything about the existence of micro-objects nor using the notion of a 'probability of *one* event'. It must be demonstrated that this way of foundation does not contradict the well-known great successes of the many-particle quantum mechanics. Such a demonstration is only possible, against all prejudices, if all steps of this demonstration will be elaborated in the most exact form, i.e. in a correct mathematical form.

This is a complicated and very hard way. It would be impossible for me to sketch out this entire development, and therefore I shall refer to the bibliography, refs. 12 to 30, where one can find the details, although there have been many improvements in the meantime which have not yet been published.

The aim of our approach is not only a foundation of the Hilbert-space structure of quantum mechanics. The aim is actually nothing else than to repeat (but in a *new* manner) what the physicists have done only intuitively to describe the atoms, electrons, protons etc., and furthermore to establish quantum mechanics as a suitable description of these micro-objects. The aim is to repeat that work of the physicists by formulating a 'physical theory' (and not by intuitive guessing) in a mathematical form. One has to reveal such structures in the mathematical description of the objectively described devices which allow one to *conclude* that *there are real micro-objects* and that *these can be described by quantum mechanics*. A method similar to that one by which one has 'theoretically' discovered Neptune as a real planet, revealing the fundamental structures of the objectively described devices (which are formulated in the mathematical form by mathematical axioms, i.e. as 'laws of nature'), yields the 'rediscovery' of the micro-objects and of quantum mechanics, describing these micro-objects completely.

If one wants to elaborate only the physically real existence of the micro-objects and the Hilbert-space structure of quantum mechanics, one needs only very general 'laws of nature', as general as the laws of thermodynamics. Therefore I have called these laws 'laws of measurement'. It is clear that one needs many more special laws of nature to construct the whole quantum mechanics in a more detailed manner. Some of these more special laws of nature can only be formulated by using group-theoretical structures.[31, 32]

Finally I want to make some remarks concerning the following question: Are there connections between the various ways of founding quantum mechanics and what do they look like?

It seems to me that there are no essential differences with respect to the fundamental thinking about quantum mechanics between Jauch (see also the contribution of Jauch to this symposium) and me; only the systematical procedures and the starting-points are different. As I see it, Jauch starts e.g. by using the conception of really existent micro-objects, whereas these real micro-objects are 'rediscovered' in the way I proceed. But in principle both theories are similar: both ways require one to compare the respective theories with the effects which are produced by the measuring devices. Concerning my approach, it was to be shown too that the effects of the 'rediscovered' micro-objects in measuring devices are consistent with my starting point, i.e. consistent with the structure of the objectively described measuring devices (see refs. 26, 27).

The other way of founding quantum mechanics, namely using a probability-logic as the basic conception, is incompatible with my approach. Although I cannot understand the notion of a probability-logic, I am able to understand the mathematical formalism of this theory. I said that these two ways are incompatible; but this does *not* mean that there is no possibility of comparing the respective 'results'. Following my approach, one also gets 'physical propositions' which concern the 'physical possibility' of events. But such propositions make no sense *before* one has written down the mathematical structure of the theory using the *usual* logic.[12] Also, following my approach, it is possible to speak about a 'probability of a physically possible event', but only *after* the theory has been formulated. This 'probability of one event' does not depend on the knowledge of a subjective observer (i.e., this probability is neither a 'subjective probability' nor a probability as one uses in a probability logic), but it does depend on the 'ensemble' selected by objectively describable methods. A detailed comparison of these two ways (and of other interpretations of quantum mechanics) is given in ref. 33.

There are many interpretations of quantum mechanics which include the consciousness of the observer as a necessary part of the measuring process. Are these interpretations compatible with my approach? This question cannot be answered by a clear 'yes' or 'no'. To me, the problem of the consciousness of the observer looks like this: The objective description of the measuring devices is compatible with the macrophysical description of the physical processes by which the results, objectively indicated by the devices, arrive at the brain. For instance, the light coming into the eyes, the electric signals transported by the nerves, etc., all these are processes which are describable in a classical manner. So there is no incompatibility between the results indicated objectively by the measuring devices and the physical description of the processes by which these results are transported to our brains.

But what about the processes in the brain itself? According to the ideas proposed by me, this is neither a problem of quantum mechanics nor of the usual macroscopic physics. The problem of the brain is a new one, which cannot be solved either by means of quantum mechanics only or by known macrophysical theories. Since the brain is a macrosystem, quantum mechanics is not able to describe the processes in the brain completely, because in my opinion quantum mechanics is not able to describe any macrosystem completely.

To illustrate the relation between quantum mechanics and a macrosystem (e.g. quantum mechanics and the brain), it might be useful to imagine a similar (but not an equivalent) situation: If you take the law of energy conservation you will find no contradictions to this law by observing experiments. On the other hand, this law alone is of course not sufficient to describe the planetary system or even a biological system. In my opinion, this holds for quantum mechanics in a similar way (please understand here the concept 'similar' in a very wide sense, see ref. 26); i.e. quantum mechanics is not a theory which completely describes either macrophysical systems or biological systems. And that is why quantum mechanics is not able to describe the brain completely.

To conclude: I do not believe quantum mechanics to be the theory of nature as a whole. Furthermore I do not believe that there will ever be a physical theory of the whole of nature. A theory of the whole of nature means for me a limit that can never be reached, but a useful limit for accelerating the development of science.

ACKNOWLEDGEMENTS

I wish to express my thanks to Miss Wegener and Dr Kanthack for many discussions concerning the formulations discussed in this paper.

BIBLIOGRAPHY

1. P. A. M. Dirac, *Die Prinzipien der Quantenmechanik*, Hirzel, Leipzig (1930).
2. J. von Neumann, *Mathematische Grundlagen der Quantenmechanik*, Springer, Berlin (1932).
3. W. Heisenberg, *Die physikalischen Prinzipien der Quantentheorie*, Hirzel, Leipzig (1941).
4. E. Wigner, *Gruppentheorie und ihre Anwendung auf die Quantenmechanik der Atomspektren*, Vieweg, Braunschweig (1931).
5. B. L. van der Waerden, *Die gruppentheoretische Methode in der Quantenmechanik*, Springer, Berlin (1932).
6. H. Weyl, *Gruppentheorie und Quantenmechanik*, Hirzel, Leipzig (1931).
7. C. F. von Weizsäcker, *Zum Weltbild der Physik*, Hirzel, Stuttgart (1958), and *Die Einheit der Natur*, Hauser, München (1971).
8. G. Ludwig, 'Die Stellung des Subjekts in der Quantentheorie', in *Veritas, Justitia, Libertas*, [Festschrift der Freien Universität zur 200-Jahrfeier der Columbia University], Berlin (1952), pp. 262–271.
9. G. Ludwig, 'Der Meßprozess', *Z. Physik* **135**, 483–511 (1953).
10. G. Ludwig, *Die Grundlagen der Quantenmechanik*, Springer-Verlag, Berlin (1954).
11. G. Ludwig, 'Gelöste und ungelöste Probleme des Meßprozesses in der Quantenmechanik', in *Werner Heisenberg und die Physik unserer Zeit*, Braunschweig (1961), pp. 150–181.
12. G. Ludwig, 'Versuch einer axiomatischen Grundlegung der Quantenmechanik und allgemeinerer physikalischer Theorien', *Z. Physik* **181**, 233–260 (1964).
13. G. Ludwig, 'An Axiomatic Foundation of Quantum Mechanics on a Nonsubjective Basis', in *Quantum Theory and Reality*, Springer-Verlag, Berlin (1967), pp. 98–104.
14. G. Ludwig, 'Attempt of an Axiomatic Foundation of Quantum Mechanics and More General Theories, II', *Commun. Math. Phys.* **4**, 331–348 (1967).
15. G. Ludwig, 'Hauptsätze über das Messen als Grundlage der Hilbertraumstruktur der Quantenmechanik', *Z. Naturforsch.* **22**a, 1303–1323 (1967).
16. G. Ludwig, 'Ein weiterer Hauptsatz über das Messen als Grundlage der Hilbertraumstruktur der Quantenmechanik', *Z. Naturforsch.* **22**a, 1324–1327 (1967).
17. G. Ludwig, 'Attempt of an Axiomatic Foundation of Quantum Mechanics and More General Theories, III', *Commun. Math. Phys.* **9**, 1–12 (1968).
18. G. Dähn, 'Attempt of an Axiomatic Foundation of Quantum Mechanics and More General Theories, IV*', *Commun. Math. Phys.* **9**, 192–211 (1968).
19. P. Stolz, 'Attempt of an Axiomatic Foundation of Quantum Mechanics and More General Theories, V*', *Commun. Math. Phys.* **11**, 303–313 (1969).
20. G. Ludwig, 'Deutung des Begriffs "physikalische Theorie" und axiomatische Grundlegung der Hilbertraumstruktur der Quantenmechanik durch Hauptsätze des Messens', *Lecture Notes in Physics*, Heft 4, Springer-Verlag, Berlin and Heidelberg (1970). An improved version of Chapter II will appear in print in *Notes in Math. Phys.* (Marburg).
21. G. Ludwig, 'The Measuring Process and an Axiomatic Foundation of Quantum Mechanics' in *Foundations of Quantum Mechanics* (1971), Rendiconti della Scuola Internazionale di Fisica 'Enrico Fermi', Il. Corso (1970), pp 287-315 (New York, 1971).
22. P. Stolz, 'Attempt of an Axiomatic Foundation of Quantum Mechanics and More General Theories, VI*' *Commun. Math. Phys.* **23**, 117–126 (1971).
23. G. Ludwig, 'A physical interpretation of an axiom within an axiomatic approach to quantum mechanics and a new formulation of this axiom as a general covering condition', *Notes in Math. Phys.* **1** (Marburg, 1971).
24. H. Neumann, 'Classical Systems and Observables in Quantum Mechanics', *Notes in Math. Phys.* **2** (Marburg, 1971).
25. G. Ludwig, 'Measurement as a process of interaction between macroscopic systems' [Lecture given by L. Kanthack at the Colloquium *The Meaning and Function of Science in Contemporary Society*, in Pennsylvania, U.S.A. (1971), to be published.]
26. G. Ludwig, 'Makroskopische Systeme und Quantenmechanik', *Notes in Math. Phys.* **5** (Marburg, 1972).
27. G. Ludwig, 'Meß- und Präparierprozesse', *Notes in Math. Phys.* **6** (Marburg, 1972).

28. G. Ludwig, 'An Improved Formulation of Some Theorems and Axioms in the Axiomatic Foundation of the Hilbert Space Structure of Quantum Mechanics', *Commun. Math. Phys.* **26**, 78–86 (1972).

29. H. Neumann, 'Classical Systems and Observables in Quantum Mechanics', *Commun. Math. Phys.* **23**, 100–116 (1971).

30. G. Ludwig, 'Transformationen von Gesamtheiten und Effekten', *Notes in Math. Phys.* **4** (Marburg, 1971).

31. H. Neumann, 'Transformation Properties of Observables', *Helv. Phys. Acta.*

32. H. Neumann, 'Classical Systems in Quantum Mechanics and their Representations in topological Spaces' Habilitationsschrift (Marburg, 1972).

33. L. Kanthack, 'Ein Vergleich verschiedener Interpretationen der Quantenmechanik, (to be published).

38

A Process Conception of Nature

David Finkelstein

The powerful conceptions of nature that have been surveyed at this Symposium incorporate two recent revolutions and yet may still be upside-down in an interesting
and suggestive sense. They employ spacetime to describe matter and process as
though spacetime were primary and process secondary. The primacy of process has
been urged by philosophers from Heraclitus to Whitehead and beyond. The dependence of our perception of spacetime upon dynamical processes was important during
the conception of special relativity, when Einstein argued operationally from the
invariance properties of Maxwell's dynamics and light signals to those of spacetime
geometry. Even theories using a deep spacetime structure recognize that the spacetime
we see is surface structure, is at least renormalized. I believe that the way has been
prepared to turn over the structure of present physics, to take process as fundamental
at the microscopic level and spacetime and matter as semimacroscopic statistical
constructs akin to temperature and entropy.

This has taken so long, I think, because we are born with an inherent spacetime
illusion. It is easy to be conscious that what we feel is at our fingertips, even when we
tap a cane, but remarkably difficult to make ourselves conscious that what we see is
at our eyes. Nor have we effectors dual to our visual receptors as we have for our
kinaesthetic. Vision is passive. The system of mechanics I would describe here is more
tactile and kinaesthetic than visual in its intuitions.

There is no great problem and no little interest in translating existing theories
into process terms, for it is an aid to forming the concept of process. One point of a
new language in physics is to suggest new hypotheses. The process language suggests
a new quantum hypothesis, which I divide into *process atomism* and *process coherence*.

In classical quantum mechanics we treat processes quite differently according to
the number of channels they involve: 1 for a creation-destruction process, 2 for a
propagation process, 3 or more for interaction processes. An atomic hypothesis of the
material variety is incorporated in the assumption that there exist discrete elementary
processes for creation–destruction: we do not permit the creation of a fraction of a
particle. But propagation is supposed to be infinitely divisible. Creation–destruction
and interaction are assigned zero duration, but propagation takes a continuously
varying time. In a process theory, to the contrary, it is more natural to assume process
atomism: *every process is a finite assembly of elementary processes*. The interre-

J. Mehra (ed.), The Physicist's Conception of Nature, 709–713. All Rights Reserved
Copyright © 1973 by D. Reidel Publishing Company, Dordrecht-Holland

lations among the elementary processes of such an assembly may be expressed by a network or graph.[1] Then a particle process is just a significant bundle of strands in such a network, and spacetime relative coordinates and proper times are statistical characterizations of the strands joining two junctions.

Assembling a world in such a system of mechanics is not too different from the design of a self-reproducing automaton.[2] We must formulate the basic process or *codon*, the combination of such processes or *code* needed to express the most general possible process (kinematics), and the natural process (dynamics) in such a way that the self-replicating processes (stationary states) in our stochastic environment are the observed elementary particle processes. The dynamical law is defined by a class D of kinematical processes, a projection in the algebra of the kinematics.

Here enters the second part of the process quantum hypothesis, process coherence. The processes of classical quantum mechanics do not form a coherent system in the sense that some of their coordinates are not subject to complementarity limitations. The momentum transfer in a process is complementary to the space coordinate increment x of the process but there is not supposed to be any quantity complementary to the duration of the process. Superpositions of Feynman amplitudes for different durations never occur. In classical quantum mechanics D is a class of high multiplicity. In a pure quantum theory D can be assumed to be a singlet projection onto one ray $|D)$. The Feynman amplitude of classical quantum mechanics and the action function of classical mechanics can be attained as successive statistical constructs from the dynamical ray $|D)$.[3]

In such a finitistic scheme there is a swift reaction between the principles of relativity and quantum theory. The linear space of the state vectors of the codon should be finite-dimensional by our atomic hypothesis, a Hilbert space by quantum principles, and carry a unitary representation of the Lorentz group according to relativity. (Not the Poincaré group: translation is expressed by combining codons in series.) But there is no finite-dimensional unitary representation of the Lorentz group.[4]

It seems possible to take this clash as seriously as that between the Galilean and Lorentz groups, and return once more to operationalism to decide the outcome. Quantum logic may be founded on two operational judgements concerning *transitions*, pairs of creation and destruction processes: the null judgement that a transition is forbidden and the universal judgement that a transition is compulsory. The null judgement may be carried out by means of an uncalibrated instrument such as a galvanometer but the universal judgement requires a rate calibration such as an ammeter or auxiliary channels for the communication of counts.

A quantum logic founded entirely upon the null judgement alone exists and has a theory resembling present quantum logic in its use of linear spaces, but its linear space lacks an intrinsic inner product. Instead all inner products appear on an equal footing. Concepts depending on the inner product such as relative probability (in contrast to absolute) may be defined only relative to an arbitrarily chosen metric. I call this logic relativistic quantum logic. It is contrasted to classical and unitary quantum logic in Table I. Its group consists of the non-singular linear and antilinear transformations

TABLE I

	Classical	Unitary Quantum	Relativistic Quantum
Logic	Distributive orthocomplemented	Orthocomplemented	–
Group	Permutation	Unitary and anti-unitary	linear and anti-linear
Algebra	Commutative *algebra	*algebra	algebra

mapping the rays of the linear space. Singling out a metric reduces this group to the unitary and anti-unitary transformation of the rays. In exactly one case the group of a relativistic quantum logic coincides with the (isochronous) Lorentz group: the binary case, where the linear space has dimension 2. Then adding a metric reduces the group to the rotation group about a fixed time axis. Indeed there is a simple well known 1–1 correspondence between time-axes in Minkowski spacetime and metrics in 2-dimensional spinor space.

Further, all the processes of unitary quantum logic that are needed for the construction of quantum assemblies, Maxwell–Boltzmann, Einstein–Bose and Fermi–Dirac, or of quantum networks, are already relativistic, do not make use of the metric of Hilbert space. They are purely linear constructions. In application to a binary quantum they are Lorentz invariant.[5]

To maintain Lorentz invariance, I abandon the unitary quantum logic for the relativistic quantum logic, regarding the metrical structure as relative, imposed from outside the system whenever a time axis would be imposed from outside the system in relativity. In the further development the metric aspect of logical structure of the system is conditional upon history, just as the pseudometric geometrical structure of spacetime is conditional in general relativity.

This identification also determines the elementary quantum process, the codons of the network. One of them must be a binary quantum[6] whose logic is described in coordinates by the calculus of relativistic spinors ψ^A. Its q numbers are the operators q_B^A on such spinors, the Pauli algebra A. All the entities that relativistic quantum logic can make of such an elementary process are networks on which may appear also the three codons whose spinors have the form ψ_A, $\psi^{\dot{A}}$, $\psi_{\dot{A}}$. The corresponding four algebras result from applying the $i$dentity, $t$ranspose, $c$omplex conjugate and $h$ermitian conjugate operation to the Pauli algebra A, so I call these four codons I, T, C and H. In the formation of dynamical laws, invariance and locality demand that the four codons link in pairs IT and CH.

Nothing but simplicity excludes the possibility that there is more than one such tetragram. As we take up each symmetry we have to choose between a phenomenological theory, ascribing the new degree of freedom to codons themselves, or a fundamental one, attributing the degree of freedom to the internal structure of networks. The most fundamental possibility is constructing the system of physics out of a single tetragram of spinors.

In such a construction as much effort goes into the dictionary expressing familiar operational concepts in this code, into the semantics of the theory, as into the kinematics and dynamics. The new theory may use many guides of the old: correspondence, operationality, invariance.

Because of their rich invariance theory, electrodynamics and gravity are natural first subjects for encoding. For electrodynamics I dissect the $GL(2, C)$ group of one of the codons, say I, into R_+, $U(1, C)$, and $SL(2, C)$, taking $U(1, C)$ as the gauge group of electromagnetism, whose generator is electric charge Q; $SL(2, C)$ as the covering group of the Lorentz group, whose generators form the angular momentum bivector S_{AB}; and R_+, the multiplicative group of the positive real numbers, leads to a conserved whole number. (Fractional representations of R_+ do not arise in a constructive procedure such as this.)

The code that suggests itself for the electron e is $. IT: IT: ... : IT .$, a process transferring unit charge and half-integer spin, and its hermitian conjugate. Each codon in the sequence is assigned a proper time $\tau/2$. Since the physical electron e has rather definite mass, e cannot be identified with just one such strand but with a coherent superposition of many of different duration, and interactions will bring more complicated networks into e.

We go from this to the Dirac equation by giving an expression for the momentum transfer $p = (p^{AB}) = p_\mu \sigma^\mu$ of the e process and the coordinate increment $x = x^\mu \sigma_\mu = x_{BA}$ in terms of the codon q numbers. The simplest choice works well enough for this purpose. The coordinate increment is taken to be a direct sum $x^\mu = \oplus \gamma^\mu \tau/2$ over all codons in the sequence, which must be taken together with its hermitian conjugate $. HC: NC: ... : HC.$ in order to go from the Pauli to the Dirac algebra. Momentum transfer is defined by expressing in terms of x those process state vectors that have definite momentum transfer p in the limit of infinite length: $|p) = \exp ikx$. With these assignments the deduction of the proper time Dirac equation is immediate.[7]

In the limit where the Dirac propagator becomes valid, so does its causal structure and translational invariance. Thus the classical Minkowski space time emerges from the underlying process theory just when classical quantum mechanics does, as a semimacroscopic statistical limit.

In the same way one deduces a proper time Maxwell equation for the double strand

$$\gamma = \frac{: IT : \quad : IT}{: TI : ... : IT .}$$

To form x we average over processes in parallel, add for processes in series.

A code for an electrodynamic interaction process suggests itself immediately, a double strand separating into two single strands, but it has not been shown to agree with quantum electrodynamics in the limit $\tau \to 0$.

The language for the description of gravitation is present from the start. This system of mechanics also suggests fundamental theories of the weak and strong interactions, but it would be speculative to present them now and they are so clear it might be superfluous as well.

The size of the chronon τ is the most pressing question for this conception of the world. Since the individual codon lacks the discrete Lorentz symmetries, its duration $\tau/2$ is comparable to the range of that interaction which lacks these symmetries most conspicuously, the weak. According to this it is not a coincidence that the least symmetric interaction has the shortest range. It senses the asymmetric codon of the world code. Using a current plausible estimate of the weak range, I deduce that present spacetime concepts break down in individual processes with durations comparable to

$$\tau \sim \hbar/40 \text{ GeV}.$$

Furthermore processes that are conventionally regarded as neutral, such as the photon, are not microscopically neutral in the present system. The double strand given for γ contains two opposing electric currents at the microscopic level. Short-range departures from neutrality should manifest themselves for all the supposedly neutral processes, the graviton and neutrino as well.

There is much that has to be done before this conception becomes a working theory. I have passed over vast areas I do not understand because only the little I begin to understand is appropriate for this volume. The theory presently lacks not merely computational power but also important concepts concerning the relation of a system to its environment. The theory cannot yet be subjected to a decisive test and its viability is still in doubt. These unexpected resemblances of the microphysical world and the genetic, Friday's footprints on a strand thought barren, are still best regarded as part of a working heuristic process.

NOTES

1. Thus the formalism is a hybrid of the relativistic continuum diagrams of Feynman and the discrete spin diagrams of Penrose.
2. See J. Neumann, *Theory of Self-Reproducing Automata* (ed. by A. W. Burks), Urbana, University of Illinois Press (1966), in which the evolution from a particle to a field theory of automata takes place. That the world too is a digital computer was suggested by R. P. Feynman.
3. J. Schwinger taught that the action principle was the beachhead for the next quantum mechanics in his 1950 classes. However neither spacetime nor fields exist in the deep structure of the physics presented here, and the connection to conventional quantum field theory must be made through Schwinger's source theory or its equivalent.
4. This famous result which seems from the present point of view so crucial for physics is due to E. P. Wigner.
5. These questions are treated under the series title of 'Spacetime Code' in *The Physical Review* (1969–).
6. C. F. von Weizsäcker reaches a similar conclusion on similar grounds. I am deeply indebted to him for discussions and encouragement in this study.
7. This deduction is due to G. Frye and L. Susskind. See D. Finkelstein, G. Frye and L. Susskind, 'Spacetime Code V' (submitted).

39

Quantum Logic and Non-Separability

Bernard d'Espagnat

Er (E. Mach) hat besonders dadurch sanierend gewirkt, dass er
deutlich machte, dass die wichtigsten physikalischen Probleme
nicht mathematisch-deduktiver Art sind, sondern solche, die sich
auf die Grundbegriffe beziehen.

A. EINSTEIN

(Letter to M. Besso, January 1948 in *A. Einstein, M. Besso –
Correspondance 1905–1955*, Hermann, Paris, 1972)

1. INTRODUCTION

The attempts at interpreting elementary quantum mechanics in terms of statements
bearing on the physical systems themselves meet with several difficulties which have
been known for a long time. The purpose of the present article is to draw attention
to the fact that at least some of these difficulties occur in a much more general theory,
which admits conventional Hilbert space quantum mechanics as a particular case.

The theory considered here is a generalization of the calculus of propositions, also
called (although improperly according to some authors) quantum logic. Initiated by
the pioneering work of Birkhoff and von Neumann[1] that calculus was mainly developed, as is well known, by Jauch and Piron[2, 3, 4, 5, 6] by Weizsäcker[7] and by Mackey[8].
Here indeed we are dealing with a generalization of the corresponding principles
since the assumption that the considered set of propositions has the properties of a
lattice will not be necessary. It is enough to assume that this set is partially ordered
and that ortho-complementation is defined on it. But obviously everything that we
have to say is compatible with the idea – most certainly quite natural and attractive –
that this set is a lattice.

Moreover, the argumentation given in this paper points – we believe – to the necessity of adding some complements to the previous works in this area. Indeed it aims
at introducing the necessary elements for an approach of a problem not yet investigated at a sufficiently deep level, namely that of the *separation of physical systems*.
A body of researches, both theoretical and experimental – some quite old[10] and some
very new[11, 12, 13, 14] in particular have shown what follows. Given two systems that
are isolated at some time (that we refer to as the present) but that have interacted in
the past, the description of the correlations that exist between these systems raises
fundamental questions. The significance of the latter as regards our understanding
of the existing theories has been greatly underestimated up to now, even in the axio-

J. Mehra (ed.), The Physicist's Conception of Nature, 714-735. All Rights Reserved
Copyright © 1973 by D. Reidel Publishing Company, Dordrecht-Holland

matical formulations that make use of advanced mathematics. As will become apparent, this narrows considerably the span of the acceptable concepts of these theories, particularly as regards the notion of separated physical systems.

Here, a review of the language of the calculus of proposition would be both boring and unnecessary. The very few technical words whose meanings are not or are only cursorily defined below are parts of the conventional vocabulary of the above mentioned theory. Their meaning can be found in the works of the authors quoted and particularly in Jauch's book, *Foundations of Quantum Mechanics*[4].

The procedure used here consists (Section 2) in giving a more precise shape to some other definitions that are also parts of the conventional language of the theory under study but whose operational meaning had, we feel, not yet been fully explained, especially in view of the questions raised by the union and separation of physical systems Within these definitions our procedure also consists in trying to state in accurate terms a few assumptions often implicitly made and that give the possibility of describing the present and the past by means of one and the same set of concepts. Using a generalization of well-known arguments, whose leading ideas are due to Einstein, Podolsky and Rosen[10], Bell[11] and Wigner[15], it consists then (Section 3) in showing that if considerable caution is not exerted as regards the fields of application of the notions of *physical systems* and of *separation* of such systems, the set of the statements thus constructed is incompatible with known facts. Indeed this necessary caution is tantamount to the need of a real change in the point of view initially taken. Investigation of the possibilities which seem to exist along this line is the subject matter of Section 4. In spite of their differences all these possibilities have non-separability in common. Section 5 investigates the full implication of this conclusion.

From a strictly scientific point of view these first five sections make up a whole, to which the sixth one adds nothing. From the point of view of man's understanding of Nature – which is the subject of this symposium – I feel on the contrary quite convinced of the real importance of the problems which I try to discuss in this last section.

2. GENERAL DEFINITIONS

2.1 PHYSICAL SYSTEMS

In quantum theory the notion of 'physical system' is a relative one. In other words there exist experimental situations that in classical physics we could analyse in terms of several systems (each having well defined properties) whereas quantum physics forbids us to do so or at least makes such analysis subject to significant qualifications. The main result of the present study will eventually be a very general confirmation of this fact. Indeed we shall have to conclude that the great relativity of the notion of a physical system is not a peculiarity of conventional quantum mechanics – defined in a Hilbert space with coefficients in the field of complex numbers – but that, on the contrary, it remains true, finally, quite independently of the formalism adopted, provided the latter is subjected to the condition that it should not lead to contradictions with some recently established facts.

However, in conventional quantum physics we *do* use, as elsewhere, the concept of system. The legitimacy of this derives from the following circumstance. If the set of experiments to be made and analysed in the future is restricted in advance to a specified one (as, in practice, is always the case) then it is often possible, at different intermediary steps, to use a description by means of physical systems, and even of localized physial systems, in order to study and predict the results related to this set of experiments. 'Possible' here means that such a procedure will not – under the said conditions – enter into conflict with the more abstract description of the intermediary steps that quantum mechanics forces upon us. But obviously this fact in itself cannot legitimate the use of the notion of system in a more absolute sense, that is to say independently of the set of proposed experiments.

Here we do not want to start from the set of assumptions of conventional quantum mechanics. We want to remain more general. In order to establish the limitations of the notion of physical system we therefore need different arguments. Our method will be that of a *reductio ad absurdum*. That is: we establish here the following convention. We shall call – quite uncritically – 'physical systems' or 'systems' for short, all the entities to which classical physics would give this name including the 'microsystems' (atoms, particles, etc.). Parenthetically it can be observed that such an attitude is at least suggested by many axiomatical formulations and that it is even explicitly chosen by those which make a systematical and unrestricted use of the notion of individual physical systems (ref. 5 is a clearcut example). With the discussion carried out in this article we shall prove that such a position cannot be maintained. For the time being however we adopt it for the sake of argument.

Two systems can differ from one another either in their general structure or in their 'individual' properties (position, speed etc.). Two systems which differ only through the latter will be said to be *of the same type*.

Just as in classical physics, two systems will be considered as isolated from each other if they do not interact through any 'force'. Non-relativistically two systems whose mutual distance is arbitrarily large at some time *t* are thus isolated from one another at time *t*. Relativistically, two 'segments of space time tubes' each of which lies entirely outside all the light cones of the points of the other one describe two systems that are isolated during the corresponding times.

2.2 Observables and Measurements

The observables are defined, on a specified type of systems, by the description of the instruments used to measure them. Several experimental procedures (to which appropriate information and rules of calculation should be added) can of course correspond to one and the same observable. We postulate that every observable can be subjected to an *ideal* (or *first kind*) measurement. Briefly speaking a first kind measurement is such that if a second measurement of the same quantity is made, immediately after the first one, on the same system, both results coincide. In what follows by *measurement* we always mean a first kind measurement.

2.3 Compatible Observables

The notion of compatibility between two observables A and B belonging to a given type, T, of systems can and should be defined operationally. Let us imagine that on each element of an ensemble E of systems of type T we make a measurement of A immediately followed by one of B which is, in turn followed by a measurement of A. If for *any* preparation of E, the result of the third measurement coincides, for each system, with that of the first we say that B is compatible with A. If B is compatible with A and A with B (reflexivity) then A and B are *compatible*.

Remark. In the case of conventional quantum mechanics the observables A and B are represented by Hamiltonian operators. If $P_j(Q_k)$ are the projectors that appear in the spectral expansion of A and B respectively, it is easily shown that the foregoing definition leads to the condition that $\langle u | P_j Q_k P_j Q_k P_j | u \rangle = \langle u | P_j Q_k P_j | u \rangle$ for any $u \in \mathcal{H}$, where $\mathcal{H}$ is the Hilbert space associated with T. It is easy to show then that this condition implies $[P_j, Q_k] = 0$ and hence $[A, B] = 0$. Reversibility always holds.

2.4 Certain Knowledge. Direct and Indirect Measurements

The validity of non-exhaustive induction is postulated: we consider that if one person made a specified measurement a great number of times, if the systems on which those measurements were made were all of the same type T and all similarly prepared and if the results obtained were all identical to one another, then this person *knows* in advance (with certainty, but we shall henceforth omit this word) that if the specified measurement is made once more on a system of type T prepared as the preceding ones, the same result will again be obtained.

The measurement considered here can be direct, that is bear on the systems themselves, or indirect, that is bear on other systems which are known to have with the former some strict correlations involving the observable of interest, and this because of their previous interactions with the said systems.

2.5 Propositions

Propositions are observables the measurement of which can only give one of two values conventionally called *yes* and *no*. Hence, a statement is a proposition only if if an instrument, by means of which its content can be verified, can be conceived. Conventional quantum mechanics associates projectors to propositions.

To any proposition a we can associate another proposition a' defined by the same experimental procedure and such that the result of the measurement of a' is said to be yes (no) whenever the result of the measurement of a is no (yes). a' is the *orthogonal complement* of a.

Remark. It is conceivable that the fact, for any statement, of being a proposition should depend on the general abilities of the considered community of observers. For example this is the case in hidden variables theories as soon as 'demons' are imagined, which would have access to the values of these variables. We shall postulate that the human community, in this respect, is endowed with well specified abilities that will be used as references.

2.6 Truth of a Proposition

(*a*) When, on the basis of measurements of the type of those considered in 2.4, whether direct or indirect, one or several people know that a measurement of a proposition *a* on a system *S* would give the result yes, were it performed, the proposition *a* is *true* on *S*.

(*b*) However, we know quite well through reference to the example of classical physics that the case considered in (*a*) is not necessarily the only one in which a proposition can be true, according to the commonly accepted language.

Indeed, on the one side we know cases in which it is possible to make a judgement on the validity of a proposition *bearing on the past*, and on the other side there exists in classical physics many a proposition that can be objectively true – in the sense of an adequation to the thing as it really is – without anybody knowing it.

Hence it is tempting to supplement the wording of (*a*) with the following two definitions.

Definition A

At time t_1 it can be asserted that a proposition *a* bearing on a system *S* will be (was) *true* at a time $t > t_1 (t < t_1)$ if the following condition is realized: it is known that if a measurement of *a* were (had been) made immediately after *t*, the *yes* result would be (would have been) obtained.

Definition B

At time t_1 it is said that a proposition *a*, bearing on a system *S* *can* be or could have been true at a time *t* (identical to or distinct from t_1) when and only when it is possible to imagine one or several persons having, at time t_1, all the information necessary to know the result that a measurement of *a* made immediately after *t* would give, or would have given, if it were, or had been performed (the *if* implies that in the case $t < t_1$, the said information should in no case proceed from the said measurement).

Instead of 'can be true' we could also say 'is true or false' or 'is trivially undecided'.

2.7 Implication

Having defined the truth of a proposition, it is now easy to define the verb 'to imply' in connection with propositions. It is said that *a* implies *b* if and only if *b* is true in all the cases in which *a* is true.

2.8 Persistent Propositions

On a given type *T* of systems, propositions having the following property often exist: if one of them is true on an isolated physical system *S* of type *T* at time *t* it remains true on *S* at later times *t'* provided that between *t* and *t'* *S* remains isolated. We call these propositions *persistent*.

In conventional quantum mechanics the persistent propositions are those whose operator (projector) commutes with the free Hamiltonian of the system); and the orthogonal complement of a persistent proposition is also a persistent proposition. Here we generalize the result and we add a new assumption:

Postulate 1

(i) The orthogonal complement of a persistent proposition is a persistent proposition.

(ii) Let a be a persistent proposition bearing on a system S. If a is true on S at a time t_1 and if S remains isolated during an interval of time (t_0, t_2) that includes t_1, then one can (at time t_1) assert that if a measurement of a had been performed at a time t lying between t_0 and t_1 the result of the measurement would have been yes.

The meaning of this postulate is discussed below (Section 5).

Consequence

Definition A then shows that proposition a was already true at time t on S.

2.9 Comments. Problems Bearing on the Individual State Concept. Causality

Definitions A and B and Postulate 1 have a particularly natural outlook in the realm of a realistic assumption bearing on everything that can be called a 'system' according to the convention stated in Section 2.1 (provisionally this includes, as we know, all the isolated microsystems, quite independently from any circumstances concerning them). But, on the other hand, these definitions and postulates do not *imply* the necessity of a realistic assumption about the systems. They can therefore be adopted without implying any postulate that would transgress the limitations set forth by the acceptance of a strictly operational viewpoint.

On the other hand, if the class of the 'systems' is supposed to include any microsystem (as the convention of Section 2.1 stipulates), then for $t < t_1$ the Definitions A and B can enter into conflict with the idea according to which a microsystem would constitute, in some way, an 'indivisible whole' with – in particular – the measuring instrument that will be used to observe it. This important point will be further examined below. For the time being we adopt the working assumption that Definitions A and B are valid.

Besides, this choice is forced upon us as soon as we require that the same conceptual schemes should be applicable to the present and to the past while maintaining the principle according to which the future cannot influence the past and keeping the usual general definition of systems given in Section 2.1. To show this let us consider again the operational definition of the truth of a proposition:

Definition d

At time t_1, it is said that a proposition a is or will be true at a time $t \geq t_1$ when it is known that a measurement of a performed immediately after that time would give with certainty the result *yes*.

And let us add to this the following postulate which can be considered as a mere formulation of the causality principle.

Postulate 2 (Causality)

One cannot change the truth values of the set of propositions that are true at a given time t on an isolated system S by merely acting on the experimental – or other – set ups with which S *will* interact in times $t' > t$. Nor can one change them a fortiori by acting, at times $t' \geqslant t$, on some *other* systems from which S is isolated at time t.

Definition d above immediately gives Definition A for the case $t \geqslant t_1$. In case $t < t_1$ (the only critical one), to assert – as Definition A implies – that if a measurement of a had been done on S immediately after t, a yes result would have been obtained, first means that an assumption is tacitly made, namely that S *did* exist as a system already at time t. This is a legitimate assumption however for it is in agreement with the extension given in Section 2.1 to the field of applicability of the word 'system'. Moreover Definition A refers to an experiment that was *not really made* (otherwise the corresponding instrument, I, would in general necessarily have changed the postulated evolution of the system). Now if an interaction (between S and some instrument such as I) taking place at some time t' immediately posterior to t *could* influence the truth values at time t of propositions refering to S, it would in fact be incorrect to define the validity of an assertion relative to S at time t by reference to an imagined measurement at t' which, in fact, is not done. Postulate 2 disposes of that objection. Under these conditions let us consider a case in which, for some reason or other, it is known, at time t_1, that if the said measurement had been done, it would have given a *yes* result. Since the systems pre-existed at t we can conceive of someone who, at time t, would have had that same knowledge on that same system. He would have applied Definition d and he would have said therefore that a is true. Hence, for coherence sake, we – at time t_1 – must also say that a *was* true.   Q.E.D.

Within the framework of the same requirements, Definition B can be justified in a very similar way. If a proposition or its orthogonal complement was true at time t in the sense of Definition d one can always of course imagine someone having, at time t_1, the information necessary to know whichever was the case. Conversely, if one can imagine, at time t_1, the existence of somebody having the information necessary to know what result a measurement of a proposition a would have given *if* it had been made at time t, then that somebody can be identified with the one who, in Definition A, is supposed to know that a yes result (for a or for a' as the case may be) would have been obtained. Hence such a person could assert, at time t_1, that either a or a' was true at time t. We can therefore say, at time t_1, that at time t, a was trivially undecided.

Finally let us state

Postulate 3

If a is a proposition about which it can be asserted, at time t_1, that it is true (that it was true at a time $t < t_1$) on a physical system S, then it is possible to assert that a is true (was true at time t) also on any $S + T$ composite system of which S constitutes a part.

Postulate 3 is an obvious truth in any realistic model since it just means what follows: if an object S has a property a, it cannot cease to have it if we consider it in conjunction with object T, and this also holds for the past. The validity of Postulate 3 in an operational scheme is discussed in Section 4.

3. SPECIAL STATEMENTS AND DISCUSSION

To study the adequacy to physics of the concepts, definitions and postulates considered above it is necessary to take a few facts into consideration. Let us systematize them by means of two statements.

Statement 1
Let U and V be two physical systems. Let $u_1, u_2, u_3 \ldots$ be propositions defined on U and similarly for $v_1, v_2, v_3 \ldots$ and V. For some sets of u_i and v_i it is possible to prepare composite systems $U + V$, U and V being isolated, such that if u_i and v_i are simultaneously measured, either the *yes* result is obtained for both or the *no* result is obtained for both, this fact being true for several i indexes, labelling distinct pairs of propositions.

Statement 2
Among the possible sets of u_i (and of v_i) considered in Statement 1 some exist that are such that each of the u_i (and each of the v_i) is a persistent proposition (see Section 2.8).

3.1 CONSEQUENCES

Let u_i and v_i satisfy Statement 2 and let us consider a composite $U + V$ physical system of the type described in Statement 1 and which remains so during a time interval (t_0, t_1). Let us assume that V remains isolated during the whole time interval (t_0, t'), with $t' > t_1$, but that at time t_1, u_1 is measured on U and the *yes* result is obtained. According to Statement 1, it is then known that if v_1 was simultaneously measured on V the *yes* result would also be obtained. According to Definition A, v_1 is therefore true on V. On the other hand V exists as a physical system (because of our convention in Section 2.1) and V is isolated during the time interval (t_0, t'). Hence, according to Postulate 1 and the consequence stated just below it, proposition v_1 was already true on V at any time t'' such that $t_0 < t'' < t_1$.

Let us now make the alternative assumption that, under the same conditions, the *no* result is obtained when u_1 is measured at time t_1. This can also be stated by asserting that a measurement of proposition u_1' is made at that time and that the *yes* result is obtained. It follows from Statement 1 that u_1' and v_1' are necessarily related to one another in the same way as u_1 and v_1 are. Hence the same argument as above applies and the conclusion is that, at time t'', v_1' was true on V.

If at time t_1, u_1 is measured, then, one of the two assumptions considered above is necessarily true (the result of a measurement on a proposition is always necessarily yes or no). Hence if we consider an enemble E of composite $U + V$ systems, all of which are of the type under study, were prepared in the same way and undergo the same treatment (namely the measurement on U described above) and moreover if we take Postulate 3 into account, we must conclude from the above that E was, at time t'', composed of two sub-ensembles, one of which being such that v_1 was true on all its elements (let this sub-ensemble be called E_1) and one of which being such that v_1' was true on its elements (let that one be called E_1').

According to Postulate 2, the truth values at t'' of the $U+V$ systems that make up E_1 and E_2 cannot be changed by a mere modification of the programme of the interactions to which these systems will be subjected after t''. In particular, these truth values are not changed if the measurement made on U at t_1 is suppressed from this programme. Nor are they changed if that measurement is replaced by a measurement of u_2. Let us consider the latter case. With regard to it we can repeat all the above argumentation merely by changing everywhere index 1 into index 2. We must therefore conclude that, at time t'', the ensemble E was also composed of two ensembles, E_2, E_2', the first one being such that v_2 was true on all its elements, the other one being such that same held for v_2'.

Thus, a consequence of the system of concepts (see Section 2.1), definitions (Definition A) and Postulates (1, 2 and 3) under investigation here, is that for any system V there are, at time t'', several propositions that are not necessarily compatible with one another (compatibility of the v_i has nowhere been assumed above) and are nevertheless true simultaneously. These propositions are v_1 and v_2 for some systems, v_1 and v_2' for some other systems etc. (Some of these classes may be empty but not all of them obviously.)

Let it be observed finally that, since *Statement 1* is assumed to hold as regards the $U+V$ systems, we can assert that at time t'', u_1 was true on all the elements of subensemble E_1: For we know from Statement 1 that if u_1 had been observed just after t'' on an element of E_1, a *yes* result would have been obtained and we can apply definition A. The same holds of course as regards the sub-ensembles E_1', E_2, E_2'. Hence a system $U+V$ on which $v_i(v_i')$ was true at t'' is also one on which $u_i(u_i')$ was true at that same time.

3.2 COMMENT AND DISCUSSION

Let us now comment on the result just obtained and let us investigate its consequences.

A first comment has to do with the notion of atomic propositions, a notion which has a considerable importance in conventional quantum mechanics.

Within the very general framework we are using, we could a priori contemplate at least two 'reasonable' definitions of the atomicity of a proposition, hereafter called Definitions 1 and 2.

Definition 1

A proposition a bearing on a system S is *atomic* if no valid assertion b, bearing on S alone, different from the absurd proposition Φ and implying a without being identical to a, exists.

Definition 2

Same as Definition 1 but with the words 'valid assertion' changed to 'proposition'.

In conventional quantum mechanics (without hidden variables) it is assumed that a proposition that is atomic on a system is so on any system of the same type. Moreover in that theory it may happen that the system $U+V$ introduced above is such that u_i, u_i', v_i, v_i' are all atomic, for a set of values of index i. This is the case for instance in an example to which we shall – after other authors[17] – often refer: U and V are two spins $\frac{1}{2}$ particles on a state of zero total spin (u_i then corresponds to the spin component of U along a direction $\mathbf{e}_i$ and v_i to that of V along $- \mathbf{e}_i$). Under such conditions the result of Section 3.1 shows what follows. With respect to any system considered at time t'' there exists one valid assertion b at least – which is either 'v_1 and v_2 are true' or 'v_1 and v_2' are true' or 'v_2' and v_2 are true' or 'v_1' and v_2' are true' – that (i) is not identical to Φ since it is true in some cases and that (ii) implies the validity of a proposition of the set v_1, v_2, v_1', v_2' defined on V (of two of them, as a matter of fact, which are either v_1 and v_2 or v_1 and v_2' etc...). If we adhere to Definition 1, this means that v_1, v_2, v_1', v_2' are not atomic, contrary to the premises. Hence we have obtained – a posteriori! – the proof that Definition 1 of atomicity is incompatible with the system of concepts, definitions and postulates under discussion.

This is not the case with Definition 2 because Assertion b is not a proposition in the sense of Definition 2.5. (Indeed, although we know that b holds for some systems V, we cannot *sort* the latter from the whole ensemble.) Nevertheless, the mere existence of assertions such as Assertion b that are shown to be true on some systems and are nevertheless distinct from propositions in the technical sense of that word is a worrying fact. Moreover we know how to introduce such assertions only when they bear on the past. Since one of our purposes is to deal with the present and with the past by means of the same conceptual system the latter remark in itself could be considered as pointing to a real difficulty concerning the application of our system of concepts, definitions and postulates to *any* theory that considers propositions bearing on the various spin states of particles with spin as atomic. However a much more general and perhaps more convincing argument is given below (Section 3.3) to show that the problem indeed exists.

It may be pointed out that in classical mechanics such difficulties do not appear. There, the atomic propositions are phase-space points, hence the orthogonal complements to atomic propositions are not atomic. In that theory if v_1 and v_2 are atomic and distinct the sub-ensemble $E_1 \cap E_2$ corresponding to both v_1 and v_2 true is indeed empty so that the assertion 'v_1 and v_2 are true' is never valid. But this emptiness is compensated by the non-emptiness of the sub-ensembles corresponding to the other possibilities.

3.3 DISCUSSION (continued)

The last observation above leads us to consider a more general viewpoint in which no assumption is made with respect to the atomicity or non-atomicity of propositions u_i, u_i', v_i, v_i'. With the help of recent results[13,14] we show in this section that even then one cannot make the system of concepts, definitions and postulates hitherto considered compatible with known facts. Just one counter example is sufficient for that purpose, and it is convenient to consider again a system of two spin $\frac{1}{2}$ particles U and V that has been produced at time t_0 in a state of zero total spin. The following argumentation is exclusively based on the results obtained in Section 3.1 as consequences of the considered system of concepts, definitions and postulates. It requires no assumption involving in any way the concept of atomicity. Nor does it require that any assumption should be made to the effect that the evolution of systems in general is governed by 'hidden' variables.*

Let e_1, e_2, e_3 be three linearly independent unit vectors in ordinary space, defining three directions. The results obtained in Section 3.1 imply that at time t'' (that is to say, before any measurement is made) any system V is such that – in particular – three propositions w_1, w_2, w_3 are simultaneously true with respect to it: where w_i is identical either to v_i (for some systems V) or to v_i' (for other systems V). When an ensemble E of N composite $U+V$ systems is considered, it is legitimate – within the framework of the system of concepts, definitions and postulates under investigation – to consider the number $n(\sigma_1, \sigma_2, \sigma_3)$ of systems $U+V$ on which it is simultaneously true that a measurement of proposition v_1 would have produced the result σ_1 if it had been done *and* that a measurement of v_2 would similarly have produced the result σ_2 *and*, finally, that a measurement of v_3 would similarly have produced σ_3 ($\sigma_i = \pm$, where $\sigma_i = +1(-1)$ corresponds to the answer *yes* (no) to v_i). The fact that V is isolated from U implies (because of postulate 2) that $n(\sigma_1, \sigma_2, \sigma_3)$ does not depend on the orientation etc. of the set-up that will 'measure' U (separability). Moreover the argumentation of Section 3.1 also leads to the conclusion that any composite system $U+V$ on which a measurement of $v_i(v_i')$ would have given the result *yes* is also a system on which a measurement of $u_i(u_i')$ would have given the result *yes* (let it be recalled that owing to the conventions made above this corresponds to opposite values of the corresponding spin components).

Now these conditions are precisely those under which a proof of Bell's inequalities[11] can be given. The principle of that proof is due to Wigner[15] and was applied by him to Bell's problem, that is to a study of the hidden variable theories. Fortunately, it can also be used here. And the proof it leads to is so simple that it is worth while describing it at full length.

Let $P(i, j)$ be the mean value on ensemble E of the observable A_iB_j, $A_k(B_k)$ being the value (in $\hbar/2$ units) of the spin component of $U(V)$ along e_k. On the other hand, let us consider the sum of two $n(\sigma_1, \sigma_2, \sigma_3)$ that differ only through the values of one of the variables σ_i; by convention let us agree to designate that sum by means of the notation consisting in a replacement of that variable by a point:

---

* It is true that the considerations of Section 3.1 led us to introduce assertions whose truth values appear as hidden variables with respect to the conceptual framework of conventional quantum mechanics. But these variables turn up here as a consequence of our system of concepts, definitions and moreover as a consequence that can be derived only for rather special composite physical systems. Its generalization to other physical systems could only come as a supplementary postulate and we do not want to introduce such a supplementary postulate into our considerations in this article.

$$n(-, +, \cdot) = n(-, +, +) + n(-, +, -).$$

It is apparent from the above that $n(-, +, \cdot)$ is the number of composite systems $U + V$ on each of which a measurement of B_2 would give $+1$ *and* a measurement of A_1 would also give $+1$. It is therefore an easy matter to express the $P(i, j)$ in terms of the $n(\sigma_1, \sigma_2, \sigma_3)$. For example:

$$NP(1, 2) = n(-, +, \cdot) + n(+, -, \cdot) - n(+, +, \cdot) - n(-, -, \cdot).$$

The explicit calculation then immediately gives the two equalities:

$$N[P(1,2) - P(1,3)] = 2[n(-, +, -) + n(+, -, +) - n(+, +, -) - n(-, -, +)]$$
$$N[1 + P(2,3)] = 2[n(+, -, +) + n(-, -, +) + n(+, +, -) + n(-, +, -)]$$

from which Bell's inequality follows:

$$|P(1,2) - P(1,3)| \leqslant 1 + P(2,3).$$

Now the values of the quantities $P(i, j)$ can be obtained from experiment so that Bell's inequality* provides a mean of testing the compatibility with the facts of our system of concepts, definitions and postulates. Moreover such experiments have been done already[13,14] with photon pairs. At their present stage these experiments cannot yet be considered entirely decisive. However the indication they give points very strongly towards a *violation* of Bell's inequality (a result already predicted by conventional quantum mechanics).** That conclusion constitutes the bulk of the proof by *reductio ad absurdum* we mentioned at the beginning of this article. It shows that the system of concepts, definitions, postulates hitherto studied is incompatible with our present experimental knowledge. This conclusion holds notwithstanding the fact that, considered separately, these concepts, definitions and postulates all look entirely 'natural'. It may therefore be used as a warning against an uncritical use of any one of these.

4. CONCEIVABLE CHANGES

The result obtained above was derived by means of an argument that, in some of its aspects, is reminiscent of the well-known Einstein, Podolsky and Rosen way of reasoning.[10] Between the two there are important differences however: contrary to these authors, we do refer here to experiments, and moreover in place of the *compatible*

* In its original form, as given here. Within the framework of hidden variable theory extensions of these inequalities have been given by several authors, but they seem to require the hidden variable postulate.
** Explicitly, the experimental articles just mentioned claim to test only the *generalizations* of Bell's inequality. However ref. 14 shows an experimental curve from which it is easy to infer that Bell's original inequality – i.e. the one written above – is also violated; although the degree of confidence of the latter inference is of course lower than that of the former. We are grateful to Prof. A. Shimony for pointing out to us that the information contained in the experimental curve of ref. 14 could be used for such a purpose.

observables, total momentum and relative coordinate used by these authors, we introduce observables u_i and v_i that are not compatible in general but still can have simultaneously a known value in some particular instances. Correspondingly, our result
applies to much more general theories than conventional quantum mechanics in
Hilbert space and is less open to philosophical controversy since our working hypotheses do not include a *realistic* assumption (the 'elements of reality') although they are
of course, compatible with it. Also, as will be apparent below, our conclusions are
different. Similarly, although the present considerations are strongly related to those
that have been developed on the subject of the 'hidden variables', neither our working
hypotheses nor our conclusions coincide exactly with those of such theories: roughly
speaking our assumptions are more general, as pointed out above. And correspondingly
the number of types of systems that can serve us as examples can happen to be more
restricted. Compared to the approaches just mentioned, ours derives from a desire
to follow the safer itinerary that pure operationalism claims to offer. Or, more precisely, it proceeds from a will to follow the rules of an operationalist methodology
that would incorporate the requirement of keeping the same conceptual schemes for
the present and for the past, without just reducing the past to the present.

On the other hand, as it stands our result is a negative one. Therefore it shows that
we must give up one at least of the concepts or postulates on the basis of which it is
derived. Let us now examine some of the possibilities existing in this respect.

4.1 Giving Up Postulate 2 (Causality)

To give up causality is clearly a possible way to escape the difficulty. On the other
hand the principle of causality in the restrictive sense (the fact that past events cannot
be changed) is so necessary to any kind of action that to give it up in a general way is
impossible. Its violation can hardly be considered except perhaps in the realm of
microphysics, or, more precisely, by a weakening of the general concept of 'system'
that would be effective exclusively for microsystems. This can be done by asserting
that a microsystem makes up an indivisible whole together with the instruments that
produce and analyse it, that it thus does not have – except by convention – properties
that are its own and that consequently it is not a 'full fledged' system. Let us call
object–instrument non-separability the principle thus introduced. In conventional
quantum mechanics the Copenhagen interpretation is based, at least in part, on considerations that are very near these ones.

4.2 Giving Up Postulate 1

A requirement was formulated above that – by thought – we should be able to apply
to the past the system of concepts and determination that we apply to the present as
soon as we imagine the same general conditions to be realized in both cases. If this
requirement is maintained then it is impossible to give up Postulate 1.

4.3 Giving Up Postulate 3

In the classical way of thinking, based on an ontology, Postulate 3 is just a statement of an obvious fact. Within the general system of concepts, definitions and postulates here considered, its validity as regards to the present follows immediately from the very definition of the truth of a proposition. With regard to the past the situation is different. Indeed let us consider a situation in which a given proposition a is persistent. To be able, at time t_1, to assert that a was true at a time $t < t_1$, on a physical system V it is sufficient, according to definition A and postulate 1, to know that at time t_1 a is true on V and that V remainded isolated during the interval of time $[t_0, t_1]$. The knowledge of the fact that, at time t_1, a is true on V can then proceed (as in the example above) from a pre-established correlation between V and some other physical system U and from a measurement carried out on U immediately before t_1. Under such conditions, however, the composite system $U + V$, contrary to the system V, does *not* remain isolated during the whole time interval $[t_0, t_1]$. Hence we cannot apply Postulate 1 to *this* system, and it is therefore impossible to make use of Definition A and to assert that a was truc at time t on $U + V$. This shows that postulate 3 is not a tautology when applied to the past. If we give up our demand that its validity should be quite general the argumentation of Section 3 is no longer possible and the difficulty we found vanishes.

However denying the validity of postulate 3 as regards past events while at the same time acknowledging the validity of the corresponding statement in relation to the present, is against our principles: for it introduces an irreducible difference between the present and the past in the manner we conceive them. If we do not accept any such difference we have to keep postulate 3.

4.4 A Refusal of the Intuitive Meaning of the Notion of System

The possibility of weakening the principle of causality discussed in Section 4.1 has led us to consider as rather natural to abandon the defence of the very general notion of system introduced in Section 2.1. However it must be observed that if such an idea is accepted, then it is by no means necessary to take all the steps considered in Section 4.1. Indeed, to invalidate the conclusion of Section 3 it is sufficient to consider that, before the time t_1 at which the measurement was made on U, the 'entities' U and V mentioned in the Statements 1 and 2 are not 'systems' in the new, more restrictive, sense that we shall henceforth give to that word. So that it is not even possible any more to speak about propositions such as the u_i and the v_i as defined on U and V respectively, as soon as some time prior to t_1 is considered.

On the other hand, of course, it still remains true that, as soon as a measurement is made on U, there exists a proposition that is true (that, if measured, would give the *yes* result on the corresponding V) independently of the distance at which that V is then from U. We must therefore concede that such a measurement acts immediately on the reality of the system, if by 'reality of a system' we understand the set of the propositions that are true on that system in the sense discussed above. Again, the non-separability comes in.

Hence we observe that under the present assumption non-separability is not the particular feature of *one* formalism or of *one* interpretation, as for example the conventional Hilbert space quantum mechanics or the Copenhagen interpretation. Unless we are prepared to consider even more drastic changes in our ways of thinking we must acknowledge that, on the contrary, the non-separability principle enters in any conceivable model that satisfies all the experimental requirements. It can be formulated by asserting that 'there are circumstances in which we cannot think of certain micro-systems as possessing as many properties as can be attributed to them in other occasions'. In our example, these circumstances are taken from the cases in which the considered system had interacted previously with other systems. And if we must believe conventional quantum mechanics, that criterion should be of general validity. If we decide to restrict the use of the word system only to the cases in which the maximum number of properties can be thought of as possessed by these systems, then we have to say that before t_1, the entities U and V introduced above are not systems.

Remark. Non-separability can also be formulated as the negation of an assumption that Einstein considered as being almost necessary and which he stated as follows:[16] 'the real factual situation of the system V is independent of what is done with the system U, which is spatially separated from the former'.

5. OUTLOOK

It must be acknowledged that in spite of the considerable coherence of the theory the basic facts of quantum physics can receive different interpretations. For instance, it would not be accurate to say that the question of determinism *versus* indeterminism is really settled by contemporary physics. It is true that a very large majority of physicists believe in the existence of micro-indeterminism. And they have very good motivations. Nevertheless, deterministic (and non-separable) theories exist[18,19] that reproduce exactly the observable predictions of the conventional formulation. Hence, indeterminism is by no means *proved*. Similarly, the principles of conventional quantum theory – as soon as the latter is stated with the necessary degree of coherence – contain in their very formulation explicit references to the concept of measurement (or of measuring instruments). It then looks as if these principles were only weakly objective, or in other words, as if they can at most claim to be valid for *any* observer (and not independently of the existence of observers). In a theory without hidden variables this result seems to be established for any finite N, where N is the number of degrees of freedom. And on the other hand the attempts at bringing back again strong objectivity (the objectivity of classical physics) in the pictures that are based on the $N = \infty$ idealization to describe the instruments did not apparently succeed in accounting realistically (that is for finite times and volumes) of the phenomena known as the 'collapse of the wave function'. But in spite of the highly convincing power of such remarks it cannot be said, here either, that an incompatibility was proved between our description of the actual facts and any conceivable strong objectivity requirement. For, again, the hidden variable theories mentioned above do represent a counter-example in that it is certainly not obvious in what sense they would lack strong objectivity.

We hope to have shown that non-separability presents a different case and that the arguments in its favour amount to much more than to a collection (be it even quite an impressive one) of mere *presumptions*. Indeed, under some disguise or other, it seems to appear as a necessary ingredient in *any* theory we care to choose, if only we adhere to a few requirements of thought that can be kept quite general. Such requirements (especially the identity of the concepts that are used to describe the present and the past) are inherent features to, in particular, any 'realistic' description. Therefore if (in the hope of giving a more firm foundation to our belief in the principle of non exhaustive induction or for any other motive) we agree to keep – or to incorporate – in our general system of concepts the one of a reality considered as the source of at least a part of man's experience (and, in that sense, prior to it) then we are forced to admit the following truth: such a reality does *not* obey the principle of separability. Hence, notwithstanding its extraordinary usefulness, the atomistic description of events and/or of micro-objects is obviously just a model.

With respect to the problem of building up an axiomatic of the proposition of physics, the foregoing conclusion should undoubtedly be a part of the preliminary requirements to be imposed. However it must be confessed that a general calculus that would take such a condition into account is, for a part, yet to be constructed. It must also be considered that, in view of all the foregoing analysis, that calculus should presumably – so that its internal coherence be guaranteed – take the form of *calculus of questions*[7]. Finally, the foregoing sequence of arguments can to some extent be reordered. An objection sometimes made to the epistemologies exclusively based on operationality takes its roots in the difficulties that these epistemologies have to cope with when they try to give the propositions referring to the past a meaning that would really refer to the latter and not merely to the present. We think that the present analysis makes this objection more precise and is at the same time a confirmation of its validity as regards any epistemology that claims to make use of propositions bearing on elementary events or systems. In other terms, it shows that with respect to the separability problem, such epistemologies are subject to limitations quite similar to those that do refer to an ontology.

6. APPENDIX

We shall not enter here into the general question of the conceptual foundations of quantum mechanics. This is much too vast a problem, and moreover we tackled it elsewhere.[9, 21] However, considering the title of the symposium, we feel it may be not entirely beside the point to enlarge to some extent the field of the foregoing discussion and to investigate the following question: it being granted that science cannot but propagate in some way or other, a view about Nature, which ones should it preferentially disseminate – or avoid to disseminate – in the public, account being taken of our present knowledge concerning the non-separability of whatever may be called 'the physical reality'?

That problem is unfortunately quite arduous. And we, physicists, are very much aware of the fact that the difficulty it raises does *not* reduce to the well-known general ambiguity that beclouds the concept 'Nature'. Indeed we have to deal with a question that is made extremely complex by specific reasons and these can be sketched (quite roughly) as follows. On the one hand we observe that the advances made in several sciences (in astrophysics and biology in particular) make it possible to account for an impressive and ever-increasing number of facts by means of the laws of physics. Indeed the assumption that these laws should ultimately account for *all* possible facts now seems far from unreasonable. And this is, in a way, a belated but concrete realization of the hope entertained by many popular thinkers of a pro-scientific tendency during the later part of the last century. Moreover it appears that one of the most fundamental instruments for this mode of explanation is quantum physics – of course – but quantum physics applied quite indirectly: as a tool merely used for determining some physical constants (binding energies etc.) that are then injected into calculations made with the help of classical physics[20]. This two-steps procedure offers, in a way, an excuse for holding fast to the classical pluralist – or, better said, *multitudinist* – conception. By this we mean a conception according to which the ultimate reality – all that really is – would essentially be constituted by an enormous number of elementary events and/or microscopic objects, each one of them being endowed with simple properties and being such that the interactions of them all – taken as local and causal – would, combined with chance, give rise to the complexity of appearances.* Considered as a tool for scientific thinking that conception is very convenient: this is why descriptions following its general lines – with eventually some more sophistication in details – are quite often made. However it then becomes tempting to shift mentally its status from that of a useful model to that of an absolute truth. In other words it then becomes tempting to adhere to what we can call a multitudinist ontology. By this we mean a metaphysics that reduces reality, both outside us and within us, to the elementary mechanisms just described. As a matter of fact it appears that a great number of scientists – particularly the non-theorists! – *do* take that step, although sometimes with minor reservations in their mind. For many of them the argument that the view is *useful in practice*, is undoubtedly important. On the other hand, since the advent of quantum mechanics it is a well known fact that this kind of uncritical realism runs very soon into considerable difficulties. In particular we know that *separability* in the sense of Einstein's claim (see Section 4) cannot be generally assumed for systems that have interacted. It is true that until now a way out still existed (hidden variables) that give to such philosophies some remnant of rational justification. With the violation of Bell's inequalities even this possibility disappears. The supporters of such a conception are therefore bound to modify their attitude. In fact they will have to recognize that in spite of its great successes (in classical physics, in molecular

---

* It is necessary to say 'multitudinist' rather than 'atomistic' since some philosophers (e.g. Russell during one period of his life) have put forward multitudinist theories whose elements were sensations instead of physical objects.

biology, etc.) the multitudinist model is quite unfit to be erected into a natural philosophy.

That evolution strongly reminds us of a fact that was once well known but had, to some extent, been forgotten: in a truly coherent realistic conception – where *the Real* is identified to what *is*, independently of our capacities and the limitations thereof – the fruitfulness of a model is not a sufficient warrant of the adequacy to the Real of the basic ideas of that model. Indeed it does not even guarantee any *approximate* adequacy of such a kind. Considered from the point of view of realism, the content of the foregoing sections only confirms once more that fundamental restriction. This it does by means of an example and with the help of a method the first aim of which is to free the argumentation from any particular mathematical formalism.

It would undoubtedly be an error to interpret the consideration above in such a way as to describe as a closed loop, as a meaningless oscillation, what in effect should be compared to a motion in spiral but still representing a real advance. Scientific research led men to the discovery of rigorous thinking: it has taught them how to go beyond infantilism and haziness in the construction of concepts and more generally, in the rational operations of the mind. Hence the contribution of Science to the maturation of man's intelligence can hardly be overstressed. And, for instance, our newly acquired knowledge of the non-separability of reality – of whatever can be described under that name at the microscopic level – is far more subtle (however incomplete it still is) than could possibly have been the intuition of it by a thinker of the foregoing centuries. But on the other hand – and, precisely, due to the appearance of new pieces of knowledge of such a kind – it would quite obviously be an error of a comparable importance (and a mark of considerable innocence) to ignore the phases and variations in outlook that such advances bring with them. Born inside a culture that had inherited a vision of the deep unity of Reality (the Great Rationalism as Merleau-Ponty called it) the scientific movement was slowly driven by the first established facts to give credence, or at least to spread among the public, the conception that we called 'multitudinist'. So much so that – again – quite a large number of scientists and of science teachers (often merely out of lack of information), instead of taking it as just a useful model, accept it, explicitly or implicitly, as a presumed universally valid description of whatever deserves the name of 'being'. But now it comes about that in spite of the considerable fruitfulness of that *model*, the corresponding conception of reality is proved at variance with experiments (provided only that certain highly probable technical hypotheses should be true, the need, even, of which will hopefully soon disappear). That conception is thus 'proved' to be false. Undoubtedly therefore we are now witnessing a decisive – even if quite progressive – change in the conceptions that any adherent to any realistic philosophy whatsoever may legitimately entertain as regards the 'real world'. This turn necessarily leads – as soon as implicit or explicit reductions of Nature to man are refused – to a conception in which a key role must somehow be given to the notion of the unity of Nature. If the concept of a Nature 'in sich' has a meaning (and this is far from being admitted by everyone), this Nature is actually – in a way – a wholeness, in the complexity of which *our* vision carves out distinct parts;

and these parts are primarily the mere reflection of our possibilities of action. More than a collection of objects or events having attributes of a type familiar to us, such a Nature appears as a *Natura Naturans* – to use a scholastic expression – somewhat similar to the God of Spinoza but even more distant from the actual phenomena. Comparable in that respect – although quite possibly in no others – to the God of Pascal, Nature conceals to us its basic unity, far behind the clouds of our *applied* rational thinking... but still lets us sometimes have a short glimpse at it.

Under these conditions it is disappointing to see that the larger number of scientists continue to hold to a naively realistic philosophy of the multitudinist variety (let us not discuss here the mere sophisticated instrumentalist or 'purely linguistic'[9] viewpoint, that have some basic arguments in their favour). It is disappointing but, *as far as research is concerned*, it is perhaps not alarming. In sciences as in arts (Fra Angelico is a good example!) a certain spirit of naïvety and innocence can be fruitful and if this happens then the conceptions animated by such a spirit should not be just dismissed as too naïve: so to speak they transcend innocence by the operation of their very innocence. As Wigner pointed out[20] the naïvely realistic conception of most of the modern scientists (theorists excluded!) fall fortunately into that class and that state of affairs will presumably last for at least one generation and perhaps even longer. From a cultural point of view, on the other hand, the preservation of such a conception – which, when all is said, is blatantly oversimplified – seems to be very dangerous. Today, great efforts – meeting with partial success – are made, by way of teaching, in order to give to vast layers of the world population a scientific knowledge and a scientific approach to various problems. Of course, the purpose of all this is first of all of a technological nature. But how could we expect from the majority of students and pupils that they should spontaneously make the distinction that we know are necessary between such conceptions as instrumentalism and realism and within the latter between the models and the truth? Particularly how could we expect this when (as is the almost general rule) their attention is never directed to this point by their instructors? Hence it is very much to be feared that the modern mass teaching of exact sciences will definitely and erroneously compromise them, in the public's mind, with the multitudinist philosophy sketched above.

Now this philosophy has two major defects. The first one is that it is so contrary to the conception of 'what is' that man tends to elaborate from his own intuition and thinking that it is more and more thrown into doubts, and that an ever increasing (and today enormous) proportion of the educated world's population has by now spontaneously ceased to give any credence to it. The growing lack of genuine interest for scientific studies is surely in part due to such a reaction; or in other words, to an instinctive demand for a culture that should be enriching and true, in a deep sense that multitudinist reductionism does not seem to have to offer. It will be a sad thing indeed if the exact sciences come to suffer from the discredit attached to a philosophy with which they should really not have much in common.

But the other defect of the said philosophy is even worse. It is that when it is really taken as such – i.e. as a supposedly true description of a Reality that would exist prior to man and independently of him – multitudinism is just simply false. Of course this is no new discovery. Most of the founders of atomic physics already stressed the incompatibility of certain aspects of multitudinism with the formalism of conventional quantum mechanics. But then, as others said, what about changing that formalism, or its interpretation at least, without changing its observable predictions? What is new is that now multitudinism as described above is contradicted by experiment *independently of the formalism*. Its falsehood as a philosophy thus seems unquestionable. Now no arguments, not even those based on fruitfulness, can justify the conscious dissemination of an *error* once it was recognized as such. To say it is merely a philosophical error would certainly be no excuse! Prejudices as regard the structure of Reality can influence quite considerably our whole mentality. Who can tell what cultural renewals might be repressed out of existence by cheating in that field?

Alas a dissemination of the kind just alluded to is very much favoured by the disinterest for general ideas that characterizes the specialists and by the resulting too piecemeal knowledge of science teachers. So much so indeed that the limitations to its nocivity is likely to come mainly – in a fully negative way! – from the growing lack of interest for exact sciences already mentioned. Could positive steps be taken? The first one would obviously be a general recognition of the existence of the danger by the community of responsible scientists (and even that is perhaps unlikely to occur!) The second one might be in the line of an effort directed towards scientific writers and commentators, so that they bring up to the public the existence of this problem, by presenting it as it really is i.e. connected with the 'quality of living' in its most intimate layers: not to the technical surroundings but to the image each of us can legitimately entertain of 'what is'. Then, and only then, would it be profitable to introduce into the scientific curriculum studies in modern epistemology that would bear on the – today fundamental – question of the comprehension and the significance of exact science. Sketchy as it is, this programme endeavours to meet needs that are deemed real, and we can only hope that they be recognized as such.

Once again let it be stressed that the importance of the perspectives opened by scientific research even in the field of pure knowledge is not disputed here. On the contrary what we want to stress is that the avenues opened by – for example – the violation of Bell's inequality extend these perspectives even farther than had been considered so far... even if they do change somewhat their direction! Obviously, to keep to the old philosophical lines, to allow the spreading among the public of fundamental views that we ourselves cannot any more consider as being really true – in spite of all their momentary fruitfulness – and to do all this under the very cover of Science would amount to a very serious falsification of the testimony of experience. Never was it so important to see to it that popularization of knowledge does not lead to anti-knowledge.

SUMMARY AND CONCLUSION

Since this symposium is on the development of the physicist's conception of nature, I wish to state very simply what impressed me most in the recent development of that subject. It is that experiments have been recently carried out that would have forced Einstein to change *his* conception of nature on a point he always considered essential.

In a discussion about his 1935 controversy with Bohr, Einstein wrote something like this: 'There is one suggestion we must absolutely hold fast to. It is that if we have two systems 1 and 2 that have once interacted and are no more interacting, the real factual situation of system 1 must not depend on what is done with system 2 which is spatially separated from the former'. For future reference let us call this the principle of separability. As is well known, a result of the Einstein, Podolsky and Rosen paper was to show quite decisively that as soon as this principle is given a completely specified meaning (this requires some care in the definitions of words) then it is *not* obeyed in quantum mechanics, as usually formulated. And for that reason they suggested that hidden variables should be introduced. Recently Bell has shown that no hidden variable theory that reproduces correctly the observable predictions of quantum mechanics can obey the principle of separability. And, this year, Clauser and Friedman have shown (experimentally, that is: quite independently of quantum mechanics) that indeed all hidden variable theories that obey the principle in question are contradicted by the data.

Hence we know for sure, contrary to Einstein's conception, that, in some respects at least, the world is non-separable. In other words the status of the concept of non-separability has changed. Previously, it was but a feature of *one* interpretation of experiments, namely the conventional Hilbert-space quantum mechanics, without hidden variables. Now we know it holds also in another quite different interpretation, the one with hidden variables, which as we all know is *consistent* even if it is not quite beautiful. And here I claimed that non-separability holds also in more abstract formulations such as the one based on the lattice of propositions (in the style of Piron and Jauch). For all these reasons I think we may safely say that non-separability is now one of the most certain general concepts in physics; more so than indeterminism of weak objectivity. This, in fact, is my real point. In other words, what I want to stress here is not the novelty of this concept but its high relative degree of certainty.

Non-separability is also culturally very important and significant for the subject of this symposium, because it shows that the general approach of Democritian atomism is a false view of nature even when applied to events. If we call 'atoms' micro-objects, having definite properties, or micro-events, then it is *we* who, so to speak, paint the distinct atoms on the canvas of non-separable reality, whatever this latter word means.

Hence we are in a circle. To state it loosely but vividly: 'the atoms make us, but we make the atoms'. If I may refer to the cartoon that Professor Eigen showed at the beginning of his lecture, this is just the problem of the hen and the egg. But the tragedy, in a way, is this: as it is popularized, and even as it is taught, science vigorously propagates the view that it goes only one way. That it is the egg that causes the hen, I mean the multiplicity of atoms that causes us and everything, in some *absolute*, *irreversible* way. To say it differently, elementary scientific teaching implicitly propagates the view (*A*) that a kind of naive realism is true; (*B*) that the ultimate reality is essentially constituted of an immense number of small elements each possessing a fixed number of definite properties, and (C) that the local and causal interactions of these elements lead to combinations that account for the complexity of the actual world. On the question that such a view of the world is but a crude model *tailored to specified purposes of ours*, an absolute silence is nearly always kept.

This mutilation of the truth throws undue discredit on the concept of *unity*, and is therefore a really serious one, culturally I mean. Moreover, it is a falsehood recognized as such by, guess, a vast majority of the participants in this symposium (I know I have not been saying anything new!). And falsehoods should not be spread. Still, that one implicitly is; and perhaps even more by scientists who are not physicists than by physicists. It seems to me that we must be aware of that situation, for if nothing is done we may fear that science will become associated, in the public mind, with a very poor and obsolete natural philosophy with which it has really nothing in common.

When extended to the system-instrument complex, non-separability constitutes the essence of the formidable problem of measurement. I think that the latter can probably *not* be solved merely by limiting the number of true observables. More precisely: there have been detailed attempts at analysing realistically the mechanism of measurement that have led me, among others, to the conclusion just mentioned. Maybe, therefore, we should say this. There are natural requirements as to a *deep* level of understanding whose scientific relevance is questioned by some people. But if these requirements are made, and many other people think they should, then we have reasons for considering the possibility of modifying quantum mechanics – perhaps, as Professor Wigner advocates, by adjunction of new variables.

REFERENCES

1. G. Birkhoff and J. Von Neumann, *Ann. Math.* **37**, 823 (1936).
2. J. M. Jauch and C. Piron, *Helv. Phys. Acta* **36**, 827 (1963).
3. C. Piron, *Helv. Phys. Acta* **37**, 439 (1964).
4. J. M. Jauch, *Foundations of Quantum Mechanics*, Addison–Wesley, Reading, Mass., U.S.A.
5. J. M. Jauch and C. Piron, *Helv. Phys. Acta* **42**, 842 (1969).
6. J. M. Jauch, in Proc. of the IL session of the Enrico Fermi Summer School on *Foundations of Quantum Mechanics*, Academic Press, New York.
7. C. F. von Weizsäcker, *Naturwiss.* **42**, 521 (1955).
8. G. Mackey, *Foundations of Quantum Mechanics*, Benjamin, New York.
9. B. d'Espagnat, *Conceptual Foundations of Quantum Mechanics*, Benjamin, Addison–Wesley, Reading, Mass., U.S.A.
10. A. Einstein, B. Podolsky and N. Rosen, *Phys. Rev.* **47**, 777 (1935).
11. J. S. Bell, *Physics* **1**, 195 (1964).
12. J. F. Clauser, M. A. Horne, A. Shimony and R. A. Holt, *Phys. Rev. Letters* **23**, 880 (1969).
13. L. R. Kasday in *Foundations of Quantum Mechanics*, Academic Press, New York.
14. S. J. Freedman and J. F. Clauser, *Phys. Rev. Letters.* **28**, 938 (1972).
15. E. Wigner, *Amer. J. Physics* **38**, 1005 (1970).
16. *A. Einstein, Philosopher-Scientist* (ed. by P. A. Schilpp), Library of Living Philosophers, Evanston, Ill., U.S.A.
17. D. Bohm, *Quantum Mechanics*, Prentice Hall, Englewood Cliffs, N. J.
18. D. Bohm, *Phys. Rev.* **85**, 166 and 180 (1952).
19. J. S. Bell, Proceedings of the 1971 Penn. State Conference (to be published).
20. E. Wigner, *Symmetries and Reflections*, Indiana University Press, Bloomington and London.
21. B. d'Espagnat, *Conceptions de la Physique contemporaine*, Hermann, Paris.

40

Physics and Philosophy*

C. F. von Weizsäcker

I was informed by Jagdish Mehra that some discussion on the question of physics and philosophy would be of interest to the participants in the symposium, and I agreed to talk about it. I am glad to express some of my views on physics and philosophy. I wondered what might be interesting and I thought that the problem could perhaps be dealt with in four steps: a discussion of objectivism which has been so much in our minds at this symposium, a mention of positivism, the philosophy of Kant, and then the philosophy of Plato, which I consider to be important in this respect.

It is quite clear that philosophy is an immense task, and one cannot do justice to it in one hour or in one evening, and one has to devote a considerable part of a lifetime to it. I shall therefore just give a few personal impressions. The arrangement which I have proposed for this talk is connected with an experience I had when I first tried to understand physics. One of my main difficulties when I tried to understand physics was that I found people were using words, and hopefully concepts, which they could use very effectively, and they could apply them very efficiently to the solution of problems, while I found it extremely hard to understand what they meant by them or what these concepts were really intended to express. So when I learned that physicists speak about space and time, about energy, about potential energy, about reality, I was at a loss to know what all this meant. On the other hand, in the beginning, I thought that they knew it very well because they could use it so well. But then I discovered that, in many cases, they knew how to use these concepts because they had learned it from their teachers, and their teachers had learned it from their teachers. I found that there had been a common usage of such concepts without a full reflection about their meaning. When I went back to the sources where these concepts had been invented, I found that to a large extent they came from the philosophers, from philosophy. They came partly from modern philosophy, especially of the seventeenth century. But the philosophy of the seventeenth century cannot be understood if one does not try to understand how it is connected with its predecessors or scholasticism. Scholasticism, in its turn, is impossible to understand if you don't know how deeply it was influenced by Aristotle and Plato. My attempt to

---

* *Editor's note:* The Symposium was almost over when numerous participants requested us to arrange a discussion on 'physics and philosophy'. At the last minute we asked Professor von Weizsäcker to give an introduction to the subject. His spontaneous remarks were deeply appreciated by all and led to a fruitful discussion.

understand anything in physics and philosophy therefore led me, I think by necessity, back to Greek philosophy.

A very good thing with the Greeks is that they do not 'refer to the Greeks'; they refer to themselves. They speak for themselves. In Aristotle and Plato, for the very first time, I had the impression that here is a philosophy which one can understand. It is my personal reaction. Others may react differently. And I can explain why, because they formed their concepts in order to explain something which they had seen themselves, and they had not inherited their concepts from somebody else.

Let us first talk about objectivism. The feeling which we all share, being physicists, is that we talk about something that exists, about objective things. Our life depends on them. For instance, we would not have assembled in Trieste if there had not been this modern technology which has been able to bring us here by car, by train, by areoplane. And this technology depends upon the understanding of the objective laws of nature. And that this is a good understanding is shown in the fact that it works. So, in this sense, we are all objectivists. It's absolutely impossible not to be an objectivist in this sense.

Then the question is: Does 'objective reality' just mean that we are able to apply all these methods? Is this not perhaps still too subjective a point of view? Well, let me insert here an anecdote. I read somewhere that, just before the American presidential election in 1908, when the Republican candidate was Taft and the Democratic candidate was Bryan, an American politician of high standing visited Mount Wilson Observatory and looked through one of the big telescopes. He looked at the Andromeda Nebula, at whatever you could see there. After having looked long through the telescope, he said, 'After all, perhaps it is not so important whether Bryan or Taft is going to be elected.' So this again is an expression of the feeling of the scientist that all our human world doesn't matter so very much as compared to the immense objectivity of the physical world.

I remember very vividly that I was deeply impressed with the beauty of the stars when I was a boy, and certainly the man of whom I just spoke was too. But I was also troubled by the question whether this beauty was not somehow in conflict with the fact that these stars are spheres of ionized gas. How can these two things go together? And it was my very immediate impression that this problem could only be resolved by seeing the beauty of the laws governing ionized gases, i.c. by the beauty of the laws of nature, which again are to be discovered and not invented. This is said in order to make the strong point that all this is not invented by us. It is discovered, and it is greater than our life, greater than we are.

On the other hand, the philosophy of objectivism seems to lead to certain difficulties and drawbacks. For instance, if we speak about matter and the laws of matter in this objective sense, the question arises: How is our own consciousness connected with all this? What does it mean that we are speaking about it? Are we strangers in the material world? Do we belong to it? What is the feeling of beauty which is so strong in the minds of great physicists? What are they sensing there? Is beauty objective, or is it subjective? Or is the subjective objective? What does all this mean?

I feel there is a difficulty in the very concept of objectivity if one confines what is called objective to the things which obey the laws of classical mechanics. This unity of nature is very difficult to understand if we try to found it on classical physics. Classical physics talks about bodies in space and time. How can life, or the human mind, be explained or understood by being reduced to bodies in space and time. If one takes the Cartesian point of view that there are two completely different substances – there is matter, the extended substance, and there is mind the thinking substance – then their connection is absolutely in the dark. So it seems that the very objectivity of classical physics makes it difficult to understand that nature, including ourselves, should be a unity.

A quite different way of understanding the unity of nature would be the idea that all in nature is somehow of the same essence, of the same structure, of the same nature, as our own consciousness. That would be some sort of a mentalist or spiritualist philosophy. That this philosophy, in turn, would make it very difficult to understand physics was clearly the view taken by classical physics. Why should a mind, or the soul, or whatever the basis of the philosophy that one could have here, obey such strange laws as the laws of mechanics? There was thus a gap in our understanding of nature in terms of classical physics, which was realized as soon as one wondered how man and all that belongs to human life was to be included in it. Therefore, for reasons which may be called philosophical, I feel that the objectivity of classical physics is some sort of half-truth. It is a very good thing, a very great achievement, but somehow it makes it more difficult than it would have seemed before to understand the fullness of reality. I am here speaking about motivation, because I think philosophy consists, to some extent, in understanding our own motivations, in knowing what we try to speak about and why we try it.

Let's see what contributions can be made to this problem partly by contemporary physics, partly by the great philosophers of the past. As for the first step, I remember very well how I met Heisenberg for the first time when I was a boy of fourteen. We happened to be in Copenhagen at the same time. Soon afterwards, in a taxi in Berlin in April 1927, he told me about the uncertainty principle. I was fourteen years old, and I was greatly moved by this new idea. I got the impression that if this was physics, one must study physics. This was the first moment when I saw that there was a hope of bringing together the two different parts: the objective world described by classical mechanics and the world of man. I didn't know how, but somehow it meant that there was a connection between the two. And this was the way in which Heisenberg himself was expressing it when he said that the sharp distinction between subject and object was no longer possible in quantum theory. Nearly all the discussion about objectivity and the other matters which we have been discussing today [in the session on Physical Description and Epistemology] is produced by this very fact, because if Heisenberg's statement is true we would have to apply concepts which would be applicable explicitly to subject *and* object, and which might illuminate this apparent distinction between matter and mind, or whatever else one may like to call these 'opposites'.

The attempt to reintroduce classical objectivity in quantum physics is partly due to the fact that the understanding and the clear conceptual formulation of the new view, which was introduced by quantum theory, is so immensely difficult. It is immensely difficult to do this in a consistent manner. But we do wish to have consistency in physics, and this induces some of us to fall back on that very system of thought to have left which is the great merit of quantum theory.

In any case, one could try to see whether there exists any philosophy which would bring subject and object together in one conceptual framework. I was deeply impressed by the philosophy of Ernst Mach in this respect when I started studying physics. Therefore I come now to positivism, because Mach is rightly counted to be one of the great men of positivism. I would say that I personally was more impressed by Mach than by any of the other positivists. Mach's idea was that one can do without any concept of a subject (of an 'I') and without any concept of a thing or an object, if one would speak of 'sensations' as the one basic reality. He called them 'elements', and he said that you could call them sensations, if you like, but then you must be very careful to understand that they are not sensations of a subject which are created by objects; rather, the ultimate reality are the sensations.

If I may use slightly more modern language, one can say that a positive source of sensations is called a thing, and a negative source of sensations is called a subject. Do sensations converge? The fact that they converge constitutes something like an 'I'. Are sensations produced? One can say that what produces them can be called an object. This was a very ingenious way of eliminating Cartesian dualism, only nobody was ever able to carry this out, and I really do not think it is a tangible theory.

One of the objections is that we speak always of things which we can have knowledge of, and if we try to formulate into language as to what this means, it means that we are not only to speak about factual sensations, about sensations which actually happen, but also about the connection between possible sensations. The law, in Mach's language, is always a connection between possible sensations and not only between actual sensations. For instance, it enables us to predict sensations. Prediction is always about the possible. An attempt to formulate clearly what might be the possible truth of Mach's idea leads us to the question: How can there be laws about the possible? This is the basic question of empiricist philosophy, and since there exists no convincing empiricist answer to this question, Mach's theory cannot be consistently carried through.

I have referred to the difficulties of empiricism in my article on the classical and quantum descriptions [in this volume]. Let me recall the main point. It is, as Hume put it very clearly, how do you know the sun will rise again tomorrow? Answer: It has risen all the past days and years and millions of years. Question: How do you know that it will rise again tomorrow? Do you see any necessity? No, there is no necessity. There is just the rule that what has happened many times in the past can be expected again in the future. How do you know this rule? Well, it has been very successful in the past. How do you know that it will apply to the future? You don't know it. There is no logical way of concluding anything about the future from the past. This has been repeated in a slightly different manner in modern times by Popper, who says that a law of nature, logically speaking, is a universal statement. And a universal statement cannot possibly be verified by enumerating all its cases, if the number of these cases is infinite. You can only falsify it. I am not going to repeat the problem of falsification, but it must be said that the real problem is not so much whether you can falsify such laws or not, but how such laws can rule in so many instances as, for instance, in quantum theory. Why should it be possible at all? It seems to be improbable from the outset that there should be a theory which you can write down in one half page of print, which applies to a thousand million cases correctly. This, I think, is the true problem.

The empiricist philosophy has not been able to answer this question. One can easily understand that it cannot possibly answer this question. This is not to say that we do not get our physics from experience. The problem is not that we should have another basis than experience. The problem is: How is experience possible? How can we understand that experience works? The modern positivist or empiricist philosophy of science has been struggling with this problem very hard and it has come into greater and greater difficulties, and this is because what it tries to achieve is impossible. They are now beginning to see it, but this is not yet a solution; it is just a problem. When I had come to the point of recognizing that this problem existed, the most natural thing was to try to find out whether there was any philosophy which had already seen this problem, and had perhaps solved it.

Now the problem had been very well articulated by David Hume. The attempt of an answer which immediately comes to our mind then is the attempt given by Kant. Kant tried to answer the problem of Hume in a manner which I would like to explain a little bit, although I personally don't consider myself to be a Kantian. In many respects I hold views which are not identical with Kant's, just because I am a man of the twentieth century, and there are many things which Kant did not know about. On the other hand, I have been giving courses and seminars on Kant now for about fifteen years, and it has been my experience that when I come back to Kant, I always find some points where I see that my objections against him were caused by the simple reason that I had not yet understood him. That's what you always experience with really good philosophers. Let me say a little bit about the consistency of Kant's ideas, because only then will we see more clearly in which respect they will not suffice for explaining the problems imposed by modern physics.

Kant's idea is that Hume's problem is to be resolved by showing that the laws which we find in experience are made possible by certain preconditions of all experience. And if we really want to understand why there should be laws at all, we should first understand that experience itself is not a trivial thing, and that many conditions must be fulfilled in order that experience should be possible. I gave an example this morning [See 'Classical and quantum descriptions' by C. F. von Weizsäcker in this volume] by saying that if there is no time, there is no experience – experience meaning to learn from the past for the future. Scientific experience, at least, means that. Since we are speaking here about scientific experience, we can be pretty sure that we can presuppose time as an element in every theory, because if there is no time there would be no experience and hence no theory.

Kant tried to do this in more detail and he did it especially in speaking of two different roots of our understanding. One is intuition, the forms of intuition; the other one is thought, the forms of thought or categories. The forms of intuition in Kant's philosophy are space and time. He takes space and time as something given, as the forms in which we must understand everything we experience. Without going into details of this theory, let me just mention one point. Many people have said that this is out of the question today: since Kant thought Euclidean geometry to be *a priori*, and he was just not aware what would happen in the nineteenth century, so we should no longer be interested in such a theory. But this is historically wrong. Kant was fully aware of the logical possibility of a geometry, different from Euclidean geometry, as found by Saccheri and Lambert, and this means that he knew that Euclid's postulate about the parallels could not be logically deduced from the other postulates. Kant's reaction to this was precisely what he expressed in his statement that all mathematics, especially geometry, rests on *synthetic* judgements a priori. A priori means that it could not be otherwise; we understand that this is true. Synthetic, because they are not analytic. Not analytic means not logically deducible. And this is just an understanding that a non-Euclidean geometry is logically possible. I would say that Kant's theory is not the naive theory which doesn't know the possibility of non-Euclidean geometry. It is an attempt to answer the various possibilities of a non-Euclidean geometry by saying that, 'All right, logically this may be possible. But we are not speaking about possible spaces, what mathematicians today call spaces, but we are speaking about that space in which our experience is made. The space in which our experience happens seems to be Euclidean. And this must be understood. This is the real problem.'

Kant's theory of mathematics is perhaps even more closely connected to really modern ideas in the theory of arithmetic, such as Brouwer's intuitionism, and Brouwer himself says that he is dependent on Kant. Brouwer's theory is that number cannot be reduced to logic, and that it can only be understood as an original intuition, the intuition of counting which is done in time. This is more or less what Kant had already said, and we can say that Kant's theory of mathematics is as modern as intuitionism, and intuitionism is the best theory of the foundations of mathematics that we have today. However, the problem of non-Euclidean geometry, in the way that it has been posed by general relativity, had not been seen by Kant, and therefore we cannot apply Kant straightforwardly to anything like this problem.

Let me mention another point in Kant. This is the categories or the forms of thought. Without going into detail, I shall just mention the points which I think are the most interesting, formulating them in a manner which is in the *Critique of Pure Reason*, but adapted to my present task. In the *Critique of Pure Reason*, Kant asks the question how the unity of consciousness, or as he says the unity of apperception (apperception being his word for consciousness), is possible *in time*? This is a strange problem, *because we are living now*. We are *not now* living in the *past*. We are *not now* living in the *future*. But our consciousness connects present, past, and future. How can it possibly do this? How can the past be present? His answer is that it is possible by means of concepts. In concepts, we can have something which goes *all through time*. His theory of the concept is that we should call concept an idea which can be a common part of many different ideas. For instance, the concept of a dog is contained in the idea of this dog that we know, and of that dog, and even in the idea of a dog which we have never seen. Kant's point is that a concept is an idea which can be contained in other ideas, that it is an idea which can apply to past experiences, to present experience, as well as to future experience. Only through concepts is it possible to unite our extended experience, extended in time. The general forms of all concepts, which Kant calls categories, are used by him to conclude that no experience would be possible if certain principles were not fulfilled, and I shall just state some of these principles.

One is the principle of substance, in which he says that there must be some underlying unchanging substratum in which changes can be observed if there is to be experience at all. This he connects in a very interesting way, with the idea that time itself cannot be observed because it is homogeneous. There must be something to represent the only reality which lasts through all time, and that is time itself. This is a difficult concept, but the interesting thing here is that Kant deduces the idea of conservation of a substance, which in the end amounts to the conservation of matter, by a consideration of the homogeneity of time. When I first understood his argument, I was immediately reminded of Noether's theorem which deduces the conservation of energy, which we know now to be identical with the conservation of matter, from the homogeneity of time, although it is not possible to draw a strictly connecting line between these two ideas.

Another idea of Kant is that the law of causality is necessary in order to have experience at all. Again, without going into details, let me remind you that Bohr's idea of the inevitability of classical concepts is really an amplification of Kant's idea. Bohr, I think, had never read Kant very carefully, but he had known Høffding very well, who was a very good Danish philosopher. Høffding himself knew Kant well, and therefore Bohr had at least a check in his conversations with Høffding whether he was on the right track or not.

In the end, Kant's theory furnishes a most interesting point. Kant's *Metaphysical Principles of Natural Science* was a hundred years after Newton's mathematical principles of natural science. I think, the title of Kant's work is given explicitly in opposition to Newton's title. In that book, Kant tries to explain how we can go from such general principles like conservation of substance, like causality, and so on, down to the real laws of nature. Then he says that some general principles can be found *a priori*. He even tries to deduce classical mechanics *a priori*, which is not quite convincing, but quite interesting. And then he says that the *special* laws of course must be found by experience.

Now if one compares that with present-day physics, one finds that this is no longer the situation, because to us there are no special laws. Of course, there are special laws which we find by experience, but if we try to reunite all of physics by quantum theory, as a general theory of change and of possible statements about change, and elementary particle physics as a general theory of all possible species of objects, then in the end we hope that at least in principle it would be possible to deduce the properties of matter from our basic laws. There would be no special laws left which would be logically independent of the basic laws. The present situation in physics is such that Kant's problem of special laws disappears. It has not yet completely disappeared, but it will disappear.

This is a point in which, in my view, most of modern methodology of science is completely on the wrong track, because there one always tries to understand how special laws can be deduced from experience. I don't think this can be done. Of course, you can find them in experience, but you cannot justify them fully from experience. You can just go in experience far enough to say that something seems to be the case, but there is no way, as was seen by Hume, for any special law to prove that because you have observed it in the past it will also hold in the future. If one makes a very precise analysis of Kant's theory, one finds that he has not justified it either, because in his theory special laws which must be deduced from experience fall under the same objection which Hume raised, and Kant did not really improve the situation.

Our problem today is not to reproduce Kant's idea. We must either renounce every understanding why physics is possible, and just accept that it happens to be so; or we must try to understand the small set of fundamental laws, which we have partially already found and partially hope to find, as the preconditions of experience, as conditions without which experience would not be possible. This, I think, is the task given to the philosophy of modern physics. Again, it is very complicated and very difficult to do, but the philosophical situation is such that either this is the task and the question is answered, or there is no philosophy of science.

Next I come to Plato. To speak about Plato is a great enterprise, and I would love to talk at length about him, but I shall be brief. Why should we talk about Plato at all? Plato, is not, like Kant, two hundred years ago but two thousand years ago, even more than that. It is quite evident that whatever he said about natural science is completely obsolete now, and we need not be interested in his idea that something corresponding to the atom is formed as so-called Platonic bodies, the tetrahedron and the cube, and so on. On the other hand, his methodological approach contains an element which is still very important; besides he is just the great philosopher of the Occident. Whitehead was completely right in saying that occidental philosophy consists in a few footnotes to Plato. So if we want to understand our philosophy, we must understand Plato anyway.

Platonic philosophy contains, as its central concept, the concept of what he calls *Idea*. To translate that into English is dangerous because Plato's *Idea* is the most objective that can exist. It is just the source of objectivity, while the idea in modern use means what we have in our minds, and this has come under the suspicion of being merely subjective. As I have said earlier [see von Weizsäcker's article on 'Classical and quantum descriptions' in this volume], I feel all this distinction of objective and subjective is somehow nonsense. Because what we call subjective, is what we know; of course, we can also speak about our errors, and they may be subjective in a sense which is not meant here. Besides, our knowledge is a fact in a real being, a human being, which we have agreed to call a subject, and I don't see what is objectionable in speaking about knowledge as well as we speak about matter. Knowledge exists as well as matter exists, and knowledge is perhaps even more important because we know about matter in the form of knowledge but we do not know about knowledge in the form of matter. It is not symmetrical. Knowledge is really the more basic concept. And knowledge means that things are the way we know them. The Modern usage of the word 'idea' is misleading because it contains all this completely unclear way of speaking about objectivity and subjectivity, which confuses everything. Everything might be far clearer if these two words were not used.

It was absolutely clear in Plato. Plato spoke about a *form*, as it is sometimes translated in modern times, as that which makes the existence and the understanding of existing things possible. How is this to be understood? A modern example, not one used by Plato himself, might help us to bring it a little bit more into our present day discussion. In one of his books, the biologist and ethologist Konrad Lorenz talks about the wild geese. He describes that he has been studying the life of these birds, and he knows that they are monogamous, that they marry just once in their life. This belongs somehow to their conditions of life; this is the way in which they can best live or are best adapted to their surroundings. But then, in studying their life in detail, he found out that in many cases they were not quite monogamous. They had some other friendships too, and even married twice or three times. Lorenz seems to have been a little bit offended by this discovery, because he was so very proud of the monogamy of the wild goose. One of the girls who cooperated with him in his institute then said, 'Well, now after all they are human.'

Well, what does this story mean? It is meant to say, and Lorenz himself used it in this manner, that what the zoologist describes is not the empirical wild goose, but the Platonic idea, the Platonic *form* of the wild goose. This is how the wild goose ought to be, not how it really is. And why so? Well, he says in Darwinist terms, in evolutionist terms, in selectionist terms, that this is the way in which it would be best adapted to its environment. It may not fulfil completely the norm, but the norm describes its optimum living conditions. This is a way of describing in biology how the concept of *form* can be reproduced. It is something which is not in any goose, which is not *within* it as Aristotle put it, but which is completely separated from any real goose; it is the norm by which we judge whether it is a good goose or not. And this is precisely the way Plato speaks.

How is this to be connected with what else we know? It can be connected in the following manner. Only under certain constant conditions of the environment will there be anything like an optimal norm for a living being. Hence we must have norms not only for how a living being should be, but also norms for the way in which anything in the material world would be. We call these norms, natural norms, and they are the higher ideas, the higher *forms* in the language of Plato.

In his simile of the cave, Plato describes people who are sitting in a cave and looking at the wall, where they only see the shadows of some things which are transported behind their backs. Then they are turned around or at least one of them is turned around completely in order to see the reality. Then he suddenly realizes how unimportant is the great art of the people, who have been sitting with him looking at the shadows. This is the art of predicting what shadow would follow the other one. They take the shadow to be the real thing. But this art is far surmounted by the understanding of one who sees the real thing. But then, he has seen only the things which are carried behind their backs in the cave. He goes into the outer world, and there he sees the shadows of things in the light of the sun, and he sees real things in the light of the sun, and then he may see the sun itself.

This comparison shows how understanding arises. The lowest level is to be able to predict what will happen to the shadows. This is science as it is understood by empiricist methodology, or this is the way elections are won. The next step is to see those things of which these shadows are shadows. This is the way of theoretical physics which tries to understand things as they are in themselves, or it is the way of real practical ethics and political theory which says how elections ought to be held. Then one goes higher up and one sees the shadows of real things in the light of the sun. This, in Plato's view, is mathematics. Mathematics is the condition of the possibility of the things which the theoretical physicist studies; it is theoretical ethics or the rules of what is really good in single cases.

From there, one must go up to see the real things themselves. This is what he called the *forms*, which are ideas like identity and difference, existence, motion, rest, etc. These are the basic ideas. And if you understand them, you must go on once more and understand what is common to all of them. What is common to all of them is what traditional philosophy calls the transcendentals, that means good and truth are one. That means One, Good, True, Being: to be one, to be good, to be understandable, that means to be true, and just to be. These are then the very highest concepts, the very highest forms without which nothing else can be understood, which are implicit in all our understanding. But because they are implicit in our understanding, we do not know of them in general. We are usually not reflecting on them, but whenever we realize that we are reflecting on anything we must finally come to the understanding that we are reflecting by means of these basic concepts. The very highest concept is *the one* which, in Plato's philosophy, is identical with the concept of the good, and this is no more a concept. You cannot even say that there 'is one', because in saying this you are saying two things: is and one.

Then you must descend again, and you must try to understand all of reality, all of what one knows, including finally even the sensual appearances, starting from this highest perspective. That is mathematical physics. Plato starts from the highest forms, and goes down to forms which are mathematical, ending up by describing models of existing things, of what we see and what we perceive by all the senses.

There are some elements in this descent which are even interesting in detail to us. In the end, he describes everything in terms of small bodies, Platonic bodies, which are the only fully symmetrical bodies that are mathematically possible in three-dimensional space. Heisenberg says that he had been greatly impressed by this, and he feels that modern physics, which is trying to reduce everything to symmetries, is somehow on a similar track. There is one additional point which I like, but it may of course be overdone. I think it can be clearly shown that the symmetries which Plato gives to his smallest bodies are only possible due to a *finitism*, because Plato says that there are smallest lines and the smallest lines are all equal. Therefore, they form triangles and, from the triangles, bodies which are regular; and they would not be regular if their sides were not equal. And they are equal because there cannot be smaller ones. That's the reason why they are equal. This is precisely the way in which we would deduce symmetries from *finitism*, and there is something in common between Plato's philosophy and our task in modern physics. How can that be? Well, it can be because it is a very simple idea. In order to understand it one does not need very much particular knowledge; one just needs an understanding of the general way of how things go together.

Human understanding, to Plato as well as to Kant, is an understanding by a finite mind. Plato's way of describing all this is to say that the world's soul, which is the basic reality in the world, and which has its appearance in the world's body, is the one reality of which the human soul is a part. The reason why we are able to understand nature, why we are able to understand our sensations, is because we are parts of that very soul which is on the basis of the existence of all that we see. Thus in symbolic language Plato explicitly formulates the principle of the unity of nature in the sense that the highest concept is unity itself. This, however, can no longer be expressed; you can just try to lead people to an understanding that this is so. Then you develop more and more of the differences, of plurality. But plurality, as d'Espagnat said, is never an ultimate truth; it is the way in which we are able to express ourselves in order to understand the details. That we are able to understand all these things is rooted in the fact that all these things are, to use the neo-Platonic phrase, emanations of *the one*, but the word emanation is not in Plato.

Let me just mention one difference, in which we completely differ from Plato. It is that to us time is the central concept and to Plato it is not. It is true that to Plato time is a very important concept, and I am not now going to discuss the differences in detail, but Plato does not have a theory of the difference of what I called fact and possibility, or what is called in the more usual language the difference between past and future. He is, of course, aware of this difference, but he does not explicitly deal with it. This means that we would have to have a theory of how all our theories have developed in time; they are not only theories *about* time, they are theories *in* time. To our philosophy, therefore, there belongs a philosophy of history and a philosophy of the history of science as necessary elements. This again is not in Plato, and cannot be in Plato. It's not even in Kant. It is just beginning to show its possibility in our era. This is one of the tasks of philosophy. If we examine this background of philosophy, of Kantian philosophy and of modern physics, then we find that there is a possibility of bringing all these things together.

Part V

Memorial Lectures

41

Recollections of Lord Rutherford*

P. L. Kapitza

I am greatly honoured to speak to you on my reminiscences of Lord Rutherford. But this is a very difficult task. At first sight I thought that to speak about the scientific achievements of so great a scientist as Rutherford would be easy. The greater the achievements of a scientist the more exactly and briefly can they be described. Rutherford created the modern study of radioactivity; he was the first to understand that it is the spontaneous disintegration of the atoms of radioactive elements. He was the first to produce the artifical disintegration of the nucleus and finally he was the first to discover that the atom has a planetary system. Each of these achievements is sufficient to make a man a great physicist. But nowadays these achievements and their fundamental values are well known not only to research students but even to schoolboys. Equally we all know the very simple and beautiful classical experiments by means of which Rutherford made his great discoveries.

You arc all well aware that from research into radioactivity there grew up an independent science which is now called nuclear physics, and of all the papers published on physical research one fifth relate to the investigation of nuclear phenomena. Both nuclear energy and the use of artificial radioactivity in science and technology are developing quickly and simultaneously. All these fields absorb the main bulk of the monetary resources spent on science and which now reaches the sum of thousands of millions of pounds, dollars and roubles. And all this for the last [over] 30 years grew out of one modest domain of physics which in the old days was called radioactivity and the father of which is justly called Rutherford. To speak of the development of nuclear technology and physics which came from the work of Rutherford and his school is very interesting and very instructive. But I am sure that such Fellows of the Royal Society as our President Professor Blackett, Sir James Chadwick, Sir John Cockcroft, Sir Charles Ellis and Sir Mark Oliphant who in the old days were the most active members of Rutherford's school and who themselves in this domain have made fundamental discoveries and researches are certainly more qualified than I to speak on these matters.

The only way in which I can satisfy your interest is to speak of Rutherford the

---

* *Editor's Note:* We had invited Professor Kapitza to give a talk on the recollections of his Cambridge years at our symposium, but unfortunately he could not attend. We are reprinting a lecture delivered by him at the Royal Society, London, 17 May 1966, and published in the *Proceedings of the Royal Society*, A **294**, 123–137 (1966). [Reproduced by permission of the Council of the Royal Society.]

man, of how I remember him during my 13 years' work in the Cavendish Laboratory, of how he worked, how he trained us young scientists and also of his relations with the scientific world. My task is therefore to draw you a portrait of a great scientist and of a great man. Frankly, this is the job of a writer and not of a scientist. If I have now decided to do so, this is mainly for the following reasons. When I look back and see myself as a young man coming to England in 1921 and starting work in the Cavendish Laboratory and, after 13 years, growing into a scientist, I feel that these years of my work were the happiest, and for all that I have been able to achieve I feel immensely grateful for the attention and kindness which Rutherford showed me, not only as a teacher but as a very kind and sympathetic man for whom I have a sincere affection and with whom I eventually became great friends.

However imperfect my recollections of Rutherford may be, this is the only way in which I can express my deep gratitude to this great and remarkable man.

As is well known, Rutherford was not only a great scientist but also a great teacher. I can recall no other scientist contemporary to Rutherford in whose laboratory so many outstanding physicists were trained. The history of science tells us that an outstanding scientist is not necessarily a great man, but a great teacher must be a great man. Therefore my task is even more difficult: I must give you a portrait not only of a scientist but of a man. I will attempt to make my portrait of Rutherford as alive as possible and for this purpose I shall illustrate my talk with episodes which I most vividly remember. From my many recollections I shall select the ones which characterize different sides of Rutherford's nature. I hope this will help you to reconstruct a lifelike picture of Rutherford in your imagination from all these fragments.

I would like to begin my recollections with a small episode which happened in 1930 in the Cavendish Laboratory. At that time a small conference was being held in Cambridge to commemorate the centenary of the birth of Maxwell, the first director of the Cavendish Laboratory. He was succeeded by Rayleigh, J. J. Thomson and Rutherford, four great physicists of the last and present century. After the official part of the meeting in which some of Maxwell's pupils talked of their reminiscences, Rutherford asked me how I liked the speeches. I answered that they were very interesting, but I was surprised that all the speakers spoke only of the positive side of Maxwell's work and personality and made a 'sugary extract' (sic) of him and I said that I would like to see Maxwell presented as a living figure with all his human traits and faults which of course every man possesses however great his genius. Rutherford as usual laughed and said that he charged me after his death to tell future generations what he was really like. Rutherford was joking and I was laughing too. And now when I try to fulfil his behest and I imagine Rutherford as I have to present him before you, I see that time has absorbed all his minor human imperfections and I can only see a great man with an astounding brain and great human qualities. How well I now understand Maxwell's pupils who spoke about him in Cambridge.

There are numerous books and articles on Rutherford as a scientist. It is widely recognized that the simplicity and clarity of his thinking, his great intuition and great temperament were very characteristic of his creative ability. Studying the works of

Rutherford in the Cavendish laboratory for high magnetic fields.

A meeting of the governing board of the Magnetic Laboratory: Rutherford (right), Cockcroft (left), Kapitza (centre).

Two views of Rutherford and Cockcroft (standing) in front of the high power impulse generator for producing high magnetic fields, taken about 1930.

Rutherford and observing how he worked I think the basic characteristics of his thinking were great independence and hence great daring.

The basic method by which science is developing consists of experimental investigations into natural phenomena and the continuous verification of the consistency of the results of our investigations with our theoretical conceptions. The progress of our knowledge of nature appears in cases when we find contradictions between theory and observation, and these contradictions, as they compel us to develop our theories, enable us to widen our knowledge of nature. The more acute these contradictions are, the more they lead us to further fundamental changes in understanding the laws of nature, on the basis of which we may use nature for our cultural development. In science, as in history, definite stages of development demand their particular kind of genius. A definite period of development requires men with corresponding mental abilities. In the history of the development of physics, as in any other experimental science, the most interesting periods arc those in which we are brought to revise our fundamental scientific conceptions. Then not only deep thinking and intuition are required from the scientist but also a daring imagination. As an illustration I shall remind you of two well known cases in the history of physics. They made a great impression on me personally. The first case concerns Franklin's creation of the study of electricity. On the basis of this study Franklin stated that electricity has a material origin and can impregnate metal and freely move in it. In his day this concept was in fundamental contradiction to the concept of the continuous nature of matter. But Franklin's view was eventually accepted as it gave a simple and complete explanation of all the electrostatic phenomena observed in his day. It is only recently, 150 years later, after J. J. Thomson discovered the electron, that Franklin's concepts were completely justified. But the most striking thing in this story is how it could have happened that Franklin, who had never before done any scientific work, could, in the course of a few years in a small remote American town when he was already a middle-aged man, find the right way by which this most important branch of science should be developed. And this happened in the middle of the eighteenth century when science was developing on the level of Newton, Huygens and Euler. How could Franklin achieve such results which were beyond the reach of professional scientists?

The other similar case in which the fundamental concepts of electricity had to be revised in the light of experiments is also well known. This is Faraday's concept of the electrical field. It is difficult to find a more revolutionary and original idea than Faraday's. He advanced the concept that electrodynamical processes must be explained by the phenomena happening in the space surrounding the conductor. I mention this case mainly because Faraday was a scientist who had no traditional scientific education, even though at that time its level was high for an average English scientist. I mention these two well known cases only to show that at a particular stage of the development of science, when new fundamental concepts have to be found, wide erudition and conventional training are not the most important characteristics of a scientist required to solve this kind of problem. It appears that in this case imagination, very concrete thinking, and most of all, daring are needed. Strict logical thinking which is so necessary in mathematics hinders the imagination of a scientist when new fundamental concepts must be found. The ability of a scientist to solve such scientific problems without showing a logical trend of thought is usually called 'intuition'. Possibly there is a way of thinking which takes place in our subconscious but the laws by which it is governed are at present unknown. If I am not mistaken, even Freud, a pioneer in the study of subconscious processes, was not aware of it. But if intuition exists as a powerful creative thinking process then doubtless Franklin and Faraday mastered it thoroughly. I am sure that Rutherford mastered it too and he has rightly been called the Faraday of our time.

When at the beginning of our century Rutherford started studying radioactivity it had already been proved experimentally that these phenomena contradicted the most fundamental law of nature, the law of conservation of energy. The explanation of radioactivity which Rutherford gave, namely the disintegration of matter, at once provided not only the key to the understanding of these phenomena but also led all investigation in the right direction. The same thing happened when Rutherford created the planetary model of the atom. At first sight this model completely contradicted laws of classical electrodynamics since in its circular motion an electron was perpetually bound to lose by radiation its kinetic energy. But the experiments of scattering the α-particles, performed by Rutherford's pupil Marsden in 1910, definitely showed the existence of a heavy nucleus in the centre of the atom and Rutherford imagined the collision of particles so clearly that even these contradictions with the fundamental laws of electrodynamics could not prevent him from establishing the planetary structure of the atom. We know that only three years later Bohr, on the basis of the developing quantum theory of light, evolved his brilliant theory of the structure of the atom which not only justified Rutherford's planetary model but also quantitatively explained the spectra of atomic radiation.

The peculiar character of Rutherford's thinking could easily be followed when talking to him on scientific topics. He liked being told about new experiments but you could easily and immediately see by his expression whether he was listening with interest or whether he was bored. You had to talk only about fundamental facts and ideas without going into the technical details in which Rutherford took no interest. I remember, when I had to bring him for approval my drawings of the impulse generator for strong magnetic fields, for politeness sake he would put them on the table before him, without noticing that they were lying upside down, and he would say to me: 'These blueprints don't interest me. Please state simply the principle on which this machine works.' He grasped the basic idea of an experiment extremely quickly, in half a word. This struck me very much, especially during my first years in Cambridge, when my knowledge of English was poor and I spoke it so badly that I could only vaguely explain my ideas, yet in spite of this Rutherford caught on very quickly and always expressed very interesting opinions.

Rutherford also liked talking about his own experiments. When he was explaining something he usually made drawings. For this purpose he kept small bits of pencil in his waistcoat pocket. He held them in a peculiar way – it always seemed to me a very inconvenient one – with the tips of his fingers and thumb. He drew with a slightly shaky hand, his drawings were always simple and consisted of a few thickly drawn lines, made by pressing hard on the pencil. More often than not the point of the pencil broke and then he would take another bit from his pocket.

A number of physicists, especially theoreticians, like to discuss science and apparently the process of argument is a way of thinking. I never heard Rutherford argue about science. Usually he gave his views on the subject very briefly, with the maximum of clarity and very directly. If anybody contradicted him he listened to the argument with interest but would not answer it and then the discussion ended.

I greatly enjoyed Rutherford's lectures. I followed the course of general physics which he gave to the undergraduates as Cavendish Professor. I did not learn much physics from this course since by that time I already possessed a fair knowledge of the subject, but from Rutherford's approach to it I learnt a great deal. Rutherford delivered his lectures with great enthusiasm. He used hardly any mathematical formulae, he used diagrams widely and accompanied his lectures with very precise but restrained gestures from which it could be seen how vividly and picturesquely Rutherford thought. I found it interesting that during the lecture he changed the topic as his thoughts, probably following some analogy, turned to a different phenomenon. This was usually connected with some new experiments made in the field of radioactivity which fascinated him and he then proceeded to speak with enthusiasm on the new subject. In this case he usually put his assistant in a difficult position by asking him to give a demonstration which was not part of the original planned version.

About the same time I also attended the lectures of J. J. Thomson in his special course on the conductivity of electricity through gas. It was interesting to notice how differently these two great scientists approached scientific problems. If Rutherford's way of thinking was inductive, then the way of thinking of J. J. Thomson was deductive.

I think it useful when training young scientists to ask them to follow a course of lectures, even an elementary one, but delivered by an eminent scientist. Listening to these lectures they will learn something that they will never find in any textbooks. In this connection I remember a conversation which I had with Sir Horace Lamb. He was telling me how he had attended Maxwell's lectures. Maxwell, he said, was not a brilliant lecturer; he usually came to lectures without any notes. When he was doing mathematics on the blackboard he often made mistakes and sometimes got muddled. From the way in which Maxwell tried to disentangle and correct his mistakes Lamb learned more than from any textbooks he ever read. Lamb told me that for him the most precious parts of Maxwell's lectures were those in which he made mistakes. No doubt the mistakes of a genius are sometimes as instructive as his achievements.

When I came to Cambridge Rutherford did no more experimental work by himself; he worked chiefly either with Chadwick or with Ellis. But in both cases he took an active part in experiments. The setting up of the apparatus was done mainly by his laboratory assistant, Crow, whom he treated rather severely. But I sometimes saw how Rutherford himself, despite his slightly shaking hands, dealt quite skilfully with the finewalled glass tubes filled with radium emanation. Although Rutherford's experiments are well known, I cannot refrain from saying a few words about them. The most attractive thing about these experiments was the clarity of setting the problem. The simplicity and directness of approach to the solution of the problem were most remarkable. From my long experience as an experimenter I have learned that the best way of correctly evaluating the capacity of a beginner as well as of a mature scientist is by his natural inclination and ability to find a simple way of solving problems. There is an excellent saying by an unknown French author which applies perfectly to Rutherford: 'La simplicité c'est la plus grande sagesse.' I should also like to quote the profound saying of a Ukrainian philosopher, Gregory Skovoroda. He was by origin a peasant and lived in the second half of the eighteenth century. His writings are most interesting but probably quite unknown in England. He said 'We must be grateful to God that He created the world in such a way that everything simple is true and everything complicated is untrue.' Rutherford's finest and simplest experiments concerned the phenomena of scattering by nuclear collisions. The methods of observation of scintillations by counters were worked out by Rutherford in collaboration with Geiger in 1908. Since then more than half a century has passed and this method and the Wilson chamber invented about the same time remain the fundamental methods for studying nuclear phenomena, and only the optical and resonance methods for determining nuclear moments have since been added. And up to now all nuclear physics possesses no experimental possibilities other than those used by Rutherford and his collaborators. The present development of nuclear physics is proceeding not by the invention of new experimental possibilities of investigating nuclear phenomena but thanks to the possibility of investigating nuclear collisions of a *larger* number of elements; and these collisions are studied in the domain of larger energies which are reached mainly by the use of powerful modern accelerators. But even now the way which leads us to the knowledge of the nucleus is still the method discovered by Rutherford, and he was the first to appreciate its fundamental value. I am referring here to the investigation of the collision of nuclei. Rutherford always liked to say 'Smash the atom!'

Even now, in the process of investigating nuclear collisions, there is one great weakness: the necessity of using statistical methods in the interpretation of experimental results. Great care is required to deduce correct general laws from limited statistical data. Someone once said about statistics: 'There are three kinds of lies: a simple lie, an impudent lie and statistics.' In fact this was said about the application of statistics to social problems. But to some extent it is true of statistics in physics. I do not think that in any other branch of physics so many mistakes and faulty discoveries were made as in the course of the interpretation of statistical data obtained from experiments on nuclear collisions. Nearly every year new particles of resonance levels are still discovered, some of which may not exist. Rutherford was well aware what danger lies concealed in the interpretation of experimental data of statistical origin, especially when the scientist anticipates definite results. Therefore Rutherford was very careful to exclude the personal element and took the following precautions during the course of these experiments: the counting of scintillations was usually done by undergraduates who did not know the purpose of the experiment; the curves were drawn by persons who did not know what results were expected. As far as I remember, Rutherford and his pupils never made a single such mistake, while in the same line of investigation a number of mistakes were made in other laboratories. I remember that in those days the most critical approach in the interpretation of statistical data was that of Chadwick on whose judgement Rutherford usually relied completely.

I did not work with Rutherford because my investigations were not connected with nuclear physics and therefore I did not see him working in his laboratory. But I know that up to the very end of his life the main bulk of his time was taken up by his personal scientific research. I expect he gave the same amount of attention and strength to directing the work of young research pupils working in the Cavendish Laboratory. The detailed guidance of scientific work he left to one of the senior scientific workers, usually Chadwick. But he himself always took an interest in the choice of problem for experiment and of the experimental approach. Until the research student began obtaining results Rutherford showed no marked interest in his work. He never bothered about detailed guidance.

He often came to the laboratory but only for a short time; just to make remarks like: 'Why don't you get a move on – when are you going to get some results?' When I started working in the Cavendish Laboratory such remarks made a great impression on me, especially as they were made in a thundering voice and with a severe expression. But eventually I found out that such utterances were automatic, maybe customary for a New Zealand farmer who when going through the fields found it useful to stimulate the workers with a few 'kind' words. That it was actually so was proved by an episode which happened a few years later in the Cavendish. One day it was necessary to break a hole through a stone wall to put through a cable needed for some experiment. The work was urgent and it happened that at that time there was a building strike and it was exceptionally difficult to find a bricklayer who would consent to work. Finally a man was found and he started work but after a while he came and said that he refused to go on. When asked why, he replied that twice a gentleman had passed by him and both times had asked him when he would finally start work and get the job done. These remarks offended the workman. When asked who this gentleman was, his description showed without a doubt that it was Rutherford. When we reproachfully pointed out to Rutherford that during a strike one should be a little careful we were surprised that Rutherford denied having said anything to the bricklayer. Obviously when he likewise grumbled at us in the laboratory for our slow work, he did it unconsciously; it was a kind of conditional reflex.

The greatest quality of Rutherford as a teacher was his ability to direct research work in the right direction, then to encourage the beginner and to give just appraisal of his achievements. What he valued most in a pupil was independent thought and originality in his work. Rutherford did his utmost to develop in his pupil an individuality. I remember how in the first years of my work in the Cavendish I once said to Rutherford: 'You know that the work of X is pretty hopeless; don't you think he's wasting his time and apparatus?' Rutherford replied that he too knew that the man was working on a hopeless problem, 'but', said Rutherford, 'it is a problem of his own and even if the work cannot be accomplished it will lead him to another original research problem which will be successful'. The future showed that Rutherford was right. As I said, Rutherford would do his utmost to develop in his pupils independence and originality of thought and as soon as a pupil showed these qualities Rutherford would pay close attention to his work. As an example of Rutherford's ability to direct the research of his pupils I remember the story, as Rutherford told it himself, of the discovery made by Moseley. In 1912 Moseley worked with Rutherford in Manchester. He was very young and Rutherford spoke of him as one of his best pupils. When Moseley came to Manchester he at once accomplished some minor research work and then eventually he came to Rutherford and told him of three different topics he would like to investigate. One of these researches was the classical work which had made Moseley's name so well known – the dependence of the wavelengths of Roentgen rays on the position of atoms in the periodical system. Rutherford at once advised Moseley to choose this work for his investigation. The future showed that Rutherford made the right choice, but he always pointed out that the idea of the experiment belonged to Moseley.

Rutherford was very particular to give credit for the exact authorship of any idea. He always did this in his lectures as well as in his published works. If anybody in the laboratory forgot to mention the author of the idea Rutherford always corrected him. He was also very particular not to give a beginner technically difficult research work. He reckoned that, even if a man was able, he needed some success to begin with. Otherwise he might be disappointed in his abilities, which could be disastrous for his future. Any success of a young research worker must be duly appreciated and must be duly acknowledged.

Once, in one of our outspoken talks, he told me that the most important thing a teacher must learn is not to be jealous of the successes of his pupils – which is not so easily done as the teacher gets older! This profound truth made a great impression on me. No doubt the greatest quality of a good teacher should be generosity. Rutherford was undoubtedly very generous and I think this is one of the main secrets which explains why so many first-class scientists came from his laboratory. There was always an atmosphere of freedom and efficiency there.

Rutherford well understood the importance that his pupils had for him. It was not merely that young research students increased the scientific productivity of the laboratory, but, as he said, 'My pupils keep me young'. This is very true, since pupils do not permit a teacher to lag behind new achievements in science. How often do we notice that when a scientist is ageing he starts opposing new ideas and underestimates the significance of new trends in science. Rutherford, with great ease and generosity, always accepted new ideas in physics like wave quantum mechanics, while a number of distinguished scientists of his generation were sceptical of the same ideas. Such conservatism is characteristic of scientists who work by themselves without having pupils to be directed and encouraged.

Rutherford was very sociable and loved talking to the scientists who came to visit him and the Cavendish Laboratory. Usually there were many such visitors. His attitude to other people's work was kind and considerate. In conversation Rutherford was very lively; he was fond of jokes and often made them himself. He laughed easily, his laughter was sincere, loud and infectious. His face was very expressive: you could see at once what mood he was in, good or bad, or whenever he was worried by anything. You always knew he was in a good temper when he good-naturedly teased the person he was talking to. The more he teased him, the more he liked him. This was particularly noticeable when he talked to Bohr or to Langevin to both of whom he was especially attached. His kindest jokes often concealed a deeper sense. I remember one occasion when he brought Professor Robert Millikan to my room in the laboratory, Rutherford said to me, 'Let me introduce you to Millikan; no doubt you know who he is. Show him your installation to produce strong magnetic fields and tell him about your experiments. But I doubt whether he will let you speak as he himself will tell you about his own experiments!' There followed loud laughter in which Millikan joined with rather less enthusiasm. Rutherford then left us, and I soon found out that his prophecy was correct.

I shall not describe the way in which Rutherford read his papers. I always liked them very much as regards both their content and their exposition. He attached great importance to the way in which his papers were presented and evidently prepared them very carefully. He taught me how to read papers to the Royal Society, and one of his instructions I still remember very clearly: 'Don't show too many slides. When it is dark in the lecture room some of the audience take the opportunity to leave!'

Rutherford's interests were not limited narrowly to physics; they were much wider. He was well read, he liked books on geography and history and liked to discuss what he had read. He absorbed all knowledge enthusiastically and always extracted the essentials.

Later on, when I became a Fellow of Trinity and used to accompany him home after dinner on Sundays, we often discussed politics. On the first day I started work in the Cavendish I was surprised to hear him saying to me that in no circumstances would he tolerate my making Communist propaganda in his laboratory. At this time this remark came quite unexpectedly. It not only surprised me, but also shocked me and to a certain extent even offended me. Undoubtedly it was a consequence of the current atmosphere of acute political struggle and was connected with the propaganda which existed in those days, only four years after the Russian revolution. Before coming to England, I was so absorbed by my research work in Russia that I was completely unaware of what was happening in Western Europe and could not appreciate the scale of the bitter political controversy which then existed. Later on when my first experimental research was published I presented Rutherford with a reprint and I made an inscription on it that this work was proof that I had come to his laboratory to do scientific work and not to make Communist propaganda. He got extremely angry with this inscription, swore and gave me the reprint back. I had foreseen this and I had another reprint in reserve with an extremely appropriate inscription with which I immediately presented him. Obviously Rutherford appreciated my foresight and the incident closed. Rutherford had a characteristically hot temper but cooled down just as quickly.

Eventually we had many conversations on political questions; we were especially concerned about the growth of fascism in Europe. Rutherford was an optimist and thought that all would soon be over. We now know that this was not the case. Rutherford, like most scientists who work in the exact sciences, had progressive political views. I involved Rutherford in some political activity on two occasions.

The first of these was connected with Langevin. In his younger days Rutherford had worked with Langevin in the same room at the Cavendish. A deep friendship developed between them. Indeed it was practically impossible not to be friendly with a man of such brilliant intelligence and exceptional moral qualities. In Paris my friends, pupils of Langevin, were greatly shocked that Langevin, undoubtedly the best French physicist, had not been elected to the French Academy as a result of his left wing political views. Langevin had taken part in a number of progressive organizations, had been the founder of the League of the Rights of Man (*ligue des droits d'hommes*) and had fought anti-semitism in the Dreyfus case. I told Rutherford of the difficulties Langevin had encountered in France and asked him whether a man who held such leftist views as Langevin could be a Foreign Member of the Royal Society. Rutherford said something I could not quite follow, then started to tell me what a really good man Langevin was, and then recalled that during the war Langevin had been very active in inventing supersonic beams propagated in water by which he had established communication between England and France across the Channel. At this point the conversion ended. I learnt later that at the next election in 1928 Langevin was elected a Foreign Member of the Royal Society and this was much earlier than his election to the French Academy.

The second example occurred much later, when Hitler started to come into power. We were very anxious about the fate of such distinguished physicists as Stern, Frank, Born and a number of others in the conditions of active and increasing anti-semitism in Germany. About this time Szillard came to England and we were faced with the question of how to get these scientists out of Germany without raising suspicion. I spoke to Rutherford and he was very willing to help, writing personal letters to these scientists, and inviting them to come to lecture in Cambridge.

Rutherford took an interest in a great variety of people, but he particularly liked people with strong personalities. When Rutherford was elected President of the Royal Society and often had to attend dinner parties with distinguished politicians, businessmen and statesmen, he was fond of telling stories afterwards about the conversations he had with them and always gave descriptions of them. I specially remember that Churchill made a great impression on Rutherford. His description of Churchill was, like all his descriptions, short and clear, and in due course I found out that it was quite correct. I well remember that Churchill in those days already regarded Hitler as a real danger to peace and called him 'a man riding a tiger'. Possibly this conversation somewhat altered Rutherford's optimistic view of the future. Rutherford's interest in understanding human psychology and his kindness to others was undoubtedly felt by them.

This explains why Rutherford's excessively direct way of speaking, which was sometimes not very tactful, was completely compensated for by his kindness and cordiality.

Of course Rutherford's correct evaluation of people and his understanding of them was due to the fact that he was a subtle psychologist. People interested him and he had the faculty of understanding them. His assessments of people were always very outspoken and direct. As in his scientific work, his description of a man was always brief and very accurate. I was always convinced that his descriptions were correct. Possibly his approach to people was also a subconscious process and could be called intuitive.

I should like to illustrate his interest in psychology with the following two episodes. In Cambridge there was a small but progressive theatre which produced Chekov's play 'Uncle Vanya'. Rutherford went to this play and was greatly taken with it. As in all Chekov's works, it deals with a psychological problem complicated by the fact that all the people in the play are highly intellectual and therefore their acceptance of life is very complex. In the play a certain retired professor comes to live on the estate of his wife. Uncle Vanya, who manages the estate, has devoted his whole life to supporting the professor. Soon Uncle Vanya finds that the professor is a fake celebrity, scholastic and pedantic in his work. Against a background of complex psychological situations Uncle Vanya fires a pistol at the professor but misses him. I remember how vividly, clearly and simply Rutherford told me this plot and his sympathy was completely on Uncle Vanya's side. The fact that Rutherford was so attracted by the play shows that he undoubtedly enjoyed disentangling complicated psychological cases of this kind.

A great impression was made on me by the following case which demonstrates Rutherford's skill in handling complicated psychological problems. I think enough time has now passed and I can tell you about this case which involved the then well known physicist, Paul Ehrenfest.

Ehrenfest was born in Austria. On one of his mountaineering excursions he met a Russian woman scientist and followed her to Russia where he married her. In Russia he published a number of outstanding theoretical works on thermodynamics. Eventually he was invited to Leiden University to take the chair of theoretical physics vacated by the great Lorentz, creator of the electronic theory of metals and one of the founders of the theory of relativity. In Leiden Ehrenfest and his house became one of the world centres of theoretical physics. Ehrenfest's main quality was his precise critical mind. He was not only a very good teacher of young scientists, who were very fond of him, but his criticisms were regarded as profound and of such high quality that leading theoretical scientists like Einstein and Bohr often came to Ehrenfest to discuss their work. Ehrenfest always noticed even the smallest contradiction or mistake. His critical remarks were made very readily, with great spirit and even sharply, but always very goodnaturedly. The quality of his criticism was greatly appreciated. Despite our difference in age we became friends and I often visited his very hospitable and very charming family and more than once was present at his scientific discussions.

Ehrenfest's exceptionally strong critical mind evidently acted as a restraint on his creative imagination and he did not succeed in producing scientific work which he himself would have considered of sufficiently high standard. In those days I did not know that in his acute nervous condition Ehrenfest suffered greatly when he could not in his work attain the level of the friends he criticized. I learnt about his feelings in the following manner. In the beginning of 1933 I received a long letter from him in which he described in detail his state of mental depression and spoke of the futility of his achievements. He had come to the conclusion that it was not worth living any more. The only way to save himself, he thought, was to leave Leiden and settle some-where away from his friends. He asked me to help him to find a chair at some small university in Canada and to ask Rutherford, who doubtless had connections in Canada, to assist him. I was, of course, very upset by this letter. We all liked Ehrenfest and all knew that his influence as a teacher and critic in the development of modern physics was colossal. I translated the letter from German into English and came to Rutherford, who had little personal acquaintance with Ehrenfest. I handed Rutherford the letter and told him that we were very worried about Ehrenfest's future as, without any doubt, the letter showed that he was mentally unbalanced; perhaps, I said, this state was only temporary and everything possible should be done to help him out of this state of depression. Rutherford said I must not worry and he would handle the case himself. I do not know what Rutherford wrote to Eherenfest but shortly afterwards I received a letter telling me that he was once again in a happy frame of mind. He said that Rutherford had explained what a great role he was playing in physics and he added that of course there was now no need for him to go to Canada. This story shows how skil-fully Rutherford dealt with a very complicated psychological case, probably better than a professional psychiatrist.

A few months later, while I was on a visit to Russia, the state of depression returned to Ehrenfest and on 25 September 1933 he committed suicide.

I should now like to recall quite a different and rather amusing case characteristic of Rutherford's attitude to the young. Once Rutherford called me into his study and I found him reading a letter and roaring with laughter. It appeared that the letter was from some Ukrainian schoolboys. They had written to say that they had organized a physics club and were proposing to continue Rutherford's fundamental work on the study of the nucleus of the atom and ask him to be an honorary member and to send them reprints of his scientific work. In the part of the letter in which they described Rutherford's achievements in nuclear physics, instead of using the correct term in physics they used a corresponding term which in slang has a different meaning. In this way the description of the structure of the atom acquired a property of the living organism. Its character is such that one does not speak about it in polite society, and it made Rutherford laugh heartily. I explained to Rutherford that the schoolboys were apparently not very well versed in English and the writing of the letter was mostly done with the use of a dictionary and the mistake was bona fide. Rutherford said that he appreciated this. He sent the boys a reply, thanking them for the honour of being elected a member of the club and promising to send them reprints.

Finally, I should like to discuss a question I have come across several times in descriptions of Rutherford's activities. The question is: did Rutherford foresee the great practical consequences which would emerge from his scientific discoveries and investigations into radioactivity?

The immense reserves of energy which are hidden in matter was understood by physicists a long time ago. The development of his view took place side by side with the development of the theory of relativity. The question which was not clear at that time was: would it eventually be possible to find technical means of making practical use of these reserves? We know now that the actual possibility of obtaining energy from nuclear collisions was becoming more and more real as nuclear phenomena were better understood. But up to the last moment it was not certain whether it would be technically possible to produce nuclear reactions with a great yield of energy. I remember only rare occasions on which I discussed this question with Rutherford and in all these conversations he expressed no interest in it. From the beginning of my acquaintance with Rutherford I noticed that he never took any interest in technical problems and I even had the impression that he was prejudiced about applied problems. Possibly this was because such problems were connected with business interests.

I am by training a chartered engineer and naturally I always took an interest in solving technical problems. During my stay in Cambridge I was approached several times to help in solving technical problems in industry. In these cases I used to take advice from Rutherford and he always said to me; 'You cannot serve God and Mammon at the same time.' Of course he was right. Once I remember Rutherford telling me about Pupin who as an able young physicist had come to Cambridge and done successful scientific work in the Cavendish Laboratory. Pupin was somewhat senior to Rutherford so they met only occasionally. Eventually Pupin turned to commercial activity in the U.S.A. and made a lot of money. Rutherford spoke disapprovingly of Pupin's activities. So I think that Rutherford's opinions on the practical applications of nuclear physics had no real value as they lay outside the scope of his interests and tastes.

In connection with Rutherford's views on industry I remember a conversation I had with him during a high table dinner at Trinity College. I do not remember how the conversation started, maybe it was under the influence of Lombroso's book, *Genius and Madness*. I was telling my neighbour that every great scientist must be to some extent a madman. Rutherford overheard this conversation and asked me, 'In your opinion, Kapitza, am I mad too?'

'Yes, Professor', I replied.

'How will you prove it?' he asked.

'Very simply', I replied. 'Maybe you remember a few days ago you mentioned to me that you had had a letter from the U.S.A., from a big American company. (I do not remember now which one it was, possibly General Electric Co.) In this letter they offered to build you a colossal laboratory in America and to pay you a fabulous salary. You only laughed at the offer and would not consider it seriously. I think you will agree with me that from the point of view of an ordinary man you acted like a madman!' Rutherford laughed and said that in all probability I was right.

The last time I saw Rutherford was in the autumn of 1934 when I went as usual to the Soviet Union to see my mother and my friends and unexpectedly was deprived of the possibility of returning to Cambridge. I did not hear his voice again, nor hear him laugh. For the next three years I had no laboratory to work in and was unable to continue my scientific work and the only scientist with whom I freely corresponded outside Russia was Rutherford. At least once every two months he wrote me long letters which I greatly valued. In these letters he gave me an account of life in Cambridge, spoke about the scientific achievements of himself and his pupils, wrote about himself, made jokes, gave good advice and invariably cheered me up in my difficult position. He understood that the important thing for me was to start my scientific work which had been interrupted for several years. It is no secret that it was only due to his intervention and help that I was able to obtain the scientific installation and apparatus of the Mond Laboratory and in three years time I was able to renew my work in the domain of low-temperature physics.

I am sure that in the course of time all Rutherford's letters will be published but even so I should like here and now to quote three short extracts which require no comment.

On 21 November 1935 he wrote:

'...I am inclined to give you a little advice, even though it may not be necessary. I think it will be important for you to get down to work on the installation of the laboratory as soon as possible, and try to train your assistants to be useful. I think you will find many of your troubles will fall from you when you are hard at work again, and I am confident that your relations with the authorities will improve at once when they see that you are working wholeheartedly to get your show going. I would not worry too much about the attitude or opinions of individuals, provided they do not interfere with your work. I daresay you will think I do not understand the situation, but I am sure that chances of your happiness in the future depend on your keeping your nose down to the grindstone in the laboratory. Too much introspection is bad for anybody!...'

On 15 May 1936 he wrote:

'...This term I have been busier than I have ever been, but as you know my temper has improved during recent years, and I am not aware that anyone has suffered from it for the last few weeks!...

'...Get down to some research even though it may not be of an epoch-making kind as soon as you can and you will feel happier. The harder the work the less time you will have for other troubles. As you know, "a reasonable number of fleas is good for a dog" – but I expect you feel you have more than the average number!'

You see what short and clear and fatherly advice he gave me. The last letter is dated 9 October 1937. He wrote in great detail about his proposed journey to India. In the last part of the letter he said: '...I am glad to say that I am feeling physically pretty fit, but I wish that life was not quite so strenuous in term time...' Ten days before his death he did not feel that it was so near.

For me the death of Rutherford meant not only the loss of a great teacher and friend; for me, as for a number of scientists, it was also the end of a whole epoch in science.

Obviously we should attribute to those years the beginning of the new period in the history of human culture which is now called the scientific-technical revolution. One of the greatest events in this revolution has been the use of atomic energy. We all know that the consequences of this revolution may be very terrible – it may destroy mankind. In 1921 Rutherford warned me not to make any Communist propaganda in his laboratory, but it now appears that just at that time he himself together with his pupils were laying the foundations for a scientific-technical revolution.

We all hope that in the end people will have sufficient wisdom to direct this scientific revolution to the benefit of humanity.

But nevertheless the year that Rutherford died there disappeared forever the happy days of free scientific work which gave us such delight in our youth. Science has lost her freedom. Science has become a productive force. She has become rich but she has become enslaved and part of her is veiled in secrecy.

I do not know whether Rutherford would continue nowadays to joke and laugh as he used to do.

42

W. Pauli's Scientific Work

Charles P. Enz

'Nur die Fülle führt zur Klarheit
Und im Abgrund wohnt die Wahrheit.'
SCHILLER

CONTENTS

1. Portrait of a Genius
2. The Old Quantum Theory. Spin and Exclusion Principle
3. The New Quantum Theory and Its Interpretation
4. Quantum Field Theory. Spin and Statistics
5. Physics of Condensed Matter. Statistical Physics
6. Neutrino, CPT-Theorem and Parity Violation
7. Aspects of Pauli's Spiritual Life

1. PORTRAIT OF A GENIUS[1]*

It was at the turn of the century, indeed in 1900, that Max Planck's idea of the energy quantum was born in Berlin. Also in 1900, on 25 April, Wolfgang Pauli was born in Vienna as the son of Wolfgang Joseph Pauli, a medical doctor, and his wife Bertha Schütz.

The chances for young Pauli to become a physicist were remarkable: Ernst Mach was his godfather whose gift for Wolfgang's christening was a silver cup with the inscription '31 Mai 1900'. Much later Pauli commented on this event in a letter dated 31 March 1953 as follows:[2]

... Among my books there is a rather dusty container, in which there is a silver cup in fin de siècle style, and in this cup there lies a card. ... This cup, of course, is a christening cup and on the card is inscribed in old-fashioned flowery type:

'Dr. E. Mach, Professor an der Universität Wien'.

It so happened that my father was very friendly with his family, and at that time was intellectually entirely under his influence. He (Mach) had thus graciously agreed to assume the role of my godfather....

Pauli's father became an associate professor at the medical school of the University of Vienna in 1907. In 1913, he was the first to lecture on physiological-chemical biology, and in 1922 he became director of the new institute for medical colloid chemistry,

---

* This article has two series of references. Superior figures as [2] refer to notes and references listed on pp. 792–796. Figures in square brackets as [2] refer to Pauli's own works which are listed on pp. 796–799.

J. Mehra (ed.), The Physicist's Conception of Nature, 766-799. All Rights Reserved
Copyright © 1973 by D. Reidel Publishing Company, Dordrecht-Holland

a field in which he became well-known by his publication of a number of important works. This was the reason why young Pauli published under the name of Wolfgang Pauli *junior* until his nomination as full professor at ETH (Eidgenössische Technische Hochschule), Zurich, in 1928.

Although in later years Pauli sometimes used, half jokingly, harsh words like 'geistige Einöde' [spiritual desert] [3] for his home town, Vienna was, at the time of Pauli's high school years, a remarkable place. In fact, Pauli's class, which graduated from Döblinger Gymnasium's humanistic section in 1918, has gone down in local history as the 'class of the geniuses'. [4] This class of 27 boys counted among its numbers two later Nobel prize winners (the second, Richard Kuhn, got the chemistry prize for 1938), two famous actors (one was Hans Thimig from the famous Viennese family of actors), three University professors, two directors of medical schools, a music historian, a politician and several industrialists. The graduation group picture, [4] however, does not show the whole class because some of the boys has already joined the Army; World War I was still haunting Europe.

But this did not prevent the class of geniuses from having its own kind of fun with school. Very typically, at the occasion of Empress Zita's birthday Pauli asked the family's girl cook to write a letter of vigorous protest to the school director complaining that Pauli's entire class was planning to demonstratively read *Die Arbeiter Zeitung* during the anniversary mass for the Empress. [5]

During the same year 1918, however, Pauli immediately attained full respectability by the submission on 22 September, still from Vienna, of his first published paper [1], on the energy components of the gravitational field. Eight months later, in his second paper [2], submitted on 4 June 1919 from Sommerfeld's institute in Munich, Pauli already acknowledges a correspondence with Hermann Weyl whose theory of gravitation was the subject of this note. The critical self-assurance for which the young Pauli became so well-known and which earned him the name of God's whip ('die Geissel Gottes') from Ehrenfest [6] is still restrained in the second paper. Indeed, in a footnote Pauli points out an error of sign in Weyl's calculation with the words: '... I should like to express, with due respect, the opinion that a small oversight has occurred in Weyl's paper.'

Algebraic signs haunted Pauli no less than any average physicist. Much later he would stand in front of the big blackboard in the lecture room 6c of the physics institute of the ETH nodding his head and murmuring 'Vorzeichen, Vorzeichen' [the signs, the signs!], while a damped noise would signal that the class had taken the opportunity to relax.

Pauli's third paper [3], submitted on 3 November 1919, again discusses the consequences of Weyl's theory and ends with a critical remark of more fundamental significance. Formulated in the best of Pauli's characterisic style, mature, precise and cautious, he observes that while Weyl's theory continually operates with the field strength in the interior of the electron it is only the force on a test particle which for the physicist is well-defined. And since there is no smaller test particle than the electron itself the notion of the electric field strength in a mathematical point seems to be an empty fiction.

And the 19-year-old Pauli continues: 'One would, of course, like to require that essentially only observable quantities be introduced into physics.' In this credo that physics should be formulated entirely in terms of observable quantities, the influence of Pauli's godfather Ernst Mach is undeniable. This credo arose to its full significance later on in the elaboration and interpretation of the new quantum theory (see Section 3). It conflicted, on the other hand, with the difficulties of renormalization theory of fields (see Section 4). With regard to the latter problem the closing sentence of Pauli's third paper [3] raised a question which is still unanswered:

Sollten wir überhaupt mit den Kontinuumstheorien für das Feld im Innern des Elektrons auf einer falschen Fährte sein?

['Is it possible that we are on the wrong track with the continuum theories for the field in the interior of the electron?'] The problem of the notion of the field and the atomicity of the electric charge bothered Pauli during all his scientific life, as will be seen in Section 7.

These first three short papers of Pauli on relativity and gravitation stand isolated; it was only thirteen years later that he came back to this field of research. However, as is well-known, this activity of Pauli culminated in a momentous work, the famous chapter on relativity in *Encyklopädie der mathematischen Wissenschaften* [4] of which Einstein wrote:[7]

No one studying this mature, grandly conceived work would believe that the author is a man of twenty-one. One wonders what to admire most, the psychological understanding for the development of ideas, the sureness of mathematical deduction, the profound physical insight, the capacity for lucid, systematical presentation, the knowledge of the literature, the complete treatment of the subject matter [or] the sureness of critical appraisal.

Although it is difficult to improve upon Einstein's eulogy of Pauli's essay on relativity, Pauli's own critical appraisal of the problem of charged particles deserves special mention. In fact in the last Section (Section 67 of ref. [4]) Pauli comes back to the questions raised in his third paper [3] concerning the concept of the electric field strength and the atomicity of electric charge: No modification of his earlier judgement was necessary.

The encyclopaedia article has withstood the test of 35 years of research so brilliantly that an English translation, updated with 'Supplementary Notes by the Author' written in 1956 [5], appeared in 1958, the year of Pauli's death. These short supplementary notes deal with some of the later developments of the theory but give in no way a complete picture. The most important omission is the problem of relativistic quantum theory, in particularly field theory, to which Pauli has significantly contributed himself (see Section 4) and which had grown to proportions beyond the scope of such supplementary notes.

Of particular interest are the supplementary notes 7 and 23. In note 7 Pauli corrected his earlier omission of the Bianchi identities. In a conversation with J. Mehra in spring 1958 Pauli commented on this omission when Mehra observed to him that the original article was already so perfect that it was surprising that he should have thought of adding supplementary notes. He replied with a wide grin, 'You know, it was not all that perfect. I had not even mentioned the Bianchi identities.'[8]

In note 23, concerning unified field theories, Pauli comments on his earlier discussion of the problem of charged particles in the following words: 'The reader of the original text of Section 67 will see that I was already at that time very doubtful regarding the possibility of explaining the atomism of matter, and particularly of electric charge, with the help of classical concepts of continuous fields alone.'

Apart from this general introduction, note 23 discusses two types of attempts at unification of fields, namely theories with unsymmetrical metric tensor and five-dimensional theories. The latter are of particular interest here because Pauli's most important research papers in general relativity deal with the 'formulation of the laws of nature in five homogeneous coordinates' [6]. In the first of these two papers published in 1933, 'Pauli gave a beautiful account of this projective geometry and its tensor analysis, which were developed from first principles and he formulated the Einstein–Maxwell equations in projective coordinates.'[9] The second paper 'dealt with the incorporation of spinors and of Dirac's equation into this geometrical structure'.[9] Bargmann comments on it: 'In my opinion, this is by far the most satisfactory exposition of spinors in general relativity – quite independent of the problem of a unified field theory.'[9] For the sake of completeness it should be mentioned that Pauli had treated the problem of spinors and the formulation of the Dirac equation in 5 dimensions in two earlier papers published in 1932 with Solomon [7].

A particular result obtained in the second paper [6] concerns an additional term in the field equation, describing an anomalous magnetic moment which, however, is too small to be observed (except, perhaps, through its influence on terrestrial magnetism, as Pauli remarks at the end of this paper). In a recent analysis of the problem, Thirring [10] has shown that there is in addition a non-vanishing, although very small, electric dipole moment due to a very peculiar violation of parity and charge conjugation in this theory.

The number of research papers in general relativity by Pauli is not large. Apart from the already mentioned publications, there is only one more, written with Einstein in Princeton in 1943, 'on the non-existence of regular stationary solutions of relativistic field equations' [8]. But Pauli always took an active interest in this field, as is evident from his contribution as president of the conference on 'Fünfzig Jahre Relativitätstheorie' [9] which was held in Berne in 1955 shortly after Einstein's death.

2. THE OLD QUANTUM THEORY. SPIN AND EXCLUSION PRINCIPLE

Pauli studied six semesters at the University of Munich where his teacher in theoretical physics was Arnold Sommerfeld. In June 1921 he obtained his doctorate with a dissertation on a particular molecular model.[11] His doctoral diploma is dated 25 July 1921 and carries the mention 'summa cum laude'.[12] His dissertation, an improved and extended version of which was published [10], is Pauli's first contribution to the old quantum theory. The difficulty of this theory was that its rules were applicable only to conditionally periodic systems which excluded systems with more than one electron. The next more complicated one-electron system beyond hydrogen-like atoms is the hydrogen molecule ion which was the subject of this work.

In the winter semester of 1921/22 Pauli was Max Born's assistant in Göttingen. During this time Born and Pauli collaborated on a paper [11] on the systematic application of astronomical perturbation theory to atomic physics.[11] A short review on perturbation theory written by Pauli appeared two years later in the *Physikalisches Handwörterbuch* [12]. Pauli spent the following summer semester in Hamburg as an assistant of W. Lenz whom he had met in Munich, where Lenz had also worked with Sommerfeld.[11]

In Göttingen, Pauli had met Niels Bohr who asked him to come to Copenhagen for one year. Pauli was much surprised by this offer, and answered 'with that certainty of which only a young man is capable: "I hardly think that the scientific demands which you will make on me will cause me any difficulty, but the learning of a foreign tongue like Danish far exceeds my abilities" ... and I went to Copenhagen in autumn 1922, where both of my contentions were shown to be wrong' [13].

After a first paper written in Copenhagen in collaboration with H. A. Kramers on the theory of band spectra [14] Pauli concentrated his efforts on the problem of the anomalous Zeeman effect. This problem appeared to him very unapproachable. When a colleague who met him strolling in Copenhagen observed that he looked unhappy Pauli answered: 'How can one look happy when he is thinking about the anomalous Zeeman effect' [13].

Pauli's first paper on the anomalous Zeeman effect [15] analyzed the multiplets in the case of strong fields that Landé had investigated for weak magnetic fields H.[13] From this analysis Pauli was able to rederive Landé's splitting factor g by using the rule that 'the sum of the energies of all states of a multiplet belonging to a given value of M remains a linear function of H, when we pass from weak to strong fields'.[14] In a second paper [16], submitted from Hamburg where he had returned in autumn 1923 as an assistant of Lenz, Pauli showed that Landé's association of the quantum numbers in the cases of weak and strong fields [15] (which are adiabatically connected) could be derived from the rules of the old quantum theory.

The third paper on the anomalous Zeeman effect [17], submitted on 2 December 1924, is of fundamental importance because it introduces, still in disguised form, the spin quantum number of the electron. Pauli starts by arguing that the then widely accepted hypothesis that closed atomic shells (in particular the K-shell) carry magnetic moment and angular momentum has to be rejected for several different reasons. The main reason was that the relativistic velocity-dependence of the electron mass leads to a dependence of the gyromagnetic ratio of the K-shell on the atomic number Z, and hence to a Z-dependence of the Zeeman-splitting, contrary to observation. Pauli then draws the remarkable conclusion that the magneto-mechanical anomaly must be due to a 'strange two-valuedness' of the quantum-theoretic properties of the outer electron which is not describable classically. The original wording of this conclusion is the following (Ref. [17], p. 385):

Die abgeschlossenen Elektronenkonfigurationen sollen nichts zum magnetischen Moment und zum Impulsmoment des Atoms beitragen. Insbesondere werden bei den Alkalien die Impulswerte des Atoms und seine Energieänderungen in einem äusseren Magnetfeld im wesentlichen als eine alleinige Wirkung des Leuchtelektrons angesehen, das auch als der Sitz der magneto-mechanischen Anomalie betrachtet wir. Die Dublettstruktur der Alkalispektren, sowie die Durchbrechung des Larmortheorems kommt gemäss diesem Standpunkt durch eine eigentümliche, klassisch nicht beschreibbare Art von Zweideutigkeit der quantentheoretischen Eigenschaften des Leuchtelektrons zustande.

The remarkable feature of this conclusion is that only four months before, in a paper submitted on 17 August 1924 [18], Pauli had proposed the hypothesis that 'in general the nucleus must possess a non-vanishing resulting angular momentum.' The association of the electron spin to this 'strange two-valuedness' thus seems compelling. Pauli explains the reasons for his hesitation in his Nobel Prize Lecture of 1946 [19] as follows: 'Although at first I strongly doubted the correctness of this idea because of its classical mechanical character, I was finally converted to it by *Thomas*' calculation[16] on the magnitude of doublet splitting.' But he continues: 'On the other hand, my earlier doubts as well as the cautious expression: "classically non-describable two-valuedness" experienced a certain verification during later developments.'

It is interesting to note that there exists a completely forgotten footnote by Pauli which explains more closely Pauli's difficulty with the 'classical mechanical character' mentioned above. This footnote, written in 1928, in fact, fully elucidates the reason why Pauli could at the same time propose the idea of the nuclear spin and reject the apparently identical idea of the electron spin. This reason is that due to the small mass of the electron the peripheral velocities of the spinning electron are much larger than the light velocity c, while nuclear masses are sufficiently large to make the peripheral velocities of rotating nuclei much smaller than c.[17] This reference is a footnote on p. 1794 of Ref. [26] (see Section 3) and reads:

In einer mehr die kinematischen Verhältnisse ins Auge fassenden Weise wird auch von einem 'rotierenden Elektron' (englisch 'spin-elektron') gesprochen. Die Vorstellung eines rotierenden materiellen Gebildes halten wir aber nicht für wesentlich und sie empfiehlt sich auch nicht wegen der Ueberlichtgeschwindigkeiten, die man dann mit in Kauf nehmen muss...

The subtle history of the electron spin has been recounted elsewhere.[18] So it is sufficient to record here the following two episodes concerning the history of nuclear spin. Sommerfeld's reaction to Pauli's proposal [18] was one of disappointment over its classical nature which he expressed on a postcard to Pauli.[19] The second episode is recounted by S. A. Goudsmit,[20] who writes: 'For a number of years, whenever I met Pauli, he would remark cryptically that he "could afford not to be quoted". It was only in the later thirties that I found out to what he referred.'

Pauli's work on the anomalous Zeeman effect culminated in the famous paper enunciating the exclusion principle [20], submitted on 16 January 1925. In an address given at a dinner in honour of Pauli's award of the Nobel prize held at the Institute for Advanced Study in Princeton on 10 December 1945 [13] Pauli said: 'The history of the discovery of the exclusion principle, for which I have received the honour of the Nobel Prize award this year, goes back to my student days in Munich.' And he continues later on: 'The series of whole numbers 2, 8, 18, 32, ..., giving the lengths of the periods in the natural system of chemical elements, was zealously discussed in Munich, including the remark of the Swedish physicist, Rydberg[21], that these numbers are of the simple form $2n^2$ if n takes on all integer values.' In his Rydberg Centennial lecture of 1955 [21] Pauli says, regarding this formula: 'This is the famous formula $2p^2$ (p integer) which Sommerfeld called "cabbalistic" in his book *Atombau und Spektrallinien* and which impressed me very much as a student.' These periods in the system of chemical elements obviously intrigued Pauli very much and in 1923 he gave his inaugural lecture as Privatdozent at the University of Hamburg on this subject (see Ref. [19], p. 133).

Of course, Rydberg's formula was immediately evident once the exclusion principle was enunciated. In the original paper [20] the latter has the following content:

Es kann niemals zwei oder mehrere äquivalente Elektronen im Atom geben, für welche in starken Feldern die Werte aller Quantenzahlen n, k_1, k_2, m_1 (oder, was dasselbe ist, n, k_1, m_1, m_2)[22] übereinstimmen. Ist ein Elektron im Atom vorhanden, für das diese Quantenzahlen (im äusseren Felde) bestimmte Werte haben, so ist dieser Zustand 'besetzt'.

Admitting that he cannot give a closer justification of this rule Pauli then goes on to discuss the consequences. And he notes:

Zunächst sehen wir, dass das Resultat von Stoner und damit die Periodenlängen 2, 8, 18, 32, ... im natürlichen System in unserer Regel unmittelbar enthalten sind.

Here reference is made to a paper by E. C. Stoner[23] which, in Pauli's own opinion, was an important step towards the discovery of the exclusion principle. He formulates Stoner's main observation as follows (Ref. [19], p. 133):

For a given value of the principal quantum number is the number of energy levels of a single electron in the alkali metal spectra in an external magnetic field the same as the number of electrons in the closed shell of the rare gases which corresponds to this principal quantum number.

The divining of the exclusion principle thus appears as a magic act, an instant vision of things falling into place as in a phase transition. And one understands the fascination Pauli had for the act of creation in science, which he analyzed much later in his life in the example of Johannes Kepler (see Section 7).

3. THE NEW QUANTUM THEORY AND
ITS INTERPRETATION

In spring 1925 Heisenberg tried by Fourier analysis to arrive at intensity formulas for the hydrogen spectrum, in the hope of being able to guess the correct quantum mechanical intensity relations. However, the Kepler problem (of the hydrogen atom) turned out to be too difficult for this task. So Heisenberg applied his ideas to the anharmonic oscillator, the result of which he communicated to Pauli in a letter on 24 June 1925. In view of the credo mentioned earlier in relation with Pauli's third paper [3], the following introduction in this letter is of particular interest:[24] 'The principle is: In the calculation of any quantities such as energy, frequency etc, only relations among essentially observable quantities should occur.' This principle, Heisenberg believed, was the basis of Einstein's relativity theory which is also recognized as the origin of the credo in Pauli's third paper [3]. Einstein, however, thought that this principle was wrong, even if he had applied it himself. His argument was that only theory decides what can be observed.[25]

The mathematical formulation of the new theory by Heisenberg[26], by Born and Jordan[27], by Dirac[28] and by Born, Jordan and Heisenberg[29] followed almost immediately. Already in October, Pauli surprised Heisenberg with the complete quantum mechanical solution of the hydrogen atom. Heisenberg writes in a letter to Pauli on 3 November 1925:[30] 'I clearly do not have to write you how glad I am about the new theory of the hydrogen atom and how much I admire that you have brought forth this theory so quickly.'

The work on the anomalous Zeeman effect described in the previous section made use of only modest mathematical tools. However it required all the logical and analytical powers of Pauli's genius. On the other hand, the paper on the hydrogen spectrum, from the point of view of the new quantum mechanics [22], required considerable mathematical intuition. The key to the solution of this purely algebraic problem was the Lenz vector[31] which is a constant of the motion of the Kepler problem. By eliminating the coordinates of the electron Pauli arrives at four algebraic equations relating the components of the Lenz vector and those of the angular momentum vector l. Choosing a representation where l_z and l^2 are diagonal he then obtains the Balmer formula.

In March 1926 there appeared Schrödinger's first communication on 'Quantisierung als Eigenwertproblem'.[32] In a letter to Jordan, dated 12 April 1926, Pauli comments on this work[33]: 'I feel that this paper is to be counted among the most important recent publications.' And he immediately goes on to show the equivalence of Schrödinger's approach with the 'Göttingen Mechanics'.[33] This major contribution of Pauli to the development of the new quantum mechanics has been brought to the attention of the physics community by B. L. van der Waerden only in September 1972.[33] It had not been published by Pauli because Schrödinger had submitted the proof of the equivalence on 18 March 1926[34], and therefore had the priority. But Pauli must have attached some importance to his independent proof since, contrary to his habit of writing his correspondence by hand, this letter is typed, and Pauli kept a signed carbon copy in a special envelope.

While the electron spin was still missing in the paper on the spectrum of the hydrogen atom (Pauli's acceptance of the spin idea occurred in March 1926, and this paper was submitted on 17 January of this year), Pauli developed the non-relativistic quantum mechanics of the magnetic electron in his second published contribution to the new quantum theory [23], submitted on 3 May 1927.

The electron spin had been discussed in matrix mechanics earlier by Heisenberg and Jordan[35] in relation to the anomalous Zeeman effect, but its integration into the Schrödinger equation met formal difficulties. They stemmed from the fact that if the wave function ψ would depend on the angle φ, which is canonically conjugate to the z-component s_z of the spin, then, due to the 'strange two-valuedness' $s_z = \pm\frac{1}{2}$, ψ would have to change sign under rotation of φ by 2π. The simple way out proposed by Pauli was to let ψ depend on s_z itself, instead of on φ. This immediately leads to a two-component wave function and to the well-known Pauli spin matrices. Pauli emphasized that his theory was to be considered as provisional since the final theory should be relativistically invariant. He could not, at that moment, realize the important role his theory would play less than a year later when Dirac derived this final theory.[36]

This was the year when Pauli came to ETH with Paul Scherrer, the experimentalist, to succeed Peter Debye. About the same time Hermann Weyl left ETH and Schrödinger left the University of Zurich. The departure of these three eminent scholars was a painful loss for Zurich. Schrödinger was succeeded by Gregor Wentzel who, together with Pauli, represented the young generation. These two brilliant young professors brought to Zurich a fresh activity in theoretical physics for which the common seminar was the adequate platform for many years. From the beginning Pauli had at his disposal at ETH a postdoctorate assistantship which became the point of departure for many brilliant careers.

Besides the published contributions to the building of the new quantum mechanics mentioned earlier, Pauli took part with Heisenberg, Bohr, Born and others in leading the 'Aufbruch in das neue Land'.[37] The clarification of the logical and epistemological content of the new theory, with its entirely new notions of intrinsic uncertainty and complementarity, was a more strenuous march of the mind than can be realized in looking back. The last journey was covered at the 5th Solvay Conference in Brussels in the autumn of 1927, which was dominated by the discussions between Bohr and Einstein, in the outcome of which Pauli had an essential part.[38]

It was also Pauli who in later years gave the most accurate formulation of the controversy between the desire 'to complete quantum mechanics in a way so as to make it into a deterministic scheme with the aid of hidden parameters'[39], and 'the interpretation of quantum mechanics based on the idea of the complementarity'.[39] He made it clear that the given physical reasons which 'have nothing to do with philosophical prejudices'[39], led him to consider the second point of view of the above alternative as the only possible one [24]. He also clarified the fruitless discussion between Born and Einstein on the same subject in three letters to Max Born dated 3 and 31 March and 15 April 1954.[40]

There exist three extensive review articles on quantum theory by Pauli, two of which had the misfortune to have been written, essentially, in 1925. Consequently they soon became obsolete and are therefore not widely known. The first, 'Quantentheorie' [25] appeared in 1926 in Geiger and Scheel's *Handbuch*; the second, 'Allgemeine Grundlagen der Quantentheorie des Atombaues' [26] was part of Müller and Pouillet's treatise of 1929. While in the first, references to the electron spin are given in footnotes added in proof, the second contains an addendum (Nachtrag) on its consequences [where the footnote quoted in the previous Section can also be found]. It is a pity to see the enormous work of detail, so characteristic of all of Pauli's reviews, wasted in these treatises.

The third review of 1933, 'Die allgemeinen Prinzipien der Wellenmechanik' [27], which also appeared in Geiger and Scheel's *Handbuch*, however, competes in fame and durability with the review on relativity [4] of twelve years before. In fact, it was reprinted in somewhat shortened form in the same year 1958 as the English translation of the relativity [5] in Flügge's *Handbuch* [28]. The deleted sections had to do with the radiation field (Sections 6 to 8 of Part B of Ref. [27]), the theory of which had in the meantime grown into an enormous literature as will be seen in the following Section.

In the reprinted part, a footnote referring to the parity-violating neutrino (see Section 6 below) has been added (apart from a slight modification) in relation to Weyl's two-component equation (Ref. [28], p. 150). Also the discussion of the states of negative energy in Dirac's theory (Section 5 of Part B of Ref. [27]) has been modified and considerably shortened. This of course is due to the fact that at the time this review was written the only known particles were the electron and the proton (the positron was discovered in 1932, see note 47 below). As a consequence there was an asymmetry between positive and negative charge which had bothered Pauli already in his third paper [3].

In view of the fact that Pauli's *Handbuch* article on wave mechanics [27] was written at ETH in Zurich, it is astonishing that there did not exist a course on wave mechanics in the physics curriculum of ETH until 1956. Clearly, Pauli gave courses on selected topics of quantum mechanics, but it was only in the winter semester of 1956/57 that he started giving a regular course on the subject [29]. These lectures were marked by Pauli's liking for 19th century mathematics symbolized by Whittaker and Watson's well-known treatise[41], which gave his 'Wellenmechanik' the particular touch of solid handicraft. Pauli had inherited this love for wanderings in the complex plane from his teacher Arnold Sommerfeld. In fact, in his obituary note for Sommerfeld [30], Pauli writes: 'After the discovery of the new quantum mechanics (1927), Sommerfeld could now usefully employ in the theory of atomic structure his old mastery of the mathematics of wave theory in the byways of the complex plane so familiar to him.' On the occasion of Sommerfeld's 70th birthday in 1938 Pauli walked these 'byways in the complex plane' in a paper on the problem of light diffraction by a wedge [31], deriving new asymptotic formulas for Sommerfeld's exact solution.

4. QUANTUM FIELD THEORY. SPIN AND STATISTICS

From the moment that quantum field theory came into existence it became Pauli's main concern for the rest of his life. With respect to his own work quantum field theory started with Dirac's two papers of 1927[42] on the interaction of radiation with atomic matter. In these papers only the transverse (radiative) part of the electromagnetic field is quantized while the Coulomb part is included in the Hamiltonian of the material system, thus violating Lorentz invariance. This problem motivated the work of Jordan and Pauli [32] which develops a relativistically invariant form of quantum electrodynamics. Although it was a preliminary step since it treated only the free field case this work paved the way of a proper formulation of quantum field theory by introducing the famous 'invariant delta function'.

In the two important papers by Heisenberg and Pauli [33], which were the first ones that Pauli submitted from ETH in Zurich in 1928, the main aim was again quantum electrodynamics. But the canonical quantization method developed in the first of the two papers is much more general and has become a standard technique in quantum field theory.

The hard problem in the general part of this paper was the proof of the relativistic invariance of the canonical commutation relations, with regard to which Pauli used to say: 'Ich warne Neugierige'.[43] In fact, the commutation relations contain the two field quantities at equal time, and to obtain these relations for arbitrary space-times separation the canonical field equations have to be used.

The application of this general formalism to electrodynamics given in the second part of the first paper, and in the second paper, lead to difficulties related to the vanishing photon mass, that is, with the gauge group. While in the first paper these difficulties were overcome by adding a term to the Lagrangian density, the second paper was formulated in the radiation gauge in which the scalar potential vanishes. This choice of the gauge gives rise to a condition on the state vectors resembling the treatment in Fermi's paper[44] which had just appeared and is therefore mentioned only in a footnote of the second paper.

The matter which interacts with the electromagnetic field is taken, in both papers, to be the then known charged particles, electrons and protons. But in the first paper both Fermi and Bose statistics are discussed, the conclusion being that both statistics work. This is the first hint to the question of the spin-statistics relation which will be discussed below.

These early efforts in quantum electrodynamics and, in particular, the 'ultraviolet' divergence problem were reviewed in the discarded sections of the *Handbuch* article of 1933 [27], mentioned earlier. In the last of these discarded sections, Pauli also comments again on the questions raised in his third paper [3] concerning the concept of the electric field strength and the atomicity of electric charge.

Another question of historical interest discussed in the first of the discarded sections of Ref. [27] had to do with zero-point energy. Pauli made a distinction between the zero-point energy of material oscillators and of radiation, thus repeating an opinion already expressed in his paper on paramagnetism of 1927 [61] (see Section 5), namely 'that material systems (e.g. crystal lattices) quite generally are distinct from the radiation with respect to the occurrence of a zero-point energy'.

From his discussions with Otto Stern during his Hamburg years, Pauli knew that the zero-point energy of material oscillators was important for the understanding of the separation of isotopes.[45] On the other hand, he also knew from a calculation made in early years that the zero-point energy of radiation would have an unreasonably large gravitational effect.[45] Hence it must be discarded, thus freeing quantum electrodynamics from its most trivial divergence problem.

Another divergence problem that turned up in 1937 through the investigation of Bloch and Nordsieck[46] was the 'infrared' one which has to do with the soft photons emitted in Bremsstrahlung. Pauli and Fierz [34] reconsidered the problem in 1938 in order to see whether the logarithmic divergence of the integrated energy loss of a charged particle in a weak external field obtained by Bloch and Nordsieck was due only to inacceptable mathematical approximations or whether it concerned a deeper physical difficulty. In giving the particle a finite extension, restricting its velocity to non-relativistic values, and neglecting the recoil by photon emission Pauli and Fierz came to the conclusion that the infrared divergence was indeed an inherent difficulty of quantum electrodynamics. But the sensitive dependence of their result on the extension of the particle left Pauli somewhat pessimistic. Too pessimistic, indeed since this divergence problem is a fact with which the physicists around accelerators have become familiar with and which does not have the fundamental aspect of its 'ultraviolet' counterpart.

Anderson's discovery of the positron in 1932[47], which confirmed Dirac's reinterpretation of the states of negative energy[48], was for Pauli the point of departure into a new direction of research in quantum field theory. It eventually culminated in the important spin-statistics relationship. Curiously, the spin-$\frac{1}{2}$ positron first led Pauli to consider the spin-0 field, in a paper written with Weisskopf [35]. This is not so surprising however, since the main point of this paper was to show that the scalar relativistic wave equation of Gordon[49] and Klein[50] also admits particles of opposite charge and identical mass, thus giving rise to pair creation and annihilation and to vacuum polarization. This result had been obtained by applying the canonical quantization method of Heisenberg and Pauli [33].

The Pauli–Weisskopf paper was published in 1934, the year of Pauli's marriage to Miss Franca Bertram, who became Pauli's devoted and understanding spouse for the rest of his life.

In the paper of Pauli and Weisskopf [35], and even more explicitly in the detailed account that Pauli gave of this work in the *Annales de l'Institut Henri Poincaré* in 1936 [36], the question whether the spin-zero field could be quantized according to the exclusion principle (anti-commutators) is raised. The answer, which is explicitly derived in Ref. [36], is that in this case it is impossible to satisfy simultaneously the relativistic invariance of the theory and the condition that the charge density commutes in different points of space (or, more generally, for space-like separations). This was the third example of the general spin-statistics relationship. The first, obviously, was that of the photons which are spin-1 bosons, and the second that of the spin-$\frac{1}{2}$ particles of Dirac whose reinterpretation of the negative energy states was only possible with the exclusion principle.

The next important step was to establish the theory of free fields of arbitrary spin. After a first attempt by Dirac[51], this was, with the acknowledged guidance of Pauli, the work of Fierz in 1939[52] who at that time was Pauli's assistant. Fierz's result was that classically the energy is positive for integral spin and indefinite for half-integral spin. Quantization then was possible according to Bose statistics in the first case and according to Fermi statistics (exclusion principle) in the second.

Pauli, in his famous spin-statistics paper [37] submitted on 19 August 1940 from Princeton, where he had gone shortly before to escape the menace of Hitlerism, could build on Fierz's result as well as on those of two earlier publications one with Fierz [38], the other with Belinfante [39]. He first gives a general 'proof of the indefinite character of the charge in case of integral, and of the energy in case of half-integral spin' (Ref. [37], Section 3). Then making use of the postulate 'that all physical quantities at finite distances exterior to the light cone ... are commutable' (Ref. [37], p. 721), he proves that 'for integral spin the quantization according to the exclusion principle is not possible' (Ref. [37], p. 722). Finally he remarks that quantization according to Bose statistics would, for half-integral spins, leave the energy indefinite which is unacceptable.

For this analysis Pauli had developed new mathematical methods concerning the classification of spinors of arbitary rank.[53] In later papers [40], [41], [42], he reviewed the particular cases of spin 0, $\frac{1}{2}$ and 1 with more conventional methods. Ref. [40] also gives the interaction describing an anomalous magnetic moment which has become known as the 'Pauli term'.

The appearance on the list of elementary particles of the mu-meson, discovered in 1937[54] and mistaken to be Yukawa's particle of the nuclear forces[55], gave field quantization a new impulse. Pauli took an active part in research on the meson theory of nuclear forces during the period until 1946 which he spent at the Institute for Advanced Study in Princeton. He published several papers, in collaboration with members of the Institute, on the strong coupling approximation. These were the papers with Dancoff [43] on the pseudoscalar case, with Kusaka [44] on the mixed pseudoscalar–vector case, and with Hu [45] on the scalar- and vector-pair theory. In autumn 1944 Pauli gave a series of lectures on meson theory and nuclear forces at MIT which subsequently were published in book form [46].

While this clearly indicates Pauli's strong interest in meson theory and nuclear forces, it was always the field-theoretic aspect, and not nuclear physics, that attracted him to this domain of research. Characteristically it was Pauli [47] who applied to meson theory the field-theoretic device of the 'λ-limiting process' of Wentzel[56] and Dirac[57], which will be discussed below in relation to an indefinite metric in Hilbert space and with Ref. [41]. This, however, was a weak coupling approximation.

However, all these efforts left Pauli somewhat dissatisfied, since on the one hand the strong coupling theories gave quantitatively wrong results for the magnetic moments of proton and neutron [43] and of the deuteron [44] and led to unstable nuclei of high charge [44] or produced an unsatisfactory neutron–proton interaction for the deuteron [45]. The weak coupling approximation, on the other hand, had the deficiency 'that if the perturbation method (development in powers of the coupling constant) is valid and if the radius a of the nucleon is supposed to be smaller than the range of the resulting nuclear forces, the coupling constant must be so small that the nuclear interaction becomes much smaller than the empirical one' (Ref. [45], p. 267).

The finite radius a of the nucleon is a characteristic feature of all these theories [43], [44], [45], [47]. More generally, the heavy particle is described by a non-relativistic source function or form factor, as was the charged particle in the work of Pauli and Fierz [34], and also like the heavy particle in Källen and Pauli's treatment of the Lee model [52] to be discussed below. In subsequent years many attempts have been made to find a relativistic generalization of the cut-off method.[58] This led to the so-called non-local field theories in which the fields are coupled at different space-time points.

Pauli took up this problem, previously discussed by C. Bloch[59] and by Kristensen and Møller[60], in a paper published in 1953 [48]. According to Pauli,

The characteristic difference between local and non-local lorentzinvariant field theories is the fact that in the latter *it is not any longer possible to define field quantities (observables) which commute with each other simultaneously in all space-like pairs of points.* Therefore, in any case, non-local theories have to be considered as an enrichment of the known mathematical possibilities for quantized lorentz-invariant field theories (Ref. [48], p. 650).

The last sentence is somewhat surprising in view of the fact that commutativity of observables for space-like separation, that is, micro-causality, was a key postulate for Pauli's spin-statistics relationship [37]. Bu it may, in a very concealed fashion, reflect the hope that this might be a possible way to answer one day his ever-present question concerning the concept of the electric field strength and the atomicity of electric charge, raised in his third paper [3]. That such a hope (if it existed at all) was far from realistic is evident from the criticism Pauli voices in Footnote 8 where he writes with regard to Bloch's form factors[59] that 'such form factors will, however, in general lead to a wrong time order of processes (acausality) even for macroscopic distances'. This violation of macro-causality is indeed a serious defect.

The year 1948 brought forth renewed endeavours in quantum field theory through the famous works of Tomonaga[61], Schwinger[62], Feynman[63] and Dyson[64]. Back at ETH Pauli joined this activity, supported by the younger people around him: Villars, Jost, then his assistant, Schafroth, succeeding Jost as assistant, and the visitors Rayski, Luttinger and Glauber. The Monday afternoon theoretical seminar in Zurich, at which Heitler and his associates from Zurich University, Fierz from Basel, and even Stueckelberg from faraway Geneva participated, was an event which attracted many prominent guests. Former Pauli assistants Kronig, Bloch, Peierls, Kemmer, Casimir, Weisskopf used to stop by. This activity is reflected in the notes on 'Feldquantisierung' [49] based on lectures given by Pauli in 1950/51 at ETH and which were widely used by the experts. The situation could not be better described than in the words of Villars[65]:

Pauli's attitude with respect to the possibilities opened up by the new approach was characteristically one of critical optimism. His criticism was primarily focused on the claim of the unambiguity of the physical predictions of the theory (after isolation of the infinities); his optimism and vivid interest due to the hope that something might actually be learned from facing the remaining difficulties rather than by claiming total success before it was actually achieved.

The problem of the 'unambiguity of the physical predictions of the theory' mentioned above was systematically analyzed in the paper by Pauli and Villars [50] on invariant regularization. In this 'formalistic', as opposed to the 'realistic' procedure (Ref. [50], p.435), the invariant Green's functions, which depend on a single mass, were replaced by modified functions depending on a mass distribution ϱ. Regularization was achieved by the condition that the zeroth and first moments of ϱ vanish. ϱ could be a continuous or a discrete distribution. Regularization by a single auxiliary mass had been used before by Feynman[66], while Stueckelberg and Rivier[67] had used an arbitrary number of masses.

The most important problem of unambiguity analyzed in the paper by Pauli and Villars was the self-energy of the photon. Wentzel[68], by formal application of Schwinger's method, had obtained a non-vanishing result, in violation of gauge invariance. With the same technique, Feynman's result[66] for the electron self-energy and Schwinger's result[69] for the magnetic moment of the electron were obtained. In addition, an older result by Pauli and Rose [51] for the finite polarization effect on a point charge could be reproduced.[70]

As Pais and Uhlenbeck had shown[71], multi-mass regularization implies an indefinite metric in Hilbert space.[72] An indefinite metric was introduced for the first time by Dirac in 1942[73] as a new method to overcome, in the words of Pauli, 'all well-known convergence difficulties of quantized field theories if it is coupled with a quite different and logically independent method due to Wentzel and improved by Dirac, the so-called λ-limiting process' (Ref. [41], p. 175). However, Pauli showed [41] that pair creation and annihilation completely invalidated this result. The reason is that for a consistent application of the λ-process the electrons must not come closer to each other than the distance λ.

In a situation similar to Dirac's an indefinite metric was introduced by Källen and Pauli [52] in their paper on the Lee model.[74] By introducing a nucleon form factor, Källen and Pauli were able to express the coupling constant renormalization of Lee in finite form which gave rise to a finite critical coupling constant g_{crit}. For a value of the renormalized coupling constant $g > g_{crit}$, the normalization factor of the V-particle state became imaginary thus leading naturally to an indefinite metric and, in addition, to an non-unitary S-matrix.

A last short note on indefinite metric in an extended Lee model where the V-particle has two complex conjugate eigenvalues was presented by Pauli at the CERN conference of 1958 [53], five months before his death. In this note Pauli concentrates 'on the possibility of a temporal description of physical phenomena, since this is necessary in order to understand causality'. The rather artificial character of this note, so untypical of Pauli, hardly conceals a certain disillusionment with quantum field theory which Pauli had arrived at towards the end of his life.

5. PHYSICS OF CONDENSED MATTER.
STATISTICAL PHYSICS

As a graduate student in Munich, Pauli wrote three short papers on magnetic and dielectric properties of gases. In the first paper [54] submitted on 18 June 1920 Pauli calculated the diamagnetic susceptibility of monatomic gases with the aid of Larmor's theorem (multi-atomic gases are excluded by the fact that Larmor's theorem is not necessarily applicable).

The result, which is known as Langevin–Pauli diamagnetism[75], was a temperature-independent expression proportional to the quadrupole moment θ along the field direction. Comparison with the sparse experimental data on helium and argon led to values of θ which were 10 times too small compared to what follows from the atomic dimensions. This led Pauli to suspect that the experimental values could be wrong. And he concludes 'Purpose of these lines is to incite further measurements which could decide this important question.' Pauli's doubts were justified since the now accepted values are indeed more than 10 times smaller.[76]

In the second paper [55], which was based on a talk given at a meeting of the natural scientists in Germany, Pauli discussed the magnetic moment as determined from the Langevin formula $\mu^2/3RT$ for the paramagnetic susceptibility.[75] He rejects the Weiss magneton (which is approximately 5 times smaller than the Bohr magneton μ_B) as a fundamental unit and explains how angular momentum quantization (according to the old quantum mechanics) leads to non-integer values of μ/μ_B. Applied to diatomic molecules he concludes by comparison with experimental data that O_2 and NO have angular momentum quantum number 2 and 1, respectively, and that the magnetic moment is perpendicular to the molecular axis.

The third paper [56], submitted on 30 July 1921, treats the analogous case of the dielectric Langevin formula where μ is the electric moment. The statistical average implied in μ is more complicated in this case. Taking the dumb-bell model of a diatomic dipolar molecule, the problem reduces to the spherical pendulum which is treated according to the old quantum mechanics. This leads to a time average of the direction cosine which is expressed in terms of phase integrals. Subsequent statistical averaging then produces a quantum mechanical moment which is 2.1471 times smaller than the classical one. No experimental data were available to Pauli to test this result.

In a sequel to this paper of 1926, Mensing and Pauli [57] rederived μ according to the new quantum mechanics which gave a result much closer to that of classical theory. In particular, in the limit of high temperatures there was an exact coincidence between quantum mechanical and classical moment. And the authors conclude

It appears, therefore, that here again, as in many other cases, the new quantum mechanics joins classical mechanics more closely with respect to statistical averages, than the old quantum theory.

These early papers on the problems of statistical physics, which are widely known, have been described here in some detail because they demonstrate Pauli's precise and eminently practical method of research in a field which is entirely different from what has been discussed so far, and because they show Pauli's very early interest in this type of problems.

In another early paper [58] written in Copenhagen, Pauli gives an interesting and not well-known derivation of Planck's radiation law, based on the energy-momentum conservation for the Compton effect.[77] He shows that in order to get Planck's law the number of Compton processes $v \to v_1$, for given initial electron momentum, must be proportional to $\varrho_v + \text{const} \cdot \varrho_v \varrho_{v_1}$, ϱ_v being the spectral density, where the bilinear term is due to fluctuations of the radiation and gives the necessary modification leading from Wien's law to Planck's law. This shows explicitly the importance of fluctuations in radiation equilibrium. Pauli then remarks that a generalization of this procedure to other systems, such as atoms, allows one also to consider optical dispersion phenomena. As observed by Bohr[78], this generalization is very closely related to Kramers' dispersion relation. Pauli also wrote a review article on the theory of black-body radiation which appeared in 1929 in Müller and Pouillet's treatise [59], although it had been completed already in 1924. The above paper [58] is described in detail in this review.

Pauli's published contributions to solid state physics are contained essentially in two early papers. The first one of 1925 [60] deals with the problem of infrared absorption of dielectric crystals, and was a communication to the German Physical Society. It is treated in the model of a linear chain with nearest neighbour anharmonic forces which are quadratic in the elongations. The result, given without comments on the calculation involved, is controversial. In fact, Peierls[79] goes so far as to say that 'The published summary of this talk is probably the only incorrect formula published under Pauli's name.' Yet Pauli's formula does not look unreasonable, at least at temperatures T above the Debye temperature where an absorption proportional to T is to be expected. As Peierls remarks, the particular sinusoidal dispersion curve of a linear chain has no non-trivial solution to the energy and wave vector conservation equations for the three-phonon process involved in this problem, and therefore the result should be zero. But at finite temperature the thermal broadening, which was not taken into account in Pauli's calculation, relaxes conservation, leading to a non-zero result.[80] Anyway, 'Pauli was not satisfied and suggested a further study of the problem'.[79] And it was indeed to Peierls, his student and subsequently his assistant at ETH, that he made this suggestion in 1929.[81]

The three-dimensional problem turned out to be complicated by the existence of different phonon branches and, in fact, had a long history to which Peierls contributed significantly.[82] Pauli's dissatisfaction, on the other hand, may well have left a deeper trace in his mind, for he did not have the happy gift of getting rid of frustrations easily. And, perhaps, his well-known remark, that 'I don't like this solid state physics... I initiated it though'[83] may well have had this secret root.

This remark of Pauli also points to a highly positive fact. The allusion that he had actually initiated solid state physics refers to his second publication in this domain, the famous paper on the Pauli paramagnetism [61], submitted on 16 December 1926. Concerning this paper Peierls remarks that 'it is probably no exaggeration to say that the modern electron theory of metals was started by Pauli's paper on the paramagnetism of an electron gas.'[84]

While the exclusion principle had already proved its success in the understanding of the atomic structure, its consequences for the statistics of identical particles were just beginning to be recognized. Fermi[85] had applied it to a gas of particles (electrons). Dirac[86] had shown that its application to the quantum mechanical solution of an ideal gas necessarily led to Fermi statistics and he argued that for a gas of material particles (as against photons), Fermi statistics, and not Bose statistics, should be applied.

Pauli adopted this point of view, though somewhat reluctantly. This is because the analogy with the gas of light quanta, which was emphasized in his paper on Planck's radiation law [58], was thus destroyed. On the other hand, he points out the difference in the significance of the zero-point energy (mentioned in the previous Section) and of the velocity in the two cases. He then gives a general derivation of Fermi statistics, including the fluctuations, by the method of the grand canonical ensemble which at that time was not yet well known (Fermi had not used it). Pauli later took over this derivation of quantum statistics almost literally into his course on statistical mechanics [62].

After an application to atoms with arbitrary angular momentum he derives in this paper [61] the paramagnetic susceptibility bearing his name. But he emphasizes that the approximation of the metallic electrons as an ideal gas would have to be refined by taking into account their interactions. He observes that due to the exclusion principle the low-temperature susceptibility is strongly reduced from the Langevin value, and is of the same order as the diamagnetic contribution, which later was derived by Landau.[87]

At the 6th Solvay Conference in Brussels in 1930, Pauli presented an exhaustive report entitled 'Les théories quantiques du magnétisme. L'électron magnétique' [63]. In the first section Pauli reviews the paramagnetism and the diamagnetism of free electrons, in particular Landau's then new theory [87] which in the report is still quoted only as an oral communication. He then discusses the exchange integral and its application by Heisenberg [88] to the theory of ferromagnetism, as well as a number of other questions related to ferromagnetism. This section is Pauli's last written contribution to solid state physics, the second section being devoted to the relativistic theory of the electron.

But Pauli's interest in statistical physics had another important side, concerning the problem of Boltzmann's H-theorem in quantum mechanics. The first paper on this problem is contained in the volume dedicated to Arnold Sommerfeld by his pupils on the occasion of his 60th birthday [64], published in 1928. Pauli sets out to show that from the point of view of wave mechanics Boltzmann's H-theorem on the increase of entropy can be given a much more general form than is possible in classical mechanics. This is due to the fact that wave mechanics is already a statistical theory which gives much greater simplicity and generality to the laws describing the energy transfer between subsystems.

Pauli derives the 'golden rule' for the probability $W_m(t)$ to find the system in a state m at time t if it was in state n at time 0. The calculation of the inverse process from m to n is more complicated and necessitates the 'hypothesis of elementary disorder', according to which the initial phases are random. This leads to the detailed balance $A_{nm} = A_{mn}$ for the transition probabilities A_{nm} and to the rate equation $\Delta W_n = = -\sum_m Z_{nm} \Delta t + \sum_m Z_{mn} \Delta t$ which later was baptized as the 'master equation' by Uhlenbeck.[89] Here $Z_{nm} = A_{nm} G_m W_n$ is the transition rate and G_m the weight of state m. The time interval Δt must be small enough so that the W_n do not vary appreciably over Δt. The H-theorem then is a simple consequence of the master equation.

Conditions for the existence of the H-theorem in quantum mechanics are investigat-

ed in a second paper by Pauli and Fierz in 1937 [65]. The derivation and the consequences of the master equation gave rise to a large number of publications [90] of which
the work of van Hove [91] is of particular importance. In fact, the weakness of Pauli's
derivation is that in order to integrate the master equation on a time scale long
compared to the interval Δt the random phase assumption has to be made at every step
Δt. Van Hove showed that the master equation can actually be derived with one initial
random phase assumption only. This of course still destroys translational invariance
in time and therefore leads to irreversibility.

It is noteworthy that this work of van Hove was among the contents of the last
course that Pauli gave at ETH. It was in the middle of this subject that the lectures
abruptly stopped on 5 December 1958, ten days before his death.

On the other hand, the last work Pauli published in a scientific journal was a paper
on phenomenological thermodynamics [66], submitted on 19 July 1957. In this paper
the chemical equilibrium of mixtures was studied with the aid of external force fields
which allowed one to do away with the cumbersome van 't Hoff boxes. This paper was
included in the second edition of Pauli's course on Thermodynamics [67].

6. NEUTRINO, CPT-THEOREM AND PARITY VIOLATION

Pauli has recounted the history of the neutrino in a conference given in the Zurich
Society of Natural Sciences in the evening of 21 January 1957 [68]. This was an exciting
day since that very afternoon Pauli had received from Telegdi in Chicago the first
results of the three experiments by Madame Wu [92], by Lederman [93] and by Telegdi [94] on
the parity violation in beta- and mu-decay. In the morning of the same day two
theoretical papers, one by Yang, Lee and Oehme [95] and the other by Lee and Yang on
the two-component theory of the neutrino [96], which had been proposed by Salam [97]
several weeks earlier [98] were in Pauli's mail. And from Villars in Geneva came the
New York Times article reporting the sensation of parity violation.[98, 99]

Understandably, Pauli was on the peak of excitement, which, however, was somewhat dampened by his embarrassment about not having believed that 'God is just
left-handed'.[98] He gave an excellent talk that evening, and at the end he broke the news
about parity violation, giving an improvised account of the problem and its importance.
(An extended version of this account is contained in Section 5 of the published version
[68] of this talk.)

Pauli's own brainchild, the neutrino, conceived 27 years earlier, had unquestionably
played a trick on him! In a letter to Madame Wu written on the afternoon of the talk
at Zurich, Pauli says[100] 'this particle neutrino, of the existence of which I am not
innocent, still persecutes me'. Its capricious left-handedness had created considerably
more publicity than its birth, which for Pauli had been the only possible way out of a
dilemma.

This dilemma was the continuous beta-spectrum of Radium E (see Fig. 1 of Ref. [68]).
Calorimetric measurements by Ellis[101] and by Lise Meitner[102] in 1930 had yielded an
energy corresponding to the average energy of the beta-spectrum, and not to its

maximum. These results ruled out Meitner's idea that the continuous character of the spectrum was due to secondary processes. In order to explain this situation Niels Bohr invoked his old idea of *statistical* energy conservation[103] in beta-decay.[104] But to Pauli's deeply rooted sense of symmetries and conservation laws such a compromise was unacceptable. The only logical alternative was to invoke a new particle so penetrating that it escaped all measuring apparatus.

The first written account of this shockingly bold idea is contained in a letter of 4 December 1930 that Pauli addressed to the 'dear radioactive ladies and gentlemen' at a physics meeting in Tübingen, which Pauli was unable to attend because of a dance in Zurich (see Ref. [68], p. 159). In this letter he characterizes this new particle as neutral spin-$\frac{1}{2}$ particle of mass smaller than 0.01 times the proton mass and calls it neutron. This name had been introduced by Rutherford in 1921[105] for a hypothetical nucleus consisting of one proton and one electron. After the discovery of the real neutron by Chadwick in 1932[106] Fermi baptized Pauli's particle 'neutrino', the little neutron, in seminars he gave in Rome (see Ref. [68], p. 162). In June 1931 Pauli gave an account of his ideas in a talk at Pasadena, and at the 7th Solvay Conference in Brussels in 1933 he discussed the entire problem of beta-decay including the neutrino hypothesis.[107]

Inspired by the discussions at the 7th Solvay Conference Fermi shortly afterwards developed his theory of the beta-decay.[108] From a comparison of the statistical factor of allowed decays at the upper end of the spectrum Fermi[108] and Perrin[109] concluded already in 1933 that the neutrino mass should be zero (see Ref. [68], p. 164).

Pauli was particularly well prepared for the theory of beta-decay by his two papers on the mathematical theory of the Dirac matrices of 1935 and 1936 [69]. But although he took an active part in the analysis of the general structure of the theory, one of the results of which was the rearrangement theorem for the permutation of the fields, he left the elaboration to his assistant Fierz.[110] Instead, Pauli looked into the field-theoretic aspect of beta-decay interaction. In a note presented to the Academy of Sciences of the U.S.S.R. in 1938, and published in Russian [70], Pauli showed that the Fermi theory of beta-decay leads to an infinite self-energy and that, therefore, the application of the perturbation theory in higher than first order (as had been done by Heisenberg[111] in his theory of cosmic showers) was unjustified. This problem has remained with field theory ever since and has led to the distinction between renormalizable and unrenormalizable interactions.[112]

The two papers on the mathematical structure of the Dirac matrices [69], on the other hand, have become of great use for the definition of the discrete symmetry operations of charge conjugation (C), parity (P) and time-reversal (T) for spin-$\frac{1}{2}$ fields. These operations had been analyzed in view of the spin-statistics problem by Schwinger[113], and in particular by Lüders.[114] Lüders made the remarkable discovery that the product CPT had more general invariance properties than each operation separately, when applied to specific interaction Hamiltonians.

Pauli looked into this problem himself, and the result was the famous paper on the CPT-theorem [71] which he dedicated to Niels Bohr's 70th birthday and which is part of the volume edited by Pauli (with the assistance of L. Rosenfeld and V. Weisskopf) for this occasion. In a dedication full of warmth and wit Pauli paraphrases Bohr's favourite verses of Schiller which were chosen to introduce this essay. He says that while trying to use a rigorous mathematical formalism 'to connect all mentioned features of the theory with the help of a richer "fullness" of plus and minus signs in an increasing "clarity" ', the epistemological analysis 'makes me aware that the final "truth" on the subject is still "dwelling in the abyss" '.

Pauli's original form of the CPT-theorem states that if the combined CPT-inversion, which he calls 'strong reflection', is properly defined for the original field quantities 'it also holds for all ordered products of them or their derivatives of finite order after application of the inversion' (Ref. [71], p. 35). The remarkable fact of the theorem is that it only supposes invariance with respect to the proper, i.e. continuous, Lorentz group and local character of the interactions, the proper spin-statistics relationship being assumed. For the proof Pauli relies heavily on the spinor analysis he had developed in his spin-statistics paper [37].

The 'final truth' of the theorem rose to full 'clarity' shortly afterwards through the work of Jost[115], who had realized at an early stage that the combined PT-inversion could be continuously connected to the identitiy by analytic continuation in the complex Lorentz group. While the CPT-theorem, together with the spin-statistics relationship, has become a corner stone of modern field theory[115], it has also played a decisive role in weak interactions for the analysis of both the P-violation of the neutrino [68] and the CP-violation of the neutral K-meson.[116]

Pauli's interest in the problem of the beta-decay was stimulated anew by two experimental events which occurred several months before the parity violation was established. On 15 June 1956 he received a telegram from F. Reines and C. Cowan from the Los Alamos nuclear reactor with the following content: 'We are happy to inform you that we have definitely detected neutrinos from fission fragments by observing inverse beta-decay of protons. Observed cross-section agrees well with expected 6×10^{-44} cm^2.' Pauli answered the same day by a night letter with the following content[117]: 'Thanks for message. Everything comes to him who knows how to wait. Pauli.'

So the existence of Pauli's brainchild was at last certified. Pauli announced the good news at the CERN symposium in July 'because otherwise everybody would ask me separately' [72]. This last excuse, of course, was superfluous, but typical for Pauli: his natural modesty did not permit him to show too openly that he was proud.

The experiment of Cowan and Reines[118] was a veritable *tour de force* since the penetration depth of neutrinos in lead is of the order of 100 light years! (see Ref. [68], p. 160). Cowan and Reines also determined upper bounds for the mass and the magnetic moment of the neutrino.[119]

The second experimental event of the year 1956, which stimulated Pauli, was Davis' negative result for the detection of the beta-transition of ^{37}Cl induced by neutrinos from a nuclear reactor.[120] The result meant that lepton charge, i.e. the number of leptons minus the number of anti-leptons, was conserved. It was in response to this result that Pauli undertook to analyze the consequences of the non-conservation of the lepton number.

Into this occupation exploded the news about parity violation, which Pauli swiftly integrated into his work. The result was the paper on the general form of the beta-decay interaction [73], submitted on 14 May 1957. The essence of this paper was that linear canonical transformations mixing right- and left-handed neutrinos and antineutrinos induced, for zero neutrino mass, 2×2 unitary unimodular transformations of the coupling constants. This group property then gave rise to four bilinear invariants of the coupling constants plus two pseudo-invariants transforming with a phase. Consequently, S-matrix elements can only depend on these six quantities which greatly simplifies the analysis of particular processes, such as the reaction of Davis, the detailed calculations of which Pauli left to his assistant.[121]

The Pauli group of transformations of the coupling constants describes nothing else than rotations for spin-$\frac{1}{2}$. This property led Gürsey[122] to its interpretation as isospin transformations. Heisenberg[123] incorporated this idea into his non-linear spinor equation which, combined with a degenerate vacuum, looked very fascinating because of the hope of being able to degenerate all elementary particles from one single equation.

For a while Pauli took an enthusiastic part in these endeavours. However, later, in spring 1958, he withdrew from this collaboration with Heisenberg because the theory had revealed too many inconsistencies. In a report published posthumously by Touschek [74] Pauli comes back once more to the problem that 'the fields of all particles are to be constructed from the minimum number of fields'. But the problem remained wide open.

7. ASPECTS OF PAULI'S SPIRITUAL LIFE

Pauli was more than just a physicist with exceptional analytical and mathematical skill: he was a natural philosopher.

Three men unquestionably left an imprint on Pauli's spirit: Ernst Mach, his godfather, Arnold Sommerfeld, his teacher, and Niels Bohr, his philosopher friend. With regard to Ernst Mach, Pauli writes in the continuation of the letter of 31 March 1953 quoted at the beginning of this essay:[2]

He evidently was a stronger personality than the catholic priest, and the result seems to be that in this way I am baptized 'antimetaphysical' instead of catholic. In any case, the card remains in the cup and in spite of my larger spiritual transformations in later years it still remains a label which I myself carry, namely: 'of antimetaphysical descent'. Indeed, Mach considered metaphysics, somewhat simplifying, to be the origin of all evil in the world – that is, psychologically speaking, as the devil himself – and that cup with the card in it remained a symbol for the 'aqua permanens' which exorcizes the evil metaphysical spirits.

This letter is written in the middle of a period of intense philosophical and psychological activity stimulated by the 'extensive and essential disccusions' with C. G. Jung in Zurich (Ref. [75], p. 167).

At that time the label 'of antimetaphysical descent' imprinted on Pauli by the spirit of Mach had faded into 'larger spiritual transformations'.[2] And the two men who had most influenced these transformations were Sommerfeld and Bohr. Theirs were the only pictures that decorated Pauli's office at ETH.[124]

Of Sommerfeld, for whom Pauli had the veneration of a timid pupil, he writes in the obituary note [30] mentioned in Section 3:

To his pupils it will remain unforgettable how in his fine sense for harmonies based on integral numbers he reconjured the spirit of Kepler.

And in the important essay on the ideas of the unconscious in relation to the sciences [76], written on the occasion of the 80th birthday of C. G. Jung on 26 July 1954, Pauli writes on p. 295:

Through my teacher A. Sommerfeld I knew very well how these Pythagorean elements appearing with Kepler are still alive today.

Here Pauli refers to an article he had dedicated to Sommerfeld's 80th birthday on 5 December 1948 [77] where already the spirit of Kepler had been invoked.

This contagious fascination for Kepler transferred by his teacher inspired Pauli to a more serious investigation of Kepler's scientific method in his remarkable essay in the volume 'Naturerklärung und Psyche' published together with C. G. Jung [75]. In this essay the 'Pythagorean elements' in Kepler's thinking are the starting point of a profound analysis of the historical, philosophical and psychological motivation of science. Pauli writes of Kepler (Ref. [75], p. 114):

To him as legitimate spiritual descendent of the Pythagoreans all beauty resides in the right proportion, for 'geometria est archetypus pulchritudinis mundi' (geometry is the primordial image of beauty). This principle of his is his strength and, at the same time, his limitation: his ideas about the regular polyhedra and the harmonic proportions clearly would not quite fit in the planetary system ....

The key word 'archetypus' mentioned explicitly in the title of the essay relates this work to Jung's psychology of the unconscious. In the essay for Jung's 80th birthday mentioned above, Pauli notes (Ref. [76], p. 295):

In search for applications of the notion of archetypus outside the modern psychology of the unconscious I first came across the historical fact that Kepler extensively and regularly used the words 'archetypus' and 'archetypalis', and this in a similar sense as Jung, namely as primordial image ['Urbild'].

With regard to Jung's work and to the importance of the unconscious in the formation of scientific ideas, Pauli gives the following motivation for his Kepler article (Ref. [75], p. 113):

My attention therefore was directed especially towards the 17th century in which the then quite new natural-scientific reasoning had grown out of the maternal soil of the magic-animistic conception of Nature as a consequence of a large spiritual effort. For the sake of illustration of the relation between archetypical conceptions and natural-scientific theories, *Johannes Kepler* (1571–1630) seemed to me particularly proper, since his ideas represent a remarkable intermediary stage between the former magic-symbolic and the modern quantitative-mathematical description of Nature.

The particular significance of the 17th century for modern science is also emphasized in a conference that Pauli gave in the Zurich Philosophical Society in February 1949 [78]. He says (Ref. [78], p. 1):

The schism of the natural sciences and mathematics as autonomous partial disciplines from an originally unified but pre-scientific natural philosophy, which occurred in the 17th century, was, on the one hand, a necessary condition for the further spiritual development of the Occident. But today the conditions for a renewed agreement of the physicists and philosophers on the epistemological foundations of the scientific description of Nature seem to me to be fulfilled.

Pauli then gives the reasons for this hope of a 'renewed agreement of the physicists and philosophers' in a formulation which he repeats almost verbally in his Kepler article (Ref. [75], p. 163–164):

The development of atomistics and quantum theory since 1900 has indeed led to the fact that physics was gradually forced to give up its proud claim to be able, in principle, to understand the *whole* world ... Just this circumstance, however, [that today we do possess natural sciences but not any more a natural-scientific conception of the world] could, as a correction of the former one-sidedness, contain the germ of a progress in direction towards a unified global conception of the world, in which the natural sciences are only one part. Herein I would like to perceive the more general significance of the idea of complementarity, which, thanks to the Danish physicist Niels Bohr, has grown out of the soil of physics.

This idea of complementarity was the focal point of the inspiration that Bohr conveyed to Pauli on their 'long and still continuing common pilgrimage since the year 1922, in which so many stations are involved'. (See the dedication in Ref. [71].) The 'soil of physics' out of which this idea has grown is the quantum mechanics which, through the indeterministic character of the uncertainty relations, leaves the physicist the freedom to choose between mutually complementary experimental setups.

With regard to this freedom Pauli says in a conference given at the meeting of the Swiss Society of Natural Sciences in Berne in 1952 (Ref. [79], p. 116):

Herewith observation acquires the character of the *irrational, unique actuality* with a not predictable result.

This irrational element built into quantum mechanics, which Pauli had already emphasized in his conference in Zurich in 1949 (Ref. [78], p. 8), was of importance to him because of its far reaching implications outside physics. In the last quoted reference he continues:

This feature of complementarity within physics leads in a natural way beyond the more limited domain of physics to analogous situations with the general conditions of the human perception.

This leads to the problem of consciousness which, in analogy to the relation between the observing apparatus and the observed system in physical experiments, implies a relation between subject and object. The passage on this relation contained in the last quoted reference impresses by the vastness of Pauli's view. And one can only agree with Arthur Koestler writing about Pauli[125]: 'He may well have possessed an even deeper knowledge of the limits of the natural sciences than most of his colleagues.'

Pauli says (Ref. [78], p. 9):

The notion of consciousness, in fact, requires a cut between subject and object, the *existence* of which is a logical necessity while, on the other hand, the *position* of the cut is, to a certain degree, arbitrary. Non-consideration of this fact gives rise to two different kinds of metaphysical extrapolations which themselves can be called mutually complementary. The one is that of the material or, more generally, physical object, the constitution of which ought to be independent of the way in which it is observed ... The complementary extrapolation is that of the Hindu-metaphysics of the pure subject of perception to which is opposed no object. Personally I have no doubt that this idea also must be recognized as untenable extrapolation. The occidental spirit cannot recognize such a notion of a trans-personal cosmic conscience to which is opposed no object, and must keep the median prescribed by the idea of complementarity.

This attitude of mediation between extremes, quite typical of Pauli, played also a role in his relationship with Jung. In the course of his work Jung was gradually forced towards a less extreme view in his psychology of the unconscious by admitting non-psychical elements. In relation to this, Koestler quotes the following sentence by Jung written in 1947 and referring to para-psychological phenomena:[126]

I frankly doubt that an exclusively psychological methodology and consideration can do justice to the phenomena in question. Not only the statement of parapsychology but also my own theoretical reflections, which I have sketched in the *Theoretische Ueberlegungen zum Wesen des Psychischen*, lead me to certain postulates which touch the domain of the atomic-physical conceptions, that is the space time continuum. Herewith the question of the trans-psychical reality is raised which is at the immediate basis of the psyche.

The influence of Pauli on Jung is obvious from these remarks. In fact, Jung openly acknowledges this influence in his contribution to the common volume *Naturerklärung und Psyche* (Ref. [75], p. 101; see also Koestler,[125] p. 110). Koestler, however, remarks:[127]

... apart from his function as auxiliary teacher in the domain of theoretical physics (which, after all, Jung made little use of) Pauli probably had only little influence on Jung's treatise.

Pauli explicitly refers to Jung's gradual change of attitude mentioned above in his essay on the occasion of Jung's 80th birthday in relation to Jung's encounter with alchemy (Ref. [76], p. 290–291):

It does not surprise us that soon after the first occurrence of this encounter with alchemy the psycho-physical problem and also the problem of the inclusion of the observer in the course of Nature assumes actuality. Indeed, in order to cope with these fundamental problems *Jung*, in 1946, applied drastic changes to the notions used by him. He does it in particular in view of the phenomena of 'extra sensory perception' (ESP)....

The psycho-physical problem mentioned in these lines has much occupied Pauli, as is clear from other passages in the same essay [76]. One reason was that dreams are examples of psycho-physical processes (see Ref. [76], p. 287). But, more generally, it was related to his hope of reaching a synthesis between Science and Psychology. In his essay on Kepler he ventures into wishful thinking when he says (Ref. [75], p. 164):

It would be most satisfactory if *physis* and *psyche* could be understood as complementary aspects of the same reality.

Admittedly, this remained an open question for Pauli.

In closing this essay the other open question has to be mentioned again which bothered Pauli during all his scientific life, and which also brings another man into the picture of Pauli's spiritual life: Albert Einstein. This open question is the one which was raised for the first time by the 19-year-old Pauli in his third paper [3] (see Section 1):

Sollten wir überhaupt mit den Kontinuumstheorien für das Feld im Innern des Elektrons auf einer falschen Fährte sein?

It was Einstein's hope of his lifetime to integrate the atomistic constitution of matter into a unified theory. He never accepted the finality of the indeterministic character of quantum theory, and his well-known argument was that 'Gott würfelt nicht', 'God does not throw dice'. This had the effect that Pauli's relation to Einstein soon became one of respectful disagreement. But Einstein's admiration for Pauli was not affected by this disagreement. In fact, at a dinner in honour of Pauli held at the Institute for Advanced Study in Princeton, when Pauli left for a visit to Europe in 1946, the old and ailing Einstein quite unexpectedly rose to give an address which was moving and fascinating in its clear simplicity. In this address Einstein designated Pauli as his successor at the Institute and called him his spiritual son, of whom he hoped that he would carry on his, i.e. Einstein's, work.[128]

In 1958, in an article on Einstein Pauli writes [80]:

Since 1927 Einstein was disappointed by the development of physics. Unyielding, he retreated into his spiritual solitude. His further work on field theory although written with the same mathematical mastery as before, seems to lack the close contact with Nature. It is doubtful whether these last theoretical attempts of Einstein will in fact find an application in physics.

This article, written in the year of Pauli's death, then takes an unexpected turn in its closing paragraph which voices a challenge to the address of the younger generation, including Pauli himself:

Had we been able to present Einstein a synthesis of his general relativity theory with quantum theory then the discussion with him would have been considerably easier. But the duality between the field and its means of measurement, although latently present in today's quantum theory of fields, is conceptually not clearly expressed. The relation of the applicability of the ordinary space-time concept in the small with the properties of the smallest physical objects, the so-called 'elementary' particles is not disclosed. Einstein's life ended with a question [posed] to the science of physics and with a behest for synthesis to us.

These remarks dramatically reflect the fact that the question which had been posed by the 19-year-old Pauli, retained its urgency to the end of his life. It is the clash between the continuity of the notion of the field and the atomicity of the electric charge (and, more generally, all the coupling constants of elementary particle interactions) which bothered Pauli to a point that he raised the question time and again throughout his work, including the introduction to his course on electrodynamics [81].

This fact is substantiated by the following passages. The first is contained in a conference given at the Zurich Philosophical Society in 1934 [82]:

But these [classical field] theories ... are unable to interpret the additional fundamental property of the [electric] charge to be atomistic .... Only a new formulation of quantum theory would be satisfactory which, just as it juxtaposes as complementary the momentum and energy conservation laws and the description in space and time, would also juxtapose the classical conservation law of charge and its atomicity in a quantum theoretical correlation by an interpretation of the numerical value of the dimensionless number $hc/(2\pi e^2) = 136.8 \pm 0.2$.

Another place where this problem surfaces, not less urgently in its explicit wording, is the closing paragraph of Pauli's Nobel lecture delivered in Stockholm on 13 December 1946 [19]:

From the point of view of logic, my report on 'Exclusion principle and quantum mechanics' has no conclusion. I believe that it will only be possible to write the conclusion if a theory will be established which will determine the value of the fine structure constant and will thus explain the atomistic structure of electricity, which is such an essential quality of all atomic sources of electric fields actually occurring in nature.

The last reference in which the fine structure constant is explicitly mentioned is in note 23 of Ref. [5], written in 1956 (see Section 1), where Pauli writes:

In this connection it should be remembered that the atomicity of electric charge has already found its expression in the specific numerical value of the fine structure constant, at theoretical understanding of which is still missing today.

The number 1/137, the fine structure constant which his teacher Sommerfeld had introduced into physics, was Pauli's link to the 'magic-symbolic' world with which he was so familiar. Pauli spent the last few days of his life in the Red Cross Hospital in Zurich, where he died on 15 December 1958. A fact which had disturbed him during these last days was that the number of his room was 137.

NOTES AND REFERENCES

1. Two biographical as well as scientific portraits of Wolfgang Pauli deserve special mention. The first is an article by R. E. Peierls in *Biographical Memoirs of Fellows of the Royal Society* (The Royal Society, London, 1959), Vol. 5, pp. 175–192. Entitled 'Wolfgang Ernst Pauli' it is the only reference which explicitly mentions Pauli's second name Ernst, which Pauli himself never used. Apart from supplementary biographical information this article also contains a selection of the well-known Pauli anecdotes. The second portrait is an article by W. Thirring in *Oesterreichs Nobelpreisträger*, edited by F. G. Smekal (Wilhelm Frick, Wien–Stuttgart–Zürich, 1961), pp. 147–157 and p. 191. While giving some additional biographical elements it contains two inaccuracies (see p. 152); in fact, Pauli became full professor (Ordinarius) at ETH in 1928 and Swiss citizen in 1948.
2. '... Unter meinen Büchern befindet sich ein etwas verstaubtes Etui, in diesem ist ein Silberbecher im Jugendstil und in diesem wiederum ist eine Karte ... Dieser Becher nun ist ein Taufbecher und auf der Karte steht in altmodisch verschnörkelten Buchstaben:

   "Dr. E. Mach, Professor an der Universität Wien".

   Es kam so, dass mein Vater sehr mit seiner Familie befreundet war, damals geistig ganz unter seinem Einfluss stand und er (Mach) sich freundlicherweise bereit erklärt hatte, die Rolle des Taufpaten bei mir zu übernehmen .... Er war wohl eine stärkere Persönlichkeit als der katholische Geistliche, und das Resultat scheint zu sein, dass ich auf diese Weise antimetaphysisch statt katholisch getauft bin. Jedenfalls bleibt die Karte im Becher und trotz meiner grösseren geistigen Wandlungen in späterer Zeit bleibt sie doch eine Etikette, die ich selber trage, nämlich: "von antimetaphysischer Herkunft". In der Tat betrachtete Mach die Metaphysik, etwas vereinfachend, als die Ursache alles Bösen auf Erden – also psychologisch gesprochen: als den Teufel schlechtweg – und jener Becher mit der Karte darin blieb ein Symbol für die "aqua permanens", welche die bösen metaphysischen Geister verscheucht. ...'
   A copy of this letter is exhibited, together with the silver cup and Mach's card, in the Pauli archive at CERN, Geneva. I am grateful to Mrs F. Pauli for having drawn my attention to these objects.
3. Pauli's remark on the occasion of Walter Thirring's acceptance of the chair of theoretical physics at Vienna University in 1958.
4. *Neue Illustrierte Wochenschau*, 46. Jahrgang, Nr. 22, Wien (29. Mai 1955); *Hamburger Abendblatt*, Jahrgang 15, Nr. 53 (3. März 1962).
5. See the second reference of note 4, the accuracy of the content of which has been confirmed to me by Mrs F. Pauli.

6. See Wolfgang Pauli, *Collected Scientific Papers* (ed. by R. Kronig and V. F. Weisskopf), Interscience, New York–London–Sydney (1964), Preface, vol. **1**, p. viii.

7. See Wolfgang Pauli, *Collected Scientific Papers*, vol. **1**, p. x, which is a translation of a review by Einstein published in *Naturwiss.* **10**, 184 (1922).

8. J. Mehra, 'Einstein, Hilbert, and the Theory of Gravitation', in this volume, see note 242 on p. 170.

9. See V. Bargmann, 'Relativity', in *Theoretical Physics in the Twentieth Century, A Memorial Volume to Wolfgang Pauli* (ed. by M. Fierz and V. F. Weisskopf), Interscience, New York–London (1960), p. 197.

10. W. Thirring, *Acta Physica Austriaca*, Supplementum 1972, and this volume.

11. See Pauli's own biographical note written in Hamburg in 1926 after having obtained the title of professor, reproduced in Wolfgang Pauli, *Collected Scientific Papers*, vol. **1**, pp. v–vi (English translation, pp. x–xi).

12. Pauli's doctor's diploma is exhibited, together with the diplomas of various honours he had obtained later, in the Pauli archive at CERN, Geneva. The following gold medals are deposited at the same place: Lorentz medal (1931), Nobel medal (1945), Franklin medal (1952), Matteucci medal (1956), Planck medal (1958).

13. A. Landé, *Z. Phys.* **15**, 189 (1923).

14. See B. L. van der Waerden, 'Exclusion Principle and Spin', in *Theoretical Physics in the Twentieth Century* [see note 9], p. 202.

15. A. Landé, *Z. Phys.* **19**, 112 (1923).

16. L. H. Thomas, *Nature* **117**, 514 (1926); *Phil. Mag.* **3**, 1 (1927). Compare also J. Frenkel, *Z. Phys.* **37**, 243 (1926).

17. An angular momentum of $\hbar/2$ leads classically to a velocity $v = \hbar/(2mr)$ for a mass m rotating at a radius r. Taking for m the electron mass m_e and for r the classical electron radius $r_e = e^2/(m_e c^2)$ leads to $v/c \cong 70$. For a nucleus the radius is at least of the same order as r_e and the mass is at least 2000 times larger than m_e so that $v/c \lesssim 0.04 \times I$ where I is the nuclear spin.

   For the electron this result was first derived by Goudsmit and Uhlenbeck at the end of September 1925. See Uhlenbeck's address delivered at Leiden in 1955 on the occasion of his acceptance of the Lorentz Professorship, quoted in B. L. van der Waerden, 'Exclusion Principle and Spin', in *Theoretical Physics in the Twentieth Century*, p. 213. I am grateful to Professors G. E. Uhlenbeck, B. L. van der Waerden and J. Mehra for interesting conversations on this point.

18. See R. Kronig, 'The Turning Point', in *Theoretical Physics in the Twentieth Century*, pp. 5–39, and B. L. van der Waerden, 'Exclusion Principle and Spin', same reference, pp. 209–216.

19. I am grateful to Professor J. Mehra for having drawn my attention to the existence of this postcard.

20. S. A. Goudsmit, *Physics Today* **14**, No. 6, 18 (1961).

21. J. R. Rydberg, *Lunds Univ. Arsskrift* **9**, No. 18 (1913). *J. Chimie phys.* **12**, 585 (1914).

22. In modern notation $k_1 = l+1$, $k_2 = j+\frac{1}{2}$, $m_1 = m_j$, $m_2 = m_j \pm \frac{1}{2}$. See B. L. van der Waerden, 'Exclusion Principle and Spin', in *Theoretical Physics in the Twentieth Century*, p. 206.

23. E. C. Stoner, *Phil. Mag.* **48**, 719 (1924).

24. Translated from B. L. van der Waerden, *Sources of Quantum Mechanics*, North-Holland, Amsterdam (1967), p. 25. See also W. Heisenberg, 'Erinnerungen an die Zeit der Entwicklung der Quantenmechanik', in *Theoretical Physics in the Twentieth Century*, p. 43.

25. See W. Heisenberg, *Der Teil und das Ganze*, Piper, München (1969), pp. 91–92. I am grateful to Professors B. L. van der Waerden and C. F. von Weizsäcker for an interesting conversation on this point.

26. W. Heisenberg, *Z. Phys.* **33**, 879 (1925).

27. M. Born and P. Jordan, *Z. Phys.* **34**, 858 (1925).

28. P. A. M. Dirac, *Proc. Roy. Soc.* **109**, 642 (1925).

29. M. Born, P. Jordan and W. Heisenberg, *Z. Phys.* **35**, 557 (1926).

30. Translated from W. Heisenberg, 'Erinnerungen an die Zeit der Entwicklung der Quantenmechanik,' in *Theoretical Physics in the Twentieth Century*, p. 43.

31. W. Lenz, *Z. Phys.* **24**, 197 (1924). For the Coulomb potential $-Ze^2/r$ Lenz' vector takes the form $(Ze^2 m)^{-1} \cdot (\mathbf{l} \times \mathbf{r}) + \mathbf{r}/r$.

32. E. Schrödinger, *Ann. Phys. (Leipzig)* **79**, 361 (1926).

33. B. L. van der Waerden, in this volume. It is interesting to note that in this letter to Jordan, Pauli derives Schrödinger's equation starting from a relativistic form which, for free particles, is not, however, the Klein–Gordon equation

$$\Box\psi = \frac{m_0{}^2 c^2}{\hbar^2}\,\psi$$

(see notes 49 and 50 below) but

$$\Box\psi = -\frac{m_0{}^2 c^2}{E^2}\frac{\partial^2\psi}{\partial t^2}.$$

Pauli must have felt that this equation looked rather strange.

34. E. Schrödinger, 'Ueber das Verhältnis der Heisenberg–Born–Jordanschen Quantenmechanik zu der meinen', *Ann. Phys. (Leipzig)* **79**, 734 (1926).
35. W. Heisenberg and P. Jordan, *Z. Phys.* **37**, 263 (1926).
36. P. A. M. Dirac, *Proc. Roy. Soc.* A **117**, 610 (1928).
37. W. Heisenberg, *Der Teil und das Ganze*, Piper, München (1969), p. 101.
38. W. Heisenberg, 'Erinnerungen an die Zeit der Entwicklung der Quantenmechanik,' in *Theoretical Physics in the Twentieth Century*, p. 47. Unfortunately, these discussions at the 5th Solvay Conference ('Electrons et Photons') have not been included in the *Collected Scientific Papers* of Pauli [see note 6]. They are listed, however, in C. P. Enz, 'Bibliography Wolfgang Pauli', in *Theoretical Physics in the Twentieth Century*, p. 306.
39. Quotations translated from Ref. [24], pp. 37 and 42.
40. Albert Einstein, Max Born, *Briefwechsel 1916–1955*, kommentiert von Max Born, Nymphenburger, München (1969), pp. 289–299.
41. E. T. Whittaker and G. N. Watson, *A Course of Modern Analysis*, Cambridge University Press (first edition, 1902).
42. P. A. M. Dirac, *Proc. Roy. Soc.* A **114**, 243, 710 (1927).
43. See G. Wentzel, 'Quantum Theory of Fields (until 1947)', in *Theoretical Physics in the Twentieth Century*, p. 51.
44. E. Fermi, *Rendiconti Acad. Lincei* **9**, 881 (1929).
45. See C. P. Enz and A. Thellung, *Helv. Phys. Acta* **33**, 839 (1960).
46. F. Bloch and A. Nordsieck, *Phys. Rev.* **52**, 54 (1937).
47. C. D. Anderson, *Phys. Rev.* **43**, 492 (1933); P. M. S. Blackett and G. P. S. Occhialini, *Proc. Roy. Soc.* A **139**, 699 (1932).
48. P. A. M. Dirac, *Proc. Roy. Soc.* A **133**, 60 (1931).
49. W. Gordon, *Z. Phys.* **40**, 117 (1926).
50. O. Klein, *Z. Phys.* **41**, 407 (1927).
51. P. A. M. Dirac, *Proc. Roy. Soc.* A **155**, 447 (1936).
52. M. Fierz, *Helv. Phys. Acta* **12**, 3 (1939).
53. See R. Jost, 'Das Pauli-Prinzip und die Lorentz-Gruppe,' in *Theoretical Physics in the Twentieth Century*, pp. 114–115.
54. S. H. Neddermeyer and C. D. Anderson, *Phys. Rev.* **51**, 884 (1937); J. C. Street and E. C. Stevenson, *Phys. Rev.* **51**, 1005 (1937).
55. H. Yukawa, *Proc. phys.-math. Soc. Japan* **17**, 48 (1935).
56. G. Wentzel, *Z. Phys.* **86**, 479, 365, (1933); **87**, 726 (1934).
57. P. A. M. Dirac, *Proc. Roy. Soc.* A **167**, 148 (1938).
58. See F. Villars, 'Regularization and Non-Singular Interactions in Quantum Field Theory,' in *Theoretical Physics in the Twentieth Century*, pp. 94–98.
59. C. Bloch, *Kgl. Danske Vidensk. Selsk. Mat.-Fys. Medd.* **26**, No. 1 (1950); **27**, No. 8 (1952).
60. P. Kristensen and C. Møller, *Kgl. Danske Vidensk. Selsk. Mat.-Fys. Medd.* **27**, No. 7 (1952).
61. S. Tomonaga, *Progr. Theor. Phys.* **1**, 27 (1946).
62. J. Schwinger, *Phys. Rev.* **74**, 1439 (1948); **75**, 651 (1949); **76**, 790 (1949).
63. R. P. Feynman, *Phys. Rev.* **76**, 749, 769 (1949).
64. F. J. Dyson, *Phys. Rev.* **75**, 486, 1736 (1949).
65. F. Villars, 'Regularization and Non-Singular Interactions in Quantum Field Theory', in *Theoretical Physics in the Twentieth Century*, p. 82.
66. R. P. Feynman, *Phys. Rev.* **74**, 1439 (1948).
67. E. C. G. Stueckelberg and D. Rivier, *Phys. Rev.* **74**, 218, 986 (1948).

68. G. Wentzel, *Phys. Rev.* **74**, 1070 (1948).

69. J. Schwinger, *Phys. Rev.* **73**, 416 (1948).

70. See F. Villars, 'Regularization and Non-Singular Interactions in Quantum Field Theory', in *Theoretical Physics in the Twentieth Century*, pp. 89–92.

71. A. Pais and G. E. Uhlenbeck, *Phys. Rev.* **79**, 145 (1950).

72. See F. Villars, 'Regularization and Non-Singular Interactions in Quantum Field Theory', in *Theoretical Physics in the Twentieth Century*, pp. 83–84.

73. P. A. M. Dirac, *Proc. Roy. Soc.* A **180**, 1 (1942).

74. T. D. Lee, *Phys. Rev.* **95**, 1329 (1954).

75. See e.g., J. H. van Vleck, *The Theory of Electric and Magnetic Susceptibilities*, Oxford (1932).

76. The experimental susceptibility of one gram atom in 10^{-6} c.g.s. units is -1.88 and 19.6 for He and Ar, respectively, as compared to 46 and 233 to 252 quoted in Ref. [54]. See *Handbook of Chemistry and Physics*, Cleveland (1962), pp. 2731, 2735.

77. A. H. Compton, *Phys. Rev.* **21**, 483 (1923); P. Debye, *Physikal. Z.* **24**, 161 (1923).

78. Niels Bohr, 'Foreword', in *Theoretical Physics in the Twentieth Century*.

79. R. E. Peierls, 'Quantum Theory of Solids', in *Theoretical Physics in the Twentieth Century*, pp. 154–155.

80. Peierls' conclusion is based on the fact that he actually had seen Pauli's notes with the detailed calculation. In this calculation Pauli made use of exact conservation (which holds true in his first order calculation) but obtained a non-zero result by error (his solution was an Umklapp process which he did not recognize as giving a zero result).

    As seen from today, the problem is complicated by the fact that the rate of the 3-phonon process responsible for the absorption simultaneously determines, and is determined by, the thermal broadening and hence is selfconsistent, i.e. non-linear. I am grateful to Professor Peierls for an illuminating conversation on this problem. I was unable to find any trace of Pauli's notes in the Pauli archive at CERN. Very likely, Pauli discarded them after Peierls had clarified the problem.

81. R. E. Peierls, *Ann. Phys. (Leipzig)* **3**, 1055 (1929). This was Peierls' doctoral thesis which he started in Spring 1929 when he came to work with Pauli. After finishing his thesis Pauli asked him to become his assistant in October 1929.

82. R. E. Peierls, 'Quantum Theory of Solids', in *Theoretical Physics in the Twentieth Century*, pp. 155–159.

83. H. B. G. Casimir, 'Pauli and the Theory of the Solid State', in *Theoretical Physics in the Twentieth Century*, p. 137.

84. R. E. Peierls, 'Quantum Theory of Solids', in *Theoretical Physics in the Twentieth Century*, p. 140. In fact, in this paper, Pauli made the first application of Fermi–Dirac statistics to the theory of metals. See A. Sommerfeld and H. Bethe, 'Elektronentheorie der Metalle', in *Handbuch der Physik* (ed. by H. Geiger and K. Scheel), 2nd ed., Springer, Berlin (1933), Vol. 24, Part 2, p. 2. Sommerfeld's paper on the electron theory of metals, based on Fermi statistics, was published only in 1928 (*Z. Phys.* **47**, 1).

85. E. Fermi, *Z. Phys.* **36**, 902 (1926).

86. P. A. M. Dirac, *Proc. Roy. Soc.* A **112**, 661 (1926).

87. L. D. Landau, *Z. Phys.* **64**, 629 (1930).

88. W. Heisenberg, *Z. Phys.* **49**, 619 (1928).

89. See A. Nordsieck, W. E. Lamb, Jr., and G. E. Uhlenbeck, *Physica* **7**, 344 (1940), p. 353. I am grateful to Professor Uhlenbeck for information on this point.

90. See, e.g., *Fundamental Problems in Statistical Mechanics* (edited by E. G. D. Cohen), North-Holland, Amsterdam (1962).

91. L. van Hove, *Physica* **21**, 517 (1955); **23**, 411 (1957).

92. C. S. Wu, E. Ambler, R. W. Hayward, D. D. Hoppes and R. P. Hudson, *Phys. Rev.* **105**, 1413 (1957).

93. R. L. Garwin, L. M. Lederman and M. Weinreich, *Phys. Rev.* **105**, 1415 (1957).

94. J. I. Friedman and V. L. Telegdi, *Phys. Rev.* **105**, 1681 (1957).

95. T. D. Lee, R. Oehme and C. N. Yang, *Phys. Rev.* **106**, 340 (1957).

96. T. D. Lee and C. N. Yang, *Phys. Rev.* **105**, 1671 (1957); see also L. D. Landau, *Nuclear Physics* **3**, 127 (1957).

97. A. Salam, *Nuovo Cimento*, **5**, 299 (1957).
98. See Pauli's own account in the letter to Weisskopf of 27 January 1957 reproduced in Wolfgang Pauli, *Collected Scientific Papers*, vol. **1**, p. xiii, translation, p. xvii.
99. The first message of parity violation was contained in a letter of John Blatt to Pauli written in Princeton, which essentially read: 'We are all very shocked by the sudden death of parity'. I am grateful to Res Jost for this information. Neither this letter nor an accompanying letter of Villars are listed in the file of the Pauli correspondence in the Pauli archive at CERN.
100. C. S. Wu, 'The Neutrino', in *Theoretical Physics in the Twentieth Century*, p. 250.
101. C. D. Ellis and W. A. Wooster, *Proc. Roy. Soc.* A **117**, 109 (1927).
102. L. Meitner and W. Orthmann, *Z. Phys.* **60**, 143 (1930).
103. See N. Bohr, H. A. Kramers and J. C. Slater, *Z. Phys.* **24**, 69 (1924) where on p. 77 the following sentence can be found: 'Diese Unabhängigkeit [der Uebergangsprozesse] reduziert nicht nur die Erhaltung der Energie zu einem statistischen Gesetz, sondern auch die Erhaltung der Bewegungsgrösse....'
104. N. Bohr, 'Chemistry and the Quantum Theory of Atomic Constitution', Faraday Lecture, *J. Chem. Soc. (London)*, p. 349 (February 1932), on p. 383.
105. See J. L. Classon, *Phil. Mag.* **42**, 596 (1921).
106. J. Chadwick, *Proc. Roy. Soc.* A **136**, 692 (1932).
107. Unfortunately, these discussions at the 7th Solvay Conferenc ('Noyaux Atomiques') have not been included in the *Collected Scientific Papers* of Pauli. They are listed, however, in C. P. Enz, 'Bibliography Wolfgang Pauli', in *Theoretical Physics in the Twentieth Century*, p. 307.
108. E. Fermi, *Ricercha Scient.* **2**, Part 12 (1933); *Z. Phys.* **88**, 161 (1934).
109. F. Perrin, *Compt. Rend. Acad. Sci. Paris* **197**, 1625 (1933).
110. M. Fierz, *Z. Phys.* **104**, 553 (1937).
111. W. Heisenberg, *Z. Phys.* **101**, 533 (1936).
112. See, e.g., H. Umezawa, *Quantum Field Theory*, North-Holland, Amsterdam (1956), Chapter XV.
113. J. Schwinger, *Phys. Rev.* **82**, 914 (1951).
114. G. Lüders, *Kgl. Danske Vidensk. Selsk. Mat.-Fys. Medd.* **28**, Nr. 5 (1954).
115. See R. Jost, 'Das Pauli-Prinzip und die Lorentz-Gruppe,' in *Theoretical Physics in the Twentieth Century*, pp. 116–136.
116. See, e.g., T. D. Lee and C. S. Wu, *Annual Rev. Nucl. Sci.* **16**, Chapter 9 (1966).
117. The sheet of paper with Pauli's handwritten text and acknowledgement of execution ('erl.[edigt] 15.6.56/15.35 h als night letter') is contained in the folder with Pauli's notes for his neutrino talk, Ref. [68], in the Pauli archive at CERN.
118. C. L. Cowan, F. Reines, F. B. Harrison, H. Kruse and A. D. Guire, *Science* **124**, 103 (1956), F. Reines and C. L. Cowan, Jr., *Nature* **178**, 446 (1956).
119. $m_\nu < 0.002 \, m_e$, $\mu_\nu < 10^{-9} \, \mu_B$, see C. L. Cowan, Jr., and F. Reines, *Phys. Rev.* **107**, 528 (1957).
120. R. Davis, *Phys. Rev.* **97**, 766 (1955), *Bull. American Phys. Soc.*, Washington Meeting (1956).
121. C. P. Enz, *Nuovo Cimento* **6**, 250 (1957).
122. F. Gürsey, *Nuovo Cimento* **7**, 411 (1958).
123. See W. Heisenberg, *Einführung in die Theorie der Elementarteilchen* (lectures edited by H. Rechenberg and K. Lagally) (1961), Chapter 6; H. P. Dürr, W. Heisenberg, H. Mitter, S. Schlieder and K. Yamazaki, *Z. Naturforsch.* **14a**, 441 (1959).
124. In the late forties Mach's picture also hung in Pauli's office in No. 4c of the physics institute of ETH. 1 am grateful to Res Jost for this information. In later years Pauli moved to room No. 3e and Mach's picture was moved to the discussion room of the institute for theoretical physics on the same floor. This and the adjacent room of Pauli's assistant were decorated by historic group pictures of physics meetings. All these photographs are now in the Pauli archive at CERN.
125. Arthur Koestler, *Die Wurzeln des Zufalls*, Scherz, Bern–München–Wien (1972), p. 98.
126. A. Koestler, *Die Wurzeln des Zufalls*, pp. 102–103.
127. A. Koestler, *Die Wurzeln des Zufalls*, pp. 107–108.
128. Private communication from Mrs F. Pauli. Unfortunately, this address has left no trace in the published literature.

REFERENCES TO W. PAULI'S WORKS

[1] 'Ueber die Energiekomponenten des Gravitationsfeldes', *Physikal. Z.* **20**, 25–27 (1919).

[2] 'Zur Theorie der Gravitation und der Elektrizität von Hermann Weyl', *Physikal. Z.* **20**, 457–467 (1919).

[3] 'Mercurperihelbewegung und Strahlenablenkung in Weyls Gravitationstheorie', *Verhandl. Deutsche Phys. Ges.* **21**, 742–750 (1919).

[4] 'Relativitätstheorie', in *Encyklopädie der mathematischen Wissenschaften*, Vol. 5, Part 2, Teubner, Leipzig (1921), pp. 539–775.

[5] *Theory of Relativity*, Pergamon, London–New York–Paris–Los Angeles (1958).

[6] 'Ueber die Formulierung der Naturgesetze mit fünf homogenen Koordinaten, Teil I: klassische Theorie', *Ann. Phys. (Leipzig)* [5] **18**, 305–336 (1933), 'Teil II: Die Dirac'schen Gleichungen für die Materiewellen', *Ann. Phys. (Leipzig)* [5] **18**, 337–372 (1933).

[7] (W. Pauli and J. Solomon) 'La théorie unitaire d'Einstein et Mayer et les équations de Dirac, I and II', *J. Phys. Radium* [7] **3**, 452–463 and 582–589 (1932).

[8] (A. Einstein and W. Pauli) 'On the non-existence of regular stationary solutions of relativistic field Equations', *Ann. of Math.* **44**, 131–137 (1943).

[9] 'Opening Talk'; 'Schlusswort des Präsidenten der Konferenz'; 'Relativitätstheorie und Wissenschaft'; in: *'Fünfzig Jahre Relativitätstheorie, Bern, 1955'*, *Helv. Phys. Acta*, Suppl. **4**, 27; 261–267; 282–286 (1956).

[10] 'Ueber das Modell des Wasserstoffmolekülions', *Ann. Phys. (Leipzig)* [4] **68**, 177–240 (1922).

[11] (M. Born and W. Pauli) 'Ueber die Quantelung gestörter mechanischer Systeme', *Z. Phys.* **10**, 137–158 (1922).

[12] 'Störungstheorie', in *Physikalisches Handwörterbuch* (ed. by A. Berliner and K. Scheel), Springer, Berlin, (1924) pp. 752–756.

[13] 'Remarks on the history of the exclusion principle', *Science* **103**, 213–215 (1946).

[14] (H. A. Kramers and W. Pauli) 'Zur Theorie der Bandenspektren', *Z. Phys.* **13**, 351–367 (1923).

[15] 'Ueber die Gesetzmässigkeiten des anomalen Zeemaneffektes', *Z. Phys.* **16**, 155–164 (1923).

[16] 'Zur Frage der Zuordnung der Komplexstrukturterme in starken und in schwachen äusseren Feldern', *Z. Phys.* **20**, 371–387 (1924).

[17] 'Ueber den Einfluss der Geschwindigkeitsabhängigkeit der Elektronenmasse auf den Zeemaneffekt', *Z. Phys.* **31**, 373–385 (1925).

[18] 'Zur Frage der theoretischen Deutung der Satelliten einiger Spektrallinien und ihrer Beeinflussung durch magnetische Felder', *Naturwiss.* **12**, 741–743 (1924).

[19] 'Exclusion Principle and Quantum Mechanics', *Prix Nobel 1946*, Stockholm (1948), p. 131–147. German translation in Ref. [68].

[20] 'Ueber den Zusammenhang des Abschlusses der Elektronengruppen im Atom mit der Komplexstruktur der Spektren', *Z. Phys.* **31**, 765–783 (1925).

[21] 'Rydberg and the periodic system of the elements', in *Proc. Rydberg Centennial Conf. Atomic, Spectroscopy, Lund, 1954*, Lund (1955), pp. 22–26. German translation in Ref. [68].

[22] 'Ueber das Wasserstoffspektrum vom Standpunkt der neuen Quantenmechanik', *Z. Phys.* **36** 336–363 (1926).

[23] 'Zur Quantenmechanik des magnetischen Elektrons', *Z. Phys.* **43**, 601–623 (1927).

[24] 'Remarques sur le problème des paramètres cachés dans la mécanique quantique et sur la théorie de l'onde pilote', in *Louis de Broglie, physicien et penseur*, Albin Michel, Paris (1953), pp. 33–42.

[25] 'Quantentheorie', in *Handbuch der Physik* (ed. by H. Geiger and K. Scheel), Springer, Berlin Vol. **23**, pp. 1–278.

[26] 'Allgemeine Grundlagen der Quantentheorie des Atombaues', in *Müller-Pouillets Lehrbuch*, Vieweg, Braunschweig (11th ed., 1929), Vol. 2, Part 2, Chap. 29, pp. 1709–1842.

[27] 'Die Allgemeinen Prinzipien der Wellenmechanik', in *Handbuch der Physik* (ed. by H. Geiger and K. Scheel), Springer, Berlin (2nd ed., 1933), Vol. **24**, Part 1, pp. 83–272.

[28] 'Die Allgemeinen Prinzipien der Wellenmechanik', in *Handbuch der Physik* (ed. by S. Flügge), Springer, Berlin (1958), Vol. 5, Part 1, pp. 1–168.

[29] *Wellenmechanik*, (ed. by F. Herlach and H. E. Knoepfel), Verein der Mathematiker und Physiker an der ETH, Zürich, (1959). English translation: *Wave Mechanics* (ed. by C. P. Enz), MIT Press, Cambridge, Mass. (1973).

[30] 'Arnold Sommerfeld', *Z. Naturforsch.* **6a**, 468 (1951). Reprinted in Ref. [68].

[31] 'On asymptotic series for functions in the theory of diffraction of light', *Phys. Rev.* **54**, 924–931 (1938).

[32] (P. Jordan and W. Pauli) 'Zur Quantenelektrodynamik ladungsfreier Felder', *Z. Phys.* **47**, 151–173 (1928).

[33] (W. Heisenberg and W. Pauli) 'Zur Quantendynamik der Wellenfelder', *Z. Phys.* **56**, 1–61 (1929); Part II: *Z. Phys.* **59**, 168–190 (1931).

[34] (W. Pauli and M. Fierz) 'Zur Theorie der Emission langwelliger Lichtquanten', *Nuovo Cimento*, **15**, 167–188 (1938).

[35] (W. Pauli and V. Weisskopf) 'Ueber die Quantisierung der skalaren relativistischen Wellengleichung', *Helv. Phys. Acta* **7**, 709–731 (1934).

[36] 'Théorie quantique relativiste des particles obéissant à la statistique de Einstein–Bose', *Ann. Inst. Henri Poincaré* **6**, 137–152 (1936).

[37] 'The connection between spin and statistics', *Phys. Rev.* **58**, 716–722 (1940).

[38] (M. Fierz and W. Pauli) 'On the relativistic wave equations for particles of arbitrary spin in an electromagnetic field', *Proc. Roy. Soc.* A **173**, 211–232 (1939).

[39] (W. Pauli and F. J. Belinfante) 'On the statistical behaviour of known and unknown elementary particles', *Physica* **7**, 177–192 (1940).

[40] 'Relativistic field theories of elementary particles', *Rev. Mod. Phys.* **13**, 203–232 (1941).

[41] 'On Dirac's new method of field quantization', *Rev. Mod. Phys.* **15**, 175–207 (1943).

[42] 'On the connection between spin and statistics', *Progr. Theor. Phys.* **5**, 526–543 (1950).

[43] (W. Pauli and S. M. Dancoff) 'The pseudoscalar meson field with strong coupling', *Phys. Rev.* **62**, 85–108 (1942).

[44] (W. Pauli and S. Kusaka) 'On the theory of a mixed pseudoscalar and a vector meson field', *Phys. Rev.* **63**, 400–416 (1943).

[45] (W. Pauli and N. Hu) 'On the strong coupling case for spin-dependent interactions in scalar- and vector-pair theories', *Rev. Mod. Phys.* **17**, 267–286 (1945).

[46] *Meson Theory of Nuclear Forces*, Interscience, New York (1946, 2nd ed. 1948).

[47] 'On applications of the λ-limiting process to the theory of the meson field', *Phys. Rev.* **64**, 332–344 (1943).

[48] 'On the Hamiltonian structure of non-local field theories', *Nuovo Cimento* **10**, 648–667 (1953).

[49] *Ausgewählte Kapitel aus der Feldquantisierung* (ed. by U. Hochstrasser and M. R. Schafroth), Verein der Mathematiker und Physiker an der ETH, Zürich (2nd ed. 1957). English translation: *Selected Topics in Field Quantization* (ed. by C. P. Enz), MIT Press, Cambridge, Mass. (1973).

[50] (W. Pauli and F. Villars) 'On the invariant regularization in relativistic quantum theory', *Rev. Mod. Phys.* **21**, 434–444 (1949).

[51] (W. Pauli and M. E. Rose) 'Remarks on the polarization effects in the positron theory', *Phys. Rev.* **49**, 462–465 (1936).

[52] (G. Källén and W. Pauli) 'On the mathematical structure of T. D. Lee's model of a renormalizable field theory', *Kgl. Danske Vidensk. Selsk. Mat.-Fys. Medd.* **30**, 3–23 (1955).

[53] 'The indefinite metric with complex roots', *Proc. Annual Internat. Conf. High Energy Physics*, CERN, Geneva, (1958), pp. 127–128.

[54] 'Theoretische Bemerkungen über den Diamagnetismus einatomiger Gase', *Z. Phys.* **2**, 201–205 (1920).

[55] 'Quantentheorie und Magneton', *Physikal. Z.* **21**, 615–617 (1920).

[56] 'Zur Theorie der Dielektrizitätskonstante zweiatomiger Dipolgase', *Z. Phys.* **6**, 319–327 (1921).

[57] (L. Mensing and W. Pauli) 'Ueber die Dielektrizitätskonstante von Dipolgasen nach der Quantenmechanik', *Physikal. Z.* **27**, 509–512 (1926).

[58] 'Ueber das thermische Gleichgewicht zwischen Strahlung und freien Elektronen,' *Z. Phys.* **18**, 272–286 (1923).

[59] 'Theorie der schwarzen Strahlung', in *Müller-Pouillet's Lehrbuch*, Vieweg, Braunschweig (11th ed., 1929), Vol. 2, Part 2, Chap. 27, pp. 1483–1553.

[60] 'Ueber die Absorption der Reststrahlen in Kristallen', *Verhandl. Deutsche Phys. Ges.* [3] **6**, 10–11 (1925).

[61] 'Ueber Gasentartung und Paramagnetismus', *Z. Phys.* **41**, 81–102 (1927).

[62] *Statistische Mechanik* (ed. by R. Schafroth), Verein der Mathematiker und Physiker an der ETH, Zürich (1951). English translation: *Statistical Mechanics* (ed. by C. P. Enz), MIT Press, Cambridge, Mass. (1973).

[63] 'Les théories quantiques du magnétisme. L'électron magnétique', in *6ème Conseil de Physique Solvay, Le Magnétisme, Bruxelles, 1930* (Paris, 1932), pp. 175–238.

[64] 'Ueber das H-Theorem vom Anwachsen der Entropie vom Standpunkt der neuen Quanten-mechanik', in *Probleme der modernen Physik, Arnold Sommerfeld zum 60. Geburtstage, gewidmet von seinen Schülern*, Hirzel, Leipzig (1928), pp. 30–45.

[65] (W. Pauli and M. Fierz) 'Ueber das H-Theorem in der Quantenmechanik', *Z. Phys.* **106**, 572–587 (1937).

[66] 'Zur Thermodynamik dissoziierter Gleichgewichtsgemische in äusseren Kraftfeldern', in *'Festschrift Jakob Ackeret'*, *Z. Angew. Math. Phys.* **9b**, 490– 497(1958).

[67] *Thermodynamik und kinetische Gastheorie* (ed. by E. Jucker), Verein der Mathematiker und Physiker an der ETH, Zürich (improved 2nd ed., 1958). English translation: *Thermodynamics and the Kinetic Theory of Gases* (ed. by C. P. Enz), MIT Press, Cambridge, Mass. (1973).

[68] 'Zur älteren und neueren Geschichte des Neutrinos', in W. Pauli, *Aufsätze und Vorträge über Physik und Erkenntnistheorie* (ed. by V. F. Weisskopf) Vieweg, Braunschweig (1961), pp. 156–180.

[69] 'Beiträge zur mathematischen Theorie der Dirac'schen Matrizen', in P. Zeeman, *Verhandelingen*, Nijhoff, Den Haag, (1935) pp. 31–43; 'Contributions mathématiques à la théorie des matrices de Dirac', *Ann. Inst. Henri Poincaré* **6**, 109–136 (1936).

[70] 'Some Basic Remarks about the Theory of β-Decay', *Izv. Akad. Nauk SSSR, Otd. Mat. i Est. Nauk*, 1938, 149–152 (In Russian, summary in German).

[71] 'Exclusion Principle, Lorentz Group and Reflection of Space-Time and Charge', in *Niels Bohr and the development of Physics*, Pergamon, London (1955) pp. 30–51.

[72] 'Announcement', in *Proc. CERN Symp. High Energy Accelerators and Pion Phys., Geneva, 1956*, Vol. 2, p. 259.

[73] 'On the conservation of the lepton charge', *Nuovo Cimento*, **6**, 204–215 (1957).

[74] (W. Pauli and B. Touschek) 'Report and Comment on F. Gürsey's "Group Structure of Elementary Particles" ', *Nuovo Cimento* **14**, 205–211 (1959).

[75] 'Der Einfluss archetypischer Vorstellungen auf die Bildung naturwissenschaftlicher Theorien bei Kepler', in *Naturerklärung und Psyche*, Rascher, Zürich (1952), pp. 109–194.

[76] 'Naturwissenschaftliche und erkenntnistheoretische Aspekte der Ideen vom Unbewussten', *Dialectica* **8**, 283–301 (1954). Reprinted in Ref. [68].

[77] 'Sommerfelds Beiträge zur Quantentheorie', *Naturwiss.* **35**, 129–132 (1948). Reprinted in Ref. [68].

[78] 'Die philosophische Bedeutung der Idee der Komplementarität', *Experientia*, **6**, 72–81 (1950). Reprinted in Ref. [68].

[79] 'Wahrscheinlichkeit und Physik', *Dialectica* **8**, 112–124 (1954). Reprinted in Ref. [68].

[80] 'Albert Einstein in der Entwicklung der Physik', *Universitas* **13**, 593–598 (1958). Reprinted in Ref. [68].

[81] *Elektrodynamik*, (ed. by A. Thellung), Verein der Mathematiker und Physiker an der ETH, Zürich (3rd ed., 1958). English translation: *Electrodynamics* (ed. by C. P. Enz), MIT Press, Cambridge, Mass. (1973).

[82] 'Raum, Zeit und Kausalität in der modernen Physik', *Scientia (Milano)* **59**, 65–76 (1936). Reprinted in Ref. [68].

43

Remarks on Enrico Fermi

S. Chandrasekhar

While I appreciate the courtesy of the organizers of this symposium in asking me to talk about Enrico Fermi, I am not altogether certain that I am the most appropriate person: there are others here in the audience who knew him much better and much longer than I did or who were associated with him in some of his more well known contributions. I am probably unique in that I was associated with him in the least known of his work; and that may be an advantage since I can reflect on Fermi's broader interest in physics and not only on those special aspects which have made him well known.

That Fermi was a great physicist requires no elaboration; and this is hardly the occasion to attempt an evaluation of his many contributions to many phases of physics. Let it suffice then to enumerate only those of his contributions that are permanently associated with his name: the Fermi transport in general relativity, the Fermi-Dirac statistics, the Fermi–Thomas atom, the Fermi resonance (in CO_2), the Fermi theory of β-decay, Fermi's work on slow neutrons, Fermi's leadership in the construction of the first self-sustaining nuclear reactor, the Fermi mechanism for the acceleration of charged particles in random magnetic fields, and the discovery of the first of the resonances in the proton–pion scattering. Many of Fermi's friends have recalled one or other of these contributions. I was associated with him in none of these, but in one of his side interests which in 1952 was the role of magnetic fields in astrophysical problems. It was apparently one of Fermi's methods to find someone who might have some knowledge of an area which he wished to learn; and get introduced to the problems of the area by conversations and discussions. And so it came about that in the fall of 1952 and the winter and spring of 1953 I saw Fermi regularly: we met for two hours every Thursday morning and we discussed a variety of astrophysical problems bearing on hydromagnetics and the origin of cosmic radiation. These were problems in which he had not worked before. Nevertheless, I was constantly amazed by the insight with which he used to penetrate to the heart of a problem. In the manner in which he reacted to new problems and new physical situations, he always gave me the impression of a musician who when presented with a new piece of music at once plays it with a perception and a discernment which often surprises even the composer. The fact, of course, was that Fermi was instantly able to bring to bear on any physical problem with which he was confronted his profound and deep feeling for physical laws: the result invariably was that the problem was illuminated and clarified.

J. Mehra (ed.), The Physicist's Conception of Nature, 800–802. All Rights Reserved
Copyright © 1973 by D. Reidel Publishing Company, Dordrecht-Holland

I believe that for most of you assembled here, Fermi is only a name that is attached to many important concepts in physics. For that reason, I should like to narrate a few incidents which might give, to those of you who did not know him, some idea as to the manner of man he was.

Let me start by describing the way he used to discuss problems with me. It often happened that during our meetings I would state some result. Sometimes he would react by saying that the result did not seem correct to him. He would then start off by saying, 'Let me assume that you are wrong; and let me argue on that basis. If I am wrong in my assumption, then at some point it must become clear that I am wrong and that you are right.' The discussion would then proceed and a point would invariably come when it became clear to both of us that Fermi was indeed right. And at that point he would change the subject. He never took pleasure in drawing attention to the fact that *he* was right. I do not know if this is the experience of others who have worked with Fermi; but certainly in my experience I have rarely known of any other scientist who showed that same degree of personal generosity in scientific discussions.

Still, there was one occasion that I got the better of him. I was once asked to talk to a seminar and when I expressed my doubts as to what I should talk about, Fermi advised, 'If I were you, I would not be technical.' And I asked, 'Do you mean if I were you, or you were me?' This baffled him.

Others with greater competence have written about Fermi's fundamental contributions to physics. But his own account of the critical moment when the effect of the slowing down of neutrons on their ability to induce nuclear transformations was discovered is perhaps worth recording.

I described to Fermi Hadamard's thesis regarding the psychology of invention in mathematics, namely, how one must distinguish four different stages: a period of conscious effort, a period of 'incubation' when various combinations are made in the subconscious mind, the moment of 'revelation' when the 'right combination' (made in the subconscious) emerges into the conscious, and finally the stage of further conscious effort. I then asked Fermi if the process of discovery in physics had any similarity. Fermi volunteered and said: 'I will tell you how I came to make the discovery which I suppose is the most important one I have made.' And he continued: 'We were working very hard on the neutron induced radioactivity and the results we were obtaining made no sense. One day, as I came to the laboratory, it occurred to me that I should examine the effect of placing a piece of lead before the incident neutrons. And instead of my usual custom, I took great pains to have the piece of lead precisely machined. I was clearly dissatisfied with something: I tried every "excuse" to postpone putting the piece of lead in its place. When finally, with some reluctance, I was going to put it in its place, I said to myself, "No: I do not want this piece of lead here; what I want is a piece of paraffin." It was just like that: with no advanced warning, no conscious, prior, reasoning. I immediately took some odd piece of paraffin I could put my hands on and placed it where the piece of lead was to have been.'

As you all know, Fermi became ill during the summer of 1954 while he was in Italy. When he returned to Chicago in the fall, all of his friends at the University were shocked to see how ill he looked. At first, the doctors could not diagnose what was wrong with Fermi. It finally became clear that the cause of the illness was either congestion of the oesophagus or cancer of the stomach and intestines. And the matter would be settled by surgery. Dr L. R. Dragstedt, who performed the surgery, told me of his conversation with Fermi the night preceding the operation. Dragstedt told him that if the problem was with the oesophagus the surgery would be complicated and would take a long time; but if it was cancer of the stomach and intestines, then they could probably do very little about it and the operation would be only of short duration. And the next day on returning from surgery, Fermi opened his eyes and noticed that he had not been in surgery for long. He turned to Dragstedt and asked him, 'Has the mitosis set in?' The answer was yes. Then he asked, 'How many more months?' And Dragstedt replied, 'Some six months.' And Fermi went back to sleep.

The following day Herbert Anderson and I went to see Fermi in the hospital. It was of course very difficult to know what to say or how to open a conversation when all of us knew what the surgery had shown. Fermi resolved the gloom by turning to me and saying, 'For a man past fifty, nothing essentially new can happen; and the loss is not as great as one might think. Now you tell me, will I be an elephant next time?'

Among the great physicists it has been my privilege to know, Fermi was unique in that he was interested, genuinely, seriously, and deeply, in all aspects of the physical world around us: all of physics interested him and he drew no lines of demarcation.

Part VI

Celebration of P.A.M. Dirac's 70th Birthday

44

The Banquet of the Symposium

In Honour of Paul Dirac

Abdus Salam: Ladies and gentlemen. As you all know, tonight's banquet is in honour of Professor P. A. M. Dirac's seventieth birthday. The first part of the ceremony tonight consists of a presentation of a commemoration volume which has been edited by Professor Wigner and myself and has been very lovingly produced by the Cambridge University Press. It is called *Aspects of Quantum Theory* and the contributors to the volume, some of whom are here tonight, are R. J. Eden and J. C. Polkinghorne, J. H. Van Vleck, Jagdish Mehra, Res Jost, Abraham Pais, A. S. Wightman, Rudolf Peierls, Werner Heisenberg, J. M. Jauch, C. Lanczos, Laurent Schwartz, Edoardo Amaldi and Nicola Cabibbo, Freeman J. Dyson, Eugene P. Wigner, and J. Strathdee and myself.

Before I request Professor Wigner to make the presentation of this volume to Professor Dirac, I shall read to you the short preface to the book:

'On the 8 August 1972 Paul Adrien Maurice Dirac will be seventy. To celebrate this occasion, some of his pupils and admirers have prepared this volume of essays. Dirac is one of the chief creators of quantum mechanics. By concentrating on just those areas of quantum theory with which he is primarily associated, we have in fact been able to range over almost all its aspects.

'Posterity will rate Dirac as one of the greatest physicists of all time. The present generation values him as one of its great teachers – teaching both through his lucid lectures as well as through his book *The Principles of Quantum Mechanics*. This exhibits a clarity and a spirit similar to those of the *Principia* written by a predecessor of his in the Lucasian Chair in Cambridge. On those privileged to know him, Dirac has left his mark, not only by his observations (which he makes rarely but which are always incisive), but even more by his human greatness. He is modest, affectionate, and sets the highest possible standards of personal and scientific integrity. He is a legend in his own lifetime and rightly so.'

On behalf of all those who have contributed to the volume, and I am sure on behalf of everyone present here tonight, we offer Dirac this volume as a token of our affection and our gratitude.

Eugene Wigner: I would like to say that in addition to our very dear friend Paul Dirac, we should remember our older friends also. I would like to propose first a toast in honour of my erstwhile teacher, at present honoured and dear friend Michel Polanyi and his wife Magda Polanyi. May they have a happy and healthy life. [*Toast.*]

J. Mehra (ed.), The Physicist's Conception of Nature, 805–819. *All Rights Reserved*
Copyright © 1973 *by D. Reidel Publishing Company, Dordrecht-Holland*

Abdus Salam and I were both wondering how we could derive some personal pleasure out of this occasion [Dirac's seventieth birthday], and we soon came to conclude that we should make some gift to Paul and Margit Dirac. Well, we thought for a while as to what should it be? But Salam soon came up with the idea that we should present this volume to Dirac, and the tray on which it was placed to Mrs Dirac. I do that now. May they have a happy life in which we all participate, and may they share the pleasures of life with their family and all of their friends.

Abdus Salam: The second presentation to Professor Dirac is on behalf of the city of Trieste and Consortia which was responsible for this beautiful building in which we have been having the symposium. Paolo Budini will make the presentation.

Paolo Budini: On behalf of the Consortia, in which both the city of Trieste and the region are represented, I have the pleasure to present to you this small token of gratitude for what you have brought to Trieste. This little statue represents Minerva, and it is supposed to be beautiful with beauty, although it is not mathematical beauty.

Abdus Salam: It is now my very great pleasure to hand over the session to our chairman, Professor Casimir.

Hendrik Casimir: Thank you very much. Ladies and gentlemen: I have been announced to be the master of ceremonies here. My notion of a master of ceremonies is of someone with a top hat who produces funny stories. This I am not going to do. It's not that I have entirely exhausted my store of anecdotes and various stories, but somehow I don't feel like it today. It is not that I am in any way unhappy. On the contrary, I feel extremely happy and grateful that it has been possible to hold this beautiful symposium, that it has been possible to meet so many old friends and to make so many new friends, that it has been possible to review many of the old problems and to see the new problems, and especially that it has been possible for us to have this banquet in honour of Dirac, whom all of us value and esteem so highly, whom all of us, I would say, without exaggeration, love so much.

Now it is my duty and my pleasure to introduce our speaker of tonight, Lord Snow. Again, that is what I shall hardly do. For how should one introduce someone whom we all know through his writings? If we are thinking of a writer we value, it's not only that we have read his books with pleasure, but we remember something of his writings. One of the characteristics of writers we admire is that they leave with us some lasting images, images of persons, of character, and perhaps a few sentences here and there which stay with us forever. But, by way of introduction, to start with an anthology of those passages and those images which I particularly cherish, or which for some reason have remained with me as valuable things, that would be just a personal presumption and not very useful to other people. I shall therefore cut short my introduction, and I shall ask Lord Snow now to be so kind as to address our gathering. Thank you.

P. A. M. Dirac receiving *Festschrift* from E. P. Wigner.
[Seated (left) Prince Torre e Tasso, (right) Lord Snow.]

H. B. G. Casimir presiding at Dirac's 70th birthday celebration and banquet.
[Seated Mrs Dirac and Abdus Salam.]

P. A. M. Dirac receiving the statuette of Minerva from P. Budini.
'Beautiful, but not mathematical beauty.'

P. A. M. Dirac receiving the
Freedom of the City from the
Mayor of Trieste.

The Classical Mind

C. P. Snow

This is a momentous occasion, but also something like a family birthday party, and I don't know which part moves me most. We are all here from many parts of the world to express our affection, our veneration, and above all our gratitude to a great man.

Looking around this enormous concourse of the conceptual intelligence, perhaps the most remarkable concourse of the conceptual intelligence that I have ever met, or ever been part of, it struck me as somewhat odd to think what would happen if a bomb went off in this room. Imagine that someone from Northern Ireland, under the mistaken impression that Paul Dirac was either a Protestant or a Catholic, it wouldn't matter much which, decided that his life should be ended. Well, the results would be of some consequence. I think the London *Times* would say, 'Setback for Theoretical Physics', a moderately understated comment in the best English manner. Since physics is a collective enterprise, of course, it wouldn't stop physics. In the rest of the world of physics, the young men would gradually get back to the position which the subject has now reached. And yet, I think, the historians would say that it did at least mark the end of a great period of physics. I, sitting between Heisenberg and Dirac, feel this extremely, and we are privileged to see all these great men here.

This subject, in its full rigour, started with very great papers nearly 50 years ago. It has not yet ended, but historians will have to pronounce on that in the future, and if life deals with us kindly, then we may see something happen again. But you are privileged, you, the partakers in this collective enterprise, are enormously privileged, for you have been taking part in what without any doubt is the greatest intellectual achievement of the mind of man. And we, who just had to watch from the outside, with a sort of vicarious delight in what was happening, still do feel, in some remote way at least that we can sit back on the sidelines and cheer.

It is wonderful that this concourse should have happened in Trieste. To anyone of a literary turn of mind, Trieste has many associations. Just down the bay is Duino, where Rainer Maria Rilke wrote his elegies. In Trieste itself, James Joyce, so they say, finished the novel. I hope that isn't true. I have a certain tenderness for that particular form of art, but it isn't a stupid statement. In the late nineteenth century, Richard Burton, one of the most remarkable of all nineteenth-century Englishmen, writer, linguist, traveller, lived out the last disappointed years of his life in Trieste. Ettore Schmitz, the son of an Austrian Jewish father and Italian mother, who became a passionate Italian nationalist, writing under the pen name of Italo Svevo, started a whole stream of novels, not quite great novels, but of a remarkably interesting line in modern literature, lived here. All that has happened in Trieste.

And we have to say then, that we are privileged to be associated with Abdus Salam's adventure in international theoretical physics which is centred here in Trieste. We know both his concern for physics and his concern for the poor of this world, what we call the third world, the developing world. I wouldn't like this evening to pass, not at least knowingly, that our consciousness of that world is there, is here and now. If we forget it, our children and grandchildren will curse us; even at best this provides a set of problems to which so far we see no answer, and which they will have to struggle through for the next fifty, sixty, or a hundred years.

And again, I would like to make one other personal tribute to my friend Jagdish Mehra. Without his initiative and imagination, his historical insight and sense, this unique symposium and concourse might not have happened at all. And historical insight is a great gift.

Now a number of you know me, and I think you know that I am not very much of a chauvinist. Nevertheless I am a little sad that this gathering isn't happening in England. After all, we have something to do with it, you know. It's sad that Dirac's seventieth birthday shouldn't have been marked in England as one of the great events of this year. Don't think that the English don't realize a great man when they see him. But they are not very good at doing anything about it. In fact, it is pleasure on one side of my face, a slightly sour grin on the other, that this hasn't happened in our own Cambridge of which he is the greatest living star.

I made a remark not long ago which I will repeat, because it was a considered remark and wasn't said lightly. I said then that 'Dirac is the greatest living Englishman.' This was carefully thought out and I meant it in a perfectly definite and explicable way. I was thinking what the world a hundred years hence will think of the England of my time. I would guess that politics, our politics, English politics, will not count very much, will not interest very much observers a hundred years hence. I suspect that is true, but in politics, it is very hard to make this kind of prediction. I suspect that even the last war would seem a minor episode in a much more important human development. I wouldn't put much money on anyone a hundred years hence to be very interested in the art of England of any kind. Perhaps Yeats will still seem a very great poet. Yeats was an Irishman and we can't lay any real claim to him. But I have no doubt that in a hundred years, in this edifice of science, one great name will be there, and that is Paul Dirac. Possibly another, and that is Francis Crick.

Remember, scientific history is rather odd. These people contributed great slabs of concrete to this enormous edifice of cumulative science. Don't think that anyone will read their papers. I should think that unlikely. But the efforts of Dirac, and probably the efforts of Crick, will also get into the bloodstream of the science of a hundred years hence. How many of you have ever read more than ten pages of the *Principia* by Paul Dirac's illustrious predecessor in the Lucasian Chair and the only Englishman who can really be mentioned in the same breath? Newton, Dirac. Darwin, Crick.

I don't think these analogies are too fanciful. The minds of Newton and Dirac seem to me to have certain resemblances. I don't think the minds of Darwin and Crick have the same resemblances, and I don't suggest to you for one moment that the temperaments of these four have anything at all in common. Newton was a singularly suspicious, disagreeable and ungenerous man. Of Paul Dirac we should all say the opposite. Darwin was a modest, retiring, apprehensive hypochondriac. I don't think Francis Crick, who is a friend of mine, would think any of those particular words applied much to him.

When I think of Newton and Dirac, I think of a sort of exercise of mine about the classical mind. A classical mind isn't the only kind of mind, but it's an exceptionally valuable one. It has certain characteristics. Well, Newton had it and he exercised it in his science, but not elsewhere. Paul Dirac has exercised it in all of his human activities.

The classical mind, it seems to me, has some of these qualities – lucidity, austerity, that is a dislike for unnecessary frills, indeed frills of any kind, but mixed with this austerity an acute aesthetic sense. And that's where people often go wrong when they are thinking about classical minds. All those I have known best, mixed with their austerity this absolutely strong and prevailing aesthetic sense. It is innate in Dirac. It is innate in all real classical minds. And another, the classical mind is candid; it's unwilling to tolerate unnecessary mystifications. Though Wittgenstein hadn't what I would regard as a classical mind except as a veneer, I think he was a passionate romantic, nevertheless, his statement 'Of what you can't speak, you must keep silent' would apply to anyone who is really gifted in this kind of sense. Finally, rationality. Rationality which doesn't despise the unreason of man, the irrational parts of our impulses. It knows what they are. But it also knows that the rational mind is the only way to explore them. Therefore, I suggest to you that the classical mind, at its best, is the supreme expression of the rational soul of man.

When I thought about this in connection with Paul Dirac, I was reminded of a very dear friend of mine, to whom I was very close, and that was the mathematician G. H. Hardy. Now mathematicians' fame doesn't usually last as long as physicists' fame, partly because they are not concerned with the natural world. Many of you perhaps will only have a school textbook knowledge of Hardy at most. Perhaps he will seem not more than a name to many of you. Nevertheless, he is someone to whom I was close, and I think he had more intellectual influence on me than anyone that I have ever known. He wouldn't have tolerated for one moment the idea that his eminence in his own field reached a peak which Paul reached in his. Let me be quite clear about that. Hardy had all the restraints of the classical mind, all its desire for a simple truth. What he used to say was that at his very best he was perhaps the fifth best mathematical artist in the world. He wouldn't have said it to me like that, because we used to talk in a curious slang derived from the game of cricket which would be incomprehensible to about ninety-nine per cent of the people here as to what it meant. And I've often thought of these two men, two of the people whom I have admired most in my entire life, and thought how very much they resembled each other. They both had the same aesthetic sense which I've referred to as one of the criteria of the classical mind. Dirac has said that, 'If you get a really beautiful mathematical equation, it must have some relation to reality. It is much more likely to be true than experiment is likely to be true,' or words to this effect. Hardy would have said exactly the same thing. That, in fact, if a mathematical theorem is beautiful then it is important. I once taxed him and said that if one would present a solution which finally proved Goldbach's theorem, and if it was ugly, would he accept it? Hardy replied, 'This is impossible. It couldn't possibly be a proof of Goldbach's theorem if it were ugly.' Hardy was usually more of a conversationalist than Dirac, and he then made this extremely terse remark characteristic of Dirac himself: 'This is impossible.'

There is another very close resemblance, I think, between Hardy and Dirac. They are both capable of this kind of anticlimactic utterance which lingers in our minds. Hardy, unlike Dirac, was something of a conversational athlete. Among his friends he delighted in the game of conversation as he delighted in all games. So he was prepared to make statements like saying, 'The excellence of a religion is inversely proportional to its number of gods.' This was usually directed at his Anglican or Catholic friends in Trinity – Catholics having four gods, churchmen having three gods, monotheists having only one god, and that was better, and you can take this obviously through to the end. That was a typical Hardy conversational gambit. Dirac is too sensitive to the feelings of others to indulge himself like that. But nevertheless, a good many of his reported remarks are not so dissimilar in tone.

Both Hardy and Dirac have another quality, which I suggested was characteristic of the classical mind. That is ultimate candour. I don't know how many of you have read Hardy's *A Mathematician's Apology* which is one of the most beautiful statements about the creative mind ever written or ever likely to be written. Written with absolute, ruthless candour. Again, Dirac hasn't made many personal statements in his life, but he has made some, and I have read a printed version of a lecture, given in Miami, which was called rather typically *The Development of Quantum Theory*, in which Dirac just for once said what parts of his life had been like, and it was said with exactly the same ruthless candour that Hardy employed. One of Dirac's main points, I remember, was that in the creative life, you have great hopes and great fears, and almost anybody tends to be overwhelmed by great fears just at the point that he ought to take the next step forward. He begins, I think, by talking about Lorentz having everything in his hands for the special theory of relativity, and not taking this one decisive step forward. Then he talks about himself, having gotten the great equation, he was satisfied with that. It seemed to fit the spins and everything else and then he did the calculations to see how it fitted the hydrogen spectrum. He took the relativistic correction to the first order and it worked, and then he says, 'I simply couldn't go further because I was scared.' Only a man of both extraordinary magnanimity and extraordinary character could have said that. I have always thought that such candour brings these two great men together.

I don't want to press the analogy too far, but I would like just to say one other thing. You will find a slight mention of Dirac in Hardy's *A Mathematician's Apology*.

In private, he said that he had known many theoretical physicists, and he had admired them beyond measure. He had a veneration for Einstein such as we all had; he thought him probably the greatest human being he'd ever met. But he said that, of all of them, Dirac was the only one, in Hardy's view, who could have been a really good pure mathematician. This had a slight edge, as Hardy in his secret heart really felt that anyone ought to do pure mathematics if he had the talent for it, and anyone who couldn't do it was really just a shade outside the sacred pale.

I've been self-indulgent here. I've been referring to a great friend of 25 years ago to try to talk about something dear to myself. But it hasn't been entirely self-indulgent. There is a certain rarefied world in which only great spirits can esteem each other properly. I can't presume to belong to that rarefied world, but I've called in aid a friend, not so great as Dirac, but a great talent, who can perhaps speak with authority.

Anyway, here we are. We've been lucky. We've seen a contribution to human thought in our own lifetime which is certainly the greatest edifice which has yet been built. There is no doubt about that. It is good for us to have had that experience. On the whole we have had a rough life. Anyone who has lived in our time has seen things which make him feel ashamed of the human species. It is not good for man to feel ashamed of the human species permanently. We want to cling on to anything which can make us feel proud. We can feel proud of this collective enterprise. We can feel proud of Paul Dirac. It is good to have lived in his time.

*   *   *

Hendrik Casimir: Thank you, Lord Snow. I would like to thank you on behalf of all of the audience for having expressed so eloquently what many of us have felt.

I should now like to call on the man who has been responsible for this symposium, and that is Jagdish Mehra. It has been said already how much he has done to bring about this meeting, but let's also say that he has been a hard taskmaster. He is a man who under the cloak of being very gentle and very kind is able to let people work very hard for him. And this is a great talent of leadership of being able to do it, and that we are here in this gathering and that it has been such a wonderful symposium, and such a wonderful experience, is I think in no small measure due to this special talent of persuasion, of being able to let people work for him in such a way that they enjoy doing it. Before giving the microphone to Jagdish Mehra, I should like to drink to his health. Jagdish, we drink to your health, happiness, and success. [*Toast.*]

And now we will see what he has to say for himself. But be careful, probably he will engage us to start working already on the next conference.

Jagdish Mehra: Thank you, Professor Casimir. Ladies and gentlemen. The occasion tonight and the gathering assisting it, are a name dropper's paradise. An Italian prince, an English peer of the realm, a British knight, numerous Nobel laureates, and intellectual eminence in such quantity as rarely comes together, are here this evening. But the true aristocracy is indeed of the mind, of the heart, and of character, and it is a prince of this aristocracy whose presence among us this evening honours all of us. May I on behalf of the sponsoring committee, lecturers and participants of the Symposium on the Development of the Physicist's Conception of Nature, a symposium which, I'm sure, will in future be referred to as the *Dirac Symposium*, extend to Paul Dirac a happy birthday and many returns thereof. Happy birthday, Paul, and all of our good wishes to you.

Hendrik Casimir: Thank you very much, Jagdish. I have now performed my share, my very modest share, of this evening's proceedings, and I shall give the microphone back to Salam.

Abdus Salam: I have only two minor tasks to perform. One is to read a cable from Professor Vladimir Fock, who sends to all the participants here his greetings. He wants us to transmit to Professor Dirac his heartiest congratulations on his seventieth birthday. The second and final task of this evening is for me most pleasant. I would like you to rise and drink to the health of the co-chairmen of the symposium, Professors Casimir and Wigner, the members of the sponsoring committee, and also to drink to Professor Dirac, which we have not done so far. [*Toast.*]
   This concludes the formal part of the evening.

Lady Peierls: The organizers of this evening's celebration have not invited any woman to give a speech. In the name of Women's Lib, I shall give a speech, although I am giving it with the best wishes of the master of ceremonies. I am delighted that so many of us made it, so many of us are still there to celebrate Professor Dirac's seventieth birthday. I hope we will celebrate other seventieth birthdays that are still to come. Tonight I also want to remember so many of those who didn't make it. I want to remember Enrico Fermi, Wolfgang Pauli, Johnny von Neumann, George Placzek, Lev Landau, George Gamov, Robert Oppenheimer, who were with us for so many years and didn't make it. In their memory, together with best wishes for all those who are still there, the lucky ones, I propose a toast. [*Toast.*]

Abdus Salam: Professor Lanczos, would you like to say a few words?

Cornelius Lanczos: Ladies and gentlemen. This honour comes to me quite unexpectedly. I am asked to give some kind of a toast. If I had known it in advance, I would have had a chance to prepare something. I have not prepared anything and therefore what I am going to say is really sincere and not something which I made up. I recently happened to be in Frankfurt to receive an honorary degree and I got a wonderful diploma in Latin, all in superlatives. My friend, who formulated this long wonderful ceremony which pointed out what a marvellous man I am, is a great Ciceronian master of the language. He told me that years ago they discussed the question whether the diploma should be in German or in Latin and then the classicists pointed out that in no language can you lie so beautifully as in Latin, and therefore they kept Latin as the language for expressing appreciation.

Abdus Salam: You are in a Latin country here, you know.

Cornelius Lanczos: In a Latin country, yes, of course. That's right. I think Italian would be a match probably for the old Latin. But English is not so bad either if you have the chance to prepare yourself.

Now, I have nothing particularly flattering to say to Professor Dirac. I am a great admirer of Dirac, although our circles are to some extent mutually exclusive, in as much as we think in different terms. But why shouldn't one think in different terms? There is the famous unified field theory on which Einstein worked so long and, in spite of my great admiration and conviction that there is such a unified field theory, it reminded me somewhat of this lovely story by the Armenian writer who lives in the United States. In the story, a man in an insane asylum plays a cello, and he plays always the same tune, and they ask him, 'Why do you play the same tune all day long?' And he says, 'Look at all these people, they have their fingers on the board going up and down all the time looking for the right tune. I have found it.' So I can only say that perhaps we are all crazy in looking for this tune which gives us the final truth, but I think the very effort of trying to find this basic tune on which everything is built up is a tremendous experience. I have the feeling that even if our circles are somewhat exclusive, but still this basic approach to nature, that the fundamental category from which we can understand everything, is beauty, is something which delights Dirac and myself.

And so I wish Dirac all the good things. I am nine years older than he is and I hope he will live, as in Jewish they say up to 'hundertundzwanzig Jahre', that's up to 120. The Jews always thought that God is envious, so in order to reach 100 one should say 120. I wish him all the good luck with 'hundertundzwanzig Jahre'.

Paul Dirac: I would just like to say that this has been a very happy occasion for me. I have met so many old friends and every one of them has been a delight to meet again. I would like to thank you all for coming on this occasion. And I would also like to thank the various people who have spoken about me. I feel that they have rather overwhelmed me with their nice remarks. They haven't talked at all about my failings, my forgetfulness, my absentmindedness.

Perhaps I might just mention a little story which illustrates to what extent absent-mindedness can lead one. This concerns Hilbert, the mathematician whom all quantum physicists know about, and he was perhaps the most absentminded man who ever lived. He was a great friend of the physicist James Franck. One day when Hilbert was walking in the street he met James Franck and he said, 'James, is your wife as mean as mine?' Well, Franck was rather taken aback by this statement and didn't quite know what to say, and he said, 'Well, what has your wife done?' And Hilbert said, 'It was only this morning that I discovered quite by accident that my wife does not give me an egg for breakfast. Heaven knows how long that has been going on for.'

Before I forget to do so, let me thank you all very much.

Abdus Salam: Professor Heisenberg, would you say a few words?

Paul Dirac: Yes, you started me off, after all.

Werner Heisenberg: When I am asked to give a speech at such a sudden moment, I cannot do anything but just tell stories. I should perhaps tell a story about our common trip from the United States to Japan. That reminds me, I should perhaps relate the two incidents in this story. One was that we had agreed to meet at the hotel at the Old Faithful geyser at Yellowstone, and I had written to Paul that it would be nice if we could look at some of the geysers going off just at the moment that they go off. When we met, I found that he had made out a very careful timetable of all the geysers which one could possibly reach at the time of their going off, and he also wrote out the distance from one geyser to the next. Thus we could walk over in such a way that by a very carefully selected trip we could actually see quite a number of them going off.

Then you know that Paul always thinks about his formulations very carefully. He does not like to answer spontaneously at once, he first thinks about things. We were on the steamer from America to Japan, and I liked to take part in the social life on the steamer and so, for instance, I took part in the dances in the evening. Paul, somehow, didn't like that too much but he would sit in a chair and look at the dances. Once I came back from a dance and took the chair beside him and he asked me, 'Heisenberg, why do you dance?' I said 'Well, when there are nice girls it is a pleasure to dance.' He thought for quite a long time about it, and after about five minutes he said, 'Heisenberg, how do you know *beforehand* that the girls are nice?'

I think this story characterizes at least one part of the mind of Dirac. About the other parts, which are certainly more important, others have said better things than I can do just now. I just want to repeat my congratulations to Paul.

Rudolf Peierls: I find it hard to add anything new to the feelings that have been expressed so well in so many ways, and I share all of them. I will also add a story about Dirac.

I can think perhaps of one that Paul no doubt will remember. A friend of ours, Hulme, was walking many years ago with Paul in Cambridge, with something rattling in his pocket, and he said apologetically, 'I am sorry about this noise. I have a bottle of aspirin in my pocket, and I took some as I had a cold, so it is normal for it to make some noise.' There was silence for a while and then, so our friend Hulme told us, Paul said, 'I suppose it makes a maximum noise when it's just half full.'

For long I tried to find out the date of this story because I had hoped that it might have preceded Dirac's hole theory and perhaps this might have been the source of it. Unfortunately, the remark was made some years after the hole theory had been developed. Well, his way of thinking seems to us different from our own, and his thoughts appear to us unexpected, but when we've heard him we find that this was obviously a right way to look at the problem. That is what has impressed us so much in his physics. I would like to add my wishes to everybody else's.

Behram Kursunoglu: I would like to relate two stories which I am sure you have never heard before.

Paul Dirac: Have you just invented them?

Behram Kursunoglu: Well, let's test it. I was walking with Professor Dirac around the lake and I counted fourteen birds on the lake. Professor Dirac said, 'Fifteen, because I saw one going under the water.' On another occasion, I had a call from Professor Wigner early one morning, about ten o'clock on Sunday. With his usual apologies Professor Wigner said, 'I shouldn't have called you really.' I said, 'Well, why did you then?' He said 'I couldn't help it. I wonder if you know Professor Dirac's telephone number?' I said 'I shall give you that cheerfully. By the way, Professor Dirac was concerned that when you come to Miami somebody should meet you at the airport, in case you get lost.' He said, 'Doesn't Dirac realize that I found my way all the way from Hungary to the United States and I couldn't get lost in Miami?' I said, 'I will relate this to Professor Dirac.' And I did, and Professor Dirac said, 'Doesn't Wigner realize that he had a lot of time then and he hasn't got that much time now?'

Hendrik Casimir: Since Dirac complained in a way that no reference had been made to any less desirable characteristics of his, I might perhaps say, that with his great precision of expression, although with great kindness, on occasion Dirac can be very severe. Oh, not rude, of course, but very severe, and very strict. I should like to relate one episode which bears it out. At one of the Copenhagen conferences there was an evening at Bohr's house, and music was being played, I believe, by Frisch, Weisskopf, and Kopfermann. Landau was there. Now Landau had many great gifts, but the love of music was not one of them. So he sat there, and since he was quite obviously bored, he started making grimaces and pulling faces, and in general making a nuisance of himself. I was also there and had been just newly married. During a pause in the music Dirac went to Landau and said kindly, 'Landau, when you don't like the music, why don't you go out of the room?' Landau answered at once in his usual aggressive way, 'Well, I wanted to go out of the room, but Mrs Casimir isn't interested in music either and I asked her to come out of the room with me and she didn't. Why didn't she go out of the room with me? Then I would have gone out of the room.' To this Dirac answered very gently and quietly, 'Obviously she preferred listening to the music to going out of the room with you.' It must have been one of the few occasions on which even Landau had no answer.

Abdus Salam: Jagdish, would you tell your Dirac story?

Jagdish Mehra: My first meeting with Paul Dirac took place in Cambridge in 1955. I had just returned to England after a couple of years with Heisenberg in Göttingen. A historian friend of mine in Cambridge, knowing of my great hero worship for Dirac, offered to take me with him to St John's College, which was also his college, and to dine at the High Table. He thought we might see Dirac there. I went with him, and true to his word, he showed me that Professor Dirac was sitting there. We sat down. The weather outside was very bad, and since in England it is always quite respectable to start a conversation with the weather, I said to Dirac, 'It is very windy, Professor.' He said nothing at all, and a few seconds later he got up and left. I was mortified, as I thought that I had somehow offended him. He went to the door, opened it, looked out, came back, sat down, and said 'Yes.'

In April 1970, when I reminded him of this, he reflected and said 'Mehra, I wonder why I did that, because I must have already known that the weather was windy when you said it, unless it had changed since I went inside.'

Abdus Salam: Professor Jauch, you have a story, I am told.

Josef Jauch: I had written a little piece in which I wanted to try to popularize some of the problems in the interpretation of quantum mechanics. It happened to be completed just at the moment when Professor Dirac and I were in Tallahassee together, and I thought that this was a wonderful opportunity to present my essay to Professor Dirac and benefit by his comments. So I left this piece with him, and a week later he brought it back to me and said, 'Thank you.' I was a little surprised that there was no more comment to it than just that. I prodded him a little bit as to what his opinion was about this piece, and he said, 'I don't like the title of it.' I asked, 'Why not?' The title, incidentally was, 'Are Quanta Real?' And then he said, 'Well, it's just like asking, is God real?' And I said, 'This is a very interesting remark, because this is just about what I wanted to communicate in this dialogue.' And then Dirac said, 'Why did it take you so many pages to say it?'

Nicholas Kemmer: Professor Casimir mentioned an occasion in Niels Bohr's house. In the early nineteen-fifties I was present at a similar occasion. We were in the great hall with the lovely picture collection and a certain lady in the company was pointing out one picture. If I were to describe it entirely devoid of artistic terms, it consisted of a lot of white ovals, with little red dots near them, which were meant to represent chickens, the bodies and the combs. And this lady was particularly impressed with this painting, which was indeed a very attractive one. Dirac was standing next to her, and she said 'Professor Dirac, what do you think of this painting?' Dirac said, 'How many chickens do you think there are on this painting?' I was the next man, and well, this seemed to be a challenge, and so I said 'Oh, I think ten.' And Paul Dirac said, 'I think there are only eight.' This was again a challenge to count and so I counted beyond the eight and pointed to another one in the corner, number nine, which however was just a white oval, without the red spot on top denoting the comb, and I said triumphantly 'Nine.' Dirac said, 'That one might be a pigeon.'

Carl Friedrich von Weizsäcker: I remember a story from the very old days in Copenhagen, in 1932 or even earlier, when Pauli was drinking tea with some others. Pauli took more sugar than others felt was really good for him, and there was a discussion about it. I am told that Dirac was called to judge the matter. After some consideration he said, 'I think that one piece of sugar is enough for Pauli.' Everybody was satisfied, and the conversation went on. After a while when they had been talking about other matters, Dirac said, 'I think one piece of sugar is enough for anybody.' Then the problem, evidently, seemed to be solved. Again, after a while, I am told, Dirac said: 'I think that cubes of sugar are made in such a way that one is enough.'

Eugene Wigner: I will tell a story in which Paul was wrong, as it is evidenced this evening. Once Polanyi, Dirac, and I had a luncheon together and we discussed questions of science, society, and all those things. All the time Dirac did not say a word. So when we went out I told Paul, 'Paul, why don't you speak up? Everybody is interested in his opinion.' And then he said something which the discussion here this evening shows was incorrect. 'There are always more people,' he said, 'willing to speak, than willing to listen.'

Appendix 1

Programme of the Symposium

Miramare, Trieste, Italy
18–25 September 1972

Monday, 18 September

MORNING:

10.00–10.30   Opening session

Structure of the universe   Chairman: G. Randers

10.30–11.30   'The universe as a whole' – Speaker: D. Sciama

11.45–12.45   'A chapter in the astrophysicist's view of the universe'
              – Speaker: S. Chandrasekhar

AFTERNOON:

Space, time and geometry   Chairman: C. Møller

  3.00–4.00   'The nature and structure of space-time' – Speaker: J. Ehlers

  4.00–5.00   Comments by G. Burbidge and discussion

EVENING LECTURE: Chairman: J. Schwinger
              'Development of the physicist's conception of nature'
              – Speaker: P. A. M. Dirac

Tuesday, 19 September

MORNING:

Space, time and geometry   Chairman: S. Chandrasekhar

  9.30–10.30   'Theory of gravitation' – Speaker: A. Trautman

10.45–11.45   'From relativity to mutability' – Speaker: J. Wheeler

11.45–12.45   Comments by C. Møller and W. Thirring and discussion

AFTERNOON:

Quantum theory   Chairman: B. L. van der Waerden

  3.00–4.00   'The wave-particle dilemma' – Speaker: L. Rosenfeld

  4.00–5.00   'From matrix mechanics and wave mechanics to unified quantum
              mechanics' – Speaker: B. L. van der Waerden

EVENING LECTURE: Chairman: E. P. Wigner
              'Some concepts in elementary particle physics' – Speaker: C. N. Yang

Wednesday, 20 September

MORNING:

Quantum theory   Chairman: Jagdish Mehra
9.00–10.00   'Development of concepts in the history of quantum theory'
             – Speaker: W. Heisenberg
10.00–11.00 'Fundamental constants and their development in time'
             – Speaker: P. A. M. Dirac
11.15–12.15 'Relativistic equations in quantum mechanics' – Speaker: E. P. Wigner
12.15–1.00   Discussion

AFTERNOON: Chairman: W. Heisenberg
3.00–4.00   'The electron: development of the first elementary particle theory'
             – Speaker: F. Rohrlich
4.00–5.00   'The development of quantum field theory' – Speaker: R. E. Peierls

EVENING LECTURE: Chairman: J. Wheeler
             'Superconductivity and superfluidity' – Speaker: H. B. G. Casimir

Thursday, 21 September

MORNING AND AFTERNOON: Sightseeing in the region

EVENING: The banquet of the symposium – in honour of P. A. M. Dirac
           Banquet speaker: Lord Snow, 'The classical mind'
           Host: Abdus Salam
           Master of Ceremonies: H. B. G. Casimir

Friday, 22 September

MORNING:

Quantum theory   Chairman: G. Wentzel
9.00–10.00   'Mathematical structure of elementary quantum mechanics'
             – Speaker: J. M. Jauch
10.00–11.00 'A report on quantum electrodynamics' – Speaker: J. Schwinger
11.15–12.15 'Progress in renormalization theory since 1949'
             – Speaker: Abdus Salam

AFTERNOON: Chairman: C. N. Yang
3.00–4.00   'Crucial experiments on discrete symmetries'
             – Speaker: V. L. Telegdi
4.00–5.00   'W. Pauli's scientific work' – Speaker: C. P. Enz

EVENING LECTURE: Chairman: R. E. Peierls
             'Approach to thermodynamic equilibrium' – Speaker: W. E. Lamb, Jr.

Saturday, 23 September

MORNING: Chairman: G. E. Uhlenbeck
Statistical description of nature
9.00–10.00   'Problems of statistical physics' – Speaker: G. E. Uhlenbeck
10.00–11.00 'Phase transitions' – Speaker: M. Kac
11.15–12.00 'Kinetic approach to non-equilibrium phenomena'
      – Speaker: E. G. D. Cohen

AFTERNOON: Chairman: M. Kac
Statistical description of nature
2.00–3.00   'Time, irreversibility and structure' – Speaker: I. Prigogine
3.00–4.15   'The origin of biological information' – Speaker: M. Eigen

EVENING: Reception for the sponsoring committee and lecturers at the Duino
Castle – Host: The Prince of Torre e Tasso

Sunday, 24 September

MORNING AND AFTERNOON: Excursion to Venice

EVENING: Concert: Trio di Trieste

Monday, 25 September

MORNING: Chairman: E. P. Wigner
Special lecture:
9.00–10.00   'The expanding earth' – Speaker: P. Jordan
Physical description and epistemology
10.00–11.00 'Classical and quantum descriptions' – Speaker: C. F. von Weizsäcker
11.15–12.30 Subject, object and measurement; completeness of description.
     Participants: J. S. Bell, L. N. Cooper, B. d'Espagnat, R. Haag,
     J. M. Jauch, G. Ludwig, I. Prigogine, C. F. von Weizsäcker

AFTERNOON: Chairman: C. F. von Weizsäcker
2.30–3.00   'Objectivity and measurement in quantum mechanics'
      – Speaker: R. Haag
3.00–4.30   Discussion (continued) on physical description, epistemology and
philosophy
     Participants: J. S. Bell, B. d'Espagnat, D. Finkelstein, J. M. Jauch,
     G. Ludwig, C. F. von Weizsäcker, E. P. Wigner

EVENING LECTURE: Chairman: Jagdish Mehra
     'Physics and philosophy' – Speaker: C. F. von Weizsäcker

Appendix 2

Participants

ABRAMSKY, J., Department of Mathematics, The University of Southampton, Southampton, England.

D'AGATA, S., Istituto di Fisica 'Guglielmo Marconi', Università degli Studi di Roma, Rome, Italy.

AGGARWAL, J. K., Department of Electrical Engineering, University of Texas, Austin, Texas, U.S.A.

AUGER, P., 12 rue Emile Faguet, Paris, France.

BARUT, A. O., Department of Physics and Astrophysics, University of Colorado, Boulder, Colorado, U.S.A.

BELL, J. S., Theory Division, C.E.R.N., Geneva, Switzerland.

BERGMANN, P. G., Department of Physics, Syracuse University, Syracuse, New York, New York, U.S.A.

BHATIA, A. B., Department of Physics, University of Alberta, Edmonton, Alberta, Canada.

BIERMANN, L., Max-Planck-Institute for Physics and Astrophysics, Munich, Fed. Rep. Germany.

BLEULER, K., Institut für Theoretische Physik, Universität Bonn, Bonn, Fed. Rep. Germany.

BOPP, F., Institut für Theoretische Physik, Universität München, Munich, Fed. Rep. Germany.

BROCKHOUSE, B., Department of Physics, McMaster University, Hamilton, Ontario, Canada.

BRUNGS, R. A., Saint-Louis University, St Louis, Missouri, U.S.A.

BUB, J., Department of Philosophy, The University of Western Ontario, London, Ontario, Canada.

BUDINI, P., International Centre for Theoretical Physics, Trieste, Italy.

BURBIDGE, G., Royal Greenwich Observatory, Hailsham, Sussex, England.

CAPRI, A. Z., Department of Physics, The University of Alberta, Edmonton, Canada.

CASIMIR, H. B. G., Philips Research Laboratories, Eindhoven, The Netherlands.

CAVALIERE, A., Laboratori Gas Ionizzati, Via Enrico Fermi, Frascati, Rome, Italy.

CHANDRASEKHAR, S., Laboratory for Astrophysics and Space Research, University of Chicago, Chicago, Illinois, U.S.A.

CHARAP, J. M., Department of Physics, Queen Mary College, University of London, London, England.

CHOQUARD, P., Laboratoire de Physique Théorique, Ecole Polytechnique Fédérale de Lausanne, Lausanne, Switzerland.

CHOW, Y., Department of Physics, University of Wisconsin–Milwaukee, Milwaukee, Wisconsin, U.S.A.

COHEN, E. G. D., The Rockefeller University, New York, New York, U.S.A.

COISH, H. R., Department of Physics, The University of Manitoba, Winnipeg, Manitoba, Canada.

COLLERAINE, A., Physics Department, Florida State University, Tallahassee, Florida, U.S.A.

CONIGLIO, A., Istituto di Fisica Teorica, Università di Napoli, Naples, Italy.

COOPER, L. N., Physics Department, Brown University, Providence, Rhode Island, U.S.A.

CORBEN, H. C., Scarborough College, University of Toronto, Toronto, Ontario, Canada.

CVIJANOVICH, G. B., Upsala College, East Orange, New Jersey, U.S.A.

DAVIES, P., Institute for Theoretical Astronomy, University of Cambridge, Cambridge, England.

DEMOPOULOS, W., Department of Philosophy, University of Western Ontario, London, Ontario, Canada.

DESER, S., Physics Department, Brandeis University, Waltham, Massachusetts, U.S.A.

DEVANATHAN, V., Department of Nuclear Physics, University of Madras, Madras, India.

DIAMESSIS, E. S., Electronics Department, N.R.C. 'Democritos', Athens, Greece.

DIRAC, P. A. M., Department of Physics, Florida State University, Tallahassee, Florida, U.S.A.

DOWKER, J. S., Department of Theoretical Physics, The Schuster Laboratory, The University of Manchester, Manchester, England.

DRIESCHNER, M., Max-Planck-Institut, Starnberg, Fed. Rep. Germany.

DÜRR, H. P., Max-Planck-Institut für Physik and Astrophysik, Munich, Fed. Rep. Germany.

EHLERS, J., Max-Planck-Institut für Physik and Astrophysik, Munich, Fed. Rep. Germany.

EIGEN, M., Max-Planck-Institut für Biophysikalische Chemie, Göttingen–Nikolausberg, Fed. Rep. Germany.

ENZ, C. P., Institut de Physique Théorique, Université de Genève, Geneva, Switzerland.

ESCH, R. J., Department of Physics, The University of British Columbia, Vancouver, Canada.

D'ESPAGNAT, B., Laboratoire de Physique Théorique et Particules Elémentaires, Université Paris-Sud, Orsay, France.

FINKELSTEIN, D., Department of Physics, Belfer Graduate School of Science, Yeshiva University, New York, New York, U.S.A.

FONDA, L., Institute of Physics, University of Trieste, Trieste, Italy.

FORD, K. W., Physics Department, University of Massachusetts, Boston, Massachusetts, U.S.A.

FRENKEL, A., Central Research Institute for Physics, Hungarian Academy of Sciences, Budapest, Hungary.

FRONSDAL, C., University of California, Physics Department, Los Angeles, California, U.S.A.

FURLAN, G., Institute of Physics, University of Trieste, Trieste, Italy.

GAGO, C., Instituto Venezolano de Investigaciones Científicas, Departamento de Física Atómica y Molecular, Caracas, Venezuela.

GALZENATI, E., Istituto di Fisica Teorica, Università di Napoli, Naples, Italy.

GAMBHIR, Y. K., International Centre for Theoretical Physics, Trieste, Italy.

GATES, D. J., Mathematics Department, The Rockefeller University, New York, New York, U.S.A.

GÉHÉNIAU, J., Faculté des Sciences, Université Libre de Bruxelles, Brussels, Belgium.

GEORGE, C., Service de Chimie Physique II, Université Libre de Bruxelles, Brussels, Belgium.

GHOSE, P., Department of Physics, Visva-Bharati University, Santiniketan, West Bengal, India.

GIANNONI, M., Institut de Physique Nucléaire, Division de Physique Théorique, Orsay, France.

GLAUBER, R. J., Lyman Laboratory of Physics, Harvard University, Cambridge, Massachusetts, U.S.A.

GRECOS, A., Service de Chimie Physique II, Université Libre de Bruxelles, Brussels, Belgium.

de GROOT, S. R., Instituut voor Theoretische Fysica, Universiteit van Amsterdam, Valckeniersstraat 65, Amsterdam, The Netherlands.

GUENIN, M., Institut de Physique Théorique, Université de Genève, Geneva, Switzerland.

GUPTA, R. P., Department of Physics, University of Allahabad, Allahabad, India.

GÜRSEY, F., Physics Department, Yale University, New Haven, Connecticut, U.S.A.

GYÖRGI, G., Central Research Institute for Physics, Hungarian Academy of Sciences, Budapest, Hungary.

HAAG, R., II. Institute for Theoretical Physics, Universität Hamburg, Hamburg, Fed. Rep. Germany.

HALPERN, F. R., Department of Physics, Revelle College, University of California, San Diego, La Jolla, California, U.S.A.

HAMENDE, A. M., International Centre for Theoretical Physics, Trieste, Italy.

HARTKÄMPER, A., Institut für Theoretische Physik, Universität Marburg, Marburg, Fed. Rep. Germany.

HEHL, F. W., Institut für Theoretische Physik, Technische Universität, Clausthal–Zellerfeld, Fed. Rep. Germany.

HEISENBERG, W., Max-Planck-Institut für Physik und Astrophysik, Munich, Fed. Rep. Germany.

HINTERMANN, A., Laboratoire de Physique Théorique, Ecole Polytechnique Fédérale de Lausanne, Lausanne, Switzerland.

HULTHÉN, L., Department of Theoretical Physics, The Royal Institute of Technology, Stockholm, Sweden.

HUND, F., Institut für Theoretische Physik, Universität Göttingen, Göttingen, Fed. Rep. Germany.

INÖNÜ, E., Physics Department, Middle East Technical University, Ankara, Turkey.

JASSELETTE, P., Service de Physique Mathématique, Université de Liège, Liège, Belgium.

JAUCH, J. M., Institut de Physique Théorique, Université de Genève, Geneva, Switzerland.

JEHLE, H., Physics Department, University of Maryland, College Park, Maryland, U.S.A.

JORDAN, P., Universität Hamburg, Hamburg, Fed. Rep. Germany.

KAC, M., The Rockefeller University, New York, New York, U.S.A.

KAFKA, P., Max-Planck-Institut für Physik und Astrophysik, Munich, Fed. Rep. Germany.

KANTHACK, L., Institut für Theoretische Physik, Universität Marburg, Marburg, Fed. Rep. Germany.

KASTELEYN, P. W., Lorentz Institute for Theoretical Physics, University of Leiden, Leiden, The Netherlands.

KEMMER, N., Department of Physics, University of Edinburgh, Edinburgh, Scotland.

KHAN, A. R., Physics Department, Forman Christian College, Lahore, Pakistan.

KLAUDER, J. R., Bell Laboratories, Murray Hill, New Jersey, U.S.A.

KURSUNOGLU, B., Center for Theoretical Studies, University of Miami, Coral Gables, Florida, U.S.A.

LAMB, W. E., JR., Physics Department, Yale University, New Haven, Connecticut, U.S.A.

LANCZOS, C., Dublin Institute for Advanced Studies, Dublin, Eire.

LEGGETT, A. J., School of Mathematical and Physical Sciences, The University of Sussex, Falmer, Brighton, Sussex, England.

LEHNSEN, J., 34 Overton Road, Scarsdale, New York, New York, U.S.A.

LELAS, J., Institute for the Philosophy of Science, Yugoslav Academy of Sciences, Zagreb, Yugoslavia.

LELAS, S., Institute for the Philosophy of Science, Yugoslav Academy of Sciences, Zagreb, Yugoslavia.

LÖWDIN, P. O., Quantum Chemistry Group, University of Uppsala, Uppsala, Sweden.

LUDWIG, G., Institut für Theoretische Physik, Universität Marburg, Marburg, Fed. Rep. Germany.

MANOUKIAN, E. B., Department of Physics, The University of Alberta, Edmonton, Alberta, Canada.

MARTIN, A., Theory Division, C.E.R.N., Geneva, Switzerland.

MATTHIAE, G., Istituto Superiore di Sanità, Laboratori di Fisica, Rome, Italy.

MAZUR, P., Lorentz Institute for Theoretical Physics, University of Leiden, Leiden, The Netherlands.

MCLELLAN, A. G., Department of Physics, University of Canterbury, Christchurch New Zealand.

MEHRA, JAGDISH, The University of Texas at Austin, Austin, Texas, U.S.A.

MELVIN, M. A., Physics Department, Temple University, Philadelphia, Pennsylvania, U.S.A.

MERCIER, R. P., Mathematics Department, The University of Southampton, Southampton, England.

MERLINI, D., Laboratoire de Physique Théorique, Ecole Polytechnique Fédérale de, Lausanne, Lausanne, Switzerland.

MESSIAH, A., Division de la Physique, CEN – Saclay, Gif-sur-Yvette, France.

MØLLER, C., NORDITA, Copenhagen, Denmark.

MOLDAUER, P. A., Argonne National Laboratory, Argonne, Illinois, U.S.A.

MÜLLER, F., Fachbereich Physik, Universität Trier–Kaiserslautern, Kaiserslautern, Fed. Rep. Germany.

MUGUR-SCHÄCHTER, M., Faculté des sciences, Université de Reims, Reims, France.

NOVARO, O., Instituto de Física, Ciencias Ciudad Universitaria, Mexico, Mexico.

OCCHIALINI, G., University of Milan, Milan, Italy.

OHANESSIAN, S., Laboratoire de Physique Théorique, Ecole Polytechnique Fédérale de Lausanne, Lausanne, Switzerland.

ÖKTEM, F., Department of Physics, Middle East Technical University, Ankara, Turkey.

OLSEN, S. O., P.O. Box 107, Lødingen, Norway.

O'RAIFEARTAIGH, L., Dublin Institute for Advanced Studies, Dublin, Eire.

PEIERLS, R. E., Department of Theoretical Physics, University of Oxford, Oxford, England.

PERES, A., The Graduate School, Technion, Israel Institute of Technology, Haifa, Israel.

PERRIN, F., Laboratoire de Physique Atomique et Moléculaire, Collège de France, Paris, France.

PIRON, C., Institut de Physique Théorique, Université de Genève, Geneva, Switzerland.

POLYANI, M., 22 Upland Park Road, Oxford, England.

POLDER, D., Philips Research Laboratories, Eindhoven, The Netherlands.

PRASSANA, A. R., Institute of Mathematical Sciences, Madras, India.

PRIGOGINE, I., Faculté des sciences, Université Libre de Bruxelles, Brussels, Belgium.

PUGH, R. E., Department of Physics, University of Toronto, Toronto, Ontario, Canada.

PULIAFITO, S., University of Mendoza, Mendoza, Argentina.

RANDERS, G., Assistant Secretary General, Scientific Affairs Division, NATO, Brussels, Belgium.

RAZMI, S., Institute of Physics, University of Islamabad, Islamabad, Pakistan.

RECHENBERG, H., Max-Planck-Institut für Physik und Astrophysik, Munich, Fed. Rep. Germany.

REES, J., Institute of Astronomy, University of Cambridge, Cambridge, England.

RIAZUDDIN, Daresbury Nuclear Physics Laboratory, Daresbury, Lancs, England.

RICHARD, J. L., Centre de Physique Théorique CNRS, Marseille, France.

ROHRLICH, F., Department of Physics, Syracuse University, Syracuse, New York, U.S.A.

ROMAN, P., Department of Physics, Boston University, Boston, Massachusetts, U.S.A.

ROMERIO, M., Institut de Physique, Université de Neuchâtel, Neuchâtel, Switzerland.

ROSENFELD, L., NORDITA, Copenhagen, Denmark.

ROSSI, M. A., Via Emilia Est. 163, Modena, Italy.

ROUVILLOIS, G., Commissariat à l'Energie Atomique, Paris, France.

ROXBURGH, I. W., Department of Mathematics, Queen Mary College, University of London, London, England.

RÜCKL, R., Pressburgerstr. 53, Munich, Fed. Rep. Germany.

RUDBERG, E., The Royal Academy of Sciences, Fack, Stockholm, Sweden.

SABATIER, P. C., Laboratoire de Physique Mathématique CNRS, Université des Sciences et Techniques, Montpellier, France.

SALAM, ABDUS, International Centre for Theoretical Physics, Miramare, Trieste, Italy.

SANDERS, J. H., Clarendon Laboratory, University of Oxford, Oxford, England.

SAURO, J. P., Southeastern Massachusetts University, North Dartmouth, Massachusetts, U.S.A.

SAXON, D. S., University of California, Los Angeles, California, U.S.A.

SCHIFF, H., Department of Physics, The University of Alberta, Edmonton, Alberta, Canada.

SCHLITT, D. W., Department of Physics, The University of Nebraska–Lincoln, Lincoln, Nebraska, U.S.A.

SCHWINGER, J., Physics Department, University of California, Los Angeles, California, U.S.A.

SCIAMA, D. W., Department of Astrophysics, University of Oxford, Oxford, England.

SIEGBAHN, K., Institute of Physics, University of Uppsala, Uppsala, Sweden.

SNOW, CHARLES PERCY, 85 Eaton Terrace, London S.W.1, England.

SOLHEIM, J. E., The Auroral Observatory, University of Tromsø, Tromsø, Norway.

SOURIAU, J. M., Physique Mathématique, Faculté des sciences de l'université, Marseille, France.

STRAUMANN, N., Institut für Theoretische Physik, Universität Zürich, Zurich, Switzerland.

TEIXEIRA, A. da F., Centro Brasileiro de Pesquisas Fisicas, Rio de Janeiro, Brasil.

TELEGDI, V. L., Enrico Fermi Institute, University of Chicago, Chicago, Illinois, U.S.A.

THIRRING, W., Institut für Theoretische Physik, Universität Wien, Vienna, Austria.

TOWNES, C. H., Department of Physics, University of California, Berkeley, California, U.S.A.

TRAUTMAN, A., Institute of Theoretical Physics, University of Warsaw, Warsaw, Poland.

UHLENBECK, G. E., The Rockefeller University, New York, New York, U.S.A.

UMEZAWA, H., Department of Physics, University of Wisconsin–Milwaukee, Milwaukee, Wisconsin, U.S.A.

URBAN, P., Institut für Theoretische Physik, Universität Graz, Graz, Austria.

VALATIN, J. G., Department of Physics, Queen Mary College, University of London, London, England.

VAN BUEREN, H. G., Department of Radioastrophysics, The Astronomical Institute at Utrecht, Utrecht, The Netherlands.

VAN DAM, H., Physics Department, University of North Carolina, Chapel Hill, North Carolina, U.S.A.

VAN DER WAERDEN, B. L., Universität Zürich, Zurich, Switzerland.

VERDE, M., Istituto di Fisica Teorica, University of Turin, Turin, Italy.

VERZAUX, P., Commissariat à l'Energie Atomique, Paris, France.

VON WEIZSÄCKER, C. F., Max-Planck-Institut, Starnberg, Fed. Rep. Germany.

VOORHEES, B. H., Department of Physics, Pars College, Tehran, Iran.

WAGNER, S., Institut für Theoretische Kern Physik, Universität Bonn, Bonn, Fed. Rep. Germany.

WALLER, I., Institute for Theoretical Physics, Uppsala University, Uppsala, Sweden.

WATAGHIN, G., University of Turin, Turin, Italy.

WEGENER, U., Institut für Theoretische Physik, Universität Marburg, Marburg, Fed. Rep. Germany.

WENTZEL, GREGOR, 77 Via Collina, Ascona, Switzerland.

WHEELER, J. A., Joseph Henry Laboratories, Princeton University, Princeton, New Jersey, U.S.A.

WIGNER, E. P., Joseph Henry Laboratories, Princeton University, Princeton, New Jersey, U.S.A.

WILLIS, J. B., Department of Mathematics, The University of Southampton, Southampton, England.

WILSON, B. G., Simon Fraser University, Burnaby, British Columbia, Canada.

YANG, C. N., Department of Physics, State University of New York, Stony Brook, Long Island, New York, U.S.A.

Index of Names